A textbook of macro and semimicro qualitative inorganic analysis

VOGEL'S TEXTBOOK OF MACRO AND SEMIMICRO QUALITATIVE INORGANIC ANALYSIS

Fifth Edition
Revised by

G. Svehla, Ph.D., D.Sc., F.R.I.C.
Reader in Analytical Chemistry,
Queen's University, Belfast

Longman London and New York

Longman Group Limited London

Associated companies, branches and representatives throughout the world

Published in the United States of America by Longman Inc., New York

First Published under the title 'A Text-book of Qualitative Chemical Analysis' 1937
Second Edition 1941
Reissue with Appendix 1943
Third Edition under the title 'A Text-book of Qualitative Chemical Analysis including Semimicro Qualitative Analysis' 1945
Fourth Edition under the title 'A Text-book of Macro and Semimicro Qualitative Inorganic Analysis' 1954
New Impression (with minor corrections) 1955
New Impression 1976
Fifth edition 1979

Library of Congress Cataloging in Publication Data
Vogel, Arthur I.
Vogel's Macro and semimicro qualitative inorganic analysis.
First-3d ed. published under title: A text-book of qualitative chemical analysis; 4th ed. published under title: A text-book of macro and semimicro qualitative inorganic analysis.
Includes index.
1. Chemistry, Analytic–Qualitative. 2. Chemistry, Inorganic. I. Svehla, G. II. Title. III. Title: Macro and semimicro qualitative inorganic analysis.
QD81.V6 1978 544 77-8290
ISBN 0-582-44367-9

Printed in Great Britain by
Richard Clay (The Chaucer Press) Ltd, Bungay, Suffolk

CONTENTS

CHAPTER V SYSTEMATIC QUALITATIVE INORGANIC ANALYSIS 395

CHAPTER VI SEMIMICRO QUALITATIVE INORGANIC ANALYSIS 461

CHAPTER VIII AN ABBREVIATED COURSE OF QUALITATIVE INORGANIC ANALYSIS 550

IX APPENDIX 566

INDEX 599

FROM PREFACE TO THE FIRST EDITION

Experience of teaching qualitative analysis over a number of years to large numbers of students has provided the nucleus around which this book has been written. The ultimate object was to provide a text-book at moderate cost which can be employed by the student continuously throughout his study of the subject.

It is the author's opinion that the theoretical basis of qualitative analysis, often neglected or very sparsely dealt with in the smaller texts, merits equally detailed treatment with the purely practical side; only in this way can the true spirit of qualitative analysis be acquired. The book accordingly opens with a long Chapter entitled 'The Theoretical Basis of Qualitative Analysis', in which most of the theoretical principles which find application in the science are discussed.

The writer would be glad to hear from teachers and others of any errors which may have escaped his notice: any suggestions whereby the book can be improved will be welcomed.

A. I. Vogel
Woolwich Polytechnic London S.E.18

CHAPTER I THE THEORETICAL BASIS OF QUALITATIVE ANALYSIS

A. CHEMICAL FORMULAE AND EQUATIONS

I.1 SYMBOLS OF THE ELEMENTS To express the composition of substances and to describe the qualitative and quantitative changes, which occur during chemical reactions in a precise, short, and straightforward way we use **chemical symbols** and **formulae**. Following the recommendations of Berzelius (1811), the symbols of chemical elements are constructed by the first letter of their international (Latin) names with, in most cases, a second letter which occurs in the same name. The first letter is a capital one. Such symbols are: O (oxygen, oxygenium) H (hydrogen, hydrogenium), C (carbon, carbonium), Ca (calcium), Cd (cadmium), Cl (chlorine, chlorinum), Cr (chromium), Cu (copper, cuprum), N (nitrogen, nitrogenium), Na (sodium, natrium), K (potassium, kalium), etc. As well as being a qualitative reference to the element, the symbol is most useful in a quantitative context. It is generally accepted that the symbol of the element represents 1 atom of the element, or, in some more specific cases, 1 grammatom. Thus C represents 1 atom of the element carbon or may represent 1 grammatom (12·011 g) of carbon. In a similar way, O represents one atom of oxygen or one grammatom (15·9994 g) of oxygen, H represents one atom of hydrogen or 1 grammatom (1·0080 g) of hydrogen etc. Names, symbols, and relative atomic masses of the elements are given in Section **IX.1**.

I.2 EMPIRICAL FORMULAE To express the composition of materials whose molecules are made up of more atoms, empirical formulae are used. These are made up of the symbols of the elements of which the substance is formed. The number of atoms of a particular element in the molecule is written as a subscript after the symbol of the element (but 1 is never written as a subscript as the symbol of the element on its own represents one atom).

Thus, the molecules of **carbon dioxide** is formed by one carbon atom and two oxygen atoms, therefore its empirical formula is CO_2. In the molecule of **water** two hydrogen atoms and one oxygen atom are present, therefore the empirical formula of water is H_2O. In the molecule of **hydrogen peroxide** on the other hand there are two hydrogen and two oxygen atoms present, its empirical formula is therefore H_2O_2.

Although there are no strict rules as to the order of symbols appearing in a formula, in the case of **inorganic** substances the symbol of the metal or that of hydrogen is generally written first followed by non-metals and finishing with oxygen. In the formulae of **organic** substances the generally accepted order is C, H, O, N, S, P.

The **determination** of the empirical formula of a compound can be made experimentally, by determining the percentage amounts of elements present in the substance using the methods of quantitative chemical analysis. At the same time the relative molecular mass of the compound has to be measured as well. From these data the empirical formula can be determined by a simple calculation. If, for some reason, it is impossible to determine the relative molecular mass the simplest (assumed) formula only can be calculated from the results of chemical analysis; the true formula might contain multiples of the atoms given in the assumed formula.

If the empirical formula of a compound is known, we can draw several conclusions about the physical and chemical characteristics of the substance. These are as follows:

(*a*) From the empirical formula of a compound we can see which elements the compound contains, and how many atoms of each element form the molecule of the compound. Thus, hydrochloric acid (HCl) contains hydrogen and chlorine; in its molecule one hydrogen and one chlorine atom are present. Sulphuric acid (H_2SO_4) consists of hydrogen, sulphur, and oxygen; in its molecule two hydrogen, one sulphur, and four oxygen atoms are present etc.

(*b*) From the empirical formula the **relative molecular mass** (molecular weight) can be determined simply by adding up the **relative atomic masses** (atomic weights) of the elements which constitute the compound. In this summation care must be taken that the relative atomic mass of a particular element is multiplied by the figure which shows the number of its atoms in the molecule. Thus, the relative molecular mass of hydrochloric acid (HCl) is calculated as follows:

$$M_r = 1{\cdot}0080 + 35{\cdot}453 = 36{\cdot}4610$$

and that of sulphuric acid (H_2SO_4) is

$$M_r = 2 \times 1{\cdot}0080 + 32{\cdot}06 + 4 \times 15{\cdot}9994 = 98{\cdot}0736$$

and so on.

(*c*) Based on the empirical formula one can easily calculate the **relative amounts of the elements present** in the compound or the **percentage composition** of the substance. For such calculations the relative atomic masses of the elements in question must be used. Thus, in hydrochloric acid (HCl) the relative amounts of the hydrogen and chlorine are

$$H:Cl = 1{\cdot}0080:35{\cdot}453 = 1{\cdot}0000:35{\cdot}172$$

and (as the relative molecular mass of hydrochloric acid is 36·461) it contains

$$100 \times \frac{1{\cdot}008}{36{\cdot}461} = 2{\cdot}76 \text{ per cent H}$$

and

$$100 \times \frac{35{\cdot}453}{36{\cdot}461} = 97{\cdot}24 \text{ per cent Cl}$$

Similarly, the relative amounts of the elements in sulphuric acid (H_2SO_4) are

$$\begin{aligned} H:S:O &= 2 \times 1{\cdot}0080:32{\cdot}06:4 \times 15{\cdot}9994 \\ &= 2{\cdot}016:32{\cdot}06:63{\cdot}9976 \\ &\qquad 1:15{\cdot}903:31{\cdot}745 \end{aligned}$$

and knowing that the relative molecular mass of sulphuric acid is 98·0763, we can calculate its percentage composition which is

$$100 \times \frac{2{\cdot}0160}{98{\cdot}0736} = 2{\cdot}06 \text{ per cent H}$$

$$100 \times \frac{32{\cdot}06}{98{\cdot}0736} = 32{\cdot}69 \text{ per cent S}$$

and

$$100 \times \frac{63{\cdot}9976}{98{\cdot}0736} = 65{\cdot}25 \text{ per cent O}$$

and so on.

(*d*) Finally, if the formula is known – which of course means that the relative molecular mass is available – we can calculate the volume of a known amount of a gaseous substance at a given temperature and pressure. If p is the pressure in atmospheres, T is the absolute temperature in degrees kelvins, M_r is the relative molecular mass of the substance in g mol^{-1} units and m is the weight of the gas in grams, the volume of the gas (v) is

$$v = \frac{mRT}{pM_r}\ \ell$$

where R is the gas constant, 0·0823 ℓ atm K^{-1} mol^{-1}. (The gas here is considered to be a perfect gas.)

I.3 VALENCY AND OXIDATION NUMBER In the understanding of the composition of compounds and the structure of their molecules the concept of valency plays an important role. When looking at the empirical formulae of various substances the question arises: are there any rules as to the number of atoms which can form stable molecules? To understand this let us examine some simple compounds containing hydrogen. Such compounds are, for example, hydrogen chloride (HCl), hydrogen bromide (HBr), hydrogen iodide (HI), water (H_2O), hydrogen sulphide (H_2S), ammonia (H_3N), phosphine (H_3P), methane (H_4C), and silane (H_4Si). By comparing these formulae one can see that one atom of some of the elements (like Cl, Br, and I) will bind one atom of hydrogen to form a stable compound, while others combine with two (O, S), three (N, P) or even four (C, Si). This number, which represents one of the most important chemical characteristics of the element, is called the **valency**. Thus, we can say that chlorine, bromine, and iodide are monovalent, oxygen and sulphur bivalent, nitrogen and phosphorus tervalent, carbon and silicon tetravalent elements and so on. Hydrogen itself is a monovalent element.

From this it seems obvious that the valency of an element can be ascertained from the composition of its compound with hydrogen. Some of the elements, for example some of the metals, do not combine with hydrogen at all. The valency of such elements can therefore be determined only in an indirect way, by examining the composition of their compounds formed with chlorine or oxygen and finding out the number of hydrogen atoms these elements replace. Thus, from the formulae of magnesium oxide (MgO) and magnesium chloride ($MgCl_2$) we can conclude that magnesium is a bivalent metal, similarly from the composition of aluminium chloride ($AlCl_3$) or aluminium oxide (Al_2O_3) it is obvious that aluminium is a tervalent metal etc.

In conclusion we can say that the valency of an element is a number which expresses how many atoms of hydrogen or other atoms equivalent to hydrogen can unite with one atom of the element in question.* If necessary the valency of the element is denoted by a roman numeral following the symbol like Cl(I), Br(I), N(III) or as a superscript, like Cl^{I}, Br^{I}, N^{III}, etc.

Some elements, like hydrogen, oxygen, or the alkali metals, seem always to have the same valency in all of their compounds. Other elements however show different valencies; thus, for example, chlorine can be mono-, tri-, penta- or heptavalent in its compounds. It is true that compounds of the same element with different valencies show different physical and chemical characteristics.

A deeper study of the composition of compounds and of the course of chemical reactions reveals that the classical concept of valency, as defined above, is not quite adequate to explain certain phenomena. Thus, for example, chlorine is monovalent both in hydrochloric acid (HCl) and in hypochlorous acid (HClO), but the marked differences in the chemical behaviour of these two acids indicate that the status of chlorine in these substances is completely different. From the theory of chemical bonding† we know that when forming hydrochloric acid, a chlorine atom takes up an electron, thus acquiring one negative charge. On the other hand, if hypochlorous acid is formed, the chlorine atom releases an electron, becoming thus a species with one positive charge. As we know, the uptake or release of electrons corresponds to reduction or oxidation (cf. Section **I.35**), we can therefore say that though chlorine is monovalent in these acids, its oxidation status is different. It is useful to define the concept of **oxidation number** and to use it instead of valency. The oxidation number is a number identical with the valency but with a sign, expressing the nature of the charge of the species in question when formed from the neutral atom. Thus, the oxidation number of chlorine in hydrochloric acid is -1, while it is $+1$ in hypochlorous acid. Similarly we can say that the oxidation number of chlorine in chlorous acid ($HClO_2$) is $+3$, in chloric acid ($HClO_3$) is $+5$, and in perchloric acid ($HClO_4$) $+7$. The concept of oxidation number will be used extensively in the present text.

I.4 STRUCTURAL FORMULAE Using the concept of valency the composition of compounds can be expressed with structural formulae. Each valency of an element can be regarded as an arm or hook, through which chemical bonds are formed. Each valency can be represented by a single line drawn outwards from the symbol of the element, like

$$\mathrm{H-}\quad \mathrm{Cl-}\quad \mathrm{O=}\quad \mathrm{N\equiv}\quad \mathrm{C\equiv}$$

The structural formulae of compounds can be expressed with lines drawn between the atoms ‡ like

* Cf. Mellor's *Modern Inorganic Chemistry*, newly revised and edited by G. D. Parkes, Longman 1967, p. 99 et f.

† Cf. Mellor op. cit., p. 155 et f.

‡ There are no restrictions about the direction of these lines (unless differentiation has to be made between stereochemical isomers). Nor is there any restriction on the distances of atoms. Structural formulae must therefore be regarded only as a step in the approximation of the true structure. A three dimensional representation with true directions and proportional distances can most adequately be made with molecular model kits.

H—Cl   H—O—H   O=C=O

NH_3 (N bonded to three H) CH_4 (C bonded to four H)

Structural formulae will be used in this text only when necessary, mainly when dealing with organic reagents. A more detailed discussion of structural formulae will not be given here; beginners should study appropriate textbooks.* Readers should be reminded that the simple hexagon

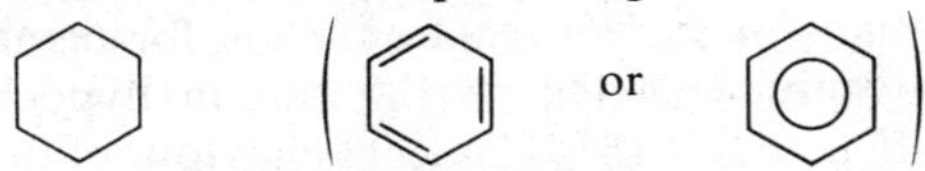

represents the benzene ring. Benzene (C_6H_6) can namely be described with the (simplified) ring formula in which double and single bonds are alternating (so-called conjugate bonds):

(Six-membered ring of C atoms, each bonded to one H, with alternating double and single C–C bonds)

All the aromatic compounds contain the benzene ring.

I.5 CHEMICAL EQUATIONS

Qualitative and quantitative relationships involved in a chemical reaction can most precisely be expressed in the form of chemical equations. These equations contain the formulae of the reacting substances on the left-hand side and the formulae of the products on the right-hand side. When writing chemical equations the following considerations must be kept in mind:

(*a*) Because of the fact that the formulae of the reacting species are on the left-hand side and those of the products are on the right, the sides generally cannot be interchanged (in this sense chemical equations are not equivalent to mathematical equations). In the cases of equilibrium reactions† when the reaction may proceed in both directions, the double arrow ($\rightleftarrows$) sign should be used instead of the equal ($=$) or single arrow ($\rightarrow$) sign.

(*b*) The individual formulae, used in the chemical reactions, must be written correctly.

(*c*) If more molecules (atoms or ions) of the same substance are involved in the reaction, an appropriate stoichiometric number has to be written in front of the formula. This number is a multiplication factor, which applies to all atoms in the formula. (Thus, for example $2Ca_3(PO_4)_2$ means that we have 6 calcium, 4 phosphorus, and 16 oxygen atoms in the equation.)

---

* Cf. Mellor's *Modern Inorganic Chemistry*, newly revised and edited by G. D. Parkes, Longman 1967, p. 155.

† Theoretically speaking, all reactions lead to an equilibrium. This equilibrium however may be shifted completely towards the formation of the products.

(*d*) A chemical equation must be written in such a way that it fulfils the law of conservation of mass, which is strictly valid for all chemical reactions. This means, that the equation should be balanced by applying proper stoichiometric numbers in such a way that the numbers of different individual atoms are the same on both sides.

(*e*) If charged species (ions or electrons) are involved in the reaction, these charges must be clearly indicated (like Fe^{3+} or Fe^{+++}) and properly balanced; the sum of charges on the left-hand side must be equal to the sum of charges on the right-hand side. The electron, as a charged particle, will be denoted by e^- in this text.

As an example let us express the equation of the reaction between calcium hydroxide and phosphoric acid.

Knowing that the products of such a reaction are calcium phosphate and water, we can write the formulae of the substances into the yet incomplete equation:

$$Ca(OH)_2 + H_3PO_4 \rightarrow Ca_3(PO_4)_2 + H_2O \qquad \text{(incomplete)}$$

(note that the sides of the equation cannot be interchanged, because the reaction will not proceed in the inverse direction). Now we try to balance the equation by applying suitable stoichiometric numbers:

$$3Ca(OH)_2 + 2H_3PO_4 = Ca_3(PO_4)_2 + 6H_2O$$

It is advisable to check the equation by counting the numbers of individual atoms on both sides. Doing so we can see that there are 3 calcium, 2 phosphorus, 12 hydrogen and 14 oxygen atoms on both sides.

It is useful to denote the physical state of the reaction partners. For this purpose the letters s, l, and g are applied for solid, liquid, and gaseous substances respectively, while the notation aq is used for species dissolved in water. These letters are used in parenthesis after the formula, e.g. AgCl(s), H_2O(l), CO_2(g), while the aq follows the formula simply, without parenthesis e.g. H_3PO_4aq. The systematic use of these notations is important only in thermodynamics, that is when the energetics of the reactions are examined. In the present text we shall use them in some cases. The formation of a precipitate will be denoted by a ↓ sign (indicating that the precipitate settles to the bottom of the solution) while the liberation of gases will be denoted by a ↑ sign. If not otherwise stated, equations will refer to reactions proceeding in dilute aqueous solutions.

Following those considerations discussed in Section **I.2.**, relative masses, mass balances, and volumes (of gaseous substances only) can be calculated on the basis of chemical equations. Such calculations are involved in all kinds of quantitative analyses based on chemical reactions.

B. AQUEOUS SOLUTIONS OF INORGANIC SUBSTANCES

I.6 ELECTROLYTES AND NON-ELECTROLYTES Quantitative inorganic analysis is based mainly on the observation of chemical reactions carried out in aqueous solutions. Other solvents are rarely employed except for special tests or operations. It is therefore important to have a general knowledge of the characteristics of aqueous solutions of inorganic substances.

A **solution** is the homogeneous product obtained when a substance (**the solute**) is dissolved in the **solvent** (water). Substances can be classified into two important groups according to their behaviour when an electric current is passed through their solution. In the first class there are those which conduct electric current; the solutions undergo chemical changes thereby. The second class is composed of materials which, when dissolved in water, do not conduct electricity and which remain unchanged. The former substances are termed **electrolytes**, and these include, with few exceptions, all inorganic substances (like acids, bases, and salts); the latter are designated **non-electrolytes**, and are exemplified by such organic materials as cane sugar, mannose, glucose, glycerine, ethanol, and urea. It must be pointed out that a substance which behaves as an electrolyte in water, e.g. sodium chloride, may not yield a conducting solution in another solvent such as ether or hexane. In the molten state most electrolytes will conduct electricity.

I.7 ELECTROLYSIS, THE NATURE OF ELECTROLYTIC CONDUCTANCE, IONS Chemically pure water practically does not conduct electricity, if however, as already stated, acids, bases, or salts are dissolved in it, the resultant solution not only conducts the electric current, but undergoes chemical changes as well. The whole process is called electrolysis.

Phenomena occurring during electrolysis can be studied in the electrolysis cell shown in Fig. I.1. The electrolyte solution is placed in a vessel, into which

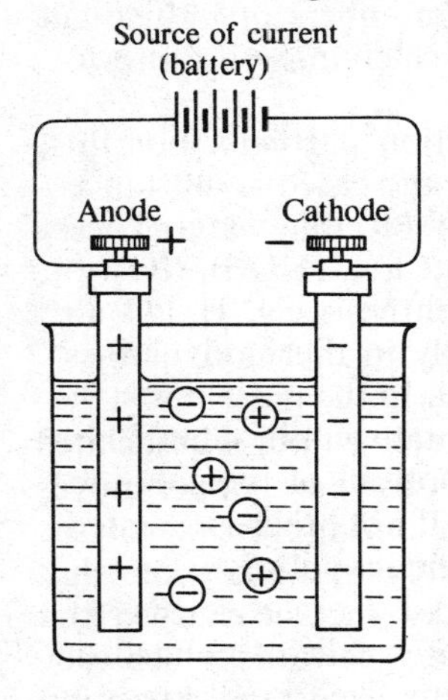

Fig I.1

two solid conductors (e.g. metals), the so called electrodes, are immersed. With the aid of a battery (or another d.c. source) a potential difference is applied between the two electrodes. The electrode with the negative charge in the electrolysis cell is called the cathode, while that with the positive charge is termed the anode.*

* It must be emphasized that the terms cathode and anode correspond to the negative and positive electrodes respectively only in electrolysis cells. According to Faraday's nomenclature, cathode is the electrode where cations lose their charge, while anions do the same on the anode. Consequently, in a battery (like the Daniell-cell) the anode is the negative and the cathode is the positive electrode.

The chemical change occurring during the course of electrolysis is observable on or in the vicinity of the electrodes. In many cases such a change is a simple decomposition. If for example a dilute solution of hydrochloric acid is electrolysed (between platinum electrodes), hydrogen gas is liberated on the cathode and chlorine on the anode; the concentration of hydrochloric acid in the solution decreases.

It is easy to demonstrate that electrolysis is always accompanied by the transport of material in an electrolysis cell. If for example the blue solution of copper sulphate and the orange solution of potassium dichromate are mixed in equimolar concentrations, a brownish solution is obtained. This solution can be placed in a U-shaped electrolysis cell and topped up with a colourless layer of dilute sulphuric acid on each side (Fig. I.2). If this solution is then electrolysed, the hitherto colourless solution next to the cathode slowly becomes blue, while

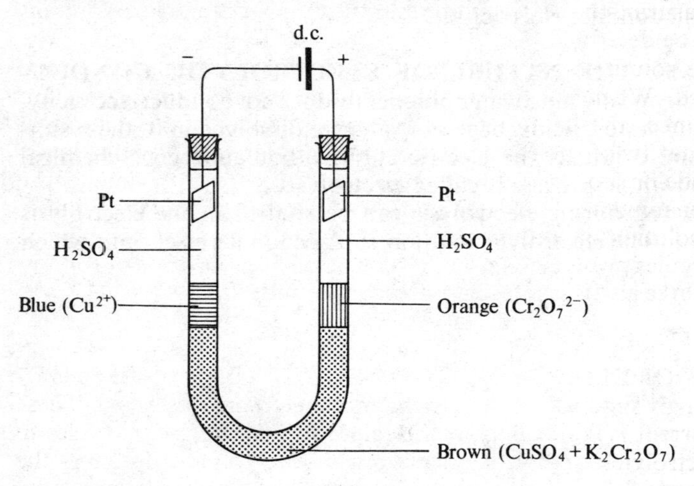

Fig. I.2

the solution next to the anode becomes orange. As the blue colour is associated with copper and the orange with dichromate, it can be said that copper moves towards the cathode and dichromate towards the anode during the electrolysis.

As such a movement can be achieved solely by electrolysis, it is obvious that those particles which move towards one of the electrodes must be charged and that this charge must be opposite to that of the electrode towards which they move. The migration of such particles is a result of the electrostatic attraction force, which is created when switching on the current. Thus the particles of hydrogen or copper, which move towards the cathode, must be positively charged, while those of chlorine or dichromate must be negatively charged. Faraday termed the charged particles in the electrolyte ions; the positively and negatively charged ions were called cations and anions respectively. It can be stated generally that solutions of electrolytes do not contain neutral molecules dispersed among the molecules of the solvent, as solutions of non-electrolytes do, but they are composed of ions. Cations and anions are present in equivalent amounts and are dispersed evenly in the solution among the molecules of the solvent; macroscopic portions of the solution therefore appear to be electrostatically neutral in all cases.

I.8 SOME PROPERTIES OF AQUEOUS SOLUTIONS It has been found experimentally that equimolecular quantities of non-electrolytes, dissolved in the same weight of solvent, will acquire identical osmotic pressures, and have the same effect upon the lowering of vapour pressure, the depression of the freezing point, and the elevation of the boiling point. Using water as a solvent, 1 mole of a non-electrolyte when dissolved in 1000 g of water lowers, for example, the freezing point of water by 1·86°C and elevates its boiling point by 0·52°C. On such a basis it is possible to determine the relative molecular mass of soluble non-electrolyte substances experimentally. When a non-electrolyte is dissolved in water, its molecules will be present as individual particles in the solution. Consequently, we can say that equal numbers of particles, present in the same amount of solution, will show identical osmotic pressure, lowering of vapour pressure, depression of the freezing point, or elevation of the boiling point. Thus, by measuring the above quantities, the number of particles present in the solution can be determined.

When electrolyte solutions are subjected to such measurements, abnormal results are obtained. When substances like sodium chloride or magnesium sulphate are examined, the depression of freezing point or the elevation of boiling point is about twice that calculated from the relative molecular mass, with calcium chloride or sodium sulphate these quantities are three times those expected. Keeping in mind what has been said above, we can say that the number of particles in the solution of sodium chloride or magnesium sulphate is twice the number of molecules present, while in the case of calcium chloride or sodium sulphate there are three particles present for each molecule.

I.9 THE THEORY OF ELECTROLYTIC DISSOCIATION In Sections **I.7** and **I.8** two, seemingly independent, experimental facts were described. These are that electric current is conducted by the migration of charged particles in the solution of electrolytes, and that in solutions of electrolyte substances the number of particles are 2, 3 . . . etc. times greater than the number of molecules dissolved. To explain these facts, Arrhenius put forward his theory of electrolytic dissociation (1887). According to this theory, the molecules of electrolytes, when dissolved in water, dissociate into charged atoms or groups of atoms, which are in fact the ions which conduct the current in electrolytes by migration. This dissociation is a reversible process; the degree of dissociation varies with the degree of dilution. At very great dilutions the dissociation is practically complete for all electrolytes.

The electrolytic dissociation (ionization) of compounds may therefore be represented by the reaction equations:

$$NaCl \rightleftarrows Na^+ + Cl^-$$
$$MgSO_4 \rightleftarrows Mg^{2+} + SO_4^{2-}$$
$$CaCl_2 \rightleftarrows Ca^{2+} + 2Cl^-$$
$$Na_2SO_4 \rightleftarrows 2Na^+ + SO_4^{2-}$$

Ions carry positive or negative charges. Since the solution is electrically neutral, the total number of positive charges must be equal to the total number of negative charges in a solution. The number of charges carried by an ion is equal to the valency of the atom or radical.

The explanation of the abnormal results obtained when measuring the depression of freezing point or elevation of boiling point is straightforward on the basis of the theory of electrolytic dissociation. In the case of sodium chloride and magnesium sulphate the measured values are twice as great as those calculated from the relative molecular mass, because both substances yield two ions per molecule when dissociated. Similarly, the depression of freezing point or elevation of boiling point of calcium chloride or sodium sulphate solutions are three times as great as of an equimolar solution of a non-electrolyte, because these substances yield three ions from each molecule when dissociating.

The phenomenon of electrolysis also receives a simple explanation on the basis of the theory of electrolytic dissociation. The conductance of electrolyte solutions is due to the fact that ions (charged particles) are present in the solution, which, when switching on the current, will start to migrate towards the electrode with opposite charge, owing to electrostatic forces. In the case of hydrochloric acid we have hydrogen and chloride ions in the solution:

$$HCl \rightleftarrows H^+ + Cl^-$$

and it is obvious that hydrogen ions will migrate towards the cathode, while chloride ions will move towards the anode. In the solution, mentioned earlier, containing copper sulphate and potassium dichromate we have the blue copper(II) ions and the orange dichromate ions present, besides the colourless potassium and sulphate ions:

$$CuSO_4 \rightleftarrows Cu^{2+} + SO_4^{2-}$$
$$K_2Cr_2O_7 \rightleftarrows 2K^+ + Cr_2O_4^{2-}$$

and this is why copper ions (together with potassium ions) moved towards the negatively charged cathode, while dichromate ions (as well as sulphate ions) moved towards the positively charged anode.

Those changes occurring on the electrodes during electrolysis can also be explained easily on the basis of the theory of electrolytic dissociation. Returning to the example of the electrolysis of hydrochloric acid, where, as said before, hydrogen ions migrate towards the cathode and chloride ions towards the anode, the electrode processes are as follows: hydrogen ions, when arriving at the cathode first take up an electron to form a neutral hydrogen atom:

$$H^+ + e^- \rightarrow H$$

Pairs of hydrogen atoms will then form hydrogen molecules, which are discharged in the form of hydrogen gas:

$$2H \rightarrow H_2(g)$$

On the anode the chloride ions release electrons, forming chlorine atoms:

$$Cl^- \rightarrow Cl + e^-$$

which again will form chlorine molecules:

$$2Cl \rightarrow Cl_2(g)$$

and are discharged in the form of chlorine gas. The electrons are taken up by the anode, and travel through the electric circuit to the cathode, where they are then taken up by hydrogen ions.

The phenomena of electrolysis are not always as simple as discussed in connection with hydrochloric acid, but it is always true that **electrons are taken up by ions on the cathode and electrons are released by ions on the anode.** It is not necessarily the cation or anion of the dissolved substance, which reacts on the electrodes, even though these ions carry the electrical current by migration. In aqueous solutions very small amounts of hydrogen and hydroxyl ions are always present due to the slight dissociation of water (cf. Sections **I.18** and **I.24**):

$$H_2O \rightleftarrows H^+ + OH^-$$

The ions of the dissolved substance and hydrogen as well as hydroxyl ions compete for discharge on the electrodes, and the successful ion is the one which needs the least energy for discharge. Using electrochemical terms we can say that under given circumstances the ion which requires a lower negative electrode potential will be discharged first on the cathode, while the one that requires a lower positive potential will be discharged on the anode. The discharge of hydroxyl ions on the anode results in the formation of oxygen gas:

$$4OH^- \rightarrow 4e^- + 2H_2O + O_2(g)$$

The competition of various ions at the electrodes for discharge may lead to various combinations. If for example sodium sulphate is electrolysed (with platinum electrodes), neither sodium nor sulphate ions ($Na_2SO_4 \rightleftarrows 2Na^+ + SO_4^{2-}$) will be discharged, but hydrogen and hydroxyl ions; the result of the electrolysis therefore is the formation of hydrogen gas on the cathode and oxygen on the anode. As hydrogen ions are removed from the vicinity of the cathode, the hydroxyl-ion concentration will surpass that of the hydrogen ions, making this part of the solution alkaline. The opposite happens around the anode, where hydrogen ions will be in excess and the solution there becomes acidic. When after the electrolysis the solution is mixed, it again becomes neutral. When electrolysing sodium chloride ($NaCl \rightleftarrows Na^+ + Cl^-$) under similar circumstances, hydrogen and chloride ions are discharged in the form of hydrogen and chlorine gas on the cathode and anode respectively. Sodium and hydroxyl ions are left behind, and the whole solution becomes alkaline. Finally, if copper sulphate ($CuSO_4 \rightleftarrows Cu^{2+} + SO_4^{2-}$) is electrolysed under the same circumstances, copper and hydroxyl ions will be discharged, the cathode being coated with a layer of copper metal, while oxygen gas is liberated on the anode. Hydrogen and sulphate ions are left behind in the solution, making the latter acidic.

In later parts of the present text we shall see that the uptake of electrons always means reduction, while the release of electrons is associated with oxidation. Briefly therefore we can say that during the course of electrolysis **reduction takes place on the cathode, while oxidation occurs on the anode.** This rule is true for any kind of electrochemical process, e.g. the same is true for the operation of electromotive cells (batteries).

I.10 DEGREE OF DISSOCIATION. STRONG AND WEAK ELECTROLYTES When discussing the theory of electrolytic dissociation, it was stated that it is a reversible process and its extent varies with concentration (and also with other physical properties, like temperature). The degree of dissociation (α) is equal to the fraction of the molecules which actually dissociate.

$$\alpha = \frac{\text{number of dissociated molecules}}{\text{total number of molecules}}$$

The value of α may vary within 0 and 1. If $\alpha = 0$, no dissociation takes place, while if $\alpha = 1$ dissociation is complete.

The degree of dissociation can be determined by various experimental methods.

The cryoscopic and ebullioscopic techniques are based on the measurement of the depression of the freezing point and the elevation of the boiling point respectively. As mentioned before, the experimental values of these were found to be higher than the theoretical ones. The ratio of these

$$\frac{\Delta\,(\text{obs})}{\Delta\,(\text{theor})} = i$$

is closely associated with the number of particles present in the solution. The value i (called van't Hoff's coefficient) gives the average number of particles formed from one molecule; as this is an average number, i is not an integer. It is always greater than unity. This number can easily be associated with the degree of dissociation. Let us consider an electrolyte which when dissociated gives rise to the formation of n ions per molecule. If 1 mole of this electrolyte is dissolved, and the degree of dissociation is α, we can calculate the total number of particles (ions plus undissociated molecules) in the following way: the number of ions (per molecule) will be $n\alpha$, while the number of undissociated molecules $1-\alpha$. The sum of these is equal to i, the van't Hoff coefficient:

$$i = n\alpha + 1 - \alpha = 1 + (n-1)\alpha$$

from which the degree of dissociation can be expressed as

$$\alpha = \frac{i-1}{n-1}$$

Thus, by calculating i from experimental data, α can be computed easily.

An important method of determining the degree of dissociation is based on the measurement of the conductivity of the electrolyte in question (**conductivity method**). This method is associated with the fact that the electric current is carried by the ions present in the solution; their relative number, which is closely connected to the degree of dissociation, will determine the conductivity of the solution. Conductivity itself is a derived quantity, as it cannot be measured as such. To determine conductivity one has to measure the **specific resistance (resistivity)** of the solution. This can be done by placing the solution in a cube-like cell of 1 cm side, two parallel faces of which are made of a conductor (platinum).* This cell can then be connected as the unknown resistance in a Wheatstone-bridge circuit, which is fed by a perfectly symmetrical (sinusoidal) alternating current at low voltage. Direct current would cause changes in the concentration of the solution owing to electrolysis. The specific resistance, ρ, is expressed in Ω cm units. The reciprocal of the specific resistance is termed

* It is not in fact necessary to use such a particular cell for the measurements; any cell of constant dimensions is suitable, provided that its 'cell constant' has been determined by a calibration procedure, using an electrolyte (e.g. potassium chloride solution), with a known specific resistance.

specific conductance or **conductivity**, κ, and is expressed in Ω^{-1} cm^{-1} units. For electrolytic solutions it is customary to define the quantity called molar conductivity, Λ. The latter is the conductance of a solution which contains 1 mole of the solute between two electrodes of indefinite size, 1 cm apart. The specific conductance and molar conductivity are connected by the relation:

$$\Lambda = \kappa V = \frac{\kappa}{c}$$

where V is the volume of the solution in cm^3 (ml), containing 1 mole of the solute, c is the concentration in mol cm^{-3}. The molar conductivity is expressed in cm^2 Ω^{-1} mol^{-1} units.

Kohlrausch discovered, in the last century, that the molar conductivity of aqueous solutions of electrolytes increases with dilution, and reaches a limiting value at very great dilutions. The increase of molar conductivity, in line with the Arrhenius theory, results from the increasing degree of dissociation; the limiting value corresponds to complete dissociation. This limiting value of the molar conductivity is denoted here by Λ_0 (the notation Λ_∞ is also used), while its value at a concentration c will be denoted by Λ_c. The degree of dissociation can be expressed as the ratio of these two molar conductivities

$$\alpha = \frac{\Lambda_c}{\Lambda_0}$$

for the given concentration (c) of the electrolyte.

The variation of molar conductivity with concentration for a number of electrolytes is shown in Table I.1. Because the conductance of solutions varies with temperature (at higher temperatures the conductance becomes higher), the temperature at which these conductances are measured must be given. Values shown on Table I.1 were measured at 25°C. It can be seen from this table that while the variation of molar conductivity of some solutions with

Table I.1 Molar conductivities of electrolytes at 25°C in cm^2 Ω^{-1} mol^{-1} units

Concentration mol ℓ^{-1}	Electrolyte						
	KCl	NaCl	HCl	NaOH	KOH	CH_3COONa	CH_3COOH
$\rightarrow 0 (= \Lambda_0)$	150·1	126·2	423·7	260·9	283·9	91·3	388·6
0·0001	149·2	125·3	—	—	—	—	—
0·0002	—	—	—	—	—	—	104·0
0·0005	148·3	124·3	422·2	246·5	270·1	89·4	64·5
0·001	147·5	123·5	421·1	244·7	268·2	88·7	48·7
0·002	146·5	122·2	419·2	242·5	266·2	87·7	35·2
0·005	144·2	119·8	414·9	238·8	262·1	85·7	22·8
0·01	141·6	117·8	410·5	234·5	258·9	83·7	16·2

concentration is slight for most of the electrolytes listed, there is a strong dependence on concentration in the case of acetic acid. The difference in behaviour can be seen better from Fig. I.3, where molar conductivities are plotted as functions of concentration, using a logarithmic scale for the latter to provide a wider range for illustration. The five substances selected for illustration represent five different groups of inorganic compounds, within each of which there is little variation, e.g. the curve for nitric acid would run very close to the curve

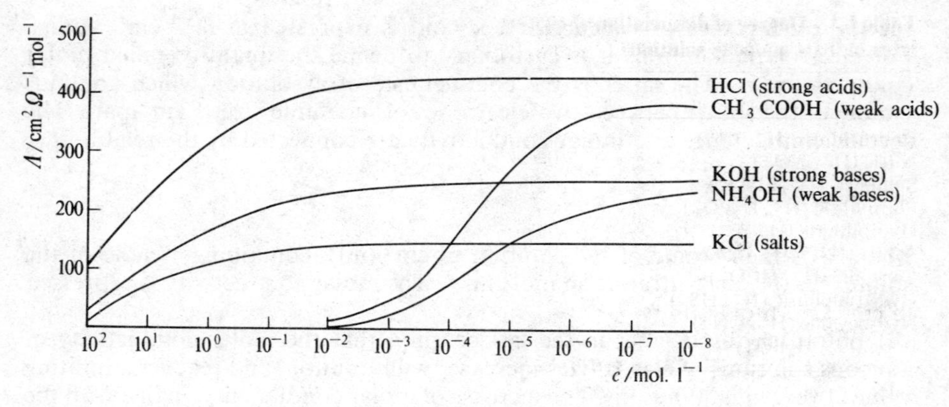

Fig. I.3

of hydrochloric acid. But if we think in terms of degrees of dissociation, we can see that there are only two groups showing different behaviour. The first group, made up of strong acids, strong bases, and salts (including those of weak acids and weak bases), is termed **strong electrolytes**. (These dissociate almost completely even at relatively low degrees of dilution 0·01M solutions), and there is little variation in the degree of dissociation at further dilution. On the other hand, **weak electrolytes** (weak acids and weak bases) start to dissociate only at very low concentrations, and the variation in the degree of dissociation is considerable at this lower concentration range.

The two methods, the cryoscopic and ebullioscopic techniques on one hand and the conductivity method on the other hand, yield strikingly similar values for the degree of dissociation, despite the substantially different principles involved in the two types of measurements. Some representative results are shown in Table I.2. It can be noted that agreement is particularly good for dilute solutions of binary electrolytes (KCl). The more concentrated the solutions, the more considerable the differences. Table I.3 shows the degree of dissociation of

Table I.2 Degree of dissociation of electrolytes, calculated from freezing point and conductivity measurements

Substance	Concentration mol ℓ$^{-1}$	α from freezing point	α from conductivity	No. of ions for one molecule, n
KCl	0·01	0·946	0·943	2
	0·02	0·915	0·924	
	0·05	0·890	0·891	
	0·10	0·862	0·864	
$BaCl_2$	0·001	0·949	0·959	3
	0·01	0·903	0·886	
	0·10	0·798	0·754	
K_2SO_4	0·001	0·939	0·957	3
	0·01	0·887	0·873	
	0·10	0·748	0·716	
$K_3[Fe(CN)_6[$	0·001	0·946	0·930	4
	0·01	0·865	0·822	
	0·10	0·715	—	

Table I.3 Degree of dissociation of electrolytes in 0·1M aqueous solutions

Acids	
Hydrochloric (H^+, Cl^-)	0·92
Nitric (H^+, NO_3^-)	0·92
Sulphuric (H^+, HSO_4^-)	0·61
Phosphoric (H^+, $H_2PO_4^-$)	0·28
Hydrofluoric (H^+, F^-)	0·085
Acetic (H^+, $CH_3.COO^-$)	0·013
Carbonic (H^+, HCO_3^-)	0·0017
Hydrosulphuric (H^+, HS^-)	0·0007
Hydrocyanic (H^+, CN^-)	0·0001
Boric (H^+, $H_2BO_3^-$)	0·0001
Salts	
Potassium chloride (K^+, Cl^-)	0·86
Sodium chloride (Na^+, Cl^-)	0·86
Potassium nitrate (K^+, Cl_3^-)	0·82
Silver nitrate (Ag^+, NO_3)	0·82
Sodium acetate (Na^+, $CH_3.COO^-$)	0·80
Barium chloride (Ba^{2+}, $2Cl^-$)	0·75
Potassium sulphate ($2K^+$, SO_4^{2-})	0·73
Sodium carbonate ($2Na^+$, CO_3^{2-})	0·70
Zinc sulphate (Zn^{2+}, SO_4^{2-})	0·40
Copper sulphate (Cu^{2+}, SO_4^{2-})	0·39
Mercuric chloride (Hg^{2+}, $2Cl^-$)	<0·01
Mercuric cyanide (Hg^{2+}, $2CN^-$)	very small
Bases	
Sodium hydroxide (Na^+, OH^-)	0·91
Potassium hydroxide (K^+, OH^-)	0·91
Barium hydroxide (Ba^{2+}, $2OH^-$)	0·81
Ammonia (NH_4^+, OH^-)	0·013

a number of electrolytes in 0·1M concentrations. From these values we can easily decide whether a particular substance is a strong or a weak electrolyte.

I.11 THE INDEPENDENT MIGRATION OF IONS. CALCULATION OF CONDUCTIVITIES FROM IONIC MOBILITIES For strong electrolytes the limiting value of the molar conductivity, Λ_0, may be determined by extending the measurements to low concentrations and then extrapolating the graph of conductivity against concentration to zero concentration. For weak electrolytes, such as acetic acid and ammonia, this method cannot be employed, since the dissociation is far from complete at the lowest concentrations at which measurements can be conveniently made ($\sim 10^{-4}$M). It is however possible to calculate these limiting conductances on the basis of the law of independent migration of ions.

As a result of prolonged and careful study of the conductance of salt solutions down to low concentrations, Kohlrausch found that the difference in molar conductivities of pairs of salts, containing similar anions and always the same two cations, is constant and independent of the nature of the anion. He found for example that the following differences of limiting molar conductivities (measured at 18°C in $cm^2\ \Omega^{-1}\ mol^{-1}$ units)

$$\Lambda_0(KCl) - \Lambda_0(NaCl) = 130{\cdot}1 - 109{\cdot}0 = 21{\cdot}1$$
$$\Lambda_0(KNO_3) - \Lambda_0(NaNO_3) = 126{\cdot}3 - 105{\cdot}3 = 21{\cdot}0$$

are very nearly equal. From these and similar results, Kohlrausch drew the conclusion that the molar conductivity of an electrolyte is made up as the sum of the conductivities of the component ions. Mathematically this can be expressed as

$$\Lambda_0 = \lambda_0^+ + \lambda_0^-$$

where λ_0^+ and λ_0^- are the limiting molar conductivities or **mobilities** of the cation and anion respectively. The ionic mobilities are computed from values of Λ_0 with the aid of **transference numbers**. These represent the current carried by the cation and anion respectively, and can be determined experimentally from the difference of concentration of electrolytes between the bulk of the solution and parts of the solution close to the cathode and anode.* Thus, for example, the transference number of chloride ion in a potassium chloride solution was found to be 0·503, while that of potassium is 0·497 (the sum of transference numbers for one particular electrolyte is by definition equal to one). The limiting value of the molar conductivity of potassium chloride solution (at 18°C) is 130·1 $cm^2\ \Omega^{-1}\ mol^{-1}$. Thus the mobility of the potassium ion is

$$\lambda_0^+(K^+) = 0{\cdot}497 \times 130{\cdot}1 = 64{\cdot}6\ cm^2\ \Omega^{-1}\ mol^{-1}$$

and that of the chloride ion is

$$\lambda_0^-(Cl^-) = 0{\cdot}503 \times 130{\cdot}1 = 65{\cdot}5\ cm^2\ \Omega^{-1}\ mol^{-1}$$

Table I.4 Limiting ionic mobilities at 18°C and 25°C in $cm^2\ \Omega^{-1}\ mol^{-1}$ units

18°C				25°C			
H^+	317·0	OH^-	174·0	H^+	348·0	OH^-	210·8
Na^+	43·5	Cl^-	65·5	Na^+	49·8	Cl^-	76·4
K^+	64·6	NO_3^-	61·8	K^+	73·4	IO_3^-	42·0
Ag^+	54·4	Br^-	67·7	Ag^+	61·9	CH_3COO^-	40·6
1/2 Ca^{2+}	52·2	I^-	66·1				
1/2 Sr^{2+}	51·7	F^-	46·8				
1/2 Ba^{2+}	55·0	ClO_3^-	55·0				
1/2 Pb^{2+}	61·6	IO_3^-	34·0				
1/2 Cd^{2+}	46·5	CH_3COO^-	32·5				
1/2 Zn^{2+}	46·0	1/2 SO_4^{2-}	68·3				
1/2 Cu^{2+}	45·9	1/2 $(COO)_2^{2-}$	61·1				

A selected number of ionic mobilities at 18°C and 25°C is shown in Table I.4. This table may be utilized for the calculation of the limiting molar conductivities of any electrolytes made up of the ions listed. Thus, for acetic acid at 25°C

$$\Lambda_0(CH_3COOH) = \lambda_0^+(H^+) + \lambda_0^-(CH_3COO^-)$$
$$= 348{\cdot}0 + 40{\cdot}6$$
$$= 388{\cdot}6\ cm^2\ \Omega^{-1}\ mol^{-1}$$

* For a more detailed discussion of transference numbers textbooks of physical chemistry should be consulted (cf.: Walter J. Moore's *Physical Chemistry*. 4th edn., Longman 1966, p. 333 et f).

The degree of dissociation can be calculated from the relation

$$\alpha = \frac{\Lambda_c}{\Lambda_0}$$

where Λ_c is the molar conductivity at the concentration c; this can be measured experimentally.

I.12 MODERN THEORY OF STRONG ELECTROLYTES The theory of electrolytic dissociation can be used to explain a great number of phenomena which are important in inorganic qualitative analysis. The theory, as put forward by Arrhenius, can be applied without much alteration as far as weak electrolytes are concerned but as further evidence – particularly of the structure of matter in the solid state – emerged, it became less and less adequate for strong electrolytes. It became clear that substances which are classified as strong electrolytes are made up of ions even in the solid (crystalline) form. In a crystal of sodium chloride, for example, there are no sodium chloride molecules present, (such molecules exist only in the sodium chloride vapour). The crystal is built up of sodium and chloride ions, arranged in a cubic lattice, one sodium ion being

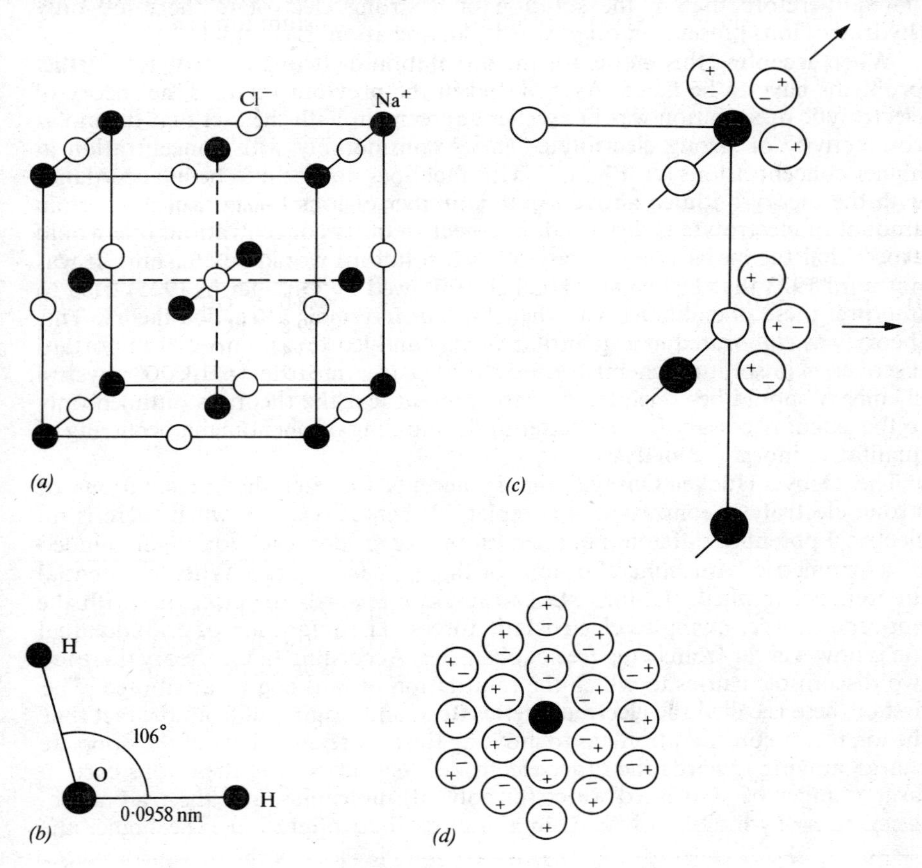

Fig. I.4

surrounded always by six chloride atoms and vice versa (see Fig. I.4a). As the ions are there in the solid state, it is incorrect to suggest that at dissolution 'molecules' dissociate into ions. The dissolution of an ionic crystal in water is a physical process.

It has been proved that in the water molecule the two hydrogen and the oxygen atoms are arranged in a triangle with a distance of 0·0958 nm between the centres of the hydrogen and oxygen atoms, and with an angle of 106° between the directions of the two hydrogen atoms (cf. Fig. I.4b). Because of this arrangement, that side of the water molecule containing the hydrogen atoms, becomes electrostatically positive, while the opposite end, where the oxygen atom is, becomes negative. Thus, the water molecule has a **dipole character**. If an ionic crystal is placed in water, these dipoles will orientate around the ions present in the outer layer of the lattice. Electrostatic force will tend to pull these ions away from the crystal (see Fig. I.4c). When an ion has been removed from the lattice, a symmetrical sphere of water molecules will orientate around it, and the whole **hydrated** ion with its sphere of water molecules will be taken away from the crystal by thermal motion. A new ion will thus be exposed to the action of water molecules, and slowly the whole crystal will dissolve. It can be stated therefore that in the solution of a strong electrolyte there are only (hydrated) ions present; in other words, 'dissociation' is complete.

When accepting this model for the dissolution of strong electrolytes, further problems have to be faced. As was said in the previous section, the theory of electrolytic dissociation was in excellent agreement with the fact that the molar conductivity of strong electrolytes varies considerably with concentration at higher concentrations (cf. Fig. I.3). This fact does not seem to be in accordance with the theory outlined above. As the number of ions is constant if a certain amount of electrolyte is dissolved, irrespective of its concentration, one would expect that the molar conductivity of such solutions would be constant. It was not until 1923 that Debye and Hückel, (followed by Onsager in 1925) tried to interpret these phenomena with their famous **interionic attraction theory**. This theory was elaborated in a quantitative way and led to a number of important discoveries in solution chemistry. For a fuller treatment textbooks of physical chemistry should be consulted.* In the present text the theory is outlined only to the extent necessary for the better understanding of phenomena occurring in qualitative inorganic analysis.

The Debye–Hückel–Onsager theory accepts the fact that in solutions of strong electrolytes ionization is complete. When at rest, i.e. when there is no electrical potential difference applied on the electrodes, each ion is surrounded by a symmetric 'atmosphere' of ions of the opposite charge. When a potential difference is applied, the ions start to migrate towards the electrode with the opposite charge, owing to electrostatic forces. The migration of an individual ion is however far from being free of obstacles. According to the theory there are two discernible causes to which the retardation of ions can be attributed. The first of these is called the **electrophoretic effect**, and originates from the fact that the ion under consideration has to move against a stream of ions of the opposite charge moving towards the other electrode. As said before, these ions carry a large number of associated water (or solvent) molecules, and the friction between these hydrated (solvated) ions retards their migration. The higher the

* Cf. Walter J. Moore's *Physical Chemistry*. 4th edn., Longman 1966, p. 359 et f.

concentration, the nearer are these ions to each other and the more pronounced is this effect. The second, the **asymmetry** or (**relaxation**) effect, is the result of the breakdown of the symmetrical atmosphere of oppositely charged ions around the ion in question. As soon as the ion starts to move towards the particular electrode, it leaves, the centre of the sphere of its ionic atmosphere, leaving behind more ions belonging to the original sphere. For a while at least an unsymmetrical distribution of ions will develop, those ions, which are 'left behind' will electrostatically attract the ion in question. As this force is exerted in just the opposite direction to that of motion, the migration of the ion is slowed down. This effect is the more pronounced the more concentrated the solution. If the electric circuit is broken, it takes some time for the arrangement of ions to become symmetrical again (in other words, for relaxation to be complete), and this time, called the **time of relaxation**, can be expressed mathematically as the function of measurable parameters of the solution. It is inversely proportional to the concentration.

The change of conductivity with dilution is therefore not attributed to changes in the degree of dissociation, as suggested by Arrhenius, but to the variation of the interionic forces outlined above. The molar conductivity, at a concentration c, can be expressed by the (simplified) equation:

$$\Lambda_c = \Lambda_0 - (A + B\Lambda_0)\sqrt{c}$$

where Λ_0 is the limiting value of the molar conductivity at zero concentration. A and B are constants (for a particular ion in a particular solvent at constant temperature) and correspond to the asymmetry and electrophoretic effects respectively. The great merits of the Debye–Hückel–Onsager theory can be judged from the fact that both A and B can be expressed with measurable parameters of the solution and with some natural constants, and that conductivities calculated with this equation agree well with the experimental values, especially if concentrations are not too high.

The ratio Λ_c/Λ_0 in the modern theory of strong electrolytes, based on complete ionization, no longer gives the degree of dissociation α for a strong electrolyte (for which it should be equal to unity); it is more proper therefore to call it the **conductivity coefficient** or **conductance ratio**. It does give the approximate degree of dissociation for weak electrolytes, but even here there are interionic forces contributing towards a lessening of the conductivity, and a correction may be applied with the aid of the Debye–Hückel–Onsager theory.

I.13 CHEMICAL EQUILIBRIUM; THE LAW OF MASS ACTION One of the most important facts about chemical reactions is that all chemical reactions are reversible. Whenever a chemical reaction is initiated, reaction products are starting to build up, and these in turn will react with each other starting a reverse reaction. After a while a dynamic equilibrium is reached; that is as many molecules (or ions) of each substance are decomposed as are formed in unit time. In some cases this equilibrium is almost completely on the side of formation of one or another substance, and the reaction thus seems to proceed until it becomes complete. In other cases it might be the experimenter's task to create the circumstances under which the reaction, which otherwise would reach an equilibrium, will become complete. This is often the case in quantitative analysis.

The conditions of chemical equilibrium can most easily be derived from the law of mass action.* This law was stated originally by Guldberg and Waage in 1867 in the following form: the velocity of a chemical reaction at constant temperature is proportional to the product of the concentrations of the reacting substances. Let us consider first a simple reversible reaction at constant temperature:

$$A + B \rightleftarrows C + D$$

The velocity with which A and B react is proportional to their concentrations, or

$$v_1 = k_1 \times [A] \times [B]$$

where k_1 is a constant known as the rate constant and the square brackets indicate the molar concentration of the substance enclosed within the brackets. Similarly, the velocity with which the reverse process occurs is given by

$$v_2 = k_2 \times [C] \times [D]$$

At equilibrium the velocities of the reverse and forward reactions are equal (the equilibrium is a dynamic and not a static one) and therefore

$$v_1 = v_2$$

or

$$k_1 \times [A] \times [B] = k_2 \times [C] \times [D]$$

By rearranging we can write

$$\frac{[C] \times [D]}{[A] \times [B]} = \frac{k_1}{k_2} = K$$

K is the **equilibrium constant** of the reaction. Its value is independent of the concentrations of the species involved; it varies slightly with temperature and pressure.

The expression may be generalized for more complex reactions. For a reversible reaction represented by the equation

$$v_A A + v_B B + v_C C + \cdots \rightleftarrows v_L L + v_M M + v_N N$$

where v_A, v_B ... etc. are the stoichiometric numbers of the reaction, the equilibrium constant can be expressed as:

$$K = \frac{[L]^{v_L} \times [M]^{v_M} \times [N]^{v_N} \ldots}{[A]^{v_A} \times [B]^{v_B} \times [C]^{v_C} \ldots}$$

Expressed in words: When equilibrium is reached in a reversible reaction at constant temperature and pressure, the product of the molecular concentrations of the resultants (the substances on the right-hand side of the equation), divided by the product of the molecular concentrations of the reactants (the substances on the left-hand side of the equation), each concentration being raised to the power equal to the number of species of that substance taking part in the reaction, is constant.

---

* It must be emphasized that the conditions of chemical equilibrium can be derived and explained most exactly on the basis of thermodynamics, that is without involving reaction rates at all. Textbooks of physical chemistry will of course contain the thermodynamical interpretation (cf. W. J. Moore's *Physical Chemistry*. 4th edn., Longman 1966, p. 167 et f.)

The expression for the equilibrium constant given above gives us the clue to the problem one often comes across in qualitative analysis: what to do in order to make a reaction complete, in other words, to shift a chemical equilibrium in a desired direction. To examine this problem, let us consider the reaction of arsenate ions with iodide. If solutions of sodium arsenate, potassium iodide, and hydrochloric acid are mixed, the solution turns yellow or brown, owing to the formation of iodine. The reaction proceeds between the various ions present, arsenite ions and water being formed simultaneously, and can be expressed with the equation

$$AsO_4^{3-} + 2I^- + 2H^+ \rightleftarrows AsO_3^{3-} + I_2 + H_2O$$

Sodium, potassium, and chloride ions, added with the reagents, do not take part in the reaction, and are therefore not included in the equation. This reaction is reversible and leads to an equilibrium. Applying the law of mass action, we can express the equilibrium constant of the reaction as

$$K = \frac{[AsO_3^{3-}] \times [I_2] \times [H_2O]}{[AsO_4^{3-}] \times [I^-]^2 \times [H^+]^2}$$

Let us suppose we want to reduce all the arsenate to arsenite, that is we want to shift the equilibrium towards the right-hand side of the equation. We can do this in several ways. If we add for example more hydrochloric acid to the solution, we can observe that the yellowish-brown colour deepens, that is more iodine is formed. The explanation of this is obvious from the expression for the equilibrium constant. When adding hydrochloric acid, we increased the hydrogen-ion concentration of the solution; thus increasing the denominator in the expression for the equilibrium constant. The equilibrium constant must remain constant, and therefore the numerator of the expression must increase as well. This can be achieved only by the increase of the individual concentrations in the numerator, which means that more arsenite, iodine, and water must be formed. In turn this means that the equilibrium has shifted towards the right-hand side. The same will happen if we add more potassium iodide to the solution. There are other ways however to achieve the same object. We can for example remove the iodine formed during the reaction by evaporation or by extraction in water-immiscible solvent. In this case the numerator of the expression decreases, and in order to keep K constant, the denominator must decrease also. This again means that more reactants are used up (and more products are formed). Generally, we can say that a **chemical equilibrium** at constant temperature and pressure *can be shifted towards the formation of the products either by adding more reactants, or by removing one of the products from the (homogeneous) equilibrium system.* In terms of reactions used in qualitative analysis this means either the addition of reagents in excess, or the removal of reaction products from the solution phase, by some means such as precipitation, evaporation, or extraction.

From the argument above it follows that opposite action will shift the equilibrium in the opposite direction. Thus, for example, adding more iodine to the equilibrium system, or removing some of the hydrogen ions with a buffer, or removing iodide ions by precipitating them with lead nitrate in the form of lead iodide will shift the equilibrium towards the formation of arsenate.

A different way of shifting equilibria towards one or another direction is based on the fact that the equilibrium constant depends on temperature and, at least in some cases, on pressure. Heating is often applied when performing qualitative analyses, though mainly in order to speed up the reactions (that is to influence kinetics) rather than to influence the conditions of equilibrium. In some cases cooling to low temperatures may achieve the object. For example, the equilibrium constant of the reaction

$$Pb^{2+} + 2I^- \rightleftarrows PbI_2(s)$$

varies with temperature in such a way that at lower temperatures the formation of lead iodide is favoured. Thus, if we want to precipitate iodide quantitatively with lead, besides adding the reagent in excess, we should perform the reaction in the cold. The equilibrium of those reactions in solutions in which some of the reactants or products are gases, may be influenced by varying the pressure above the solution. Calcium carbonate precipitate for example can be dissolved by introducing carbon dioxide gas in a closed vessel until the pressure in the vessel increases to a few atmospheres, when the equilibrium

$$CaCO_3(s) + CO_2(g) + H_2O \rightleftarrows Ca^{2+} + 2HCO_3^-$$

shifts towards the formation of calcium hydrogen carbonate. For the same reason, hydrogen sulphide gas is more effective if added at a slightly increased pressure, than when bubbled through a solution in an open vessel. If, on the other hand, the product of the reaction is a gas, the equilibrium can be shifted easily towards the formation of the products by removing the gas at a reduced pressure.

I.14 ACTIVITY AND ACTIVITY COEFFICIENTS In our deduction of the law of mass action we used the concentrations of species as variables, and deduced that the value of the equilibrium constant is independent of the concentrations themselves. More thorough investigations however showed that this statement is only approximately true for dilute solutions (the approximation being the better, the more dilute are the solutions), and in more concentrated solutions it is not correct at all. Similar discrepancies arise when other thermodynamic quantities, notably electrode potentials or chemical free energies are dealt with. To overcome these difficulties, and still to retain the simple expressions derived for such quantities, G. N. Lewis introduced a new thermodynamic quantity, termed **activity**, which when applied instead of concentrations in these thermodynamic functions, provides an exact fit with experimental results. This quantity has the same dimensions as concentration. The activity, a_A, of a species A is proportional to its actual concentration [A], and can be expressed as

$$a_A = f_A \times [A]$$

Here f_A is the **activity coefficient**, a dimensionless quantity, which varies with concentration. For the simple equilibrium reaction, mentioned in Section **I.13**

$$A + B \rightleftarrows C + D$$

the equilibrium constant can be expressed more precisely as

$$K = \frac{a_C \times a_D}{a_A \times a_B} = \frac{f_C[C] \times f_D[D]}{f_A[A] \times f_B[B]} = \frac{f_C \times f_D}{f_A \times f_B} \times \frac{[C] \times [D]}{[A] \times [B]}$$

Activities, and thus activity coefficients, must be raised to appropriate powers, just like concentrations, if the stoichiometric numbers differ from 1. Thus, for the general reversible reaction, dealt with in Section **I.13**:

$$\nu_A A + \nu_B B + \nu_C C + \cdots \rightleftarrows \nu_L L + \nu_M M + \nu_N N + \cdots$$

the equilibrium constant should be expressed more precisely as

$$K = \frac{a_L^{\nu_L} \times a_M^{\nu_M} \times a_N^{\nu_N} \times \cdots}{a_A^{\nu_A} \times a_B^{\nu_B} \times a_C^{\nu_C} \times \cdots}$$
$$= \frac{(f_L[L])^{\nu_L} \times (f_M[M])^{\nu_M} \times (f_N[N])^{\nu_N} \times \cdots}{(f_A[A])^{\nu_A} \times (f_B[B])^{\nu_B} \times (f_C[C])^{\nu_C} \times \cdots}$$
$$= \frac{f_L^{\nu_L} \times f_M^{\nu_M} \times f_N^{\nu_N} \times \cdots}{f_A^{\nu_A} \times f_B^{\nu_B} \times f_C^{\nu_C} \times \cdots} \times \frac{[L]^{\nu_L} \times [M]^{\nu_M} \times [N]^{\nu_N} \times \cdots}{[A]^{\nu_A} \times [B]^{\nu_B} \times [C]^{\nu_C} \times \cdots}$$

The activity coefficient varies with concentration. This variation is rather complex; the activity coefficient of a particular ion being dependent upon the concentration of all ionic species present in the solution. As a measure of the latter, Lewis and Randall (1921) introduced the quantity called **ionic strength**, I, and defined it as the half sum of the products of the concentration of each ion multiplied by the square of its charge. With mathematical symbols this can be expressed as

$$I = \tfrac{1}{2} \Sigma c_i z_i^2$$

where c_i is the concentration of the ith component, and z_i is its charge. Thus, if a solution is 0·1 molar for nitric acid and 0·2 molar for barium nitrate, the concentrations of each ion being

$$c_{H^+} = 0{\cdot}1 \text{ mol } \ell^{-1}$$
$$c_{Ba^{2+}} = 0{\cdot}2 \text{ mol } \ell^{-1}$$
$$c_{NO_3^-} = 0{\cdot}3 \text{ mol } \ell^{-1}$$

and the charges

$$z_{H^+} = 1$$
$$z_{Ba^{2+}} = 2$$
$$z_{NO_3^-} = 1$$

for the same ions respectively, the ionic strength of the solution will be

$$I = \tfrac{1}{2}(c_{H^+} z_{H^+}^2 + c_{Ba^{2+}} z_{Ba^{2+}}^2 + c_{NO_3^-} z_{NO_3^-}^2)$$
$$= \tfrac{1}{2}(0{\cdot}1 \times 1 + 0{\cdot}2 \times 4 + 0{\cdot}3 \times 1) = 0{\cdot}6$$

The correlation between activity coefficient and ionic strength can be deduced from the quantitative relationships of the Debye–Hückel–Onsager theory. Without giving details of this deduction* it is interesting to quote the final result:

$$\log f_i = -0{\cdot}43 e^3 N^2 \sqrt{\frac{2\pi\rho_0}{1000 R^3 \varepsilon^3 T^3}} \times z_i^2 \sqrt{I}$$

In this expression e is the charge of the electron, N the Avogadro constant, R the gas constant, ρ_0 the density of the solvent, ε the dielectric constant of the solvent,

* For details textbooks of physical chemistry should be consulted. Cf. W. J. Moore's *Physical Chemistry*. 4th edn., Longman 1966, p. 354 et f.

and T the absolute temperature. Though the expression is rather complicated, it is worth noting that it contains natural constants and some easily measurable physical quantities. This shows the great merit of the Debye–Hückel–Onsager theory: it is able to provide a quantitative picture of the characteristics of electrolyte solutions.

Inserting the adequate values of constants and physical quantities for dilute aqueous solutions at room temperatures (T = 298K) the expression can be

Table I.5 Mean activity coefficients of various electrolytes

Molar concentration	0·001	0·01	0·05	0·1	0·2	0·5	1·0	2·0
HCl	0·966	0·904	0·830	0·796	0·767	0·758	0·809	1·01
HBr	0·966	0·906	0·838	0·805	0·782	0·790	0·871	1·17
HNO_3	0·965	0·902	0·823	0·785	0·748	0·715	0·720	0·78
HIO_3	0·96	0·86	0·69	0·58	0·46	0·29	0·19	0·10
H_2SO_4	0·830	0·544	0·340	0·265	0·209	0·154	0·130	0·12
NaOH	—	—	0·82	—	0·73	0·69	0·68	0·70
KOH	—	0·90	0·82	0·80	—	0·73	0·76	0·89
$Ba(OH)_2$	—	0·712	0·526	0·443	0·370	—	—	—
$AgNO_3$	—	0·90	0·79	0·72	0·64	0·51	0·40	0·28
$Al(NO_3)_3$	—	—	—	0·20	0·16	0·14	0·19	0·45
$BaCl_2$	0·88	0·72	0·56	0·49	0·44	0·39	0·39	0·44
$Ba(NO_3)_2$	0·88	0·71	0·52	0·43	0·34	—	—	—
$CaCl_2$	0·89	0·73	0·57	0·52	0·48	0·52	0·71	—
$Ca(NO_3)_2$	0·88	0·71	0·54	0·48	0·42	0·38	0·35	0·35
$CdCl_2$	0·76	0·47	0·28	0·21	0·15	0·09	0·06	—
$CdSO_4$	0·73	0·40	0·21	0·17	0·11	0·07	0·05	0·04
$CuCl_2$	0·89	0·72	0·58	0·52	0·47	0·42	0·43	0·51
$CuSO_4$	0·74	0·41	0·21	0·16	0·11	0·07	0·05	—
$FeCl_2$	0·89	0·75	0·62	0·58	0·55	0·59	0·67	—
KF	—	0·93	0·88	0·85	0·81	0·74	0·71	0·70
KCl	0·965	0·901	0·815	0·769	0·719	0·651	0·606	0·576
KBr	0·965	0·903	0·822	0·777	0·728	0·665	0·625	0·602
KI	0·965	0·905	0·84	0·80	0·76	0·71	0·68	0·69
$KClO_3$	0·967	0·907	0·813	0·755	—	—	—	—
$KClO_4$	0·965	0·895	0·788	—	—	—	—	—
K_2SO_4	0·89	0·71	0·52	0·43	0·36	—	—	—
$K_4Fe(CN)_6$	—	—	0·19	0·14	0·11	0·67	—	—
LiBr	0·966	0·909	0·842	0·810	0·784	0·783	0·848	1·06
$Mg(NO_3)_2$	0·88	0·71	0·55	0·51	0·46	0·44	0·50	0·69
$MgSO_4$	—	0·40	0·22	0·18	0·13	0·09	0·06	0·05
NH_4Cl	0·96	0·88	0·79	0·74	0·69	0·62	0·57	—
NH_4Br	0·96	0·87	0·78	0·73	0·68	0·62	0·57	—
NH_4I	0·96	0·89	0·80	0·76	0·71	0·65	0·60	—
NH_4NO_3	0·96	0·88	0·78	0·73	0·66	0·56	0·47	—
$(NH_4)_2SO_4$	0·87	0·67	0·48	0·40	0·32	0·22	0·16	—
NaF	—	0·90	0·81	0·75	0·69	0·62	—	—
NaCl	0·966	0·904	0·823	0·780	0·730	0·68	0·66	0·67
NaBr	0·966	0·914	0·844	0·800	0·740	0·695	0·686	0·734
NaI	0·97	0·91	0·86	0·83	0·81	0·78	0·80	0·95
$NaNO_3$	0·966	0·90	0·82	0·77	0·70	0·62	0·55	0·48
Na_2SO_4	0·89	0·71	0·53	0·45	0·36	0·27	0·20	—
$NaClO_4$	0·97	0·90	0·82	0·77	0·72	0·64	0·58	—
$Pb(NO_3)_2$	0·88	0·69	0·46	0·37	0·27	0·17	0·11	—
$ZnCl_2$	0·88	0·71	0·56	0·50	0·45	0·38	0·33	—
$ZnSO_4$	0·70	0·39	—	0·15	0·11	0·07	0·05	0·04

simplified to

$$\log f_i = -0{\cdot}509 z_i^2 \sqrt{I}$$

For the mean activity coefficient of a salt the expression

$$\log f = -0{\cdot}509 z_+ z_- \sqrt{I}$$

is valid, where z_+ and z_- are the charges of the cation and anion respectively. The expression is applicable to solutions of low ionic strengths (up to $I = 0{\cdot}01$) in the strictest sense.

A number of mean activity coefficients are collected in Table I.5. Activity coefficients generally first decrease with increasing concentrations, then, after passing through a minimum, rise again, often exceeding the value 1. This is illustrated well on the diagrams of Fig. I.5 where activity coefficients of some electrolyte solutions are plotted against the square root of concentration.

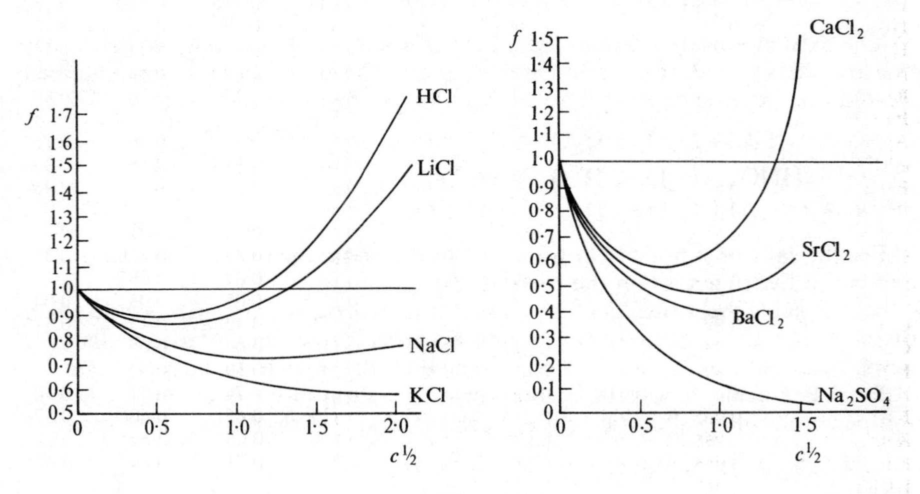

Fig. I.5

Activity coefficients of non-ionized molecules do not differ appreciably from unity. In dilute solutions of weak electrolytes the differences between activities and concentrations (calculated from the degree of dissociation) is negligible.

From all that has been said about activity and activity coefficients, it is apparent that whenever precise results are to be expected, activities should be used when expressing equilibrium constants or other thermodynamic functions. In the present text however we shall be using simply concentrations. For the dilute solutions of strong and weak electrolytes that are mainly used in qualitative analysis, errors introduced into calculations are not considerable.

C. CLASSICAL THEORY OF ACID-BASE REACTIONS

I.15 ACIDS, BASES, AND SALTS Inorganic substances can be classified into three important groups: acids, bases, and salts.

An **acid** is most simply defined as a substance which, when dissolved in water, undergoes dissociation with the formation of hydrogen ions as the only positive ions. Some acids and their dissociation products are as follows:

$$HCl \rightleftarrows H^+ + Cl^-$$
hydrochloric acid — chloride ion

$$HNO_3 \rightleftarrows H^+ + NO_3^-$$
nitric acid — nitrate ion

$$CH_3COOH \rightleftarrows H^+ + CH_3COO^-$$
acetic acid — acetate ion

Actually, hydrogen ions (protons) do not exist in aqueous solutions. Each proton combines with one water molecule by coordination with a free pair of electrons on the oxygen of water, and hydronium ions are formed:

$$H^+ + H_2O \rightarrow H_3O^+$$

The existence of hydronium ions, both in solutions and in the solid state has been proved by modern experimental methods. The above dissociation reactions should therefore be expressed as the reaction of the acids with water:

$$HCl + H_2O \rightleftarrows H_3O^+ + Cl^-$$
$$HNO_3 + H_2O \rightleftarrows H_3O^+ + NO_3^-$$
$$CH_3COOH + H_2O \rightleftarrows H_3O^+ + CH_3COO^-$$

For the sake of simplicity however we shall denote the hydronium ion by H^+ and call it hydrogen ion in the present text.

All the acids mentioned so far produce one hydrogen ion per molecule when dissociating; these are termed **monobasic acids**. Other monobasic acids are: perchloric acid ($HClO_4$), hydrobromic acid (HBr), hydriodic acid (HI) etc.

Polybasic acids dissociate in more steps, yielding more than one hydrogen ion per molecule, Sulphuric acid is a dibasic acid and dissociates in two steps:

$$H_2SO_4 \rightleftarrows H^+ + HSO_4^-$$
$$HSO_4^- \rightleftarrows H^+ + SO_4^{2-}$$

yielding hydrogen sulphate ions and sulphate ions after the first and second step respectively. Phosphoric acid is tribasic:

$$H_3PO_4 \rightleftarrows H^+ + H_2PO_4^-$$
$$H_2PO_4^- \rightleftarrows H^+ + HPO_4^{2-}$$
$$HPO_4^{2-} \rightleftarrows H^+ + PO_4^{3-}$$

The ions formed after the first, second, and third dissociation step, are termed dihydrogen phosphate, (mono)hydrogen phosphate, and phosphate ions respectively.

The degree of dissociation differs from acid to acid. **Strong acids** dissociate almost completely at medium dilutions (cf. Section **I.10**), these are therefore strong electrolytes. Strong acids are: hydrochloric, nitric, perchloric acid, etc. Sulphuric acid is a strong acid as far as the first dissociation step is concerned, but the degree of dissociation in the second step is smaller. **Weak acids** dissociate only slightly at medium or even low concentrations (at which, for example, they are applied as analytical reagents). Weak acids are therefore weak electrolytes.

Acetic acid is a typical weak acid; other weak acids are boric acid (H_3BO_3), even as regards the first dissociation step, carbonic acid (H_2CO_3) etc. Phosphoric acid can be termed as a medium strong acid on the basis of the degree of the first dissociation; the degree of the second dissociation is smaller, and smallest is that of the third dissociation. There is however, no sharp division between these classes. As we will see later (cf. Section **I.16**) it is possible to express the strength of acids and bases quantitatively.

A **base** can be most simply defined as a substance which, when dissolved in water, undergoes dissociation with the formation of hydroxyl ions as the only negative ions. Soluble metal hydroxides, like sodium hydroxide or potassium hydroxide are almost completely dissociated in dilute aqueous solutions:

$$NaOH \rightleftarrows Na^+ + OH^-$$
$$KOH \rightleftarrows K^+ + OH^-$$

These are therefore **strong bases**. Aqueous ammonia solution, on the other hand, is a **weak base**. When dissolved in water, ammonia forms ammonium hydroxide, which dissociates to ammonium and hydroxide ions:

$$NH_3 + H_2O \rightleftarrows NH_4OH \rightleftarrows NH_4^+ + OH^-$$

It is however more correct to write the reaction as

$$NH_3 + H_2O \rightleftarrows NH_4^+ + OH^-$$

Strong bases are therefore strong electrolytes, while weak bases are weak electrolytes. There is however no sharp division between these classes, and, as in the case of acids, it is possible to express the strength of bases quantitatively.

According to the historic definition, **salts** are the products of reactions between acids and bases. Such processes are called **neutralization** reactions. This definition is correct in the sense that if equivalent amounts of pure acids and bases are mixed, and the solution is evaporated, a crystalline substance remains, which has the characteristics neither of an acid nor of a base. These substances were termed salts by the early chemists. If reaction equations are expressed as the interaction of molecules,

$$\underset{\text{acid}}{HCl} + \underset{\text{base}}{NaOH} \rightarrow \underset{\text{salt}}{NaCl} + H_2O$$

the formation of the salt seems to be the result of a genuine chemical process. In fact however this explanation is incorrect. We know that both the (strong) acid and the (strong) base as well as the salt (cf. Section **I.10**) are almost completely dissociated in the solution, viz.

$$HCl \rightleftarrows H^+ + Cl^-$$
$$NaOH \rightleftarrows Na^+ + OH^-$$
$$NaCl \rightleftarrows Na^+ + Cl^-$$

while the water, also formed in the process, is almost completely undissociated. It is more correct therefore to express the neutralization reaction as the chemical combination of ions:

$$H^+ + Cl^- + Na^+ + OH^- \rightarrow Na^+ + Cl^- + H_2O$$

In this equation, Na^+ and Cl^- ions appear on both sides. As nothing has therefore happened to these ions, the equation can be simplified to

$$H^+ + OH^- \rightarrow H_2O$$

showing that the essence of any acid-base reaction (in aqueous solution) is the formation of water. This is indicated by the fact, among others, that the heat of neutralization is approximately the same (56·9 kJ) for the reaction of one mole of any monovalent strong acid and base. The salt in the solid state is built up of ions, arranged in a regular pattern in the crystal lattice. Sodium chloride, for instance, is built up of sodium ions and chloride ions so arranged that each ion is surrounded symmetrically by six ions of the opposite sign; the crystal lattice is held together by electrostatic forces due to the charges of the ions (cf. Fig. I.4).

Amphoteric substances, or ampholytes, are able to engage in neutralization reactions both with acids and bases (more precisely, both with hydrogen and hydroxyl ions). Aluminium hydroxide, for example, reacts with strong acids, when it dissolves and aluminium ions are formed:

$$Al(OH)_3(s) + 3H^+ \rightarrow Al^{3+} + 3H_2O$$

In this reaction aluminium hydroxide acts as a base. On the other hand, aluminium hydroxide can also be dissolved in sodium hydroxide:

$$Al(OH)_3(s) + OH^- \rightarrow [Al(OH)_4]^-$$

when tetrahydroxoaluminate ions are formed. In this reaction aluminium hydroxide behaves as an acid. The amphoteric behaviour of certain metal hydroxides is often utilized in qualitative inorganic analysis, notably in the separation of the cations of the third group.

I.16 ACID-BASE DISSOCIATION EQUILIBRIA. STRENGTH OF ACIDS AND BASES The dissociation of an acid (or a base) is a reversible process to which the law of mass action can be applied. The dissociation of acetic acid, for example, yields hydrogen and acetate ions:

$$CH_3COOH \rightleftarrows CH_3COO^- + H^+$$

Applying the law of mass action to this reversible process, we can express the equilibrium constant as

$$K = \frac{[H^+][CH_3COO^-]}{[CH_3COOH]}$$

The constant K is termed the **dissociation equilibrium constant** or simply the **dissociation constant**. Its value for acetic acid is $1{\cdot}76 \times 10^{-5}$ at 25°C.

In general, if the dissociation of a **monobasic acid** HA takes place according to the equilibrium

$$HA \rightleftarrows H^+ + A^-$$

the dissociation equilibrium constant can be expressed as

$$K = \frac{[H^+] \times [A^-]}{[HA]}$$

The stronger the acid, the more it dissociates, hence the greater is the value of K, the dissociation equilibrium constant.

Dibasic acids dissociate in two steps, and both dissociation equilibria can be characterized by separate dissociation equilibrium constants. The dissociation of the dibasic acid H_2A can be represented by the following two equilibria:

$$H_2A \rightleftarrows H^+ + HA^-$$
$$HA^- \rightleftarrows H^+ + A^{2-}$$

Applying the law of mass action to these processes we can express the two dissociation equilibrium constants as

$$K_1 = \frac{[H^+] \times [HA^-]}{[H_2A]}$$

and

$$K_2 = \frac{[H^+] \times [A^{2-}]}{[HA^-]}$$

K_1 and K_2 are termed first and second dissociation equilibrium constants respectively. It must be noted that $K_1 > K_2$, that is the first dissociation step is always more complete than the second.

A tribasic acid H_3A dissociates in three steps:

$$H_3A \rightleftarrows H^+ + H_2A^-$$
$$H_2A^- \rightleftarrows H^+ + HA^{2-}$$
$$HA^{2-} \rightleftarrows H^+ + A^{3-}$$

and the three dissociation equilibrium constants are

$$K_1 = \frac{[H^+][H_2A]}{[H_3A]},$$
$$K_2 = \frac{[H^+][HA^{2-}]}{[H_2A^-]}$$
$$K_3 = \frac{[H^+][A^{3-}]}{[HA^{2-}]}$$

for the first, second, and third steps respectively. It must again be noted that $K_1 > K_2 > K_3$, that is the first dissociation step is the most complete, while the third is the least complete.

Similar considerations can be applied to **bases**. Ammonium hydroxide (i.e. the aqueous solution of ammonia) dissociates according to the equation:

$$NH_4OH \rightleftarrows NH_4^+ + OH^-$$

The dissociation equilibrium constant can be expressed as

$$K = \frac{[NH_4^+] \times [OH^-]}{[NH_4OH]}$$

The actual value of this dissociation constant is $1{\cdot}79 \times 10^{-5}$ (at 25°C). Generally, if a monovalent base BOH* dissociates as

$$BOH \rightleftarrows [B^+]+[OH^-]$$

the dissociation equilibrium constant can be expressed as

$$K = \frac{[B^+]\times[OH^-]}{[BOH]}$$

It can be said again that the stronger the base the better it dissociates, and therefore the larger the value of the dissociation equilibrium constant.

The exponent of the dissociation equilibrium constant, called *pK*, is defined by the equations

$$pK = -\log K = \log \frac{1}{K}$$

Its value is often quoted instead of that of *K*. The usefulness of *pK* will become apparent when dealing with the hydrogen-ion exponent or *p*H.

We saw already that the value of the dissociation constant is correlated with the degree of dissociation, and so with the strength of the acid or base. The degree of dissociation depends on the concentration, and therefore it cannot be used to characterize the strength of the acid or base without stating the circumstances under which it is measured. The value of the dissociation equilibrium constant, on the other hand, is independent of the concentration (more precisely of the activity) of the acid, and therefore provides the most adequate quantitative measure of the strength of the acid or base. A selected list of *K* and *pK* values is given in Table I.6. Accurate values for strong acids do not appear in the table, because their dissociation constants are so large that they cannot be measured reliably.

The values of dissociation constants can be used with advantage when calculating the concentrations of various species (notably the hydrogen-ion concentration) in the solution. A few examples of such calculations are given below:

Example 1 Calculate the hydrogen-ion concentration in a 0·01M solution of acetic acid.

The dissociation of acetic acid takes place according to the equilibrium:

$$CH_3COOH \rightleftarrows H^+ + CH_3COO^-$$

the dissociation equilibrium constant being

$$K = \frac{[H^+]\times[CH_3COO^-]}{[CH_3COOH]} = 1{\cdot}75\times10^{-5}$$

* Organic amines behave like monovalent weak bases. Their behaviour can be explained along similar lines as the basic character of ammonia. The general formula of a monoamine is $R{-}NH_2$ (where R is a monovalent organic radical), showing that one hydrogen of the ammonia is replaced by the radical R. When dissolved in water, amines hydrolyse and dissociate as

$$RNH_2 + H_2O \rightleftarrows RNH_3OH \rightleftarrows RNH_3^+ + OH^-$$

and the law of mass action can be applied to this dissociation just like for that of ammonia. For more detailed account see I. L. Finar's *Organic Chemistry*, Vol. 1. *The Fundamental Principles*. 5th edn., Longman 1967, p. 343 et f.

Table I.6 Dissociation constants of acids and bases

Acid	°C	Dissociation step	K	pK
Monobasic acids				
HCl	25	1	$\sim 10^7$	~ -7
HBr	25	1	$\sim 10^9$	~ -9
HI	25	1	$\sim 3 \times 10^9$	~ -9.48
HF	25	1	$6{\cdot}7 \times 10^{-4}$	3·17
HCN	18	1	$4{\cdot}79 \times 10^{-10}$	3·32
HCNO	25	1	$2{\cdot}2 \times 10^{-4}$	3·66
HCNS	25	1	$1{\cdot}42 \times 10^{-1}$	0·85
HClO	15	1	$3{\cdot}2 \times 10^{-8}$	7·49
$HClO_2$	25	1	$4{\cdot}9 \times 10^{-3}$	2·31
HIO	25	1	2×10^{-10}	9·70
HNO_2	20	1	7×10^{-4}	3·15
HNO_3	30	1	22	−1·34
CH_3COOH	20	1	$1{\cdot}75 \times 10^{-5}$	4·76
HCOOH	20	1	$1{\cdot}77 \times 10^{-4}$	3·75
CH_2Cl—COOH	20	1	$1{\cdot}39 \times 10^{-3}$	2·86
$CHCl_2$—COOH	25	1	$5{\cdot}1 \times 10^{-2}$	1·29
C_6H_5OH	20	1	$1{\cdot}05 \times 10^{-10}$	9·98
C_6H_5COOH	20	1	$6{\cdot}24 \times 10^{-5}$	4·20
C_2H_5COOH	20	1	$1{\cdot}34 \times 10^{-5}$	4·87
Dibasic acids				
H_2CO_3	25	1	$4{\cdot}31 \times 10^{-7}$	6·37
	25	2	$5{\cdot}61 \times 10^{-11}$	10·25
H_2S	20	1	$9{\cdot}1 \times 10^{-8}$	7·04
	20	2	$1{\cdot}2 \times 10^{-15}$	14·92
H_2SO_3	18	1	$1{\cdot}66 \times 10^{-2}$	1·78
	18	2	$1{\cdot}02 \times 10^{-2}$	1·99
H_2SO_4	20	1	$\sim 4 \times 10^{-1}$	0·4
		2	$1{\cdot}27 \times 10^{-2}$	1·9
$(COOH)_2$	20	1	$2{\cdot}4 \times 10^{-2}$	1·62
	20	2	$5{\cdot}4 \times 10^{-5}$	4·27
$C_4H_6O_6$	20	1	$9{\cdot}04 \times 10^{-4}$	3·04
(tartaric acid)		2	$4{\cdot}25 \times 10^{-5}$	4·37
Tribasic acids				
H_3AsO_4	18	1	$5{\cdot}62 \times 10^{-3}$	2·25
	18	2	$1{\cdot}70 \times 10^{-7}$	6·77
	18	3	$2{\cdot}95 \times 10^{-12}$	11·53
H_3BO_3	20	1	$5{\cdot}27 \times 10^{-10}$	9·28
	20	2	$1{\cdot}8 \times 10^{-13}$	12·74
	20	3	$1{\cdot}6 \times 10^{-14}$	13·80
H_3PO_4	20	1	$7{\cdot}46 \times 10^{-3}$	2·13
	20	2	$6{\cdot}12 \times 10^{-8}$	7·21
	20	3	$4{\cdot}8 \times 10^{-13}$	12·32
$C_6H_8O_7$	20	1	$7{\cdot}21 \times 10^{-4}$	3·14
(citric acid)	20	2	$1{\cdot}70 \times 10^{-5}$	4·77
	20	3	$4{\cdot}09 \times 10^{-5}$	4·39

Table I.6 Dissociation constants of acids and bases

Acid	°C	Dissociation step	K	pK
Bases				
NaOH	25	1	~4	−0·60
LiOH	25	1	$6{\cdot}65\times10^{-1}$	0·18
NH_4OH	20	1	$1{\cdot}71\times10^{-5}$	4·77
$Ca(OH)_2$	25	1	4×10^{-2}	1·40
	25	2	$3{\cdot}74\times10^{-3}$	2·43
$Mg(OH)_2$	25	2	$2{\cdot}6\times10^{-3}$	2·58
$CH_3—NH_2$	20	1	$4{\cdot}17\times10^{-4}$	3·38
$(CH_3)_2{=}NH$	20	1	$5{\cdot}69\times10^{-4}$	3·24
$(CH_3)_3{\equiv}N$	20	1	$5{\cdot}75\times10^{-5}$	4·24
$C_2H_5—NH_2$	20	1	$3{\cdot}02\times10^{-4}$	3·52
$(C_2H_5)_2{=}NH$	25	1	$8{\cdot}57\times10^{-4}$	3·07
$(C_2H_5)_3{\equiv}N$	25	1	$5{\cdot}6\times10^{-4}$	3·25
$C_6H_5—NH_2$ (aniline)	20	1	4×10^{-10}	9·40
C_5H_5N (pyridine)	20	1	$1{\cdot}15\times10^{-9}$	8·94
C_9H_7N (quinoline)	20	1	$5{\cdot}9\times10^{-10}$	9·23

Neglecting the small amounts of hydrogen ions originating from the dissociation of water (cf. Section **I.18**), we can say that all hydrogen ions originate from the dissociation of acetic acid. Hence the hydrogen-ion concentration is equal to the concentration of acetate ions:

$$[H^+] = [CH_3COO^-]$$

Some of the acetic acid in the solution will remain undissociated, while some molecules dissociate. The total concentration c (0·01M) of the acid is therefore the sum of the concentration of undissociated acetic acid and that of acetate ions:

$$c = [CH_3COOH]+[CH_3COO^-] = 0{\cdot}01$$

These equations can be combined into

$$K = \frac{[H^+]^2}{c-[H^+]}$$

Rearranging and expressing $[H^+]$ we obtain

$$[H^+] = \frac{-K+\sqrt{K^2+4cK}}{2} \qquad \text{(i)}$$

Inserting $K = 1{\cdot}75\times10^{-5}$ and $c = 0{\cdot}01$ we have

$$[H^+] = \frac{-1{\cdot}75\times10^{-5}+\sqrt{3{\cdot}06\times10^{-10}+7\times10^{-7}}}{2} = 4{\cdot}10\times10^{-4}\ \text{mol}\ \ell^{-1}$$

(The second root of equation (i) with a minus sign in front of the square root leads to a negative concentration value, which has no physical meaning.)

From this example we can see that in a 0·01M solution of acetic acid only about 4% of the molecules are dissociated.

Example 2 Calculate the concentrations of the ions HS^- and S^{2-} in a saturated solution of hydrogen sulphide.

A saturated aqueous solution of hydrogen sulphide (at 20°C and 1 atm pressure) is about 0·1M, (the precise figure is 0·1075 mol ℓ^{-1}). The dissociation constants of hydrogen sulphide are

$$K_1 = \frac{[H^+]\times[HS^-]}{[H_2S]} = 8{\cdot}73\times10^{-7} \qquad \text{(i)}$$

and

$$K_2 = \frac{[H^+]\times[S^{2-}]}{[HS^-]} = 3{\cdot}63\times10^{-12} \qquad \text{(ii)}$$

(for 20°C). As the second dissociation constant is very small, the value of $[S^{2-}]$ is exceedingly small. Thus only the first ionization step may be taken into consideration, when the correlation

$$[H^+] = [HS^-] \qquad \text{(iii)}$$

holds. Because of the small degree of even the first ionization, the total concentration (0·1 mol ℓ^{-1}) can be regarded as equal to the concentration of undissociated hydrogen sulphide:

$$[H_2S] = 0{\cdot}1 \qquad \text{(iv)}$$

Combining equations (i), (iii), and (iv) we have

$$[HS^-] = \sqrt{K_1[H_2S]} = \sqrt{8{\cdot}73\times10^{-7}\times0{\cdot}1} = 2{\cdot}95\times10^{-4}$$

and the combination of (ii) and (iii) yields the value of $[S^{2-}]$:

$$[S^{2-}] = K_2\frac{[HS^-]}{[H^+]} = K_2 = 3{\cdot}63\times10^{-12}$$

If one multiplies equations (i) and (ii) together and transposes

$$[S^{2-}] = \frac{3{\cdot}17\times10^{-18}}{[H^+]^2}$$

one finds that the concentration of sulphide ions is inversely proportional to the square of hydrogen-ion concentration. Thus, by adjusting the hydrogen-ion concentration by adding an acid or a base to a solution, the concentration of sulphide ions can be adjusted to a predetermined, preferential value. This principle is used in the separation of metal ions of the 2nd and 3rd groups.

I.17 EXPERIMENTAL DETERMINATION OF THE DISSOCIATION EQUILIBRIUM CONSTANT. OSTWALD'S DILUTION LAW The dissociation equilibrium constant and the degree of dissociation at a given concentration are interlinked. To find this correlation let us consider the dissociation of a weak monobasic acid. The dissociation reaction can be written as

$$HA \rightleftarrows H^+ + A^-$$

with the dissociation equilibrium constant

$$K = \frac{[H^+]\times[A^-]}{[HA]} \qquad \text{(i)}$$

The total concentration of the (undissociated plus dissociated) acid is c, thus the correlation

$$c = [HA]+[A^-] \qquad \text{(ii)}$$

holds. The degree of dissociation is α. The concentration of hydrogen ions and that of the dissociated anion will be equal, and can be expressed as

$$[H^+] = [A^-] = c\alpha \qquad \text{(iii)}$$

Combining equations (i), (ii), and (iii) we can write

$$K = \frac{c\alpha \times c\alpha}{c-c\alpha} = \frac{c\alpha^2}{1-\alpha}$$

or, using the notation V for the dilution of the solution

$$V = \frac{1}{c} \qquad \text{(in } \ell \text{ mol}^{-1} \text{ units)}$$

the equilibrium constant can be written as

$$K = \frac{\alpha^2}{V(1-\alpha)}$$

If c or V is known and α is determined by one of the experimental methods mentioned in Section **I.10**, K can be calculated by these equations. These equations are often referred to as **Ostwald's dilution law**, as they express the correlation between dilution and the degree of dissociation. As the latter is proportional to the molar conductivity of the solution, the above correlation describes the particular shapes of the conductivity curves shown in Fig. I.3.

The way in which dissociation constants are obtained from experimental data is illustrated in Table I.7, in which the dissociation equilibrium constant of acetic acid is computed from molar conductivities. The average value

Table I.7 Calculation of the dissociation equilibrium constant of acetic acid from measured values of molar conductivity

Concentration $\times 10^5$	Λ	α	$K\times 10^5$
1·873	102·5	0·264	1·78
5·160	65·95	0·170	1·76
9·400	50·60	0·130	1·83
24·78	31·94	0·080	1·82
38·86	25·78	0·066	1·83
56·74	21·48	0·055	1·84
68·71	19·58	0·050	1·84
92·16	16·99	0·044	1·84
112·2	15·41	0·040	1·84
0	388·6	—	—

($1{\cdot}82 \times 10^{-5}$) of the equilibrium constant agrees well with the true value ($1{\cdot}78 \times 10^{-5}$ at 25°C).

I.18 THE DISSOCIATION AND IONIC PRODUCT OF WATER

Kohlrausch and Heidweiller (1894) found, after careful experimental studies, that the purest water possesses a small, but definite conductance. Water must therefore be slightly ionized in accordance with the dissociation equilibrium:

$$H_2O \rightleftarrows H^+ + OH^-$$

Applying the law of mass action to this dissociation, we can express the equilibrium constant as

$$K = \frac{[H^+] \times [OH^-]}{[H_2O]}$$

From the experimental values obtained for the conductance of water the value of K can be determined; this was found to be $1{\cdot}82 \times 10^{-16}$ at 25°C. This low value indicates that the degree of dissociation is negligible; all the water can therefore in practice be regarded as undissociated. Thus the concentration of water (relative molecular mass = 18) is constant, and can be expressed as

$$[H_2O] = \frac{1000}{18} = 55{\cdot}6 \text{ mol } \ell^{-1}$$

We can therefore collect the constants to one side of the equation and can write

$$K_w = [H^+] \times [OH^-] = 1{\cdot}82 \times 10^{-16} \times 55{\cdot}6 = 1{\cdot}01 \times 10^{-14} \quad \text{(at 25°C)}$$

the new constant, K_w is called the **ionic product of water**. Its value is dependent on temperature (cf. Table I.8); for room temperature the value

Table I.8 The ionic product of water at various temperatures

Temperature (°C)	$K_w \times 10^{14}$	Temperature (°C)	$K_w \times 10^{14}$
0	0·12	35	2·09
5	0·19	40	2·92
10	0·29	45	4·02
15	0·45	50	5·48
20	0·68	55	7·30
25	1·01	60	9·62
30	1·47	—	—

$$K_w = 10^{-14}$$

is generally accepted and used.

The importance of the ionic product of water lies in the fact that its value can be regarded as constant not only in pure water, but also in diluted aqueous solutions, such as occur in the course of qualitative inorganic analysis. This means that if, for example, an acid is dissolved in water, (which, when dissociating, produces hydrogen ions), the concentration of hydrogen ions can increase only at the expense of hydroxyl-ion concentration. If, on the other hand, a base is dissolved, the hydroxyl-ion concentration increases and hydrogen-ion concentration decreases.

We can define the term neutral solution more precisely along these lines. A solution is neutral if it contains equal concentrations of hydrogen and hydroxyl ions; that is if

$$[H^+] = [OH^-]$$

In a **neutral solution** therefore

$$[H^+] = [OH^-] = \sqrt{K_w} = 10^{-7} \text{ mol } \ell^{-1}$$

In an acid solution the hydrogen-ion concentration exceeds this value, while in an alkaline solution the reverse is true. Thus

in an **acid solution** $[H^+] > [OH^-]$ and $[H^+] > 10^{-7}$
in an **alkaline solution** $[H^+] < [OH^-]$ and $[H^+] < 10^{-7}$

In all cases the acidity or alkalinity of the solution can be expressed in quantitative terms by the magnitude of the hydrogen-ion (or hydroxyl-ion) concentration. It is sufficient to use only one of these for any solution; knowing one we can calculate the other using the equation

$$[H^+] = \frac{10^{-14}}{[OH^-]}$$

In a M solution of a strong monobasic acid (supposing that the dissociation is complete) the hydrogen-ion concentration is 1 mol ℓ^{-1}. On the other hand, in a M solution of a strong monovalent base the hydroxyl-ion concentration is 1 mol ℓ^{-1}, thus the hydrogen-ion concentration is 10^{-14} mol ℓ^{-1}. The hydrogen-ion concentration of most of the aqueous solutions dealt with in chemical analysis (other than concentrated acids, used mainly for dissolution of samples), lies between these values.

I.19 THE HYDROGEN-ION EXPONENT (*p*H) In the practice of chemical analysis one frequently deals with low hydrogen-ion concentrations. To avoid the cumbersome practice of writing out such figures with factors of negative powers of 10, Sörensen (1909) introduced the **hydrogen-ion exponent** or *p*H, defined by the relationship:

$$p\text{H} = -\log [H^+] = \log \frac{1}{[H^+]} \quad \text{or} \quad [H^+] = 10^{-p\text{H}}$$

Thus, the quantity *p*H is equal to the logarithm of the hydrogen-ion concentration* with negative sign, or the logarithm of the reciprocal hydrogen-ion concentration. It is very convenient to express the acidity or alkalinity of a solution by its *p*H. From the considerations of Section **I.18** it follows that the *p*Hs of aqueous solutions will in most cases lie between the values of 0 and 14. In a 1M solution of a strong monobasic acid

$$p\text{H} = -\log 1 = 0$$

while the *p*H of a 1M strong monovalent base is

$$p\text{H} = -\log 10^{-14} = 14$$

* more precisely: hydrogen-ion activity, i.e.

$$p\text{H} = -\log a_{\text{H}^+} \quad \text{or} \quad a_{\text{H}^+} = 10^{-p\text{H}}$$

If a solution is neutral,

$$pH = -\log 10^{-7} = 7$$

From the above definition it follows that

for an acid solution $pH < 7$
for an alkaline solution $pH > 7$

The term pOH is sometimes used in an analogous way for the hydroxyl-ion exponent, that is

$$pOH = -\log[OH^-] = \log\frac{1}{[OH^-]} \quad \text{or} \quad [OH^-] = 10^{-pOH}$$

For any aqueous solution the correlation

$$pH + pOH = 14$$

holds.

Figure I.6 will serve as a useful mnemonic for the relation between $[H^+]$, pH, $[OH^-]$, and pOH in acid and alkaline solutions.

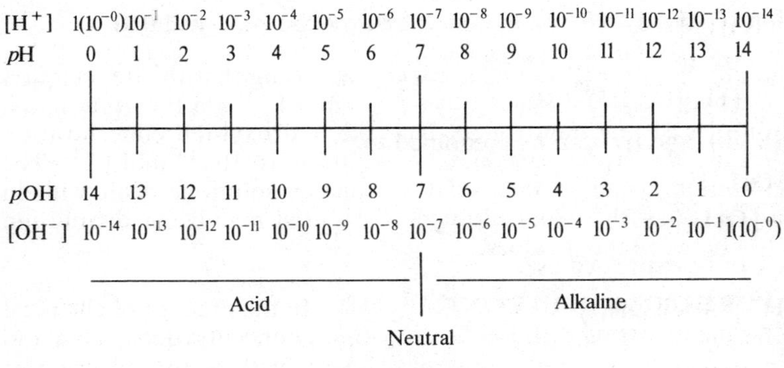

Fig. I.6

Example 3 Calculate the pH of a solution of 0·01M acetic acid (cf. *Example* 1 in Section **I.16**).

In *Example* 1 we found that the hydrogen-ion concentration is

$$[H^+] = 4{\cdot}10 \times 10^{-4} \text{ mol } \ell^{-1}$$

Thus,

$$pH = -\log(4{\cdot}10 \times 10^{-4}) = -(\log 4{\cdot}10 + \log 10^{-4}) = -(0{\cdot}61 - 4) = 3{\cdot}39$$

Example 4 Calculate the molarity of an ammonia solution with a pH of 10·81.

The dissociation of ammonium hydroxide takes place according to the equilibrium

$$NH_4OH \rightleftarrows NH_4^+ + OH^- \qquad \text{(i)}$$

with the dissociation equilibrium constant (cf. Table I.6):

$$K_b = \frac{[NH_4^+] \times [OH^-]}{[NH_4OH]} = 1{\cdot}71 \times 10^{-5} \qquad \text{(ii)}$$

In a solution of pure ammonia the concentrations of ammonium and hydroxyl ions are equal (cf. (i) above)

$$[NH_4^+] = [OH^-] \qquad \text{(iii)}$$

The unknown concentration, c, of ammonia is equal to the sum of the concentrations of the ammonium ions and the undissociated ammonia

$$c = [NH_4^+]+[NH_4OH] \qquad \text{(iv)}$$

Finally, we have the correlation

$$[H^+]\times[OH^-] = 10^{-14} \qquad \text{(v)}$$

valid for all dilute aqueous solutions.

From the definition of pH

$$pH = -\log[H^-] \qquad \text{(vi)}$$

we can calculate the hydrogen-ion concentration in the following way:

$$pH = 10{\cdot}81 = -\log[H^+] = -(0{\cdot}19-11)$$
$$[H^+] = 1{\cdot}55\times10^{-11}$$

From equation (v) the hydroxyl-ion concentration can be calculated

$$[OH^-] = \frac{10^{-14}}{[H^+]} = \frac{10^{-14}}{1{\cdot}55\times10^{-11}} = 6{\cdot}45\times10^{-4}\ \text{mol}\ \ell^{-1} \qquad \text{(vii)}$$

Equations (ii), (iii), and (iv) can be combined to

$$K_b = \frac{[OH^-]^2}{c-[OH^-]}$$

from which c can be expressed as

$$c = \frac{[OH^-]^2+K_b[OH^-]}{K_b}$$
$$= \frac{[OH^-]^2}{K_b} + [OH^-]$$
$$= \frac{(6{\cdot}54\times10^{-4})^2}{1{\cdot}71\times10^{-5}}+6{\cdot}45\times10^{-4}$$
$$= 2{\cdot}495\times10^{-2} \approx 2{\cdot}5\times10^{-2}$$

Thus the solution is 0·025 molar.

The method using the exponent rather than the quantity itself has been found useful for expressing other small numerical quantities which arise in qualitative analysis. Such quantities include (cf. Section **I.16**) (i) dissociation constants and (ii) other ionic concentrations.

(i) For any acid with a dissociation constant K_a

$$pK_a = -\log K_a = \log\frac{1}{K_a}$$

Similarly, for any base with a dissociation constant K_b

$$pK_b = -\log K_b = \log \frac{1}{K_b}$$

Example 5 The dissociation constant of acetic acid is $1{\cdot}75 \times 10^{-5}$ (at 20°C). Calculate its pK_a value.

$$\begin{aligned} pK_a = -\log K_a &= -\log 1{\cdot}75 \times 10^{-5} \\ &= -(\log 1{\cdot}75 \times \log 10^{-5}) \\ &= -(0{\cdot}24 - 5) = 4{\cdot}75 \end{aligned}$$

Example 6 The pK_b value of dimethylamine is 3·24 (at 20°C). Calculate the K_b dissociation constant.

$$\begin{aligned} pK_b = -\log K_b &= 3{\cdot}24 = 4 - 0{\cdot}76 = -(0{\cdot}76 - 4) \\ \log K_b &= 0{\cdot}76 - 4 \\ K_b &= \text{num}\log(0{\cdot}76 - 4) = 5{\cdot}76 \times 10^{-4} \end{aligned}$$

The correct value (cf. Table I.6) is 5.69×10^{-4}. The difference comes from the fact, that pK values are usually rounded off to the third significant figure. (The accurate value of pK_b would be 3·24489.)

(ii) For any ion I of concentration [I]

$$pI = -\log[I] = \log \frac{1}{[I]}$$

Example 7 Find the pNa value of a solution in which $[Na^+] = 8 \times 10^{-5}$ mol ℓ^{-1}.

$$\begin{aligned} pNa = -\log[Na^+] &= -(\log 8 \times 10^{-5}) = -(\log 8 + \log 10^{-5}) \\ &= -(0{\cdot}90 - 5) = 4{\cdot}10. \end{aligned}$$

I.20 HYDROLYSIS When salts are dissolved in water, the solution is not always neutral in reaction. The reason for this phenomenon is that some of the salts interact with water; hence it is termed **hydrolysis**. As a result, hydrogen or hydroxyl ions remain in the solution in excess, the solution itself becoming acid or basic respectively.

In order to understand the phenomenon of hydrolysis better, it is useful to examine the behaviour of four categories of salts separately. All existing salts fall into one of the following categories:

I. Those derived from strong acids and strong bases, e.g. potassium chloride.
II. Those derived from weak acids and strong bases, e.g. sodium acetate.
III. Those derived from strong acids and weak bases, e.g. ammonium chloride.
IV. Those derived from weak acids and weak bases, e.g. ammonium acetate.

These groups behave differently with respect to hydrolysis.

I. Salts of strong acids and strong bases, when dissolved in water, show a neutral reaction, as neither the anion nor the cation combines with hydrogen or hydroxyl ions respectively to form sparingly dissociated products. The dissociation equilibrium of water

$$H_2O \rightleftarrows H^+ + OH^-$$

is therefore not disturbed. The concentration of hydrogen ions in the solution equals that of hydroxyl ions; thus, the solution reacts **neutral**.

II. Salts of weak acids and strong bases, when dissolved in water, produce a solution which reacts **alkaline**. The reason for this is that the anion combines with hydrogen ions to form a sparingly dissociated weak acid, leaving hydroxyl ions behind. In a solution of sodium acetate for example the following two equilibria exist:

$$H_2O \rightleftarrows H^+ + OH^-$$

$$CH_3COO^- + H^+ \rightleftarrows CH_3COOH$$

Thus, the hydrogen ions, formed from the dissociation of water, will partly combine with acetate ions. The two equations can therefore be added, which results in the overall **hydrolysis equilibrium**:

$$CH_3COO^- + H_2O \rightleftarrows CH_3COOH + OH^-$$

In the solution, hydroxyl ions will be in excess over hydrogen ions, and the solution will react alkaline.

In general, if the salt of a monobasic weak acid HA is dissolved in water, the A^- anion will combine with hydrogen ions to form the undissociated acid:

$$A^- + H^+ \rightleftarrows HA \qquad \text{(i)}$$

The hydrogen ions are produced by the dissociation of water:

$$H_2O \rightleftarrows H^+ + OH^- \qquad \text{(ii)}$$

Although the dissociation of pure water is almost negligible, when dissolving the salt, more and more water molecules become ionized, because the removal of hydrogen ions from equilibrium (ii) will, according to the law of mass action, shift the equilibrium to the right. The two equilibria can be combined by addition into

$$A^- + H_2O \rightleftarrows HA + OH^- \qquad \text{(iii)}$$

The equilibrium constant of this process is called the **hydrolysis constant**, and can be written as:

$$K_h = \frac{[HA^-][OH^-]}{[A^-]} \qquad \text{(iv)}$$

The concentration of water, as in the case of its dissociation equilibrium, can be regarded as constant and can therefore be involved in the value of the K_h hydrolysis constant. The greater the value of K_h, the greater is the degree of hydrolysis and the more alkaline the solution.

Values of the hydrolysis constant need not be measured and tabulated separately, because they are correlated to the dissociation constants of the weak acid (cf. equation (i)) and the ionization constant of water (cf. equation (ii)):

$$K_a = \frac{[H^+][A^-]}{[HA]} \qquad \text{(v)}$$

and

$$K_w = [H^+][OH^-] \qquad \text{(vi)}$$

Dividing (vi) by (v) we have

$$\frac{K_w}{K_a} = K_h = \frac{[HA][OH^-]}{[A^-]} \tag{vii}$$

Thus, the hydrolysis constant is equal to the ratio of the ionization constants of water and of the acid.

The degree of hydrolysis can be defined as the fraction of each mole of the anion hydrolysed in the equilibrium. If c is the total concentration of the anion (i.e. of the salt) and x is the degree of hydrolysis, the actual concentrations of the species involved in the hydrolysis equilibrium are as follows:

$$[OH^-] = xc$$
$$[HA] = xc$$
$$[A^-] = c - cx = c(1-x)$$

With these values the hydrolysis constant can be expressed as

$$K_h = \frac{cx^2}{1-x} \tag{viii}$$

By rearranging equation (viii) we can express the degree of hydrolysis as

$$x = -\frac{K_h}{2c} + \sqrt{\frac{K_h^2}{4c^2} + \frac{K_h}{c}} \tag{ix}$$

If x is small (2–5 per cent), equation (viii) reduces to

$$K_h = x^2c$$

from which

$$x = \sqrt{\frac{K_h}{c}} \tag{ixa}$$

The **hydrogen-ion concentration** of a solution obtained by dissolving c moles of a salt per litre can be calculated easily. From the stoichiometry of equilibrium (iii) it follows that, if the amount of hydroxyl ions originating from the self-dissociation of water is negligible, the hydroxyl-ion concentration and the concentration of the undissociated acid are equal:

$$[OH] = [HA] \tag{x}$$

If the degree of hydrolysis is not too great, the total concentration, c, of the salt is equal to the concentration of the anion:

$$[A^-] = c \tag{xi}$$

Combining equations (iv), (x), and (xi) we can express the hydrolysis constant as

$$K_h = \frac{[OH^-]^2}{c} \tag{xii}$$

Now we can combine this expression with expressions (v) and (vi), when the hydrogen-ion concentration of the solution can be expressed as follows:

$$[H^+] = \sqrt{\frac{K_w K_a}{c}} = 10^{-7}\sqrt{\frac{K_a}{c}} \tag{xiii}$$

It is easier to memorize the expression of pH, obtained by logarithmization of equation (xiii):

$$pH = 7+\tfrac{1}{2}pK_a+\tfrac{1}{2}\log c \qquad \text{(xiv)}$$

Example 8 Calculate the hydrolysis constant, the degree of hydrolysis, and the hydrogen-ion concentration and pH of a 0·1M solution of sodium acetate at room temperature.

From Table I.6 $K_a = 1{\cdot}75\times10^{-5}$, $pK_a = 4{\cdot}76$.

The hydrolysis constant can be calculated from equation (vii):

$$K_h = \frac{K_w}{K_a} = \frac{10^{-14}}{1{\cdot}75\times10^{-5}} = 5{\cdot}72\times10^{-10}$$

The degree of hydrolysis is calculated from equation (ix):

$$x = -\frac{K_h}{2c}+\sqrt{\frac{K_h^2}{4c^2}+\frac{K_h}{c}}$$

$$= -\frac{5{\cdot}72\times10^{-10}}{2\times0{\cdot}1}+\sqrt{\frac{3{\cdot}28\times10^{-19}}{4\times0{\cdot}01}+\frac{5{\cdot}72\times10^{-10}}{0{\cdot}1}}$$

$$x = 7{\cdot}56\times10^{-5}$$

Thus the degree of hydrolysis is slight, only 0·0075 per cent. The same result is obtained therefore from the simplified expression (ixa):

$$x = \sqrt{\frac{K_h}{c}} = \sqrt{\frac{5{\cdot}72\times10^{-10}}{0{\cdot}1}} = 7{\cdot}56\times10^{-5}$$

The hydrogen-ion concentration is obtained from equation (xiii):

$$H^+ = 10^{-7}\sqrt{\frac{K_a}{c}} = 10^{-7}\sqrt{\frac{1{\cdot}75\times10^{-5}}{0{\cdot}1}} = 1{\cdot}32\times10^{-9}$$

The pH of the solution is

$$pH = -\log 1{\cdot}32\times10^{-9} = -(0{\cdot}12-9) = 8{\cdot}88$$

The same result is obtained from equation (xiv):

$$pH = 7+\tfrac{1}{2}pK_a+\tfrac{1}{2}\log c = 7+\frac{4{\cdot}76}{2}-\tfrac{1}{2} = 8{\cdot}88$$

Example 9 Calculate the hydrolysis constant, the degree of hydrolysis, and the pH of a 0·1M solution of sodium sulphide.

From Table I.6 the two ionization constants for hydrogen sulphide are $K_{a1} = 9{\cdot}1\times10^{-8}$; $pK_{a1} = 7{\cdot}04$ and $K_{a2} = 1{\cdot}2\times10^{-15}$; $pK_{a2} = 14{\cdot}92$.

The hydrolysis of sulphide ions take place in 2 steps:

$$S^{2-}+H_2O \rightleftarrows HS^-+OH^- \qquad (K_{h1})$$

$$HS^-+H_2O \rightleftarrows H_2S+OH^- \qquad (K_{h2})$$

The two hydrolysis constants for the two steps are calculated from equation (vii):

$$K_{h1} = \frac{K_w}{K_{a2}} = \frac{10^{-14}}{1{\cdot}2\times10^{-15}} = 8{\cdot}33$$

and

$$K_{h2} = \frac{K_w}{K_{a1}} = \frac{10^{-14}}{9{\cdot}1 \times 10^{-8}} = 1{\cdot}10 \times 10^{-7}$$

Since $K_{h1} \gg K_{h2}$ the second hydrolysis step can be neglected in the calculations.
The degree of hydrolysis is calculated from equation (ix):

$$x = \frac{K_{h1}}{2c} + \sqrt{\frac{K_{h1}^2}{4c^2} + \frac{K_{h1}}{c}} = -\frac{8{\cdot}33}{2 \times 0{\cdot}1} + \sqrt{\frac{69{\cdot}44}{0{\cdot}04} + \frac{8{\cdot}33}{0{\cdot}1}}$$

$$x = 0{\cdot}988 \text{ or } 99 \text{ per cent}$$

Because of this high degree of hydrolysis, the simplified equation (ixa) cannot be used any more.

Example 10 Calculate the degree of hydrolysis and the pH of a 0·1M solution of (*a*) sodium carbonate and (*b*) sodium hydrogen carbonate.

From Table I.6 for H_2CO_3 $K_{a1} = 4{\cdot}31 \times 10^{-7}$; $pK_{a1} = 6{\cdot}37$ and $K_{a2} = 5{\cdot}61 \times 10^{-11}$; $pK_{a2} = 10{\cdot}25$.

(*a*) The hydrolysis of the carbonate ion takes place in two stages:

$$CO_3^{2-} + H_2O \rightleftharpoons HCO_3^- + OH^- \qquad (K_{h1})$$

and

$$HCO_3^- + H_2O \rightleftharpoons H_2CO_3 + OH^- \qquad (K_{h2})$$

The two hydrolysis constants for the two steps are calculated from equation (vii):

$$K_{h1} = \frac{K_w}{K_{a2}} = \frac{10^{-14}}{5{\cdot}61 \times 10^{-11}} = 1{\cdot}79 \times 10^{-4}$$

and

$$K_{h2} = \frac{K_w}{K_{a1}} = \frac{10^{-14}}{4{\cdot}31 \times 10^{-7}} = 2{\cdot}32 \times 10^{-8}$$

Again $K_{h1} \gg K_{h2}$ meaning that the second hydrolysis step can be neglected in the calculations. The degree of hydrolysis can be calculated from equation (ix):

$$x = -\frac{K_{h1}}{2c} + \sqrt{\frac{K_{h1}^2}{4c^2} + \frac{K_{h1}}{c}}$$

$$= -\frac{1{\cdot}79 \times 10^{-4}}{0{\cdot}2} + \sqrt{\frac{3{\cdot}20 \times 10^{-8}}{4 \times 0{\cdot}01} + \frac{1{\cdot}79 \times 10^{-4}}{0{\cdot}1}}$$

$$x = 4{\cdot}22 \times 10^{-2} \text{ or } 4{\cdot}22 \text{ per cent}$$

The approximate expression (ixa) yields a figure close to the above:

$$x = \sqrt{\frac{K_{h1}}{c}} = \sqrt{\frac{1{\cdot}79 \times 10^{-4}}{0{\cdot}1}} = 4{\cdot}23 \times 10^{-2} \text{ or } 4{\cdot}23 \text{ per cent}$$

The pH of the solution is calculated from equation (xiv):

$$pH = 7 + \tfrac{1}{2}pK_{a2} + \tfrac{1}{2}\log c = 7 + \frac{10{\cdot}25}{2} - \tfrac{1}{2} = 11{\cdot}63$$

The solution is again strongly alkaline

(*b*) The hydrolysis of the hydrogen carbonate ion proceeds as

$$HCO_3^- + H_2O \rightleftarrows H_2CO_3 + OH^- \qquad \text{(xv)}$$

The hydrolysis constant has already been calculated (cf. *Example* 10 (*a*)):

$$K_{h2} = 2{\cdot}32 \times 10^{-8}$$

the degree of hydrolysis can be calculated from equation (ix):

$$x = -\frac{K_{h2}}{2c} + \sqrt{\frac{K_{h2}^2}{4c^2} + \frac{K_{h2}}{c}}$$

$$= -\frac{2{\cdot}32 \times 10^{-8}}{2 \times 0{\cdot}1} + \sqrt{\frac{5{\cdot}39 \times 10^{-16}}{4 \times 0{\cdot}01} + \frac{2{\cdot}32 \times 10^{-8}}{0{\cdot}1}}$$

$x = 4{\cdot}82 \times 10^{-4}$ or $0{\cdot}048$ per cent

The same result is obtained from the approximate expression (ixa):

$$x = \sqrt{\frac{K_{h2}}{c}} = \sqrt{\frac{2{\cdot}32 \times 10^{-8}}{0{\cdot}1}} = 4{\cdot}82 \times 10^{-4}$$

The calculation of pH needs more consideration, as apart from the hydrolysis (equation xv), the dissociation equilibria:

$$HCO_3^- \rightleftarrows CO_3^{2-} + H^+ \qquad \text{(xvi)}$$

$$H_2O \rightleftarrows H^+ + OH^- \qquad \text{(xvii)}$$

and

$$H_2CO_3 \rightleftarrows HCO_3^- + H^+ \qquad \text{(xviii)}$$

produce hydrogen and hydroxyl ions also. A mathematically correct treatment, taking all these equilibria into consideration, yields a complex expression for the hydrogen-ion concentration, which can only be solved by successive approximation. An approximate expression can be obtained by combining equilibria (xv), (xvi), and (xvii) to obtain

$$2HCO_3^- \rightleftarrows H_2CO_3 + CO_3^{2-}$$

showing that in the solution the concentrations of carbonic acid and carbonate ions are equal:

$$[H_2CO_3] = [CO_3^{2-}] \qquad \text{(xix)}$$

Combining the expressions for the primary and secondary dissociation constants of carbonic acid we obtain:

$$K_{a1} \times K_{a2} \frac{[H^+][HCO_3^-]}{[H_2CO_3]} \times \frac{[H^+][CO_3^{2-}]}{[HCO_3^-]}$$

$$= \frac{[H^+]^2[CO_3^{2-}]}{[H_2CO_3]} = [H^+]^2 \qquad \text{(cf. xix)}$$

and thus the hydrogen-ion concentration

$$[H^+] = \sqrt{K_{a1}K_{a2}} = \sqrt{4{\cdot}31 \times 10^{-7} \times 5{\cdot}61 \times 10^{-11}} = 4{\cdot}95 \times 10^{-9} \text{ mol } \ell^{-1}$$

For the pH we obtain

$$pH = -\log(4{\cdot}95\times10^{-9}) = -(0{\cdot}69-9) = 8{\cdot}31$$

The experimental value, obtained by glass electrode measurements is pH = 8·18, indicating that the error introduced by the simplified way of calculation together with the error due to the use of concentrations instead of activities is very small indeed.

III. Salts of strong acids and weak bases, when dissolved in water, produce a solution, which reacts **acidic**. The M^+ cation of the salt reacts with hydroxyl ions, produced by the dissociation of water, forming the weak base MOH and leaving hydrogen ions behind:

$$H_2O \rightleftarrows H^+ + OH^-$$

$$M^+ + OH^- \rightleftarrows MOH$$

The overall **hydrolysis equilibrium** can be written as

$$M^+ + H_2O \rightleftarrows MOH + H^+ \qquad \text{(i)}$$

Since hydrogen ions are formed in this reaction, the solution will become acid. The **hydrolysis constant** can be defined as

$$K_h = \frac{[MOH][H^+]}{[M^+]} = \frac{K_w}{K_b} \qquad \text{(ii)}$$

showing that the hydrolysis constant is equal to the ratio of the ionization constants of water and the weak base.

The **degree of hydrolysis** (x) for the salt of a monovalent weak base can again be correlated to the hydrolysis constant as

$$K_h = \frac{x^2c}{1-x} \qquad \text{(iii)}$$

from which the degree of hydrolysis can be expressed as

$$x = \frac{K_h}{2c} + \sqrt{\frac{K_h^2}{4c^2} + \frac{K_h}{c}} \qquad \text{(iv)}$$

If x is small (2–5 per cent), this reduces to

$$x = \sqrt{\frac{K_h}{c}} \qquad \text{(iva)}$$

In these equations c represents the concentration of the salt. The **hydrogen-ion concentration** can be obtained from the equation of the hydrolysis constant (ii) because, according to the stoichiometry of equation (i), the concentration of the undissociated weak base is equal to that of the hydrogen ions:

$$[MOH] = [H^+] \qquad \text{(v)}$$

In this assumption we neglected the small concentration of hydrogen ions originating from the dissociation of water. We can also say that, provided that the degree of hydrolysis is not too great, the concentration of the cation M^+ is equal to the total concentration of the salt:

$$[M^+] = c \qquad \text{(vi)}$$

Combining equations (ii), (v), and (vi) we can express the hydrogen-ion concentration as:

$$[H^+] = \sqrt{\frac{K_w}{K_b}c} = 10^{-7}\sqrt{\frac{c}{K_b}} \qquad \text{(vii)}$$

The pH of the solution is

$$pH = 7-\tfrac{1}{2}pK_b-\tfrac{1}{2}\log c \qquad \text{(viii)}$$

Example 11 Calculate the degree of hydrolysis and the pH of a 0·1M solution of ammonium chloride.

From Table I.6 the dissociation constant of ammonium hydroxide $K_b = 1{\cdot}71\times10^{-5}$, $pK_b = 4{\cdot}77$.

The hydrolysis equilibrium can be written as

$$NH_4^+ + H_2O \rightleftarrows NH_4OH + H^+$$

The hydrolysis constant

$$K_h = \frac{K_w}{K_b} = \frac{10^{-14}}{1{\cdot}71\times10^{-5}} = 5{\cdot}86\times10^{-10}$$

As the value of this constant is small, the degree of hydrolysis can be calculated from the approximate formula (iva):

$$x = \sqrt{\frac{K_h}{c}} = \sqrt{\frac{5{\cdot}86\times10^{-10}}{0{\cdot}1}}$$

$$= 7{\cdot}66\times10^{-5} \text{ or } 0{\cdot}0077 \text{ per cent}$$

The pH of the solution can be calculated from equation (viii):

$$pH = 7-\tfrac{1}{2}pK_b-\tfrac{1}{2}\log c = 7-\frac{4{\cdot}77}{2}+\tfrac{1}{2} = 5{\cdot}17$$

IV. Salts of weak acids and weak bases, when dissolved in water, undergo hydrolysis in a rather complex way. The hydrolysis of the cation leads to the formation of the undissociated weak base.

$$M^+ + H_2O \rightleftarrows MOH + H^+ \qquad \text{(i)}$$

while the hydrolysis of the anion yields the weak acid:

$$A^- + H_2O \rightleftarrows HA + OH^- \qquad \text{(ii)}$$

The hydrogen and hydroxyl ions formed in these processes recombine in part to form water:

$$H^+ + OH^- \rightleftarrows H_2O \qquad \text{(iii)}$$

These equations may however not be added together, unless the dissociation constants of the acid and the base happen to be equal. Depending on the relative values of these dissociation constants, three things may happen:

If $K_a > K_b$ (the acid is stronger than the base), the concentration of hydrogen ions will exceed that of hydroxyl ions, the solution will become acid.

If $K_a < K_b$ (the base is stronger than the acid), the reverse will happen and the solution becomes alkaline.

If $K_a = K_b$ (the acid and the base are equally weak), the two concentrations will be equal, and the solution will be neutral.

Such is the case of ammonium acetate, as the dissociation constants of acetic acid ($K_a = 1{\cdot}75 \times 10^{-5}$) and of ammonium hydroxide ($K_b = 1{\cdot}71 \times 10^{-5}$) are practically equal. In this case the hydrolysis can be described by the equation:

$$NH_4^+ + CH_3COO^- + H_2O \rightleftarrows NH_4OH + CH_3COOH$$

which, in fact, is the sum of the three equilibria:

$$NH_4^+ + H_2O \rightleftarrows NH_4OH + H^+$$
$$CH_3COO^- + H_2O \rightleftarrows CH_3COOH + OH^-$$
$$H^+ + OH^- \rightleftarrows H_2O$$

These three equilibria correspond to the general equations (i), (ii), and (iii) in that order respectively.

In general, equations (i) and (ii) can be added, when the overall hydrolysis equilibrium can be expressed as

$$M^+ + A^+ + 2H_2O \rightleftarrows MOH + HA + H^+ + OH^- \qquad \text{(iv)}$$

The **hydrolysis constant** can be expressed as:

$$K_h = \frac{[MOH][HA][H^+][OH^-]}{[H^+][A^-]} = \frac{K_w}{K_a K_b}$$

The **degree of hydrolysis** is different for the anion and for the cation (unless the two dissociation constants are equal).

The calculation of **hydrogen-ion concentration** is rather difficult, because all the equilibria prevailing in the solution have to be taken into consideration. The equations defining the equilibrium constants

$$K_a = \frac{[H^+][A^-]}{[HA]} \qquad \text{(v)}$$

$$K_b = \frac{[M^+][OH^-]}{[MOH]} \qquad \text{(vi)}$$

$$K_w = [H^+][OH^-] \qquad \text{(vii)}$$

contain altogether six unknown concentrations; another three equations must be found therefore to solve the problem. One of these can be derived from the fact that, because of electroneutrality, the sum of the concentrations of cations and anions in the solution must be equal (the so-called 'charge balance' condition):

$$[H^+] + [M^+] = [OH^-] + [A^-] \qquad \text{(viii)}$$

The total concentration of the salt, c, can be expressed in two ways. First, it is equal to the sum of concentrations of the anion and the undissociated acid:

$$c = [A^-] + [HA] \qquad \text{(ix)}$$

Second, it is also equal to the sum of the concentrations of the cation and of the undissociated base:

$$c = [M^+] + [MOH] \qquad \text{(x)}$$

Combining the six equations (v)–(x), we can express the hydrogen–ion concentration as

$$[H^+] = K_a\left(\frac{c}{[H^+] + \dfrac{K_b[H^+]c}{K_w + K_b[H^+]} - \dfrac{K_w}{[H^+]}} - 1\right) \qquad \text{(xi)}$$

This equation is implicit for $[H^+]$. To solve it, we have to find an approximate value of $[H^+]$, use this in the right-hand side of equation (xi), solve for $[H^+]$, and use this new value in a successive approximation. As a first approximation the value

$$[H^+]^- = \sqrt{K_w \frac{K_a}{K_b}}$$

may be used, especially if K_a and K_b do not differ too considerably.

I.21 BUFFER SOLUTIONS In the practice of inorganic qualitative (and quantitative) analysis it is often necessary to adjust the hydrogen-ion concentration to a certain value before a test, and to maintain this hydrogen-ion concentration during the course of the analysis. If a strongly acid (*p*H 0–2) or strongly alkaline (*p*H 12–14) medium is necessary, the addition of sufficient amounts of a strong acid or a strong base can achieve this task. If however the *p*H of the solutions has to be kept between say 2 and 12, the above method will not help.

Let us consider for example a case, when we need to maintain a *p*H of 4 in a solution during our analytical operations. We may add hydrochloric acid to the (originally neutral) solution in such an amount, that the concentration of the free acid should be 0·0001M in the final mixture. Given a solution of say 10 ml, this means that we must have 0·036 mg free hydrochloric acid present. This is a very small amount, and can be easily changed by reaction with traces of alkaline dissolved from the glass or by ammonia traces, present in the atmosphere of the laboratory. Similarly, solutions containing small amounts of alkali hydroxides are sensitive to carbon dioxide present in the air. It is impossible therefore to maintain the *p*H in a weakly acid, neutral, or weakly alkaline solution simply by the addition of a calculated amount of a strong acid or base.

Let us consider now a mixture of a weak acid and its salt, such as a mixture of acetic acid and sodium acetate. In such a solution the sodium acetate, like any other salt, is almost completely dissociated. The dissociation of acetic acid

$$CH_3COOH \rightleftarrows CH_3COO^- + H^+$$

is however almost negligible, because the presence of large amounts of acetate ions (which originate from the dissociation of sodium acetate), will shift the equilibrium towards the formation of undissociated acetic acid (that is towards the left in the above equation). This solution will possess a certain *p*H, and this *p*H will be kept remarkably well even if considerable amounts of acids or bases are added. If hydrogen ions (that is a strong acid) are added, these will combine with the acetate ions in the solution to form undissociated acetic acid:

$$CH_3COO^- + H^+ \rightarrow CH_3COOH$$

the hydrogen-ion concentration of the solution remains therefore virtually unchanged; all that happened is that the amount of acetate ions decreased while the amount of undissociated acetic acid increased. If, on the other hand, hydroxyl ions are added, these will react with the acetic acid:

$$CH_3COOH + OH^- \rightarrow CH_3COO^- + H_2O$$

again the hydrogen- (and hydroxyl-) ion concentration will not change considerably; only the number of acetate ions will increase while the amount of acetic acid will decrease. Such solutions show therefore a certain resistance towards both acids and alkalis, hence they are termed **buffer solutions**. A buffer solution can also be prepared by dissolving a weak base and its salt together. A mixture of ammonium hydroxide and ammonium chloride shows resistance again hydrogen ions, because the latter react with the (undissociated) ammonium hydroxide:

$$NH_4OH + H^+ \rightarrow NH_4^+ + H_2O$$

while the resistance against hydroxyl ions is based on the formation of the undissociated base from ammonium ions (which originate from the salt):

$$NH_4^+ + OH^- \rightarrow NH_4OH$$

In general, buffer solutions contain a mixture of a weak acid and its salt or a weak base and its salt. The hydrogen-ion concentration can be calculated from considerations of the chemical equilibrium which exists in such solutions. Considering a buffer made up of a weak acid and its salt, the dissociation equilibrium

$$HA \rightleftarrows H^+ + A^-$$

exists in the solution. The equilibrium constant can be expressed as

$$K_a = \frac{[H^+][A^-]}{[HA]}$$

from which the hydrogen-ion concentration can be expressed as

$$[H^+] = K_a \frac{[HA]}{[A^-]} \qquad \text{(i)}$$

The free acid present is almost completely undissociated, because the presence of large amounts of the anion A^-, which originates from the salt. The total concentration of the acid, c, will therefore be (approximately) equal to the concentration of the undissociated acid,

$$c_a \approx [HA] \qquad \text{(ii)}$$

For the same reason the total concentration of the salt, c, will be (approximately) equal to the concentration of the anion:

$$c_s \approx [A^-] \qquad \text{(iii)}$$

Combining equations (i), (ii), and (iii) we can express the hydrogen-ion concentration as

$$[H^+] = K_a \frac{c_a}{c_s} \qquad \text{(iv)}$$

or the pH as

$$pH = pK_a + \log \frac{c_s}{c_a} \qquad \text{(v)}$$

Similarly, if the buffer is made up from a weak base MOH and its salt, containing the cation M^+, the dissociation equilibrium which prevails in such a solution is

$$MOH \rightleftarrows M^+ + OH^-$$

for which the dissociation equilibrium constant can be expressed as

$$K_b = \frac{[M^+][OH^-]}{[MOH]} \qquad \text{(vi)}$$

With similar considerations, we can write for the total concentration of the base, c, and for the concentration of the salt, c, that

$$c_b \approx [MOH] \qquad \text{(vii)}$$

and

$$c_s \approx [M^+] \qquad \text{(viii)}$$

Finally, we know that in any aqueous solution the ionic product of water (cf. Section **I.18**):

$$K_w = [H^+][OH^-] = 10^{-14} \qquad \text{(ix)}$$

By combining equations (vi), (vii), (viii), and (ix) we can express the hydrogen-ion concentration of such a buffer as

$$[H^+] = \frac{K_w}{K_b} \times \frac{c_s}{c_b} \qquad \text{(x)}$$

or the pH as

$$pH = 14 - pK_b - \log \frac{c_s}{c_b} \qquad \text{(xi)}$$

where $14 = -\log K_w = pK_w$.

Example 12 Calculate the hydrogen-ion concentration and *pH* of a solution prepared by mixing equal volumes of 0·1M acetic acid and 0·2M sodium acetate.

We use equation (iv):

$$[H^+] = K_a \frac{c_a}{c_s}$$

From Table I.6 $K_a = 1{\cdot}75 \times 10^{-5}$, $c_a = 0{\cdot}05 \text{ mol } \ell^{-1}$ and $c_s = 0{\cdot}1 \text{ mol } \ell^{-1}$ (note that both original concentrations were halved when the solutions were mixed). Thus

$$[H^+] = 1{\cdot}75 \times 10^{-5} \times \frac{0{\cdot}05}{0{\cdot}1} = 8{\cdot}75 \times 10^{-6}$$

and $pH = -\log(8{\cdot}75 \times 10^{-6}) = -(0{\cdot}94 - 6) = 5{\cdot}04$.

Example 13 We want to prepare 100 ml of a buffer with pH = 10. We have 50 ml 0·4M ammonia solution. How much ammonium chloride must be added and dissolved before diluting the solution to 100 ml?

We use equation (xi):

$$pH = 14 - pK_b - \log \frac{c_s}{v_b}$$

Here pH = 10, pK_b = 4·77 (from Table I.6), and c_b = 0·2 (that is if 50 ml 0·4 molar solution is diluted to 100 ml it becomes 0·2 molar). Rearranging and inserting the above values we obtain

$$\log c_s = 14 - 4{\cdot}77 + \log 0{\cdot}4 - 10 = -1{\cdot}17$$

$$c_s = \text{num.log}(-1{\cdot}17) = \text{num.log}(0{\cdot}83 - 2) = 6{\cdot}76 \times 10^{-2} \text{ mol } \ell^{-1}$$

The relative molecular mass of NH_4Cl being 53·49, the weight of salt needed for 1 ℓ is $6{\cdot}76 \times 10^{-2} \times 53{\cdot}49 = 3{\cdot}59$ g. For 100 ml we need one-tenth of this, that is 0·359 g.

There are several buffer systems which can easily be prepared and used in the laboratory. Compositions of some of these, covering the pH range from 1·5 to 11, are shown in Table I.9.

Table I.9 Composition of buffer solutions

A. Standard buffer solutions The following standards are suitable for the calibration of pH meters and for other purposes which require an accurate knowledge of pH.

Solution	***p*H at**		
	12°C	25°C	38°C
0·1M $KHC_2O_4.H_2C_2O_4.2H_2O$	—	1·48	1·50
0·1M HCl + 0·09M KCl	—	2·07	2·08
Saturated solution of potassium hydrogen tartrate, $KHC_4H_4O_6$	—	3·57	—
0·05M potassium hydrogen phthalate, $KHC_8H_4O_4$	4·000 (15°C)	4·005	4·015
0·1M CH_3COOH + 0·1M CH_3COONa	4·65	4·64	4·65
0·025M KH_2PO_4 + 0·025M $Na_2HPO_4.12H_2O$	—	6·85	6·84
0·05M $Na_2B_4O_7.12H_2O$	—	9·18	9·07

Solutions of known pH for colorimetric determinations are conveniently prepared by mixing appropriate volumes of certain standard solutions. The compositions of a number of typical buffer solutions are given below.

B. Solutions for the *p*H range 1·40–2·20 at 25°C (German and Vogel, 1937).

X ml of 0·1M *p*-toluenesulphonic acid monohydrate (19·012 g ℓ$^{-1}$) and *Y* ml of 0·1M sodium *p*-toluenesulphonate (19·406 g ℓ$^{-1}$), diluted to 100·0 ml.

X (ml)	*Y* (ml)	*p*H	*X* (ml)	*Y* (ml)	*p*H
48·9	1·1	1·40	13·2	36·8	1·90
37·2	12·8	1·50	10·0	40·0	2·00
27·4	22·6	1·60	7·6	42·4	2·10
19·0	31·0	1·70	4·4	45·6	2·20
16·6	33·4	1·80			

Table I.9 Composition of buffer solutions

C. Solutions for the *p*H range 2·2–8·0 (McIlvaine, 1921).

20·00 ml mixtures of *X* ml of 0·2M Na_2HPO_4 and *Y* ml of 0·1M citric acid.

X (ml) Na_2HPO_4	*Y* (ml) **Citric acid**	*p*H	*X* (ml) Na_2HPO_4	*Y* (ml) **Citric acid**	*p*H
0·40	19·60	2·2	10·72	9·28	5·2
1·24	18·76	2·4	11·15	8·85	5·4
2·18	17·82	2·6	11·60	8·40	5·6
3·17	16·83	2·8	12·09	7·91	5·8
4·11	15·89	3·0	12·63	7·37	6·0
4·94	15·06	3·2	13·22	6·78	6·2
5·70	14·30	3·4	13·85	6·15	6·4
6·44	13·56	3·6	14·55	5·45	6·6
7·10	12·90	3·8	15·45	4·55	6·8
7·71	12·29	4·0	16·47	3·53	7·0
8·28	11·72	4·2	17·39	2·61	7·2
8·82	11·18	4·4	18·17	1·83	7·4
9·25	10·65	4·6	18·73	1·27	7·6
9·86	10·14	4·8	19·15	0·85	7·8
10·30	9·70	5·0	19·45	0·55	8·0

D. Solutions for the *p*H ranges 2·2–3·8, 4·0–6·2, 5·8–8·0, 7·8–10·0 at 20°C (Clark and Lubs, 1916).

(A) *p*H 2·2–3·8. 50 ml 0·2M KHphthalate + *P* ml 0·2M HCl, diluted to 200 ml
(B) *p*H 4·0–6·2. 50 ml 0·2M KHphthalate + *Q* ml 0·2M NaOH, diluted to 200 ml
(C) *p*H 5·8–8·0. 50 ml 0·2M KH_2PO_4 + *R* ml 0·2M NaOH, diluted to 200 ml
(D) *p*H 7·8–10·0. 50 ml 0·2M H_3BO_3 and 0·2M KCl* + *S* ml 0·2M NaOH, diluted to 200 ml

A		**B**		**C**		**D**	
P (ml) HCl	*p*H	*Q* (ml) NaOH	*p*H	*R* (ml) NaOH	*p*H	*S* (ml) NaOH	*p*H
46·60	2·2	0·40	4·0	3·66	5·8	2·65	7·8
39·60	2·4	3·65	4·2	5·64	6·0	4·00	8·0
33·00	2·6	7·35	4·4	8·55	6·2	5·90	8·2
26·50	2·8	12·00	4·6	12·60	6·4	8·55	8·4
20·40	3·0	17·50	4·8	17·74	6·6	12·00	8·6
14·80	3·2	23·65	5·0	23·60	6·8	16·40	8·8
9·65	3·4	29·75	5·2	29·54	7·0	21·40	9·0
6·00	3·6	35·25	5·4	34·90	7·2	26·70	9·2
2·65	3·8	39·70	5·6	39·34	7·4	32·00	9·4
—	—	43·10	5·8	42·74	7·6	36·85	9·6
—	—	45·40	6·0	45·17	7·8	40·80	9·8
—	—	47·00	6·2	46·85	8·0	43·90	10·0

* That is a solution containing 12·369 g H_3BO_2 and 14·911 g KCl per litre.

E. Solutions for the *p*H range 2·6–12·0 at 18°C – universal buffer mixture (Johnson and Lindsey, 1939) A mixture of 6·008 g of A.R. citric acid, 3·893 g of A.R. potassium dihydrogen phosphate, 1·769 g of A.R. boric acid and 5·266 g of pure diethylbarbituric acid is dissolved in water and made up to 1 litre. The *p*H values of mixtures of 100 ml of this solution with various volumes (*X*) of 0·2M sodium hydroxide solution (free from carbonate) are tabulated below.

Table I.9 Composition of buffer solutions

*p*H	*X* (ml)	*p*H	*X* (ml)	*p*H	*X* (ml)
2·6	2·0	5·8	36·5	9·0	72·7
2·8	4·3	6·0	38·9	9·2	74·0
3·0	6·4	6·2	41·2	9·4	74·9
3·2	8·3	6·4	43·5	9·6	77·6
3·4	10·1	6·6	46·0	9·8	79·3
3·6	11·8	6·8	48·3	10·0	80·8
3·8	13·7	7·0	50·6	10·2	82·0
4·0	15·5	7·2	52·9	10·4	82·9
4·2	17·6	7·4	55·8	10·6	83·9
4·4	19·9	7·6	58·6	10·8	84·9
4·6	22·4	7·8	61·7	11·0	86·0
4·8	24·8	8·0	63·7	11·2	87·7
5·0	27·1	8·2	65·6	11·4	89·7
5·2	29·5	8·4	67·5	11·6	92·0
5·4	31·8	8·6	69·3	11·8	95·0
5·6	34·2	8·8	71·0	12·0	99·6

I.22 THE EXPERIMENTAL DETERMINATION OF *p*H In some instances it may be important to determine the *p*H of the solution experimentally. According to the accuracy we need and the instrumentation available we can have a choice of several techniques. A few of these will be discussed here.

A. ***The use of indicators and indicator test papers*** An indicator is a substance which varies in colour according to the hydrogen-ion concentration. It is generally a weak organic acid or weak base employed in a very dilute solution. The undissociated indicator acid or base has a different colour to the dissociated product. In the case of an indicator acid, HInd, dissociation takes place according to the equilibrium

$$\mathrm{HInd} \rightleftarrows \mathrm{H^+} + \mathrm{Ind^-}$$

The colour of the indicator anion, $\mathrm{Ind^-}$, is different from the indicator acid. If the solution to which the indicator is added is acid, that is it contains large amounts of hydrogen ions, the above equilibrium will be shifted towards the left, that is the colour of the undissociated indicator acid becomes visible. If however the solution becomes alkaline, that is hydrogen ions are removed, the equilibrium will be shifted towards the formation of the indicator anion, and the colour of the solution changes. The colour change takes place in a narrow, but definite range of *p*H. Table I.10 summarizes the colour changes and the *p*H ranges of indicators within which these colour changes take place. If we possess a set of such indicator solutions, we can easily determine the approximate *p*H of a test solution. On a small strip of filter paper or on a spot-test plate we place a drop of the indicator and then add a drop of the test solution, and observe the colour. If for example we find that under such circumstances thymol blue shows a yellow (alkaline) colour, while methyl orange a red (acid) one, we can be sure that the *p*H of the solution is between 2·8 and 3·1.

Some of the indicators listed in Table I.10 may be mixed together to obtain a so-called 'universal' indicator, and with such the approximate *p*H of the solution can be determined with one single test. Such a 'universal' indicator may

Table I.10 Colour changes and *p*H range of some indicators

Indicator	Chemical name	Colour in acid solution	Colour in alkaline solution	*p*H range
Brilliant cresyl blue (acid)	Amino-diethylamino-methyl diphenazonium chloride	Red-orange	Blue	0·0–1·0
α-Naphthol-benzein (acid)		Colourless	Yellow	0·0–0·8
Methyl violet	Pentamethyl *p*-rosaniline hydrochloride	Yellow	Blue-green	0·0–1·8
Cresol red (acid)	*o*-Cresolsulphone-phthalein	Red	Yellow	1·2–2·8
Thymol blue (acid)	Thymol-sulphone-phthalein	Red	Yellow	1·2–2·8
Meta cresol purple	*m*-Cresolsulphone-phthalein	Red	Yellow	1·2–2·8
Bromophenol blue	Tetrabromophenol-sulphone phthalein	Yellow	Blue	2·8–4·6
Methyl orange	Dimethylamino-azo-benzene-sodium sulphonate	Red	Yellow	3·1–4·4
Congo red	Diphenyl-bis-azo-α-naphthylamine-4-sulphonic acid	Violet	Red	3·0–5·0
Bromocresol green	Tetrabromo-*m*-cresol-sulphone-phthalein	Yellow	Blue	3·8–5·4
Methyl red	*o*-Carboxybenzene-azo-dimethylaniline	Red	Yellow	4·2–6·3
Chlorophenol red	Dichlorophenol-sulphone-phthalein	Yellow	Red	4·8–6·4
Azolitmin (litmus)		Red	Blue	5·0–8·0
Bromothymol blue	Dibromo-thymol-sulphone-phthalein	Yellow	Blue	6·0–7·6
Diphenol purple	*o*-Hydroxy-diphenyl-sulphone-phthalein	Yellow	Violet	7·0–8·6
Cresol red (base)	*o*-Cresol-sulphone-phthalein	Yellow	Red	7·2–8·8
α-Naphthol-phthalein	α-Naphthol-phthalein	Yellow	Blue	7·3–8·7
Thymol blue (base)	Thymol-sulphone-phthalein	Yellow	Blue	8·0–9·6
α-Naphthol-benzein (base)		Yellow	Blue-green	8·2–10·0
Phenolphthalein		Colourless	Red	8·3–10·0
Thymolphthalein		Colourless	Blue	9·3–10·5
Brilliant cresyl blue (base)	(See above)	Blue	Yellow	10·8–12·0

be prepared, after Bogen, by dissolving 0·2 g phenolphthalein, 0·4 g methyl red, 0·6 g dimethylazobenzene, 0·8 g bromothymol blue, and 1 g thymol blue in 1 ℓ absolute ethanol. The solution must be neutralized by adding a few drops of dilute sodium hydroxide solution until its colour turns to pure yellow. According to the *p*H of the solution this 'universal' indicator shows different colours, the approximate *p*H values with their corresponding colours are given in the following small table:

*p*H	2	4	6	8	10	12
colour	red	orange	yellow	green	blue	purple

Small strips of filter paper may be impregnated with this solution and dried. Such an indicator test paper may conveniently be stored for longer times. For a test one strip of this paper should be dipped into the solution, and the colour examined.

Firms manufacturing and selling chemicals, normally market wide-range universal *p*H test papers. The composition of the indicator mixture is generally not disclosed, but a convenient colour chart is supplied with the paper strips, by the aid of which the approximate *p*H can easily be determined by comparing the colour of the paper with that shown on the chart. With a single paper the approximate *p*H of a test solution may be determined with an accuracy of 0·5–1·0 *p*H units, within the *p*H range of 1–11. Some firms also market series of papers, by which the *p*H can be determined with an accuracy of 0·1–0·2 *p*H unit. The appropriate narrow-range test paper to use must be selected by a preliminary test with a wide-range paper.

On the other hand, in some cases we may only have simply to test whether the solution is acid or alkaline. For this test a strip of litmus paper may be used. In acid solutions litmus turns to red, while in alkaline ones it shows a blue colour. The transition *p*H range lies around *p*H = 7.

***B. The colorimetric determination of* pH** The principle outlined under Section **I.22.A** can be made more precise by using known amounts of buffers and indicator solutions and comparing the colour of the test solution with a set of reference standards under identical experimental circumstances. First, the approximate *p*H of the test solution is determined by one of the methods described in Section **I.22.A.** Then a series of buffer solutions is prepared (cf. Section **I.21**, Table I.9), differing successively in *p*H by about 0·2 and covering the range around the approximate value. Equal volumes, say 5 or 10 ml, of the buffer solutions are placed in test tubes of colourless glass, having approximately the same dimensions, and a small, identical quantity of a suitable indicator for the particular *p*H range is added to each tube. A series of different colours corresponding to the different *p*H values is thus obtained. An equal volume of the test solution is then treated with an equal volume of indicator to that used for the buffer solutions, and the resulting colour is compared with that of the individual coloured reference standards. When a complete (or almost com-

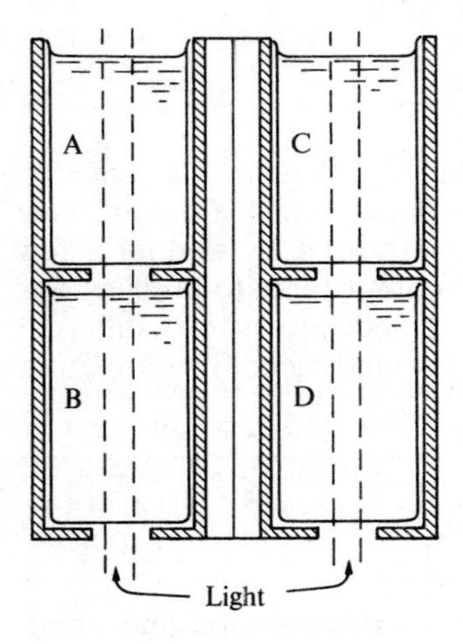

Fig. I.7

plete) match is found, the test solution and the corresponding buffer solution have the same *p*H ±0·2 unit. For matching the colours the buffer solutions may be arranged in the holes of a test tube stand in order of increasing *p*H, the test solution is then moved from hole to hole until the best colour match is obtained. Special stands with white background and standards for making the comparison are available commercially (e.g. from The British Drug Houses Ltd.). The commercial standards, prepared from buffer solutions, are not permanent and must be checked every few months.

For turbid or slightly coloured solutions, the simple comparison method described above, can no longer be applied. The interference due to the coloured substance can be eliminated in a simple way by a device introduced by H. Walpole (1916), shown in Fig. I.7. A, B, C, and D are glass cylinders with plain bottoms standing in a box, which is painted dull black in the inside. A contains the solution to be tested with the indicator, B contains an equal volume of water, C contains the solution of known *p*H with the indicator, while D contains the same volume of the solution to be tested as was originally added to A, but indicator is *not* added to D. Viewing through the two pairs of tubes from above, the colour of the test solution is compensated for. When making the measurements, only tube C has to be removed and replaced by another standard for better matching.

The preparation of reference standards in this procedure is a tedious task and may require considerable time. Time can be saved by applying what is called a **permanent colour standard method**, which requires a special device, the so-called **comparator**. The Lovibond comparator,* shown on Fig. I.8 employs nine permanent glass colour standards, fitted on a revolving disc. The device is fitted

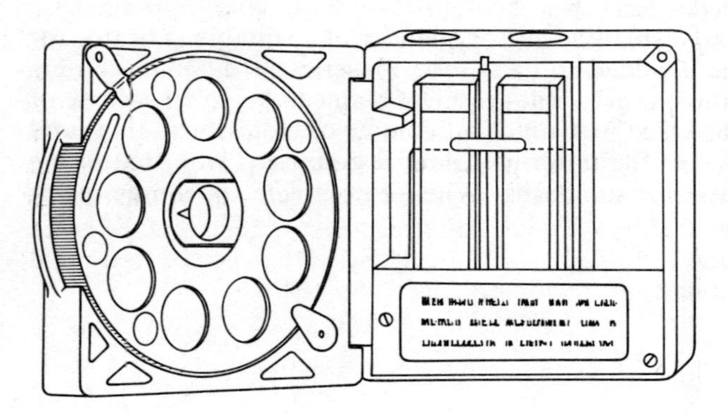

Fig. I.8

with low compartments to receive small test tubes or rectangular glass cells. There is also an opal glass screen against which colours can be compared. The disc can revolve in the comparator, and each colour standard passes in turn in front of an aperture through which the solution in the cell (or cells) can be observed. As the disc revolves, the *p*H of the colour standard visible in the

* Manufactured by The Tintometer Ltd., Milford, Salisbury, England. A similar apparatus is marketed by Hellige, Inc., of Long Island City 1, N.Y. U.S.A., this utilizes Merck's (U.S.A.) indicators. The glass discs in the two instruments are not interchangeable.

aperture appears in a special recess. The Lovibond comparator is employed with B.D.H. indicators. The colour discs available include cresol red (acid and base range), thymol blue (acid and base range), bromophenol blue, bromocresol blue, bromocresol green, methyl red, chlorophenol red, bromocresol purple, bromothymol blue, phenol red, diphenol purple, cresol red, thymol blue, and the B.D.H. 'universal' indicator. The *p*H ranges of these indicators are listed in Table I.10.

A determination of the approximate *p*H of the solution is made with a 'universal' or 'wide range' indicator or with an indicator test paper (see under Section **I.22.A**) and then the suitable disc is selected and inserted into the comparator. A specified amount (with the Lovibond comparator 10 ml) of the unknown solution is placed in the glass test tube or cell, the appropriate quantity of indicator (normally 0·5 ml) is added and the colour is matched against the glass disc. Provision is made in the apparatus for the application of the Walpole technique by the insertion of a 'blank' containing the solution. It is claimed that results accurate to 0·2 *p*H unit can be achieved.

C. The potentiometric determination of* pH The most advanced and precise method of the measurement of *p*H is based on the measurement of the electromotive force (e.m.f.) of an electrochemical cell, which contains the solution of the unknown *p*H as electrolyte, and two electrodes. The electrodes are connected to the terminals of an electronic voltmeter, most often called simply a *p*H-meter. If properly calibrated with a suitable buffer of a known *p*H, the *p*H of the unknown solution can be read directly from the scale.

The e.m.f. of an electrochemical cell can be regarded as the absolute value of the difference of the electrode potentials of the two electrodes.† The two electrodes applied in building the electrochemical cell have different roles in the measurement, and must be chosen adequately. One of the electrodes, termed the **indicator electrode** acquires a potential which depends on the *p*H of the solution. In practice the glass electrode is used as the indicator electrode. The second electrode, on the other hand, has to have a constant potential, independent of the *p*H of the solution, to which the potential of the indicator electrode therefore can be compared in various solutions, hence the term **reference electrode** is applied for this second electrode. In *p*H measurements the (saturated) calomel electrode is applied as an indicator electrode.

The measured e.m.f. of the cell can thus be expressed as

$$\text{e.m.f.} = |E_{gl} - E_{cal}|$$

Here E_{cal} is the electrode potential of the calomel electrode, which is constant.

$$E_{cal} = \text{const}$$

The potential of the saturated calomel electrode is +0·246 V at 25°C (measured against the standard hydrogen electrode). E_{gl}, the potential of the glass electrode, on the other hand, depends on the *p*H of the solution. For the *p*H region 2–11

* The full understanding of this section implies some knowledge of electrode potentials, which is discussed later in this book (cf. Section **I.39**). The treatment in this section is factual, and aims to describe the necessary knowledge required for a proper measurement of *p*H.

† For a detailed discussion of electrochemical cells textbooks of physical chemistry should be consulted, e.g. W. J. Moore's *Physical Chemistry*. 4th edn., Longman 1966, p. 379 et f.

(where the accurate determination is most important) the *p*H-dependence of the potential of the glass electrode can be expressed as

$$E_{gl} = E_{gl}^0 - 0{\cdot}059\,p\mathrm{H}$$

Here E_{gl}^0 is the standard potential of the glass electrode. This quantity varies from specimen to specimen, it depends also on the age and on the pretreatment of the electrode. Within one set of measurements it can be regarded as constant. If we adapt the usual calibration process, described below, it is not necessary to measure the standard potential and to deduct the potential of the calomel electrode from the results, as the *p*H can be read directly from the *p*H-meter.

The **glass electrode** (Fig. I.9) contains the *p*H-sensitive glass in the form of a small bulb, which is fused to an ordinary glass tube. The *p*H-sensitive glass is made by manufacturers according to various specifications. The composition of the most important glasses used in glass electrodes are listed in Table I.11.

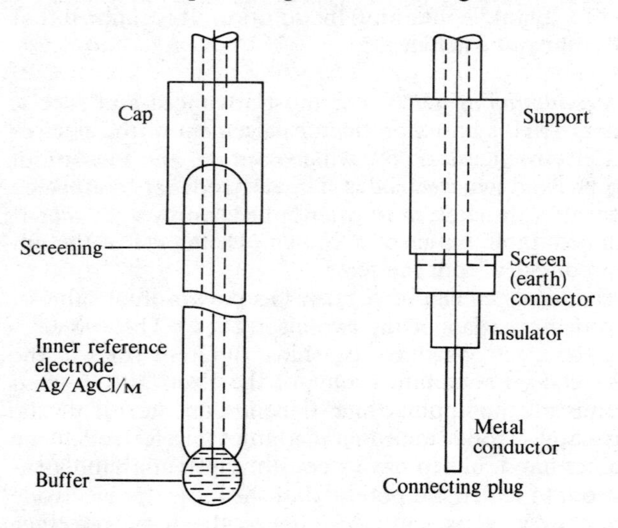

Fig. I.9

Table I.11 Composition of glasses used in the manufacture of glass electrodes*

	Li_2O	Na_2O	Cs_2O	CaO	BaO	La_2O_3	SiO_2
Dole glass	—	21·4	—	6·4	—	—	72·2
Perley glass	28	—	3	—	—	4	65
Lithium–barium glass	24	—	—	—	8	—	68

* B. Csákváry, Z. Boksay and G. Bouquet: *Anal. Chim. Acta*, **56** (1971) p. 279.

The bulb is filled with an acid solution or with an acid buffer, which is connected to the circuit by a platinum wire. Usually there is an internal reference electrode (a silver–silver chloride electrode) included in the circuit, placed somewhere between the bulb and the top of the glass tube. This internal reference electrode is switched in series with the wire leading to the electrolyte in the bulb, and is connected to the input of the *p*H-meter. The role of the internal reference electrode is to protect the glass electrode from an accidental loading by electricity. It is non-polarizable and has, just like the calomel electrode, a constant potential. The glass bulb itself is made of a very thin glass and is therefore very delicate; it must be handled with the greatest care. The proper operation of the glass electrode requires that the electrode glass itself should be wet and in a 'swollen' state; glass electrodes therefore must be kept always dipped in water or in dilute acid. If the glass electrode is left to dry out, it will not give reproducible readings on the *p*H-meter. (In such cases the electrode has to be soaked in 0·1M hydrochloric acid for 1–2 days, when its response usually returns.) Though the swollen glass of the electrode is capable of conducting, it represents a high resistance in the circuit. The resistance of a glass electrode is usually about 10^8–10^9 Ω, which means that the current in the circuit is extremely low. (The current must also be low with electrodes of low resistances, to avoid polarization.) The cable leading from the glass electrode is therefore screened; the electrical signal being passed through the inside lead, while the screening cable is switched, in most cases together with the input of the calomel electrode, to the instrument body and through this it is earthed. As was said before, the glass electrode is suitable for the accurate measurement of *p*H within the range 2 to 11. Below this *p*H, (at high hydrogen-ion concentrations), a rather high so-called diffusion potential is superimposed on the measured e.m.f. This varies considerably with the hydrogen-ion concentration itself, and therefore reliable results cannot be obtained even with the most careful calibration. At *p*H values above 11 the so-called alkaline error of the glass electrode occurs, making its response non-linear to *p*H. Over *p*H 2 and below *p*H 11 the glass electrode operates reliably. As a rule, each measurement should be preceded by a calibration with a buffer, the *p*H of which should stand as near to the *p*H of the test solution as possible.

The bulb of a new glass electrode is sometimes coated with a wax layer for protection. This should be removed by dipping the electrode into an organic solvent (specified in the instruction leaflet), and then soaking the electrode in dilute hydrochloric acid for a few days. When not in use, the electrode should be kept in distilled water or in dilute hydrochloric acid.

A **calomel electrode** is basically a mercury electrode, the electrode potential of which depends solely on the concentration of mercury(I) (Hg_2^{2+}) ions in the solution with which it is in contact. The concentration of mercury(I) ions is kept constant (though low) by adding mercury(I) chloride precipitate (Hg_2Cl_2, calomel) to the solution, and by applying a large concentration of potassium chloride. In the saturated calomel electrode a saturated solution of potassium chloride is applied; saturation is maintained by keeping undissolved crystals of potassium chloride in the solution. At constant temperature the chloride-ion concentration is constant, this means that the concentration of mercury(I) ions remains constant (cf. Section **I.26**), and thus the electrode potential remains constant too. As long as both calomel and potassium chloride are present in solid form, this concentration of mercury(I) ions will remain constant even if a considerable current passes through the electrode. This electrode is therefore non-polarizable. The potential of a saturated calomel electrode at 25°C is +0·246 V against the standard hydrogen electrode.

A simple form of calomel electrode, suitable for elementary work, is shown in Fig. I.10. It can be made of a simple reagent bottle, into which a rubber stopper with two bores is placed. Through one of the holes a glass tube is fitted,

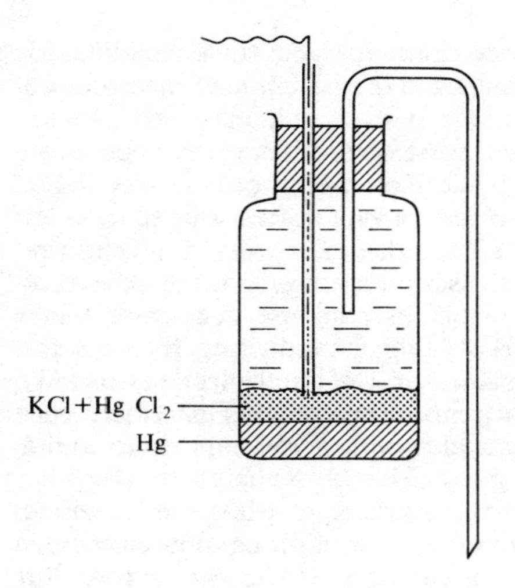

Fig. I.10

with a platinum wire fused through its bottom, which is in direct contact with the mercury in the bottom of the electrode vessel. The platinum wire is then connected to the circuit. The mercury must be of the purest available quality, if possible trebly distilled. Over the mercury a freshly precipitated and carefully washed layer of calomel must be placed. Washing can be done by shaking the precipitate with distilled water and decantation; the procedure being repeated 8–10 times. A liberal amount of analytical grade solid potassium chloride should be added and the vessel filled with saturated potassium chloride solution. Then a small tube, bent in U-shape (see Fig. I.10) should be filled with a hot solution of concentrated potassium chloride, to 100 ml of which 0·5 g of agar has been added. When cooling the solution freezes into the tube but remains conductive, thus enabling electrical contact between the electrode and the test solution. On cooling the level of solution decreases because of contraction, it should therefore either be topped up, or the empty parts of the U-tube cut away. This salt bridge then has to be fitted into the second hole of the rubber stopper. The electrode is now ready for use. The wire leading from the electrode to the *p*H-meter need not be screened. The end of the salt bridge should be kept dipped into concentrated potassium chloride solution when storing.

The ***p*H-meter** is an electronic voltmeter with a high input resistance. (The input resistance of a good *p*H-meter is in the region of 10^{12}–10^{13} Ω.) Both valve and transistorized instruments are in use. They generally operate from the mains, and contain their own power supply circuit with a rectifier. Cheaper instruments contain a differential amplifier, the d.c. input signal being amplified directly in the instrument. More expensive instruments convert the d.c. signal, coming from the measuring cell, into an a.c. signal, which is then amplified, the d.c. component filtered, and finally the amplified signal is rectified. With both instruments the amplified signal is then displayed on a meter, calibrated in *p*H units (and in most cases also in millivolts). A third type of electronic *p*H-meter is also known; with this the electric signal coming from the cell is compensated by turning a potentiometer knob until a galvanometer shows zero deflection. Because of the low currents which circulate in the cell, such an instrument needs also an amplifier between the galvanometer and the potentiometric circuit. On such instruments the *p*H is read from the position of the potentiometer knob.

When **measuring the *p*H** the instrument has to be switched on first, and a sufficient time, ranging from a few minutes to half an hour, must be allowed until complete thermal and electrical equilibrium is achieved. Then, (but not always) the 'zero' knob must be adjusted until the meter shows a deflection given in the instruction manual (generally 0 to 7 on the *p*H-scale). The 'temperature selector' must be set to the room temperature. Now a suitable buffer is chosen, with *p*H nearest to the expected *p*H of the test solution. The glass and calomel electrodes are dipped into the buffer, and the electrodes connected to the relevant input terminals. Usually, the input terminal of the glass electrode looks somewhat special to accommodate a plug with a screened cable, and is marked 'glass' or 'indicator'. The input of the calomel electrode, on the other hand, is generally an ordinary banana socket, marked 'reference' or 'calomel'. (In any case the instruction manual should be followed or an experienced person consulted.) The 'range selector' switch should then be set from 'zero' (or 'standby') position to the range which incorporates the *p*H of the buffer, and the 'buffer adjustment' knob operated until the meter deflects to a position on the *p*H-scale identical to the *p*H of the buffer. The 'selector' switch is then switched to 'zero' position, the electrodes are taken out of the buffer, rinsed carefully with distilled water, and immersed in the test solution. The 'selector' is again set to the same position as before, and the *p*H of the test solution read from the scale.

With *p*H-meters based on the principle of compensation, the operations are similar to those mentioned above, but the potentiometer knob (with the *p*H-scale) is set to a position corresponding to the *p*H of the buffer, and the galvanometer zeroed with the 'buffer adjustment' knob. When the test solution is measured, the galvanometer is zeroed with the potentiometer knob, and the *p*H of the solution read from its scale.

When the measurement is finished, the 'selector' must be switched to 'zero' position, and the electrodes rinsed with distilled water and stored away. The glass electrode must be kept in water or dilute hydrochloric acid, while the salt bridge of the calomel electrode should be left dipped into concentrated potassium chloride. When finishing for the day, the *p*H-meter is switched off, otherwise it should be left on, rather than switched on and off frequently.

D. THE BRØNSTED–LOWRY THEORY OF ACIDS AND BASES

I.23 DEFINITION OF ACIDS AND BASES The classical concepts of acids and bases, as outlined in Sections **I.15–I.22** are sufficient to explain most of the acid-base phenomena encountered in qualitative inorganic analysis carried out in aqueous solutions. Nevertheless this theory has limitations, which become most apparent if acid-base phenomena in non-aqueous solutions have to be interpreted. In the classical acid-base theory two ions, the hydrogen ion (that is the proton) and the hydroxyl ion are given special roles. It was, however, pointed out that while the proton has indeed exceptional properties, to which acid-base function can be attributed, the hydroxyl ion possesses no exceptional qualities entitling it to a specific role in acid-base reactions. This point can be illustrated with some experimental facts. It was found for example that perchloric acid acts as an acid not only in water, but also in glacial acetic acid or liquid ammonia as solvents. So does hydrochloric acid. It is reasonable to suggest therefore that the proton (the only common ion present in both acids) is responsible for their acid character. Sodium hydroxide, while it acts as a strong base in water, shows no special base characteristics in the other solvents (though it reacts readily with glacial acetic acid). In glacial acetic acid, on the other hand, sodium acetate, shows properties of a true base, while sodium amide ($NaNH_2$) takes up such a role in liquid ammonia. Other experimental facts indicate that in glacial acetic acid all soluble acetates, and in liquid ammonia all soluble amides possess base properties. However none of the three ions, hydroxyl, acetate, or amide (NH_2^-), can be singled out as solely responsible for base behaviour.

Such considerations led to a more general definition of acids and bases, which was proposed independently by J. N. Brønsted and T. M. Lowry in 1923. They defined **acid** as any substance (in either the molecular or the ionic state) which donates protons (H^+), and a **base** as any substance (molecular or ionic) which accepts protons. Denoting the acid by A and the base by B, the acid-base equilibrium can be expressed as

$$A \rightleftarrows B + H^+$$

Such an equilibrium system is termed a **conjugate (or corresponding) acid-base system.** A and B are termed a conjugate acid-base pair. It is important to realize that the symbol H^+ in this definition represents the bare proton (unsolvated hydrogen ion), and hence the new definition is in no way connected to any solvent. The equation expresses a hypothetical scheme for defining the acid and base – it can be regarded as a 'half reaction' which takes place only if the proton, released by the acid, is taken up by another base.

Some acid-base systems are as follows:

$$\begin{aligned}
\text{Acid} &\rightleftarrows \text{Base} + H^+ \\
HCl &\rightleftarrows Cl^- + H^+ \\
HNO_3 &\rightleftarrows NO_3^- + H^+ \\
H_2SO_4 &\rightleftarrows HSO_4^- + H^+ \\
HSO_4^- &\rightleftarrows SO_4^{2-} + H^+ \\
CH_3COOH &\rightleftarrows CH_3COO^- + H^+ \\
H_3PO_4 &\rightleftarrows H_2PO_4^- + H^+ \\
H_2PO_4^- &\rightleftarrows HPO_4^{2-} + H^+ \\
HPO_4^{2-} &\rightleftarrows PO_4^{3-} + H^+ \\
NH_4^+ &\rightleftarrows NH_3 + H^+ \\
NH_3 &\rightleftarrows NH_2^- + H^+ \\
H_3O^+ &\rightleftarrows H_2O + H^+ \\
H_2O &\rightleftarrows OH^- + H^+
\end{aligned}$$

From these examples it can be seen that according to the Brønsted–Lowry theory, acids can be:

(*a*) uncharged molecules known as acids in the classical acid-base theory, like HCl, HNO_3, H_2SO_4, CH_3COOH, H_3PO_4 etc.

(*b*) anions, like HSO_4^-, $H_2PO_4^-$, HPO_4^{2-} etc.

(*c*) cations, like NH_4^+, H_3O^+ etc.

According to this theory, bases are substances which are able to accept protons (and not, as in the classical acid-base theory, those, which produce hydroxyl or any other ion). The following are included:

(*a*) uncharged molecules, like NH_3 and H_2O etc.

(*b*) anions, like Cl^-, NO_3^-, NH_2^-, OH^- etc.

It is important to note that those substances (alkali hydroxides) which are, according to the classical acid-base theory, strong bases are in fact not forming uncharged molecules, but are invariably ionic in nature even in the solid state. Thus, the formula NaOH is illogical, the form Na^+, OH^- or $Na^+ + OH^-$ would really express the composition of sodium hydroxide. The basic nature of these strong bases is due to the OH^- ions which are present in the solid state or aqueous solution.

Some substance (like HSO_4^-, $H_2PO_4^{2-}$, HPO_4^{2-}, NH_3, H_2O etc.) can function both as acids and bases, depending on the circumstances. These substances are called amphoteric electrolytes or **ampholytes**.

As already pointed out, the equation

$$A \rightleftarrows B + H^+$$

does not represent a reaction which can take place on its own; the free proton, the product of such a dissociation, because of its small size and the intense electric field surrounding it, will have a great affinity for other molecules, especially those with unshared electrons, and therefore cannot exist as such to any appreciable extent in solution. The free proton is therefore taken up by a base of a second acid-base system. Thus, for example, A_1 produces a proton according to the equation:

$$A_1 \rightarrow B_1 + H^+$$

this proton is taken up by B_2, forming an acid A_2

$$B_2 + H^+ \rightarrow A_2$$

As these two reactions can proceed only simultaneously (and never on their own), it is more proper to express these together in one equation as

$$A_1 + B_2 \rightarrow B_1 + A_2$$

Generally, an acid-base reaction can be written as

$$\text{Acid}_1 + \text{Base}_2 \rightarrow \text{Base}_1 + \text{Acid}_2$$

These equations represent a transfer of a proton from A_1 (Acid_1) to B_2 (Base_2). Reactions between acids and bases are hence termed **protolytic reactions**. All these reactions lead to equilibrium, in some cases the equilibrium may be shifted almost completely in one or another direction. The overall direction of these reactions depends on the relative strengths of acids and bases involved in these systems.

In the classical acid-base theory various types of acid-base reactions (like dissociation, neutralization and hydrolysis) had to be postulated to interpret experimental facts. The great advantage of the Brønsted–Lowry theory is that all these different types of reactions can be interpreted commonly as simple protolytic reactions. Moreover, the theory can easily be extended to acid-base reactions in non aqueous solvents, where the classical acid-base theory has proved to be less adaptable.

Some examples of protolytic reactions are collected below:

$$Acid_1 + Base_2 \rightleftarrows Acid_2 + Base_1$$

$$HCl + H_2O \rightleftarrows H_3O^+ + Cl^- \qquad (i)$$

$$CH_3COOH + H_2O \rightleftarrows H_3O^+ + CH_3COO^- \qquad (ii)$$

$$H_2SO_4 + H_2O \rightleftarrows H_3O^+ + HSO_4^- \qquad (iii)$$

$$HSO_4^- + H_2O \rightleftarrows H_3O^+ + SO_4^{2-} \qquad (iv)$$

$$H_3O^+ + OH^- \rightleftarrows H_2O + H_2O \qquad (v)$$

$$CH_3COOH + NH_3 \rightleftarrows NH_4^+ + CH_3COO^- \qquad (vi)$$

$$H_2O + CH_3COO^- \rightleftarrows CH_3COOH + OH^- \qquad (vii)$$

$$NH_4^+ + H_2O \rightleftarrows H_3O^+ + NH_3 \qquad (viii)$$

$$H_2O + HPO_4^{2-} \rightleftarrows H_2PO_4^- + OH^- \qquad (ix)$$

$$H_2O + H_2O \rightleftarrows H_3O^+ + OH^- \qquad (x)$$

Reactions (i) to (iv) represent 'dissociations' of acids, reaction (v) is the common reaction, called 'neutralization', of strong acids with strong bases, reaction (vi) describes the neutralization reaction between acetic acid and ammonia which takes place in the absence of water, reactions (vii) to (ix) represent 'hydrolysis' reactions, while reaction (x), which is the same as reaction (v) but in the opposite direction, describes the 'dissociation' (or, more properly, the **autoprotolysis**) of water. Some of these reactions will be discussed in more detail in subsequent chapters.

I.24 THE PROTOLYSIS OF ACIDS. STRENGTHS OF ACIDS AND BASES It is of interest to examine the processes which take place when an acid is dissolved in a solvent, first of all in water. According to the Brønsted–Lowry theory this dissolution is accompanied by a protolytic reaction, in which the solvent (water) acts as a base. To elucidate these processes, let us examine what happens if a strong acid (hydrochloric acid) and a weak acid (acetic acid) undergo protolysis.

Hydrogen chloride in the gaseous or pure liquid state does not conduct electricity, and possesses all the properties of a covalent compound. When the gas is dissolved in water, the resulting solution is found to be an excellent conductor of electricity, and therefore contains a high concentration of ions. Evidently water, behaving as a base, has reacted with hydrogen chloride to form hydronium and chloride ions:

$$HCl + H_2O \rightleftarrows H_3O^+ + Cl^-$$

From the original acid (HCl) and base (H_2O) a new acid (H_3O^+) and a new base (Cl^-) have been formed. This equilibrium is completely shifted towards the right; all the hydrogen chloride is transformed into hydronium ions. Similar conclusions can be drawn for other strong acids (like HNO_3, H_2SO_4, $HClO_4$); when dissolved in water, their protolysis yields hydronium ions. Of the two acids, (the strong acid and H_3O^+), involved in each protolytic reaction, hydronium ion is the weaker acid. Water as a solvent has thus a *levelling* effect on strong acids; each strong acid is levelled to the strength of hydronium ions.

When acetic acid is dissolved in water, the resulting solution has a relatively low conductivity indicating that the concentration of ions is relatively low. The reaction

$$CH_3COOH + H_2O \rightleftarrows H_3O^+ + CH_3COO^-$$

proceeds only slightly towards the right. Thus, hydrochloric acid is a stronger acid than acetic acid, or, what is equivalent to the former statement, the acetate ion is a stronger base than the chloride ion. The strength of an acid thus depends upon the readiness with which the solvent can take up protons as compared with the anion of the acid. An acid, like hydrogen chloride, which gives up H^+ readily to the solvent to yield a solution with a high concentration of H_3O^+ is termed a **strong acid**. An acid, like acetic acid, which gives up its protons less readily, affording a solution with a relatively low concentration of H_3O^+ is called a **weak acid**. It is clear also, that if the acid is strong, its conjugate base must be weak and vice versa: if the acid is weak, the conjugate base is strong, *i.e.* possesses a powerful tendency to combine with H^+.

The strength of acids can be measured and compared by the value of their protolysis equilibrium constant. For the protolysis of acetic acid this equilibrium constant can be expressed as

$$K_a = \frac{[H_3O^+][CH_3COO^-]}{[CH_3COOH]}$$

the expression being identical to that of the **ionization constant**, defined and described in Section **I.16**. Ionization constants of acids are listed in Table I.6.

The protolysis of acids in water can be described by the general equation:

$$\text{Acid} + H_2O \rightleftarrows H_3O^+ + \text{Base}$$

and the protolysis constant (or ionization constant) can be expressed in general terms as

$$K_a = \frac{[H_3O^+][\text{Base}]}{[\text{Acid}]}$$

The higher the ionization constant, the stronger the acid, and consequently the weaker the base. Thus the value of K_a is at the same time a measure of the strength of the base; there is no need to define a base ionization constant separately.

Base dissociation constants, listed in Table I.6, are related to the protolysis constant of their conjugate acid through the equation

$$K_a = \frac{K_w}{K_b}$$

This expression can easily be derived from the case of ammonia. In the view of the Brønsted–Lowry theory the dissociation of ammonium hydroxide is more properly the reaction of ammonia with water.*

$$NH_3 + H_2O \rightleftarrows NH_4^+ + OH^-$$

* This statement does not contradict the fact that ammonium hydroxide does really exist; this has been proved beyond doubt by various physicochemical measurements. Cf. Mellor's *Modern Inorganic Chemistry*, revised and edited by G. D. Parkes, 6th edn., Longman 1967, p. 434 et f.

the K_b base dissociation constant for this process can be expressed by

$$K_b = \frac{[NH_4^+][OH^-]}{[NH_3]} \tag{i}$$

The protolysis of the ammonium ion, on the other hand, can be described as

$$NH_4^+ + H_2O \rightleftarrows NH_3 + H_3O^+$$

with the protolysis constant

$$K_a = \frac{[NH_3][H_3O^+]}{[NH_4^+]} \tag{ii}$$

The ionization constant (or autoprotolysis constant) of water (cf. Section **I.18**), is

$$K_w = [H_3O^+][OH^-] \tag{iii}$$

Combining the three expressions (i), (ii), and (iii) the correlation

$$K_a = \frac{K_w}{K_b}$$

can easily be proved.

I.25 INTERPRETATION OF OTHER ACID-BASE REACTIONS WITH THE BRØNSTED–LOWRY THEORY As already outlined, the great advantage of the Brønsted–Lowry theory lies in the fact that any type of acid-base reaction can be interpreted with the simple reaction scheme

$$\text{Acid}_1 + \text{Base}_2 \rightleftarrows \text{Base}_1 + \text{Acid}_2$$

The following examples serve to elucidate the matter:

Neutralization reactions between strong acids and metal hydroxides in aqueous solutions are in fact reactions between the hydronium ion and the hydroxide ion:

$$H_3O^+ + OH^- \rightleftarrows H_2O + H_2O$$
$$\text{Acid}_1 + \text{Base}_2 \rightleftarrows \text{Base}_1 + \text{Acid}_2$$

Neutralization reactions may proceed in the absence of water; in such a case the 'undissociated' acid reacts directly with hydroxyl ions, which are present in the solid phase. Such reactions have little if any practical importance in qualitative analysis.

Displacement reactions, like the reaction of acetate ions with a strong acid, are easy to understand. The stronger acid (H_3O^+) reacts with the conjugate base (CH_3COO^-) of the weaker acid (CH_3COOH), and the conjugate base (H_2O) of the stronger acid is formed:

$$H_3O^+ + CH_3COO^- \rightleftarrows H_2O + CH_3COOH$$
$$\text{Acid}_1 + \text{Base}_2 \rightleftarrows \text{Base}_1 + \text{Acid}_2$$

The displacement of a weak base (NH_3) with a stronger base (OH^-) from its salt can be explained also:

$$OH^- + NH_4^+ \rightleftarrows NH_3 + H_2O$$
$$\text{Base}_1 + \text{Acid}_2 \rightleftarrows \text{Base}_2 + \text{Acid}_1$$

Hydrolysis is an equilibrium between two conjugate acid-base pairs, in which water can play the part of a weak acid or a weak base. In the hydrolysis of acetate ions water acts as an acid:

$$CH_3COO^- + H_2O \rightleftarrows CH_3COOH + OH^-$$
$$\text{Base}_1 + \text{Acid}_2 \rightleftarrows \text{Acid}_1 + \text{Base}_2$$

while in the hydrolysis of the ammonium ion it acts as a weak base:

$$NH_4^+ + H_2O \rightleftarrows NH_3 + H_3O^+$$
$$\text{Acid}_1 + \text{Base}_2 \rightleftarrows \text{Base}_1 + \text{Acid}_2$$

The hydrolysis of heavy metal ions can also be explained easily, keeping in mind that these heavy metal ions are in fact aquacomplexes (like $[Cu(H_2O)_4]^{2+}$ $[Al(H_2O)_4]^{3+}$ etc.), and these ions are conjugate acids of the corresponding metal hydroxides. The first step of the hydrolysis of the aluminium ion can be explained, for example, by the acid-base reaction

$$[Al(H_2O)_4]^{3+} + H_2O \rightleftarrows [Al(H_2O)_3OH]^{2+} + H_3O^+$$
$$\text{Acid}_1 + \text{Base}_2 \rightleftarrows \text{Base}_1 + \text{Acid}_2$$

This hydrolysis may proceed further until aluminium hydroxide, is formed.

The dissociation (more properly, the autoprotolysis) of water, is in fact the reversal of the process of neutralization, in which one molecule of water plays the role of an acid, the other that of a base:

$$H_2O + H_2O \rightleftarrows H_3O^+ + OH^-$$
$$\text{Acid}_1 + \text{Base}_2 \rightleftarrows \text{Acid}_2 + \text{Base}_1$$

The quantitative treatment of these equilibria is formally similar to those described in Sections **I.15–I.22** of this chapter, and will not be repeated here. Results and expressions are indeed identical if aqueous solutions are dealt with. The great advantage of the Brønsted–Lowry theory is that it can be adapted easily for acid-base reactions in any protic (that is, proton-containing) solvents.

E. PRECIPITATION REACTIONS

I.26 SOLUBILITY OF PRECIPITATES A large number of reactions employed in qualitative inorganic analysis involve the formation of precipitates. A precipitate is a substance which separates as a solid phase out of the solution. The precipitate may be crystalline or colloidal, and can be removed from the solution by filtration or by centrifuging. A precipitate is formed if the solution becomes oversaturated with the particular substance. The solubility (S) of a precipitate is by definition equal to the molar concentration of the saturated solution. Solubility depends on various circumstances, like temperature, pressure, concentration of other materials in the solution, and on the composition of the solvent.

The variation of solubility with pressure has little practical importance in qualitative inorganic analysis, as all operations are carried out in open vessels at atmospheric pressure; slight variations of the latter do not have marked influence on the solubility. More important is the variation of the solubility with temperature. In general it can be said, that solubilities of precipitates increase with temperature, though in exceptional cases (like calcium sulphate) the opposite is true. The rate of increase of solubility with temperature varies, in some cases it is marginal, in other cases considerable. The variation of solubility with temperature can, in some cases, serve as the basis of separation. The separation of lead from silver and mercury(I) ions can be achieved, for example, by precipitating the three ions first in the form of chlorides, followed by treating the mixture with hot water. The latter will dissolve lead chloride, but will leave silver and mercury(I) chlorides practically undissolved. After filtration of the hot solution, lead ions will be found in the filtrate and can be identified by characteristic reactions.

The variation of solubility with the composition of the solvent has some importance in inorganic qualitative analysis. Though most of the tests are carried out in aqueous solutions, in some cases it is advantageous to apply other substances (like alcohols, ethers, etc.) as solvents. The separation of alkali metals can for example be achieved by the selective extraction of their salts by various solvents. In other cases the reagent used in the test is dissolved in a non-aqueous solvent, and the addition of the reagent to the test solution in fact changes the composition of the medium.

Solubility depends also on the nature and concentration of other substances, mainly ions, in the mixture. There is a marked difference between the effect of the so-called common ions and of the foreign ions. A **common ion** is an ion which is also a constituent of the precipitate. With silver chloride for example, both silver and chloride ions are common ions, but all other ions are foreign. It can be said in general, that the solubility of a precipitate decreases considerably if one of the common ions is present in excess though this effect might be counterbalanced by the formation of a soluble complex with the excess of the common ion. The solubility of silver cyanide, for example, can be suppressed by adding silver ions in excess to the solution. If, on the other hand, cyanide ions are added in excess, first the solubility decreases slightly, but when larger amounts of cyanide are added, the precipitate dissolves completely owing to the formation of dicyanoargentate $[Ag(CN)_2]^-$ complex ion. In the presence of a **foreign ion**, the solubility of a precipitate increases, but this increase is generally slight, unless a chemical reaction (like complex formation or an acid-base reaction) takes place between the precipitate and the foreign ion, when the increase of solubility is more marked. Because of the importance of the effects of common and foreign ions on the solubility of precipitates in qualitative inorganic analysis, these will be dealt with in more detail in subsequent sections.

I.27 SOLUBILITY PRODUCT The saturated solution of a salt, which contains also an excess of the undissolved substance, is an equilibrium system to which the law of mass action can be applied. If, for example, silver chloride precipitate is in equilibrium with its saturated solution, the following equilibrium is established:

$$AgCl \rightleftarrows Ag^+ + Cl^-$$

This is a heterogeneous equilibrium, as the AgCl is in the solid phase, while the

Ag^+ and Cl^- ions are in the dissolved phase. The equilibrium constant can be written as

$$K = \frac{[Ag^+][Cl^-]}{[AgCl]}$$

The concentration of silver chloride in the solid phase is invariable and therefore can be included into a new constant K_s, termed the **solubility product**:

$$K_s = [Ag^+][Cl^-]$$

Thus in a saturated solution of silver chloride, at constant temperature (and pressure) the product of concentration of silver and chloride ions is constant.

What has been said for silver chloride can be generalized. For the saturated solution of an electrolyte $A_{\nu_A}B_{\nu_B}$ which ionizes into $\nu_A A^{m+}$ and $\nu_B B^{n-}$ ions:

$$A_{\nu_A}B_{\nu_B} \rightleftarrows \nu_A A^{m+} + \nu_B B^{n-}$$

the solubility product (K_s) can be expressed as

$$K_s = [A^{m+}]^{\nu_A} \times [B^{n-}]^{\nu_B}$$

Thus it can be stated that, in a saturated solution of a sparingly soluble electrolyte, the product of concentrations of the constituent ions for any given temperature is constant, the ion concentration being raised to powers equal to the respective numbers of ions of each kind produced by the dissociation of one molecule of the electrolyte. This principle was stated first by W. Nernst in 1889.

The ion concentrations in the expression of the solubility product are to be given in mol ℓ^{-1} units. The unit of K_s itself is therefore (mol $\ell^{-1})^{\nu_A+\nu_B}$.

In order to explain many of the precipitation reactions in qualitative inorganic analysis, values of solubility products of precipitates are useful. Some of the most important values are collected in Table I.12. The values were selected from the most trustworthy sources in the literature. The values of solubility products are determined by various means, and the student is referred to textbooks of physical chemistry for a description of these methods. Many of these constants are obtained by indirect means, such as measurements of electrical conductivity, the e.m.f. of cells, or from thermodynamic calculations, using data obtained by calorimetry. The various methods however, do not always give consistent results, and this may be attributed to various causes including the following. In some cases the physical structure, and hence the solubility, of the precipitate at the time of precipitation is not the same as that of an old or stabilized precipitate; this may be due to the process known as 'ripening', which is a sort of recrystallization, or it may be due to a real change of crystal structure. Thus, for nickel sulphide three forms (α, β, and γ) have been reported with solubility products of 3×10^{-21}, 1×10^{-26} and 2×10^{-28} respectively; another source gives the value of $1{\cdot}4 \times 10^{-24}$. The α-form is said to be that of the freshly precipitated substance: the other forms are produced on standing. For cadmium sulphide a value of $1{\cdot}4 \times 10^{-28}$ has been computed from thermal and other data (Latimer, 1938), whilst direct determination leads to a solubility product of $5{\cdot}5 \times 10^{-25}$ (Belcher, 1949).

The solubility product relation explains the fact that the solubility of a substance decreases considerably if a reagent containing a common ion with the substance is added. Because the concentration of the common ion is high, that of the other ion must become low in the saturated solution of the substance;

Table I.12 Solubility products of precipitates at room temperature

Substance	Solubility product	Substance	Solubility product
AgBr	$7{\cdot}7\times10^{-13}$	FeS	$4{\cdot}0\times10^{-19}$
$AgBrO_3$	$5{\cdot}0\times10^{-5}$	Hg_2Br_2	$5{\cdot}2\times10^{-23}$
AgCNS	$1{\cdot}2\times10^{-12}$	Hg_2Cl_2	$3{\cdot}5\times10^{-18}$
AgCl	$1{\cdot}5\times10^{-10}$	Hg_2I_2	$1{\cdot}2\times10^{-28}$
$Ag_2C_2O_4$	$5{\cdot}0\times10^{-12}$	Hg_2S	1×10^{-45}
Ag_2CrO_4	$2{\cdot}4\times10^{-12}$	HgS	4×10^{-54}
AgI	$0{\cdot}9\times10^{-16}$	$K_2[PtCl_6]$	$1{\cdot}1\times10^{-5}$
$AgIO_3$	$2{\cdot}0\times10^{-8}$	$MgCO_3$	$1{\cdot}0\times10^{-5}$
Ag_3PO_4	$1{\cdot}8\times10^{-18}$	MgC_2O_4	$8{\cdot}6\times10^{-5}$
Ag_2S	$1{\cdot}6\times10^{-49}$	MgF_2	$7{\cdot}0\times10^{-9}$
Ag_2SO_4	$7{\cdot}7\times10^{-5}$	$Mg(NH_4)PO_4$	$2{\cdot}5\times10^{-13}$
$Al(OH)_3$	$8{\cdot}5\times10^{-23}$	$Mg(OH)_2$	$3{\cdot}4\times10^{-11}$
$BaCO_3$	$8{\cdot}1\times10^{-9}$	$Mn(OH)_2$	$4{\cdot}0\times10^{-14}$
BaC_2O_4	$1{\cdot}7\times10^{-7}$	MnS	$1{\cdot}4\times10^{-15}$
$BaCrO_4$	$1{\cdot}6\times10^{-10}$	$Ni(OH)_2$	$8{\cdot}7\times10^{-19}$
$BaSO_4$	$9{\cdot}2\times10^{-11}$	NiS	$1{\cdot}4\times10^{-24}$
Bi_2S_3	$1{\cdot}6\times10^{-72}$	$PbBr_2$	$7{\cdot}9\times10^{-5}$
$CaCO_3$	$4{\cdot}8\times10^{-9}$	$PbCl_2$	$2{\cdot}4\times10^{-4}$
CaC_2O_4	$2{\cdot}6\times10^{-9}$	$PbCO_3$	$3{\cdot}3\times10^{-14}$
CaF_2	$3{\cdot}2\times10^{-11}$	$PbCrO_4$	$1{\cdot}8\times10^{-14}$
$CaSO_4$	$2{\cdot}3\times10^{-4}$	PbF_2	$3{\cdot}7\times10^{-8}$
CdS	$1{\cdot}4\times10^{-28}$	PbI_2	$8{\cdot}7\times10^{-9}$
$Co(OH)_2$	$1{\cdot}6\times10^{-18}$	$Pb_3(PO_4)_2$	$1{\cdot}5\times10^{-32}$
$Co(OH)_3$	$2{\cdot}5\times10^{-43}$	PbS	5×10^{-29}
CoS	3×10^{-26}	$PbSO_4$	$2{\cdot}2\times10^{-8}$
$Cr(OH)_3$	$2{\cdot}9\times10^{-29}$	$SrCO_3$	$1{\cdot}6\times10^{-9}$
CuBr	$1{\cdot}6\times10^{-11}$	SrC_2O_4	$5{\cdot}0\times10^{-8}$
CuCl	$1{\cdot}0\times10^{-6}$	$SrSO_4$	$2{\cdot}8\times10^{-7}$
CuI	$5{\cdot}0\times10^{-12}$	TlCl	$1{\cdot}5\times10^{-4}$
CuS	1×10^{-44}	TlI	$2{\cdot}8\times10^{-8}$
Cu_2S	2×10^{-47}	Tl_2S	1×10^{-22}
CuSCN	$1{\cdot}6\times10^{-11}$	$Zn(OH)_2$	1×10^{-17}
$Fe(OH)_2$	$4{\cdot}8\times10^{-16}$	ZnS	1×10^{-23}
$Fe(OH)_3$	$3{\cdot}8\times10^{-38}$		

The dimension of the solubility product is $(\text{mol}\ \ell^{-1})^{\nu_A+\nu_B}$, the individual ion concentrations therefore are always expressed in mol ℓ^{-1} units.

the excess of the substance will therefore be precipitated. If therefore one ion has to be removed from the solution by precipitation, the reagent must be applied in excess. Too great excess of the reagent may however do more harm than good, as it may increase the solubility of the precipitate because of complex formation.

The effect of foreign ions on the solubility of precipitates is just the opposite; the solubility increases slightly in the presence of foreign ions.

To explain the effect of foreign ions on the solubility of precipitates, one has to bear in mind that the solubility product relation, in the strictest sense, has to be expressed in terms of activities. For the saturated solution of the electrolyte $A_{\nu_A}B_{\nu_B}$, which ionizes into $\nu_A A^{m+}$ and $\nu_B B^{n-}$ ions

$$A_{\nu_A}B_{\nu_B} \rightleftarrows \nu_A A^{m+} + \nu_B B^{n-}$$

the solubility product (K_s) must be expressed as

$$K_s = a_{A^{m+}}^{\nu A} \times a_{B^{n-}}^{\nu B} = f_{A^{m+}}^{\nu A} \times f_{B^{m+}}^{\nu B} \times [A^{m+}]^{\nu A} \times [B^{n-}]^{\nu B}$$

The activity coefficients $f_{A^{m+}}$ and $f_{B^{n-}}$ depend however on the concentration of all ions (that is common and foreign ions) in the solution. The higher the total concentration of the ions in the solution, the higher the ionic strength, consequently the lower the activity coefficients (cf. Section **I.14**). As the solubility product must remain constant, the concentrations $[A^{m+}]$ and $[B^{n-}]$ must increase to counterbalance the decrease of the activity coefficients; hence the increase in solubility.

The graphs on Fig. I.11 illustrate the effects of common and foreign ions

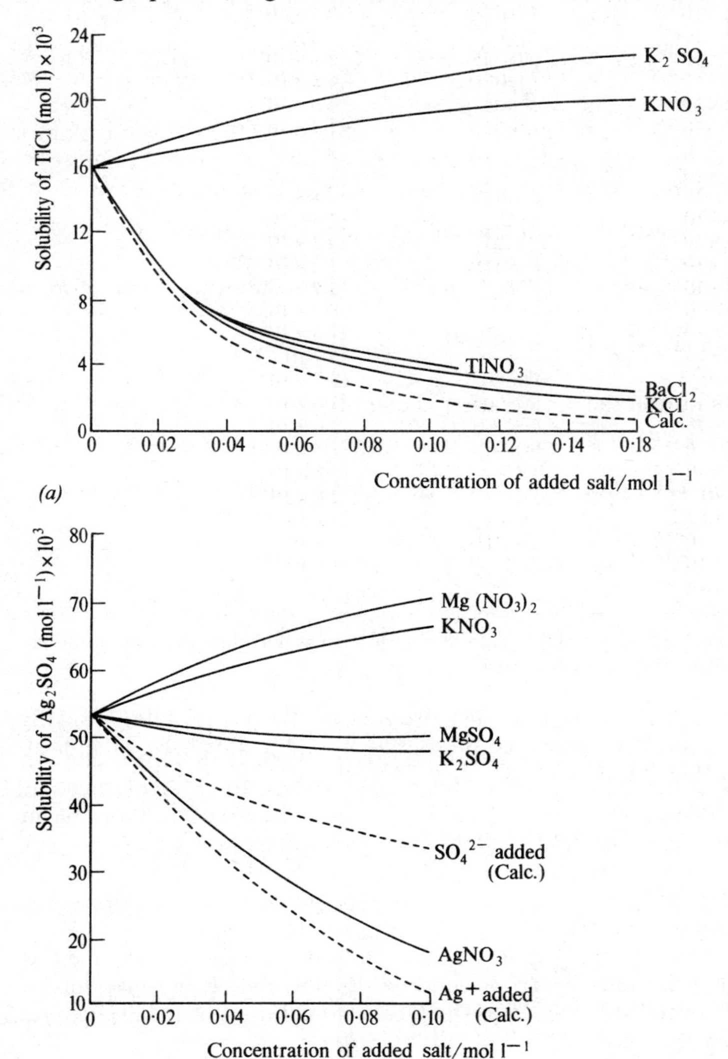

Fig. I.11

more quantitatively. In the case of TlCl the three salts with common ions decrease the solubility of the salt considerably, though somewhat less than the solubility product predicts (dotted line), because of the simultaneous decrease of the activity coefficient (the so-called **salt effect**). The two salts with no common ions, on the other hand, increase the solubility, the divalent sulphate exerting the greater effect. This is quite predictable, as in the expression (cf. Section **I.14**)

$$\log f_i = -Az_i^2\sqrt{I}$$

the charge of the ion, z, has an emphasized role. In the case of Ag_2SO_4 the excess of $AgNO_3$ decreases the solubility somewhat less than simple theory (which does not take activity coefficients into consideration) predicts; $MgSO_4$ and K_2SO_4 decrease the solubility only slightly, whilst KNO_3 and $Mg(NO_3)_2$ markedly increase the solubility with the divalent magnesium ion causing the greater increase. The effects of $MgSO_4$ and K_2SO_4 are obviously the results of simultaneous common-ion and salt effects.

The following examples may help the student to understand the subject more fully. Note that in these examples activities are not taken into consideration; solubility products are everywhere expressed in terms of concentrations.

Example 14 A saturated solution of silver chloride contains 0·0015 g of dissolved substance in 1 litre. Calculate the solubility product.

The relative molecular mass of AgCl is 143·3. The solubility (S) therefore is

$$S = \frac{0{\cdot}0015}{143{\cdot}3} = 1{\cdot}045\times10^{-5}\ \text{mol}\ \ell^{-1}$$

In the saturated solution the dissociation is complete:

$$AgCl \rightleftarrows Ag^+ + Cl^-$$

Thus, one mole of AgCl produces 1 mole each of Ag^+ and Cl^-. Hence

$$[Ag^+] = 1{\cdot}045\times10^{-5}\ \text{mol}\ \ell^{-1}$$
$$[Cl^-] = 1{\cdot}045\times10^{-5}\ \text{mol}\ \ell^{-1}$$

and

$$K_s = [Ag^+]\times[Cl^-] = 1{\cdot}045\times10^{-5}\times1{\cdot}045\times10^{-5}$$
$$= 1{\cdot}1\times10^{-10}\ (\text{mol}\ \ell^{-1})^2$$

Example 15 Calculate the solubility product of silver chromate, knowing that 1 litre of the saturated solution contains $3{\cdot}57\times10^{-2}$ g of dissolved material. The relative molecular mass of Ag_2CrO_4 is 331·7, hence the solubility

$$S = \frac{3{\cdot}57\times10^{-2}}{331{\cdot}7} = 1{\cdot}076\times10^{-4}\ \text{mol}\ \ell^{-1}$$

The dissociation

$$Ag_2CrO_4 \rightleftarrows 2Ag^+ + CrO_4^{2-}$$

is complete; 1 mole of Ag_2CrO_4 yields 2 moles of Ag^+ and 1 mole of CrO_4^{2-}. Thus, the concentrations of the two ions are as follows:

$$[Ag^+] = 2S = 2{\cdot}152\times10^{-4}$$
$$[CrO_4^{2-}] = S = 1{\cdot}076\times10^{-4}$$

The solubility product

$$K_s = [Ag^+]^2 \times [CrO_4^{2-}] = (2{\cdot}152 \times 10^{-4})^2 \times 1{\cdot}076 \times 10^{-4}$$
$$= 5{\cdot}0 \times 10^{-12}\ (\text{mol}\ \ell^{-1})^3$$

Example 16 The solubility product of lead phosphate is $1{\cdot}5 \times 10^{-32}$. Calculate the concentration of its saturated solution in g ℓ^{-1} units.

The dissociation equation is

$$Pb_3(PO_4)_2 \rightleftarrows 3Pb^{2+} + 2PO_4^{3-}$$

If S is the solubility (in mol ℓ^{-1}), we have

$$[Pb^{3+}] = 3\,S$$

and

$$[PO_4^{3-}] = 2\,S$$

or

$$K_s = 1{\cdot}5 \times 10^{-32} = [Pb^{2+}]^3 \times [PO_4^{3-}]^2 = (3\,S)^3 \times (2\,S)^2$$

or

$$1{\cdot}5 \times 10^{-32} = 108\,S^5$$

Thus

$$S = \sqrt[5]{\frac{1{\cdot}5 \times 10^{-32}}{108}} = 1{\cdot}68 \times 10^{-7}\ \text{mol}\ \ell^{-1}$$

The relative molecular mass of $Pb_3(PO_4)_2$ is 811·5. Thus the amount of substance dissolved per litre (m) is

$$m = 811{\cdot}5 \times 1{\cdot}68 \times 10^{-7} = 1{\cdot}37 \times 10^{-4}\ \text{g}\ \ell^{-1}$$

Example 17 Given that the solubility product of magnesium hydroxide is $3{\cdot}4 \times 10^{-11}$, calculate the concentration of hydroxyl ions in a saturated aqueous solution. The dissociation

$$Mg(OH)_2 \rightleftarrows Mg^{2+} + 2OH^-$$

being complete, we may put $[Mg^{2+}] = x$; then $[OH^-] = 2x$. The solubility product can thus be written as

$$K_s = [Mg^{2+}] \times [OH^-]^2 = x(2x)^2 = 4x^3 = 3{\cdot}4 \times 10^{-11}$$

Hence

$$x = \sqrt[3]{\frac{3{\cdot}4 \times 10^{-11}}{4}} = 2{\cdot}04 \times 10^{-4}\ \text{mol}\ \ell^{-1}$$

and as $[OH^-] = 2x$, we can calculate

$$[OH^-] = 2 \times 2{\cdot}04 \times 10^{-4} = 4{\cdot}08 \times 10^{-4}\ \text{mol}\ \ell^{-1}$$

Example 18 What is the concentration of silver ions (in mol ℓ^{-1} units) remaining in a solution of $AgNO_3$ after the addition of HCl to make the final chloride-ion concentration 0·05 molar?

The solubility product of AgCl is

$$K_s = 1{\cdot}5 \times 10^{-10} = [Ag^+] \times [Cl^-]$$

In the final solution $[Cl^-] = 5 \times 10^{-2}$ mol ℓ^{-1}. Thus

$$[Ag^+] = \frac{K_s}{[Cl^-]} = \frac{1{\cdot}5 \times 10^{-10}}{5 \times 10^{-2}} = 3 \times 10^{-9} \text{ mol } \ell^{-1}$$

Example 19 To 100 ml of a solution, which contains $8{\cdot}29 \times 10^{-3}$ g lead ions, 100 ml of 10^{-3}M sulphuric acid is added. How much lead remains in the solution unprecipitated?

When mixing the reagents, $PbSO_4$ is precipitated:

$$Pb^{2+} + SO_4^{2-} \rightleftarrows PbSO_4$$

The solubility product of $PbSO_4$ is $2{\cdot}2 \times 10^{-8}$ and the relative atomic mass of Pb is 207·2.

1 litre of the same solution would contain $8{\cdot}29 \times 10^{-2}$ g of Pb^{2+}; the molar concentration of Pb^{2+} in the original solution is

$$[Pb^{2+}]_{or} = \frac{8{\cdot}29 \times 10^{-2}}{207{\cdot}2} = 4 \times 10^{-4} \text{ mol } \ell^{-1}$$

while that of sulphate ions is, as given in the question

$$[SO_4^{2-}]_{or} = 10^{-3} \text{ mol } \ell^{-1}$$

At the instant of mixing, these concentrations are halved, because each solution is diluted to twice its original volume. At the same time precipitation occurs. If x mol of Pb^{2+} is precipitated, this will carry again x mol of SO_4^{2-} into the precipitate. Thus, when equilibrium is reached, the concentrations of these ions can be expressed as

$$[Pb^{2+}] = \frac{4 \times 10^{-4}}{2} - x = 2 \times 10^{-4} - x \text{ mol } \ell^{-1}$$

and

$$[SO_4^{2-}] = \frac{10^{-3}}{2} - x = 5 \times 10^{-4} - x \text{ mol } \ell^{-1}$$

The solubility product

$$K_s = [Pb^{2+}][SO_4^{2-}]$$

can therefore be expressed as

$$2{\cdot}2 \times 10^{-8} = (2 \times 10^{-4} - x)(5 \times 10^{-4} - x) \qquad \text{(i)}$$

Rearranging the above equation we obtain

$$x^2 - 7 \times 10^{-4} x + 7{\cdot}8 \times 10^{-8} = 0$$

From which x can be expressed as

$$x = \frac{7 \times 10^{-4} \pm \sqrt{49 \times 10^{-8} - 4 \times 7 \times 10^{-8}}}{2}$$

which yields the two roots, $x_1 = 5{\cdot}61 \times 10^{-4}$ and $x_2 = 1{\cdot}4 \times 10^{-4}$. The two roots both fulfil equation (i), but of these x_1 has obviously no physical meaning, as it would yield negative concentration values for both ions in the expression of the solubility product. The value $x_2 = 1{\cdot}4 \times 10^{-4}$ is therefore the one which has to be taken into consideration. The concentration of lead ions in the final solution thus becomes

$$[Pb^{2+}] = 2\times10^{-4} - 1{\cdot}4\times10^{-4} = 6\times10^{-5}\ \text{mol}\ \ell^{-1}$$

In 200 ml of solution we have one-fifth of this number of moles, i.e. $1{\cdot}2\times10^{-5}$ mol Pb^{2+}. Multiplying by the relative atomic mass we can calculate the amount of lead which was left unprecipitated:

$$m_{Pb^{2+}} = 202{\cdot}7\times1{\cdot}2\times10^{-5} = 2{\cdot}43\times10^{-3}\ \text{g}$$

showing that under such circumstances about one-third of the original amount of lead ($8{\cdot}29\times10^{-3}$ g) remains dissolved.

I.28 APPLICATIONS OF THE SOLUBILITY PRODUCT RELATION

In spite of its limitations (as outlined in the previous section) the solubility product relation is of great value in qualitative analysis, since with its aid it is possible not only to explain but also to predict precipitation reactions. The solubility product is in reality an ultimate value which is attained by the ionic product when equilibrium has been established between the solid phase of the slightly soluble salt and the solution. If conditions are such that the ionic product is different from the solubility product, the system will seek to adjust itself in such a manner that the ionic product attains the value of the solubility product. Thus, if the ionic product is arbitrarily made greater than the solubility product, for example by the addition of another salt with a common ion, the adjustment of the system results in the precipitation of the solid salt. Conversely, if the ionic product is made smaller than the solubility product, as, for instance, by diminishing the concentration of one of the ions, equilibrium in the system is attained by some of the solid salt passing into solution.

As an example of the formation of a precipitate, let us consider the case of silver chloride. The solubility product is

$$K_s = [Ag^+]\times[Cl^-] = 1{\cdot}5\times10^{-10}$$

Let us suppose that to a solution which is 0·1 molar in silver ions we add enough potassium chloride to produce momentarily a chloride concentration 0·01 molar. The ionic product in such a case would be $0{\cdot}1\times0{\cdot}01 = 10^{-3}$. As $10^{-3} > 1{\cdot}5\times10^{-10}$, equilibrium will not exist and precipitation of silver chloride will take place

$$Ag^+ + Cl^- \rightleftarrows AgCl$$

until the value of ionic product has been reduced to that of the solubility product, i.e. until $[Ag^+]\times[Cl^-] = 1{\cdot}5\times10^{-10}$. At this point equilibrium is reached (cf. Section **I.13**), that is the rate of formation of silver chloride precipitate equals the rate of its dissolution. The actual ionic concentrations can easily be calculated (cf. *Example* 19). Such a solution is now saturated for silver chloride. If then we add either a soluble chloride or a silver salt in small quantity, a slight further precipitation of silver chloride takes place, until equilibrium is reached again and so on. It should be pointed out, that the solubility product defines a state of equilibrium, but does not provide information about the rate at which this equilibrium is established. The rate of formation of precipitates will be discussed in a separate section (cf. Section **I.29**).

Attention must also be drawn to the fact that complete precipitation of a sparingly soluble electrolyte is impossible, because no matter how much the concentration of one ion is arbitrarily increased (and there are physical limitations in this respect too), the concentration of the other ion cannot be decreased to zero since the solubility product has a constant value. The concentration of the ion can of course be reduced to a very small value indeed: in *Example* 18 the silver-ion concentration is as low as 3×10^{-9} mol ℓ^{-1} (or $3 \times 10^{-9} \times 107{\cdot}87 = 3{\cdot}236 \times 10^{-7}$ g ℓ^{-1}) which is negligible for most practical purposes. In practice, it is found that, after a certain point, a further excess of precipitant does not materially increase the weight of precipitate. Indeed, a large excess of the precipitant may cause some of the precipitate to dissolve, either as a result of increased salt effect (see Section **I.27**) or as a result of complex ion formation (for more details, see Section **I.31**). Some results of Forbes (1911), collected in Table I.13, on the effect of larger amounts of sodium chloride on the solubility

Table I.13 The effect of sodium chloride on the solubility of silver chloride

NaCl present mol ℓ^{-1}	**Dissolved Ag^+** mol ℓ^{-1}
0·933	$8{\cdot}6 \times 10^{-5}$
1·433	$1{\cdot}84 \times 10^{-4}$
2·272	$5{\cdot}74 \times 10^{-4}$
3·000	$1{\cdot}19 \times 10^{-3}$
4·170	$3{\cdot}34 \times 10^{-3}$
5·039	$6{\cdot}04 \times 10^{-3}$

of silver chloride illustrate this point. These results show why only a moderate excess of the reagent should be used when carrying out precipitation reactions.

On the basis of this general discussion, we can now consider some direct applications of the solubility product principle to quantitative inorganic analysis.

Precipitation of sulphides Hydrogen sulphide gas is a frequently used reagent in qualitative inorganic analysis. When hydrogen sulphide gas is passed into a solution, metal sulphides are precipitated. For this precipitation the rule mentioned above can be applied: precipitation may take place only if the product of concentrations of metal ions and sulphide ions (taken at proper powers) exceed the value of the solubility product. While the concentration of metal ions usually does fall into the range of $1–10^{-3}$ mol ℓ^{-1}, the concentration of sulphide ion may vary considerably, and can easily be selected by the adjustment of the *p*H of the solution to a suitable value.

This variation of the sulphide ion concentration with *p*H is due to the fact that hydrogen sulphide is a weak acid itself, with two dissociation steps:

$$H_2S \rightleftarrows H^+ + HS^-$$

with

$$K_1 = \frac{[H^+][HS^-]}{[H_2S]} = 9{\cdot}1 \times 10^{-8}$$

and

$$HS^- \rightleftarrows H^+ + S^{2-}$$

with

$$K_2 = \frac{[H^+][S^{2-}]}{[HS^-]} = 1{\cdot}2 \times 10^{-15}$$

Multiplying the two equations we obtain

$$\frac{[H^+]^2[S^{2-}]}{[H_2S]} = K_1K_2 = 1{\cdot}09 \times 10^{-22} \approx 10^{-22}$$

At room temperature (25°C) and atmospheric pressure the saturated aqueous solution of hydrogen sulphide is almost exactly 0·1 molar. As the substance is a weak acid, its dissociation may be ignored, and the value $[H_2S] = 0{\cdot}1$ inserted into the above equation:

$$\frac{[H^+]^2[S^{2-}]}{0{\cdot}1} = 10^{-22}$$

The expression can be rearranged to give

$$[S^{2-}] = \frac{10^{-23}}{[H^+]^2} \qquad \text{(i)}$$

This equation shows the correlation between hydrogen-ion concentration and the concentration of sulphide ions. It can be seen that the sulphide-ion concentration is inversely proportional to the square of hydrogen-ion concentration. In strongly acid solutions ($[H^+] = 1$) the sulphide-ion concentration may not be greater than 10^{-23} mol ℓ^{-1}. Under such circumstances only the most insoluble sulphides can be precipitated. In a neutral solution ($[H^+] = 10^{-7}$) the sulphide-ion concentration rises to 10^{-9} mol ℓ^{-1}, enabling the precipitation of metal sulphides with higher solubility products.

Equation (i) can be simplified further if we introduce the quantity *p*S, the sulphide ion exponent. Its definition is analogous to that of *p*H:

$$pS = -\log[S^{2-}]$$

Using such a notation, equation (i) becomes

$$pS = 23 - 2pH$$

This equation can easily be memorized and used for quick calculations. The equation is strictly valid for the *p*H range 0–8; over *p*H = 8 the dissociation of hydrogen sulphide cannot be disregarded any more, and therefore the simple treatment outlined above cannot be used. With proper mathematical treatment the sulphide ion exponent even for *p*H above 8 can be calculated; results of such calculations are summarized on the graph of Fig. I.12. This graph can be used if predictions on the precipitation of sulphides are required. This is illustrated in the following examples.

Example 20 Given is a solution containing 0·1M $CuSO_4$ and 0·1M $MnSO_4$. What happens if (*a*) the solution is acidified to achieve *p*H = 0 and saturated with hydrogen sulphide gas, and (*b*) if ammonium sulphide solution is added, which adjusts the *p*H to 10? The solubility products of CuS and MnS are

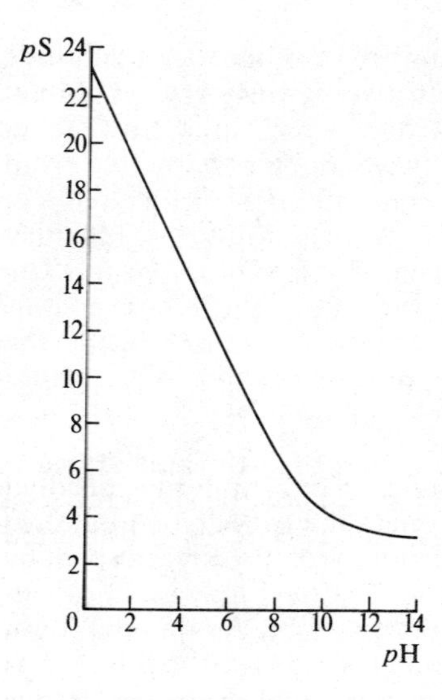

Fig. I.12

1×10^{-44} and $1{\cdot}4 \times 10^{-15}$ respectively (cf. Table I.12).

(*a*) From Fig. I.12, at pH $= 0$ the value of pS is 23, that is $[S^{2-}] = 10^{-23}$ mol ℓ^{-1}. The metal-ion concentrations being 10^{-1} mol ℓ^{-1} in both cases, the product of ion concentrations is 10^{-29} for both ions. Because $10^{-24} > 1 \times 10^{-44}$, copper sulphide will be precipitated, while as $10^{-24} <$ $1{\cdot}4 \times 10^{-15}$ manganese sulphide will not be precipitated at all. It is possible therefore to separate copper and manganese at pH $= 0$.

(*b*) Using Fig. I.12, we find at pH $= 10$ the value 4 for pS. This corresponds to $[S^{2-}] = 10^{-4}$ mol ℓ^{-1}. The product of ion concentrations is 10^{-5} for both metal ions. As $10^{-5} > 1{\cdot}4 \times 10^{-15} > 1 \times 10^{-44}$, both CuS and MuS will be precipitated under such conditions.

Example 21 Given a 0·01M solution of $ZnCl_2$, what is the lowest pH at which ZnS can be precipitated?

From Table I.12 the solubility product of ZnS is taken as 1×10^{-23}. Thus

$$[Zn^{2+}][S^{2-}] = 10^{-23}$$

and $[Zn^{2+}] = 10^{-2}$, the sulphide-ion concentration in the saturated solution is

$$[S^{2-}] = \frac{10^{-23}}{10^{-2}} = 10^{-21}$$

and pS $= 21$. From the equation

$$pS = 23 - 2\,pH$$

the minimum value of pH at which precipitation occurs is

$$pH = \frac{23-pS}{2} = \frac{23-21}{2} = 1$$

In fact, if we want to precipitate ZnS quantitatively, the pH must be raised even higher. At pH 4 to 5, that is from a solution containing an acetate buffer, ZnS will precipitate easily.

If similar calculations are carried out for a number of other metal sulphide precipitates it is easy to classify these metals into two distinct groups. Metal ions like Ag^+, Pb^{2+}, Hg_2^{2+}, Bi^{3+}, Cu^{2+}, Cd^{2+}, Sn^{2+}, As^{3+} and Sb^{3+} form sulphides under virtually any circumstances e.g. they can be precipitated from strongly acid (pH = 0) solutions. Other metal ions, like Fe^{2+}, Fe^{3+}, Ni^{2+}, Co^{2+}, Mn^{2+}, and Zn^{2+} cannot be precipitated from acid solutions, but they will form sulphides in neutral or even slightly acid (buffered) solutions. The difference is used in the analytical classification of these ions; the first set of ions mentioned form the so-called first and second groups of cations, while the second set are members of the third group. The separation of these ions is based on the same phenomenon.

Precipitation and dissolution of metal hydroxides The solubility product principle can also be applied to the formation of metal hydroxide precipitates; these are also made use of in qualitative inorganic analysis. Precipitates will be formed only if the concentrations of the metal and hydroxyl ions are momentarily higher than those permitted by the solubility product. As the metal-ion concentration in actual samples does not vary much ($10^{-1}-10^{-3}$ mol ℓ^{-1} is the usual range), it is the hydroxyl-ion concentration which has the decisive role in the formation of such precipitates. Because of the fact that in aqueous solutions the product of hydrogen- and hydroxyl-ion concentrations is strictly constant ($K_w = 10^{-14}$ at 25°C, cf. Section **I.18**), the formation of a metal-hydroxide precipitate depends mainly on the pH of the solution. Using the solubility product principle, it is possible to calculate the (minimum) pH required for the precipitation of a metal hydroxide.

Example 22 Calculate the pH (*a*) at which the precipitation of $Fe(OH)_3$ begins from a 0·01M solution of $FeCl_3$, and (*b*) the pH at which the concentration of Fe^{3+} ions in the solution does not exceed 10^{-5}M, that is to say, when the precipitation is practically complete. The value of the solubility product (cf. Table I.12) is

$$K_s = [Fe^{3+}][OH^-]^3 = 3{\cdot}8 \times 10^{-38}$$

(*a*) With $[Fe^{3+}] = 10^{-2}$, the hydroxyl-ion concentration

$$[OH^-] = \sqrt[3]{\frac{K_s}{[Fe^{3+}]}} = \sqrt[3]{\frac{3{\cdot}8 \times 10^{-38}}{10^{-2}}} = 1{\cdot}56 \times 10^{-12}$$

The hydrogen-ion concentration is

$$[H^+] = \frac{K_w}{[OH^-]} = \frac{10^{-14}}{1{\cdot}56 \times 10^{-12}} = 6{\cdot}41 \times 10^{-3}$$

and the pH

$$pH = -\log[H^+] = -\log(6{\cdot}41 \times 10^{-3}) = 2{\cdot}19$$

Thus, $Fe(OH)_3$ will start to precipitate at pH = 2·19.

(*b*) With $[Fe^{3+}] = 10^{-5}$, the hydroxyl-ion concentration

$$[OH^-] = \sqrt[3]{\frac{K_s}{[Fe^{3+}]}} = \sqrt[3]{\frac{3{\cdot}8 \times 10^{-38}}{10^{-5}}} = 1{\cdot}56 \times 10^{-11}$$

The hydrogen-ion concentration

$$[H^+] = \frac{K_w}{[OH^-]} = \frac{10^{-14}}{1{\cdot}56 \times 10^{-11}} = 6{\cdot}41 \times 10^{-4}$$

and the pH

$$pH = -\log[H^+] = -\log(6{\cdot}41 \times 10^{-4}) = 3{\cdot}19$$

Thus at pH = 3·19 $Fe(OH)_3$ is completely precipitated.

Results of similar calculations, made on several metal-hydroxide precipitates, are summarized on the graphs of Fig. I.13. (Erdey, 1963). The shaded area in each graph shows the pH region in which the precipitate is formed; the areas left white correspond to circumstances under which the ions are in the dissolved phase. The upper ends of the oblique lines, drawn as boundaries, correspond to solutions containing 10^{-2} mol ℓ^{-1} metal ions, these are therefore the pH values at which precipitation begins. The lower ends, on the other hand, mark

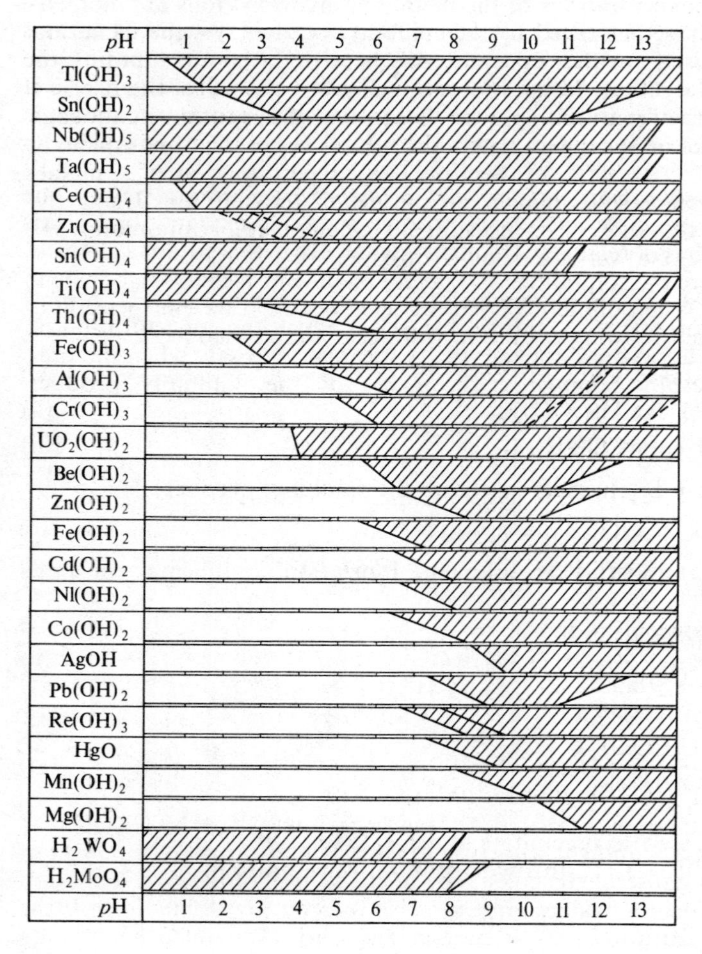

Fig. I.13

the pH at which precipitation becomes complete (i.e. where the concentration of metal ions in the solution decreases below 10^{-5} mol ℓ^{-1}). The figure also shows the dissolution pH values of amphoteric hydroxides in alkalis. Some of the hydroxides (like AgOH or $Cu(OH)_2$) may dissolve in ammonia solution at even lower pH values owing to the formation of ammine complexes, these pH values however are not shown on the diagram. Dashed lines correspond to irreversible precipitation and dissolution, or to uncertain values in general.

From the diagram important conclusions may be drawn, and even separation schemes may be devised. If for example we have a solution which contains Fe^{3+}, Ni^{2+} and Zn^{2+} ions, the following method could be used for their separation: first the pH is adjusted to 4, when $Fe(OH)_3$ is precipitated. The precipitate is filtered, washed, and the filtrate treated with an excess of sodium hydroxide. By adjusting the pH to 12–13 in this way, $Ni(OH)_2$ is precipitated, but the $Zn(OH)_2$, formed during the addition of the reagent, is redissolved. The $Ni(OH)_2$ precipitate can then be filtered and washed. The three ions are thus separated.

It is useful to state that there are complete separation systems of cations devised, based entirely on the separation of metal hydroxides, omitting the use of sulphides altogether. These however will not be discussed here. As it will be seen later, the separation of ions within the third group of cations, is almost entirely based on the differences in the solubilities of their hydroxides.

Solubility of sparingly soluble salts of weak acids in strong mineral acids The solubility product principle enables us to give a simple explanation of this phenomenon, which is of relatively frequent occurrence in quantitative analysis. Typical examples are the solubilities of calcium oxalate or barium carbonate in hydrochloric acid. When dilute hydrochloric acid is added to a suspension of calcium oxalate, the following equilibria will occur simultaneously:

$$Ca(COO)_2(s) \rightleftarrows Ca^{2+} + (COO)_2^{2-} \quad \text{(i)}$$

$$(COO)_2^{2-} + H^+ \rightleftarrows H(COO)_2^- \quad \text{(ii)}$$

$$H(COO)_2^- + H^+ \rightleftarrows (COOH)_2 \quad \text{(iii)}$$

Equilibrium (i) can be characterized with the solubility product:

$$K_s = [Ca^{2+}] \times [(COO)_2^{2-}] = 2{\cdot}6 \times 10^{-9} \quad \text{(iv)}$$

For equilibria (ii) and (iii) the following ionization equilibrium constants will be valid

$$K_{a2} = \frac{[H^+][(COO)_2^{2-}]}{[H(COO)_2^-]} = 5{\cdot}4 \times 10^{-5} \quad \text{(v)}$$

and

$$K_{a1} = \frac{[H^+][H(COO)_2^-]}{[(COOH)_2]} = 2{\cdot}4 \times 10^{-2} \quad \text{(vi)}$$

It is evident from equation (v) that if $[H^+]$ is large, it will decrease $[(COO)_2^{2-}]$ by forming $H(COO)_2^-$ (and subsequently $(COOH)_2$ in a saturated solution of calcium oxalate. Because of the constancy of the solubility product (equation (iv)) some of the calcium oxalate must dissolve to produce oxalate ions. If the hydrogen-ion concentration is high enough, the whole amount of precipitate may dissolve.

Similarly, if hydrochloric acid is added to a solution containing barium carbonate precipitate, the following equilibria will exist in the solution:

$$BaCO_3(s) \rightleftarrows Ba^{2+} + CO_3^{2-} \qquad (vii)$$

$$H^+ + CO_3^{2-} \rightleftarrows HCO_3^- \qquad (viii)$$

$$H^+ + HCO_3^- \rightleftarrows H_2CO_3 \rightleftarrows H_2O + CO_2(g) \qquad (ix)$$

These equilibria can be characterized with the solubility product

$$K_s = [Ba^{2+}] \times [CO_3^{2-}] = 8{\cdot}1 \times 10^{-9} \qquad (x)$$

and with the ionization constants of carbonic acid

$$K_1 = \frac{[H^+][HCO_3^-]}{[H_2CO_3]} = 4{\cdot}31 \times 10^{-7} \qquad (xi)$$

and

$$K_2 = \frac{[H^+][CO_3^{2-}]}{[HCO_3^-]} = 5{\cdot}61 \times 10^{-11} \qquad (xii)$$

Now, because of the low values of the two ionization constants of carbonic acid, hydrogen ions will immediately combine with carbonate ions present in the solution (owing to process vii) forming first hydrogen carbonate (equation viii), then ultimately carbonic acid (equation ix). This decomposes to water and carbon dioxide, and (unless pumped back into the solution at higher pressure) the latter is removed from the system by bubbling out of the solution. If there are enough hydrogen ions added, the whole set of equilibria gets shifted towards the right-hand sides, ending up in the complete decomposition and dissolution of barium carbonate. Any other metal carbonate undergoes similar decomposition when treated with an acid. Even an acid as weak as acetic acid will decompose carbonates.

Fractional precipitation The calculation as to which of two sparingly soluble salts will be precipitated under given experimental conditions may be also made with the aid of the solubility product principle. An example of great practical importance is the Mohr method for the estimation of halides. In this process a solution of chloride ions is titrated with a standard solution of silver nitrate, a small quantity of potassium chromate being added to serve as an indicator. Here two sparingly soluble salts may be formed, viz. silver chloride (a white precipitate) and silver chromate (which is red):

$$Ag^+ + Cl^- \rightleftarrows AgCl(s)$$

and

$$2Ag^+ + CrO_4^{2-} \rightleftarrows Ag_2CrO_4(s)$$

The solubility products of these precipitates are as follows (cf. Table I.12).

$$K_s(AgCl) = [Ag^+] \times [Cl^-] = 1{\cdot}5 \times 10^{-10} \qquad (i)$$

$$K_s(Ag_2CrO_4) = [Ag^+]^2 \times [CrO_4^{2-}] = 2{\cdot}4 \times 10^{-12} \qquad (ii)$$

In the saturated solution of the mixture of both precipitates these solubility equilibria will prevail simultaneously. From the two equations we can therefore obtain

$$\frac{[Cl^-]^2}{[CrO_4^{2-}]} = \frac{(1{\cdot}5\times10^{-10})^2}{2{\cdot}4\times10^{-12}} = \frac{1}{1{\cdot}1\times10^8} \tag{iii}$$

This expression shows that under equilibrium conditions the concentration of chromate ions in the solution is always much greater than that of the chloride ions. If therefore to a mixture of chloride and chromate ions, silver ions are added, these will combine with chloride ions, forming silver chloride precipitate until the concentration of chloride ions in the solution decreases to such an extent, that the ratio expressed in equation (iii) is achieved. From then onwards the two precipitates will be formed simultaneously. If a 0·1M solution of sodium chloride is titrated with silver nitrate in the presence of 0·002M potassium chromate, the concentration of chloride ions at which silver chromate starts to precipitate can be expressed from equation (iii):

$$[Cl^-] = \sqrt{\frac{1}{1{\cdot}1\times10^8}[CrO_4^{2-}]} = \sqrt{\frac{2\times10^{-3}}{1{\cdot}1\times10^8}} = 4{\cdot}26\times10^{-6}\text{M}$$

Therefore practically all chloride ions are removed from the solution before silver chromate is formed.

Another example of some practical interest is the precipitation with sulphate of a solution containing strontium and barium ions. The solubility products of the precipitates $SrSO_4$ and $BaSO_4$ are as follows (cf. Table I.12):

$$K_s(SrSO_4) = [Sr^{2+}]\times[SO_4^{2-}] = 2{\cdot}8\times10^{-7} \tag{iv}$$

$$K_s(BaSO_4) = [Ba^{2+}]\times[SO_4^{2-}] = 9{\cdot}2\times10^{-11} \tag{v}$$

These two expressions and also the relation

$$\frac{[Sr^{2+}]}{[Ba^{2+}]} = \frac{2{\cdot}8\times10^{-7}}{9{\cdot}2\times10^{-11}} \approx \frac{3000}{1} \tag{vi}$$

must be satisfied for the equilibrium between the two precipitates. It follows therefore that if in a solution $[Sr^{2+}] > 3000[Ba^{2+}]$ and sulphate ions are added, $SrSO_4$ will be precipitated until the ratio given in equation (vi) is reached; from then onwards the two ions will precipitate simultaneously. On the other hand, if originally $[Sr^{2+}] < 3000[Ba^{2+}]$ and sulphate ions are added again, first $BaSO_4$ will be precipitated until the ratio $[Sr^{2+}]/[Ba^{2+}] = 3000$ is reached; from then onwards both precipitates will be formed together. In the supernatant liquid the ratio expressed in equation (vi) will be maintained.

Similar calculations may be carried out for the pairs $SrSO_4$ and $CaSO_4$, and $BaSO_4$ and $CaSO_4$; the results have an important bearing on the separation of Group IV cations.

As previously pointed out, the values of many solubility products are not known with any great accuracy: for this reason considerable caution must be exercised in predicting whether or not a given ion can be separated from one or more ions on the basis of solubility product equations, particularly where there is some doubt as to the exact magnitude of the solubility product.

I.29 MORPHOLOGICAL STRUCTURE AND PURITY OF PRECIPITATES Precipitation is probably the most frequently used method in the practice of qualitative analysis. The occurrence of a precipitate as the result of the addition of a certain reagent may be used as a test for a given ion. In such cases we simply observe whether the precipitate formed has the proper colour and general appearance and sometimes carry out tests with further reagents, observing their effects on the precipitate. Precipitates may however be produced for the sake of separation. Doing so, a suitable reagent is added, which forms precipitate(s) with only one or a few of the ions present. After the addition of a suitable amount of the reagent the precipitate is filtered and washed. Some of the ions remain dissolved, while others are to be found in the precipitate. In order to achieve as far as possible a quantitative separation, the precipitate must be easily filterable and must be free of contamination.

The ease with which a precipitate can be filtered and washed depends largely upon the morphological structure of the precipitate, that is the shape and size of its crystals. Obviously, the larger the crystals that are formed during the course of precipitation, the easier they can be filtered, and very probably (though not necessarily) the quicker they will settle out of the solution, which again will assist filtration. The shape of crystals is also important. Simple structures, like cubes, octahedrons, or needles, are advantageous, because it is easy to wash them after filtration. Crystals with more complex structures, containing dents and cavities, will retain some of the mother liquid, even after thorough washing. With precipitates made up of such crystals quantitative separations are less likely to be achieved.

The size of crystals formed during the course of precipitation depends mainly on two important factors: these are the rate of nucleation and the rate of crystal growth.

The **rate of nucleation** can be expressed by the number of nuclei formed in unit time. If the rate of nucleation is high, a great number of crystals are formed, but none of these will grow too large, i.e. a precipitate made up of small particles is formed. The rate of nucleation depends on the degree of **supersaturation** of the solution. Experience has shown that the formation of crystals from a homogeneous solution often does not start at the concentration of ions allowed by the solubility product, but is delayed until the concentration of the solute is much higher than that of the saturated solution. Such a supersaturated solution may stay for quite a long time in this metastable state; often special procedures are required (shaking, scratching the wall of the vessel with a glass rod, inoculation with a crystal) to promote crystallization. The higher the degree of supersaturation, the greater is the probability of forming new nuclei, thus the higher the rate of nucleation.

The **rate of crystal growth** is the other factor which influences the size of crystals formed during the course of precipitation. If this rate is high, large crystals are formed. The rate of crystal growth depends also on the degree of supersaturation. It is advisable however, to create circumstances under which supersaturation is moderate, allowing only a relatively small number of nuclei to be formed, which in turn can grow into large crystals.

The morphological structure of precipitates can often be improved by after-treatment. It is known* that the solubility of very small particles is considerably greater than that of a larger crystal of the same substance. If a mixture made of the mother liquor and the precipitate is simply allowed to stand for a longer time, the small particles will slowly dissolve in the mother liquor, while the larger particles will in turn grow; thus a recrystallization occurs. This process

* Cf. L. Erdey's *Gravimetric Analysis*, Vol. I, Pergamon Press 1963, p. 81 et f.

of **ageing or ripening** can be considerably accelerated by keeping such mixtures at higher temperatures, e.g. allowing them to stand on a water bath. The process of ageing precipitates in this way is often called **digestion**. After such treatment the precipitate becomes more easily filterable and washable; this is why this step is included in almost all procedures of gravimetric analysis. On the other hand, ageing of a precipitate might have an unwanted effect. A freshly obtained precipitate might, for example, be soluble in acids or bases, while if aged, it becomes resistant to such agents. When dealing with the reactions of ions attention will be drawn to such cases.

When a precipitate is separating from a solution it is not always perfectly pure; it may contain various amounts of impurities, dependent upon the nature of the precipitate and conditions of precipitation. The contamination of the precipitate by substances which are normally soluble in the mother liquor is termed **coprecipitation**. Two important processes may be distinguished which cause coprecipitation. The first is the **adsorption** of foreign particles on the surface of a growing crystal, while the second is the **occlusion** of foreign particles during the process of crystal growth. Adsorption, in general, is greatest with gelatinous precipitates and least for those of pronounced macrocrystalline character.

Some precipitates are deposited slowly and the solution is in the state of supersaturation for a considerable time. Thus, when calcium oxalate is precipitated in the presence of larger amounts of magnesium ions, the precipitate is practically pure at first, but if it is allowed to remain in contact with the solution, magnesium oxalate forms slowly (and the presence of calcium oxalate precipitate tends to accelerate this process). Thus, the calcium oxalate precipitate becomes contaminated owing to **post-precipitation** of magnesium oxalate. Post-precipitation often occurs with sparingly soluble substances which tend to form supersaturated solutions, they usually have an ion in common with the primary precipitate. Another typical example is the precipitation of copper or mercury(II) sulphide in dilute acid solution, which become contaminated, if zinc ions are present, by post-precipitation of zinc sulphide. Zinc ions alone may not be precipitated with sulphide ions under identical circumstances.

I.30 THE COLLOIDAL STATE It sometimes happens in qualitative analysis that a substance does not appear as a precipitate when the reactants are present in such concentrations that the solubility product of the substance is greatly exceeded and precautions are taken against supersaturation of the resulting solution. Thus when hydrogen sulphide is passed into a cooled solution of arsenic(III) oxide, no precipitate is discernible when one looks through the resulting mixture. The solution, however, has acquired a deep-yellow colour and when viewed by reflected light shows a marked opalescence. If a powerful beam of light is passed through the solution and the latter viewed through a microscope at right angles to the incident light, a scattering of light (bright spots of light against a dark background) is observed, evidently due to the light reflected by the particles in suspension in the solution. The scattering of light is called the **Tyndall effect**, whilst the device suitable for viewing the Tyndall beam in a microscope is termed the **ultramicroscope**. True solutions, i.e. those with particles of molecular dimensions do not exhibit a Tyndall effect. Clearly the reaction has taken place forming arsenic(III) sulphide but particles are in such a fine state of division that they do not appear as a precipitate. They are in fact in the **colloidal state** or in **colloid solution**. Other colloidal solutions which may be encountered in qualitative analysis include iron(III), chromium(III), and aluminium hydroxides, copper(II), manganese(II), and nickel sulphides, silver chloride and silicic acid.

Further examination of the colloidal solution of arsenic(III) sulphide brings to light other peculiar properties. When an attempt is made to filter the solution, the particles are found to pass right through the filter paper. Also, when the colloidal solution is allowed to stand for a long time, no appreciable settling takes place; no precipitation takes place upon shaking with solid arsenic(III) sulphide, thus ruling out the possibility of supersaturation. The addition of, say, aluminium sulphate solution however brings about immediate precipitation of arsenic(III) sulphide, although there is no apparent reaction between the Al^{3+} or SO_4^{2-} and any ion in the solution. Potassium chloride solution produces the same effect but considerably more of it must be added. Any electrolyte, in fact, causes precipitation, i.e. **coagulation or flocculation** of the colloidal material. Heating the solution also favours coagulation. It is evident that the colloidal state must be avoided in qualitative analysis, and a more detailed account of the phenomenon will therefore be given.

The colloidal state of matter is distinguished by a certain range of particle size, as a consequence of which certain characteristic properties become apparent. Colloidal properties are in general exhibited by substances of particle size ranging between 0·2 μm and 5 nm (2×10^{-7} and 5×10^{-9} m). Ordinary filter paper will retain particles up to a diameter of 10–20 μm ($1–2 \times 10^{-5}$ m), so that colloidal solutions, just like true solutions, pass through an ordinary filter paper (the size of ions is of the order of 0·1 nm $= 10^{-10}$ m). The limit of vision under the microscope is about 5–10 nm ($5–10 \times 10^{-9}$ m). Colloidal solutions are therefore not true solutions. Close examination shows that they are not homogeneous, but consist of suspension of solid or liquid particles in a liquid. Such a mixture is known as a **disperse system**; the liquid (usually water in qualitative analysis) is called the **dispersion medium** and the colloid the **disperse phase**.

An important consequence of the smallness of the size of the particles in a colloidal solution is that the ratio of the surface to the volume is extremely large. Phenomena which depend upon the size of the surface, such as adsorption, will therefore play an important part. The effect of particle size upon the area of the surface will be apparent from the following example. The total surface area of 1 ml of material in the form of a cube of 1 cm side is 6 cm^2. When it is divided into cubes of 10^{-6} cm (10^{-8} m) size (which approximates to many colloidal systems), the total surface area of the same volume of material is 6×10^6 cm^2.

Although colloidal particles cannot be separated from those of molecular dimensions by the use of ordinary filter papers – the best quantitative filter papers retain particles larger than about 1 μm in diameter – separation can be effected by the use of special devices. The procedure known as **dialysis** utilizes the fact that substances in true solution, provided the molecules are not too large, can pass through membranes of parchment or collodion, while colloidal particles are retained. The separation can also be effected by **ultra-filtration**. Filter papers are impregnated with collodion, or with gelatin subsequently hardened by formaldehyde, thereby making the pores small enough to retain particles of colloidal dimensions. The ultimate size of the pores depends upon the particular paper used and upon the concentration of the solution employed to impregnate it. The solution is poured on the filter and the passage of the liquid is accelerated by suction or by pressure. It may be mentioned that other factors (e.g. rates of diffusion and adsorption) in addition to pore size determine whether or not particles of a given size will pass through an ultra-filter.

Colloidal systems, in which a liquid is the dispersion medium, are often termed **sols** to distinguish them from true solutions: the nature of the liquid is indicated by the use of a prefix, e.g. aquasols, alcosols, etc. The solid produced upon coagulation or flocculation of a sol was originally described as a gel, but the name is now generally restricted to those cases in which the whole system sets to a semi-solid state without any supernatant liquid initially. Some authors employ the word gel to include gelatinous precipitates, such as aluminium and iron(III) hydroxides, formed from sols, whereas others refer to them as **coagels**. The process of dispersing a flocculated solid or gel (or coagel) to form a colloidal solution is called **peptization**.

Colloidal solutions may be divided roughly into two main groups, designated as **lyophobic** (Greek: solvent hating) and **lyophilic** (Greek: solvent loving); when water is the dispersion medium, the terms hydrophobic and hydrophilic are employed. The chief properties of each class are summarized in Table I.14, but it must be emphasized that the distinction is not an absolute one since some, particularly sols of metallic hydroxides, exhibit intermediate properties.

Table I.14 Some properties of colloid systems

Hydrophobic sols	Hydrophilic sols
1. The viscosity of the sols is similar to that of the medium. Examples: sols of the metals, silver halides, metallic hydroxides, and barium sulphate.	**1.** The viscosity is very much higher than that of the medium; they set to jelly-like masses, often termed gels (or coagels). Examples: sols of silicic acid, tin(IV), gelatin, starches, and proteins.
2. A comparatively minute quantity of an electrolyte results in flocculation. The change is, in general, irreversible; water has no effect upon the flocculant.	**2.** Small quantities of electrolytes have little effect: large amounts may cause precipitation, 'salting out'. The change is, in general, reversible upon the addition of water.
3. The particles, ordinarily, have an electric charge of definite sign, which can be changed only by special methods. The particles migrate in one direction in an electric field (cataphoresis or electrophoresis).	**3.** The particles change their charge readily, e.g. they are positively charged in acid medium and negatively charged in a basic medium. Uncharged particles are also known. The particles may migrate in either direction or not at all in an electric field.
4. The ultramicroscope reveals bright particles in vigorous motion (Brownian movement).	**4.** Only a diffuse light is exhibited in the ultramicroscope.
5. The surface tension is similar to that of water.	**5.** The surface tension is often lower than that of water: foams are often produced readily.

Colloid particles in solution exhibit the phenomenon called the **Brownian motion**, which can be observed by the ultramicroscope. When placed into an electric field, they usually migrate towards one of the electrodes, showing that the colloid particles possess a distinct charge. This charge can be both positive and negative. Negatively charged colloids include the sols of metals, sulphur, metallic sulphides, silicic acid, stannic acid, silver halides, gums, starch, and certain acid dyestuffs, whilst sols of metallic hydroxides and of certain basic dyestuffs usually carry a positive charge.

The stability of a colloidal solution is intimately associated with the electrical charge on the particles. Thus in the formation of an arsenic(III) sulphide sol by precipitation with hydrogen sulphide in faintly acid solution, sulphide ions are primarily adsorbed (since every precipitate has a tendency to adsorb its own ions), and in order to maintain the electroneutrality of the solution, an equivalent quantity of hydrogen ions is secondarily adsorbed. Thus, an electrical double layer is created around each particle, with the positive side facing the solution. As a consequence, the colloid particles repel each other, preventing the formation of larger (microscopic or macroscopic) particles. If this double layer is destroyed, the colloid coagulates. This can be achieved, for example, by adding larger amounts of an electrolyte to the solution (**salting-out effect**). The ions of the electrolyte, being present in large concentrations, interfere with the formation of a spherical electrical double layer around the particles, which are therefore not inhibited further from coagulation. It appears that ions of opposite charge to those primarily adsorbed on the surface are necessary for coagulation. The minimum amount of electrolyte necessary to cause flocculation is called the **flocculation** or **coagulation value**. It has been found that the latter depends upon the valency of the ions of the opposite charge to that on the colloidal particle, the higher the valency the smaller the coagulation value; the nature of the ions has some influence also. If two sols of opposite charge are mixed (e.g. iron(III) hydroxide and arsenic(III) sulphide), mutual coagulation usually occurs because of the neutralization of charges.

The above remarks apply largely to hydrophobic sols. Hydrophilic sols are generally much more difficult to coagulate than hydrophobic sols. If a small amount of a hydrophilic sol (preferably, but not necessarily, with the same charge), e.g. gelatin, is added to a hydrophobic sol, e.g. of gold, then the latter appears to be strongly protected against the flocculating action of electrolytes. It may be that the particles of the lyophilic sol are adsorbed by the lyophobic sol and impart their properties to the latter. The hydrophilic substance is known as a **protective colloid**. An explanation is thus provided of the relative stability of the otherwise unstable gold (or other noble metal) sols produced by the addition of a little gelatin or of the salts of protalbic or lysalbic acids (the latter are obtained by the alkaline hydrolysis of albumin). For this reason organic matter, which might form a protective colloid, must generally be destroyed before proceeding with an analysis.

During the coagulation of a colloid by an electrolyte, the ion of opposite sign to that of the colloid is adsorbed to a varying degree on the surface; the higher the valency of the ion, the more strongly is it adsorbed. In all cases, the precipitate will be contaminated by surface adsorption. Upon washing the precipitate with water, part of the adsorbed electrolyte is removed, and a new difficulty may arise. The electrolyte concentration in the supernatant liquid may fall below the coagulation value, and the precipitate will pass into colloidal solution again. It can be prevented by washing the precipitate with a suitable electrolyte.

The adsorptive properties of colloids find some application in analysis, e.g. in the removal of phosphates by tin(IV) hydroxide oxide in the presence of nitric acid (Section **V.13**); and in the formation of coloured lakes from colloidal metallic hydroxides and certain soluble dyes (see Section **III.23** and **III.35** for aluminium and magnesium respectively). There is, however, some evidence that lake formation may be partly chemical in character.

Precipitates obtained from dilute or very concentrated solutions are often in the form of very fine crystals. These fine precipitates will generally become filterable if allowed to stand for some time in contact with the mother liquor, preferably, if the solubility permits, near the boiling point of the solution. The addition of macerated filter paper is beneficial in assisting the filtration of colloidal precipitates. The macerated filter paper increases the speed of filtration by retaining part of the precipitate and thus preventing the clogging of the pores of the filter paper.

F. COMPLEXATION REACTIONS

I.31 THE FORMATION OF COMPLEXES In the practice of qualitative inorganic analysis ample use is made of reactions which lead to the formation of complexes. A complex ion (or molecule) comprises a **central atom (ion)** and a number of **ligands** closely attached to the former. The relative amounts of these components in a stable complex seems to follow a well-defined stoichiometry, although this cannot be interpreted within the classical concept of valency. The central atom can be characterized by the **coordination number**, an integer figure, which shows the number of (monodentate) ligands which may form a stable complex with one central atom. In most cases the coordination number is 6 (as in the case of Fe^{2+}, Fe^{3+}, Zn^{2+}, Cr^{3+}, Co^{3+}, Ni^{2+}, Cd^{2+}) sometimes 4 (Cu^{2+}, Cu^{+}, Pt^{2+}), but the numbers 2(Ag^{+}) and 8 (some of the ions in the platinum group) do occur.

The coordination number represents the number of spaces available around the central atom or ion in the so-called coordination sphere, each of which can be occupied by one (monodentate) ligand. The arrangement of ligands around the central ion is symmetrical. Thus, a complex with a central atom of a coordination number of 6, comprises the central ion, in the centre of an octahedron, while the six ligands occupy the spaces defined by the vertices of the octahedron. To the coordination number 4 a tetrahedric symmetry normally corresponds, although a planar (or nearly planar) arrangement, where the central ion is in the centre of a square and the four ions occupy the four corners of the latter, is common as well.

Simple inorganic ions and molecules like NH_3, CN^-, Cl^-, H_2O form monodentate ligands, that is one ion or molecule occupies one of the spaces available around the central ion in the coordination sphere, but bidentate (like the dipyridyl ion), tridentate, and also tetradentate ligands are known. Complexes made of polydentate ligands are often called **chelates**, the name originating from the Greek word for the claw of the crab, which bites into an object like the polydentate ligand 'catches' the central ion. The formation of chelate complexes is used extensively in quantitative chemical analysis (complexometric titrations).*

Formulae and names of some complex ions are as follows:

$[Fe(CN)_6]^{4-}$ hexacyanoferrate(II)

---

* Cf. A. I. Vogel's *A Text-Book of Quantitative Inorganic Analysis*. 3rd edn., Longman 1966, p. 415 et f.

$[Fe(CN)_6]^{3-}$	hexacyanoferrate(III)
$[Cu(NH_3)_4]^{2+}$	tetramminecuprate(II)
$[Cu(CN)_4]^{3-}$	tetracyanocuprate(I)
$[Co(H_2O)_6]^{3+}$	hexaquocobaltate(III)
$[Ag(CN)_2]^-$	dicyanoargentate(I)
$[Ag(S_2O_3)_2]^{3-}$	dithiosulphatoargentate(I)

From these examples the rules of nomenclature are apparent. The central atom (like Fe, Cu, Co, Ag) is followed by the formula of the ligand (CN, NH_3, H_2O, S_2O_3), with the stoichiometric index number, (which, in the case of monodentate ligands is equal to the coordination number). The formula is placed inside square brackets, and the charge of the ion is shown outside the brackets in the usual way. When expressing concentrations of complexes, brackets of the type { } will be used to avoid confusion. In the name of the ion, first the (Greek) number, then the name of the ligand is expressed, followed by the name of the central atom and its oxidation number (valency).

The classical rules of valency do not apply for complex ions. To explain the particularities of chemical bonding in complex ions, various theories have been developed. As early as 1893, A. Werner suggested that, apart from normal valencies, elements possess secondary valencies which are used when complex ions are formed. He attributed directions to these secondary valencies, and thereby could explain the existence of stereoisomers, which were prepared in great numbers at that time. Later G. N. Lewis (1916), when describing his theory of chemical bonds based on the formation of electron pairs, explained the formation of complexes by the donation of a whole electron pair by an atom of the ligand to the central atom. This so-called **dative bond** is sometimes denoted by an arrow, showing the direction of donation of electrons. In the structural formula of the tetramminecuprate(II) ion

$$\left[\begin{array}{ccc} & NH_3 & \\ & \downarrow & \\ H_3N \rightarrow & Cu & \leftarrow NH_3 \\ & \uparrow & \\ & NH_3 & \end{array}\right]^{2+}$$

the arrows indicate that an electron pair is donated by each nitrogen to the copper ion. Although the Lewis theory offers a comprehensive explanation of chemical structures in relatively simple terms, a deeper understanding of the nature of the chemical bond necessitated the formulation of new theories. Among these the **ligand field theory** explains the formation of complexes on the basis of an electrostatic field created by the coordinated ligand around the inner sphere of the central atom. This ligand field causes the splitting of the energy levels of the *d*-orbitals of the central atom, which in turn produces the energy responsible for the stabilization of the complex (ligand field stabilization energy). For a more detailed study of the ligand field theory, appropriate text-books should be consulted.*

* Cf. W. J. Moore's *Physical Chemistry*. 4th edn., Longman 1962, p. 545 et f.

The charge on a complex ion is the sum of the charges on the ions which form the complex:

$$Ag^{+} + 2CN^{-} \rightarrow [Ag(CN)_2]^{-}$$
$$Cu^{2+} + 4CN^{-} \rightarrow [Cu(CN)_4]^{2-}$$
$$Fe^{2+} + 6CN^{-} \rightarrow [Fe(CN)_6]^{4-}$$
$$Fe^{3+} + 6CN^{-} \rightarrow [Fe(CN)_6]^{3-}$$

If neutral molecules are involved as ligands in the formation of complexes, the charge on the complex ion remains the same as that on the central atom:

$$Ag^{+} + 2NH_3 \rightarrow [Ag(NH_3)_2]^{+}$$
$$Ni^{2+} + 6NH_3 \rightarrow [Ni(NH_3)_6]^{2+}$$

Complexes with mixed ligands may have quite different charges:

$$Co^{3+} + 4NH_3 + 2NO_2^{-} \rightarrow [Co(NH_3)_4(NO_2)_2]^{+} \quad \text{(positive)}$$
$$Co^{3+} + 3NH_3 + 3NO_2^{-} \rightarrow [Co(NH_3)_3(NO_2)_3] \quad \text{(neutral)}$$
$$Co^{3+} + 2NH_3 + 4NO_2^{-} \rightarrow [Co(NH_3)_2(NO_2)_4]^{-} \quad \text{(negative)}$$

The formation of complexes in qualitative inorganic analysis is often observed and is used for separation or identification. One of the most common phenomena occurring when complex ions are formed is **a change of colour** in the solution. Some examples are:

$$\underset{\text{blue}}{Cu^{2+}} + 4NH_3 \rightarrow \underset{\text{deep, dark blue}}{[Cu(NH_3)_4]^{2+}}$$

$$\underset{\text{pale green}}{Fe^{2+}} + 6CN^{-} \rightarrow \underset{\text{yellow}}{[Fe(CN)_6]^{4-}}$$

$$\underset{\text{green}}{Ni^{2+}} + 6NH_3 \rightarrow \underset{\text{blue}}{[Ni(NH_3)_6]^{2+}}$$

$$\underset{\text{yellow}}{Fe^{3+}} + 6F^{-} \rightarrow \underset{\text{colourless}}{[FeF_6]^{3-}}$$

Another important phenomenon often observed when complexes are formed, is **an increase of solubility**; many precipitates may dissolve because of complex formation:

$$AgCl(s) + 2NH_3 \rightarrow [Ag(NH_3)_2]^{+} + Cl^{-}$$
$$AgI(s) + 2S_2O_3^{2-} \rightarrow [Ag(S_2O_3)_2]^{3-} + I^{-}$$

Complex formation is responsible for the dissolution of precipitates in excess of the reagent:

$$AgCN(s) + CN^{-} \rightarrow [Ag(CN)_2]^{-}$$
$$BiI_3(s) + I^{-} \rightarrow [BiI_4]^{-}$$

There are differences in the stabilities of complexes. Thus, copper(II) sulphide can be precipitated from a solution of copper ions with hydrogen sulphide:

$$Cu^{2+} + H_2S \rightarrow CuS(s) + 2H^{+}$$

The same precipitate is formed if hydrogen sulphide gas is introduced into the dark-blue solution of tetramminocuprate(II) ions:

$$[Cu(NH_3)_4]^{2+} + H_2S \rightarrow CuS(s) + 2NH_4^+ + 2NH_3$$

From the colourless solution of tetracyanocuprate(I) $[Cu(CN)_4]^{3-}$ ions however, hydrogen sulphide does not precipitate copper sulphide, indicating that the tetracyano complex is more stable than the tetrammine complex. On the other hand, cadmium(II) ions form both tetrammine $[Cd(NH_3)_4]^{2+}$ and tetracyano $[Cd(CN)_4]^{2-}$ complexes, but hydrogen sulphide gas can precipitate the yellow cadmium sulphide from both solutions, even though cadmium sulphide is more soluble than the copper(I) sulphide precipitate (cf. Table I.12). This fact indicates that the tetracyanocadmiate(II) $[Cd(CN)_4]^{2-}$ complex is less stable than the tetracyanocuprate(I) $[Cu(CN)_4]^{3-}$

I.32 THE STABILITY OF COMPLEXES In the previous section hints were made about the differences in stabilities of various complexes. In order to be able to make more quantitative statements and comparisons, a suitable way has to be found to express the stability of complexes. The problem in many ways is similar to that of expressing the relative strength of acids and bases. This was done on the basis of their dissociation constants (cf. Section **I.16**), obtained by applying the law of mass action to these dissociation equilibria. A similar principle can be applied for complexes.

Let us consider the dicyanoargentate(I) $[Ag(CN)_2]^-$ complex. This ion dissociates to form silver and cyanide ions:

$$[Ag(CN)_2]^- \rightleftarrows Ag^+ + 2CN^-$$

The fact that such a dissociation takes place can be proved easily by experiments. Silver ions (which are the product of this dissociation) can be precipitated by hydrogen sulphide gas as silver sulphide Ag_2S, and also silver metal can be deposited on the cathode from such a solution by electrolysis (note that the dicyanoargentate ion, with its negative charge moves towards the anode when electrolysed). By applying the law of mass action to the dissociation mentioned above, we can express the dissociation constant or instability constant as

$$K = \frac{[Ag^+] \times [CN^-]^2}{\{[Ag(CN)_2]\}}$$

The constant has a value of $1{\cdot}0 \times 10^{-21}$ at room temperature. By inspection of this expression it must be evident that if cyanide ions are present in excess, the silver ion concentration in the solution must be very small. The lower the value of the instability constant, the more stable is the complex and vice versa. A selected list of instability constants, (all of which have importance in qualitative inorganic analysis) is shown in Table I.15.

It is interesting to compare these values and to predict what happens if, to a solution which contains the complex ion, a reagent is added which, under normal circumstances, would form a precipitate with the central ion. It is obvious that the higher the value of the instability constant, the higher the concentration of free central ion (metal ion) in the solution, and therefore the more probable it is that the product of ion concentrations in the solution will exceed the value of the solubility product of the precipitate and hence the precipitate

Table I.15 Instability constants of complex ions

Equilibrium	Instability constant
$[Ag(NH_3)_2]^+ \rightleftharpoons Ag^+ + 2NH_3$	$6{\cdot}8 \times 10^{-3}$
$[Ag(S_2O_3)_2]^{3-} \rightleftharpoons Ag^+ + 2S_2O_3^{2-}$	$1{\cdot}0 \times 10^{-18}$
$[Ag(CN)_2]^- \rightleftharpoons Ag^+ + 2CN^-$	$1{\cdot}0 \times 10^{-21}$
$[Cu(CN)_4]^{3-} \rightleftharpoons Cu^+ + 4CN^-$	$5{\cdot}0 \times 10^{-28}$
$[Cu(NH_3)_4]^{2+} \rightleftharpoons Cu^{2+} + 4NH_3$	$4{\cdot}6 \times 10^{-14}$
$[Cd(NH_3)_4]^{2+} \rightleftharpoons Cd^{2+} + 4NH_3$	$2{\cdot}5 \times 10^{-7}$
$[Cd(CN)_4]^{2-} \rightleftharpoons Cd^{2+} + 4CN^-$	$1{\cdot}4 \times 10^{-17}$
$[CdI_4]^{2-} \rightleftharpoons Cd^{2+} + 4I^-$	5×10^{-7}
$[HgCl_4]^{2-} \rightleftharpoons Hg^{2+} + 4Cl^-$	$6{\cdot}0 \times 10^{-17}$
$[HgI_4]^{2-} \rightleftharpoons Hg^{2+} + 4I^-$	$5{\cdot}0 \times 10^{-31}$
$[Hg(CN)_4]^{2-} \rightleftharpoons Hg^{2+} + 4CN^-$	$4{\cdot}0 \times 10^{-42}$
$[Hg(SCN)_4]^{2-} \rightleftharpoons Hg^{2+} + 4SCN^-$	$1{\cdot}0 \times 10^{-22}$
$[Co(NH_3)_6]^{3+} \rightleftharpoons Co^{3+} + 6NH_3$	$2{\cdot}2 \times 10^{-34}$
$[Co(NH_3)_6]^{2+} \rightleftharpoons Co^{2+} + 6NH_3$	$1{\cdot}3 \times 10^{-5}$
$[I_3]^- \rightleftharpoons I^- + I_2$	$1{\cdot}4 \times 10^{-2}$
$[Fe(SCN)]^{2+} \rightleftharpoons Fe^{3+} + SCN^-$	$3{\cdot}3 \times 10^{-2}$
$[Zn(NH_3)_4]^{2+} \rightleftharpoons Zn^{2+} + 4NH_3$	$2{\cdot}6 \times 10^{-10}$
$[Zn(CN)_4]^{2-} \rightleftharpoons Zn^{2+} + 4CN^-$	2×10^{-17}

will start to form. The lower this solubility product, the more probable that the precipitate will in fact be formed.

Equally an assessment can be made of the possibility of being able to dissolve an existing precipitate with a certain complexing agent. Obviously, the more stable the complex, the more probable it is that the precipitate will dissolve. On the other hand, the less soluble the precipitate, the more difficult it will be to find a suitable complexing agent to dissolve it.

Firm predictions can easily be made on the basis of simple calculations. The following examples illustrate the way in which these calculations are made:

Example 23 A solution contains tetracyanocuprate(I) $[Cu(CN)_4]^{3-}$ and tetracyanocadmiate(II) $[Cd(CN)_4]^{2-}$ ions, both in 0·5M concentration. The solution has a *p*H of 9 and contains 0·1 mol ℓ^{-1} free cyanide ions. Can copper(I) sulphide Cu_2 and/or cadmium sulphide CdS be precipitated from this solution by introducing hydrogen sulphide gas?

From Table I.12 we have the solubility products

$$K_s(Cu_2S) = 2 \times 10^{-47}$$
$$K_s(CdS) = 1{\cdot}4 \times 10^{-28}$$

and from Table I.15 we take the values of the following instability constants:

$$K_1 = \frac{[Cu^+][CN^-]^4}{\{[Cu(CN)_4]^{3-}\}} = 5 \times 10^{-28} \qquad \text{(i)}$$

and

$$K_2 = \frac{[Cd^{2+}][CN^-]^4}{\{[Cd(CN)_4]^{2-}\}} = 1{\cdot}4 \times 10^{-17} \qquad \text{(ii)}$$

We have to calculate the concentrations of the various species present in the solution. The hydrogen-ion concentration being 10^{-9} mol ℓ^{-1}, the sulphide ion concentration can be expressed as (cf. Section **I.28**):

$$[S^{2-}] = \frac{10^{-23}}{[H^+]^2} = \frac{10^{-23}}{(10^{-9})^2} = 10^{-5} \text{ mol } \ell^{-1}$$

Because of the low values of the instability constants, the complexes are practically undissociated, thus the concentration of both complex ions are 0·5 mol ℓ^{-1}. The concentration of cyanide ions being 10^{-1} mol ℓ^{-1}, the concentrations of the free metal ions can be expressed from (i) and (ii) as

$$[Cu^+] = \frac{K_1 \times \{[Cu(CN)_4]^{3-}\}}{[CN^-]^4} = \frac{5 \times 10^{-28} \times 0{\cdot}5}{(10^-)^4}$$
$$= 2{\cdot}5 \times 10^{-24} \text{ mol } \ell^{-1}$$

and

$$[Cd^{2+}] = \frac{K_2 \times \{[Cd(CN)_4]^{2-}\}}{[CN^-]^4} = \frac{1{\cdot}4 \times 10^{-17} \times 0{\cdot}5}{(10^{-1})^4}$$
$$= 7 \times 10^{-14} \text{ mol } \ell^{-1}$$

Now we compare the products of concentrations with the solubility products. For copper(I) ions we have

$$[Cu^+]^2 \times [S^{2-}] = (2{\cdot}5 \times 10^{-24})^2 \times 10^{-5} = 6{\cdot}25 \times 10^{-53}$$

As $6{\cdot}25 \times 10^{-53} < K_s(Cu_2S)$, it is obvious that copper(I) sulphide will not be precipitated under such circumstances. On the other hand, for cadmium ions we have

$$[Cd^{2+}] \times [S^{2-}] = 7 \times 10^{-14} \times 10^{-5} = 7 \times 10^{-19}$$

As $7 \times 10^{-19} > K_s(CdS)$ the concentrations of the ionic species are higher than permitted by the solubility product, therefore cadmium(II) sulphide will be precipitated from such a solution.

This difference in the behaviour of copper and cadmium ions is utilized for the separation of copper and cadmium. First, excess ammonia is added to the solution, when the tetrammine complexes of copper(II) and cadmium are formed (and hydroxides of other ions may be precipitated). Then potassium cyanide is added to the solution, when the tetracyano complexes are formed, and at the same time copper(II) ions are reduced to copper(I). The deep-blue colour of the tetrammine cuprate(II) ions (which serves as a test for copper) disappears, and a colourless solution is obtained. If hydrogen sulphide gas is now introduced, the yellow precipitate of cadmium(II) sulphide is formed; by this the presence of cadmium ions is proved. By filtering the mixture the separation of copper and cadmium is achieved.

Example 24 What happens if, to a mixture which contains 0·1432 g silver chloride and 0·2348 g silver iodide, (*a*) ammonia and (*b*) potassium cyanide solution is added? The final volume of the solution is 100 ml, and the concentrations of free ammonia and free potassium cyanide are 2 mol ℓ^{-1} and 0·05 mol ℓ^{-1} respectively.

By comparing the relative molecular masses of AgCl (143·2) and AgI (234·8) with the actual weights we can see that the amount of each precipitate present is 10^{-3} mol; if therefore they dissolve completely in the 100 ml solution, the concentrations of chloride and iodide ions will be 10^{-2} mol ℓ^{-1}. From the stoichiometry of the dissociation reactions, which lead to the formation of $[Ag(NH_3)_2]^+$ and $[Ag(CN)_2]^-$ complex ions, it follows that their concentrations will also be 10^{-2} mol ℓ^{-1} in these solutions. The instability constants of these complex ions are taken from Table I.15; they are

$$K_1 = \frac{[Ag^+][NH_3]^2}{\{[Ag(NH_3)_2]^+\}} = 6{\cdot}8 \times 10^{-8} \tag{iii}$$

and

$$K_2 = \frac{[Ag^+][CN^-]^2}{\{[Ag(CN)_2]^-\}} = 1 \times 10^{-21} \tag{iv}$$

From Table I.12 we take the solubility products of AgCl and AgI:

$$K_s(AgCl) = 1{\cdot}5 \times 10^{-10}$$

and

$$K_s(AgI) = 0{\cdot}9 \times 10^{-16}$$

With these data in hand, we can solve the problems.

(*a*) If ammonia is added, we have the concentrations

$$[NH_3] = 2 \text{ mol } \ell^{-1}$$
$$\{[Ag(NH_3)_2]^+\} = 10^{-2} \text{ mol } \ell^{-1}$$

The concentration of free silver ions can be calculated from equation (iii)

$$[Ag^+] = \frac{K_1\{[Ag(NH_3)_2]^+\}}{[NH_3]^2} = \frac{6{\cdot}8 \times 10^{-8} \times 10^{-2}}{2^2} = 1{\cdot}7 \times 10^{-10} \text{ mol } \ell^{-1}$$

If the precipitates were completely dissolved, the concentrations of chloride and iodide ions would be

$$[Cl^-] = 10^{-2} \text{ mol } \ell^{-1}$$
$$[I^-] = 10^{-2} \text{ mol } \ell^{-1}$$

Comparing the products of concentrations with the solubility products, we have

$$[Ag^+] \times [Cl^-] = 1{\cdot}7 \times 10^{-10} \times 10^{-2} = 1{\cdot}7 \times 10^{-12} < K_s(AgCl)$$

Thus the silver chloride precipitate will dissolve in ammonia. On the other hand

$$[Ag^+] \times [I^-] = 1{\cdot}7 \times 10^{-10} \times 10^{-2} = 1{\cdot}7 \times 10^{-12} > K_s(AgI)$$

This result shows that silver iodide will not dissolve in ammonia (though some of the silver ions will be complexed). These results are in good agreement with experimental facts.

(*b*) A similar procedure can be adapted for the case of cyanide. Here

$$[CN^-] = 5 \times 10^{-2} \text{ mol } \ell^{-1}$$
$$\{[Ag(CN)_2]^-\} = 10^{-2} \text{ mol } \ell^{-1}$$
$$[Cl^-] = 10^{-2} \text{ mol } \ell^{-1}$$
$$[I^-] = 10^{-2} \text{ mol } \ell^{-1}$$

From equation (iv) the concentration of free silver ions can be calculated:

$$[Ag^+] = \frac{K_2\{[Ag(CN)_2]^-\}}{[CN^-]^2} = \frac{10^{-21}\times 10^{-2}}{(5\times 10^{-2})^2} = 4\times 10^{-21}\ \text{mol}\ \ell^{-1}$$

Comparing the products of ion concentrations with the solubility products

$$[Ag^+]\times[Cl^-] = 4\times 10^{-21}\times 10^{-2} = 4\times 10^{-23} < K_s(AgCl)$$

$$[Ag^+]\times[I^-] = 4\times 10^{-21}\times 10^{-2} = 4\times 10^{-23} < K_s(AgI)$$

we can see that these concentrations are less than 'allowed' by the solubility product for a saturated solution of these silver halides. Consequently, both precipitates will dissolve in potassium cyanide. Again, this reasoning can be verified by experiment.

I.33 THE APPLICATION OF COMPLEXES IN QUALITATIVE INORGANIC ANALYSIS The formation of complexes has two important fields of application in inorganic qualitative analysis:

(a) Specific tests for ions Some reactions, leading to the formation of complexes, can be applied as tests for ions. Thus, a very sensitive and specific reaction for copper is the test with ammonia, when the dark-blue tetramminocuprate ions are formed:

$$\underset{\text{blue}}{Cu^{2+}} + 4NH_3 \rightleftarrows \underset{\text{dark blue}}{[Cu(NH_3)_4]^{2+}}$$

the only other ion which gives a somewhat similar reaction is nickel, which forms a hexamminenickelate(II) $[Ni(NH_3)_6]^{2+}$ ion. With some experience however copper and nickel can be distinguished from each other.

Another important application is the test for iron(III) ions with thiocyanate. In slightly acid medium a deep-red colouration is formed, owing to the stepwise formation of a number of complexes:

$$Fe^{3+} + SCN^- \rightleftarrows [FeSCN]^{2+}$$

$$[FeSCN]^{2+} + SCN^- \rightleftarrows [Fe(SCN)_2]^+$$

$$[Fe(SCN)_2]^+ + SCN^- \rightleftarrows [Fe(SCN)_3]$$

$$[Fe(SCN)_3] + SCN^- \rightleftarrows [Fe(SCN)_4]^-$$

$$[Fe(SCN)_4]^- + SCN^- \rightleftarrows [Fe(SCN)_5]^{2-}$$

$$[Fe(SCN)_5]^{2-} + SCN^- \rightleftarrows [Fe(SCN)_6]^{3-}$$

of these $[Fe(SCN)_3]$ is a non-electrolyte; it can be readily extracted into ether or amyl alcohol. The complexes with positive charges are cations and migrate towards the cathode if electrolysed, while those with negative charges are anions, moving towards the anode when electrolysed. This reaction is specific for iron(III) ions; even iron(II) ions do not react with thiocyanate. The test in fact is often used to test for iron(III) in the presence of iron(II) ions.

Some complexes are precipitates, like the bright-red precipitate formed between nickel(II) ions and dimethylglyoxime:

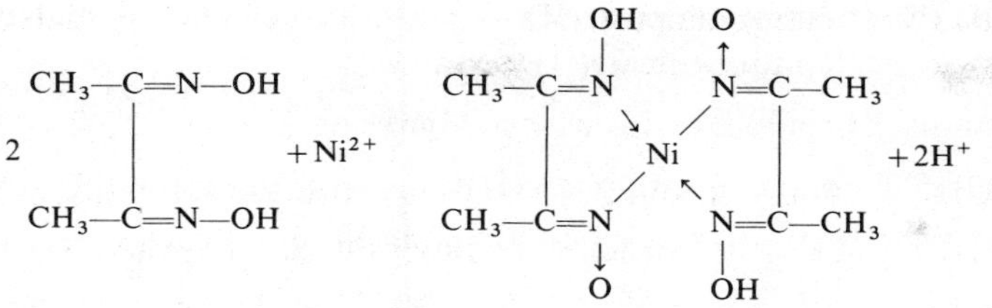

$$2\ \begin{array}{l} CH_3{-}C{=}N{-}OH \\ \quad\ \ | \\ CH_3{-}C{=}N{-}OH \end{array} + Ni^{2+} \rightarrow \begin{array}{ccccc} & & OH & & O \\ & & | & & \uparrow \\ CH_3{-}C{=}N & & & & N{=}C{-}CH_3 \\ | & \searrow & & \swarrow & | \\ & & Ni & & \\ | & \nearrow & & \nwarrow & | \\ CH_3{-}C{=}N & & & & N{=}C{-}CH_3 \\ & & \downarrow & & | \\ & & O & & OH \end{array} + 2H^+$$

(the arrows represent coordinative bonds). This reaction is specific and sensitive for nickel, if carried out under proper experimental circumstances (cf. Section **III.27**).

(b) Masking When testing for a specific ion with a reagent, interferences may occur owing to the presence of other ions in the solution, which also react with the reagent. In some cases it is possible to prevent this interference with the addition of reagents, so-called masking agents, which form stable complexes with the interfering ions. There is no need for the physical separation of the ions involved, and therefore the time for the test can be cut considerably. A classical example of masking has already been mentioned: for the test of cadmium with hydrogen sulphide copper can be masked with cyanide ions (cf. Section **I.31** and also *Example* 23 in Section **I.32**). Another example of the use of masking is the addition of organic reagents containing hydroxyl groups (like tartaric or citric acid) to solutions containing iron(III) or chromium(III) ions to prevent the precipitation of their hydroxides. Such solutions may then be made alkaline without the danger of these metals being hydrolysed, and tests for other ions can be made.

Masking may also be achieved by dissolving precipitates or by the selective dissolution of a precipitate from a mixture. Thus, when testing for lead in the presence of silver, we may produce a mixture of silver and lead chloride precipitates:

$$Ag^+ + Cl^- \rightarrow AgCl(s)$$
$$Pb^{2+} + 2Cl^- \rightarrow PbCl_2(s)$$

If ammonia is added, silver chloride dissolves in the form of the diammine-argentate ion:

$$AgCl(s) + 2NH_3 \rightarrow [Ag(NH_3)_2]^+ + Cl^-$$

while lead chloride (mixed with some lead hydroxide) remains as a white precipitate. In this way, without any further test, the presence of lead can be confirmed.

I.34 THE MOST IMPORTANT TYPES OF COMPLEXES APPLIED IN QUALITATIVE ANALYSIS In qualitative inorganic analysis complexes (both ions and molecules) are often encountered. Some important examples of these are:

(a) Aquocomplexes Most common ions exist in aqueous solution (and some also in the crystalline state) in the form of aquocomplexes. Such ions are

$[Ni(H_2O)_6]^{2+}$ hexaquonickelate(II)
$[Al(H_2O)_6]^{3+}$ hexaquoaluminato

$[Cu(H_2O)_4]^{2+}$ tetraquocuprate(II)
$[Zn(H_2O)_4]^{2+}$ tetraquozincate(II)

Some of the anions, like sulphate, form aquocomplexes as well:

$[SO_4(H_2O)]^{2-}$ monoaquosulphate(II)

The hydronium ion H_3O^+ is in fact an aquocomplex itself, and could be written as $[H(H_2O)]^+$.

Note that the formula of solid copper sulphate pentahydrate for example should be written precisely as $[Cu(H_2O)_4][SO_4(H_2O)]$. The usual formula, $CuSO_4.5H_2O$ does not account for the fact that there are two different types of water molecules (copper–water and sulphate–water) in the crystal structure. This can easily be proved. On heating, first four molecules of water are released from crystalline copper sulphate, at around 120°C, while the fifth molecule can only be removed at a much higher temperature, 240°C.

In spite of the fact that these aquocomplexes exist, we normally ignore the coordinated water molecules in formulae and equations. Thus, instead of using the formulae mentioned, we shall write simply Ni^{2+}, Al^{3+}, Cu^{2+}, Zn^{2+}, SO_4^{2-}, and H^+ in the text, unless the formation or decomposition of the aquocomplex plays a specific role in the chemical reaction.

(b) Ammine complexes These have already been mentioned. They are formed if excess ammonia is added to the solution of certain metal ions. Such complexes are

$[Ag(NH_3)_2]^+$ diammineargentate(I)
$[Cu(NH_3)_4]^{2+}$ tetramminecuprate(II)
$[Co(NH_3)_6]^{2+}$ hexamminecobaltate(II)

These ions exist only at high (>8) *p*H; the addition of mineral acids decomposes them.

(c) Hydroxocomplexes (amphoteric hydroxides) Certain metal hydroxide precipitates, like zinc hydroxide $Zn(OH)_2$, may be dissolved either by acids or by bases, that is they display both acid and base character. For this reason these precipitates are often termed **amphoteric hydroxides**. While their dissolution in acid results in the formation of the aquocomplex of the metal which, in turn, is normally regarded as the simple metal ion (like Zn^{2+}), the dissolution in excess base is in fact due to the formation of hydroxocomplexes, as in the process

$$Zu(OH)_2 + 2OH^- \rightleftarrows [Zn(OH)_4]^{2-}$$

The tetrahydroxozincate(II) ion is sometimes (mainly in older texts) represented as the zincate anion, ZnO_2^{2-}. Similar soluble hydroxocomplexes are as follows:

$[Pb(OH)_4]^{2-}$ tetrahydroxoplumbate(II)
$[Sn(OH)_4]^{2-}$ tetrahydroxostannate(II)
[but $[Sn(OH)_6]^{2-}$ hexahydroxostannate(IV)]
$[Al(OH)_4]^-$ tetrahydroxoaluminate

In fact, some of these are mixed aquo-hydroxo complexes, and the proper formulae of the tetrahydroxocomplexes are $[Pb(H_2O)_2(OH)_4]^{2-}$,

$[Sn(H_2O)_2(OH)_4]^{2-}$, and $[Al(H_2O)_2(OH)_4]^-$ respectively.

(d) Halide complexes. Halide ions are often coordinated to metal ions, forming halide complexes. If, for example, an excess of hydrochloric acid is added to a solution which contains iron(III) ions (in a suitably high concentration), the solution turns yellow. This colour change (or deepening of colour) is due to the formation of hexachloroferrate(III) $[FeCl_6]^{3-}$ ions. Silver chloride precipitate may be dissolved in concentrated hydrochloric acid, when dichloroargentate(I) $[AgCl_2]^-$ ions are formed. An excess of potassium iodide dissolves the black bismuth iodide BiI_3 precipitate under the formation of the tetraiodobismuthate(III) $[BiI_4]^-$ ion. Especially stable are some of the fluoride complexes like hexafluoroaluminate $[AlF_6]^{3-}$, the colourless hexafluoroferrate(III) $[FeF_6]^{3-}$ and the hexafluorozirconate(IV) $[ZrF_6]^{2-}$ ions. For this reason, fluorine is often used as a masking agent both in qualitative and quantitative analysis.

(e) Cyanide and thiocyanate complexes Cyanide ions form stable complexes with a number of metals. Such complexes are

$[Ag(CN)_2]^-$	dicyanoargentate
$[Cu(CN)_4]^{3-}$	tetracyanocuprate(I)
$[Fe(CN)_6]^{4-}$	hexacyanoferrate(II)
$[Fe(CN)_6]^{3-}$	hexacyanoferrate(III)

Note that in $[Cu(CN)_4]^{3-}$ copper is monovalent.

Cyanide is often used as a masking agent. In *Example* 23 its use for masking copper for the identification of cadmium has been discussed.

Thiocyanate is used in some cases for the detection of ions. Its reaction with iron(III) ion is characteristic and can be used for detecting both ions. The deep-red colour observed is due to the formation of a number of thiocyanatoferrate(III) ions and also of the chargeless molecule $[Fe(SCN)_3]$. The blue tetrathiocyanatocobaltate(II) $[Co(SCN)_4]^{2-}$ complex is sometimes used for the detection of cobalt.

(f) Chelate complexes The ligands in complexes listed under (*a*) to (*e*) are all monodentate. Polydentate ligands, on the other hand, are quite common and form very stable complexes. These are termed chelates or chelate complexes. Oxalate is probably the simplest bidentate ligand, forming chelate complexes such as

$[Fe(C_2O_4)_3]^{3-}$	trioxalatoferrate(III) ion
$[Sn(C_2O_4)_3]^{2-}$	trioxalatostannate(IV) ion

Oxyacids, like citric or tartaric acids, and polyols, like saccharose are also used, mainly as masking agents, in qualitative analysis. The action of some specific reagents, like α-α'-bipyridyl for iron(II) and dimethylglyoxime for nickel(II), is also based on the formation of chelate complexes. In quantitative analysis the formation of chelates is frequently utilized (complexometric titrations).*

* Cf. A. I. Vogel's *A Text-Book of Quantitative Inorganic Analysis.* 3rd edn., Longman 1966, p. 415 et f.

G. OXIDATION-REDUCTION REACTIONS

I.35 OXIDATION AND REDUCTION All the reactions mentioned in the previous sections were ion-combination reactions, where the oxidation number (valency) of the reacting species did not change. There are however a number of reactions in which the state of oxidation changes, accompanied by the interchange of electrons between the reactants. These are called **oxidation-reduction reactions** or, in short, **redox reactions**.

Historically speaking the term oxidation was applied to processes where oxygen was taken up by a substance. In turn, reduction was considered to be a process in which oxygen was removed from a compound. Later on, the uptake of hydrogen was also called reduction so the loss of hydrogen had to be called oxidation. Again, other reactions, in which neither oxygen nor hydrogen take part, had to be classified as oxidation or reduction until the most general definition of oxidation and reduction, based on the release or uptake of electrons, was arrived at. Before trying to define more precisely what these terms mean, let us examine a few of these reactions.*

(*a*) The reaction between iron(III) and tin(II) ions leads to the formation of iron(II) and tin(IV):

$$2Fe^{3+} + Sn^{2+} \rightarrow 2Fe^{2+} + Sn^{4+}$$

If the reaction is carried out in the presence of hydrochloric acid, the disappearance of the yellow colour (characteristic for Fe^{3+}) can easily be observed. In this reaction Fe^{3+} is **reduced** to Fe^{2+} and Sn^{2+} **oxidized** to Sn^{4+}. What in fact happens is that Sn^{2+} donates electrons to Fe^{3+}, thus an electron transfer takes place.

(*b*) If a piece of iron (e.g. a nail) is dropped into the solution of copper sulphate, it gets coated with red copper metal, while the presence of iron(II) can be identified in the solution. The reaction which takes place is

$$Fe + Cu^{2+} \rightarrow Fe^{2+} + Cu$$

In this case the iron metal donates electrons to copper(II) ions. Fe becomes oxidized to Fe^{2+} and Cu^{2+} reduced to Cu.

(*c*) The dissolution of zinc in hydrochloric acid is also an oxidation-reduction reaction:

$$Zn + 2H^+ \rightarrow Zn^{2+} + H_2$$

Electrons are taken up by H^+ from Zn; the chargeless hydrogen atoms combine to H_2 molecules and are removed from the solution. Here Zn is oxidized to Zn^{2+} and H^+ is reduced to H_2.

(*d*) In acid medium, bromate ions are capable of oxidizing iodide to iodine, themselves being reduced to bromide:

$$BrO_3^- + 6H^+ + 6I^- \rightarrow Br^- + 3I_2 + 3H_2O$$

It is not so easy to follow the transfer of electrons in this case, because an acid-base reaction (the neutralization of H^+ to H_2O) is superimposed on the redox

* In this section only such oxidation-reduction reactions are dealt with as have importance in qualitative analysis. Other processes, with technological or historical importance, such as combustion or extraction of metals are not treated here as these fall outside the scope of this book.

step. It can be seen however that six iodide ions lose six electrons, which in turn are taken up by a single bromate ion.

(*e*) Even more complicated is the oxidation of hydrogen peroxide to oxygen and water by permanganate, which in turn is reduced to manganese(II):

$$2MnO_4^- + 5H_2O_2 + 6H^+ \rightarrow 2Mn^{2+} + 5O_2 + 8H_2O$$

A more detailed examination (cf. Section **I.36**) shows that altogether ten electrons are donated by (five molecules of) hydrogen peroxide to (two) permanganate ions in this process.

Looking at these examples, we can draw a few general conclusions and can define oxidation and reduction in the following ways:

(i) Oxidation is a process which results in the loss of one or more electrons by substances (atoms, ions, or molecules). When an element is being oxidized, its oxidation state changes to more positive values. An oxidizing agent is one that gains electrons, and is reduced during the process. This definition of oxidation is quite general, it therefore applies also to processes in the solid, molten, or gaseous states.

(ii) Reduction is, on the other hand, a process which results in the gain of one or more electrons by substances (atoms, ions, or molecules). When an element is being reduced, its oxidation state changes to more negative (or less positive) values. A reducing agent is accordingly one that loses electrons and becomes oxidized during the process. This definition of reduction is again quite general and applies also to processes in the solid, molten, or gaseous states.

(iii) From all the examples quoted it can be seen that oxidation and reduction always proceed simultaneously. This is quite obvious, because the electron(s) released by a substance must be taken up by another one. If we talk about the oxidation of one substance, we must keep in mind that at the same time the reduction of another substance also takes place. It is logical therefore to talk about oxidation-reduction reactions (or redox reactions) when referring to processes which involve transfer of charges.

I.36 REDOX SYSTEMS (HALF-CELLS) Although all oxidation-reduction reactions are based on the transfer of electrons, this cannot always be seen immediately from the reaction equations. These processes can be better understood if they are split into two separate steps, the oxidation of one substance and the reduction of another one. Let us look into the examples quoted in the previous section.

(*a*) The reaction between iron(III) and tin(II) ions

$$2Fe^{3+} + Sn^{2+} \rightarrow 2Fe^{2+} + Sn^{4+} \qquad \text{(i)}$$

is made up of the reduction of iron(III) ions

$$2Fe^{3+} + 2e^- \rightarrow 2Fe^{2+} \qquad \text{(ii)}$$

and the oxidation of tin(II) ions

$$Sn^{2+} \rightarrow Sn^{4+} + 2e^- \qquad \text{(iii)}$$

In these steps it is necessary to write down the exact number of electrons which are released or taken up in order to balance the charges. It is easy to see from these steps what actually happens if the reaction proceeds: the electrons released by Sn^{2+} are taken up by Fe^{3+}. It can also be seen that equation (i) is the sum of (ii) and (iii), but the electrons are cancelled out during the summation.

(*b*) The reaction between iron metal and copper ions

$$Fe + Cu^{2+} \rightarrow Fe^{2+} + Cu$$

consists of the reduction of Cu^{2+}

$$Cu^{2+} + 2e^- \rightarrow Cu$$

and of the oxidation of Fe

$$Fe \rightarrow Fe^{2+} + 2e^-$$

The two electrons released by Fe are taken up by Cu^{2+} in this process.

(*c*) The dissolution of zinc in acids

$$Zn + 2H^+ \rightarrow Zn^{2+} + H_2$$

involves the reduction of H^+

$$2H^+ + 2e^- \rightarrow H_2$$

and the oxidation of Zn

$$Zn \rightarrow Zn^{2+} + 2e^-$$

Again, the two electrons released by Zn are taken up by H^+.

(*d*) In the reaction between bromate and iodide

$$BrO_3^- + 6H^+ + 6I^- \rightarrow Br^- + 3I_2 + 3H_2O$$

we have the reduction of bromate

$$BrO_3^- + 6H^+ + 6e^- \rightarrow Br^- + 3H_2O$$

and the oxidation of iodide

$$6I^- \rightarrow 3I_2 + 6e^-$$

and the electrons released by the iodide are taken up by bromate ions.

(*e*) Finally, the reaction between permanganate and hydrogen peroxide in an acid medium

$$2MnO_4^- + 5H_2O_2 + 6H^+ \rightarrow 2Mn^{2+} + 5O_2 + 8H_2O$$

is made up of the reduction of permanganate

$$2MnO_4^- + 16H^+ + 10e^- \rightarrow 2Mn^{2+} + 8H_2O$$

and of the oxidation of hydrogen peroxide

$$5H_2O_2 \rightarrow 5O_2 + 10H^+ + 10e^-$$

The electrons released by H_2O_2 are taken up by MnO_4^-.

In general therefore, each oxidation-reduction reaction can be regarded as the sum of an oxidation and a reduction step. It has to be emphasized that these individual steps cannot proceed alone; each oxidation step must be accompanied by a reduction and vice versa. These individual reduction or oxidation steps, which involve the release or uptake of electrons are often called **half-cell reactions** (or simply half-cells) because from combinations of them galvanic cells (batteries) can be built up. The latter aspect of oxidation-reduction reactions will be dealt with later (cf. Section **I.39**).

All the oxidation-reduction reactions used in examples (*a*) to (*e*) proceed in one definite direction; e.g. Fe^{3+} can be reduced by Sn^{2+}, but the opposite process, the oxidation of Fe^{2+} by Sn^{4+} will not take place. That is why the single arrow was used in all the reactions, including the half-cell processes as well. If however we examine one half-cell reaction on its own, we can say that normally it is reversible.* Thus, while Fe^{3+} can be reduced (e.g. by Sn^{2+}) to Fe^{2+}, it is also true that with a suitable agent (e.g. MnO_4^-) Fe^{2+} can be oxidized to Fe^{3+}. It is quite logical to express these half-cell reactions as chemical equilibria, which also involve electrons, as

$$Fe^{3+} + e^- \rightleftharpoons Fe^{2+} \quad \text{(i)}$$

and also

$$Sn^{4+} + 2e^- \rightleftharpoons Sn^{2+} \quad \text{(ii)}$$

$$Fe^{2+} + 2e^- \rightleftharpoons Fe \quad \text{(iii)}$$

$$Cu^{2+} + 2e^- \rightleftharpoons Cu \quad \text{(iv)}$$

$$Zn^{2+} + 2e^- \rightleftharpoons Zn \quad \text{(v)}$$

$$H^+ + e^- \rightleftharpoons \tfrac{1}{2}H_2 \quad \text{(vi)}$$

$$I_2 + 2e^- \rightleftharpoons 2I^- \quad \text{(vii)}$$

$$BrO_3^- + 6H^+ + 6e^- \rightleftharpoons Br^- + 3H_2O \quad \text{(viii)}$$

$$MnO_4^- + 8H^+ + 5e^- \rightleftharpoons Mn^{2+} + 4H_2O \quad \text{(ix)}$$

$$O_2 + 2H^+ + 2e^- \rightleftharpoons H_2O_2 \quad \text{(x)}$$

The substances which are involved in such an equilibrium form a **redox system.** Thus, we can speak about the iron(III)–iron(II) or about the tin(IV)–tin(II) or the permanganate–manganese(II) system and so on. In a redox system therefore, an oxidized and a reduced form of a substance are in equilibrium, in which electrons (and in some cases protons) are exchanged. For practical purposes we will classify these redox systems in two categories.

(i) **Simple redox systems** are those in which, between the oxidized and reduced forms of the substance, only electrons are exchanged. Of the systems mentioned above (i) to (vii) fall into this category. Such systems can generally be described by the following equilibrium:

$$a\,\mathrm{Ox} + ne \rightleftharpoons b\,\mathrm{Red}$$

Here Ox and Red represent the oxidized and reduced form of the substance respectively, *a* and *b* are stoichiometric numbers, while *n* is the number of electrons exchanged. If the numbers of moles on the two sides of the equilibrium are equal (that is $a = b$) we have a **homogeneous** redox system like those (i) to (v), in other cases as (vi) and (vii) it is called **inhomogeneous**. In the simplest cases $a = b = 1$, when the system can be written as

$$\mathrm{Ox} + ne^- \rightleftharpoons \mathrm{Red}$$

(ii) **Combined redox and acid-base systems** involve not only the exchange of electrons, but also protons (hydronium ions) are exchanged, as in any acid-

* It must be emphasized however, that not all half-cell reactions are thermodynamically speaking reversible (e.g. the oxidation of thiosulphate to tetrathionate: $2S_2O_3^{2-} \rightarrow S_4O_6^{2-} + 2e$).

base system. Such are the systems (viii) to (x) among the examples given above.

These systems are indeed combinations of redox and acid-base steps. Let us consider for example the permanganate–manganese(II) system. The electrons which are released are taken up by manganese atoms. The oxidation number (valency) of manganese in permanganate being +7, the exchange of electrons could be expressed as

$$Mn^{7+} + 5e^- \rightleftarrows Mn^{2+} \qquad \text{(xi)}$$

which indeed represents the pure redox step. Mn^{7+} ions are however unstable in water; they hydrolyse in the acid-base step to form permanganate ions:

$$Mn^{7+} + 4H_2O \rightleftarrows MnO_4^- + 8H^+ \qquad \text{(xii)}$$

The combination of these two steps yields the equation quoted under (ix):

$$MnO_4^- + 8H^+ + 5e^- \rightleftarrows Mn^{2+} + 4H_2O \qquad \text{(ix)}$$

The MnO_4^-–Mn^{2+} system is homogeneous, because from one permanganate ion, one manganese(II) ion is formed; there is no change in the number of moles An inhomogeneous combined redox and acid-base system is, for example, the dichromate–chromium(III) system:

$$Cr_2O_7^{2-} + 14H^+ + 6e^- \rightleftarrows 2Cr^{3+} + 7H_2O \qquad \text{(xiii)}$$

This system is considered to be inhomogeneous, because when $Cr_2O_7^{2-}$ is reduced to Cr^{3+} there is an increase in the number of moles containing chromium.

It must be emphasized that redox systems do not necessarily consist of ions. Although most inorganic systems are made up wholly or partly of ions, a number of organic redox systems are in fact equilibria between molecules. The dehydroascorbic acid ($C_6H_6O_6$) and ascorbic acid ($C_6H_8O_6$) redox system for example

$$C_6H_6O_6 + 2H^+ + 2e^- \rightleftarrows C_6H_8O_6$$

does not contain ions (other than hydrogen). Many other organic redox systems (like the quinone–hydroquinone system) follow a similar pattern, as there are 2 protons and 2 electrons involved. Many of the organic redox systems are in fact irreversible; the use of double arrows in the equations in such cases is not really justified.

I.37 BALANCING OXIDATION-REDUCTION EQUATIONS In Section **I.5** the ways in which chemical equations can be balanced were outlined, bearing in mind that whoever studies qualitative inorganic analysis, has already studied elementary chemistry and would therefore be quite familiar with this procedure. Balancing of oxidation-reduction equations however is a quite difficult task. Unless one knows the number of electrons exchanged in the process, one can easily be mistaken. Let us examine, for example, the reaction between permanganate ions and hydrogen peroxide in acid medium. Even if we know that the products of the reaction are manganese(II) ions, oxygen gas, and water we might end up with a wrong equation. The following equations

$$2MnO_4^- + H_2O_2 + 6H^+ \dashrightarrow 2Mn^{2+} + 3O_2 + 4H_2O$$
$$2MnO_4^- + 3H_2O_2 + 6H^+ \dashrightarrow 2Mn^{2+} + 4O_2 + 6H_2O$$

$$2MnO_4^- + 5H_2O_2 + 6H^+ \longrightarrow 2Mn^{2+} + 5O_2 + 8H_2O$$
$$2MnO_4^- + 7H_2O_2 + 6H^+ \dashrightarrow 2Mn^{2+} + 6O_2 + 10H_2O$$
etc.

are formally all correct: the numbers of atoms and charges are equal on both sides in all these equations. In fact, only one of these (the third) is truly correct, and expresses the stoichiometry of this process adequately. If we examine the oxidation and reduction steps separately, only the third equation produces equal numbers of electrons in these steps.

In order to balance oxidation-reduction equations, we must therefore find out how many electrons are released by the reducing agent and taken up by the oxidizing agent. This can easily be done if the half-cell reaction equations of the redox systems involved are known. In the above example, if we write up the two half-cell equations:

$$H_2O_2 \rightleftarrows O_2 + 2H^+ + 2e^-$$

and

$$MnO_4^- + 8H^+ + 5e^- \rightleftarrows Mn^{2+} + 4H_2O$$

we can see at once that 5 molecules of hydrogen peroxide will release 10 electrons, which in turn will be taken up by 2 permanganate ions. From this ratio the equation then can be balanced quite easily.

In general, balancing of oxidation-reduction equations should be made by taking the following steps:

1. Ascertain the products of the reaction.
2. Express the half cell reaction equations of the reduction and oxidation steps involved.
3. Multiply each half-cell equation by a factor so that both equations contain the number of electrons.
4. Finally, add these equations and cancel out substances which appear on both sides of the resulting equation.

The following examples may serve as illustrations:

Example 25 Describe the reaction which proceeds between Fe^{3+} and Sn^{2+}.

1. We must know that the products are Fe^{2+} and Sn^{4+}.
2. The half-cell reactions are as follows:

$$Fe^{3+} + e^- \rightleftarrows Fe^{2+} \qquad \text{(i)}$$
$$Sn^{2+} \rightleftarrows Sn^{4+} + 2e^- \qquad \text{(ii)}$$

3. If we multiply (i) by 2, we can add the two equations:

$$2Fe^{3+} + 2e^- + Sn^{2+} \rightarrow 2Fe^{2+} + Sn^{4+} + 2e^-$$

which can be simplified to

$$2Fe^{3+} + Sn^{2+} \rightarrow 2Fe^{2+} + Sn^{4+}$$

Example 26 Bromate ions can be reduced by iodide in acid medium. Express the reaction equation.

1. We know that the products of this reaction are bromide ions, iodine, and, probably, water.
2. Thus, we express the half-cell reactions as

$$BrO_3^- + 6H^+ + 6e^- \rightarrow Br^- + 3H_2O \qquad \text{(iii)}$$

and

$$2I^- \rightarrow I_2 + 2e^- \qquad \text{(iv)}$$

3.–4. Summing (iii) + 3 × (iv) we obtain:

$$BrO_3^- + 6H^+ + 6e^- + 6I^- \rightarrow Br^- + 3H_2O + 3I_2 + 6e^-$$

which after simplification becomes:

$$BrO_3^- + 6H^+ + 6I^- \rightarrow Br^- + 3I_2 + 3H_2O$$

Example 27 Cadmium sulphide precipitate can be dissolved in hot nitric acid. Let us try to express the equation of this reaction.

1. When nitric acid acts as an oxidizing agent, nitrogen monoxide (NO) is formed. From the cadmium sulphide, sulphur is formed (unless the acid is too concentrated and hot), and cadmium ions remain dissolved. Again water is probably formed.
2. The half-cell reaction of the nitric acid–nitrogen oxide system is

$$HNO_3 + 3H^+ + 3e^- \rightarrow NO\uparrow + 2H_2O \qquad \text{(v)}$$

We can discuss the oxidation of cadmium sulphide in two steps. First the dissociation of the precipitate takes place:

$$CdS\downarrow \rightleftarrows Cd^{2+} + S^{2-} \qquad \text{(vi)}$$

(This dissociation becomes complete, as the S^{2-} ions are removed continuously from the solution by reaction vii.) This is followed by the oxidation of S^{2-}:

$$S^{2-} \rightarrow S\downarrow + 2e^- \qquad \text{(vii)}$$

3.–4. The sum of 2 × (v) + 3 × (vi) and 3 × (vii) provides equal numbers of electrons on both sides:

$$2HNO_3 + 6H^+ + 6e^- + 3CdS\downarrow + 3S^{2-}$$
$$\rightarrow 2NO\uparrow + 4H_2O + 3Cd^{2+} + 3S^{2-} + 3S\downarrow + 6e^-$$

After simplification the equation becomes

$$2HNO_3 + 6H^+ + 3CdS\downarrow \rightarrow 2NO\uparrow + 3Cd^{2+} + 3S\downarrow + 4H_2O$$

Example 28 In slightly acid, almost neutral solution, permanganate ions are capable of oxidizing manganese(II) ions. Express the reaction equation.

1. The product of the reaction is manganese dioxide precipitate.
2. The half-cell reaction of the reduction of permanganate is a three-electron process:

$$MnO_4^- + 4H^+ + 3e^- \rightarrow MnO_2\downarrow + 2H_2O \qquad \text{(viii)}$$

while at the oxidation of manganese(II) 2 electrons are liberated:

$$Mn^{2+} + 2H_2O \rightarrow MnO_2\downarrow + 4H^+ + 2e^- \qquad \text{(ix)}$$

3.–4. Combining 2 × (viii) + 3 × (ix) we obtain 6 electrons on both sides:

$$2MnO_4^- + 8H^+ + 6e^- + 3Mn^{2+} + 6H_2O$$
$$\rightarrow 2MnO_2\downarrow + 4H_2O + 3MnO_2\downarrow + 12H^+ + 6e^-$$

This equation can be simplified to

$$2MnO_4^- + 3Mn^{2+} + 2H_2O \rightarrow 5MnO_2 + 4H^+$$

It is worth while noting that hydrogen ions are formed during the reaction, and if these build up to a higher concentration they may reverse the process. In order to make the reaction complete, the solution must be buffered. The reaction is utilized for the titrimetric determination of manganese by the so-called Volhard–Wolff method, when zinc oxide is used as a buffer.

Example 29 Glycerol $CH_2(OH)—CH(OH)—CH_2(OH)$ or simply $C_3H_8O_3$ can slowly be oxidized by dichromate ions in hot acid solution. Express the equation of the reaction.

1. The products of the reaction are: CO_2, H_2O as well as Cr^{3+} ions.
2. The half-cell equation of the reduction of dichromate is

$$Cr_2O_7^{2-} + 14H^+ + 6e^- \rightarrow 2Cr^{3+} + 7H_2O \qquad \text{(x)}$$

while oxidation of each glycerol molecule produces 14 electrons (note that this reaction is not reversible):

$$C_3H_8O_3 + 3H_2O \rightarrow 3CO_2\uparrow + 14H^+ + 14e^- \qquad \text{(xi)}$$

3. As (x) involves 6 electrons and (xi) involves 14, we have to find their lowest common multiple, which is 42.
4. Thus, $7\times(x)+3\times(xi)$ yields 42 electrons on both sides:

$$7Cr_2O_7^{2-} + 98H^+ + 42e^- + 3C_3H_8O_3 + 9H_2O$$
$$\rightarrow 14Cr^{3+} + 49H_2O + 9CO_2\uparrow + 42H^+ + 42e^-$$

One must not be puzzled by the unusually high stoichiometric numbers. After simplification we get a much less complicated equation:

$$7Cr_2O_7^{2-} + 3C_3H_8O_3 + 56H^+ \rightarrow 14Cr^{3+} + 9CO_2\uparrow + 40H_2O$$

Example 30 The oxidation of thiocyanate ions by permanganate in acid solution is applied in quantitative analysis. Express the reaction equation.

1. The products of the reaction are SO_4^{2-}, CO_2, N_2, Mn^{2+}, and H_2O.
2. The half-cell equation of the reduction of permanganate is well known:

$$MnO_4^- + 8H^+ + 5e^- \rightarrow Mn^{2+} + 4H_2O \qquad \text{(xii)}$$

The oxidation of thiocyanate is a rather complex process, which may be dealt with in two separate steps. First, sulphate and cyanide ions are formed:

$$SCN^- + 4H_2O \rightarrow SO_4^{2-} + CN^- + 8H^+ + 6e^- \qquad \text{(xiii)}$$

In a second step cyanide ions are oxidized to nitrogen and carbon dioxide:

$$2CN^- + 4H_2O \rightarrow 2CO_2\uparrow + N_2\uparrow + 8H^+ + 10e^- \qquad \text{(xiv)}$$

3. and 4. A combination of 22 (xii) + 10 (xiii) + 5 (xiv) provides equal numbers of electrons on both sides:

$$22MnO_4^- + 176H^+ + 110e^- + 10SCN^- + 40H_2O + 10CN^- + 20H_2O$$
$$\rightarrow 22Mn^{2+} + 88H_2O + 10SO_4^{2-} + 10CN^- + 80H^+ + 60e^-$$
$$+ 10CO_2\uparrow + 5N_2\uparrow + 40H^+ + 50e^-$$

On simplifying the final equation becomes

$$22MnO_4^- + 10SCN^- + 56H^+ \rightarrow 22Mn^{2+} + 10SO_4^{2-} + 10CO_2\uparrow + 5N_2\uparrow + 28H_2O$$

It is very important that users of this book should acquire a routine knowledge of balancing oxidation-reduction equations. Further practice may be obtained by combining various oxidizing and reducing agents mentioned in *Examples* 25 to 30, for which the half-cell reactions can easily be checked from the numbered equations. A careful study of Section **I.38** which follows might also help.

I.38 IMPORTANT OXIDIZING AND REDUCING AGENTS A large number of oxidation-reduction reactions will be listed among the reactions used for the identification of ions. For reference purposes a number of such reactions have been collected in the present chapter, grouped around substances which are used in practice as oxidizing and reducing agents.

Potassium permanganate, $KMnO_4$ This is a dark-brown solid, which produces a violet solution when dissolved in water, which is characteristic for permanganate ions. Potassium permanganate is a strong oxidizing agent, which acts differently according to the *p*H of the medium.

(*a*) **In acid solution** permanganate ions are reduced by a five-electron process, when the oxidation number of manganese changes from +7 to +2:

$$MnO_4^- + 8H^+ + 5e^- \rightarrow Mn^{2+} + 4H_2O$$

Some important oxidations carried out with permanganate ions are as follows:

$$MnO_4^- + 5Fe^{2+} + 8H^+ \rightarrow Mn^{2+} + 5Fe^{3+} + 4H_2O$$

$$2MnO_4^- + 10I^- + 16H^+ \rightarrow 2Mn^{2+} + 5I_2 + 8H_2O$$

$$2MnO_4^- + 5H_2S + 6H^+ \rightarrow 2Mn^{2+} + 5S\downarrow + 8H_2O$$

(*b*) **In neutral or slightly alkaline** solution permanganate is reduced to manganese dioxide, when in a three-electron process the oxidation number of manganese changes from +7 to +4:

$$MnO_4^- + 4H^+ + 3e^- \rightarrow MnO_2\downarrow + 2H_2O$$

MnO_2 is a darkish-brown precipitate. As an example the oxidation of manganese(II) salts can serve:

$$2MnO_4^- + 3Mn^{2+} + 2H_2O \rightarrow 5MnO_2\downarrow + 4H^+$$

As H^+ ions are formed, which might reverse the reaction, buffers have to be used. The reaction is utilized for the titrimetric determination of manganese, when ZnO is normally used as buffer.

(*c*) **In strongly alkaline solutions** (at *p*H 13 or over) permanganate can be reduced to manganate in a one-electron process:

$$MnO_4^- + e^- \rightarrow MnO_4^{2-}$$

The oxidation number of manganese in manganate is +6. MnO_4^{2-} ions exhibit a characteristic green colour. When permanganate is heated with alkalis, such a reduction takes place and oxygen is formed:

$$4MnO_4^- + 4OH^- \rightarrow 4MnO_4^{2-} + 2H_2O + O_2\uparrow$$

Potassium dichromate $K_2Cr_2O_7$ This strong oxidizing agent is an orange-red solid, which produces an orange solution in water. In strongly acid solutions dichromate ions are reduced to chromium(III):

$$Cr_2O_7^{2-} + 14H^+ + 6e^- \rightarrow 2Cr^{3+} + 7H_2O$$

The oxidation number of Cr changes from +6 to +3. The solution becomes light green, the colour originating from Cr^{3+} ions. Some important oxidations with dichromate are as follows:

$$Cr_2O_7^{2-} + 6Fe^{2+} + 14H^+ \rightarrow 2Cr^{3+} + 6Fe^{3+} + 7H_2O$$

$$Cr_2O_7^{2-} + 6I^- + 14H^+ \rightarrow 2Cr^{3+} + 3I_2 + 7H_2O$$

$$Cr_2O_7^{2-} + 3Sn^{2+} + 14H^+ \rightarrow 2Cr^{3+} + 3Sn^{4+} + 7H_2O$$

$$\underset{\text{(formaldehyde)}}{Cr_2O_7^{2-} + 3HCHO} + 8H^+ \rightarrow 2Cr^{3+} + \underset{\text{(formic acid)}}{3HCOOH} + 4H_2O$$

Nitric acid HNO_3 The oxidizing action of nitric acid depends on the concentration of the acid and the temperature of the solution. Normally, nitrogen oxide is formed in a three-electron process:

$$HNO_3 + 3H^+ + 3e^- \rightarrow NO + 2H_2O$$

The NO gas is colourless, but readily reacts with atmospheric oxygen, when the reddish-brown nitrogen dioxide is formed:

$$2NO + O_2 \rightarrow 2NO_2 \tag{i}$$

It is instructive to carry out such an oxidation (e.g. the dissolution of iron metal) in a flask with a narrow neck. Pouring not too concentrated nitric acid (e.g. a 1+1 mixture of concentrated acid with H_2O) on iron filings and heating the mixture, one can see that the dissolution produces a colourless gas inside the flask:

$$Fe + HNO_3 + 3H^+ \rightarrow Fe^{3+} + NO\uparrow + 2H_2O$$

Over the neck of the flask, the gas, exposed to atmospheric oxygen turns brownish red, because reaction (i) proceeds.

Concentrated, or half-concentrated HNO_3 is mostly used to dissolve metals and precipitates. Such reactions are:

$$3Ag + HNO_3 + 3H^+ \rightarrow 3Ag^+ + NO\uparrow + 2H_2O$$

$$3CuS + 2HNO_3 + 6H^+ \rightarrow 3Cu^{2+} + 3S\downarrow + 2NO\uparrow + 4H_2O$$

Under the conditions in which these reactions are carried out nitric acid is far from being completely dissociated, the formula HNO_3 is therefore used in the above equations

The Halogens, Cl_2, Br_2, and I_2 The action of halogens is dependent upon the conversion of electrically neutral halogen molecules into halogen ions by accepting electrons:

$$Cl_2 + 2e^- \rightarrow 2Cl^-$$

$$Br_2 + 2e^- \rightarrow 2Br^-$$

$$I_2 + 2e^- \rightarrow 2I^-$$

The oxidizing power of halogens decrease with increasing relative atomic mass. Iodine is a mild oxidant, while the iodide ion often acts as reducing agent. Some oxidations with halogens, used in qualitative analysis, are as follows:

$$Cl_2 + 2Fe^{2+} \rightarrow 2Cl^- + 2Fe^{3+}$$
$$Br_2 + AsO_3^{3-} + H_2O \rightarrow AsO_4^{3-} + 2Br^- + 2H^+$$
$$I_2 + 2S_2O_3^{2-} \rightarrow S_4O_6^{2-} + 2I^-$$

Aqua regia . The mixture of three volume of concentrated HCl and one volume of concentrated HNO_3, called 'aqua regia' or the kingly water, is a strong oxidizing agent, which is able to oxidize (and to dissolve) noble metals like gold and platinum. Its action is based on the formation of chlorine:

$$HNO_3 + 3HCl \rightarrow NOCl\uparrow + Cl_2\uparrow + 2H_2O$$

This equation is somewhat simplified; in fact there are more products formed during the process. Nitrosyl chloride, NOCl, is one of the products which can easily be identified. The oxidative action of chlorine is based on the process described in the previous section. The dissolution of gold can be expressed with the equation:

$$3HNO_3 + 9HCl + 2Au \rightarrow 3NOCl + 6Cl^- + 2Au^{3+} + 6H_2O$$

Hydrogen peroxide, H_2O_2 Although often quoted as a strong oxidizing agent, hydrogen peroxide may act both as an oxidizing and as a reducing agent. Its oxidizing action is based on a two-electron process, which results in the formation of water:

$$H_2O_2 + 2H^+ + 2e^- \rightarrow 2H_2O$$

As a reducing agent, hydrogen peroxide releases 2 electrons and oxygen gas is formed:

$$H_2O_2 \rightarrow O_2 + 2H^+ + 2e^-$$

Its role in redox reactions depends on the relative oxidizing or reducing strength of the reaction partner, and also on the pH of the solution.

Oxidations with hydrogen peroxide in acid medium are, for example,

$$H_2O_2 + 2H^+ + 2I^- \rightarrow I_2 + 2H_2O$$
$$H_2O_2 + 2H^+ + 2Fe^{2+} \rightarrow 2Fe^{3+} + 2H_2O$$

It can however act as oxidant in alkaline medium too. An alkaline solution containing chromium(III) in the form of tetrahydroxochromate(III) $[Cr(OH)_4]^-$ can be oxidized to chromate(VI) with H_2O_2:

$$3H_2O_2 + 2[Cr(OH)_4]^- \rightarrow 2CrO_4^{2-} + 2H^+ + 6H_2O$$

Reductions with hydrogen peroxide can also be achieved both in acid (i) and (ii) and alkaline medium (iii):

$$5H_2O_2 + 2MnO_4^- + 6H^+ \rightarrow 5O_2\uparrow + 2Mn^{2+} + 8H_2O \quad \text{(i)}$$
$$3H_2O_2 + 2Au^{3+} \rightarrow 2Au + 3O_2\uparrow + 6H^+ \quad \text{(ii)}$$
$$H_2O_2 + 2[Fe(CN)_6]^{3-} \rightarrow 2[Fe(CN)_6]^{4-} + 2H^+ + O_2\uparrow \quad \text{(iii)}$$

Sulphur dioxide, SO_2, and sulphurous acid H_2SO_3 Sulphur dioxide gas, when dissolved in water, forms sulphurous acid. It is a strong reducing agent, its action is based on the transformation of sulphite ion to sulphate. The oxidation number of sulphur changes from +4 to +6, hence 2 electrons are released during the process:

$$SO_3^{2-} + H_2O \rightarrow SO_4^{2-} + 2H^+ + 2e^-$$

The reagent is sometimes employed by adding sodium sulphite Na_2SO_3 to the acidified solution to be reduced. Some reductions with SO_3^{2-} are as follows:

$$SO_3^{2-} + 2Fe^{3+} + H_2O \rightarrow SO_4^{2-} + 2Fe^{2+} + 2H^+$$
$$SO_3^{2-} + I_2 + H_2O \rightarrow SO_4^{2-} + 2I^- + 2H^+$$
$$3SO_3^{2-} + Cr_2O_7^{2-} + 8H^+ \rightarrow 3SO_4^{2-} + Cr^{3+} + 4H_2O$$
$$SO_3^{2-} + AsO_4^{3-} \rightarrow SO_4^{2-} + AsO_3^{3-}$$

Hydrogen sulphide H_2S Hydrogen sulphide gas or its saturated solution is used as a precipitant in qualitative inorganic analysis. If oxidizing ions, like $Cr_2O_7^{2-}$, MnO_4^-, Fe^{3+}, AsO_4^{3-} or substances like HNO_3 or Cl_2 are present, it undergoes oxidation, when elementary S is formed:

$$H_2S \rightarrow S\downarrow + 2H^+ + 2e^-$$

The sulphide precipitates formed will therefore also contain some free sulphur. The reaction equations are as follows:

$$3H_2S + Cr_2O_7^{2-} + 8H^+ \rightarrow 3S\downarrow + 2Cr^{3+} + 7H_2O$$
$$5H_2S + 2MnO_4^- + 6H^+ \rightarrow 5S\downarrow + 2Mn^{2+} + 8H_2O$$
$$H_2S + 2Fe^{3+} \rightarrow S\downarrow + 2Fe^{2+} + 2H^+$$
$$H_2S + Cl_2 \rightarrow S\downarrow + 2Cl^- + 2H^+$$
$$3H_2S + 2HNO_3 \rightarrow 3S\downarrow + 2NO\uparrow + 4H_2O$$

Hydriodic acid HI (the iodide ion I^-) Iodide ions reduce a number of substances, themselves being oxidized to iodine:

$$2I^- \rightarrow I_2 + 2e^-$$

The oxidation number of iodine changes from −1 to 0. Iodide ions are mostly added in the form of potassium iodide KI. Reductions with I^- are e.g.:

$$6I^- + BrO_3^- + 6H^+ \rightarrow 3I_2 + Br^- + 3H_2O$$
$$5I^- + IO_3^- + 6H^+ \rightarrow 3I_2 + 3H_2O$$
$$2I^- + Cl_2 \rightarrow I_2 + 2Cl^-$$
$$6I^- + Cr_2O_7^{2-} + 14H^+ \rightarrow 3I_2 + 2Cr^{3+} + 7H_2O$$
$$10I^- + 2MnO_4^- + 16H^+ \rightarrow 5I_2 + 2Mn^{2+} + 8H_2O$$

If a solution of potassium iodide is acidified with concentrated hydrochloric acid and the solution is left exposed to air, it slowly turns to yellow and later brown, because of oxidation by atmospheric oxygen:

$$4I^- + O_2 + 4H^+ \rightarrow 2I_2 + 2H_2O$$

Tin(II) chloride $\boldsymbol{SnCl_2}$ Tin(II) ions are strong reducing agents. When oxidized to tin(IV) the oxidation number of tin increases from +2 to +4, corresponding to the release of 2 electrons:

$$Sn^{2+} \rightarrow Sn^{4+} + 2e^-$$

Some reductions with tin(II) are

$$Sn^{2+} + 2HgCl_2 \rightarrow Sn^{4+} + Hg_2Cl_2 + 2Cl^-$$
$$Sn^{2+} + Hg_2Cl_2 \rightarrow Sn^{4+} + 2Hg + 2Cl^-$$
$$Sn^{2+} + Cl_2 \rightarrow Sn^{4+} + 2Cl^-$$
$$Sn^{2+} + Fe^{3+} \rightarrow Sn^{4+} + 2Fe^{2+}$$

Solutions of $SnCl_2$ do not keep well because atmospheric oxygen oxidizes the tin(II) ion:

$$2Sn^{2+} + O_2\uparrow + 4H^+ \rightarrow 2Sn^{4+} + 2H_2O$$

Metals like zinc, iron, and aluminium These are often used as reducing agents. Their action is due to the formation of ions, normally of their lowest oxidation state:

$$Zn \rightarrow Zn^{2+} + 2e^-$$
$$Fe \rightarrow Fe^{2+} + 2e^-$$
$$Al \rightarrow Al^{3+} + 3e^-$$

Zinc can be used for reduction both in acid and alkaline medium:

$$3Zn\downarrow + 2Sb^{3+} \rightarrow 2Sb\downarrow + 3Zn^{2+}$$
$$4Zn\downarrow + NO_3^- + 7OH^- + 6H_2O \rightarrow 4[Zn(OH)_4]^{2-} + NH_3$$
$$Zn\downarrow + NO_3^- + 2H^+ \rightarrow Zn^{2+} + NO_2^- + H_2O$$
$$Fe\downarrow + Cu^{2+} \rightarrow Cu\downarrow + Fe^{2+}$$
$$Fe\downarrow + Sn^{4+} \rightarrow Sn^{2+} + Fe^{2+}$$

The reactions in which metals dissolve in acids or alkalis are also reductions of the dissolving agents, as:

$$Zn\downarrow + 2H^+ \rightarrow Zn^{2+} + H_2\uparrow$$
$$Fe\downarrow + 2H^+ \rightarrow Fe^{2+} + H_2\uparrow$$
$$2Al\downarrow + 6H^+ \rightarrow 2Al^{3+} + 3H_2\uparrow$$
$$Zn\downarrow + 2OH^- + 2H_2O \rightarrow [Zn(OH)_4]^{2-} + H_2\uparrow$$
$$2Al\downarrow + 2OH^- + 6H_2O \rightarrow 2[Al(OH)_4]^- + 3H_2\uparrow$$

I.39 REDOX REACTIONS IN GALVANIC CELLS When discussing oxidation-reduction reactions we have not mentioned ways in which the directions of such reactions can be predicted. In other words, discussions in the previous chapters were aimed at understanding *how* oxidation-reduction reactions proceed, but there was no mention of *why* they take place. In this and the next few sections the problem will be dealt with in some detail.

The direction of chemical reactions can always be predicted from thermodynamical data. Thus, if the Gibbs free energy change of a reaction is calculated, we can definitely state whether a given chemical reaction may proceed or not. To perform such calculations a good working knowledge of thermodynamics is needed, which however is not expected from the readers of this book. In the previous chapters therefore the problem of the directions of chemical reactions was dealt with on the basis of the equilibrium constant. From the value of the equilibrium constant one can easily make semiquantitative estimations, e.g. if the value of such a constant is high or low, the equilibrium between the reactants and products is shifted to one or another extreme, meaning that the reaction in fact will proceed in one or another direction. Such a treatment has been used when dealing with acid-base, precipitation, and complexation reactions.

Although the law of mass action is equally valid for oxidation-reduction processes, and therefore conclusions as to the direction of reactions may be drawn from the knowledge of equilibrium constants, traditionally a different approach is used for such processes. This has both historical and practical reasons. As pointed out in the previous sections, in oxidation-reduction processes electrons are transferred from one species to another. This transfer may occur directly, i.e. one ion collides with another and during this the electron is passed on from one ion to the other. It is possible, however, to pass these electrons through electrodes and leads from one ion to the other. A suitable device in which this can be achieved is a **galvanic cell**, one of which is shown in Fig. I.14. A galvanic cell consists of two **half-cells**, each made up of an electrode and an electrolyte. The two electrolytes are connected with a salt bridge and, if

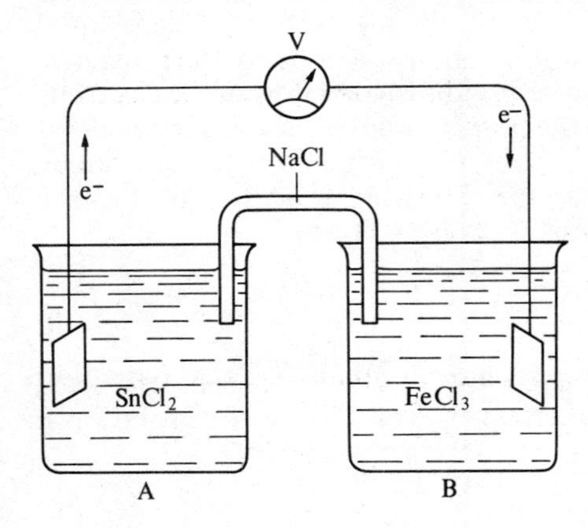

Fig. I.14

the electrodes are connected by wires, electrons will flow in the direction indicated. The movement of electrons in the lead means that an electrical current is flowing. Because of their practical importance, galvanic cells were extensively studied before theories of redox reactions were formulated. For this reason, interpretation of redox reactions is traditionally based on phenomena occurring in galvanic cells, and this tradition is observed in this text also.

The direction of this electron flow in the cell is strongly associated with the direction of the chemical reaction(s) involved in the process. Electrically speaking, the direction of electron flow depends on the sign of the potential difference between the electrodes; electrons will flow from the negative electrode through the lead towards the positive electrode. The magnitudes of electrode potentials are therefore of primary importance when trying to interpret oxidation-reduction processes in a quantitative way.

Let us examine the operation of a few galvanic cells.

(a) We have already dealt with the reduction of iron(III) ions with tin(II), which leads to the formation of iron(II) and tin(IV) ions:

$$2Fe^{3+} + Sn^{2+} \rightarrow 2Fe^{2+} + Sn^{4+} \qquad \text{(i)}$$

If solutions of iron(III) chloride and tin(II) chloride are mixed, this reaction proceeds instantaneously.

The same reaction proceeds in the galvanic cell shown in Fig. I.14. The solutions of tin(II) chloride and iron(III) chloride, each acidified with dilute hydrochloric acid to increase conductivity, are placed in separate beakers A and B, and the two solutions are connected by means of a 'salt bridge' containing sodium chloride. The latter consists of an inverted U-tube filled with a solution of a conducting electrolyte, such as potassium chloride, and stoppered at each end with a plug of cotton wool to arrest mechanical flow. It connects the two solutions while preventing mixing. The electrolyte in a solution in the salt bridge is always selected so that it does not react chemically with either of the solutions which it connects. Platinum foil electrodes are introduced into each of the solutions, and are connected to a voltmeter V of a high internal resistance. When the circuit is closed, the voltmeter shows a deflection corresponding to the difference of the voltages of the two electrodes. If the resistance of the meter is so high that no current can flow in the circuit, the measured voltage is equal to the **electromotive force** or **e.m.f.** of the cell. If, on the other hand, the resistance of the circuit is low, a current will flow, corresponding to the flow of the electrons from the negative electrode (A) towards the positive one (B). If the current flows for a while, tin(IV) ions can be detected in solution A while iron(II) ions can be found in solution B. This indicates that the following processes took place during the operation of the cell:

In solution A:

$$Sn^{2+} \rightarrow Sn^{4+} + 2e^- \qquad \text{(ii)}$$

and the two electrons are taken up by the electrode. These are then conducted to the other electrode where they are taken up by iron(III) ions. In solution B therefore the reaction

$$2Fe^{3+} + 2e^- \rightarrow 2Fe^{2+} \qquad \text{(iii)}$$

will proceed. The sum of equations (ii) and (iii) being equal to (i) we can see that the basis of the operation of this galvanic cell was an oxidation-reduction process, which would proceed normally if the reactants were mixed. The basis difference between the two processes is that the reactants (Fe^{3+} and Sn^{2+}) in the galvanic cell are separated from each other.

(*b*) If a piece of zinc is dipped into a solution of copper sulphate, its surface becomes coated with copper metal and the presence of zinc ions in the solution can be detected. The chemical reaction which takes place can be described by the following equation:

$$Cu^{2+} + Zn\downarrow \rightarrow Cu\downarrow + Zn^{2+} \qquad \text{(iv)}$$

In this process, electrons donated by zinc atoms were taken up by copper ions.

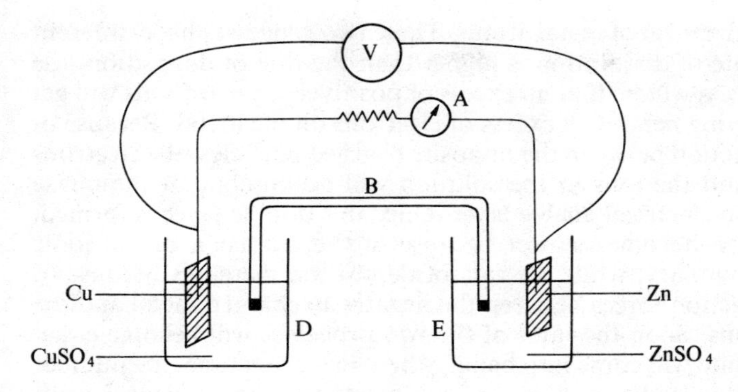

Fig. I.15

The same process takes place in the **Daniell cell**, which is shown in Fig. I.15. In vessel D a copper foil, immersed in a solution of copper sulphate, forms the +ve pole of the cell, while in vessel E the zinc foil, dipped into zinc sulphate, is the −ve pole. The role and construction of the salt bridge B is the same as in the previous cell. The voltmeter V measures the e.m.f. of the cell (the measured voltage being equal to the e.m.f. only if there is practically no current flowing in the circuit). If the electrodes are connected through a resistor, current will flow which can be measured on the ammeter A. If the cell has been operating for a while, it is possible to detect that the weight of the zinc electrode has decreased, while the weight of the copper electrode has increased; at the same time the concentration of zinc ions in the vessel E has increased, and that of copper ions in vessel D has decreased. Thus, the chemical reactions which took place in vessels E and D respectively were

$$Zn \rightarrow Zn^{2+} + 2e^- \qquad \text{(v)}$$

and

$$Cu^{2+} + 2e^- \rightarrow Cu \qquad \text{(vi)}$$

Note that the sum of equations (v) and (vi) equals (iv), meaning that the chemical processes in both cases were the same.

I.40 ELECTRODE POTENTIALS When a galvanic cell is constructed, a potential difference is measurable between the two electrodes. If the flow of current is negligible, this potential difference is equal to the electromotive force (e.m.f.) of the cell. The latter can be regarded as the absolute value of the difference of two individual **electrode potentials**, E_1 and E_2.

$$\text{e.m.f.} = |E_1 - E_2|$$

These electrode potentials are potential differences themselves, which are formed between the electrode (solid phase) and the electrolyte (liquid phase). Their occurrence can be most easily interpreted by the formation of double layers on the phase boundaries. If a piece of metal is immersed in a solution which contains its own ions (e.g. Zn in a solution of $ZnSO_4$), two processes will immediately start. First, the atoms of the outside layer of the metal will dissolve, leaving electrons on the metal itself, and slowly diffuse into the solution as metal ions. Second, metal ions from the solution will take up electrons from the metal and get deposited in the form of metal atoms. These two processes have different initial rates. If the rate of dissolution is higher than the rate of deposition, the net result of this process will be that an excess of positively charged ions will get into the solution, leaving behind an excess of electrons on the metal. Because of the electrostatic attraction between the opposite charged particles, the electrons in the metal phase and the ions in the solution will accumulate at the phase boundary, forming an **electrical double layer**. Once this double layer is formed, the rate of dissolution becomes slower because of the repulsion of the ionic layer at the phase boundary, while the rate of deposition increases because of the electrostatic attraction forces between the negatively charged metal and the positively charged ions. Soon the rates of the two processes will become equal and an equilibrium state will come into being, when, in a given time, the number of ions discharged equals the number of ions produced. As a result a well-defined potential difference will develop between the metal and solution, and the metal will acquire a negative potential with respect to the solution.

If, on the other hand, the initial rate of deposition is higher than the initial rate of dissolution, the electrical double layer will be formed just in the opposite sense, and as a result the metal becomes positive with respect to the solution. This is the case with the copper electrode in the Daniell cell.

The potential difference established between a metal and a solution of its salt will depend on the nature of metal itself and on the concentration of the ions in the solution. For a reversible metal electrode with the electrode reaction

$$\mathrm{Me} \rightarrow \mathrm{Me}^{n+} + ne^-$$

the E electrode potential can be expressed as

$$E = E^\ominus + \frac{RT}{nF} \ln a_{\mathrm{Me}^{n+}} \approx E^\ominus + \frac{RT}{nF} \ln[\mathrm{Me}^{n+}]$$

where the activity $a_{\mathrm{Me}^{n+}}$ can in most practical cases be replaced by the concentration of the metal $[\mathrm{Me}^{n+}]$. This equation was first deduced by Nernst in 1888, and is therefore called the **Nernst equation**. In the equation R is the gas constant (expressed in suitable units, e.g. $R = 8{\cdot}314\ \mathrm{J\ K^{-1}\ mol^{-1}}$) F is the Faraday number ($F = 9{\cdot}6487 \times 10^4\ \mathrm{C\ mol^{-1}}$), T is the absolute temperature (K). $E^\ominus$ is the **standard potential**, a constant, which is characteristic for the metal in question.

The electrodes just referred to are reversible with respect to the metallic ion, that is to a cation. It is possible to construct electrodes which are reversible with respect to an anion. Thus, when silver, in contact with solid silver chloride, is immersed into a solution of potassium chloride, the potential will depend on the concentration of the chloride ion, and the electrode will be reversible to this ion. The calomel electrode, described in Section **I.22** is also reversible to chloride ions.

It is not possible to measure the potential difference between the solution and the electrode, because in order to do this the solution must be connected to a conductor, i.e. a piece of another metal must be dipped into it. On the phase boundary another electrical double layer will be formed and in fact another, unknown electrode potential is developed. It is impossible therefore to measure **absolute electrode potentials**, only their differences. As seen before, the e.m.f. of a cell can be measured relatively easily, and this e.m.f. is the algebraic difference of the two electrode potentials. Building up cells from two electrodes, of which one is always the same, we can determine the relative values of electrode potentials, which can be used then for practical purposes. All that has to be done is to select a suitable **standard reference electrode**, to which all electrode potentials can be related. In practice the standard reference electrode used for comparative purposes is the **standard hydrogen electrode**. This is a reversible hydrogen electrode, with hydrogen gas of $1{\cdot}0133 \times 10^5$ Pa (= 1 atm) pressure being in equilibrium with a solution of hydrogen ions of unit activity. The potential of this electrode is taken arbitrarily as zero. All electrode potentials are then calculated on this **hydrogen scale**.

A standard hydrogen electrode can easily be built from a platinum foil, coated by platinum black by an electrolytic process, and immersed in a solution of hydrochloric acid containing hydrogen ions of unit activity (a mixture of 1000 g water and 1·184 mol hydrogen chloride can be used in practice). Hydrogen gas at a pressure of 1 atm is passed over the foil. A convenient form of the standard hydrogen electrode is shown on Fig. I.16. The gas is introduced

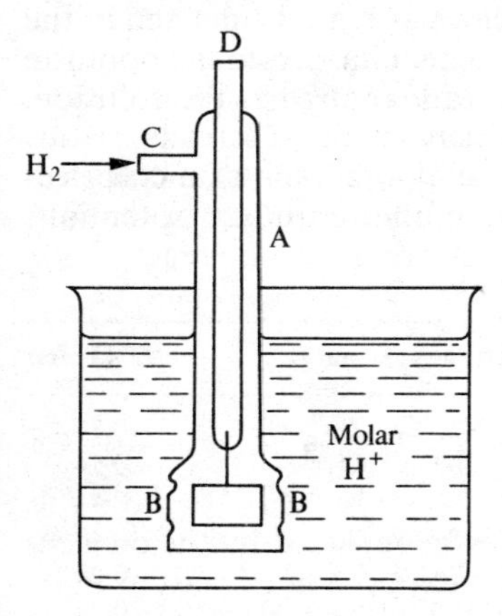

Fig. I.16

through side tube C and escapes through openings B in the surrounding gas tube A; the foil is thus kept saturated with the gas. The hydrogen gas, used for this purpose, must be meticulously purified, e.g. by bubbling it through solutions of $KMnO_4$ and $AgNO_3$. Connection between the platinum foil sealed in tube D and an outer circuit is made with mercury in D. The platinum black has the remarkable property of adsorbing large quantities of hydrogen, and it permits the change from the gaseous form into the ionic form and the reverse without hindrance:

$$2H^+ + 2e^- \rightleftarrows H_2$$

This electrode therefore behaves as if composed entirely of hydrogen, that is as a hydrogen electrode. By connecting this standard hydrogen electrode through a salt bridge to an electrode of an unknown potential, a galvanic cell is obtained, and the measured e.m.f. will be equal to the electrode potential of the unknown electrode, its sign will be equal to the polarity of the electrode in question in this cell. When using the standard hydrogen electrode as a reference electrode time must be allowed for the system to reach equilibrium; normally 30–60 min should elapse before the final measurement is taken.

Because of slowness of response and elaborate equipment needed for handling hydrogen gas, the standard hydrogen electrode is only occasionally used in practice as a reference electrode. Instead, other electrodes are used, such as the calomel electrode (cf. Section **I.22**) or the silver–silver chloride electrode. These are easy to manipulate and their electrode potentials are constant, having been determined once and for all by direct reference to the standard hydrogen electrode. For more details on such electrodes, textbooks of physical chemistry should be consulted.*

From the Nernst equation we can see that the electrode potential of a metal electrode, immersed in a solution of its ions, depends on the concentration (more precisely, activity) of these ions. If the activity of ions in the solution is unity (1 mol ℓ^{-1}), the expression becomes,

$$E = E^{\circ}$$

thus the electrode potential becomes equal to the standard electrode potential itself. The **standard electrode potential** of a metal can therefore be defined as the e.m.f. produced when a half-cell consisting of the element immersed in a solution of its ions possessing unit activity is coupled with a standard hydrogen electrode. The sign of the potential is the same as the polarity of the electrode in this combination. Table I.16 contains values of standard potentials of metal electrodes. In this table metals are arranged in the order of their standard potentials,

Table I.16 Standard potentials of metal electrodes at 25°C

Electrode reaction	E° (Volts)	Electrode reaction	E° (Volts)
$Li^{+}+e^{-} \rightleftarrows Li$	−3·04	$Zn^{2+}+2e^{-} \rightleftarrows Zn$	−0·76
$K^{+}+e^{-} \rightleftarrows K$	−2·92	$Cr^{3+}+3e^{-} \rightleftarrows Cr$	−0·74
$Ba^{2+}+2e^{-} \rightleftarrows Ba$	−2·90	$Fe^{2+}+2e^{-} \rightleftarrows Fe$	−0·44
$Sr^{2+}+2e^{-} \rightleftarrows Sr$	−2·89	$Cd^{2+}+2e^{-} \rightleftarrows Cd$	−0·40
$Ca^{2+}+2e^{-} \rightleftarrows Ca$	−2·87	$Co^{2+}+2e^{-} \rightleftarrows Co$	−0·28
$Na^{+}+e^{-} \rightleftarrows Na$	−2·71	$Ni^{2+}+2e^{-} \rightleftarrows Ni$	−0·25
$Ce^{3+}+3e^{-} \rightleftarrows Ce$	−2·48	$Sn^{2+}+2e^{-} \rightleftarrows Sn$	−0·14
$Mg^{2+}+2e^{-} \rightleftarrows Mg$	−2·37	$Pb^{2+}+2e^{-} \rightleftharpoons Pb$	−0·13
$Th^{4+}+4e^{-} \rightleftarrows Th$	−1·90	$2H^{+}+2e^{-} \rightleftarrows H_2$ (Pt)	0·00
$Be^{2+}+2e^{-} \rightleftarrows Be$	−1·85	$Cu^{2+}+2e^{-} \rightleftarrows Cu$	+0·34
$V^{3+}+3e^{-} \rightleftarrows V$	−1·80	$Hg_2^{2+}+2e^{-} \rightleftarrows 2Hg$	+0·79
$Al^{3+}+3e^{-} \rightleftarrows Al$	−1·66	$Ag^{+}+e^{-} \rightleftarrows Ag$	+0·80
$Mn^{2+}+2e^{-} \rightleftarrows Mn$	−1·18	$Pd^{2+}+2e^{-} \rightleftarrows Pd$	+0·99
		$Au^{3+}+3e^{-} \rightleftarrows Au$	+1·50

starting with negative values and finishing with positive potentials. What is obtained is the so-called **electrochemical series** of the metals. The more negative the potential of a metal, the greater the tendency of the metal to pass into the ionic state and vice versa. A metal with a more negative potential will displace any other metal below it in the series from solutions of its salts. Thus, magnesium, aluminium, zinc, and iron will all displace copper from solutions, lead will displace copper, mercury or silver, copper will displace silver and mercury and so on. Metals with negative standard potentials displace hydrogen and can

* See e.g. W. J. Moore's *Physical Chemistry*. 4th edn., Longman 1966, p. 379 et f.

therefore be dissolved in acids with the evolution of hydrogen; those with positive standard potentials can be dissolved only in oxidizing acids (like HNO_3).

The standard electrode potential is a quantitative expression of the readiness of the element to lose electrons. It is therefore a measure of the strength of a metal as a reducing agent. The more negative the electrode potential of a metal, the more powerful its action as a reductant.

I.41 OXIDATION-REDUCTION POTENTIALS In the previous section metal electrodes were dealt with, and it was shown that the equilibrium between the metal ion and the metal

$$Me^{n+} + ne^- \rightleftarrows Me \qquad \text{(i)}$$

gives rise to a potential difference between the electrode and the solution. We have also seen how the electrode potential, with the aid of a reference electrode, can be measured and expressed on the hydrogen scale.

Equation (i) above represents a redox half-cell with the metal ion as the oxidized form and the metal itself as the reduced form. It is quite logical therefore to conclude that not only those redox systems which involve a solid metal in the half-cell reaction will give rise to an electrode potential but that any oxidation-reduction system can be characterized with a sort of potential. The question, of course, is how such a potential can be measured. When discussing the standard hydrogen electrode we have already seen such a system; the reduced form, hydrogen, was adsorbed on a platinum electrode (with an exceptionally high surface, secured by the electrolytically deposited platinum black). The platinum electrode, which in fact acted as an inert electrode (i.e. a mere sonde), was able to take up or provide electrons released or accepted by the hydrogen gas or hydrogen ions respectively. Experience has shown that an inert electrode, like platinum, gold (or in some cases mercury), is capable of measuring potentials originating from oxidation-reaction equilibria. These are often called **oxidation-reduction potentials**. As it will be clear later, the electrode potentials discussed in Section **I.40** are themselves oxidation-reduction potentials, representing a special, but from the practical point of view very important, class.

If we have a solution in which both iron(II) and iron(III) ions are present, we can construct a half-cell by immersing a platinum foil as an electrode into it, and connecting the platinum electrode to the electrical circuit. The platinum should not be coated with platinum black in this case as we do not want any adsorption to take place on the surface, a so-called **bright platinum electrode** must therefore be used. With a suitable salt bridge we can connect this solution to another half-cell (e.g. a standard hydrogen or calomel electrode) and the e.m.f. of this cell can be measured. The potential of the half-cell corresponds to the half-cell equilibrium:

$$Fe^{3+} + e^- \rightleftarrows Fe^{2+}$$

Although oxidation-reduction potentials can be treated in a general way, (as in most textbooks of physical chemistry and electrochemistry), here we shall first classify redox systems and then deal separately with the potentials within each class. This not only suits the beginner but has been proved most useful for the study of inorganic qualitative and quantitative analysis. The four classes dealt with here are (a) metal electrodes, (b) simple redox systems, (c) combined redox and acid-base systems and (d) gas electrodes.

(a) Metal electrodes The half-cell reaction on which their operation is based, can be written as

$$Me^{n+} + ne^- \rightleftharpoons Me$$

and the electrode potential can be calculated from the expression:

$$E = E^{\ominus} + \frac{RT}{nF} \ln a_{Me^{n+}} \approx E^{\ominus} + \frac{RT}{nF} \ln [Me^{n+}]$$

Standard electrode potentials are collected in Table I.16, and these electrodes were discussed in details in Section **I.40**.

(b) Simple redox systems The half-cell reaction can be symbolized in general terms as

$$Ox + ne^- \rightleftharpoons Red$$

Ox represents the oxidized form and Red the reduced form of the redox couple. Such systems are

$$Fe^{3+} + e^- \rightleftharpoons Fe^{2+}$$
$$I_2 + 2e^- \rightleftharpoons 2I^-$$
$$Sn^{4} + 2e^- \rightleftharpoons Sn^{2+} \qquad \text{etc.}$$

Both the oxidized and reduced forms are in the dissolved phase. The oxidation-reduction potential can be measured with an inert electrode (platinum, gold, or in some cases mercury). The relation between ion concentrations (activities) and the oxidation-reduction potential can be expressed by the Nernst equation as

$$E = E^{\ominus} + \frac{RT}{nF} \ln \frac{a_{Ox}}{a_{Red}} \approx E^{\ominus} + \frac{RT}{nF} \ln \frac{[Ox]}{[Red]}$$

The $E^{\ominus}$ **standard potential** can be measured in a solution containing the oxidized and reduced form in **equal (molar) concentrations (activities)**. Thus, the standard oxidation-reduction potential of the iron(III)–iron(II) system could be measured in a solution containing 0·1M $FeCl_3$ and 0·1M $FeCl_2$ or 0·01M $FeCl_3$ and 0·01M $FeCl_2$ etc. Standard oxidation-reduction potentials of simplex redox systems are shown in Table I.17. The more positive this standard potential, the stronger an oxidant is the oxidized form; the more negative the standard potential, the stronger a reductant is the reduced form. The implications of these statements will be discussed in more detail in Section **I.43**.

(c) Combined redox and acid-base systems The half-cell reaction of such systems (cf. Section **I.36**) can be written as

$$Ox + mH^+ + ne^- \rightleftharpoons Red + \frac{m}{2} H_2O$$

Ox again represents the oxidized and Red the reduced form of the system. Such half-cells are

$$MnO_4^- + 8H^+ + 5e^- \rightleftharpoons Mn^{2+} + 4H_2O$$
$$AsO_4^{3-} + 2H^+ + 2e^- \rightleftharpoons AsO_3^{3-} + H_2O$$
$$Cr_2O_7^{2-} + 14H^+ + 6e^- \rightleftharpoons 2Cr^{3+} + 7H_2O$$
$$BrO_3^- + 6H^+ + 6e^- \rightleftharpoons Br^- + 3H_2O \qquad \text{etc.}$$

Table I.17 Standard oxidation-reduction potentials of simple redox systems at 25°C

Redox system		$E°$ (Volts)
$Au^{3+} + 2e^-$	$\rightleftharpoons Au^+$	+1·29
$Ce^{4+} + e^-$	$\rightleftharpoons Ce^{3+}$	+1·62
$Co^{3+} + e^-$	$\rightleftharpoons Co^{2+}$	+1·84
$Cr^{3+} + e^-$	$\rightleftharpoons Cr^{2+}$	−0·41
$Cu^{2+} + e^-$	$\rightleftharpoons Cu^+$	+0·16
$Fe^{3+} + e^-$	$\rightleftharpoons Fe^{2+}$	+0·76
$[Fe(CN)_6]^{3-} + e^-$	$\rightleftharpoons [Fe(CN)_6]^{4-}$	+0·36
$2Hg^{2+} + 2e^-$	$\rightleftharpoons Hg_2^{2+}$	+0·91
$Mn^{3+} + e^-$	$\rightleftharpoons Mn^{2+}$	+1·51
$MnO_4^- + e^-$	$\rightleftharpoons MnO_4^{2-}$	+0·56
$Pb^{4+} + 2e^-$	$\rightleftharpoons Pb^{2+}$	+1·69
$S_2O_8^{2-} + 2e^-$	$\rightleftharpoons 2SO_4^{2-}$	+2·05
$S_4O_6^{2-} + 2e^-$	$\rightleftharpoons 2S_2O_3^{2-}$	+0·17
$Sn^{4+} + 2e^-$	$\rightleftharpoons Sn^{2+}$	+0·15
$Ti^{4+} + e^-$	$\rightleftharpoons Ti^{3+}$	+0·10
$H_2 + 2e^-$	$\rightleftharpoons 2H^+$	±0·00
$Br_2 + 2e^-$	$\rightleftharpoons 2Br^-$	+1·01
$Cl_2 + 2e^-$	$\rightleftharpoons 2Cl^-$	+1·36
$I_2 + 2e^-$	$\rightleftharpoons 2I^-$	+0·52

The oxidation-reduction potential can again be measured with a platinum or gold (inert) electrode. The correlation between concentrations (activities) and the oxidation-reduction potential can be expressed with the Nerst equation as

$$E = E^{\ominus} + \frac{RT}{nF} \ln \frac{a_{\text{Ox}} a_{\text{H}^+}^m}{a_{\text{Red}}} \approx E^{\ominus} + \frac{RT}{nF} \ln \frac{[\text{Ox}][\text{H}^+]^m}{[\text{Red}]}$$

From this expression it can be seen clearly that the oxidation-reduction potential of such systems depends as well on the hydrogen-ion concentration(pH) of the solution. The $E^{\ominus}$ **standard oxidation-reduction potential** can be measured in a solution containing the oxidized and reduced forms in equal concentrations and 1 mol ℓ^{-1} hydrogen ions (that is, at pH 0). Thus, for example, the standard potential of the permanganate–manganese(II) system could be measured in a solution containing 0·1M $KMnO_4$ and 0·1M $MnSO_4$ (or 0·01M $KMnO_4$ and 0·01M $MnSO_4$ etc.) at pH 0. Standard oxidation-reduction potentials of combined redox and acid-base systems are shown in Table I.18. The dependence of the oxidation-reduction potential on the pH must always be kept in mind when trying to make deductions from the values of this Table. Predictions as to the direction of reactions can again be made on the basis of the values of oxidation-reduction potentials (cf. *Example* 33 in Section **I.42** and also see Section **I.43**).

Inhomogeneous redox systems The forms of Nernst equation quoted in Section **I.41**(a), (b), and (c) are, strictly speaking, valid only for homogeneous redox systems, where there is no change in the number of molecules (or ions) when the substance is reduced or oxidized. For *inhomogeneous* systems, where this is not the case, general equations would be too complex to quote, but the

Table I.18 Standard oxidation-reduction potentials of combined redox and acid-base systems at 25°C

Redox system		$E°$ (Volts)
$AsO_4^{3-} + 2H^+ + 2e^-$	$\rightleftharpoons AsO_3^{3-} + H_2O$	+0·56
$BrO_3^- + 6H^+ + 6e^-$	$\rightleftharpoons Br^- + 3H_2O$	+1·42
$ClO_3^- + 6H^+ + 6e^-$	$\rightleftharpoons Cl^- + 3H_2O$	+1·45
$ClO_4^- + 8H^+ + 8e^-$	$\rightleftharpoons Cl^- + 4H_2O$	+1·34
$Cr_2O_7^{2-} + 14H^+ + 6e^-$	$\rightleftharpoons 2Cr^{3+} + 7H_2O$	+1·36
$H_2O_2 + 2H^+ + 2e^-$	$\rightleftharpoons 2H_2O$	+1·77
$IO_3^- + 6H^+ + 6e^-$	$\rightleftharpoons I^- + 3H_2O$	+1·09
$MnO_4^- + 4H^+ + 3e^-$	$\rightleftharpoons MnO_2(s) + 2H_2O$	+1·70
$MnO_4^- + 8H^+ + 5e^-$	$\rightleftharpoons Mn^{2+} + 4H_2O$	+1·52
$SO_4^{2-} + 2H^+ + 2e^-$	$\rightleftharpoons SO_3^{2-} + H_2O$	+0·20
$SeO_4^{2-} + 2H^+ + 2e^-$	$\rightleftharpoons SeO_3^{2-} + H_2O$	+1·15
$O_2 + 4H^+ + 4e^-$	$\rightleftharpoons 2H_2O$	+1·23

potential can be easily expressed if the half-cell reaction is known. The rule is that in the argument of the logarithm in the Nernst equation the concentration must be taken at powers corresponding to their stoichiometrical numbers. Such inhomogeneous systems are for example

$$I_2 + 2e^- \rightleftharpoons 2I^-$$

$$2Hg^{2+} + 2e^- \rightleftharpoons Hg_2^{2+}$$

$$Cr_2O_7^{2-} + 14H^+ + 6e^- \rightleftharpoons 2Cr^{3+} + 7H_2O$$

The oxidation-reduction potentials of these systems can be expressed as

$$E = E^{\ominus} - \frac{RT}{2F} \ln \frac{[I^-]^2}{[I_2]}$$

$$E = E^{\ominus} - \frac{RT}{2F} \ln \frac{[Hg_2^{2+}]}{[Hg^{2+}]^2}$$

and

$$E = E^{\ominus} - \frac{RT}{6F} \ln \frac{[Cr^{3+}]^2}{[Cr_2O_7^{2-}][H^+]^{14}}$$

respectively. Standard potentials, $E^{\ominus}$, for these systems can be taken from Tables I.17 and I.18.

(d) Gas electrodes Gaseous substances, when donating or accepting electrons, may act as electrodes. As the gases themselves do not conduct electricity, a suitable inert electrode (e.g. platinum or graphite) must be used as a link to the electrical circuit. The solid inert electrode may act as the catalyst of the ionization process, as we have already seen with the hydrogen electrode.

The potential of gas electrodes can easily be calculated with a suitable form of the Nernst equation. This will be elucidated in connection with three electrodes of practical importance.

(i) **The hydrogen electrode** operates on the basis of the half-cell reaction:

$$2H^+ + 2e^- \rightleftharpoons H_2$$

This can be regarded as a simple redox system; its potential can be expressed as

$$E = E^{\ominus\prime} - \frac{RT}{2F}\ln\frac{a_{H_2}}{a^2_{H^+}} \approx E^{\ominus\prime} - \frac{RT}{2F}\ln\frac{[H_2]}{[H^+]^2}$$

In this expression the activity (or concentration) of hydrogen gas cannot easily be measured. It can however be replaced by its pressure, as, for a given temperature, the activity (or concentration) is proportional to the pressure.

$$a_{H_2} \approx [H_2] = k\,p_{H_2}$$

The proportionality factor, k, being constant, the expression

$$E^{\ominus\prime} - \frac{RT}{2F}\ln k = E^{\ominus}$$

can be used to define the standard potential, $E^{\ominus}$. Thus, the potential of the hydrogen electrode can be expressed as

$$E = E^{\ominus} - \frac{RT}{2F}\ln\frac{p_{H_2}}{a^2_{H^+}} \approx E^{\ominus} - \frac{RT}{2F}\ln\frac{p_{H_2}}{[H^+]^2}$$

Thus the potential of the hydrogen electrode depends on the hydrogen-ion concentration in the solution and on the pressure of hydrogen gas above the solution. The standard potential, $E^{\ominus}$, can be measured in a system where the activity of hydrogen ions is unity in the solution and the pressure of hydrogen gas over the solution is 1 atm. We have seen that the standard potential, $E^{\ominus}$, of such an electrode is by definition 0.

(ii) The operation of the **oxygen electrode** is based on the half-cell reaction

$$O_2 + 4H^+ + 4e^- \rightleftarrows 2H_2O$$

Applying the Nernst equation for such a process, which is a combined redox and acid-base reaction, we can write

$$E = E^{\ominus\prime} - \frac{RT}{4F}\ln\frac{[H_2O]^2}{[O_2][H^+]^4} = E^{\ominus} + \frac{RT}{4F}\ln\left(p_{O_2}[H^+]^4\right)$$

The final expression is arrived at in a similar way to that for the hydrogen electrode. The concentration of water is regarded as constant in dilute aqueous solutions. The value of $E^{\ominus}$, measurable at pH = 0 with p_{O_2} = 1 atm (where p_{O_2} is pressure of oxygen over the solution) is +1·23 V. It is useful and easy to memorize the standard potentials of the hydrogen electrode ($E^{\ominus}$= 0 V) and of the oxygen electrode ($E^{\ominus}$ = 1·23 V, one, two, three). The importance of these quantities will become apparent later (cf. Section **I.43**).

(iii) The **chlorine electrode** operates on the basis of the half-cell reaction

$$Cl_2 + 2e^- \rightleftarrows 2Cl^-$$

Its electrode potential can be expressed as

$$E = E^{\ominus\prime} - \frac{RT}{2F}\ln\frac{[Cl^-]}{[Cl_2]} = E^{\ominus} - \frac{RT}{2F}\ln\frac{[Cl^-]^2}{p_{Cl_2}}$$

The value of the $E^{\ominus}$ standard potential is +1·36 V.

The overall conclusions we can draw from the examples mentioned are that the potential of gas electrodes depends on the concentration of the particular ion in the solution which is involved in the half-cell reaction and on the pressure of the gas in the electrode system, which, in other words, is the pressure of gas measurable above the solution.

I.42 CALCULATIONS BASED ON THE NEAREST EQUATION Values of standard potentials are normally available for room temperatures (25°C), like those shown in Tables I.16 to I.18. Calculations based on the Nernst equations are therefore usually restricted to this temperature.* The prelogarithmic factor of the Nernst equation can thus be simplified, and 10-based logarithms can be introduced. With $R = 8{\cdot}3143\ \mathrm{J\,K^{-1}\,mol^{-1}}$, $T = 298{\cdot}15\ \mathrm{K}$, $F = 9{\cdot}6487 \times 10^4\mathrm{C\,mol^{-1}}$, and $\ln 10 = 2{\cdot}0326$ the constant $\frac{RT}{F} \ln 10 = 0{\cdot}0592$ can be obtained. Thus the potential of a metal electrode becomes

$$E = E^{\circ} + \frac{0{\cdot}0592}{n} \log [\mathrm{Me}^{n+}]$$

and the same prelogarithmic factor can be used in all other forms of the Nernst equation.

The following examples may illustrate the ways in which oxidation-reduction potentials can be calculated.

Example 31 Calculate the e.m.f. of a Daniell cell, made up by immersing a copper foil into a 0·15M $CuSO_4$ solution and a zinc rod into 0·25M $ZnSO_4$ solution and linking the two half-cells together.

According to Table 16 the values of the standard potentials are as follows:

$$\text{for} \quad Cu^{2+} + 2e^- \rightleftarrows Cu \qquad E^{\circ}_{Cu} = 0{\cdot}34\ \mathrm{V}$$

and

$$\text{for} \quad Zn^{2+} + 2e^- \rightleftarrows Zn \qquad E^{\circ}_{Zn} = -0{\cdot}76\ \mathrm{V}$$

The e.m.f. of the cell can be expressed as

$$\text{e.m.f.} = |E_{Cu} - E_{Zn}|$$

Thus first the two electrode potentials have to be calculated. These are

$$E_{Cu} = E^{\circ}_{Cu} + \frac{0{\cdot}0592}{2} \log[Cu^{2+}]$$

$$= 0{\cdot}34 + \frac{0{\cdot}0592}{2} \log 0{\cdot}15 = 0{\cdot}316\ \mathrm{V}$$

and

$$E_{Zn} = E^{\circ}_{Zn} + \frac{0{\cdot}0592}{2} \log[Zn^{2+}]$$

$$= -0{\cdot}76 + \frac{0{\cdot}0592}{2} \log 0{\cdot}25 = -0{\cdot}778\ \mathrm{V}$$

---

* When calculating potentials for other than room temperature, it is not enough simply to insert the appropriate temperature into the prelogarithmic factor of the Nernst equation, because the standard potential, E°, varies with temperature also. The correlation between electrode potential and temperature is given by the appropriate form of the *Gibbs–Helmholtz equation* (cf. S. Glasstone's *Textbook of Physical Chemistry*. 2nd edn., MacMillan, London, 1964, p. 924).

and the e.m.f.:

$$\text{e.m.f.} = |E_{Cu} - E_{Zn}| = |0{\cdot}316 - (-0{\cdot}778)| = 1{\cdot}094 \text{ V}$$

Example 32 A solution contains 0·05M $FeCl_2$ and 0·15M $FeCl_3$. What is the oxidation-reduction potential measurable in the solution?

From Table I.17 the standard potential for the half-cell

$$Fe^{3+} + e^- \rightleftarrows Fe^{2+}$$

is taken; $E^\ominus = 0{\cdot}76$ V.

The appropriate form of the Nernst equation ($n = 1$):

$$E = E^\ominus - 0{\cdot}0592 \log \frac{[Fe^{2+}]}{[Fe^{3+}]} = 0{\cdot}76 - 0{\cdot}0592 \log \frac{0{\cdot}05}{0{\cdot}15} = 0{\cdot}788 \text{ V}$$

Example 33 20 ml 0·02M $KMnO_4$, 10 ml 0·5M H_2SO_4 and 5 ml 0·1M $FeSO_4$ solutions are mixed and diluted with water to 100 ml. Calculate the oxidation-reduction potential of the solution.

From the reaction equation

$$MnO_4^- + 5Fe^{2+} + 8H^+ \rightarrow 5Fe^{3+} + Mn^{2+} + 4H_2O$$

it can be seen that only a portion of the $KMnO_4$ will be reduced; the solution therefore will contain both MnO_4^- and Mn^{2+} ions (while practically all the Fe^{2+} disappears and equivalent amounts of Fe^{3+} are formed). Thus, the oxidation-reduction potential will be that of the permanganate–manganese(II) system:

$$MnO_4^- + 8H^+ + 5e^- \rightleftarrows Mn^{2+} + 4H_2O$$

for this the standard potential is (cf. Table I.18) $E^\ominus = 1.52$.

To calculate the potential the Nernst equation can be written as

$$E = E^\ominus - \frac{0{\cdot}0592}{5} \log \frac{[Mn^{2+}]}{[MnO_4^-][H^+]^8}$$

Now we have to calculate the concentrations in the argument of the logarithm. Disregarding the slight decrease of the hydrogen-ion concentration* due to the reaction, we can calculate $[H^+]$ as

$$[H^+] = \frac{2 \times 10 \times 0{\cdot}5}{100} = 0{\cdot}1 \text{ mol } \ell^{-1}$$

To calculate $[MnO_4^-]$ and $[Mn^{2+}]$ is more complicated but in fact we do not need their actual values; it is sufficient to calculate the ratio $[Mn^{2+}]/[MnO_4^-]$. This can be done easily. From the stoichiometry of the reaction it follows that, of the original 20 ml 0·02M $KMnO_4$, 15 ml is left unreacted while the equivalent of 5 ml is reduced to Mn^{2+}. Thus the ratio can be expressed as

$$\frac{[Mn^{2+}]}{[MnO_4^-]} = \frac{5}{15} = \frac{1}{3}$$

Now the oxidation-reduction potential can be calculated:

$$E = 1{\cdot}52 - \frac{0{\cdot}0592}{5} \log \frac{1}{3(0{\cdot}1)^8} = 1{\cdot}431 \text{ V}$$

---

* The amount of hydrogen ions, used up in this reaction, is, according to the stoichiometry of the reaction, 8×10^{-4} mol.

Example 34 20 ml 0·1M NaCl and 2 ml 0·1M $AgNO_3$ are mixed and diluted to 100 ml. Calculate the potential of a silver electrode dipped into this solution.

The potential of a silver electrode can be expressed by the Nernst equation as

$$E = E^{\ominus} + 0{\cdot}059 \log [Ag^+]$$

$E^{\ominus}$ for the half-cell $Ag^+ + e^- \rightleftarrows Ag$ is 0·80 V (cf. Table I.16). To calculate the electrode potential we have to calculate $[Ag^+]$. This concentration will be extremely low, because practically all the silver is precipitated by the sodium chloride:

$$Ag^+ + Cl^- \rightleftarrows AgCl\downarrow$$

For this precipitate the solubility product can be taken from Table I.12:

$$K_s = [Ag^+][Cl^-] = 1{\cdot}5 \times 10^{-10}$$

This expression can be combined with the Nernst equation to yield

$$E = E^{\ominus} + 0{\cdot}059 \log \frac{K_s}{[Cl^-]}$$

The concentration of chloride ions can be calculated quite easily. Of the 20 ml 0·1M NaCl the equivalent of 2 ml has been removed by precipitation; thus

$$[Cl^-] = \frac{18}{100} 0{\cdot}1 = 0{\cdot}018 \text{ mol } \ell^{-1}$$

The potential can now be calculated:

$$E = 0{\cdot}80 + 0{\cdot}0592 \log \left(\frac{10^{-10}}{0{\cdot}018} \right) = 0{\cdot}313 \text{ V}$$

Example 35 What is the potential of the oxygen electrode in a solution having a *p*H of 8?

The potential of this electrode (cf. Section **I.41d**) can be expressed as

$$E = E^{\ominus} + \frac{0{\cdot}0592}{4} \log (p_{O_2}[H^+]^4)$$

Here $E^{\ominus} = 1{\cdot}23$ V. For pH $= 8$ we have $[H^+] = 10^{-8}$. The pressure of oxygen over the solution is

$$p_{O_2} = 0{\cdot}21 \text{ atm}$$

under atmospheric conditions. Thus, the potential is,

$$E = 1{\cdot}23 + \frac{0{\cdot}0592}{4} \log \{0{\cdot}21(10^{-8})^4\} = 0{\cdot}748 \text{ V}$$

I.43 CONCLUSIONS DRAWN FROM THE TABLES OF OXIDATION-REDUCTION POTENTIALS From the values of oxidation-reduction potentials we can easily find out whether a particular oxidation-reduction reaction is feasible or not. We have already seen the rules that govern the displacement of metals by one another, and the feasibility of dissolving metals in acid with the liberation of hydrogen. Those conclusions can now be extended and generalized. It can be said that *the more positive the oxidation-reduction* *potential of a redox system, the stronger an oxidant is its oxidized form and vice versa*: the more negative its oxidation-reduction potential, the stronger a reducing agent is its reduced form. The direction of an oxidation-reduction reaction can also easily be predicted; the oxidized form of the system with a more positive potential will oxidize the reduced form of a system with the more negative potential, and never the other way round. From Table I.17 we can, for example, quote the following two systems:

$$Fe^{3+} + e^- \rightleftarrows Fe^{2+} \qquad E^\ominus = +0{\cdot}76 \text{ V}$$

$$Sn^{4+} + 2e^- \rightleftarrows Sn^{2+} \qquad E^\ominus = +0{\cdot}15 \text{ V}$$

Because of the relative values of the oxidation-reduction potentials, iron(III) ions will oxidize tin(II) ions, the reaction

$$2Fe^{3+} + Sn^{2+} \rightarrow 2Fe^{2+} + Sn^{4+}$$

can proceed *only* in this direction, and never in the opposite direction; that is, iron(II) ions can never reduce tin(IV).

Similarly, from the Tables I.17 and I.18 we can see, for example, that permanganate ions (in acid medium) can oxidize chloride, bromide, iodide, iron(II), and hexacyanoferrate(II) ions, also that iron(III) ions may oxidize arsenite or iodide ions but never chromium(III) or chloride ions etc. It must be emphasized that the standard potentials are to be used only as a rough guide; the direction of a reaction will depend on the actual values of oxidation-reduction potentials. These, if the concentrations of the species are known, can be calculated easily by means of the Nernst equation.

The concentrations of the various species must be taken into consideration especially if combined redox and acid-base systems are involved. From the data, taken from Table I.17 and I.18 for example:

$$AsO_4^{3-} + 2H^+ + 2e^- \rightleftarrows AsO_3^{3-} + H_2O \qquad E^\ominus = 0{\cdot}56 \text{ V}$$

$$I_2 + 2e^- \rightleftarrows 2I^- \qquad E^\ominus = 0{\cdot}52 \text{ V}$$

one could draw the conclusion that arsenate ions will oxidize iodide:

$$AsO_4^{3-} + 2H^+ + 2I^- \rightarrow I_2 + AsO_3^{3-} + H_2O$$

but the reaction cannot go in the opposite direction. This in fact is true only if the solution is strongly acid ($p\text{H} \leqq 0$). The oxidation-reduction potential of the arsenate–arsenite system depends on the pH:

$$E = E^\ominus - \frac{0{\cdot}0592}{2} \log \frac{[AsO_3^{3-}]}{[AsO_4^{3-}][H^+]^2}$$

$$= E^\ominus - \frac{0{\cdot}0592}{2} \log \frac{[AsO_3^{3-}]}{[AsO_4^{3-}]} - 0{\cdot}0592\, pH$$

At $p\text{H} = 6$ the potential of a solution containing arsenate and arsenite ions at equal concentrations decreases to $+0{\cdot}20$ V. Under such circumstances therefore the opposite reaction will occur:

$$I_2 + AsO_3^{3-} + H_2O \rightarrow AsO_4^{3-} + 2H^+ + 2I^-$$

Both reactions are indeed used in qualitative inorganic analysis (cf. Section **III.12**, reaction 5 and Section **III.13**, reaction 5).

I.44 EQUILIBRIUM CONSTANT OF OXIDATION-REDUCTION REACTIONS The law of mass action being valid for oxidation-reduction reactions, the feasibility of such reactions can most properly be decided on the basis of their equilibrium constant. The equilibrium constant of oxidation-reduction reactions is, in turn, strongly related to the differences in the standard oxidation-reduction potentials of the systems involved. The proper explanation of this correlation needs a good working knowledge of chemical thermodynamics. In the following factual treatment this is assumed; for a more detailed understanding the reader should consult textbooks of physical chemistry.*

Let us consider the general form of a redox reaction

$$a\mathrm{Ox}_1 + b\mathrm{Red}_2 + \cdots \overset{(ne^-)}{\rightleftarrows} c\mathrm{Ox}_2 + d\mathrm{Red}_1 + \cdots \qquad \text{(i)}$$

Here the subscripts 1 and 2 refer to the individual redox systems. The equilibrium constant of such a reaction (Section **I.13**), expressed with concentrations is

$$K = \frac{[\mathrm{Ox}_2]^c[\mathrm{Red}_1]^d \ldots}{[\mathrm{Ox}_1]^a[\mathrm{Red}_2]^b \ldots} \qquad \text{(ii)}$$

This equilibrium constant is related to the standard free energy change, ΔG°, of this reaction:

$$-\Delta G^\circ = RT \ln K \qquad \text{(iii)}$$

The negative sign originates from the sign conventions usually adopted in chemical thermodynamics. This standard free energy change is equal to the electrical work done by a galvanic cell built from the two half-cells which are involved in reaction (i). If a moles of Ox_1 and b moles of Red_2 are used up in this reaction, the electrical work done is

$$nF(E_1^\circ - E_2^\circ) = -\Delta G^\circ \qquad \text{(iv)}$$

where E_1° and E_2° are the standard oxidation-reduction potentials of systems 1 and 2 respectively, and n is the number of electrons exchanged during the reaction (more precisely, the number of Faradays passing through the electrical circuit if a moles of Ox_1 and b moles of Red_2 are used up). Equations (iii) and (iv) can be combined to

$$\log K = \frac{F \log_{10} \mathrm{e}}{RT} n(E_1^\circ - E_2^\circ)$$

For room temperature ($T = 298{\cdot}15$ K), with $\log \mathrm{e} = 0{\cdot}4343$, $F = 9{\cdot}6487 . 10^4$ C mol^{-1} and R = 8·314 J K^{-4} mol^{-1}, we can obtain the expression

$$\log K = 16{\cdot}905n(E_1^\circ - E_2^\circ) \qquad \text{(v)}$$

which can be used for practical calculations. A few examples may illustrate how such calculations should be carried out.

Example 36 Calculate the equilibrium constant of the reaction

$$2\mathrm{Fe}^{3+} + \mathrm{Sn}^{2+} \overset{(2e^-)}{\rightleftarrows} 2\mathrm{Fe}^{2+} + \mathrm{Sn}^{4+}$$

---

* See e.g. W. J. Moore's *Physical Chemistry*. 4th edn., Longman 1966, pp. 174 and 385.

The equilibrium constant can be written as

$$K = \frac{[Fe^{2+}]^2[Sn^{4+}]}{[Fe^{3+}]^2[Sn^{2+}]}$$

Here $n = 2$, $E_1^\ominus = +0{\cdot}76$ V and $E_2^\ominus = +0{\cdot}15$ V (cf. Table I.17). From expression (v) we obtain

$$\log K = 16{\cdot}905 \times 2 \times (0{\cdot}76 - 0{\cdot}15) = 20{\cdot}62$$

or

$$K = 10^{20{\cdot}62} = 4{\cdot}21 \times 10^{20}$$

Example 37 Calculate the equilibrium constant of the reaction

$$MnO_4^- + 8H^+ + 5Fe^{2+} \overset{(5e^-)}{\rightleftarrows} Mn^{2+} + 5Fe^{3+} + 4H_2O$$

For this reaction

$$K = \frac{[Mn^{2+}][Fe^{3+}]^5}{[MnO_4^-][H^+]^8[Fe^{2+}]^5} \qquad \text{(vi)}$$

Here $n = 5$, $E_1^\ominus = 1{\cdot}52$ V, $E_2^\ominus = 0{\cdot}76$ V (cf. Tables I.17 and I.18). Using equation (v) we have

$$\log K = 16{\cdot}905 \times 5 \times (1{\cdot}52 - 0{\cdot}76) = 64{\cdot}24$$

or

$$K = 10^{64{\cdot}24} = 1{\cdot}73 \times 10^{64}$$

Using this example we can show the correlation between standard oxidation-reduction potentials *without* applying the thermodynamical concepts mentioned above. The oxidation-reduction potential of the system

$$MnO_4^- + 8H^+ + 5e^- \rightleftarrows Mn^{2+} + 4H_2O$$

can be expressed as

$$E_1 = E_1^\ominus - \frac{0{\cdot}0592}{5} \log \frac{[Mn^{2+}]}{[MnO_4^-][H^+]^8} \qquad \text{(vii)}$$

and that of the system

$$Fe^{3+} + e^- \rightleftarrows Fe^{2+}$$

by

$$E_2 = E_2^\ominus - 0{\cdot}0592 \log \frac{[Fe^{2+}]}{[Fe^{3+}]} \qquad \text{(viii)}$$

Equation (viii) can be transformed easily to

$$E_2 = E_2^\ominus - \frac{0{\cdot}0592}{5} \log \frac{[Fe^{2+}]^5}{[Fe^{3+}]^5} \qquad \text{(ix)}$$

If the reaction reaches equilibrium, the oxidation-reduction potentials of both systems are equal:

$$E_1 = E_2 \qquad \text{(x)}$$

Equations (vii), (ix), and (x) can then be combined to give

$$E_1^\ominus - E_2^\ominus = \frac{0{\cdot}592}{5} \log \frac{[Mn^{2+}][Fe^{3+}]^5}{[MnO_4^-][H^+]^8[Fe^{2+}]^5} \qquad \text{(xi)}$$

As the concentrations in the argument of the logarithm are equilibrium concentrations, equations (vi) and (xi) can be combined, leading to

$$\log K = 16{\cdot}905 \times 5 \times (E_1^{\ominus} - E_2^{\ominus})$$

which is identical with expression (v) deduced earlier ($n = 5$ for this particular reaction).

From such examples it becomes apparent that the greater the difference between the standard oxidation-reduction potentials, the higher the value of the equilibrium constant, that is the reactions become the more complete. In practice, a difference of 0·3 V for $n = 1$ secures a value for K greater than 10^5, which means that in practical terms the reaction will take place quantitatively. If, on the other hand, the difference of standard potentials, as defined by equations (i) and (v) is negative, the reaction is not feasible; in fact it will proceed in the opposite direction.

H. SOLVENT EXTRACTION

I.45 THE DISTRIBUTION OR PARTITION LAW It is a well-known fact that certain substances are more soluble in some solvents than in others. Thus iodine is very much more soluble in carbon disulphide, chloroform, or carbon tetrachloride than it is in water. Furthermore, when certain liquids such as carbon disulphide and water, and also ether and water, are shaken together in a vessel and the mixture allowed to stand, the two liquids separate out into two layers. Such liquids are said to be immiscible (carbon disulphide and water) or partially miscible (ether and water), according as to whether they are almost insoluble or partially soluble in one another. If iodine is shaken with a mixture of carbon disulphide and water and then allowed to settle, the iodine will be found to be distributed between the two solvents. A state of equilibrium exists between the solution of iodine in carbon disulphide and the solution of iodine in water. It has been found that when the amount of iodine is varied, the ratio of the concentrations is constant at any given temperature. That is:

$$\frac{\text{Concentration of iodine in carbon disulphide}}{\text{Concentration of iodine in water}} = \frac{c_2}{c_1} = K_d$$

The constant K_d is known as the **partition**, or **distribution, coefficient**. Some experimental results are collected in Table I.19. It is important to note that the ratio c_2/c_1 is constant only when the dissolved substance has the same relative molecular mass in both solvents. The **distribution** or **partition law** may be formulated thus: when a solute distributes itself between two immiscible solvents there exists for each molecular species, at a given temperature, a constant ratio of distribution between the two solvents, and this distribution ratio is independent of any other molecular species which may be present. The value of the ratio varies with the nature of the two solvents, the nature of the solute, and the temperature.

The removal of a solute from an aqueous solution by a water-immiscible solvent is called **solvent extraction**. This technique is often applied for separations.

Table I.19 The distribution of iodine between water and organic solvents at 25°C

Concentration of iodine		K_d
in the organic phase mol ℓ^{-1}	in the aqueous phase mol ℓ^{-1}	
Org. phase: carbon disulphide		
1·387	$3{\cdot}231 \times 10^{-3}$	429
1·017	$2{\cdot}522 \times 10^{-3}$	403
0·520	$1{\cdot}261 \times 10^{-3}$	412
0·323	$0{\cdot}788 \times 10^{-3}$	410
	Average:	413·5
Org. phase: chloroform		
$0{\cdot}338 \times 10^{-1}$	$0{\cdot}25 \times 10^{-3}$	135·2
$1{\cdot}54 \times 10^{-1}$	$1{\cdot}20 \times 10^{-3}$	128·3
$2{\cdot}32 \times 10^{-1}$	$1{\cdot}84 \times 10^{-3}$	126·1
$3{\cdot}21 \times 10^{-1}$	$2{\cdot}42 \times 10^{-3}$	132·6
	Average:	130·6
Org. phase: carbon tetrachloride		
$1{\cdot}01 \times 10^{-2}$	$1{\cdot}332 \times 10^{-4}$	75·8
$1{\cdot}64 \times 10^{-2}$	$2{\cdot}151 \times 10^{-4}$	76·3
$4{\cdot}10 \times 10^{-2}$	$5{\cdot}130 \times 10^{-4}$	80·0
$6{\cdot}01 \times 10^{-2}$	$7{\cdot}391 \times 10^{-4}$	81·3
$7{\cdot}84 \times 10^{-2}$	$9{\cdot}448 \times 10^{-4}$	82·6
$12{\cdot}17 \times 10^{-2}$	$14{\cdot}38 \times 10^{-4}$	84·1
	Average:	80·1

I.46 THE APPLICATION OF SOLVENT EXTRACTION IN QUALITATIVE ANALYSIS A few examples of the application of solvent extraction in qualitative analysis are as follows:

(a) Removal of bromine and of iodine from aqueous solution When an aqueous solution of iodine is shaken with carbon disulphide, the concentration of iodine in the resulting carbon disulphide layer is about 400 times that in water. The carbon disulphide layer may be removed with the aid of a separatory funnel and the process repeated. In this way the concentration of iodine in the aqueous solution may be reduced to a very small value, although theoretically it cannot be reduced completely to zero. The following calculation will illustrate the point.

Example 38 Ten milligrams iodine are suspended in 12 ml water, and shaken with 2 ml CCl_4 until equilibrium is reached. Calculate the weight of iodine remaining in the aqueous layer.

Let x be the weight (in milligrams) of iodine which remains in the aqueous phase. Its concentration will become

$$[I_2]_{aq} = \frac{x}{253{\cdot}8 \times 12} \text{ in mol } \ell^{-1} \text{ (or mmol ml}^{-1}\text{)}$$

units. (253·8 = 2 × 126·9 is the relative molecular mass of iodine).

In the CCl_4 10-x mg iodine will be found, its concentration being

$$[I_2]_{CCl_4} = \frac{(10-x)}{253{\cdot}8 \times 2} \text{ mol } \ell^{-1}$$

From Table I.19 we have $K_d = 80{\cdot}1$ for the distribution coefficient:

$$K_d = \frac{[I_2]_{CCl_4}}{[I_2]_{aq}} = \frac{\dfrac{(10-x)}{253{\cdot}8 \times 2}}{\dfrac{x}{253{\cdot}8 \times 12}} = \frac{\dfrac{10-x}{2}}{\dfrac{x}{12}} = 80{\cdot}1$$

from this $x = 0{\cdot}70$ mg.

If the CCl_4 layer is withdrawn with the aid of a separatory funnel and the residual aqueous layer is shaken with a second 2 ml batch fresh CCl_4, the quantity of iodine still remaining in the aqueous layer, y, can be computed from the equation

$$K_d = 80{\cdot}1 = \frac{\dfrac{0{\cdot}70-y}{2}}{\dfrac{y}{12}}$$

from which $y = 0{\cdot}052$ mg.

It can be shown that after a third extraction 3.62×10^{-3} mg and after a fourth one only $2{\cdot}1 \times 10^{-5}$ mg iodine is left behind.

If instead of three successive extractions with 2 ml portions of CCl_4, the original 10 ml of the aqueous suspension were treated with a single 6 ml extraction CCl_4, the weight of iodine remaining in the aqueous layer would be reduced only to 0·24 mg, as can be shown by a calculation similar to the above. This is a simple illustration of the fact that, in performing extractions, it is more efficient and also more economical to carry out a number of successive extractions with small portions of the solvent rather than a single extraction with a large quantity.

Use is made of the partition principle in the detection of bromides, of iodides, and in the detection of bromides and iodides in the presence of each other.

(b) Various tests in qualitative analysis (i) Chromium pentoxide is more soluble in amyl alcohol (or in ether) than in water; by shaking the dilute aqueous solution with amyl alcohol (or with ether), a concentrated solution in the latter solvent is obtained, and the presence of chromate or of hydrogen peroxide is indicated by the blue colour.

(ii) The compound ammonium tetrathiocyanatocobaltate, with the formula $(NH_4)_2[Co(CNS)_4]$, produced by the action of a concentrated solution of ammonium thiocyanate upon a cobalt(II) ion is more soluble in amyl alcohol than in water; the blue colouration of the amyl alcohol layer, due to the formation of a concentrated solution of this compound, is a sensitive and characteristic test for cobalt.

(c) Study of hydrolysis In the hydrolysis of a salt of a weak base and a strong acid or of a weak acid and a strong base, there is an equilibrium between the salt, the free acid, and the free base. The hydrolysis, for our present purpose, may be written:

$$\text{Salt} + \text{Water} \rightleftharpoons \text{Acid} + \text{Base}$$

The concentration of the weak acid or of the weak base can be determined by distribution between water and another solvent, such as benzene or chloroform; the partition coefficient of the acid or base between the water and the other solvent must, of course, be known. The degree of hydrolysis may then be calculated from the concentration of the salt and the determined concentration of the weak acid or base. An example of such a salt is aniline hydrochloride. This is partially hydrolysed into aniline and hydrogen chloride. On shaking the aqueous solution with benzene the aniline will distribute itself between the water and benzene in the ratio of the distribution coefficient. The initial concentration of aniline hydrochloride is known, the concentration of the free aniline in the aqueous solution can be computed from that found in the benzene solution, and from this the total concentration of aniline, produced by hydrolysis, is deduced. Sufficient data are then available for the calculation of the degree of hydrolysis.

(d) The determination of the constitution of complex halide ions Iodine is much more soluble in an aqueous solution of potassium iodide than it is in water; this is due to the formation of tri-iodide ions, I_3^-. The following equilibrium exists in such a solution:

$$I_2 + I^- \rightleftharpoons I_3^-$$

If the solution is titrated with standard sodium thiosulphate solution, the total concentration of the iodine, both as free I_2 and combined as I_3^-, is obtained, since, as soon as some iodine is removed by interaction with the thiosulphate, a fresh amount of iodine is liberated from the tri-iodide in order to maintain the equilibrium. If, however, the solution is shaken with carbon tetrachloride, in which iodine alone is appreciably soluble, then the iodine in the organic layer is in equilibrium with the free iodine in the aqueous solution. By determining the concentration of the iodine in the carbon tetrachloride solution, the concentration of the free iodine in the aqueous solution can be calculated from the known distribution coefficient, and therefrom the total concentration of the free iodine present at equilibrium. Subtracting this from the total iodine, the concentration of the combined iodine (as I_3^-) is obtained; by subtracting the latter value from the initial concentration of potassium iodine the concentration of the free KI is deduced. The equilibrium constant:

$$K = \frac{[I^-] \times [I_2]}{[I_3^-]}$$

is then computed.

A similar method has been used for the investigation of the equilibrium between bromine and bromides:

$$Br_2 + Br^- \rightleftharpoons Br_3^-$$

Distribution measurements have also been used to prove the existence of the tetramminecuprate(II) ion, $[Cu(NH_3)_4]^{2+}$, in an aqueous ammoniacal solution of copper sulphate, the partition of the free ammonia being studied between chloroform and water:

$$[Cu(NH_3)_4]^{2+} \rightleftarrows Cu^{2+} + 4NH_3$$

CHAPTER II EXPERIMENTAL TECHNIQUES OF QUALITATIVE INORGANIC ANALYSIS

II.1 INTRODUCTION Before the student attempts to carry out the analytical reactions of the various cations and anions detailed in Chapters III and IV, he should be familiar with the operations commonly employed in qualitative analysis, that is with the laboratory technique involved. It is assumed that the student has had some training in elementary practical chemistry; he should be familiar with such operations as solution, evaporation, crystallization, distillation, precipitation, filtration, decantation, bending of glass tubes, preparation of ignition tubes, boring of corks, and construction of a wash bottle. These will therefore be either very briefly discussed or not described at all in the following pages.

Qualitative analysis may be carried out on various scales. In **macro analysis** the quantity of the substance employed is 0·5–1 gram and the volume of solution taken for the analysis is about 20 ml. In what is usually termed **semimicro analysis**, the quantity used for analysis is reduced by a factor of 0·1–0·05, i.e. to about 0·05 gram and the volume of solution to about 1 ml. For **micro analysis** the factor is of the order of 0·01 or less. There is no sharp line of demarcation between semimicro and micro analysis: the former has been called centigram analysis and the latter milligram analysis, but these terms indicate only very approximately the amounts used in the analysis. It will be noted that only the scale of the operations has been reduced; the concentrations of the ions remain unchanged. Special experimental techniques have been developed for handling the smaller volumes and amounts of precipitate, and these will be described in some detail. For routine analysis by students, the choice lies between macro and semimicro analysis. There are many advantages in adopting the semimicro technique; these include:

(i) Reduced consumption of chemicals with a considerable saving in the laboratory budget.

(ii) The greater speed of the analysis, due to working with smaller quantities of materials and the saving of time in carrying out the various standard operations of filtration, washing, evaporation, saturation, with hydrogen sulphide, etc.

(iii) Increased sharpness of separation, e.g. washing of precipitates can be carried out rapidly and efficiently when a centrifuge replaces a filter.

(iv) The amount of hydrogen sulphide used is considerably reduced.

(v) Much space is saved both on the reagent shelves and more especially in the lockers provided immediately below the bench for the housing of the individual student's apparatus; this latter merit may be turned to good use by reducing the size of the bench lockers considerably and thus effectively increasing the accommodation of the laboratory.

(vi) The desirability of securing a training in the manipulation of small amounts of material.

For these, and also other, reasons many laboratories now employ semimicro analysis, particularly for the elementary courses. Both macro and semimicro procedures will be given separately in this book in order that the requirements of all types of students may be met. Nevertheless, when the semimicro technique is adopted, students are recommended to read the sections dealing with macro technique. It may be said that when the general technique of semimicro analysis has been mastered and appreciated, no serious difficulty should be encountered in adapting a macro procedure to the semimicro scale. Apart from drop reactions, few applications of the micro technique will be described in the text.

Qualitative analysis utilizes two kinds of tests, dry reactions and wet reactions. The former are applicable to solid substances and the latter to substances in solution. Most of the dry reactions to be described can be used with only minor modifications for semimicro analysis. Dry tests appear to have lost their popularity in certain quarters; they do, however, often provide useful information in a comparatively short time and a knowledge as to how they are carried out is desirable for all students of qualitative analysis. Different techniques are employed for wet reactions in macro, semimicro, and micro analysis.

II.2 DRY REACTIONS A number of useful tests can be carried out in the dry, that is without dissolving the sample. Instructions for such operations are given below.

1. ***Heating*** The substance is placed in a small ignition tube (bulb tube), prepared from soft glass tubing, and heated in a Bunsen flame, gently at first and then more strongly. Small test-tubes, 60–70 mm × 7–8 mm, which are readily obtainable and are cheap, may also be employed. Sublimation may take place, or the material may melt or may decompose with an attendant change in colour, or a gas may be evolved which can be recognized by certain characteristic properties.

2. ***Blowpipe tests*** A luminous Bunsen flame (air holes completely closed), about 5 cm long, is employed for these tests. A **reducing flame** is produced by placing the nozzle of a mouth blowpipe just outside the flame, and blowing gently so as to cause the inner cone to play on the substance under examination. An **oxidizing flame** is obtained by holding the nozzle of the blowpipe about one-third within the flame and blowing somewhat more vigorously in a direction parallel with the burner top; the extreme tip of the flame is allowed to play upon the substance. Figure II.1 illustrates the oxidizing and reducing flames.

The tests are carried out upon a clean charcoal block in which a small cavity has been made with a penknife or with a small coin. A little of the substance is placed in the cavity and heated in the oxidizing flame. Crystalline salts break into smaller pieces; burning indicates the presence of an oxidizing agent (nitrate, nitrite, chlorate, etc.). More frequently the powdered substance is mixed with twice its bulk of anhydrous sodium carbonate or, preferably, with 'fusion mixture' (an equimolecular mixture of sodium and potassium carbon-

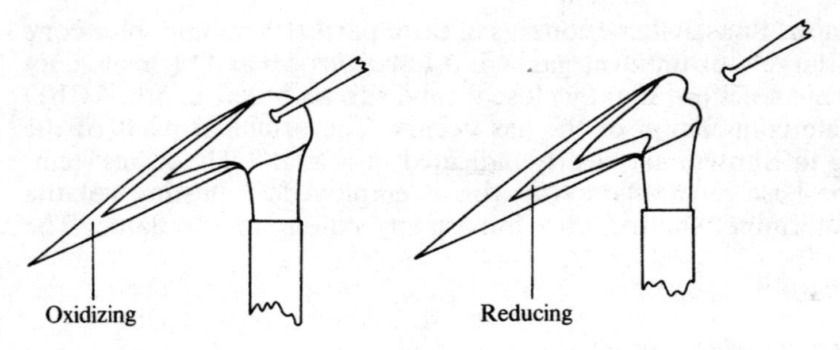

Fig. II.1

ates; this has a lower melting point than sodium carbonate alone) in the reducing flame. The initial reaction consists in the formation of the carbonates of the cations present and of the alkali salts of the anions. The alkali salts are largely adsorbed by the porous charcoal, and the carbonates are, for the most part, decomposed into the oxides and carbon dioxide. The oxides of the metals may further decompose, or be reduced to the metals, or they may remain unchanged. The final products of the reaction are therefore either the metals alone, metals and their oxides, or oxides. The oxides of the noble metals (silver and gold) are decomposed, without the aid of the charcoal, to the metal, which is often obtained as a globule, and oxygen. The oxides of lead, copper, bismuth, antimony, tin, iron, nickel, and cobalt are reduced either to a fused metallic globule (lead, bismuth, tin, and antimony) or to a sintered mass (copper) or to glistening metallic fragments (iron, nickel, and cobalt). The oxides of cadmium, arsenic, and zinc are readily reduced to the metal, but these are so volatile that they vaporize and are carried from the reducing to the oxidizing zone of the flame, where they are converted into difficulty volatile oxides. The oxides thus formed are deposited as an incrustation round the cavity of the charcoal block. Zinc yields an incrustation which is yellow while hot and white when cold; that of cadmium is brown and is moderately volatile; that of arsenic is white and is accompanied by a garlic odour due to the volatilization of the arsenic. A characteristic incrustation accompanies the globules of lead, bismuth, and antimony.

The oxides of aluminium, calcium, strontium, barium, and magnesium are not reduced by charcoal; they are infusible and glow brightly when strongly heated. If the white residue or white incrustation left on a charcoal block is treated with a drop of cobalt nitrate solution and again heated, a bright-blue colour, which probably consists of either a compound or a solid solution of cobaltous and aluminium oxides (Thenard's blue) indicates the presence of aluminium;* a pale-green colour, probably of similar composition (Rinmann's green), is indicative of zinc oxide; and a pale pink mass is formed when magnesium oxide is present.

3. ***Flame tests*** In order to understand the operations involved in the flame colour tests and the various bead tests to be described subsequently, it is necessary to have some knowledge of the structure of the non-luminous Bunsen flame (Fig. II.2).

---

* A blue colour is also given by phosphates, arsenates, silicates, or borates.

The non-luminous Bunsen flame consists of three parts: (i) an inner blue cone ADB consisting largely of unburnt gas; (ii) a luminous tip at D (this is only visible when the air holes are slightly closed); and (iii) an outer mantle ACBD in which complete combustion of the gas occurs. The principal parts of the flame, according to Bunsen, are clearly indicated in Fig. II.2 The lowest temperature is at the base of the flame (*a*); this is employed for testing volatile substances to determine whether they impart any colour to the flame. The

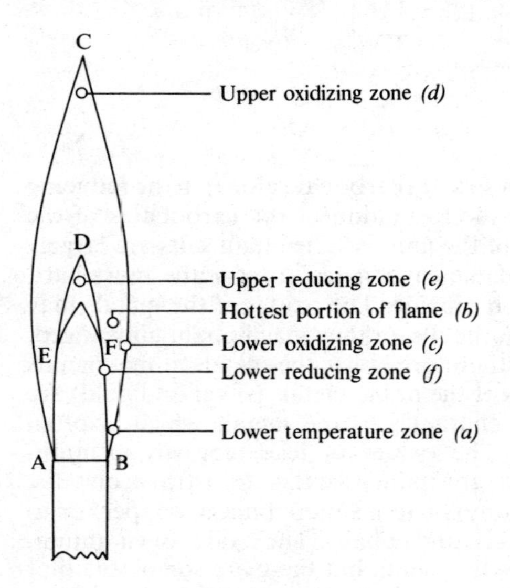

Fig. II.2

hottest part of the flame is the fusion zone at *b* and lies at about one-third of the height of the flame and approximately equidistant from the outside and inside of the mantle; it is employed for testing the fusibility of substances, and also, in conjunction with *a*, in testing the relative volatilities of substances or of a mixture of substances. The **lower oxidizing zone** (*c*) is situated on the outer border of *b* and may be used for the oxidation of substances dissolved in beads of borax, sodium carbonate, or microcosmic salt. The **upper oxidizing zone** (*d*) consists of the non-luminous tip of the flame; here a large excess of oxygen is present and the flame is not so hot as at *c*. It may be used for all oxidation processes in which the highest temperature is not required. The **upper reducing zone** (*e*) is at the tip of the inner blue cone and is rich in incandescent carbon; it is especially useful for reducing oxide incrustations to the metal. The **lower reducing zone** (*f*) is situated in the inner edge of the mantle next to the blue cone and it is here that the reducing gases mix with the oxygen of the air; it is a less powerful reducing zone than *e*, and may be employed for the reduction of fused borax and similar beads.

We can now return to the flame tests. Compounds of certain metals are volatilized in the non-luminous Bunsen flame and impart characteristic colours to the flame. The chlorides are among the most volatile compounds, and these are prepared *in situ* by mixing the compound with a little concentrated hydrochloric acid before carrying out the tests. The procedure is as follows. A thin platinum wire* about 5 cm long and 0·03–0·05 mm diameter, fused into the end of a short piece of glass tubing or glass rod which serves as a handle, is employed. This is first thoroughly cleaned by dipping it into concentrated hydrochloric acid contained in a watch glass and then heating it in the fusion zone (b) of the Bunsen flame; the wire is clean when it imparts no colour to the flame. The wire is dipped into concentrated hydrochloric acid on a watch glass, then into a little of the substance being investigated so that a little adheres to the wire. It is then introduced into the lower oxidizing zone (c), and the colour imparted to the flame observed. Less volatile substances are heated in the fusion zone (b); in this way it is possible to make use of the difference in volatilities for the separation of the constituents of a mixture.

A table showing the colours imparted to the flame by salts of different metals is given in Section **V.2**(3). Carry out flame tests with the chlorides of sodium, potassium, calcium, strontium, and barium and record the colours you observe. Repeat the test with a mixture of sodium and potassium chlorides. The yellow colouration due to the sodium masks that of the potassium. View the flame through two thicknesses of cobalt glass; the yellow sodium colour is absorbed and the potassium flame appears crimson.

Potassium chloride is much more volatile than the chlorides of the alkaline earth metals. It is therefore possible to detect potassium in the lower oxidizing flame and the calcium, strontium, and barium in the fusion zone.

After all the tests, the platinum wire should be cleaned with concentrated hydrochloric acid. It is a good plan to store the wire permanently in the acid. A cork is selected that just fits into a test-tube, and a hole is bored through the cork through which the glass holder of the platinum wire is passed. The test-tube is about half filled with concentrated hydrochloric acid so that when the cork is placed in position, the platinum wire is immersed in the acid.

A platinum wire sometimes acquires a deposit which is removed with difficulty by hydrochloric acid and heat. It is then best to employ fused potassium hydrogen sulphate. A coating of potassium hydrogen sulphate is made to adhere to the wire by drawing the hot wire across a piece of the solid salt. Upon passing the wire slowly through a flame, the bead of potassium pyrosulphate which forms travels along the wire, dissolving the contaminating deposits. When cool, the bead is readily dislodged. Any small residue of pyrosulphate dissolves at once in water, whilst the last traces are usually removed by a single moistening with concentrated hydrochloric acid, followed by heating. The resulting bright clean platinum wire imparts no colour to the flame.

4. ***Spectroscopic tests. Flame spectra*** The only worthwhile way to employ flame tests in analysis is to resolve the light into its component tints and to identify the cations present by their characteristic sets of tints. The instrument employed to resolve light into its component colours is called a **spectroscope**. A simple form is shown in Fig. II.3. It consists of a collimator A which throws a beam of parallel rays on the prism B, mounted on a turntable; the telescope C through which the spectrum is observed; and a tube D, which contains a scale of reference lines which may be superposed upon the spectrum. The spectroscope

---

* If a platinum wire is not available, a short length of chromel (or nichrome) wire, bent into a small loop at one end and inserted into a cork (to serve as a handle) at the other, may be used. This is not as satisfactory as platinum wire and is not recommended.

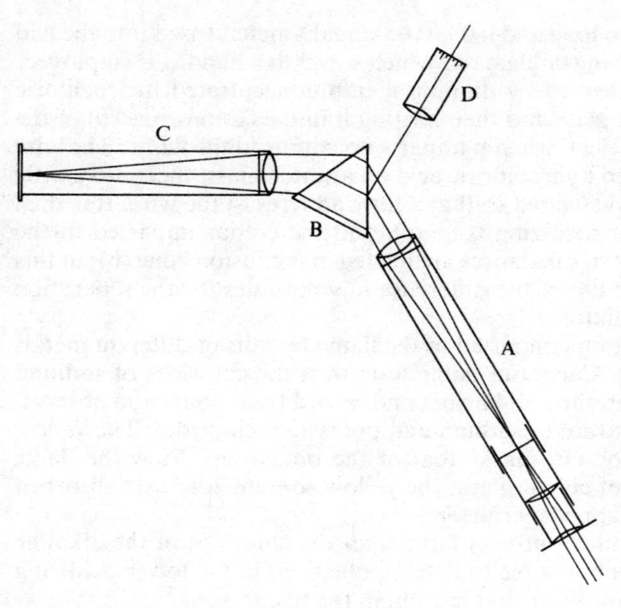

Fig. II.3

is calibrated by observing the spectra of known substances, such as sodium chloride, potassium chloride, thallium chloride, and lithium chloride. The conspicuous lines are located on a graph drawn with wavelengths as ordinates and scale divisions as abscissae. The wavelength curve may then be employed in obtaining the wavelength of all intermediate positions and also in establishing the identity of the component elements of a mixture.

To adjust the simple table spectroscope described above (which is always mounted on a rigid stand), a lighted Bunsen burner is placed in front of the collimator A at a distance of about 10 cm from the slit. Some sodium chloride is introduced by means of a clean platinum wire into the lower part of the flame, and the tube containing the adjustable slit rotated until the sodium line, as seen through the telescope C, is in a vertical position. (If available, it is more convenient to employ an electric discharge 'sodium lamp', such as is marketed by the General Electric Company: this constitutes a high-intensity sodium light source.) The sodium line is then sharply focused by suitably adjusting the sliding tubes of the collimator and the telescope. Finally, the scale D is illuminated by placing a small electric lamp in front of it, and the scale sharply focused. The slit should also be made narrow in order that the position of the lines on the scale can be noted accurately.

A smaller, relatively inexpensive, and more compact instrument, which is more useful for routine tests in qualitative analysis, is the **direct vision spectroscope** with comparison prism, shown in Fig. II.4.

The light from the 'flame' source passes through the central axis of the instrument through the slit, which is adjustable by a milled knob at the side. When the comparison prism is interposed, half the length of the slit is covered and thus light from a source in a position at right angles to the axis of the instrument will fall on one-half of the slit adjacent to the direct light which enters the other half.

This light passes through an achromatic objective lens and enters a train of five prisms of the 60° type, three being of crown glass and the alternate two of flint glass. The train of prisms gives an angular dispersion of about 11° between the red and the blue ends of the spectrum. The resulting spectrum, which can be focused by means of a sliding tube adjustment, is observed through the window. There is a subsidiary tube adjacent to the main tube: the former contains a graticule on a glass disc, which is illuminated either from the same source of light as that being observed or from a small subsidiary source (e.g. a flash lamp bulb). It is focused by means of a lens system comprising two achromatic combinations between which is a right angle prism. This prism turns the beam of light so that it falls on the face of the end prism (of the train of five prisms) and is reflected into the observer's eye, where it is seen superimposed upon the spectrum. An adjusting screw is provided to alter the position of the right angle

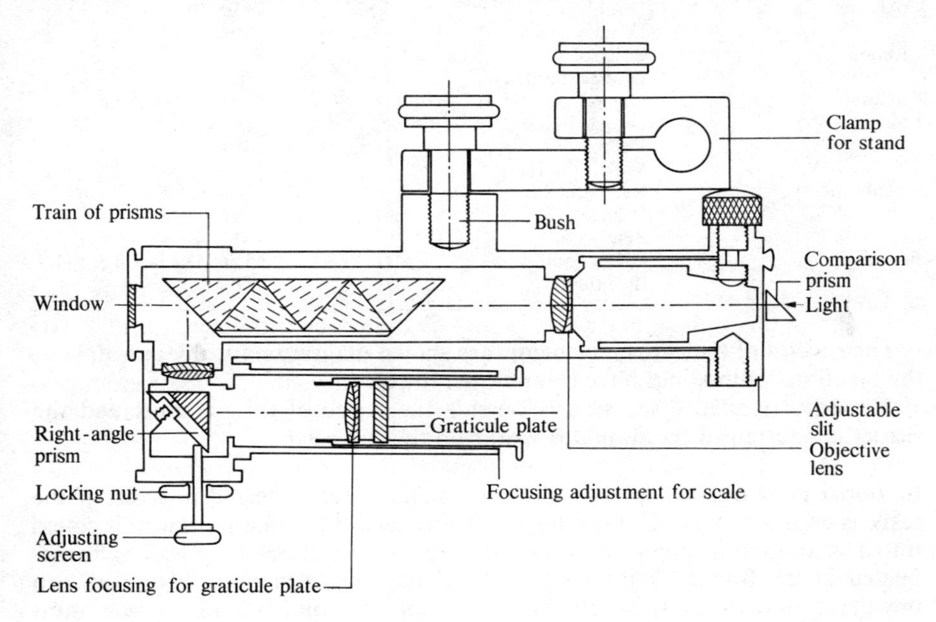

Fig. II.4

prism in order to adjust the scale relative to the spectrum. The scale is calibrated directly into divisions of 10 nanometers (or 100 Å in older instruments) and has also an indication mark at the D-line: it is 'calibrated' by means of a sodium source, and the adjusting screw is locked into position by means of a locking nut. The instrument can be mounted on a special stand.

If a sodium compound is introduced into the colourless Bunsen flame, it colours it yellow; if the light is examined by means of a spectroscope, a bright yellow line is visible. By narrowing the slit, two fine yellow lines may be seen. The mean wavelength corresponding to these two lines is $5{\cdot}893 \times 10^{-7}$ m. [Wavelengths are generally expressed in nanometers (nm). 1 nm $= 10^{-9}$ m. The old units of Angstrom (Å, 10^{-10} m) and millimicron (mμ $= 10^{-9}$ m, identical to nm) are now obsolete.] The mean wavelength of the two sodium lines is therefore 589·3 nm. The elements which are usually identified by the spectroscope in qualitative analysis are: sodium, potassium, lithium, thallium and, less frequently, because of the comparative complexity of their spectra, calcium, strontium, and barium. The wavelengths of the brightest lines, visible through a good-quality direct vision spectroscope, are collected in Table II.1. As already stated, the spectra of the alkaline earth metals are relatively complex and consist of a number of fine lines; the wavelengths of the brightest of these are given. If the resolution of the spectroscope is small, they will appear as bands.

Table II.1 Commonly occurring spectrum lines

Element	Description of line(s)	Wavelength in nm
Sodium	Double yellow	589·0, 589·6
Potassium	Double red	766·5, 769·9
	Double violet	404·4, 404·7
Lithium	Red	670·8
	Orange (faint)	610·3
Thallium	Green	535·0
Calcium	Orange band	618·2–620·3
	Yellowish-green	555·4
	Violet (faint)	422·7
Strontium	Red band	674·4, 662·8
	Orange	606·0
	Blue	460·7
Barium	Green band	553·6, 534·7, 524·3, 513·7
	Blue (faint)	487·4

The spectra of the various elements are shown diagrammatically in Fig. II.5; the positions of the lines have been drawn to scale.

A more extended discussion is outside the scope of this volume, and the reader is referred to the standard works on the subject.*

5. ***Borax bead tests*** A platinum wire, similar to that referred to under flame tests, is used for the borax bead tests. The free end of the platinum wire is coiled into a small loop through which an ordinary match will barely pass. The loop is heated in the Bunsen flame until it is red hot and then quickly dipped into powdered borax $Na_2B_4O_7$, $10H_2O$. The adhering solid is held in the hottest part of the flame; the salt swells up as it loses its water of crystallization and shrinks upon the loop forming a colourless, transparent, glass-like bead consisting of a mixture of sodium metaborate and boric anhydride.†

$$Na_2B_4O_7 = 2NaBO_2 + B_2O_3$$

The bead is moistened and dipped into the finely powdered substance so that a minute amount of it adheres to the bead. It is important to employ a minute

* See for example: L. H. Ahrens and R. S. Taylor: *Spectrochemical Analysis*. 2nd edn., Addison-Wesley 1961 or M. Slavin: *Emission Spectrochemical Analysis*, Wiley 1971.

† Some authors do not recommend the use of a loop on the platinum wire as it is considered that too large a surface of the platinum is thereby exposed. According to their procedure, the alternate dipping into borax and heating is repeated until a bead 1·5–2 mm diameter is obtained. The danger of the bead falling off is reduced by holding the wire horizontally. It is the author's experience that the loop method is far more satisfactory, especially in the hands of beginners, and is less time-consuming.

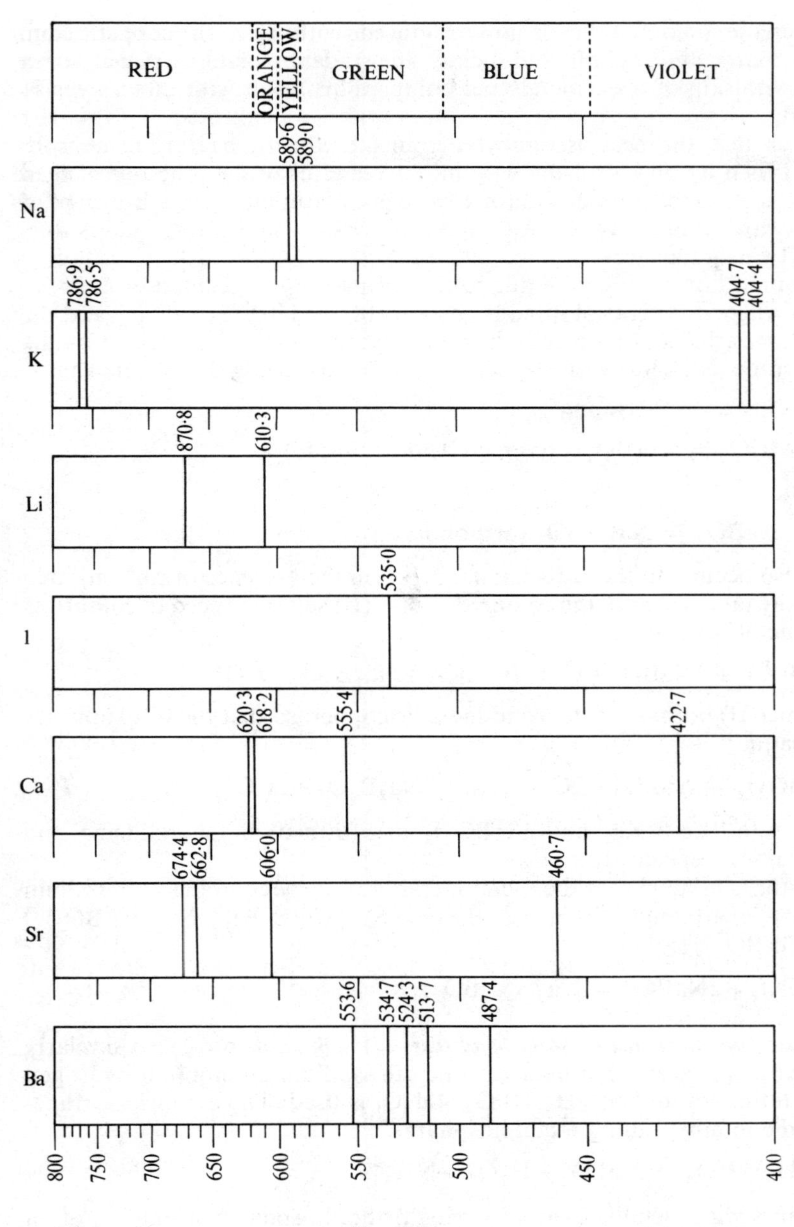

Fig. II.5

amount of substance as otherwise the bead will become dark and opaque in the subsequent heating. The bead and adhering substance are first heated in the lower reducing flame, allowed to cool and the colour observed. They are then heated in the lower oxidizing flame, allowed to cool and the colour observed again.

Characteristic coloured beads are produced with salts of copper, iron, chromium, manganese, cobalt, and nickel. The student should carry out borax bead tests with salts of these metals and compare his result with those given in Chapter III.

After each test, the bead is removed from the wire by heating it again to fusion, and then jerking it off the wire into a vessel of water. The borax bead also provides an excellent method for cleaning a platinum wire; a borax bead is run backwards and forwards along the wire by suitably heating, and is then shaken off by a sudden jerk.

The coloured borax beads are due to the formation of coloured borates; in those cases where different coloured beads are obtained in the oxidizing and the reducing flames, borates corresponding to varying stages of oxidation of the metal are produced. Thus with copper salts in the oxidizing flame, one has:

$$Na_2B_4O_7 = 2NaBO_2+B_2O_3$$

$$CuO+B_2O_3 = Cu(BO_2)_2 \text{ (copper(II) metaborate)}$$

The reaction:

$$CuO+NaBO_2 = NaCuBO_3 \text{ (orthoborate)}$$

probably also occurs. In the reducing flame (i.e. in the presence of carbon), two reactions may take place: (i) the coloured copper(II) salt is reduced to colourless copper(I) metaborate:

$$2Cu(BO_2)_2+2NaBO_2+C = 2CuBO_2+Na_2B_4O_7+CO\uparrow$$

(ii) the copper(II) borate is reduced to metallic copper, so that the bead appears red and opaque:

$$2Cu(BO_2)_2+4NaBO_2+2C = 2Cu+2Na_2B_4O_7+2CO\uparrow$$

With iron salts, $Fe(BO_2)_2$ and $Fe(BO_2)_3$ are formed in the reducing and oxidizing flames respectively.

Some authors assume that the metal metaborate may combine with sodium metaborate to give complex borates of the type $Na_2[Cu(BO_2)_4]$, $Na_2[Ni(BO_2)_4]$ and $Na_2[Co(BO_2)_4]$:

$$Cu(BO_2)_2+2NaBO_2 = Na_2[Cu(BO_2)_4]$$

6. ***Phosphate (or microcosmic salt) bead tests*** The bead is produced similarly to the borax head except that microcosmic salt, sodium ammonium hydrogen phosphate tetrahydrate $Na(NH_4)HPO_4.4H_2O$, is used. The colourless, transparent bead contains sodium metaphosphate:

$$Na(NH_4)PHO_4 = NaPO_3+H_2O\uparrow+NH_3\uparrow$$

This combines with metallic oxides forming orthophosphates, which are often coloured. Thus a blue phosphate bead is obtained with cobalt salts:

$$NaPO_3+CoO = NaCoPO_4$$

The sodium metaphosphate glass exhibits little tendency to combine with acidic oxides. Silica, in particular, is not dissolved by the phosphate bead. When a silicate is strongly heated in the bead, silica is liberated and this remains suspended in the bead in the form of a semi-translucent mass; the so-called silica 'skeleton' is seen in the bead during and after fusion. This reaction is employed for the detection of silicates:

$$CaSiO_3 + NaPO_3 = NaCaPO_4 + SiO_2$$

It must, however, be pointed out that many silicates dissolve completely in the bead so that the absence of a silica 'skeleton' does not conclusively prove that a silicate is not present.

In general, it may be stated that the borax beads are more viscous than the phosphate beads. They accordingly adhere better to the platinum wire loop. The colours of the phosphates, which are generally similar to those of the borax beads, are usually more pronounced. The various colours of the phosphate beads are collected in the following table.

Table II.2 Microcosmic salt bead tests

Oxidizing flame	**Reducing flame**	**Metal**
Green when hot, blue when cold.	Colourless when hot, red when cold.	Copper
Yellowish- or reddish-brown when hot, yellow when cold.	Yellow when hot, colourless to green when cold.	Iron
Green, hot and cold.	Green, hot and cold.	Chromium
Violet, hot and cold.	Colourless, hot and cold.	Manganese
Blue, hot and cold.	Blue, hot and cold.	Cobalt
Brown, hot and cold.	Grey when cold.	Nickel
Yellow, hot and cold.	Green when cold.	Vanadium
Yellow when hot, yellow-green when cold.	Green, hot and cold.	Uranium
Pale yellow when hot, colourless when cold	Green when hot, blue* when cold.	Tungsten
Colourless, hot and cold.	Yellow when hot, violet* when cold.	Titanium

* Blood red when fused with a trace of iron(II) sulphate.

7. ***Sodium carbonate bead tests*** The sodium carbonate bead is prepared by fusing a small quantity of sodium carbonate on a platinum wire loop in the Bunsen flame; a white, opaque head is produced. If this is moistened, dipped into a little potassium nitrate and then into a small quantity of a manganese compound, and the whole heated in the oxidizing flame, a green bead of sodium manganate is formed:

$$MnO + Na_2CO_3 + O_2 = Na_2MnO_4 + CO_2\uparrow$$

A yellow bead is obtained with chromium compounds, due to the production of sodium chromate:

$$2Cr_2O_3 + 4Na_2CO_3 + 3O_2 = 4Na_2CrO_4 + 4CO_2\uparrow$$

II.3 WET REACTIONS These tests are made with substances in solution. A reaction is known to take place (*a*) by the formation of a precipitate, (*b*) by the evolution of a gas, or (*c*) by a change of colour. The majority of the reactions of qualitative analysis are carried out in the wet way and details of these are given in later chapters. The following notes upon the methods to be adopted in carrying out the tests will be found of value and should be carefully studied.

1. ***Test-tubes*** The best size for general use is 15×2 cm with 25 ml total capacity. It is useful to remember that 10 ml liquid fills a test-tube of this size to a depth of about 5·5 cm. Smaller test-tubes are sometimes used for special tests. For heating moderate volumes of liquids a somewhat larger tube, about $18 \times 2{\cdot}5$ cm, the so-called 'boiling tube' is recommended. A test-tube brush should be available for cleaning the tubes.

2. ***Beakers*** Those of 50, 100, and 250 ml capacity and of the Griffin form are the most useful in qualitative analysis. Clock glasses of the appropriate size should be provided. For evaporations and chemical reactions which are likely to become vigorous, the clock glass should be supported on the rim of the beaker, by means of V-shaped glass rods.

3. ***Conical or Erlenmeyer flasks*** These should be of 50, 100, and 250 ml capacity, and are useful for decompositions and evaporations. The introduction of a funnel, whose stem has been cut off, prevents loss of liquid through the neck of the flask and permits the escape of steam.

4. ***Stirring rods*** A length of glass rod, of about 4 mm diameter, is cut into suitable lengths and the ends rounded in the Bunsen flame. The rods should be about 20 cm long for use with test-tubes and 8–10 cm long for work with basins and small beakers. Open glass tubes must not be used as stirring rods. A rod pointed at one end, prepared by heating a glass rod in the flame, drawing it out when soft as in the preparation of a glass jet and then cutting it into two, is employed for piercing the apex of a filter paper to enable one to transfer the contents of a filter paper by means of a stream of water from a wash bottle into another vessel.

A rubber-tipped glass rod or 'policeman' is employed for removing any solid from the sides of glass vessels. A stirring rod of polythene (polyethylene) with a thin fan-shaped paddle on each end is available commercially and functions as a satisfactory 'policeman' at laboratory temperature: it can be bent into any form.

5. ***Wash bottle*** This may consist of a 500 ml flat-bottomed flask, and the stopper carrying the two tubes should be preferably of rubber (cf. Fig. II.19*a*). It is recommended that the wash bottle be kept ready for use filled with hot water as it is usual to wash precipitates with hot water; this runs through the filter paper rapidly and has a greater solvent power than cold water, so that less is required for efficient washing. Asbestos string or cloth should be wound round the neck of the flask in order to protect the hand.

A rubber bulb may be attached to the short tube as in Fig. II.19*b*: this form of wash bottle is highly convenient and is more hygienic than the type requiring blowing by the mouth.

Polyethylene wash bottles (Fig. II.19*d*) can be operated by pressing the elastic walls of the bottle. The pressure pushes the liquid out through the tube. The flow rate can easily be regulated by changing the pressure exerted by the hand. When set aside the walls flatten out again and air is sucked into the bottle through the tube. As the air bubbles through the liquid, the latter will always be saturated with dissolved oxygen and this might interfere with some of the tests. These flasks cannot be heated, and although they will take hot liquids, it is very inconvenient to use them because one has to grip the hot walls.

6. ***Precipitation*** When excess of a reagent is to be used in the formation of a precipitate, this does not mean that an excessive amount should be employed. In most cases, unless specifically stated, only a moderate amount over that required to bring about the reaction is necessary. This is usually best detected by filtering a little of the mixture and testing the filtrate with the reagent; if no further precipitation occurs, a sufficient excess of the reagent has been added. It should always be borne in mind that a large excess of the precipitating agent may lead to the formation of complex ions and consequent partial solution of the precipitate; furthermore, an unnecessary excess of the reagent is wasteful and may lead to complications at a subsequent stage of the analysis. When studying the reactions of ions the concentrations of the reagents are known and it is possible to judge the required volume of the reagent by quick mental calculation.

7. ***Precipitation with hydrogen sulphide*** This operation is of such importance in qualitative analysis that it merits a detailed discussion. One method, which is sometimes employed, consists in passing a stream of gas in the form of bubbles through the solution contained in an open beaker, test-tube, or conical flask; this procedure is sometimes termed the 'bubbling' method. The efficiency of the method is low, particularly in acid solution; absorption of the gas takes place at the surface of the bubbles and, since the gas is absorbed slowly, most of it escapes into the air of the fume chamber and is wasted. It must be remembered that the gas is highly poisonous. The 'bubbling' method is not recommended and should not be used for macro analysis. The most satisfactory procedure (the 'pressure' method) is best described with the aid of Fig. II.6. The solution is contained in a small conical flask A, which is provided with a stopper and lead-in tube; B is a wash bottle containing water and serves to remove any hydrochloric acid spray that might be carried over from the Kipp's apparatus in the gas stream; C is a stopcock which controls the flow of gas from the generator, whilst stopcock D provides an additional control* for the gas flow. The conical flask is connected to the wash bottle by a short length of rubber tubing. The stopper is first loosened in the neck of the flask and the gas stream turned on (C first, followed by D) so as to displace most of the air in the flask: this will take not more than about 30 seconds. With the gas flowing, the stopper is inserted tightly and the flask is gently shaken with a rotary motion; splashing of the liquid on to the hydrogen sulphide delivery tube should be avoided. In order to ensure that all the air has been expelled, it is advisable to loosen the stopper in A again to sweep out the gas and then to stopper the flask tightly. Passage of the gas is continued with gentle rotation of the flask until the bubbling of the gas in B has almost ceased.† At this point the solution in A should be saturated with hydrogen sulphide and precipitation of the sulphides should be complete: this will normally require only a few minutes. Complete precipitation should be tested for by separating the precipitate by filtration, and repeating the procedure with the filtrate until the hydrogen sulphide produces

---

* Stopcock D is optional; it prevents diffusion of air into the wash bottle and thus ensures an almost immediate supply of hydrogen sulphide.

† It must be borne in mind that the gas may contain a small proportion of hydrogen due to the usually present in the commercial iron(II) sulphide. Displacement of the gas in the flask (by loosening the stopper) when the bubbling has diminished considerably will ensure complete precipitation.

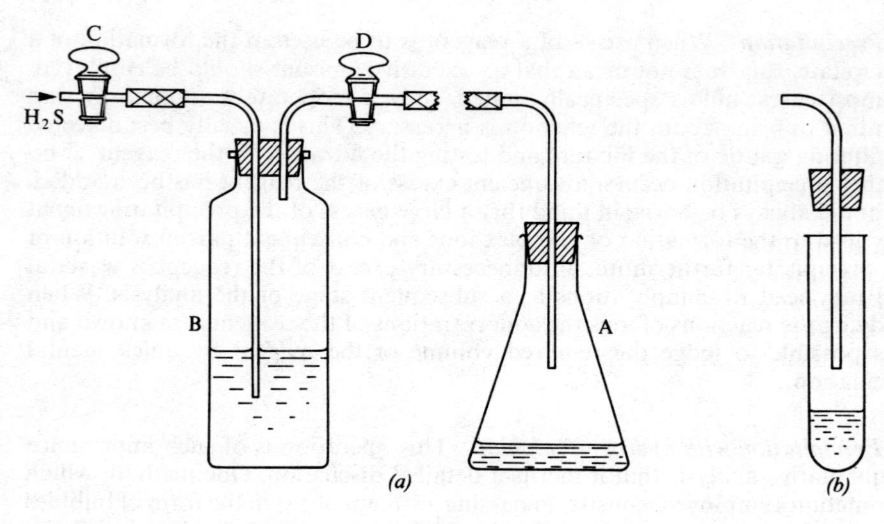

Fig. II.6

no further precipitate. Occasionally a test-tube replaces the conical flask; it is shown (with the stopper) in (*b*). The delivery tube must be thoroughly cleaned after each precipitation. The advantages of the 'pressure' method are: (i) a large surface of the liquid is presented to the gas and (ii) it prevents the escape of large amounts of unused gas.

Alternatively, a saturated aqueous solution of hydrogen sulphide can be used as a reagent. This can most easily be prepared in the bottle B of the apparatus shown in Fig. II.6. Such a reagent can most conveniently be used in teaching laboratories or classroom demonstrations when studying the reactions of ions. For a quantitative precipitation of sulphides (e.g. for separation of metals) the use of hydrogen sulphide gas is however recommended.

8. ***Filtration*** The purpose of filtration is, of course, to separate the mother liquor and excess of reagent from the precipitate. A moderately fine-textured filter paper is generally employed. The size of the filter paper is controlled by the quantity of precipitate and not by the volume of the solution. The upper edge of the filter paper should be about 1 cm from the upper rim of the glass funnel. It should never be more than about two-thirds full of the solution. Liquids containing precipitates should be heated before filtration except in special cases, for example that of lead chloride which is markedly more soluble in hot than cold water. Gelatinous precipitates, which usually clog the pores of the filter paper and thus considerably reduce the rate of filtration, may be filtered through fluted filter paper or through a pad of filter papers resting on the plate of a Buchner funnel (Fig. II.9*b* or *c*); this procedure may be used when the quantity of the precipitate is large and it is to be discarded. The best method is to add a little filter paper pulp to the solution and then to filter in the normal manner. The filter paper pulp may consist of (*a*) filter paper clippings, ashless grade, (*b*) a filtration 'accelerator' or 'ashless tablet' (Whatman), or (*c*) 'dry-dispersed', ash-free, analytical filter paper pulp (Schleicher and Schuell, U.S.A.). All the various forms of pure filter paper pulp increase the speed of filtration by re-

taining part of the precipitate and thus preventing the clogging of the pores of the filter paper.

When a precipitate tends to pass through the filter paper, it is often a good plan to add an ammonium salt, such as ammonium chloride or nitrate, to the solution; this will help to prevent the formation of colloidal solutions. The addition of a Whatman filtration 'accelerator' may also be advantageous.

A precipitate may be **washed** by decantation, as much as possible being retained in the vessel during the first two or three washings, and the precipitate then transferred to the filter paper. This procedure is unnecessary for coarse, crystalline, easy filterable precipitates as the washing can be carried out directly on the filter paper. This is best done by directing a stream of water from a wash bottle first around the upper rim of the filter and following this down in a spiral towards the precipitate in the apex; the filter is filled about one-half to two-thirds full at each washing. The completion of washing i.e. the removal of the precipitating agent is tested for by chemical means; thus if a chloride is to be removed, silver nitrate solution is used. If the solution is to be tested for acidity or alkalinity, a drop of the thoroughly stirred solution, removed upon the end of a glass rod, is placed in contact with a small strip of 'neutral' litmus paper or of 'wide range' or 'universal' test paper (Section **I.22**) on a watch glass. Other test papers are employed similarly.

9. ***Removal of the precipitate from the filter*** If the precipitate is bulky, sufficient amounts for examination can be removed with the aid of a small nickel or stainless steel spatula. If the amount of precipitate is small, one or two methods may be employed. In the first, a small hole is pierced in the base of the filter paper with a pointed glass rod and the precipitate washed into a test-tube or a small beaker with a stream of water from the wash bottle. In the second, the filter paper is removed from the funnel, opened out on a clock glass, and scraped with a spatula.

It is frequently necessary to dissolve a precipitate completely. This is most readily done by pouring the solvent, preferably whilst hot, on to the filter and repeating the process, if necessary, until all the precipitate has passed into solution. If it is desired to maintain a small volume of the liquid, the filtrate may be poured repeatedly through the filter until all the precipitate has passed into solution. When only a small quantity of the precipitate is available, the filter paper and precipitate may be heated with the solvent and filtered.

10. ***Aids to filtration*** The simplest device is to use a funnel with a long stem, or better to attach a narrow-bored glass tube, about 45 cm long and bent as shown in Fig. II.7, to the funnel by means of rubber tubing. The lower end of the tube or of the funnel should touch the side of the vessel in which the filtrate is being collected in order to avoid splashing. The speed of filtration depends *inter alia* upon the length of the water column.

Where large quantities of liquids and/or precipitates are to be handled, or if rapid filtration is desired, filtration under diminished pressure is employed: a metal or glass water pump may be used to provide the reduced pressure. A filter flask of 250–500 ml capacity is fitted with a two-holed rubber bung; a long glass tube is passed through one hole and a short glass tube, carrying a glass stopcock at its upper end, through the other hole. The side arm of the flask is connected by means of thick-walled rubber tubing ('pressure' tubing) to another flask, into

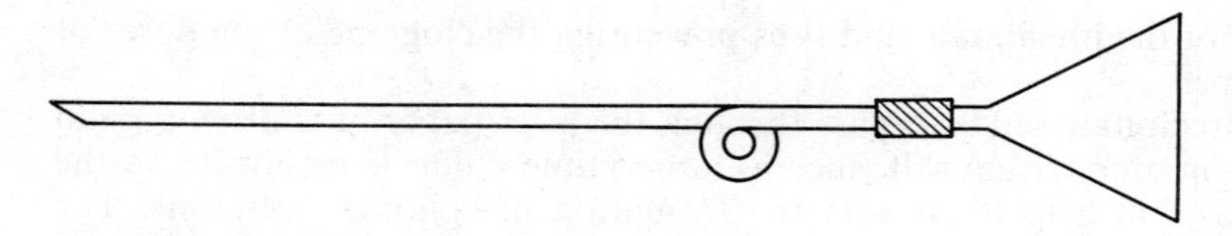

Fig. II.7

the mouth of which a glass funnel is fitted by means of a rubber bung. (Fig. II.9*a*). Upon applying suction to the filter paper fitted into the funnel in the usual way, it will be punctured or sucked through, particularly when the volume of the liquid in the funnel is small. To surmount the difficulty, the filter paper must be supported in the funnel. For this purpose either a Whatman filter cone (No. 51) made of a specially hardened filter paper, or a Schleicher and Schuell (U.S.A.) filter paper support (No. 123), made from a muslin-type material which will not retard filtration, may be used. Both types of support are folded with the filter paper to form the normal type of cone (Fig. II.8). After the filter paper has been supported in the funnel, filtration may be carried out in the usual manner under the partial vacuum created by the pump, the stopcock T being closed. When filtration is complete, the stopcock T is opened, air thereby entering the apparatus which thus attains atmospheric pressure; the filter funnel may now be removed from the filter flask.

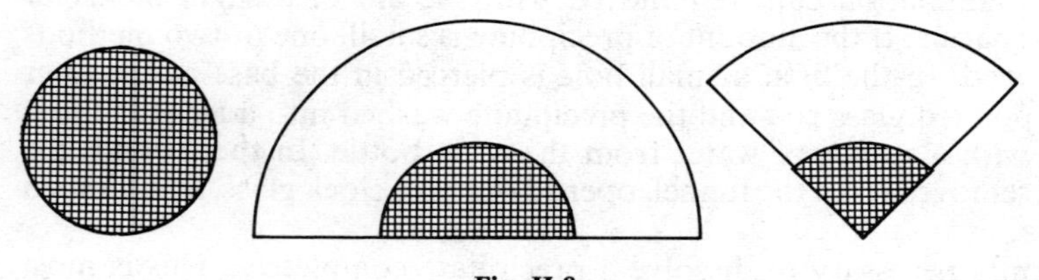

Fig. II.8

For a large quantity of precipitate, a small **Buchner funnel** (*b* in Fig. II.9, shown enlarged for the sake of clarity) is employed. This consists of a porcelain funnel in which a perforated plate is incorporated. Two thicknesses of well-fitting filter paper cover the plate. The Buchner funnel is fitted into the filter flask by means of a cork. When the volume of liquid is small, it may be collected in a test-tube placed inside the filter flask. The **Jena 'slit sieve'** funnel,* shown in *c*, is essentially a transparent Buchner funnel; its great advantage over the porcelain Buchner funnel is that it is easy to see whether the funnel is perfectly clean.

Strongly acidic or alkaline solutions cannot be filtered through ordinary filter paper. They may be filtered through a small pad of glass wool or of asbestos placed in the apex of a glass funnel. A more convenient method, applicable to strongly acidic and mildly basic solutions, is to employ a **sintered glass funnel** (Fig. II.9*d*); the filter plate, which is available in various porosities, is fused into a resistance glass (borosilicate) funnel. Filtration is carried out under reduced pressure exactly as with a Buchner funnel.

* A Pyrex 'slit sieve' funnel, with 65 mm disc, is available commercially.

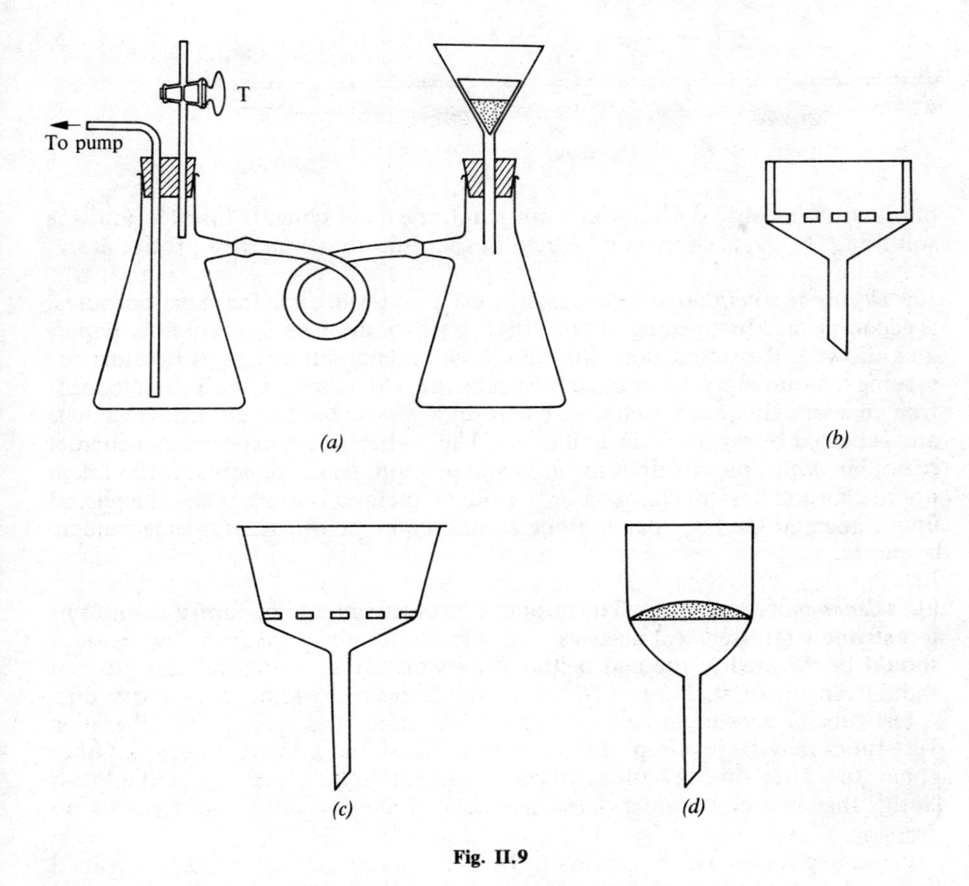

Fig. II.9

11. ***Evaporation*** The analytical procedure may specify evaporation to a smaller volume or evaporation to dryness. Both operations can be conveniently carried out in a porcelain evaporating dish or casserole; the capacity of the vessel should be as small as possible for the amount of liquid being reduced in volume. The most rapid evaporation is achieved by heating the dish directly on a wire gauze. For many purposes a water bath (a beaker half-filled with water maintained at the boiling point is quite suitable) will serve as a source of heat; the rate of evaporation will of course be slower than by direct heating with a flame. Should corrosive fumes be evolved during the evaporation, the process must be carried out in the fume cupboard. When evaporating to dryness, it is frequently desirable, in order to minimize spattering and bumping, to remove the dish whilst there is still a little liquid left; the heat capacity of the evaporating dish is usually sufficient to complete the operation without further heating.

The reduction in volume of a solution may also be accomplished by direct heating in a small beaker over a wire gauze or by heating in a wide test-tube ('boiling-tube'), held in a holder, by a free flame; in the latter case care must be taken that the liquid does not bump violently. A useful **anti-bumping device**, applicable to solutions from which gases (hydrogen sulphide, sulphur dioxide, etc.) are to be removed by boiling, is shown in Fig. II.10. It consists of a length

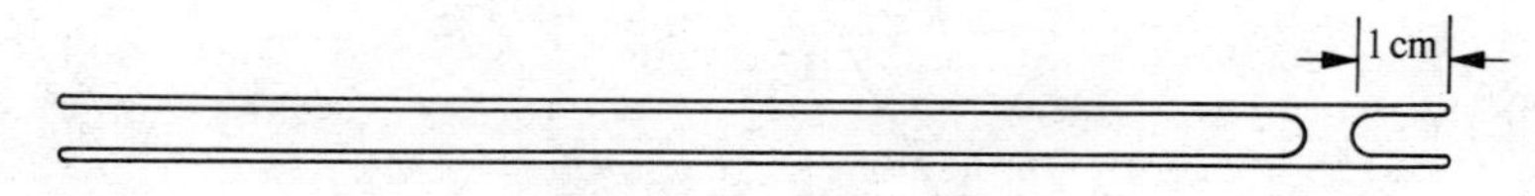

Fig. II.10

of glass tubing sealed off about 1 cm from one end, which is inserted into the solution. The device must not be used in solutions that contain a precipitate.

12. ***Drying of precipitates*** Partial drying, which is sufficient for many purposes, is accomplished by opening out the filter, laying it upon several dry filter papers and allowing them to absorb the water. More complete drying is obtained by placing the funnel containing the filter paper in a 'drying cone' (a hollow tinned-iron cone or cylinder), which rests either upon a sand bath or upon a wire gauze and is heated by means of a small flame. The funnel is thus exposed to a current of hot air, which rapidly dries the filter and precipitate. Great care must be taken not to char the filter paper. A safer but slower method is to place the funnel and filter paper, or the filter paper alone resting upon a clock glass, inside a steam oven.

13. ***Cleaning of apparatus*** The importance of using clean apparatus cannot be too strongly stressed. All glassware should be put away clean. A few minutes should be devoted at the end of the day's work to 'cleaning up'; the student should remember that wet dirt is very much easier to remove than dry dirt. A test-tube brush should be used to clean test-tubes and other glass apparatus. Test-tubes may be inverted in the test-tube stand and allowed to drain. Other apparatus, after rinsing with distilled water, should be wiped dry with a 'glass cloth', that is a cloth which has been washed at least once and contains no dressing.

Glass apparatus which appears to be particularly dirty or greasy is cleaned by soaking in chromosulphuric acid (concentrated sulphuric acid containing about 100 g of potassium dichromate per litre), followed by a liberal washing with tap water, and then with distilled water.

14. ***Some working hints*** (*a*) Always work in a tidy, systematic manner. Remember a tidy bench is indicative of a methodical mind. A string duster is useful to wipe up liquids spilt upon the bench. All glass and porcelain apparatus must be scrupulously clean.

(*b*) Reagent bottles and their stoppers should not be put upon the bench. They should be returned to their correct places upon the shelves immediately after use. If a reagent bottle is empty, it should be returned to the store-room for filling.

(*c*) When carrying out a test which depends upon the formation of a precipitate, make sure that both the solution to be tested and the reagent are absolutely free from suspended particles. If this is not the case, filter the solutions first.

(*d*) Do not waste gas or chemicals. The size of the Bunsen flame should be no larger than is absolutely necessary. It should be extinguished when no longer required. Avoid using unnecessary excess of reagents. Reagents should always be added portion-wise.

(*e*) Pay particular attention to the disposal of waste. Neither strong acids nor strong alkalis should be thrown into the sink; they must be largely diluted first, and the sink flushed with much water. Solids (corks, filter paper, etc.) should be placed in the special boxes provided for them in the laboratory. On no account may they be thrown into the sink.

(*f*) All operations involving (i) the passage of hydrogen sulphide into a solution, (ii) the evaporation of concentrated acids, (iii) the evaporation of solutions for the removal of ammonium salts, and (iv) the evolution of poisonous or disagreeable vapours or gases, must be conducted in the fume chamber.

(*g*) All results, whether positive, negative, or inconclusive, must be recorded neatly in a notebook at the time they are made. The writing up of experiments should not be postponed until after the student has left the laboratory. Apart from inaccuracies which may thus creep in, the habit of performing experiments and recording them immediately is one that should be developed from the very outset.

(*h*) If the analysis is incomplete at the end of the laboratory period, label all solutions and precipitates clearly. It is a good plan to cover these with filter paper to prevent the entrance of dust, etc.

II.4 SEMIMICRO APPARATUS AND SEMIMICRO ANALYTICAL OPERATIONS The essential technique of semimicro analysis does not differ very greatly from that of macro analysis. Since volumes of the order of 1 ml are dealt with, the scale of the apparatus is reduced; it may be said at once that as soon as the simple technique has been acquired and mastered, the student will find it just as easy to manipulate these small volumes and quantities as to work with larger volumes and quantities in ordinary test-tubes (150 × 20 mm) and related apparatus. The various operations occupy less time and the consumption of chemicals and glassware is reduced considerably; these two factors are of great importance when time and money are limited. Particular care must be directed to having both the apparatus and the working bench scrupulously clean.

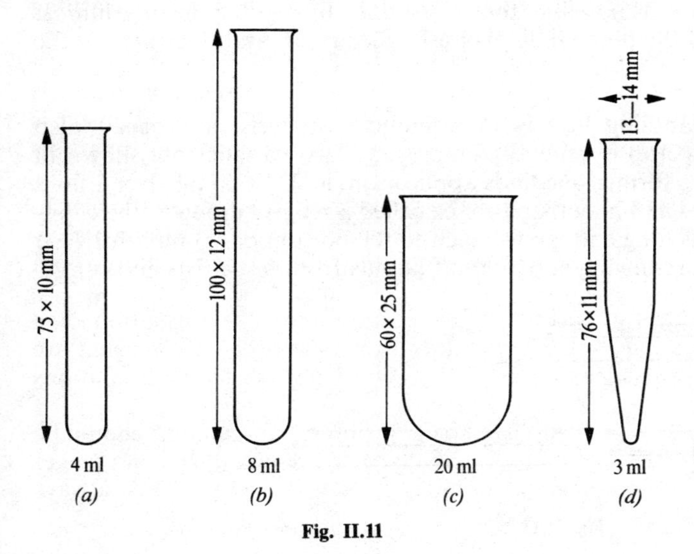

Fig. II.11

1. ***Test-tubes and centrifuge tubes*** Small Pyrex test-tubes (usually 75 × 10 mm, 4 ml, sometimes 100 × 12 mm, 8 ml) are used for reactions which do not require boiling. When a precipitate is to be separated by centrifuging, a test-tube with a tapered bottom, known as a **centrifuge tube** (Fig. II.11*d*) is generally employed; here, also, the contents cannot be boiled as 'bumping' will occur. Various sizes are available; the 3 ml centrifuge tube is the most widely used and will be adopted as standard throughout this book. For rapid concentration of a solution by means of a free flame, the **semimicro boiling tube** (60 × 25 mm, Pyrex; Fig. II.11*c*) will be found convenient.

2. ***Stirring rods*** Solutions do not mix readily in semimicro test-tubes and centrifuge tubes; mixing is effected by means of stirring rods. These can readily be made by cutting 2 mm diameter glass rod into 12 cm lengths. A handle may be formed, if desired, by heating about 1 cm from the end and bending it back at an angle of 45° (see Fig. II.12*b*). The sharp edges are fire-polished by heating

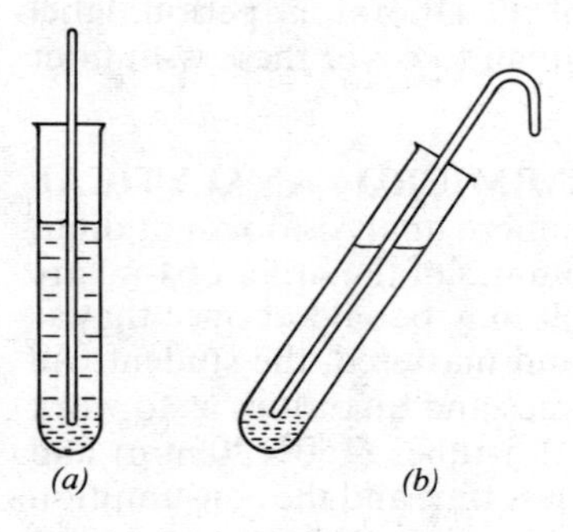

Fig. II.12

momentarily in a flame. In washing a precipitate with water or other liquid, it is essential to stir the precipitate so that every particle is brought in contact with as large a volume of liquid as possible: this is best done by holding the tube almost horizontal, to spread the precipitate over a large surface, and then stirring the suspension.

3. ***Droppers*** For handling liquids in semimicro analysis, a **dropper** (also termed a dropper pipette) is generally employed. Two varieties are shown in Fig. II.13*a* and *b*. The former one finds application, in 30 or 60 ml (1 or 2 fluid ounce) reagent bottles and may therefore be called a **reagent dropper**; the capillary of the latter (*b*) is long enough to reach to the bottom of a 3 ml centrifuge tube, and is used for removing supernatant liquids from test-tubes and centri-

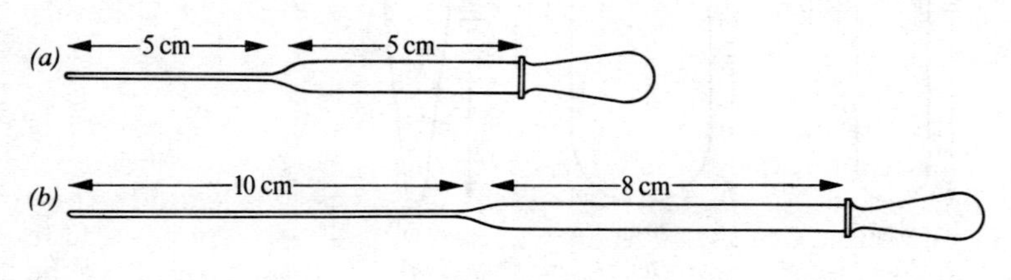

Fig. II.13

fuge tubes and for the quantitative addition of reagents. Dropper *b* will be referred to as a **capillary dropper**.

Whilst both types of dropper may be purchased, it is quite a simple operation (and excellent practice) to make them from glass tubing. To make a capillary dropper, take a piece of glass tubing about 20 cm long and of 7 mm bore and heat it near the middle of a Bunsen or blowpipe flame, rotating it slowly all the time. When the centre is soft, allow the glass to thicken slightly by continuing the heating and rotating whilst exerting a gentle inward pressure from the ends. Remove the tube from the flame and slowly draw out the softened portion to form a fairly thick-walled capillary about 20 cm long and 2 mm diameter. When cold, cut the capillary at the mid-point with a file. Fire-polish the capillary ends by rotating in a flame for a moment or two. In order that a rubber bulb (teat) may fit securely on the wide end, heat it cautiously in a flame until just soft, and whilst rotating slowly open it up with a file or glass-worker's triangular 'reamer'; alternatively, the softened end of the tube may be quickly pressed down on an asbestos or uralite sheet or upon some other inert surface. A reagent dropper is made in a similar manner except that it is unnecessary to appreciably thicken the middle of the tube before drawing out.

Before use the droppers must be calibrated, i.e. the volume of the drop delivered must be known. Introduce some distilled water into the clean dropper by dipping the capillary end into some distilled water in a beaker and compressing and then releasing the rubber teat or bulb. Hold the dropper vertically over a clean dry 5 ml measuring cylinder, and gently press the rubber bulb. Count the number of drops until the meniscus reaches the 2 ml mark. Repeat the calibration until two results are obtained which do not differ by more than 2 drops. Calculate the volume of a single drop. The dropper should deliver between 30 and 40 drops per ml. Attach a small label to the upper part of the dropper giving the number of drops per ml.

The standard commercial form of **medicine dropper**, with a tip of 1·5 mm inside diameter and 3 mm outside diameter, delivers drops of dilute aqueous solutions about 0·05 ml in volume, i.e. about 20 drops per ml. This dropper is somewhat more robust than that shown in Fig. II.13*a*, as it has a shorter and thicker capillary: this may be an advantage for elementary students, but the size of the drop (*c*. 20 per ml) may be slightly too large when working with volumes of the order of 1 ml. However, if this dropper is used, it should be calibrated as described in the previous paragraph.

It must be remembered that the volume of the drop delivered by a dropper pipette depends upon the density, surface tension, etc., of the liquid. If the dropper delivers 20 drops of distilled water, the number of drops per ml of other liquids will be very approximately as follows: dilute aqueous solutions, 20–22; concentrated hydrochloric acid, 23–24; concentrated nitric acid, 36–37; concentrated sulphuric acid, 36–37; acetic acid, 63; and concentrated ammonia solution, 24–25.

4. ***Reagent bottles and reagents*** A **semimicro reagent bottle** may be easily constructed by inserting a reagent dropper through a cork or rubber stopper that fits a 30 or 60 ml bottle – as in Fig. II.14*a*. These dropping bottles (Fig. II.14*b*) may be purchased and are inexpensive; the stoppers of these bottles are usually made of a hard rubber or plastic composition, and this as well as the rubber teat (or bulb) are attacked by concentrated inorganic acids. A dropping

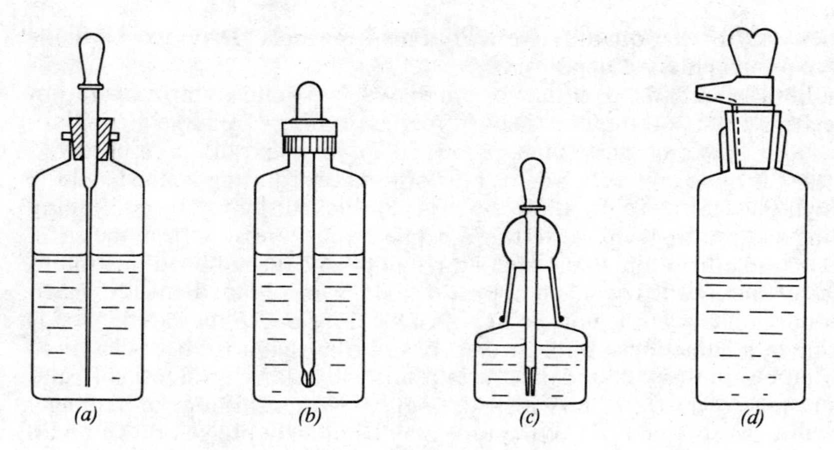

Fig. II.14

bottle of 30 ml capacity with an interchangeable glass cap (Fig. II.14*c*) is also marketed. The bottles *a* and *b* cannot be used for concentrated acids and other corrosive liquids because of their action upon the stoppers. The simplest containers for these corrosive liquids are 30 or 60 ml T.K. dropping bottles (Fig. II.14*d*).

Each student should be provided with a solid wooden stand housing a set of reagents in 30 ml or, preferably, 60 ml bottles. The following are used in the author's laboratory. Some may prefer to have the Ammonia solution, conc. in a T.K. bottle, and the dilute mineral acids in the reagent bottles fitted with reagent droppers. Others may like to have the Hydrochloric acid, dil. of 2·5–3M strength. These are questions of personal preference, and the decision will rest with the teacher.

Table II.3 Reagents for semimicro work

Reagent Bottles Fitted with Reagent Droppers (Fig. II.14*a* or *b*)			
Sodium hydroxide	2M	Acetic acid, dil.	2M
Ammonium sulphide	M	Ammonia solution, conc.	15M
Potassium hydroxide	2M	Ammonia solution, dil.	2M
Barium chloride	0·25M	Potassium hexacyanoferrate(II)	0·025M
Silver nitrate	0·1M	Potassium chromate	0·1M
Iron(III) chloride	0·5M	Ammonium carbonate	M
T.K. Dropper Bottles			
Hydrochloric acid, conc.	12M	Hydrochloric acid, dil.	2M
Sulphuric acid, conc.	18M	Sulphuric acid, dil.	M
Nitric acid, conc.	16M	Nitric acid, dil.	2M

The other reagents, which are used less frequently, are kept in 60 or 125 ml dropping bottles (30 ml for expensive or unstable reagents) on the reagent shelf (**side shelf** or **side rack reagents**); further details of these will be found in the Appendix. Two sets of these bottles should be available in each laboratory. When using these side shelf reagents, great care should be taken that the droppers do not come into contact with the test solutions, thus contaminating the reagents. If accidental contact should be made, the droppers must be thoroughly rinsed with distilled water and then dried. Under no circumstances should the capillary end of the dropper be dipped into any foreign solution.

5. ***The centrifuge*** The separation of a precipitate from a supernatant liquid is carried out with the aid of a centrifuge. This is an apparatus for the separation of two substances of different density by the application of centrifugal force which may be several times that of gravity. In practice, the liquid containing the suspended precipitate is placed in a semimicro centrifuge tube. The tube and its contents, and a similar tube containing an equal weight of water are placed in diagonally opposite buckets of the centrifuge, and the cover is placed in position; upon rotation for a short time and after allowing the buckets to come to rest and removing the cover, it will be found that the precipitate has separated at the bottom of the tube. This operation (**centrifugation**) replaces filtration in macro analysis. The supernatant liquid can be readily removed by means of a capillary dropper; the clear liquid may be called the centrifugate or 'centrate'.

The advantages of centrifugation are: (i) speed, (ii) the precipitate is concentrated into a small volume so that small precipitates are observed readily and their relative magnitudes estimated, (iii) the washing of the precipitate can be carried out rapidly and efficiently, and (iv) concentrated acids, bases, and other corrosive liquids can be manipulated easily.

The **theory** of the centrifuge is given below. The rate of settling r_s (cm s^{-1}) of spherical particles of density d_p and of radius a (cm) in a medium of dynamic viscosity η (poise) and of density d_m is given by Stokes' law:

$$r_s = \frac{2a^2g(d_p - d_m)}{9\eta} \qquad \text{(i)}$$

where g is the acceleration due to gravity (981 cm s^{-2}). It is evident from the equation that the rate of settling is increased by: an increase in the size of the particles a; an increase in difference between the density of the particles d_p and that of the medium d_m; a decrease in the viscosity of the medium; and an increase in the acceleration due to gravity g. Fine crystalline particles tend to increase in size when allowed to stand in the liquid in which they are precipitated,

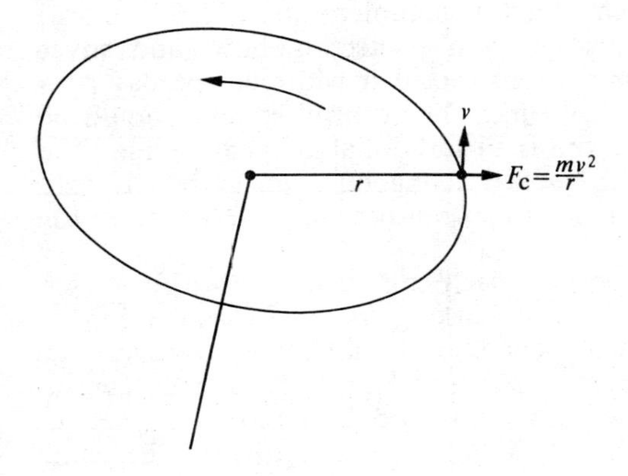

Fig. II.15

particularly when the solution is maintained at elevated temperatures and shaken occasionally. The coagulation of colloidal and gelatinous precipitates is also accelerated by high temperatures and stirring, and also by the addition of certain electrolytes. Moreover, an increase in temperature reduces the viscosity and also increases the difference $(d_p - d_m)$. This explains why centrifuge tubes or test-tubes in which precipitates are formed are usually placed in a hot water bath and shaken from time to time before separation of the mother liquor. The effect of these factors is, however, comparatively small. To accelerate greatly the rate of settling, the force on the particle g (acceleration due to gravity) must be changed. This is readily and conveniently achieved with the aid of a centrifuge.

The relative rates of settling in a centrifuge tube and in a stationary test-tube may be derived as follows. The centrifugal force F_c upon a particle of effective mass m (grams) moving in a circle of diameter r (cm) at n revolutions per second is given by:

$$F_c = ma = \frac{mv^2}{r} = \frac{m(2\pi rn)^2}{r} = 4\pi^2 mrn^2,$$

where a is the acceleration (cm s^{-2}), v is the velocity (cm s^{-2}), and $\pi = 3{\cdot}1416$. The gravitational force F_g on a particle of mass m is given by:

$$F_g = m.g$$

their ratio

$$\frac{F_c}{F_g} = \frac{m.4\pi^2 rn^2}{mg} = \frac{4\pi^2 rn^2}{g}$$

Expressing rotations per second n in rotations per minute N, and substituting for π, we have:

$$\frac{F_c}{F_g} = \frac{4\pi^2 rN^2}{981 \times 60^2} = 1{\cdot}118 \times 10^{-5} rN^2 \qquad \text{(ii)}$$

Thus a centrifuge with a radius of 10 cm and a speed of 2000 revolutions per minute has a comparative centrifugal force of $1{\cdot}118 \times 10^{-5} \times 10 \times (2000)^2 = 447$, say, 450 times the force of gravity. A precipitate which settles in, say, 10 minutes by the force of gravity alone will settle in $10 \times 60/450$ or in 1·3 seconds in such a centrifuge. The great advantage of a centrifuge is thus manifest.

Several types of centrifuge are available for semimicro analysis. These are:

A. A small 2-tube hand centrifuge with protecting bowl and cover (Fig. II.16 and Fig. II.33); if properly constructed it will give speeds up to 2–3000 rev min^{-1} with 3 ml centrifuge tubes. The central spindle should be provided with a locking screw or nut: this is an additional safeguard against the possibility of the head carrying the buckets flying off – an extremely rare occurrence. The hand-driven centrifuge is inexpensive and is satisfactory for elementary courses.

B. An inexpensive constant speed, electrically-driven centrifuge (see Figs II.32 and II.33) is marketed and has a working speed of 1450 rev min^{-1}. It is supplied with a dual purpose head. The buckets can either 'swing out' to the horizontal position or, by means of a rubber adapter, the buckets can be held at a fixed angle of 45° (M.S.E.) or 51° (International); in the latter case, the instrument acts as an 'angle' centrifuge. This is an alternative to the hand centrifuge.

C. A variable speed, electrically-driven centrifuge (see Fig. II.33) with speed indicator is the ideal instrument for centrifugation, but is relatively expensive. It is recommended that at least one of these should be available in every laboratory for demonstration purposes.

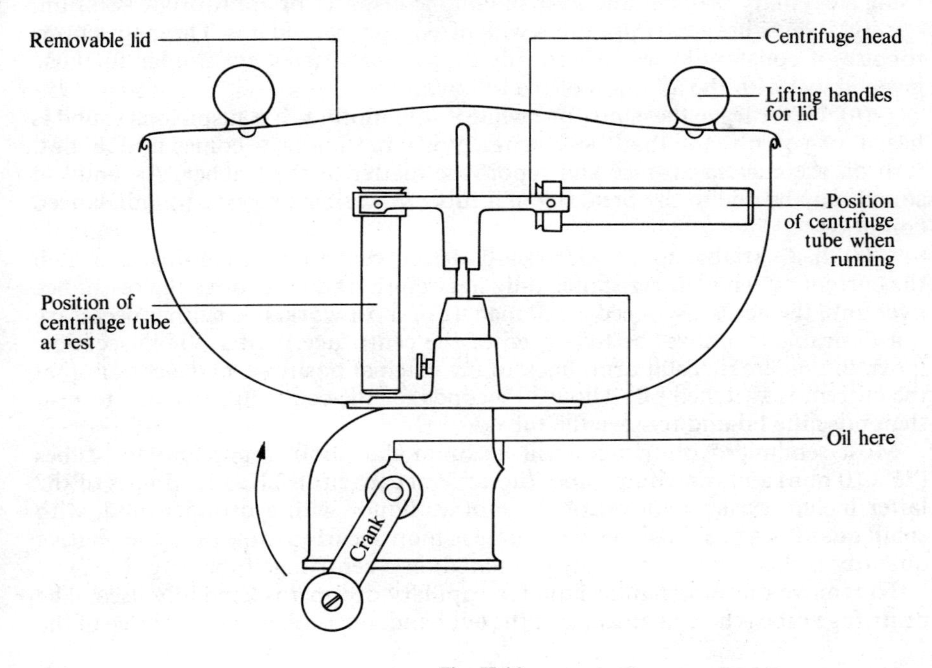

Fig. II.16

When using a hand-driven centrifuge, the following points should be borne in mind:

(*a*) The two tubes should have approximately the same size and weight.

(*b*) The tube should not be filled beyond 1 cm from the top. Spilling may corrode the buckets and produce an unbalanced head.

(*c*) Before centrifuging a precipitate contained in a centrifuge tube, prepare a balancing tube by adding sufficient distilled water from a dropper to an empty tube of the same capacity until the liquid levels in both tubes are approximately the same.

(*d*) Insert the tubes in diametrically opposite positions in the centrifuge; the head (sometimes known as a rotor) will then be balanced and vibration will be reduced to a minimum. Fix the cover in place.

(*e*) Start the centrifuge slowly and smoothly, and bring it to the maximum speed with a few turns of the handle. Maintain the maximum speed for 30–45 seconds, and then allow the centrifuge to come to rest of its own accord by releasing the handle. Do not attempt to retard the speed of the centrifuge with the hand. A little practice will enable one to judge the exact time required to pack the precipitate tightly at the bottom of the tube. It is of the utmost importance to avoid strains or vibrations as these may result in stirring up the mixture and may damage the apparatus.

(*f*) Before commencing a centrifugation, see whether any particles are floating on the surface of the liquid or adhering to the side of the tube. Surface tension effects prevent surface particles from settling readily. Agitate the surface with a stirring rod if necessary, and wash down the side of the centrifuge tube using a capillary dropper and a small volume of water or appropriate solution.

(*g*) Never use centrifuge tubes with broken or cracked lips. The instructions for use of constant speed, electrically-driven centrifuges are similar to those given above with the addition of the following:

(*h*) Never leave the centrifuge while it is in motion. If a suspicious sound is heard, or you observe that the instrument is vibrating or becomes unduly hot, turn off the current at once and report the matter to the teacher. An unusual sound may be due to the breaking of a tube; vibration suggests an unbalanced condition.

(*i*) If a variable speed, electrically-driven centrifuge is employed, switch the current on with the resistance fully in circuit; gradually move the resistance over until the necessary speed is attained. (For most work, it is neither necessary nor desirable to utilize the full speed of the centrifuge.) After 30–45 seconds, move the slider (rheostat arm) back to the original position; make certain that the current is switched off. Allow 30 seconds for the centrifuge to come to rest, then raise the lid and remove the tubes.

Most semimicro centrifuges will accommodate both semimicro test-tubes (75×10 mm) and centrifuge tubes (up to 5 ml capacity). The advantages of the latter include easier removal of the mother liquor with a dropper, and, with small quantities of solids, the precipitate is more clearly visible (and the relative quantity is therefore more easily estimated) in a centrifuge tube (Fig. II.17).

To remove the supernatant liquid, a capillary dropper is generally used. The centrifuge tube is held at an angle in the left hand, the rubber teat or nipple of the

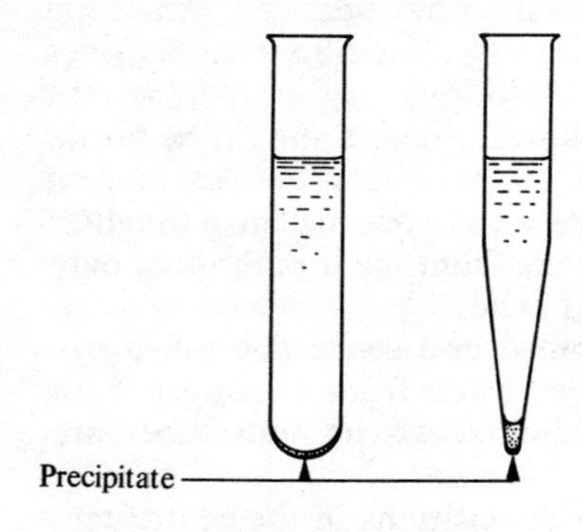

Fig. II.17

capillary dropper, held in the right hand, is compressed to expel the air and the capillary end is lowered into the tube until it is just below the liquid (Fig. II.18). As the pressure is very slowly released the liquid rises in the dropper and the latter is lowered further into the liquid until all the liquid is removed. Great care should be taken as the capillary approaches the bottom of the centrifuge tube that its tip does not touch the precipitate. The solution in the dropper should be perfectly clear; it can be transferred to another vessel by merely compressing the rubber bulb. In difficult cases, a little cotton wool may be inserted

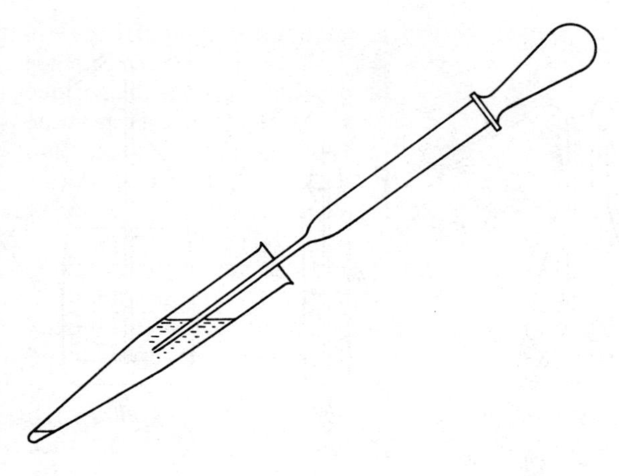

Fig. II.18

in the tip of the dropper and allowed to protrude about 2 mm below the glass tip; any excess of cotton wool should be cut off with scissors.

6. ***Washing the precipitates*** It is essential to wash all precipitates in order to remove the small amount of solution present in the precipitate, otherwise it will be contaminated with the ions present in the centrifugate. It is best to wash the precipitate at least twice, and to combine the first washing with the centrifugate. The wash liquid is a solvent which does not dissolve the precipitate but dilutes the quantity of mother liquor adhering to it. The wash liquid is usually water, but may be water containing a small amount of the precipitant (common ion effect) or a dilute solution of an electrolyte (such as an ammonium salt) since water sometimes tends to produce colloidal solutions, i.e. to peptize the precipitate.

To wash a precipitate in a centrifuge tube, 5–10 drops of water or other reagent are added and the mixture thoroughly stirred (stirring rod or platinum wire); the centrifuge tube is then counterbalanced against another similar tube containing water to the same level and centrifuged. The supernatant liquid is removed by a capillary dropper, and the washing is repeated at least once.

7. ***Wash bottles*** For most work in semimicro analysis a 30 or 60 ml glass stoppered bottle is a suitable container for distilled water; the latter is handled with a reagent dropper. Alternatively, a bottle carrying its own dropper (Fig. II.14*a* or *b*) may be used. A small conical flask (25 or 50 ml) may be used for hot water. For those who prefer wash bottles, various types are available (Fig. II.19): *a* is a 100 or 250 ml flat-bottomed flask with a jet of 0·5–1 mm diameter, and is mouth-operated; *b* is a hand-operated wash bottle (flask, 125 ml; rubber bulb, 50 ml); *c* is a Pyrex 50 ml graduated wash bottle, and *d* is a polythene bottle, from which the wash solution can be obtained by squeezing.

8. ***Transferring of precipitates*** In some cases precipitates can be transferred from semimicro test-tubes with a small spatula (two convenient types in nickel or monel metal are shown in Fig. II.20). This operation is usually difficult,

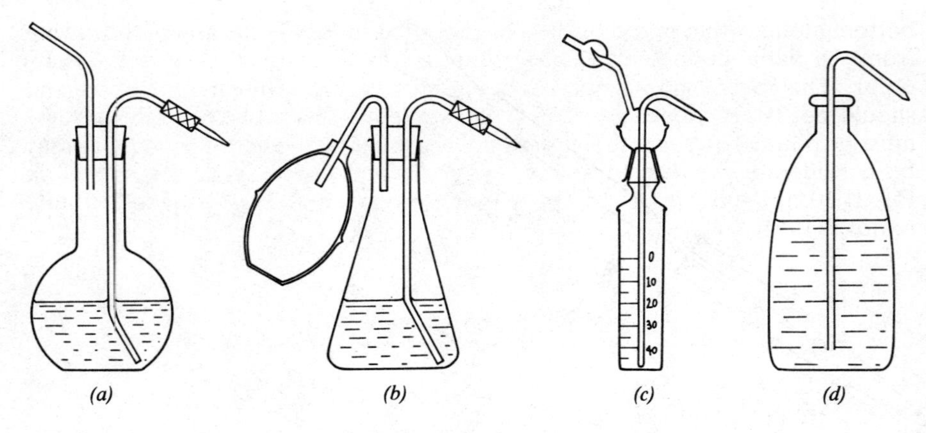

Fig. II.19

particularly for centrifuge tubes. Indeed, in semimicro analysis it is rarely necessary to transfer actual precipitates from one vessel to another. If, for some reason, transfer of the precipitate is essential, a wash liquid or the reagent itself is added, the mixture vigorously stirred and the resulting suspension transferred to a reagent dropper, and the contents of the latter ejected into the other vessel; if required, the liquid is removed by centrifugation.

If the precipitate in a test-tube is to be treated with a reagent in an evaporating dish or crucible, the reagent is added first, the precipitate brought into suspension by agitation with a stirring rod, and the suspension is then poured into the open dish or crucible. The test-tube may be washed by holding it in an almost vertical (upside down) position with its mouth over the receptacle and directing a fine stream of solution or water from a capillary dropper on to the sides of the test-tube.

9. ***Heating of solutions*** Solutions in semimicro centrifuge tubes cannot be heated over a free flame owing to the serious danger of 'bumping' (and consequent loss of part or all of the liquid) in such narrow tubes. The 'bumping' or spattering of hot solutions may often be dangerous and lead to serious burns if the solution contains strong acids or bases. Similar remarks apply to semimicro test-tubes. However, by heating the side of the test-tube (8 ml) and not the

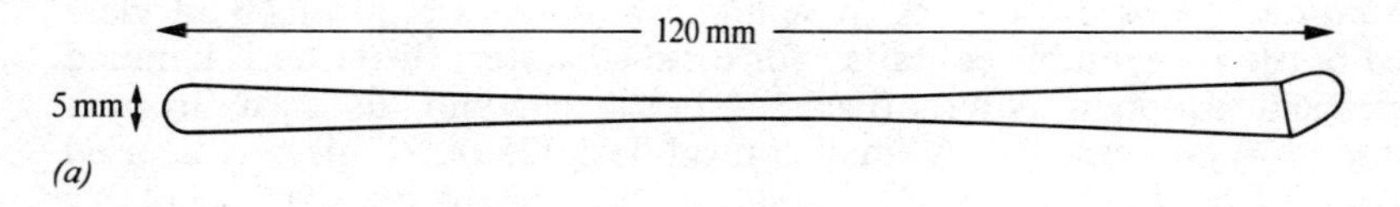

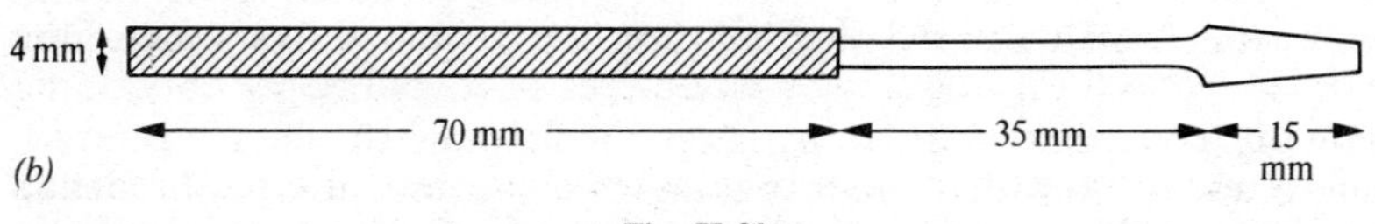

Fig. II.20

bottom alone with a micro burner, as indicated in Fig. II.21, and withdrawing from the flame periodically and shaking gently, 'bumping' does not usually occur. The latter heating operation requires very careful manipulation and should not be attempted by elementary students. The mouth of the test-tube must be pointed away from the students nearby. The danger of 'bumping' may be considerably reduced by employing the anti-bumping device shown in Fig. II.10; the tube should be about 1 cm longer than the test-tube to facilitate removal.

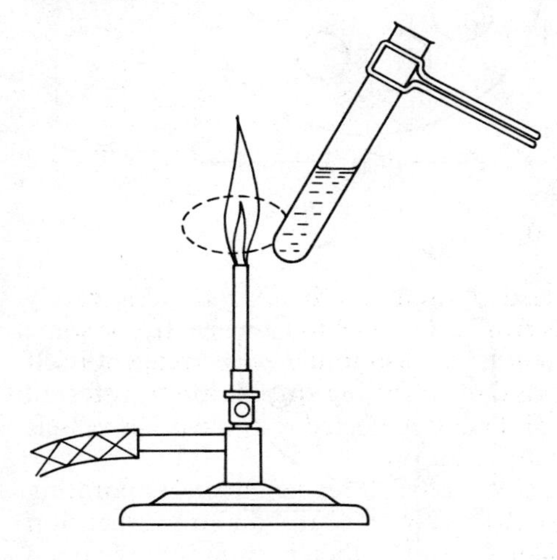

Fig. II.21

On the whole it is better to resort to safer methods of heating. The simplest procedure is to employ a small water bath. This may consist of a 250 ml Pyrex beaker, three-quarters filled with water and covered with a lead or galvanized iron plate (Fig. II.22) drilled with two holes to accommodate a test-tube and a centrifuge tube. It is a good plan to wind a thin rubber band

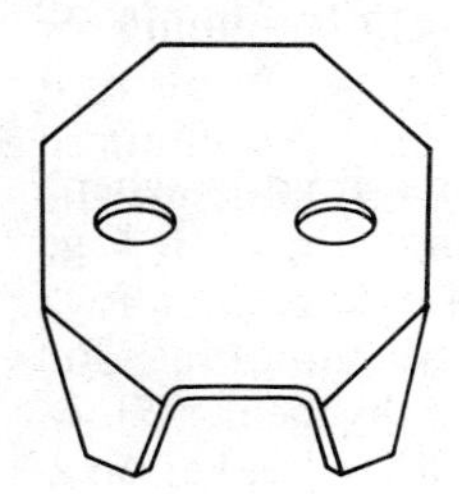

Fig. II.22

about 5 mm from the top of the tube; this will facilitate the removal of the hot tube from the water bath without burning the fingers and, furthermore, the rubber band can be used for attaching small pieces of folded paper containing notes of contents, etc.

A more elaborate arrangement which will, however, meet the requirements of several students, is Barber's water bath rack (Fig. II.23). The dimensions of a

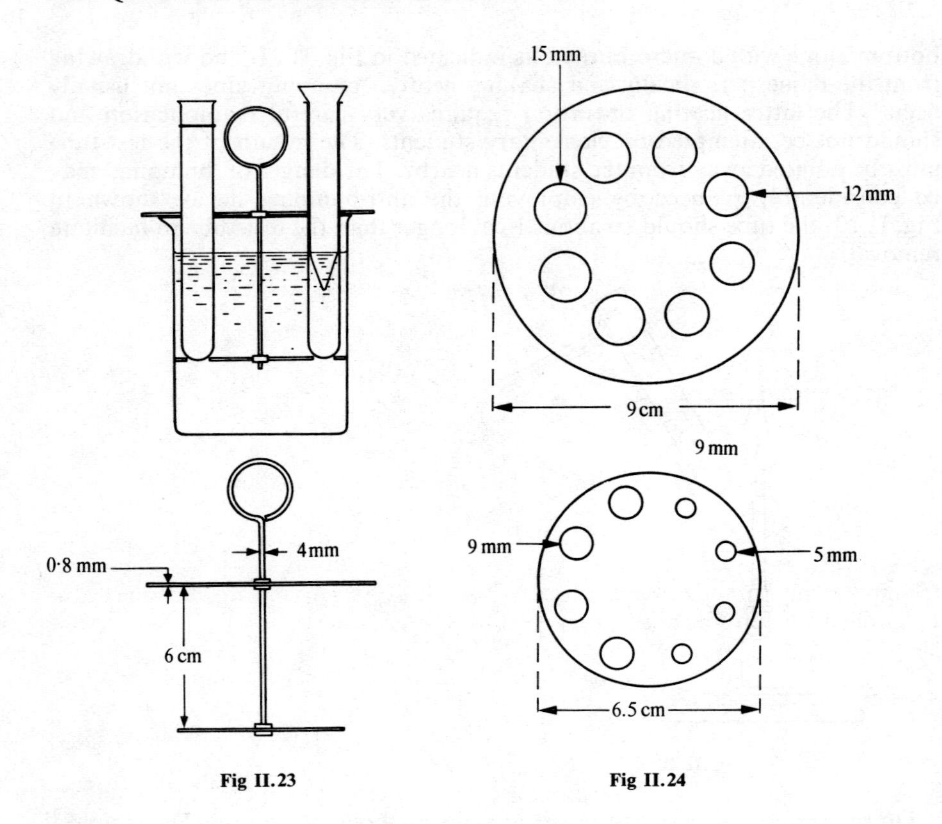

Fig II.23

Fig II.24

rack of suitable size are given in Fig. II.24. This rack will accommodate four centrifuge tubes and four semimicro test-tubes. The apparatus is constructed of monel metal, stainless steel, a plastic material which is unaffected by water at 100°C, or of brass which is subsequently tinned. The brass may be tinned by boiling with 20 per cent sodium hydroxide solution containing a few lumps of metallic tin.

10. ***Evaporations*** Where rapid concentration of a liquid is required or where volatile gases must be expelled rapidly, the semimicro boiling tube (*c* in Fig. II.11) may be employed. Two useful holders, constructed of a light metal alloy, are shown in Fig. II.25; *b* is to be preferred as the boiling tube cannot fall out by mere pressure on the holder at the point where it is usually held.* Slow evaporation may be achieved by heating in a test-tube, crucible, or beaker on a water bath.

If evaporation to dryness is required, a small casserole (*c.* 6 ml) or crucible (3–8 ml) may be employed. This may be placed in an air bath consisting of a 30 ml nickel crucible and supported thereon by an asbestos or uralite ring (Fig. II.26*a*) and heated with a semimicro burner. Alternatively, a small Pyrex beaker (say, of 50 or 100 ml capacity) may be used with a silica or nichrome

* A small wooden clothes-peg (spring type) may also be used for holding semimicro test-tubes.

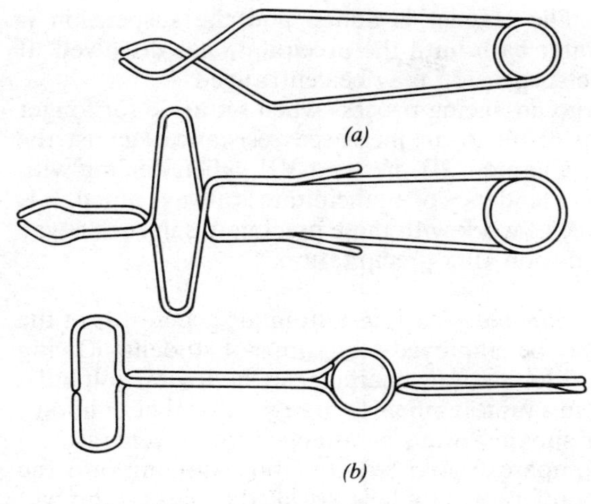

Fig. II.25

triangle to support the crucible or casserole (Fig. II.26*b*): this device may also be applied for evaporations in semimicro beakers.

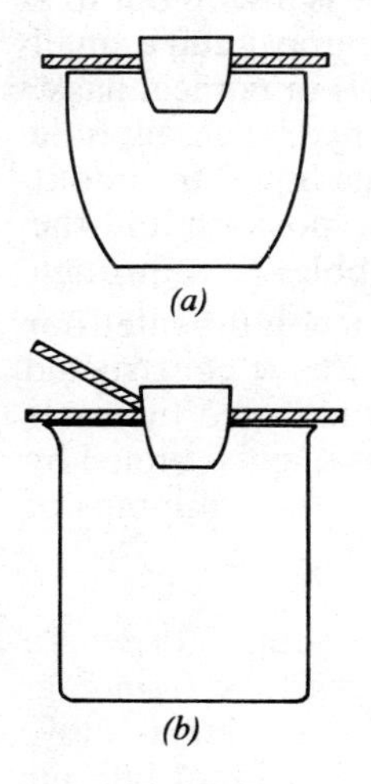

Fig. II.26

Evaporation to dryness may also be accomplished by direct intermittent heating with a micro burner of a crucible (supported on a nichrome or silica triangle) or of a semimicro beaker (supported on a wire gauze). A little practice is required in order to achieve regular boiling by intermittent heating with a flame and also to avoid 'bumping' and spattering; too hot a flame should not be used. In many cases the flame may be removed whilst a little liquid remains; the heat capacity of the vessel usually suffices to complete the evaporation without further heating. If corrosive fumes are evolved, the operation should be conducted in the fume cupboard.

11. ***Dissolving of precipitates*** The reagent is added and the suspension is warmed, if necessary, on the water bath until the precipitate has dissolved. If only partial solution occurs, the suspension may be centrifuged.

Some precipitates may undergo an ageing process when set aside for longer times (e.g. overnight). This can result in an increased resistance against the reagent used for dissolution. In Chapters III, IV, and VII, when dealing with the particulars of reactions, such a tendency of a precipitate is always noted. It is advisable to arrange the timetable of work with these precipitates in such a way that dissolution can be attempted soon after precipitation.

12. ***Precipitations with hydrogen sulphide*** Various automatic generators of the Kipp type are marketed, and may be employed for groups of students. Owing to the highly poisonous and very obnoxious character of hydrogen sulphide, these generators are always kept in a fume cupboard (draught chamber or hood.) A wash bottle containing water should always be attached to the generator in order to remove acid spray (compare Fig. II.6*a*). The tube dipping into the liquid in the wash bottle should preferably be a heavy-walled capillary; this will give a better control of the gas flow and will also help to prolong the life of the charge in the generator.

Precipitation may be carried out in a centrifuge tube, but a semimicro test-tube or 10 ml conical flask is generally preferred; for a large volume of solution, a 25 ml Erlenmeyer flask may be necessary. The delivery tube is drawn out to a thick-walled capillary (1–2 mm in diameter) and carries at the upper end a small rubber stopper which fits the semimicro test-tube* (Fig. II.27) or conical flask. The perfectly clean delivery tube is connected to the source of hydrogen sulphide as in Fig. II.6, and a slow stream of gas is passed through the liquid for about 30 seconds in order to expel the air, the cork is pushed into position and the passage of hydrogen sulphide is continued until very few bubbles pass through the liquid; the vessel is shaken gently from time to time. The tap in the gas

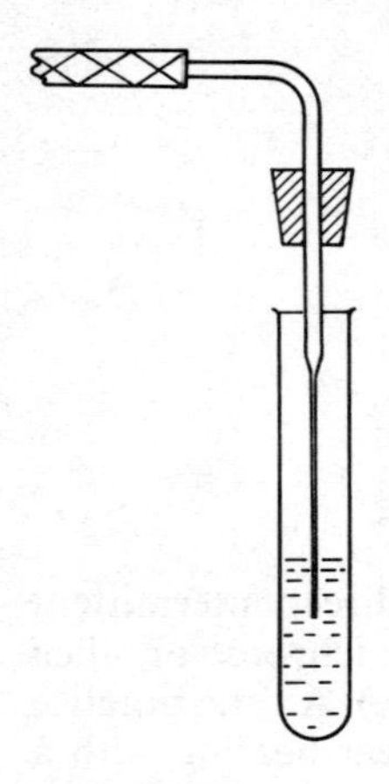

Fig. II.27

* For a 3 ml centrifuge tube or a 4 ml test-tube, the top part of the rubber teat (bulb) from a dropper makes a satisfactory stopper: a small hole is made in the rubber bulb and the delivery tube is carefully pushed through it.

generator is then turned off; the delivery tube is disconnected and immediately rinsed with distilled water. If the liquid must be warmed, the stopper is loosened, the vessel placed in the water bath for a few minutes, and the gas introduced again.

If an individual generator is desired, a Pyrex test-tube (150 × 20 mm) charged with 'Aitch-tu-ess' or an equivalent mixture (essentially an intimate mixture of a solid paraffin hydrocarbon, sulphur, and asbestos) is utilized: this yields hydrogen sulphide when heated by a micro burner and the evolution of gas ceases (but not usually abruptly) when the source of heat is removed. The test-tube type of generator is depicted in Fig. II.28. The small hole (vent) is covered

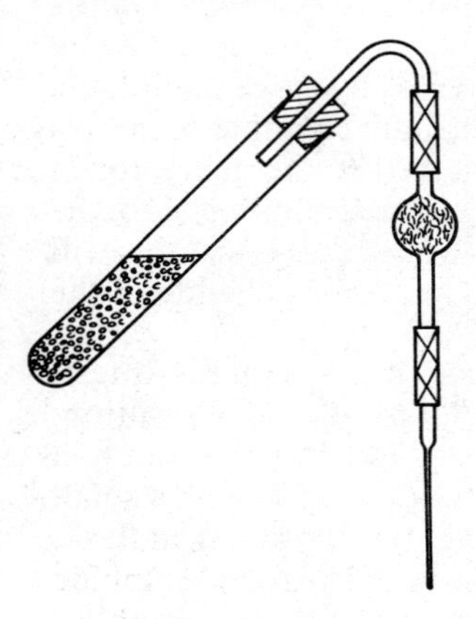

Fig. II.28

with the finger during the passage of the gas. Since the hydrogen sulphide evolution does not cease abruptly with the removal of the flame, an absorption bottle (Fig. II.29) is generally employed to absorb the unused gas. It consists of a 60 ml bottle containing 10 per cent sodium hydroxide solution: the lower end of the longer, 6 mm, tube is just above the surface of the solution. When saturation of the solution with the gas is completed, the capillary pipette and cotton wool bulb are removed and the absorption bottle is attached. The method is not strongly recommended because 'pressure' precipitation (cf. Section **II.3**) is difficult and, if attempted, introduces an element of danger.

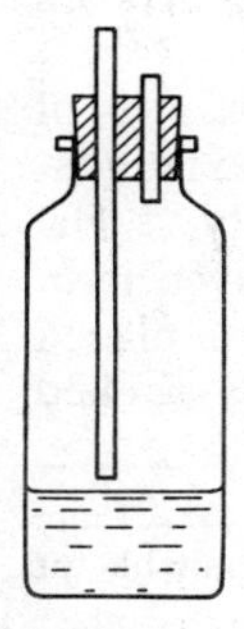

Fig. II.29

13. ***Identification of gases*** Many anions (e.g. carbonate, sulphide, sulphite, thiosulphate, and hypochlorite) are usually identified by the volatile decomposition products obtained with the appropriate reagents. Suitable apparatuses for this purpose are shown in Fig. II.30. The simplest form (*a*) consists of a semimicro test-tube with the accompanying 'filter tube':* a strip of test paper (or of filter paper moistened with the necessary reagent) about 3–4 mm wide is suspended in the 'filter tube'. In those cases where spray is likely to affect the test paper, a loose plug of cotton wool should be placed at the narrow end of the 'filter tube'. Apparatus *b* is employed when the test reagent is a liquid. A

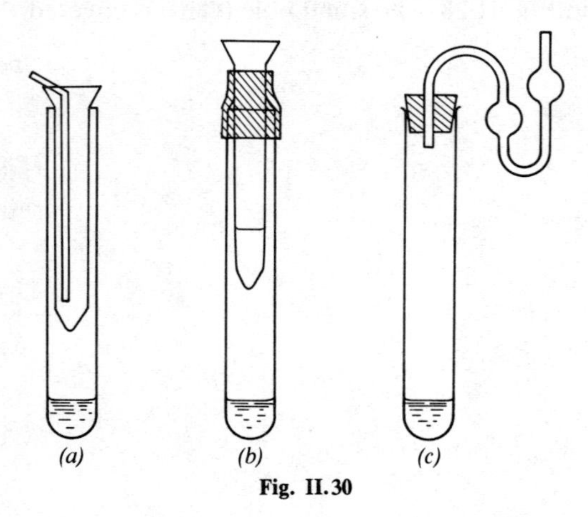

Fig. II.30

short length of wide-bore rubber tubing is fitted over the mouth of a semimicro test-tube with about 1 cm protruding over the upper edge. The chemical reaction is started in the test-tube and a 'filter tube' containing a tightly-packed plug of cotton wool is inserted through the rubber collar, the filter plug again packed down with a 4 mm glass rod, and 0·5–1 ml of the reagent introduced. An alternative apparatus for liquid reagents is shown in *c*: it is a type of absorption pipette and is attached to the 4 ml test-tube by a rubber stopper or a tightly-fitting paraffined cork. A drop or two of the liquid reagent is introduced into the absorption tube. The double bulb ensures that all the evolved gas reacts, and it also prevents the test reagent from being sucked back into the reaction mixture. All the above apparatus, *a–c*, may be warmed by placing in the hot water rack (e.g. Fig. II.23). They will meet all normal requirements for testing gases evolved in reactions in qualitative analysis.

When the amounts of evolved gas are likely to be small, the apparatus of Fig. II.31 may be used; all the evolved gas may be swept through the reagent by a stream of air introduced by a rubber bulb of about 100 ml capacity.** The approximate dimensions of the essential part of the apparatus are given in *a*, whilst the complete assembly is depicted in *b*. The sample under test is placed in **B**, the test reagent is introduced into the 'filter tube' A over a tightly-packed

* Test-tubes (75 × 10 mm and 100 × 12 mm) together with the matching 'filter tubes' (55 × 7 mm and 80 × 9 mm) are standard products; they are inexpensive and therefore eminently suitable for elementary and other students. The smaller size is satisfactory for most purposes.

** The bulb of a commercial rectum or ear syringe is satisfactory.

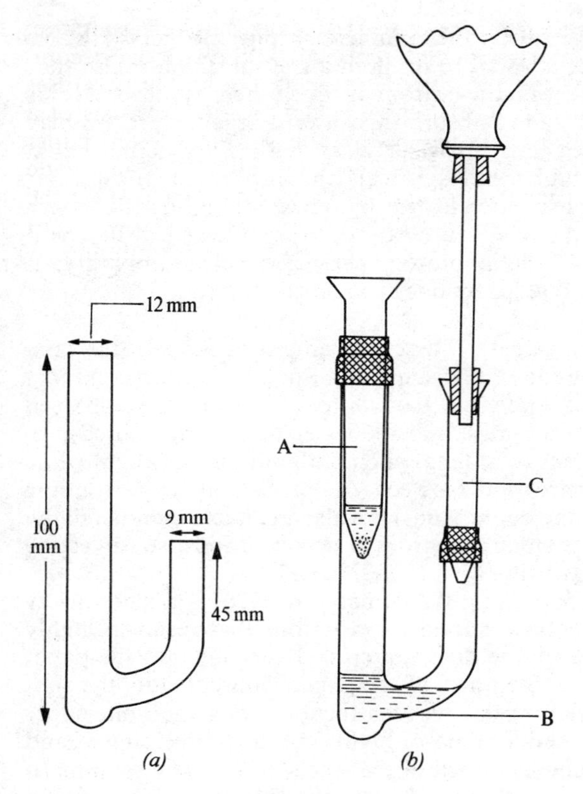

Fig. II.31

plug of cotton wool (or other medium), and the acid or other liquid reagent, is added through the 'filter tube' C. The rubber bulb is inserted into C, and by depressing it gently air is forced through the apparatus, thus sweeping out the gases through the test reagent in A. The apparatus may be warmed by placing it in a hot water bath.

14. ***Cleaning of apparatus*** It is essential to keep all apparatus scrupulously clean if trustworthy results are to be obtained. All apparatus must be thoroughly cleaned with chromosulphuric acid (concentrated sulphuric acid containing about 100 g of potassium dichromate per litre) or with a brush and cleaning powder. The apparatus is then rinsed several times with tap water, and repeatedly with distilled water. Special brushes (Fig. II.32) are available for semi-

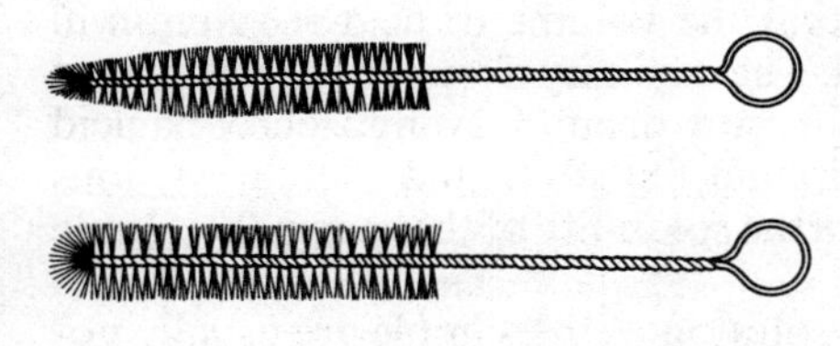

Fig. II.32

micro test-tubes and centrifuge tubes; the commercial 'pipe cleaners' are also satisfactory. The tubes may be allowed to drain in a special stand or else they may be inverted in a small beaker on the bottom of which there are several folds of filter paper or a filter paper pad to absorb the water. Larger apparatus may be allowed to drain on a clean linen towel or glass cloth. Droppers are best cleaned by first removing the rubber bulbs or teats and allowing distilled water to run through the tubes; the teats are cleaned by repeatedly filling them with distilled water and emptying them. When clean, they are allowed to dry on a linen glass cloth. At the end of the laboratory period, the clean apparatus is placed in a box with cover, so that it remains clean until required.

15. ***Spot plates. Drop-reaction paper*** These are employed chiefly for confirmatory tests (see Section **II.6** for a full discussion). Spot plates with a number of circular cavities are marketed, either black or white. The former are employed for white or light-coloured precipitates, and the latter for dark-coloured precipitates. Transparent spot plates (e.g. Jena) and combination black and white spot plates (with line of demarcation between the black and white running exactly through the centres of the depressions) are also available commercially. Spot plates are useful for mixing small quantities of reagents, and also for testing the *p*H of a solution colorimetrically.

Drop-reaction or spot-test paper (e.g. Whatman, No. 120) is a soft variety of pure, highly porous paper which is used for reactions that result in highly coloured precipitates. The precipitate does not spread very far into the paper because of the filtering action of the pores of the paper; consequently the spot test technique is employed in highly sensitive identification tests, the white paper serving as an excellent background for dark or highly coloured precipitates, and incidentally providing a permanent record of the test.

Semi-quantitative results may be obtained by the use of the Yagoda 'confined-spot-test' papers (Schleicher and Schuell, U.S.A., Nos. 211Y and 597Y): these are prepared with a chemically inert, water-repelling barrier which constricts the spot reaction to a uniform area of fixed dimensions.

16. ***Calculation of the volume of precipitating reagents*** This type of calculation is instructive, for it will assist the student to appreciate fully the significance of the quantities of reagent employed in semimicro analysis. Let us assume that a sample contains 1 mg of silver ions and 2 mg of mercury(I) ions, and that the precipitating agent is 2M hydrochloric acid. To calculate the volume of the precipitating agent required, the number of moles of the ion or ions to be precipitated and the number of moles per ml of the precipitating agent must be ascertained. In the present sample there will be $(1 \times 10^{-3})/107{\cdot}9 = 9{\cdot}3 \times 10^{-6}$ mol silver ion and $(2 \times 10^{-3})/200{\cdot}6 = 1 \times 10^{-5}$ mol mercury(I) ion, i.e. the total amount of the cations is $1{\cdot}93 \times 10^{-5}$ mol. Now 1 ml of 2M hydrochloric acid contains 2×10^{-3} mol chloride ion, so that the volume of acid required will be $(1{\cdot}93 \times 10^{-5})/(2 \times 10^{-3}) = 0{\cdot}00965$ ml. The capillary dropper delivers a drop of about 0·03 ml (30–40 drops per ml), hence 1 drop of 2M hydrochloric acid contains sufficient chloride ions to precipitate the silver and mercury(I) ions and also to provide a large excess to reduce the solubility of the sparingly soluble chlorides (common ion effect).

In practice, the weights of the ions in a solution of the sample are usually not known, so that exact calculations cannot be made. However, if the weight of sample employed is known, the maximum amount of precipitating agent required can be easily calculated. This is one of the reasons for taking a known weight (weighed to the nearest milligram) of the sample for analysis. Let us consider an actual sample. We may assume that 50 mg of the solid sample (or a solution containing the same amount of dissolved material) is taken for analysis, of which about 35 mg constitutes a Group I cation. Simple stoichiometric considerations reveal that among such ions lead, with the relative atomic mass of 207·2 requires the largest amount of hydrochloric acid for complete precipitation. Thus, if we base our calculation on lead, we cannot err on the wrong side. The 35 mg of lead constitutes $35 \times 10^{-3}/207{\cdot}2 = 1{\cdot}7 \times 10^{-4}$ mol, for which $3{\cdot}4 \times 10^{-4}$ mol hydrochloric acid is needed. If a 2M solution of hydrochloric acid is available, we require $3{\cdot}4 \times 10^{-4}/2 = 1{\cdot}7 \times 10^{-4}$ ℓ or 0·17 ml of the acid. This volume equals approximately 6 drops; and the addition of 8–10 drops will suffice for complete precipitation, taking into account the common ion effect as well.

17. ***Some practical hints*** (Cf. Section **II.3**, 14)

(*a*) Upon commencing work, arrange the more common and frequently used apparatus in an orderly manner on your bench. Each item of apparatus should have a definite place so that it can be found readily when required. All apparatus should have been cleaned during the previous laboratory period.

(*b*) Weigh the sample for analysis (if a solid) to the nearest milligram: 50 mg is a suitable quantity.

(*c*) Read the laboratory directions carefully and be certain that you understand the purpose of each operation – addition of reagents, etc. Examine the label on the bottle before adding the reagent. Serious errors leading to a considerable loss of time and, possibly, personal injury may result from the use of the wrong reagent. Return each reagent bottle to its proper place immediately after use.

(*d*) When transferring a liquid reagent with a reagent dropper always hold the dropper just above the mouth of the vessel and allow the reagent to 'drop' into the vessel. Do not allow the dropper tip to touch anything outside the reagent bottle; the possible introduction of impurities is thus avoided. Similar remarks apply to the use of T.K. dropper bottles.

(*e*) Never dip your own dropper into a reagent. Pour a little of the reagent (e.g. a corrosive liquid) into a small clean vessel (test-tube, crucible, beaker, etc.) and introduce your dropper into this. Never return the reagent to the bottle; it is better to waste a little of the reagent than to take the risk of contaminating the whole supply.

(*f*) Do not introduce your spatula into a reagent bottle to remove a little solid. Pour or shake a little of the solid on to a clean, dry watch glass, and use this. Do not return the solid reagent to the stock bottle. Try to estimate your requirements and pour out only the amount necessary.

(*g*) All operations resulting in the production of fumes (acid vapours, volatile ammonium salts, etc.) or of poisonous or disagreeable gases (hydrogen sulphide, chlorine, sulphur dioxide, etc.) must be performed in the fume chamber.

(*h*) Record your observations briefly in your note-book in ink immediately after each operation has been completed.

(*i*) Keep your droppers scrupulously clean. Never place them on the bench. Rinse the droppers several times with distilled water after use. At the end of each laboratory period, remove the rubber teat or cap and rinse it thoroughly.

(*j*) During the course of the work place dirty centrifuge tubes, test-tubes, etc., in a definite place, preferably in a beaker, and wash them at convenient intervals. This task can often be done while waiting for a solution to evaporate or for a precipitate to dissolve while being heated on a water bath.

(*k*) Adequately label all solutions and precipitates which must be carried over to the next laboratory period.

(*l*) When in difficulty, or if you suspect any apparatus (e.g. the centrifuge) is not functioning efficiently, consult the teacher.

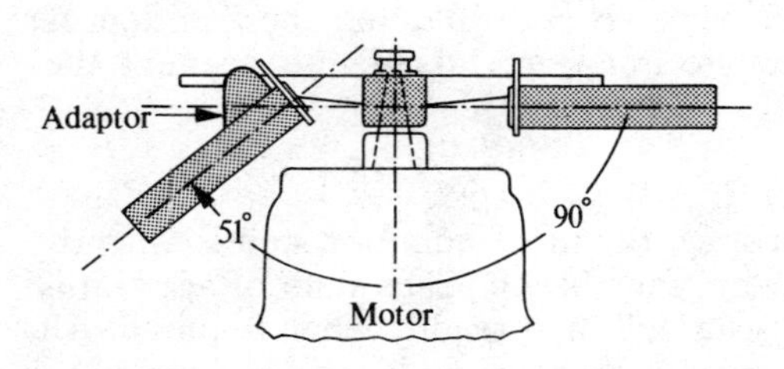

Fig. II.33

18. ***Semimicro apparatus*** The apparatus suggested for each student is listed below. (The liquid reagents recommended for each student are given in Section **II.4**(4).)

1 wash bottle, polythene (Fig. II.19*d*)
1 beaker, Griffin form, 250 ml
1 hot water rack constructed of tinned copper (Figs. II.23 and II.24) or
1 lead cover (Fig. II.22) for water bath
1 beaker, 5 ml
1 beaker, 10 ml
1 conical flask, 100 ml, and one 50 ml rubber bulb (for wash bottle, Fig. II.19)
2 conical flasks, 10 ml
1 conical flask, 25 ml
6 test-tubes, 75 × 10 mm,* 4 ml, with rim
2 test-tubes, 75 × 10 mm,* 4 ml, without rim
2 'filter tubes', 55 × 7 mm* (Fig. II.30)
1 gas absorption pipette (with 75 × 10 mm test-tube and rubber stopper, Fig. II.30)
4 centrifuge tubes, 3 ml (Fig. II.11*d*)
2 semimicro boiling tubes, 60 × 25 mm, 20 ml
1 wooden stand (to house test-tubes, 'filter tubes', gas absorption pipette, conical flasks, etc.)

* Semimicro test-tubes (100 × 12·5 mm, 8 ml) and the appropriate 'filter tubes' (80 × 9 mm) are marketed, and these may find application in analysis. The smaller 4 ml test-tubes will generally suffice: their great advantage is that they can be used directly in a semimicro centrifuge in the buckets provided for the 3 ml centrifuge tubes.

2 medicine droppers, complete with rubber teats (bulbs)
2 reagent droppers (Fig. II.13*a*)
2 capillary droppers (Fig. II.13*b*)
1 stand for droppers
2 anti-bump tubes (Fig. II.10)
1 crucible, porcelain, 3 ml (23 × 15 mm)
1 crucible, porcelain, 6 ml (28 × 20 mm)
1 crucible, porcelain, 8 ml (32 × 19 mm)
3 rubber stoppers (one 1 × 2 cm, two 0·5 × 1·5 cm, to fit 25 ml conical flask, test-tube, and gas absorption pipette)
30 cm glass tubing, 4 mm outside diameter (for H_2S apparatus)
10 cm rubber tubing, 3 mm (for H_2S apparatus)
5 cm rubber tubing, 5 mm (for 'filter tubes')
30 cm glass rod, 3 mm (for stirring rods, Fig. II.12)
1 measuring cylinder, 5 ml
1 watch glass, 3·5 cm diameter
2 cobalt glasses, 3 × 3 cm
2 microscope slides
1 platinum wire (5 cm of 0·3 mm diameter)
1 forceps, 10 cm
1 semimicro spatula (Fig. II.20*a* or *b*)
1 semimicro test-tube holder (Fig. II.25)
1 semimicro test-tube brush (Fig. II.31)
1 pipe cleaner
1 spot plate (6 cavities)
1 wide-mouthed bottle, 25 ml filled with cotton wool
1 wide-mouthed bottle, 25 ml filled with strips (2 × 2 cm) of drop reaction paper
1 dropping bottle, labelled DISTILLED WATER
1 packet blue litmus paper
1 packet red litmus paper
1 triangular file (small)
1 tripod and wire gauze (for water bath)
1 retort stand and one iron ring 7·5 cm
1 wire gauze (for ring)
1 triangle, silica
1 triangle, nichrome
1 burner, Bunsen or Tirrill
1 semimicro burner

II.5 MICRO APPARATUS AND MICROANALYTICAL OPERATIONS

In micro analysis the scale of operations is reduced by a factor of 0·01 as compared with macro analysis. Thus whereas in macro analysis the weights and volumes for analysis are 0·5–1 g and about 10 ml, and in semimicro analysis 50 mg and 1 ml respectively, in micro analysis the corresponding quantities are about 5 mg and 0·1 ml. Micro analysis is sometimes termed milligram analysis to indicate the order of weight of the sample employed. It must be pointed out that whilst the weight of the sample for analysis has been reduced, the ratio of weight to volume has been retained and in consequence the concentration of the individual ions, and other species, is maintained. A special technique must be used for handling such small quantities of materials. There is no sharp line of demarcation between semimicro and micro analysis and much of the technique described for the former can, with suitable modifications to allow for the reduction in scale by about one-tenth, be utilized for the latter. Some of the modifications, involving comparatively simple apparatus, will be described. No attempt will be made to deal with operations centred round the microscope (magnification up to 250) as the specialized technique is outside the scope of this volume.*

The small amounts of material obtained after the usual systematic separations can be detected, in many cases, by what is commonly called **spot analysis**, i.e. analysis which utilizes spots of solutions (about 0·05 ml or smaller) or a fraction of milligram of solids. Spot analysis has been developed as a result of the researches of numerous chemists: the names of Tananaeff, Krumholz, Wenger, van Niewenburg, Gutzeit and, particularly, Feigl and their collaborators must be mentioned in this connection. In general, spot reactions are preferable to tests which depend upon the formation and recognition of crystals under the microscope, in that they are easier and quicker to carry out, less susceptible to slight variations of experimental conditions, and can be interpreted more readily. Incomplete schemes of qualitative inorganic analysis have been proposed which are based largely upon spot tests: these cannot, however, be regarded as entirely satisfactory for very few spot tests are specific for particular ions and also the adoption of such schemes will not, in the long run, help in the development of micro qualitative analysis. Furthermore, such schemes, when considered from the point of view of training of students in the theory and practice of analysis, are pedagogically unsound. It seems to the writer that the greatest potential progress lies in the use of the common macro procedures (or simple modifications of them) to effect the preliminary separations for which the specialized micro technique is adopted, followed by the utilization of spot tests after the group or other separation has been effected. Hence the following pages will contain an account of the methods which can be used for performing macro operations on a micro scale and also a discussion of the technique of spot analysis.

Micro centrifuge tubes (Fig. II.34) of 0·5–2 ml capacity† replace test-tubes, beakers and flasks for most operations. Two types of centrifuge tubes are shown: *b* is particularly useful when very small amounts of precipitate are being handled. Centrifuge tubes are conveniently supported in a rack consisting of a wooden block provided with 6 to 12 holes, evenly spaced, of 1·5 cm diameter and 1·3 cm deep.

Solutions are separated from precipitates by centrifuging. Semimicro centrifuges (Section **II.4.**5), either hand-operated or electrically-driven, can be used. Adapters are provided inside the buckets (baskets) in order to accommodate micro centrifuge tubes with narrow pen ends.

Precipitations are usually carried out in micro centrifuge tubes. After centrifuging, the precipitate collects in the bottom of the tube. The supernatant liquid

---

* For a detailed account, see E. M. Chamot and C. W. Mason: *Handbook of Chemical Microscopy*, Volume I (1938) and Volume II (1940) (J. Wiley; Chapman and Hall); also A. A. Benedetti-Pichler: *Introduction to the Microtechnique of Inorganic Analysis* (1942) (J. Wiley; Chapman and Hall.)

† The dimensions given ensure that the centrifuge tubes fit into the buckets of a semimicro centrifuge. The dimensions for other capacities are: Type a: 0·5 ml, 60 × 6·3 mm (ext.); 2 ml, 68 × 11·25 mm (ext.); 3 ml, 76 × 11·25 mm (ext.). Type b: 0·5 ml, 63 mm, 8·5 mm (ext.), 2·5 mm (int.).

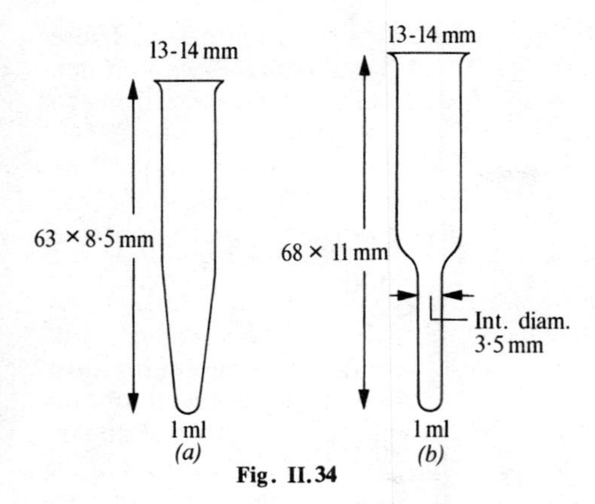

Fig. II.34

may be removed either by a capillary dropper (Fig. II.13) or by means of a **transfer capillary pipette**. The latter consists of a thin glass tube (internal diameter about 2 mm: this can be prepared from wider tubing) 20 to 25 cm in length with one end drawn out in a micro flame to a tip with a fine opening. The correct method of transferring the liquid to the capillary pipette will be evident from Fig. II.35. The centrifuge cone is held in the right hand and the capillary pipette is pushed slowly towards the precipitate so that the point of the capillary remains just below the surface of the liquid. As the liquid rises in the pipette, the latter is gradually lowered, always keeping the tip just below the

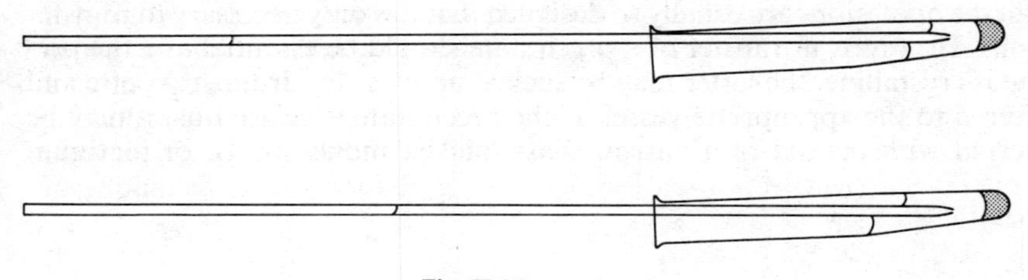

Fig. II.35

surface of the liquid until the entire solution is in the pipette and the tip is about 1 mm above the precipitate. The pipette is removed and the liquid blown or drained out into a clean dry centrifuge tube.

Another useful method for **transferring the centrifugate** to, say, another centrifuge tube will be evident upon reference to Fig. II.36. The siphon is made of thermometer capillary and is attached to the capillary pipette by means of a short length of rubber tubing of 1 mm bore. The small hole A in the side of the tube permits perfect control of the vacuum; the side arm of the small test-tube is situated near the bottom so as to reduce spattering of the liquid in the centrifuge tube inside the test-tube when the vacuum is released. Only gentle suction is applied and the opening A is closed with the finger; upon removing the finger, the action of the capillary siphon ceases immediately.

For the **washing of precipitates** the wash solution is added directly to the precipitate in the centrifuge tube and stirred thoroughly either by a platinum wire or by means of a micro stirrer, such as is shown in Fig. II.37; the latter can

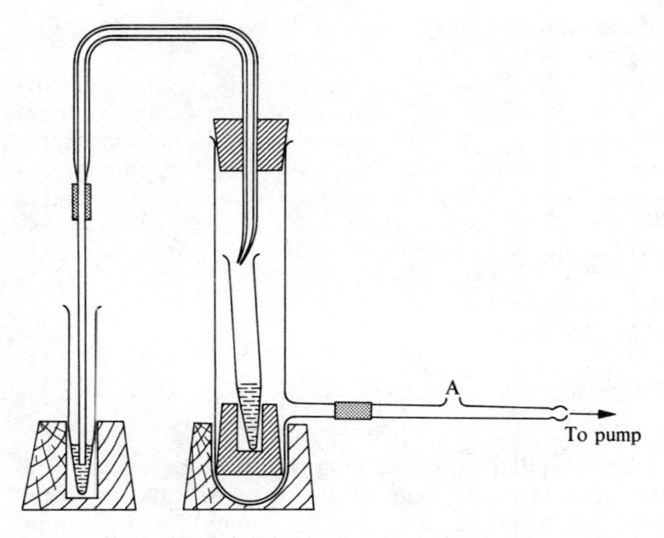

Fig. II.36

readily be made from thin glass rod. The mixture is then centrifuged, and the clear solution removed by a transfer capillary pipette as already described. It may be necessary to repeat this operation two or three times to ensure complete washing.

The **transfer of precipitates** is comparatively rare in micro qualitative analysis. Most of the operations are usually so designed that it is only necessary to transfer solutions. However, if transfer of a precipitate should be essential and the precipitate is crystalline, the latter may be sucked up by a dry dropper pipette and transferred to the appropriate vessel. If the precipitate is gelatinous, it may be transferred with the aid of a narrow glass, nickel, monel metal, or platinum

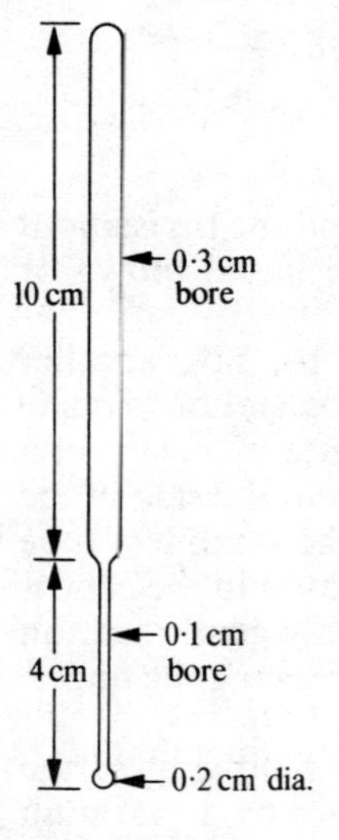

Fig. II.37

spatula. The centrifuge tube must be of type *a* (Fig. II.34) if most of the precipitate is to be removed.

The **heating of solutions** in centrifuge tubes is best carried out by supporting them in a suitable stand (compare Figs. II.22–23) and heating on a water bath. When higher temperatures are required, as for evaporation, the liquid is transferred to a micro beaker or micro crucible; this is supported by means of a nichrome wire triangle as indicated in Fig. II.38. Micro beakers may be heated

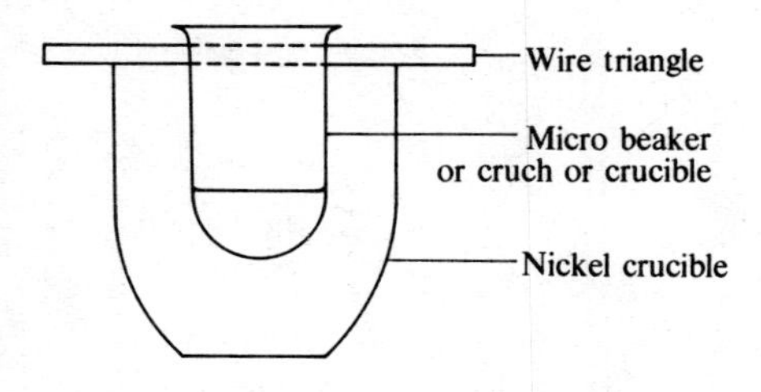

Fig. II.38

by means of the device shown in Fig. II.39, which is laid on a water bath.

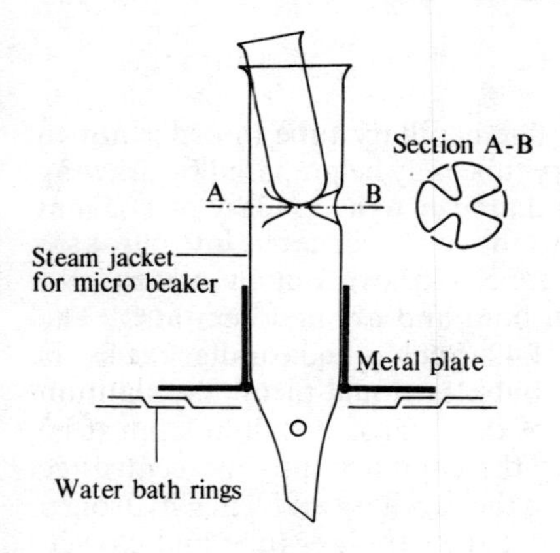

Fig. II.39

Another valuable method for concentrating solutions or evaporating to dryness directly in a centrifuge tube consists of conducting the operation on a water bath in a stream of filtered air supplied through a capillary tube fixed just above the surface of the liquid. The experimental details will be evident by reference to Fig. II.40.

Micro centrifuge tubes are cleaned with a feather, 'pipe cleaner' or small test-tube brush (cf. Fig. II.31). They are then filled with distilled water and emptied by suction as in Fig. II.41. After suction has commenced and the liquid removed, the tube is filled several times with distilled water without removing the suction device. Dropper pipettes are cleaned by repeatedly filling and emptying them with distilled water, finally separating rubber bulb and glass tube, and rinsing both with distilled water from a wash bottle. A transfer capillary pipette is cleaned by blowing a stream of water through it.

The **passage of hydrogen sulphide into a solution in a micro centrifuge tube**

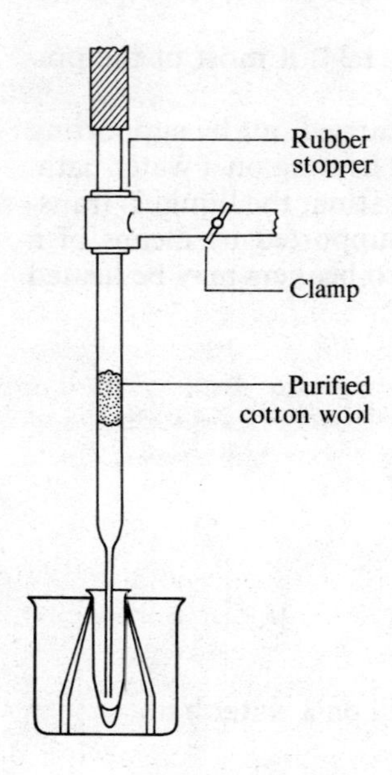

Fig. II.40

is carried out by leading the gas through a fine capillary tube in order not to blow the solution out of the tube. The delivery tube may be prepared by drawing out part of a length of glass tubing of 6 mm diameter to a capillary of 1–2 mm bore and 10–20 cm long. A plug of pure cotton wool is inserted into the wide part of the tubing, and then the capillary tube is drawn out by means of a micro burner to a finer tube of 0·3–0·5 mm bore and about 10 cm long. The complete arrangement is illustrated in Fig. II.42. Such a fine capillary delivers a stream of very small bubbles of gas; large bubbles would throw the solution out of the micro centrifuge tube. The flow of the hydrogen sulphide must be commenced before introducing the point of the capillary into the centrifuge cone. If this is not done, the solution will rise in the capillary and when hydrogen sulphide is admitted, a precipitate will form in the capillary tube and clog it.

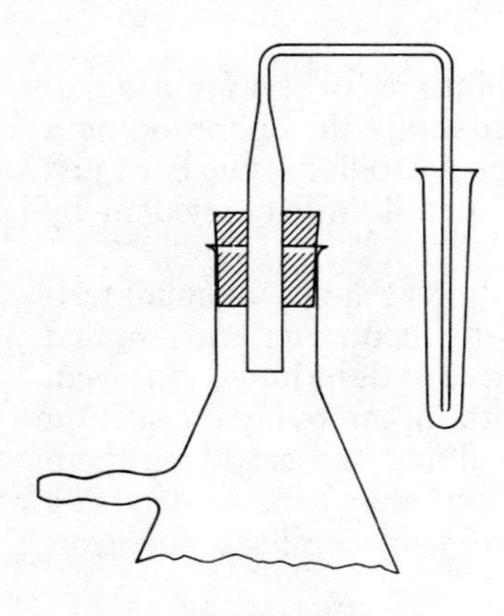

Fig. II.41

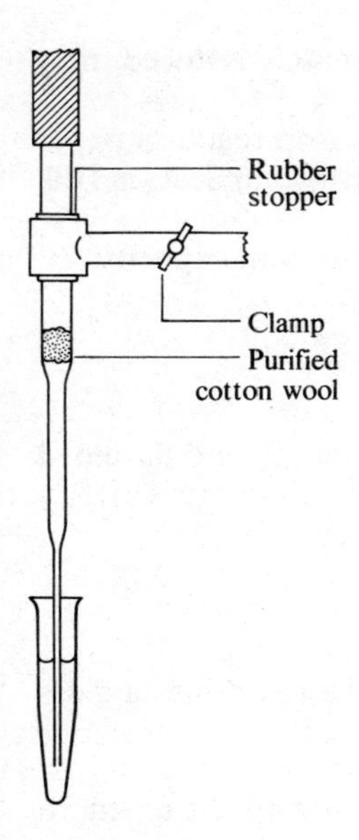

Fig. II.42

The point of saturation of the solution is indicated by an increase in the size of the bubbles; this is usually after about two minutes.

The **identification of gases** obtained in the reactions for anions may be conducted in an Erlenmeyer (or conical) flask of 5 or 10 ml capacity; it is provided with a rubber stopper which carries at its lower end a 'wedge point' pin, made preferably of nickel or monel metal (Fig. II.43). A small strip of impregnated test paper is placed on the loop of the pin. The stoppered flask, containing the reaction mixture and test paper, is placed on a water bath for five minutes. An

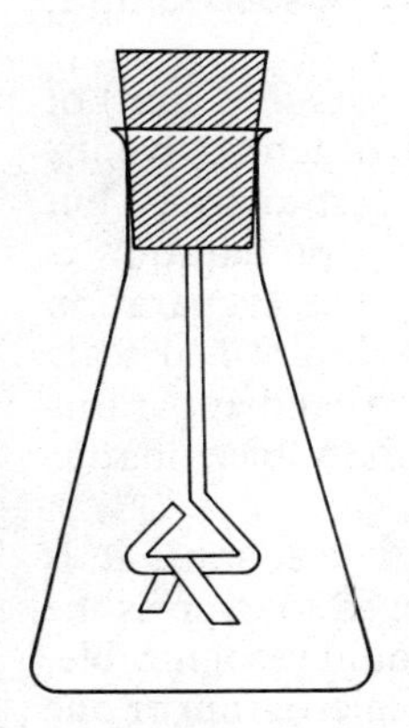

Fig. II.43

improved apparatus consists of a small flask provided with a ground glass stopper on to which a glass hook is fused (compare Fig. II.55); the test paper is suspended from the glass hook. If the evolved gas is to be passed through a liquid reagent, the apparatus of Fig. II.29*b* or *c*, proportionately reduced, may be employed.

Confirmatory tests for ions may be carried out either on drop reaction paper or upon a spot plate. The technique of spot tests is fully described in Section **II.6**.

The following **micro apparatus*** will be found useful:

Micro porcelain, silica, and platinum crucibles of 0·5 to 2 ml capacity
Micro beakers (5 ml) and micro conical flasks (5 ml)
Micro centrifuge tubes (0·5, 1·0 and 2·0 ml)
Micro test-tubes (40–50 × 8 mm)
Micro volumetric flasks (1, 2, and 5 ml)
Micro nickel, monel, or platinum spatula, 7–10 cm in length and flattened at one end
Micro burner
Micro agate pestle and mortar
Magnifying lens, 5 × or 10 ×
A pair of small forceps
A small platinum spoon, capacity 0·5–1 ml, with handle fused into a glass tube. (This may be used for fusions.)

II.6 SPOT TEST ANALYSIS The term 'spot reaction' is applied to micro and semimicro tests for compounds or for ions. In these chemical tests manipulation with drops (macro, semimicro, and micro) play an important part. Spot reactions may be carried out by any of the following processes:

(i) By bringing together one drop of the test solution and of the reagent on porous or non-porous surfaces (paper, glass, or porcelain).

(ii) By placing a drop of the test solution on an appropriate medium (e.g. filter paper) impregnated with the necessary reagents.

(iii) By subjecting a strip of reagent paper or a drop of the reagent to the action of the gases liberated from a drop of the test solution or from a minute quantity of the solid substance.

(iv) By placing a drop of the reagent on a small quantity of the solid sample, including residues obtained by evaporation or ignition.

(v) By adding a drop of the reagent to a small volume (say, 0·5–2 ml) of the test solution and then extracting the reaction products with organic solvents.

The actual 'spotting' is the fundamental operation in spot test analysis, but it is not always the only manipulation involved. Preliminary preparation is usually necessary to produce the correct reaction conditions. The preparation may involve some of the operations of macro analysis on a diminished scale (compare Section **II.5**), but it may also utilize certain operations and apparatus peculiar to spot test analysis. An account of the latter forms the subject matter of the present section.

Before dealing with the apparatus required for spot test reactions, it is necessary to define clearly the various terms which are employed to express the sensitivity of a test. The **limit of identification** is the smallest amount recognizable, and is usually expressed in **micrograms** (μg) or **gamma** (γ), one microgram or one gamma being one-thousandth part of a milligram or one-millionth part of a gram:

$$1\ \mu g = 1\gamma = 0{\cdot}001\ mg = 10^{-6}\ g$$

* All glass apparatus must be of resistance (e.g. Pyrex) glass.

Throughout this text the term **sensitivity** will be employed synonymously with limit of identification. The **concentration limit** is the greatest dilution in which the test gives positive results; it is expressed as a ratio of substance to solvent or solution. For these two terms to be comparable, a standard size drop must be used in performing the test. Throughout this book, unless otherwise stated, sensitivity will be expressed in terms of a standard drop of 0·05 ml.

The removal and addition of drops of test and reagent solutions is most simply carried out by using glass tubing, about 20 cm long and 3 mm external diameter; drops from these tubes have an approximate volume of 0·05 ml. The capillary dropper (Fig. II.13*b*) may also be employed. A useful glass pipette, about 20 cm long, may be made from 4 mm tubing and drawn out at one end in the flame (Fig. II.44); the drawn-out ends of these pipettes may be made of varying bores. A liberal supply of glass tubes and pipettes should always be kept at hand. They may be stored in a beaker about 10 cm high with the constricted end downwards and resting upon a pad of pure cotton wool; the beaker and pipettes can be protected against dust by covering with a sheet of polythene. Pipettes which are used frequently may be supported horizontally on a stand constructed of thin glass rod. After use, they should be immersed in beakers filled with distilled water: interchanges are thus prevented and subsequent thorough cleaning is facilitated.

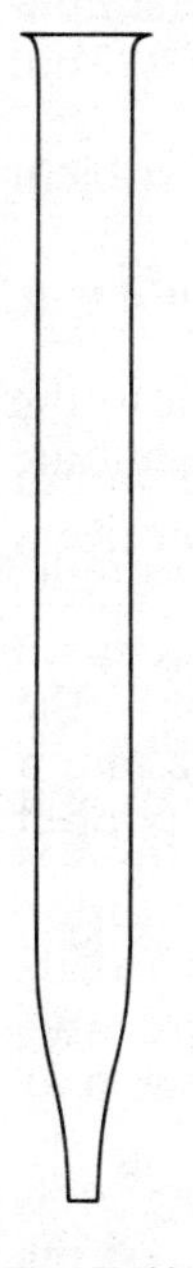

Fig. II.44

Very small and even-sized drops can be obtained by means of **platinum wire loops**; the size of the loop can be varied and by calibrating the various loops (by weighing the drops delivered), the amount of liquid delivered from each loop is known fairly accurately. A number of loops are made by bending platinum wire of suitable thickness; the wires should be attached in the usual manner to lengths of glass rod or tubing to act as handles. They are kept in

Pyrex test-tubes fitted with corks or rubber stoppers and labelled with particulars of the size of drop delivered. It must be pointed out that new smooth platinum wire allows liquids to drop off too readily, and hence it is essential to roughen it by dipping into chloroplatinic acid solution, followed by heating to glowing in a flame; this should be repeated several times. Micro burettes sometimes find application for the delivery of drops.

Reagent solutions can be added from dropping bottles of 25–30 ml capacity (see Section **II.4.4**). A stock bottle for water and for solutions which do not deteriorate on keeping, is shown in Fig. II.45; it permits facile addition in drops.

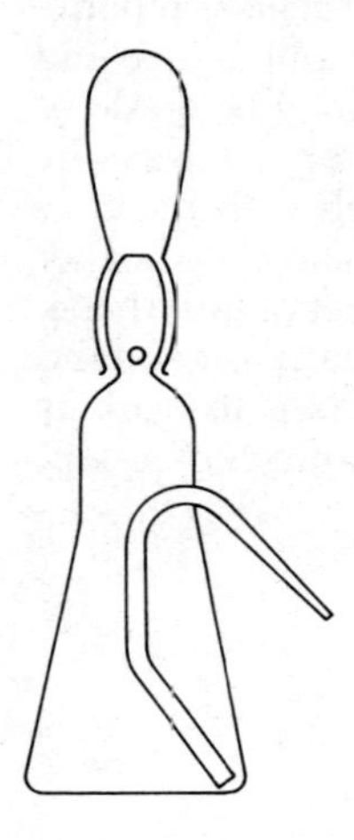

Fig. II.45

This stock bottle is constructed from a Pyrex flask into which a tube with a capillary end is fused; a small rubber bulb is placed over the drawn-out neck, which has a small hole to admit air.

Digestion of solid samples with acid or solvent may be performed in small crucibles heated on a metal hot plate, or in an air bath (Fig. II.38), or in the glass apparatus illustrated in Fig. II.46. The last-named is heated over a micro burner, and then rotated so that supernatant liquid or solution may be poured off drop-wise without danger or loss.

Spot tests may be performed in a number of ways: on a spot plate, in a micro

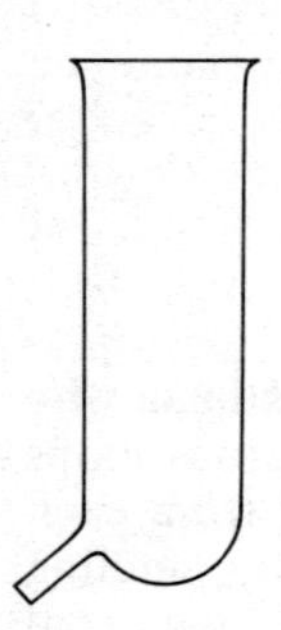

Fig. II.46

crucible, test tube, or centrifuge tube, or on filter paper. Gas reactions are carried out in special apparatus.

The commercial **spot plates** are made from glazed porcelain and usually contain 6 to 12 depressions of equal size that hold 0·5 to 1 ml of liquid. It is advisable, however, to have several spot plates with depressions of different sizes. The white porcelain background enables very small colour changes to be seen in reactions that give coloured products; the colour changes are more readily perceived by comparison with blank tests in adjacent cavities of the spot plate. Where light-coloured or colourless precipitates or turbidities are formed, it is better to employ black spot plates. Transparent spot plates of resistance glass (e.g. Jena) are also available; these may be placed upon glossy paper of suitable colour. The drops of test solution and reagent brought together on a spot plate must always be mixed thoroughly; a glass stirrer (Fig. II.37) or a platinum wire may be employed.

Traces of turbidity and of colour are also readily distinguished in micro test-tubes (50×8 mm) or in micro centrifuge tubes. As a general rule, these vessels are employed in testing dilute solutions so as to obtain a sufficient depth of colour. The liquid in a micro centrifuge tube or in a test-tube may be warmed in a special stand immersed in a water bath (compare Figs. II.22–23) or in the apparatus depicted in Fig. II.47. The latter is constructed of thin aluminium or

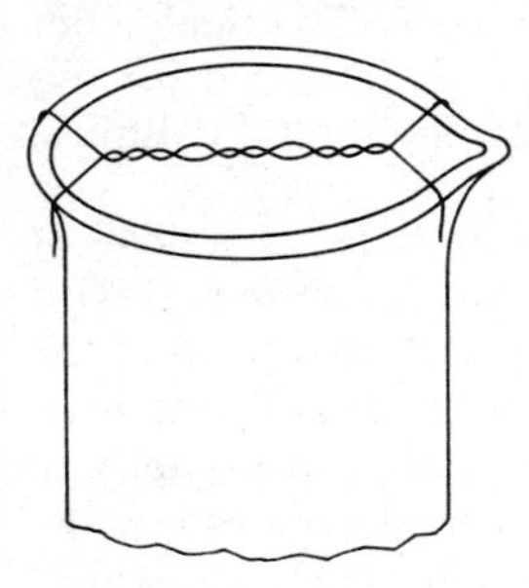

Fig. II.47

nickel wire; the tubes will slip through the openings and rest on their collars. The wire holder is arranged to fit over a small beaker, which can be filled with water at the appropriate temperature.

For heating to higher temperatures (>100°C) micro porcelain crucibles may be employed: they are immersed in an air bath (Fig. II.38) and the latter heated by a micro burner, or they can be heated directly on an asbestos mat. Small silica watch glasses also find application for evaporations.

It must be emphasized that all glass and porcelain apparatus, including spot plates and crucibles, must be kept scrupulously clean. It is a good plan to wash all apparatus (particularly spot plates) immediately after use. Glassware and porcelain crucibles are best cleaned by immersion in a chromic acid–sulphuric acid mixture or in a mixture of concentrated sulphuric acid and hydrogen peroxide, followed by washing with a liberal quantity of distilled water, and drying. The use of chromic acid mixture is not recommended for spot plates.

The great merit of glass and porcelain apparatus is that it can be employed with any strength of acid and base: it is also preferred when weakly coloured compounds (especially yellow) are produced or when the test depends upon slight colour differences. Filter paper, however, cannot be used with strongly acidic solutions for the latter cause it to tear, whilst strongly basic solutions produce a swelling of the paper. Nevertheless, for many purposes and especially for those dependent upon the application of capillary phenomena, spot reactions carried out on filter paper possess advantages over those in glass or porcelain: the tests generally have greater sensitivity and a permanent (or semi-permanent) record of the experiment is obtained.

Spot reactions upon filter paper are usually performed with Whatman drop reaction paper No. 120, but in some cases Whatman No. 3 is utilized: the Schleicher and Schuell (U.S.A.) equivalents are Nos. 601 and 598. These papers possess the desirable property of rapidly absorbing the drops without too much spreading, as is the case with thinner papers. Although impurities have been reduced to minimum values, these papers may contain traces of iron and phosphate: spot reactions for these are better made with quantitative filter paper (Whatman No. 42 or, preferably, the hardened variety No. 542). The paper should be cut into strips 6×2 cm or 2×2 cm, and stored in petri dishes or in vessels with tightly-fitting stoppers.

Spot test papers are marketed in which the spot reaction is confined to a uniform area of fixed dimensions, produced by surrounding the area by a chemically inert, water-repelling barrier. These papers are developed by H. Yagoda and may be employed for semi-quantitative work. The Schleicher and Schuell (U.S.A.) No. 211Y 'confined spot test' paper is intended for use with single drops of solution, and No. 597Y for somewhat larger volumes. They are often referred to as Yagoda test papers.

Spot reactions on paper do not always involve interaction between a drop of test solution and one of the reagent. Sometimes the paper is impregnated with the reagent and the dry **impregnated reagent paper** is spotted with a drop of the solution. Special care must be taken in the choice of the impregnating reagent. Organic reagents, that are only slightly soluble in water but dissolve readily in alcohol or other organic solvents, find extensive application. Water-soluble salts of the alkali metals are frequently not very stable in paper. This difficulty can often be surmounted by the use of sparingly soluble salts of other metals. In this way the concentration of the reactive ion can be regulated automatically by the proper selection of the impregnating salt, and the specificity of the test can be greatly improved by thus restricting the number of possible reactions. Thus potassium xanthate (see under Molybdenum, Section **VII.6**) has little value as an impregnating agent since it decomposes rapidly and is useless after a few days. When, however, cadmium xanthate is used, a paper is obtained which gives sensitive reactions only with copper and molybdenum and will keep for months. Similarly the colourless zinc hexacyanoferrate(II) offers parallel advantages as a source of hexacyanoferrate(II) ions: it provides a highly sensitive test for iron(II) ions. A further example is paper impregnated with zinc, cadmium, or antimony sulphides: such papers are stable, each with its maximum sulphide-ion concentration (controlled by its solubility product) and hence only those metallic sulphides are precipitated whose solubility products are sufficiently low. Antimony(III) sulphide paper precipitates only silver, copper, and mercury in the presence of lead, cadmium, tin, iron, nickel, cobalt and zinc. It might be expected that the use of 'insoluble' reagents would decrease the reaction rate. This retardation is not significant when paper is the medium because of the fine state of division and the great surface available. Reduction in sensitivity becomes appreciable only when the solubility products of the reagent and the reaction product approach the same order of magnitude. This is naturally avoided in the selection of reagents.

Filter paper may be impregnated with reagents by the following methods:

(i) For reagents that are soluble in water or in organic solvents, strips of filter paper are bathed in the solutions contained in beakers or in dishes. Care must be taken that the strips do not stick to the sides of the vessel or to one another, as this will prevent a uniform impregnation. The immersion should last about 20 minutes and the solution should be stirred frequently or the vessel gently rotated to produce a swirling of the solution. The strips are removed from the bath, allowed to drain, pinned to a string (stretched horizontally), and allowed to dry in the air. Uniform drying is of great importance.

Alternatively, the reagent may be sprayed on to the filter paper. The all-glass spray shown in Fig. II.48 (not drawn to scale) gives excellent results. A rubber

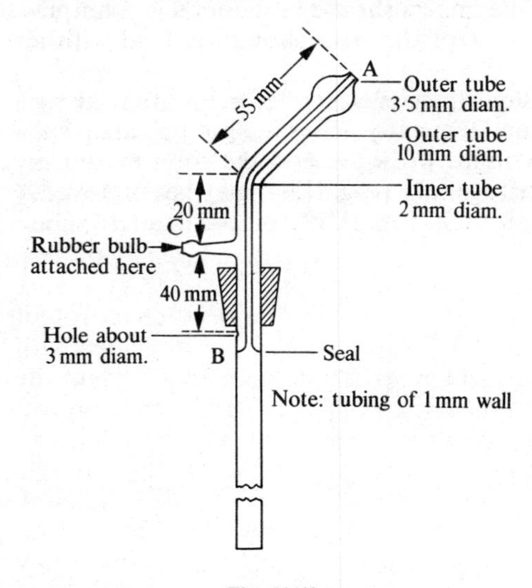

Fig. 11.48

bulb is attached at C; the cork is fitted into a boiling tube or small flask charged with the impregnating solution. The paper is sprayed first on one side and then on the other.

(ii) For reagents that are precipitated on the paper, the strips are soaked rapidly and uniformly with the solution of one of the reactants, dried and then immersed similarly in a solution of the precipitant. The excess reagents are then removed by washing, and the strips dried. The best conditions (concentration of solutions, order in which applied, etc.) must be determined by experiment. In preparing highly impregnated paper, the precipitation should never be made with concentrated solutions as this may lead to an inhomogeneous precipitation and the reagent will tend to fall off the paper after it is washed and dried. It is essential to carry out the soaking and precipitation separately with dilute solutions and to dry the paper between the individual precipitations. Sometimes it is preferable to use a reagent in the gaseous form, e.g. hydrogen sulphide for sulphides and ammonia for hydroxides; there is then no danger of washing away the precipitate.

The reactions are carried out by adding a drop of the test solution from a capillary pipette, etc., to the centre of the horizontal reagent paper resting across a porcelain crucible or similar vessel; an unhindered capillary spreading follows and a circular spot results. With an impregnated reagent paper, the resulting change in colour may occur almost at once, or it may develop after the application of a further reagent. It is usually best not to place a drop of the test solution on the paper but to allow it to run slowly from a capillary tip (0·2–1 mm diameter) by touching the tip on the paper. The test drop then enters over a minute area, precipitation or adsorption of the reaction product occurs in the immediate surrounding reagion, where it remains fixed in the fibres whilst the clear liquid spreads radially outwards by capillarity. A concentration of the coloured product, which would otherwise be spread over the whole area originally wetted by the test drop, is obtained, thus rendering minute quantities distinctly visible. A greater sensitivity is thus obtained than by adding a free drop of the test solution. Contact with the fingers should be avoided in manipulation with drop-reaction papers; a corner of the strip should be held with a pair of clean forceps.

The problem of **separating solid and liquid phases** either before or after taking a sample drop or two of the test solution frequently arises in spot test analysis. When there is a comparatively large volume of liquid and the solid matter is required, centrifugation in a micro centrifuge tube (Fig. II.34) may be employed. Alternatively, a micro sintered glass filter tube (Fig. II.49), placed in a test-tube

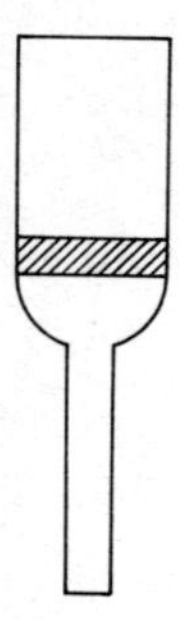

Fig. II.49

of suitable size, may be subjected to centrifugation: this device simplifies the washing of a precipitate. If the solid is not required, the liquid may be collected in a capillary pipette by sucking through a small pad of purified cotton wool placed in the capillary end; upon removing the cotton wool and wiping the pipette, the liquid may be delivered clear and free from suspended matter.

A useful **filter pipette** is shown in Fig. II.50. It is constructed of tubing of 6 mm diameter. A rubber bulb is attached to the short arm A; the arm B is ground flat, whilst the arm C is drawn out to a fine capillary; a short piece of rubber tubing is fitted over the top of B. For filtering, a disc of filter paper of the same diameter as the outside diameter of the tube (cut out from filter paper by means of a sharp cork borer or by a hand punch) is placed on the flat ground surface of B, the tube F placed upon it and then held in position by sliding the rubber tubing just far enough over the paper to hold it when the tube F is removed. The filter pipette may be used either by placing a drop of the solution

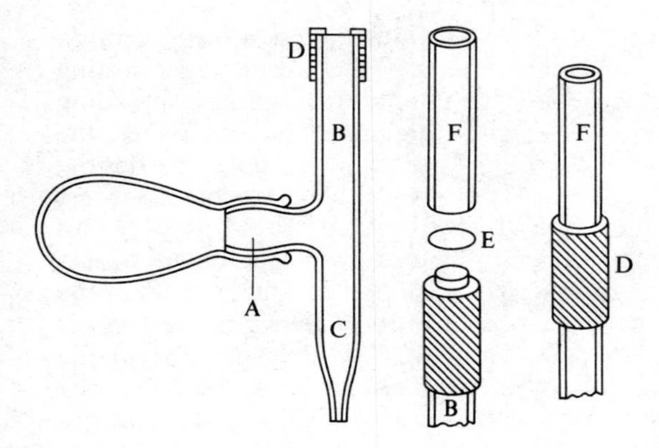

Fig. II.50

on the filter disc or by immersing the tube end B into the crucible, test-tube, or receptacle containing the solution to be filtered. The bulb is squeezed by the thumb and middle finger, and the dropper point closed with the forefinger thus allowing the solution to be drawn through the paper when the bulb is released. To release the drops of filtered liquid thus obtained, the filter pipette is inverted over the spot plate, in an inclined position with the bulb uppermost. Manipulation of the bulb again forces the liquid in the tip on to the spot plate. The precipitate on the paper can be withdrawn for any further treatment by simply sliding the rubber tubing D down over the arm B.

Another method involves the use of an **Emich filter stick** fitted through a rubber stopper into a thick-walled suction tube; the filtrate is collected in a micro test-tube (Fig. II.51). The filter stick has a small pad of purified asbestos above the constriction.

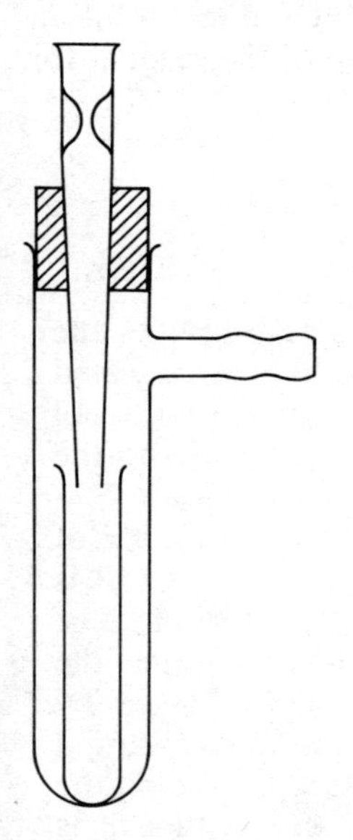

Fig. II.51

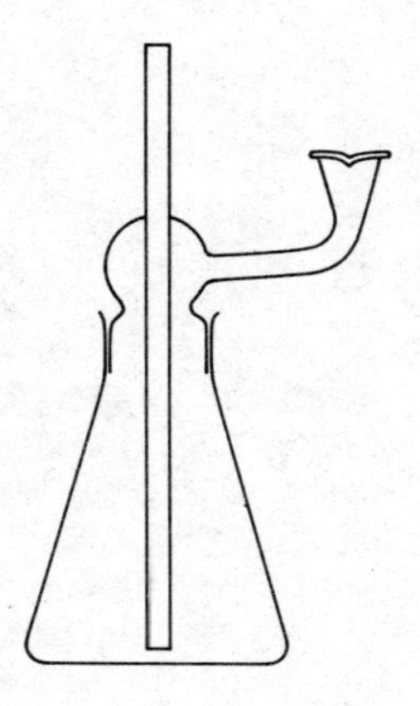

Fig. II.52

The apparatus illustrated in Fig. II.52 may be employed when the filter paper (or drop reaction paper) must be **heated in steam**; the filter paper is placed on the side arm support. By charging the flask with hydrogen sulphide solution, ammonia solution, chlorine or bromine water, the apparatus can be used for treating the filter paper with the respective gases or vapours.

Fusion and solution of a melt may be conducted either in a platinum wire loop or in a platinum spoon (0·5–1 ml capacity) attached to a heavy platinum wire and fused into a glass holder.

Gas reactions may be performed in specially devised apparatus. Thus in testing for carbonates, sulphides, etc., it is required to absorb the gas liberated in a drop of water or reagent solution. The apparatus is shown in Fig. II.53, and consists of a micro test-tube of about 1 ml capacity, which can be closed with a small ground glass stopper fused to a glass knob. The reagent and test solution or test solid are placed in the bottom of the tube, and a drop of the reagent for

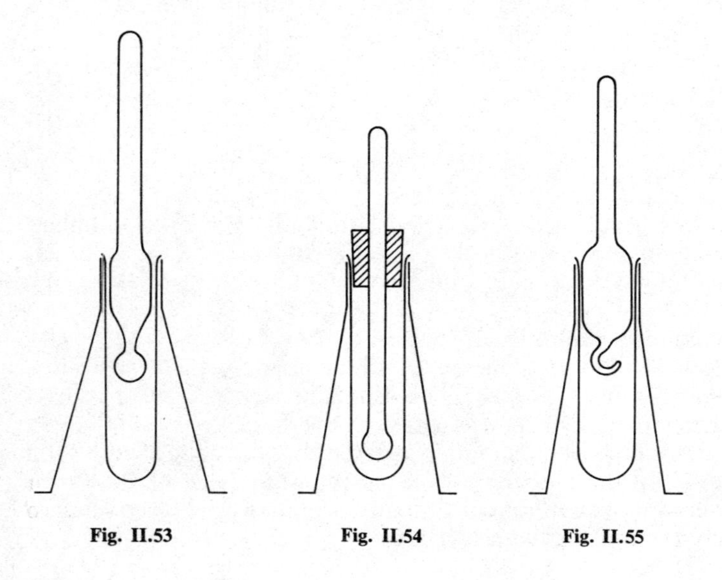

Fig. II.53 **Fig. II.54** **Fig. II.55**

the gas is suspended on the knob of the stopper. The gas is evolved in the tube, if necessary, by gently warming, and is absorbed by the reagent on the knob. Since the apparatus is closed, no gas can escape, and if sufficient time is allowed it is absorbed quantitatively by the reagent. A drop of water may replace the reagent on the stopper; the gas is dissolved, the drop may be washed on to a spot plate or into a micro crucible and treated with the reagent. The apparatus, shown in Fig. II.54, which is closed by a rubber stopper, is sometimes preferable, particularly when minute quantities of gas are concerned; the glass tube, blown into a small bulb at the lower end, may be raised or lowered at will, whilst the change of colour and reaction products may be rendered more easily visible by filling the bulb with gypsum or magnesia powder. In some reactions, e.g. in testing for ammonia, it may be desirable to suspend a small strip of reagent paper from a glass hook fused to the stopper as in Fig. II.55. When a particular gas has to be identified in the presence of other gases, the apparatus shown in Fig. II.56 should be used; here the stopper for the micro test-tube consists of a small glass funnel on top of which the impregnated filter paper is laid in order to absorb the gas. The impregnated filter paper permits the passage of other gases and only retains the gas to be tested by the formation of a non-volatile compound that can be identified by means of a spot test. Another useful apparatus is shown in Fig. II.57; it consists of a micro test-tube into which is

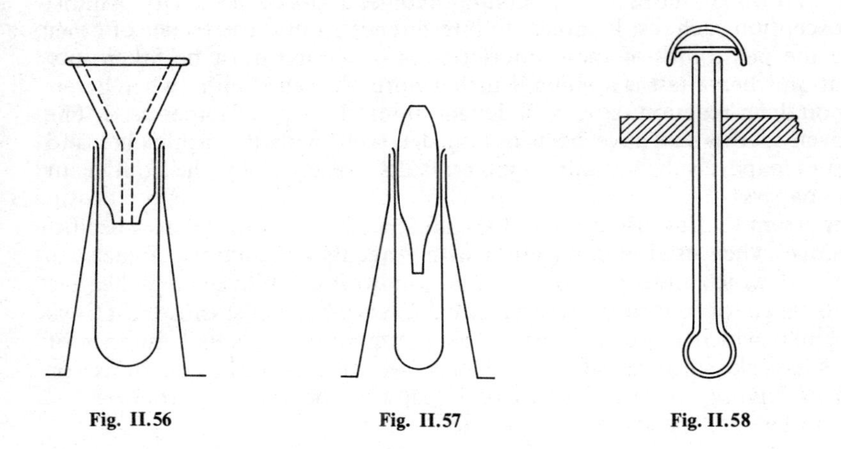

Fig. II.56 **Fig. II.57** **Fig. II.58**

placed a loosely-fitting glass tube narrowed at both ends. The lower capillary end is filled to a height of about 1 mm with a suitable reagent solution; if the gas liberated forms a coloured compound with the reagent, it can easily be seen in the capillary.

Where **high temperatures or glowing** are necessary for the evolution of the gas, a simple hard glass tube supported in a circular hole in an asbestos or 'uralite' plate (Fig. II.58) may be used. The open end of the tube should be covered by a small piece of reagent paper and kept in position by means of a glass cap.

Micro distillation is sometimes required, e.g. in the chromyl chloride test for a chloride (see Section **IV.14**). The apparatus depicted in Fig. II.59 is suitable for the distillation of very small quantities of a mixture. A micro crucible or a micro centrifuge tube may be employed as a receiver.

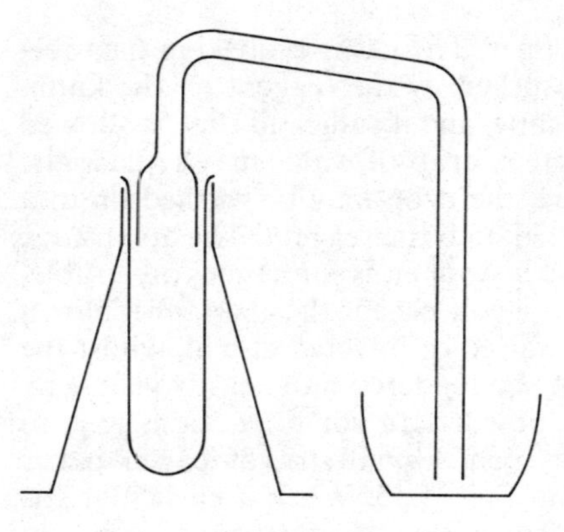

Fig. II.59

In Chapters III, IV and VII the experimental details are given for the detection of a number of ions by spot tests. The sensitivities given are, as a general rule, for a solution containing only the ion in question. It must be remembered that this is the most favourable case, and that in actual practice the presence of other ions usually necessitates a modification of the procedure which is frequently indicated, and which, more often than not, involves a loss of sensitivity. Almost without exception each test is subject to interference from the presence of other ions, and the possibility of these interferences occurring must be taken into consideration when a test is applied. Furthermore, the sensitivities when determined upon drop-reaction paper will depend upon the type of paper used. The figures given in the text have been obtained largely with the Schleicher and Schuell spot paper: substantially similar results are given by the equivalent Whatman papers.

It is important to draw attention to the difference between the terms 'specific' and 'selective' when used in connection with reagents or reactions. Reactions (and reagents), which under the experimental conditions employed are indicative or one substance (or ion) only are designated as specific, whilst those reactions (and reagents) which are characteristic of a comparatively small number of substances are classified as selective. Hence we may describe reactions (or reagents) as having varying degrees of selectivity; however, a reaction (or reagent) can be only specific or not specific.

CHAPTER III REACTIONS OF THE CATIONS

III.1 CLASSIFICATION OF CATIONS (METAL IONS) INTO ANALYTICAL GROUPS For the purpose of systematic qualitative analysis, cations are classified into five groups on the basis of their behaviour against some reagents. By the systematic use of these so-called group reagents we can decide about the presence or absence of groups of cations, and can also separate these groups for further examination. Systematic qualitative analysis by separations will be dealt with extensively in Chapter V, but the reactions of cations will be dealt with here according to the order defined by this group system. Apart from being the traditional way of presenting the material, it makes the study of these reactions easier because ions of analogous behaviour are dealt with within one group.

The group reagents used for the classification of most common cations are hydrochloric acid, hydrogen sulphide, ammonium sulphide, and ammonium carbonate. Classification is based on whether a cation reacts with these reagents by the formation of precipitates or not. It can therefore be said that **classification** of the most common **cations is based on the differences of solubilities of their chlorides, sulphides, and carbonates.**

The five groups of cations and the characteristics of these groups are as follows:

Group I Cations of this group form precipitates with dilute hydrochloric acid. Ions of this group are **lead**, **mercury(I)**, and **silver**.

Group II The cations of this group do not react with hydrochloric acid, but form precipitates with hydrogen sulphide in dilute mineral acid medium. Ions of this group are **mercury(II)**, **copper**, **bismuth**, **cadmium**, **arsenic(III)**, **arsenic(V)**, **antimony(III)**, **antimony(V)**, **tin(II)**, and **tin(III)(IV)**. The first four form the sub-group IIa and the last six the sub-group IIb. While sulphides of cations in Group IIa are insoluble in ammonium polysulphide, those of cations in Group IIb are soluble.

Group III Cations of this group do not react either with dilute hydrochloric acid, or with hydrogen sulphide in dilute mineral acid medium. However they form precipitates with ammonium sulphide in neutral or ammoniacal medium. Cations of this group are **cobalt(II)**, **nickel(II)**, **iron(II)**, **iron(III)**, **chromium(III)**, **aluminium**, **zinc**, and **manganese(II)**.

Group IV Cations of this group do not react with the reagents of Groups I, II, and III. They form precipitates with ammonium carbonate in the presence of ammonium chloride in neutral or slightly acidic medium. Cations of this group are: **calcium**, **strontium**, and **barium**.

Some systems of group classification exclude ammonium chloride besides ammonium carbonate as the group reagent; in such cases magnesium must also be included in this group. Since, however, in the course of systematic analysis considerable amounts of ammonium chloride will be present when the cations of the fourth group are to be precipitated, it is more logical not to include magnesium into Group IV.

Group V Common cations, which do not react with reagents of the previous groups, form the last group of cations, which include **magnesium**, **sodium**, **potassium**, **ammonium**, **lithium**, and **hydrogen ions**.

This group system of cations can be extended to include less common ions as well. Classification of these ions, together with their reactions will be given in Chapter VII.

III.2 NOTES ON THE STUDY OF THE REACTIONS OF IONS When studying the reactions of ions, experimental techniques described in Chapter II should be applied. The reactions can be studied both in macro and semimicro scale, and the majority of the reactions can be applied as a spot test as well. Hints on the preparation of reagents are given in the Appendix of this book. The reagents are listed there in alphabetical order, with notes on their stability. Most reagents are poisonous to some extent, and should therefore be handled with care. Those reagents which are exceptionally poisonous or hazardous must be specially labelled and must be used with utmost care. In the list of reactions these reagents will be marked as (**POISON**) or (**HAZARD**). One should not use these reagents when working alone in a laboratory; the supervisor or a colleague should always be notified before using them.

The concentration of reagents is in most cases chosen to be molar, meaning that it is easy to calculate the relative volumes of the reactant and the reagent needed to complete the reaction. It is not advisable to add the calculated amount of reagent at once to the solution (cf. Chapter II), but the final amount should be equal or more than the equivalent. In some cases it is impossible or impractical to prepare a M reagent; thus 0·5 or even 0·1M reagents have to be used sometimes. It is easy to predict the volume of a particular reagent needed to complete the reaction from the concentrations. Acids and bases are applied mostly in 2M concentrations in order to avoid unnecessary dilution of the mixture.

Taking notes when studying these reactions is absolutely necessary for a student. A logical, clear way of making notes is essential. Although it would be wrong to copy the text of this book, it is important to note (*a*) the reagent, and any special experimental circumstances applied when performing the test, (*b*) the changes observed, and (*c*) the equation of the reaction or some other explanation of what had happened. A useful way of making notes is the following. The left page of an open notebook is divided into two equal sections by a vertical line. The left column can be headed 'TEST' and should contain a brief description of the test itself, including the reagent and experimental circumstances. The second column (still on the left-hand page), headed

'OBSERVATION' should contain the visible change which occurred when carrying out the test. Finally, the entire right-hand page should be reserved for 'EXPLANATION', where the reaction equation can be entered. A typical page of a notebook containing some of the reactions of lead(II) ions is shown on Table III.1. It is also advisable to summarize reactions within one group in the form of tables, as shown on Table III.2 for the first group of cations.

The column 'TEST' should be made up before making the actual experiments. When writing this up, the student will be able to devise his experiments and to make best use of the time available in the laboratory. The 'OBSERVATION' column should be filled when actually making the experiments, while the 'EXPLANATION' page should be made up after leaving the laboratory. Finally, the reaction tables should be made up, when the reactions of the particular group have been studied and explained. This systematic way of study enables us to devote precious laboratory time entirely to experiments, and, by dealing with the one particular reaction altogether four times, helps the student actually to learn the subject as well.

III.3 FIRST GROUP OF CATIONS: LEAD(II), MERCURY(I), AND SILVER(I)

Group reagent: dilute (2M) hydrochloric acid.

Group reaction: white precipitate of lead chloride $PbCl_2$, mercury(I) chloride Hg_2Cl_2, and silver chloride AgCl.

Cations of the first group form insoluble chlorides. Lead chloride, however, is slightly soluble in water and therefore lead is never completely precipitated when adding dilute hydrochloric acid to a sample; the rest of the lead ions are quantitatively precipitated with hydrogen sulphide in acidic medium together with the cations of the second group.

Nitrates of these cations are very soluble. Among sulphates lead sulphate is practically insoluble, while silver sulphate dissolves to a much greater extent. The solubility of mercury(I) sulphate lies in between. Bromides and iodides are

Table III.1 One page of a laboratory notebook

Test Group 1 Pb^{2+}	Observation	Explanation
1. HCl	white ppt.	$Pb^{2+}+2Cl^- \rightarrow PbCl_2\downarrow$
$+NH_3$	no change	no ammine complexes (but $Pb(OH)_2\downarrow$)
+hot water	dissolves	33·4 g $PbCl_2$ dissolves per litre at 100°C
2. $H_2S(+HCl)$	black ppt.	$Pb^{2+}H_2S \rightarrow PbS\downarrow+2H^+$
+conc. HNO_3	white ppt.	$3PbS\downarrow+8HNO_3 \rightarrow 3Pb^{2+}+2NO\uparrow+4H_2O+3S\downarrow$
+boiling	white ppt. (different)	$S\downarrow+2HNO_3 \rightarrow SO_4^{2-}+2H^++2NO\uparrow$ $Pb^{2+}+SO_4^{2-} \rightarrow PbSO_4\downarrow$
3. NH_3	white ppt.	$Pb^{2+}+2NH_3+2H_2O \rightarrow Pb(OH)_2\downarrow+2NH_4^+$
+excess	no change	Pb^{2+} does not form ammine complexes
4. NaOH	white ppt.	$Pb^{2+}+2OH^- \rightarrow Pb(OH)_2\downarrow$
+excess	dissolves	$Pb(OH)_2\downarrow+2OH^- \rightleftarrows [Pb(OH)_4]^{2-}$ $Pb(OH)_2$: amphoteric
5. KI	yellow ppt.	$Pb^{2+}+2I^- \rightarrow PbI_2\downarrow$
+excess	no change	no iodo complexes

Table III.2 Tabulating reactions of Group I cations

	Pb^{2+}	Hg_2^{2+}	Ag^+
HCl	white $PbCl_2\downarrow$	white $Hg_2Cl_2\downarrow$	white $AgCl\downarrow$
+NH	no change	black $Hg\downarrow + HgNH_2Cl\downarrow$	dissolves $[Ag(NH_3)_2]^+$
+hot water	dissolves	no change	no change
$H_2S(+HCl)$	black $PbS\downarrow$	black $Hg\downarrow + HgS\downarrow$	black $Ag_2S\downarrow$
+ccHNO_3, boiling	white $PbSO_4\downarrow$	white $Hg_2(NO_3)_2S\downarrow$	dissolves Ag^+
NH_3, small amounts	white $Pb(OH)_2\downarrow$	black $Hg\downarrow + HgO{,}HgNH_2NO_3\downarrow$	brown $Ag_2O\downarrow$
+excess	no change	no change	dissolves $[Ag(NH_3)_2]^+$
NaOH, small amounts	white $Pb(OH)_2\downarrow$	black, $Hg_2O\downarrow$	brown, $Ag_2O\downarrow$
+excess	dissolves $[Pb(OH)_4]^{2-}$	no change	no change
KI, small amounts	yellow $PbI_2\downarrow$	green $Hg_2I_2\downarrow$	yellow $AgI\downarrow$
+excess	no change	grey $Hg\downarrow + [HgI_4]^{2-}$	no change
K_2CrO_4	yellow $PbCrO_4\downarrow$	red $Hg_2CrO_4\downarrow$	red $Ag_2CrO_4\downarrow$
$+NH_3$	no change	black $Hg\downarrow + HgNH_2NO_3\downarrow$	dissolves $[Ag(NH_3)_2]^+$
KCN, small amounts	white $Pb(CN)_2\downarrow$	black $Hg\downarrow + Hg(CN)_2$	white $AgCN\downarrow$
+excess	no change	no change	dissolves $[Ag(CN)_2]^-$
Na_2CO_3	white $PbO.PbCO_3\downarrow$	yellowish-white $Hg_2CO_3\downarrow$	yellowish-white $Ag_2CO_3\downarrow$
+boiling	no change	black $Hg\downarrow + HgO\downarrow$	brown $Ag_2O\downarrow$
Na_2HPO_4	white $Pb_3(PO_4)_2\downarrow$	white $Hg_2HPO_4\downarrow$	yellow $Ag_3PO_4\downarrow$
Specific reaction	Benzidine ($+Br_2$) blue colour	Diphenyl carbazide violet colour	*p*-dimethylamino-benzylidene-rhodanine ($+HNO_3$) violet colour

also insoluble, though precipitation of lead halides is incomplete, and the precipitates dissolve quite easily in hot water. Sulphides are insoluble. Acetates are more soluble, though silver acetate might be precipitated from more concentrated solutions. Hydroxides and carbonates are precipitated with an equivalent amount of the reagent, an excess however might act in various ways. There are differences in their behaviour towards ammonia as well.

III.4 LEAD, Pb (A_r: 207·19) Lead is a bluish-grey metal with a high density (11·48 g ml^{-1} at room temperature). It readily dissolves in medium concentrated nitric acid (8M), and nitrogen oxide is formed also:

$$3Pb + 8HNO_3 \rightarrow 3Pb^{2+} + 6NO_3^- + 2NO\uparrow + 4H_2O$$

The colourless nitrogen oxide gas, when mixed with air, is oxidized to red nitrogen dioxide:

$$2NO\uparrow(\text{colourless}) + O_2\uparrow \rightarrow 2NO_2\uparrow(\text{red})$$

With concentrated nitric acid a protective film of lead nitrate is formed on the surface of the metal and prevents further dissolution. Dilute hydrochloric or sulphuric acid have little effect owing to the formation of insoluble lead chloride or sulphate on the surface.

Reactions of lead(II) ions A solution of lead nitrate (0·25M) or lead acetate (0·25M) can be used for the study of these reactions.

1. Dilute hydrochloric acid (or soluble chlorides): a white precipitate in cold and not too dilute solution:

$$Pb^{2+} + 2Cl^- \rightleftharpoons PbCl_2\downarrow$$

The precipitate is soluble in hot water (33·4 g ℓ$^{-1}$ at 100°C while only 9·9 g ℓ$^{-1}$ at 20°C), but separates again in long, needle-like crystals when cooling. It is also soluble in concentrated hydrochloric acid or concentrated potassium chloride when the tetrachloroplumbate(II) ion is formed:

$$PbCl_2\downarrow + 2Cl^- \rightarrow [PbCl_4]^{2-}$$

If the precipitate is washed by decantation and dilute ammonia is added, no visible change occurs [difference from mercury(I) or silver ions], though a precipitate-exchange reaction takes place and lead hydroxide is formed:

$$PbCl_2\downarrow + 2NH_3 + 2H_2O \rightarrow Pb(OH)_2\downarrow + 2NH_4^+ + 2Cl^-$$

2. Hydrogen sulphide in neutral or dilute acid medium: black precipitate of lead sulphide:

$$Pb2^+ + H_2S \rightarrow PbS\downarrow + 2H^+$$

Precipitation is incomplete if strong mineral acids are present in more than 2M concentration. Because hydrogen ions are formed in the above reaction, it is advisable to buffer the mixture with sodium acetate.

Introducing hydrogen sulphide gas into a mixture which contains white lead chloride precipitate, the latter is converted into (black) lead sulphide in a precipitate-exchange reaction:

$$PbCl_2\downarrow + H_2S \rightarrow PbS\downarrow + 2H^+ + 2Cl^-$$

If the test is carried out in the presence of larger amounts of chloride [potassium chloride (saturated)], initially a red precipitate of lead sulpho-chloride is formed when introducing hydrogen sulphide gas:

$$2Pb^{2+} + H_2S + 2Cl^- \rightarrow Pb_2SCl_2\downarrow + 2H^+$$

This however decomposes on dilution (a) or on further addition of hydrogen sulphide (b) and black lead sulphide precipitate is formed:

$$Pb_2SCl_2\downarrow \rightarrow PbS\downarrow + PbCl_2\downarrow \quad (a)$$

$$Pb_2SCl_2\downarrow + H_2S \rightarrow 2PbS\downarrow + 2Cl^- \quad (b)$$

Lead sulphide precipitate decomposes when concentrated nitric acid is added, and white, finely divided elementary sulphur is precipitated:

$$3PbS\downarrow + 8HNO_3 \rightarrow 3Pb^{2+} + 6NO_3^- + 3S\downarrow + 2NO\uparrow + 4H_2O$$

If the mixture is boiled, sulphur is oxidized by nitric acid to sulphate (a), which forms immediately white lead sulphate precipitate (b) with the lead ions in the solution:

$$S\downarrow + 2HNO_3 \rightarrow SO_4^{2-} + 2H^+ + 2NO\uparrow \quad (a)$$

$$Pb^{2+} + SO_4^{2-} \rightarrow PbSO_4\downarrow \quad (b)$$

Boiling lead sulphide with hydrogen peroxide (3%), the black precipitate turns white owing to the formation of lead sulphate:

$$PbS\downarrow + 4H_2O_2 \rightarrow PbSO_4\downarrow + 4H_2O$$

The great insolubility of lead sulphide in water ($4{\cdot}9 \times 10^{-11}$ g ℓ^{-1}) explains why hydrogen sulphide is such a sensitive reagent for the detection of lead, and why it can be detected in the filtrate from the separation of the sparingly soluble lead chloride in dilute hydrochloric acid.

Note: Hydrogen sulphide is a highly poisonous gas, and all operations with the gas must be conducted in the fume chamber. Every precaution must be observed to prevent the escape of hydrogen sulphide into the air of the laboratory.

3. *Ammonia solution*: white precipitate of lead hydroxide

$$Pb^{2+} + 2NH_3 + 2H_2O \rightarrow Pb(OH)_2\downarrow + 2NH_4^+$$

The precipitate is insoluble in excess reagent.

4. *Sodium hydroxide*: white precipitate of lead hydroxide

$$Pb^{2+} + 2OH^- \rightarrow Pb(OH)_2\downarrow$$

The precipitate dissolves in excess reagent, when tetrahydroxoplumbate(II) ions are formed:

$$Pb(OH)_2\downarrow + 2OH^- \rightarrow [Pb(OH)_4]^{2-}$$

Thus, lead hydroxide has an amphoteric character.

Hydrogen peroxide (a) or ammonium peroxodisulphate (b), when added to a solution of tetrahydroxoplumbate(II) forms a black precipitate of lead dioxide by oxidizing bivalent lead to the tetravelent state:

$$[Pb(OH)_4]^{2-} + H_2O_2 \rightarrow PbO_2\downarrow + 2H_2O + 2OH^- \qquad \text{(a)}$$

$$[Pb(OH)_4]^{2-} + S_2O_8^{2-} \rightarrow PbO_2\downarrow + 2H_2O + 2SO_4^{2-} \qquad \text{(b)}$$

5. *Dilute sulphuric acid (or soluble sulphates)*: white precipitate of lead sulphate:

$$Pb^{2+} + SO_4^{2-} \rightarrow PbSO_4\downarrow$$

The precipitate is insoluble in excess reagent. Hot, concentrated sulphuric acid dissolves the precipitate owing to formation of lead hydrogen sulphate:

$$PbSO_4\downarrow + H_2SO_4 \rightarrow Pb^{2+} + 2HSO_4^-$$

Solubility is much lower in the presence of ethanol.

Lead sulphate precipitate is soluble in more concentrated solutions of ammonium acetate (10M) (a) or ammonium tartrate (6M) (b) in the presence of ammonia, when tetraacetateplumbate(II) and ditartratoplumbate(II) ions are formed:

$$PbSO_4\downarrow + 4CH_3COO^- \rightarrow [Pb(CH_3COO)_4]^{2-} + SO_4^{2-} \qquad \text{(a)}$$

$$PbSO_4\downarrow + 2C_4H_4O_6^{2-} \rightarrow [Pb(C_4H_4O_6)_2]^{2-} + SO_4^{2-} \qquad \text{(b)}$$

The stabilities of these complexes are not very great; chromate ions, for example, can precipitate lead chromate from their solution.

When boiled with sodium carbonate the lead sulphate is transformed into lead carbonate in a precipitate-exchange reaction:

$$PbSO_4\downarrow + CO_3^{2-} \rightarrow PbCO_3\downarrow + SO_4^{2-}$$

By washing the precipitate by decantation with hot water, sulphate ions can be removed and the precipitate will dissolve in dilute nitric acid

$$PbCO_3\downarrow + 2H^+ \rightarrow Pb^{2+} + H_2O + CO_2\uparrow$$

6. *Potassium chromate in neutral, acetic acid or ammonia solution:* yellow precipitate of lead chromate

$$Pb^{2+} + CrO_4^{2-} \rightarrow PbCrO_4\downarrow$$

Nitric acid (a) or sodium hydroxide (b) dissolve the precipitate:

$$2PbCrO_4\downarrow + 2H^+ \rightleftarrows 2Pb^{2+} + Cr_2O_7^{2-} + 2H_2O \quad \text{(a)}$$

$$PbCrO_4\downarrow + 4OH^- \rightleftarrows [Pb(OH)_4]^{2-} + CrO_4^{2-} \quad \text{(b)}$$

Both reactions are reversible; by buffering the solution with ammonia or acetic acid respectively, lead chromate precipitates again.

7. *Potassium iodide:* yellow precipitate of lead iodide

$$Pb^{2+} + 2I^- \rightarrow PbI_2\downarrow$$

The precipitate is moderately soluble in boiling water to yield a colourless solution, from which it separates on cooling in golden yellow plates.

An excess of a more concentrated (6M) solution of the reagent dissolves the precipitate and tetraiodoplumbate(II) ions are formed:

$$PbI_2\downarrow + 2I^- \rightleftarrows [PbI_4]^{2-}$$

The reaction is reversible; on diluting with water the precipitate reappears.

8. *Sodium sulphite in neutral solution:* white precipitate of lead sulphite

$$Pb^{2+} + SO_3^{2-} \rightarrow PbSO_3\downarrow$$

The precipitate is less soluble than lead sulphate, though it can be dissolved by both dilute nitric acid (a) and sodium hydroxide (b).

$$PbSO_3\downarrow + 2H^+ \rightarrow Pb^{2+} + H_2O + SO_2\uparrow \quad \text{(a)}$$

$$PbSO_3\downarrow + 4OH^- \rightarrow [Pb(OH)_4]^{2-} + SO_3^{2-} \quad \text{(b)}$$

9. *Sodium carbonate:* white precipitate of a mixture of lead carbonate and lead hydroxide

$$2Pb^{2+} + 2CO_3^{2-} + H_2O \rightarrow Pb(OH)_2\downarrow + PbCO_3\downarrow + CO_2\uparrow$$

On boiling no visible change takes place [difference from mercury(I) and silver(I) ions]. The precipitate dissolves in dilute nitric acid and even in acetic acid and CO_2 gas is liberated:

$$Pb(OH)_2\downarrow + PbCO_3\downarrow + 4H^+ \rightarrow 2Pb^{2+} + 3H_2O + CO_2\uparrow$$

10. *Disodium hydrogen phosphate:* white precipitate of lead phosphate

$$3Pb^{2+} + 2HPO_4^{2-} \rightleftarrows Pb_3(PO_4)_2\downarrow + 2H^+$$

The reaction is reversible; strong acids (nitric acid) dissolve the precipitate. The precipitate is also soluble in sodium hydroxide.

11. Potassium cyanide (**POISON**): white precipitate of lead cyanide

$$Pb^{2+} + 2CN^- \rightarrow Pb(CN)_2\downarrow$$

which is insoluble in the excess of the reagent. This reaction can be used to distinguish lead(II) ions from mercury(I) and silver(I), which react in different ways.

12. Tetramethyldiamino-diphenylmethane (or 'tetrabase') (0·5 %)

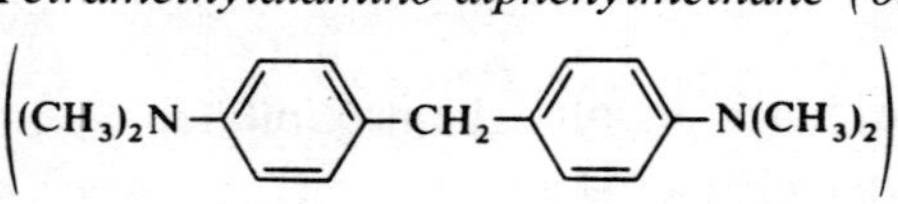

$(CH_3)_2N$ — CH_2 — $N(CH_3)_2$:

a blue oxidation product {hydrol: —CH_2 → —CH(OH)} is formed under the conditions given below.

Place 1 ml test solution in a 5 ml centrifuge tube, add 1 ml 2M potassium hydroxide and 0·5–1 ml 3 per cent hydrogen peroxide solution. Allow to stand for 5 minutes. Separate the precipitate by centrifugation, and wash once with cold water. Add 2 ml reagent, shake and centrifuge. The supernatant liquid is coloured blue.

The ions of bismuth, cerium, manganese, thallium, cobalt, and nickel give a similar reaction: iron and large quantities of copper interfere.

Concentration limit: 1 in 10,000.

13. Benzidine (0·05 %) (DANGER: THE REAGENT IS CARCINOGENIC)

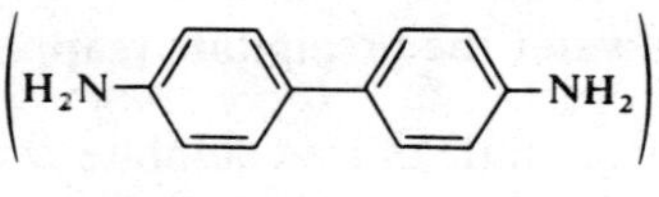

H_2N — NH_2 :

the so-called 'benzidine blue' is produced upon oxidation with lead dioxide. The ions of bismuth, cerium, manganese, cobalt, nickel, silver, and thallium give a similar reaction, but by performing the test in an alkaline extract (i.e. with a tetrahydroxoplumbate(II) solution), only thallium interferes. Oxidation is conveniently carried out by sodium hypobromite; the excess of the latter is destroyed by ammonia $[2NH_3 + 3OBr^- \rightarrow N_2\uparrow + 3Br^- + 3H_2O]$.

Place a drop of the test solution upon drop-reaction paper, and treat successively with 2 drops 3M sodium hydroxide and 1 drop saturated bromine water. Add 2 drops 1:1 ammonia solution; remove the excess ammonia by waving the paper over a small flame. Add 2 drops reagent: a blue colour develops.

Sensitivity: 1 μg Pb. *Concentration limit:* 1 in 50,000.

14. Gallocyanine (1 %)

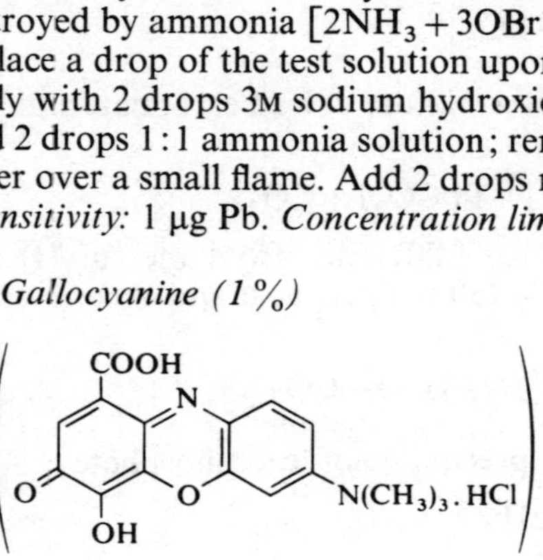

COOH, N, O, O, OH, $N(CH_3)_3 . HCl$:

deep-violet precipitate of unknown composition. The test is applicable to finely divided lead sulphate precipitated on filter paper.

Place a drop of the test solution upon drop-reaction paper, followed by a drop each of 1 per cent aqueous pyridine and the gallocyanine reagent (blue). Remove the excess of the reagent by placing several filter papers beneath the drop-reaction paper and adding drops of the pyridine solution to the spot until the wash liquid percolating through is colourless; move the filter papers to a fresh position after each addition of pyridine. A deep violet spot is produced.

Sensitivity: 1–6 μg Pb. *Concentration limit:* 1 in 50,000.

In the presence of silver, bismuth, cadmium, or copper, proceed as follows. Transfer a drop of the test solution to a drop-reaction paper and add a drop of M sulphuric acid to fix the lead as lead sulphate. Remove the soluble sulphates of the other metals by washing with about 3 drops of M sulphuric acid, followed by a little 96 per cent ethanol. Dry the paper on a water bath, and then apply the test as detailed above.

15. Diphenylthiocarbazone (or Dithizone) (0·005 %)

$$\left(SC \begin{matrix} \diagup NH.NHC_6H_5 \\ \diagdown N{=}NC_6H_5 \end{matrix} \right):$$

brick-red complex salt in neutral, ammoniacal, alkaline, or alkalicyanide solution.

Place 1 ml of the neutral or faintly alkaline solution in a micro test-tube, introduce a few small crystals of potassium cyanide, and then 2 drops of the reagent. Shake for 30 seconds. The green colour of the reagent changes to red.

Sensitivity: 0·1 μg Pb (in neutral solution). *Concentration limit:* 1 in 1,250,000.

Heavy metals (silver, mercury, copper, cadmium, antimony, nickel, and zinc, etc.) interfere, but this effect may be eliminated by conducting the reaction in the presence of much alkali cyanide: excess of alkali hydroxide is also required for zinc. The reaction is extremely sensitive, but it is not very selective.

The reagent is prepared by dissolving 2–5 mg dithizone in 100 ml carbon tetrachloride or chloroform. It does not keep well.

16. Dry tests **a.** Blowpipe test. When a lead salt is heated with alkali carbonate upon charcoal, a malleable bead of lead, (which is soft and will mark paper), surrounded with a yellow incrustation of lead monoxide is obtained.

b. Flame test. Pale blue (inconclusive).

III.5 MERCURY, Hg (A_r: 200·59) – MERCURY(I) Mercury is a silver-white, liquid metal at ordinary temperatures and has a density of 13·534 g ml^{-1} at 25°C. It is unaffected when treated with hydrochloric or dilute sulphuric acid (2M), but reacts readily with nitric acid. Cold, medium concentrated (8M) nitric acid with an excess of mercury yields mercury(I) ions:

$$6Hg + 8HNO_3 \rightarrow 3Hg_2^{2+} + 2NO\uparrow + 6NO_3^- + 4H_2O$$

with an excess of hot concentrated nitric acid mercury(II) ions are formed:

$$3Hg+8HNO_3 \rightarrow 3Hg^{2+}+2NO\uparrow+6NO_3^-+4H_2O$$

Hot, concentrated sulphuric acid dissolves mercury as well. The product is mercury(I) ion if mercury is in excess

$$2Hg+2H_2SO_4 \rightarrow Hg_2^{2+}+SO_4^{2-}+SO_2\uparrow+2H_2O$$

while if the acid is in excess, mercury(II) ions are formed:

$$Hg+2H_2SO_4 \rightarrow Hg^{2+}+SO_4^{2-}+SO_2\uparrow+2H_2O$$

The two ions, mercury(I) and mercury(II) behave quite differently against reagents used in qualitative analysis, and hence belong to two different analytical groups. Mercury(I) ions belong to the first group of cations, their reactions will therefore be treated here. Mercury(II) ions, on the other hand, are in the second cation group; their reactions will therefore be dealt with later, together with the other members of that group.

Reactions of mercury(I) ions A solution of mercury(I) nitrate (0·05M) can be used for the study of these reactions.

1. *Dilute hydrochloric acid or soluble chlorides:* white precipitate of mercury(I) chloride (calomel)

$$Hg_2^{2+}+2Cl^- \rightarrow Hg_2Cl_2\downarrow$$

The precipitate is insoluble in dilute acids.

Ammonia solution converts the precipitate into a mixture of mercury(II) amidochloride and mercury metal, both insoluble precipitates:

$$Hg_2Cl_2+2NH_3 \rightarrow Hg\downarrow+Hg(NH_2)Cl\downarrow+NH_4^++Cl^-$$

the reaction involves disproportionation, mercury(I) is converted partly to mercury(II) and partly to mercury metal. This reaction can be used to differentiate mercury(I) ions from lead(II) and silver(I).

The mercury(II) amidochloride is a white precipitate, but the finely divided mercury makes it shiny black. The name calomel, coming from Greek (*καον μελασ* = nice black) refers to this characteristic of the originally white mercury(I) chloride precipitate.

Mercury(I) chloride dissolves in aqua regia, forming undissociated but soluble mercury(II) chloride:

$$3Hg_2Cl_2\downarrow+2HNO_3+6HCl \rightarrow 3HgCl_2+2NO\uparrow+4H_2O$$

2. *Hydrogen sulphide in neutral or dilute acid medium:* black precipitate, which is a mixture of mercury(II) sulphide and mercury metal

$$Hg_2^{2+}+H_2S \rightarrow Hg\downarrow+HgS\downarrow+2H^+$$

Owing to the extremely low solubility product of mercury(II) sulphide the reaction is very sensitive.

Sodium sulphide (colourless), dissolves the mercury(II) sulphide (but leaves mercury metal) and a disulphomercurate(II) complex is formed:

$$HgS+S^{2-} \rightarrow [HgS_2]^{2-}$$

After removing the mercury metal by filtration, black mercury(II) sulphide can again be precipitated by acidification with dilute mineral acids:

$$[HgS_2]^{2-} + 2H^+ \rightarrow HgS\downarrow + H_2S\uparrow$$

Sodium disulphide (yellow) dissolves both mercury and mercury(II) sulphide:

$$HgS\downarrow + Hg\downarrow + 3S_2^{2-} \rightarrow 2[HgS_2]^{2-} + S_3^{2-}$$

This rather complicated reaction can be understood more easily by breaking it down into the following steps:

First mercury is oxidized by the disulphide, yielding mercury(II) sulphide and (mono)sulphide ions:

$$Hg\downarrow + S_2^{2-} \rightarrow HgS\downarrow + S^{2-} \qquad \text{(a)}$$

Mercury(II) sulphide then dissolves in the (mono) sulphide formed in the previous reaction

$$HgS\downarrow + S^{2-} \rightarrow [HgS_2]^{2-} \qquad \text{(b)}$$

Mercury(II) sulphide, which was originally present in the precipitate, reacts with disulphide ions yielding disulphomercurate(II) and trisulphide ions:

$$HgS + 2S_2^{2-} \rightarrow HgS_2^{2-} + S_3^{2-} \qquad \text{(c)}$$

Combining the reactions (a), (b) and (c) together we obtain the reaction described above.

Aqua regia dissolves the precipitate, yielding undissociated mercury(II) chloride and sulphur:

$$12HCl + 4HNO_3 + 3Hg\downarrow + 3HgS\downarrow = 6HgCl_2 + 3S\downarrow + 4NO\uparrow + 8H_2O$$

This reaction can be understood as the sum of the following steps:

When making up aqua regia chlorine atoms are formed:

$$3HCl + HNO_3 \rightarrow 3Cl + NO\uparrow + 2H_2O \qquad \text{(a)}$$

These react partly with mercury, forming mercury(II) chloride:

$$Hg\downarrow + 2Cl \rightarrow HgCl_2 \qquad \text{(b)}$$

Another part of chlorine reacts with mercury(II) sulphide

$$HgS\downarrow + 2Cl \rightarrow HgCl_2 + S\downarrow \qquad \text{(c)}$$

Combination of 4(a) + 3(b) + 3(c) yields the equation

$$12HCl + 4HNO_3 + 3Hg\downarrow + 3HgS\downarrow = 6HgCl_2 + 3S\downarrow + 4NO\uparrow + 8H_2O$$

When heated with aqua regia, sulphur is oxidized to sulphuric acid and the solution becomes clear:

$$S\downarrow + 6HCl + 2HNO_3 \rightarrow S_4^{2-} + 6Cl^- + 8H^+ + 2NO\uparrow$$

3. *Ammonia solution:* black precipitate which is a mixture of mercury metal and basic mercury(II) amidonitrate, (which itself is a white precipitate)

$$2Hg_2^{2+} + NO_3^- + 4NH_3 + H_2O \rightarrow HgO.Hg\begin{matrix} \diagup NH_2 \\ \diagdown NO_3 \end{matrix}\downarrow + 2Hg\downarrow + 3NH_4^+$$

This reaction can be used to differentiate between mercury(I) and mercury(II) ions.

4. *Sodium hydroxide:* black precipitate of mercury(I) oxide

$$Hg_2^{2+} + 2OH^- \rightarrow Hg_2O\downarrow + H_2O$$

The precipitate is insoluble in excess reagent, but dissolves readily in dilute nitric acid.

When boiling, the colour of the precipitate turns to grey, owing to disproportionation, when mercury(II) oxide and mercury metal are formed:

$$Hg_2O\downarrow \rightarrow HgO\downarrow + Hg\downarrow$$

5. *Potassium chromate in hot solution:* red crystalline precipitate of mercury(I) chromate

$$Hg_2^{2+} + CrO_4^{2-} \rightarrow Hg_2CrO_4\downarrow$$

If the test is carried out in cold, a brown amorphous precipitate is formed with an undefined composition. When heating the precipitate turns to red, crystalline mercury(I) chromate.

Sodium hydroxide turns the precipitate into black mercury(I) oxide:

$$Hg_2CrO_4\downarrow + 2OH^- \rightarrow Hg_2O\downarrow + CrO_4^{2-} + H_2O$$

6. *Potassium iodide, added slowly in cold solution:* green precipitate of mercury(I) iodide

$$Hg_2^{2+} + 2I^- \rightarrow Hg_2I_2\downarrow$$

If excess reagent is added a disproportionation reaction takes place, soluble tetraiodomercurate(II) ions and a black precipitate of finely divided mercury being formed:

$$Hg_2I_2\downarrow + 2I^- \rightarrow [HgI_4]^{2-} + Hg\downarrow$$

When boiling the mercury(I) iodide precipitate with water, disproportionation again takes place, and a mixture of red mercury(II) iodide precipitate and finely distributed black mercury is formed:

$$Hg_2I_2\downarrow \rightarrow HgI_2\downarrow + Hg\downarrow$$

7. *Sodium carbonate in cold solution:* yellow precipitate of mercury(I) carbonate:

$$Hg_2^{2+} + CO_3^{2-} \rightarrow Hg_2CO_3\downarrow$$

The precipitate turns slowly to blackish grey, when mercury(II) oxide and mercury are formed:

$$Hg_2CO_3\downarrow \rightarrow HgO\downarrow + Hg\downarrow + CO_2\uparrow$$

The decomposition can be speeded up by heating the mixture.

8. *Disodium hydrogen phosphate:* white precipitate of mercury(I) hydrogen phosphate:

$$Hg_2^{2+} + HPO_4^{2-} \rightarrow Hg_2HPO_4\downarrow$$

9. Potassium cyanide (**POISON**) produces mercury(II) cyanide solution and mercury precipitate:

$$Hg_2^{2+} + 2CN^- \rightarrow Hg\downarrow + Hg(CN)_2$$

Mercury(II) cyanide, though soluble, is practically undissociated.

10. Tin(II) chloride reduces mercury(I) ions to mercury metal, which appears in the form of a greyish-black precipitate:

$$Hg_2^{2+} + Sn^{2+} \rightarrow 2Hg\downarrow + Sn^{4+}$$

Mercury(II) ions react in a similar way.

11. Potassium nitrite reduces mercury metal from a solution of mercury(I) ions in cold, in the form of a greyish-black precipitate:

$$Hg_2^{2+} + NO_2^- + H_2O \rightarrow 2Hg\downarrow + NO_3^- + 2H^+$$

Under similar circumstances mercury(II) ions do not react. The spot test technique is as follows. Place a drop of the faintly acid test solution upon drop-reaction paper and add a drop of 50 per cent potassium nitrite solution. A black (or dark grey) spot is produced. The test is highly selective. Coloured ions yield a brown colouration which may be washed away, leaving the black spot.

12. Glossy copper sheet or copper coin If a drop of mercury(I) nitrate is placed on a glossy copper surface, a deposit of mercury metal is formed:

$$Cu + Hg_2^{2+} \rightarrow Cu^{2+} + 2Hg\downarrow$$

Rinsing, drying, and rubbing the surface with a dry cloth, a glittery, silverish spot is obtained. Heating the spot in a Bunsen-flame, mercury evaporates and the red copper surface becomes visible again. Mercury(II) solutions react in a similar way.

13. Aluminium sheet If a drop of mercury(I) nitrate is placed on a clean aluminium surface, aluminium amalgam is formed and aluminium ions pass into solution:

$$3Hg_2^{2+} + 2Al \rightarrow 2Al^{3+} + 6Hg\downarrow$$

The aluminium which is dissolved in the amalgam is oxidized rapidly by the oxygen of the air and a voluminous precipitate of aluminium hydroxide is formed. The remaining mercury amalgamates a further batch of aluminium, which again is oxidized, thus considerable amounts of aluminium get corroded.

14 Diphenylcarbazide (1% in alcohol)

$$\left(\begin{array}{l} \quad NH-NH \\ \; / \\ C=O \\ \; \backslash \\ \quad NH-NH \end{array} \right):$$

forms a violet-coloured compound with mercury(I) or mercury(II) ions, the composition of which is not quite understood. In the presence of 0·2M nitric acid the test is selective for mercury. Under these circumstances the sensitivity is 1 μg Hg_2^{2+} or Hg^{2+} with a concentration limit of 1 in 5×10^4.

Drop test: Impregnate a piece of filter paper with the freshly prepared reagent. Add 1 drop 0·4M nitric acid, and on top of the latter one drop of the test solution. In the presence of mercury a violet colour is observable. The test is most sensitive if the filter paper is left to dry at room temperature.

15. Dry test. All compounds of mercury when heated with a large excess (7–8 times the bulk) of anhydrous sodium carbonate in a small dry test-tube yield a grey mirror, consisting of fine drops of mercury, in the upper part of the tube. The globules coalesce when they are rubbed with a glass rod.

Note: Mercury vapour is extremely poisonous, and not more than 0·1 gram of the substance should be used in the test.

III.6 SILVER, Ag (A_r: 107·868) Silver is a white, malleable, and ductile metal. It has a high density (10·5 g ml^{-1}) and melts at 960·5°C. It is insoluble in hydrochloric, dilute sulphuric (M) or dilute nitric (2M) acid. In more concentrated nitric acid (8M) (a) or in hot, concentrated sulphuric acid (b) it dissolves:

$$6Ag + 8HNO_3 \rightarrow 6Ag^+ + 2NO\uparrow + 6NO_3^- + 4H_2O \quad \text{(a)}$$

$$2Ag + 2H_2SO_4 \rightarrow 2Ag^+ + SO_4^{2-} + SO_2\uparrow + 2H_2O \quad \text{(b)}$$

Silver forms monovalent ion in solution, which is colourless. Silver(II) compounds are unstable, but play an important role in silver-catalysed oxidation-reduction processes. Silver nitrate is readily soluble in water, silver acetate, nitrite and sulphate are less soluble, while all the other silver compounds are practically insoluble. Silver complexes are however soluble. Silver halides are sensitive to light; these characteristics are widely utilized in photography.

Reactions of silver(I) ions A solution of silver nitrate (0·1M) can be used to study these reactions.

1. Dilute hydrochloric acid (or soluble chlorides): white precipitate of silver chloride

$$Ag^+ + Cl^- \rightarrow AgCl\downarrow$$

With concentrated hydrochloric acid precipitation does not occur. Decanting the liquid from over the precipitate, it dissolves in concentrated hydrochloric acid, when a dichloroargentate complex is formed:

$$AgCl\downarrow + Cl^- \rightleftarrows [AgCl_2]^-$$

By diluting with water, the equilibrium shifts back to the left and the precipitate reappears.

Dilute ammonia solution dissolves the precipitate, when diammineargentate complex ion is formed:

$$AgCl\downarrow + 2NH_3 \rightarrow [Ag(NH_3)_2]^+ + Cl^-$$

Dilute nitric acid or hydrochloric acid neutralizes the excess ammonia, and the precipitate reappears because the equilibrium is shifted back towards the left.

Potassium cyanide (**POISON**) dissolves the precipitate with formation of the dicyanoargentate complex:

$$AgCl\downarrow + 2CN^- \rightarrow [Ag(CN)_2]^- + Cl^-$$

The safest way to study this reaction is as follows: decant the liquid from the precipitate, and wash it 2–3 times with water by decantation. Then apply the reagent.

Sodium thiosulphate dissolves the precipitate with the formation of dithiosulphatoargentate complex:

$$AgCl\downarrow + 2S_2O_3^{2-} \rightarrow [Ag(S_2O_3)_2]^{3-} + Cl^-$$

This reaction takes place when fixing photographic negatives or positive prints after development.

Sunlight or ultraviolet irradiation decomposes the silver chloride precipitate, which turns to greyish or black owing to the formation of silver metal:

$$2AgCl\downarrow \xrightarrow{(h\nu)} 2Ag\downarrow + Cl_2\uparrow$$

The reaction is slow and the actual reaction mechanism is very complicated. Other silver halides show similar behaviour. Photography is based on these reactions. In the camera these processes are only initiated; the photographic material has to be 'developed' to complete the reaction. Greyish or black silver particles appear on places irradiated by light; a 'negative' image of the object is therefore obtained. The excess of silver halide has to be removed (to make the developed negative insensitive to light) by fixation.

2. *Hydrogen sulphide* (gas or saturated aqueous solution) in neutral or acidic medium: black precipitate of silver sulphide

$$2Ag^+ + H_2S \rightarrow Ag_2S\downarrow + 2H^+$$

Hot concentrated nitric acid decomposes the silver sulphide, and sulphur remains in the form of a white precipitate:

$$3Ag_2S\downarrow + 8HNO_3 \rightarrow S\downarrow + 2NO\uparrow + 6Ag^+ + 6NO_3^- + 4H_2O$$

The reaction can be understood better if written in two steps:

$$3Ag_2S\downarrow + 2HNO_3 \rightarrow S\downarrow + 2NO\uparrow + 3Ag_2O\downarrow + H_2O$$

$$3Ag_2O\downarrow + 6HNO_3 \rightarrow 6Ag^+ + 6NO_3^- + 3H_2O$$

If the mixture is heated with concentrated nitric acid for a considerable time, sulphur is oxidized to sulphate and the precipitate disappears:

$$S\downarrow + 2HNO_3 \rightarrow SO_4^{2-} + 2NO\uparrow + 2H^+$$

The precipitate is insoluble in ammonium sulphide, ammonium polysulphide, ammonia, potassium cyanide, or sodium thiosulphate. Silver sulphide can be precipitated from solutions containing diammine-, dicyanato- or dithiosulphatoargentate complexes with hydrogen sulphide.

3. *Ammonia solution:* brown precipitate of silver oxide

$$2Ag^+ + 2NH_3 + H_2O \rightarrow Ag_2O\downarrow + 2NH_4^+$$

The reaction reaches an equilibrium and therefore precipitation is incomplete at any stage. (If ammonium nitrate is present in the original solution or the solution is strongly acidic no precipitation occurs.) The precipitate dissolves in excess of the reagent, and diammineargentate complex ions are formed:

$$Ag_2O\downarrow + 4NH_3 + H_2O \rightarrow 2[Ag(NH_3)_2]^+ + 2OH^-$$

The solution should be disposed of quickly, because when set aside silver nitride Ag_3N precipitate is formed, which explodes readily even in a wet form.

4. *Sodium hydroxide:* brown precipitate of silver oxide:

$$2Ag^+ + 2OH^- \rightarrow Ag_2O\downarrow + H_2O$$

A well-washed suspension of the precipitate shows a slight alkaline reaction owing to the hydrolysis equilibrium:

$$Ag_2O\downarrow + H_2O \rightleftharpoons 2Ag(OH)_2\downarrow \rightleftharpoons 2Ag^+ + 2OH^-$$

The precipitate is insoluble in excess reagent.

The precipitate dissolves in ammonia solution (a) and in nitric acid (b):

$$Ag_2O\downarrow + 4NH_3 + H_2O \rightarrow 2[Ag(NH_3)_2]^+ + 2OH^- \quad \text{(a)}$$

$$Ag_2O\downarrow + 2H^+ \rightarrow 2Ag^+ + H_2O \quad \text{(b)}$$

5. *Potassium iodide:* yellow precipitate of silver iodide

$$Ag^+ + I^- \rightarrow AgI\downarrow$$

The precipitate is insoluble in dilute or concentrated ammonia, but dissolves readily in potassium cyanide (**POISON**) (a) and in sodium thiosulphate(b):

$$AgI + 2CN^- \rightarrow [Ag(CN)_2]^- + I^- \quad \text{(a)}$$

$$AgI + 2S_2O_3^{2-} \rightarrow [Ag(S_2O_3)_2]^{3-} + I^- \quad \text{(b)}$$

6. *Potassium chromate in neutral solution:* red precipitate of silver chromate

$$2Ag^+ + CrO_4^{2-} \rightarrow Ag_2CrO_4\downarrow$$

Spot test: place a drop of the test solution on a watch glass or on a spot plate, add a drop of ammonium carbonate solution and stir (this renders any mercury(I) or lead ions unreactive by precipitation as the highly insoluble carbonates). Remove one drop of the clear liquid and place it on drop-reaction paper together with a drop of the potassium chromate reagent. A red ring of silver chromate is obtained.

The reaction can be used for microscopic test, when a piece of potassium chromate crystal has to be dropped into the test solution. The formation of needle-like red crystals of silver chromate can be observed distinctly.

The precipitate is soluble in dilute nitric acid (a) and in ammonia solution (b):

$$2Ag_2CrO_4\downarrow + 2H^+ \rightleftharpoons 4Ag^+ + Cr_2O_7^{2-} + H_2O \quad \text{(a)}$$

$$Ag_2CrO_4\downarrow + 4NH_3 \rightarrow 2[Ag(NH_3)_2]^+ + CrO_4^{2-} \quad \text{(b)}$$

The acidified solution turns to orange because of the formation of dichromate ions in reaction (a).

7. Potassium cyanide (**POISON**) when added dropwise to a neutral solution of silver nitrate: white precipitate of silver cyanide:

$$Ag^+ + CN^- \rightarrow AgCN\downarrow$$

When potassium cyanide is added in excess, the precipitate disappears owing to the formation of dicyanoargentate ions:

$$AgCN\downarrow + CN^- \rightarrow [Ag(CN)_2]^-$$

8. Sodium carbonate: yellowish-white precipitate of silver carbonate:

$$2Ag^+ + CO_3^{2-} \rightarrow Ag_2CO_3\downarrow$$

When heating, the precipitate decomposes and brown silver oxide precipitate is formed:

$$Ag_2CO_3\downarrow \rightarrow Ag_2O\downarrow + CO_2\uparrow$$

Nitric acid (a) and ammonia solution (b) dissolve the precipitate

$$Ag_2CO_3\downarrow + 2H^+ \rightarrow 2Ag^+ + CO_2\uparrow + H_2O \quad \text{(a)}$$

$$Ag_2CO_3\downarrow + 4NH_3 \rightarrow 2[Ag(NH_3)_2]^+ + CO_3^{2-} \quad \text{(b)}$$

Carbon dioxide gas is evolved in reaction (a).

9. Disodium hydrogen phosphate in neutral solution: yellow precipitate of silver phosphate:

$$3Ag^+ + HPO_4^{2-} \rightarrow Ag_3PO_4\downarrow + H^+$$

Nitric acid (a) and ammonia solution (b) dissolve the precipitate:

$$Ag_3PO_4\downarrow + 3H^+ \rightarrow 3Ag^+ + H_3PO_4 \quad \text{(a)}$$

$$Ag_3PO_4\downarrow + 6NH_3 \rightarrow 3[Ag(NH_3)_2]^+ + PO_4^{3-} \quad \text{(b)}$$

Phosphoric acid, formed in reaction (a) is a medium-strong acid, which is only slightly dissociated if nitric acid is present in excess.

10. Hydrazine sulphate (saturated): when added to a solution of diammine-argentate ions, forms finely divided silver metal, while gaseous nitrogen is evolving:

$$4[Ag(NH_3)_2]^+ + H_2N{-}NH_2.H_2SO_4 \rightarrow$$
$$\rightarrow 4Ag\downarrow + N_2\uparrow + 6NH_4^+ + 2NH_3 + SO_4^{2-}$$

If the vessel in which the reaction is carried out is clean, silver adheres to the glass walls forming an attractive mirror.

Procedure: Fill two-thirds of a test-tube with chromosulphuric acid (concentrated), and set aside overnight. Next day empty the test-tube, rinse cautiously with running cold water, then with distilled water. Into this test-tube pour 2 ml silver nitrate (0·1M) and 2 ml distilled water. Then add dilute ammonia (2M) dropwise, mixing the solution vigorously by shaking, until the last traces of the silver oxide precipitate disappear. Then add 2 ml saturated hydrazine sulphate solution and shake the mixture vigorously. The silver mirror forms within a few seconds. The solution should be discarded after the test (cf.

reaction 3). The silver mirror can be removed most easily by dissolving it in nitric acid (8M).

11. p-*Dimethylaminobenzylidene–rhodanine* (in short: rhodanine reagent, 0·3% solution in acetone): reddish-violet precipitate in slightly acidic solutions:

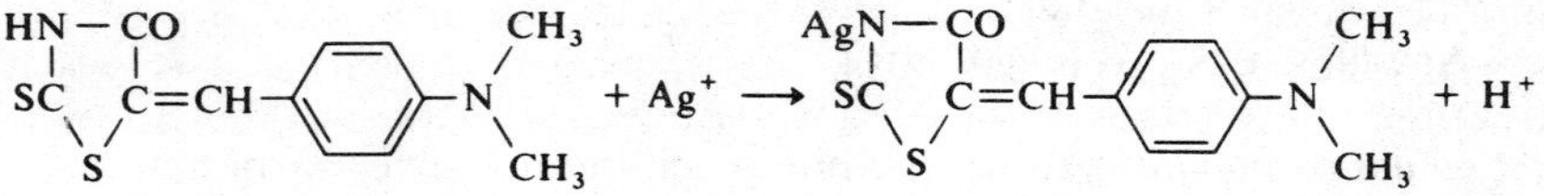

```
HN—CO                    CH3          AgN—CO                    CH3
 |   |                  /               |   |                  /
SC   C=CH—⟨benzene⟩—N        + Ag+ →  SC   C=CH—⟨benzene⟩—N        + H+
  \ /                   \               \ /                   \
   S                     CH3             S                     CH3
```

Mercury, copper, gold, platinum, and palladium salts form similar compounds and therefore interfere.

Spot test: To 1 drop of the test solution add 1 drop nitric acid (2M), then 1 drop of the reagent. A red-violet precipitate or stain is formed if silver ions are present. Alternatively, the test may be performed on a spot plate, or in a semimicro test tube; in the latter case the excess of the reagent is extracted with diethyl ether or amyl alcohol, when violet specks of the silver complex will be visible under the yellow solvent layer.

In the presence of mercury, gold, platinum, or palladium first add 1 drop potassium cyanide (10%, **HIGHLY POISONOUS**) solution to the test solution then follow the procedure given above.

To detect silver in a mixture of lead chloride, mercury(I) chloride, and silver chloride) (Group I), the mixture is treated with 10 per cent potassium cyanide solution whereby mercury(II) cyanide, mercury, and dicyanoargentate $[Ag(CN_2]^-$ are formed: after filtration (or centrifugation), a little of the clear filtrate is treated on a spot plate with a drop of the reagent and 2 drops nitric acid (2M). A red colouration is formed in the presence of Ag in weakly acid solution.

12. Dry test (blowpipe test) When a silver salt is heated with alkali carbonate on charcoal, a white malleable bead without an incrustation of the oxide results; this is readily soluble in nitric acid. The solution is immediately precipitated by dilute hydrochloric acid, but not by very dilute sulphuric acid (difference from lead).

III.7 SECOND GROUP OF CATIONS: MERCURY(II), LEAD(II), BISMUTH(III), COPPER(II), CADMIUM(II), ARSENIC(III) AND (V), ANTIMONY(III) AND (V), AND TIN (II) AND (IV)

Group reagent: hydrogen sulphide (gas or saturated aqueous solution).

Group reaction: precipitates of different colours; mercury(II) sulphide HgS (black), lead(II) sulphide PbS (black), copper(II) sulphide CuS (black), cadmium sulphide CdS (yellow), bismuth(III) sulphide Bi_2S_3 (brown), arsenic(III) sulphide As_2S_3 (yellow), arsenic(V) sulphide (yellow), antimony(III) sulphide Sb_2S_3 (orange), antimony(V) sulphide (orange), tin(II) sulphide SnS (brown), and tin(IV) sulphide SnS_2 (yellow).

Cations of the second group are traditionally divided into two sub-groups; the copper sub-group and the arsenic sub-group. The basis of this division is the solubility of the sulphide precipitates in ammonium polysulphide. While sulphides of the copper sub-group are insoluble in this reagent, those of the arsenic sub-group do dissolve under the formation of thiosalts.

The copper sub-group consists of mercury(II), lead(II), bismuth(III), copper(II), and cadmium(II). Although the bulk of lead(II) ions are precipitated with dilute hydrochloric acid together with other ions of Group I, this precipitation is rather incomplete owing to the relatively high solubility of lead(II) chloride. In the course of systematic analysis therefore lead ions will still be present when the precipitation of the second group of cations is the task. Reactions of lead(II) ions were already described with those of the cations of the first group (see Section **III.4**).

The chlorides, nitrates, and sulphates of the cations of the copper sub-group are quite soluble in water. The sulphides, hydroxides, and carbonates are insoluble. Some of the cations of the copper sub-group (mercury(II), copper(II), and cadmium(II)) tend to form complexes (ammonia, cyanide ions, etc.).

The arsenic sub-group consists of the ions arsenic(III), arsenic(V), antimony(III), antimony(V), tin(II) and tin(IV). These ions have amphoteric character: their oxides form salts both with acids and bases. Thus, arsenic(III) oxide can be dissolved in hydrochloric acid, (6M) and arsenic(III) cations are formed:

$$As_2O_3 + 6HCl \rightarrow 2As^{3+} + 6Cl^- + 3H_2O$$

At the same time arsenic(III) oxide dissolves in sodium hydroxide (2M), when arsenite anions are formed:

$$As_2O_3 + 6OH^- \rightarrow 2AsO_3^{3-} + 3H_2O$$

The dissolution of sulphides in ammonium polysulphide can be regarded as the formation of thiosalts from anhydrous thioacids. Thus the dissolution of arsenic(III) sulphide (anhydrous thioacid) in ammonium sulphide (anhydrous thiobase), yields the formation of ammonium- and thio-arsenite ions (ammonium thioarsenite: a thiosalt):

$$As_2S_3\downarrow + 3S^{2-} \rightarrow 2AsS_3^{3-}$$

All the sulphides of the arsenic sub-group dissolve in (colourless) ammonium sulphide except tin(II) sulphide; to dissolve the latter, ammonium polysulphide is needed, which acts partly as an oxidizing agent, thiostannate ions being formed:

$$SnS\downarrow + S_2^{2-} \rightarrow SnS_3^{2-}$$

Note that while tin is bivalent in the tin(II) sulphide precipitate, it is tetravalent in the thiostannate ion.

Arsenic(III), antimony(III), and tin(II) ions can be oxidized to arsenic(V), antimony(V), and tin(IV) ions respectively. On the other hand, the latter three can be reduced by proper reducing agents. The oxidation-reduction potentials of the arsenic(V)–arsenic(III) and antimony(V)–antimony(III) systems vary with *p*H, therefore the oxidation or reduction of the relevant ions can be assisted by choosing an appropriate *p*H for the reaction.

III.8 MERCURY, Hg (A_r: 200·59) – MERCURY(II) The most important physical and chemical properties of the metal were described in Section **III.5**.

Reactions of mercury(II) ions The reactions of mercury(II) ions can be studied with a dilute solution of mercury(II) nitrate (0·05M).

1. Hydrogen sulphide (gas or saturated aqueous solution): in the presence of dilute hydrochloric acid, initially a white precipitate of mercury(II) chlorosulphide (a), which decomposes when further amounts of hydrogen sulphide are added and finally a black precipitate of mercury(II) sulphide is formed (b).

$$3Hg^{2+} + 2Cl^- + 2H_2S \rightarrow Hg_3S_2Cl_2\downarrow + 4H^+ \quad (a)$$

$$Hg_3S_2Cl_2\downarrow + H_2S \rightarrow 3HgS\downarrow + 2H^+ + 2Cl^- \quad (b)$$

Mercury(II) sulphide is one of the least soluble precipitates known ($K_s = 4 \times 10^{-54}$).

The precipitate is insoluble in water, hot dilute nitric acid, alkali hydroxides, or (colourless) ammonium sulphide.

Sodium sulphide (2M) dissolves the precipitate when the disulphomercurate(II) complex ion is formed:

$$HgS\downarrow + S^{2-} \rightarrow [HgS_2]^{2-}$$

Adding ammonium chloride to the solution, mercury(II) sulphide precipitates again.

Aqua regia dissolves the precipitate:

$$3HgS\downarrow + 6HCl + 2HNO_3 \rightarrow 3HgCl_2 + 3S\downarrow + 2NO\uparrow + 4H_2O$$

Mercury(II) chloride is practically undissociated under these circumstances. Sulphur remains as a white precipitate, which however dissolves readily if the solution is heated, to form sulphuric acid:

$$2HNO_3 + S\downarrow \rightarrow SO_4^{2-} + 2H^+ + 2NO\uparrow$$

2. Ammonia solution: white precipitate with a mixed composition; essentially it consists of mercury(II) oxide and mercury(II) amidonitrate:

$$2Hg^{2+} + NO_3^- + 4NH_3 + H_2O \rightarrow HgO.Hg(NH_2)NO_3\downarrow + 3NH_4^+$$

The salt, like most of the mercury compounds, sublimes at atmospheric pressure.

3. Sodium hydroxide when added in small amounts: brownish-red precipitate with varying composition; if added in stoichiometric amounts the precipitate turns to yellow when mercury(II) oxide is formed:

$$Hg^{2+} + 2OH^- \rightarrow HgO\downarrow + H_2O$$

The precipitate is insoluble in excess sodium hydroxide. Acids dissolve the precipitate readily.

This reaction is characteristic for mercury(II) ions, and can be used to differentiate mercury(II) from mercury(I).

4. Potassium iodide when added slowly to the solution: red precipitate of mercury(II) iodide:

$$Hg^{2+} + 2I^- \rightarrow HgI_2\downarrow$$

The precipitate dissolves in excess reagent, when colourless tetraiodomercurate(II) ions are formed:

$$HgI_2 + 2I^- \rightarrow [HgI_4]^{2-}$$

An alkaline solution of potassium tetraiodomercurate(II) serves as a selective and sensitive reagent for ammonium ions (Nessler's-reagent cf. Section **III.38**, reaction 2).

5. *Potassium cyanide* (**POISON**): does not cause any change in dilute solutions (difference from other ions of the copper sub-group).

6. *Tin(II) chloride*: when added in moderate amounts: white, silky precipitate of mercury(I) chloride (calomel):

$$2Hg^{2+} + Sn^{2+} + 2Cl^- \rightarrow Hg_2Cl_2\downarrow + Sn^{4+}$$

This reaction is widely used to remove the excess of tin(II) ions, used for prior reduction, in oxidation-reduction titrations.

If more reagent is added, mercury(I) chloride is further reduced and black precipitate of mercury is formed:

$$Hg_2Cl_2\downarrow + Sn^{2+} \rightarrow 2Hg\downarrow + Sn^{4+} + 2Cl^-$$

Spot test in the presence of aniline: Treat a drop of the test solution on a filter paper or a spot plate with a drop of tin(II) chloride solution and a drop of aniline. A brown or black stain of mercury metal is produced.

Aniline adjusts the pH of the solution to an appropriate value, at which antimony does not interfere. Bismuth and copper also have no effect; silver, gold, and molybdenum do interfere.

7. *Copper sheet or coin* reduces mercury(II) ions to the metal:

$$Cu + Hg^{2+} \rightarrow Cu^{2+} + Hg\downarrow$$

For practical hints for the test see Section **III.5**, reaction 12.

8. *Diencuprato(II) sulphate* reagent in the presence of potassium iodide*: dark blue-violet precipitate in neutral or ammoniacal solution. Tetraiodomercurate(II) ions are first produced:

$$Hg^{2+} + 4I^- \rightarrow [HgI_4]^{2-}$$

these react with the complex diencuprate ions, forming diencuprato(II)-tetraiodomercurate(II) precipitate:

$$[Cu(en)_2]^{2+} + [HgI_4]^{2-} \rightarrow [Cu(en)_2][HgI_4]\downarrow$$

The reaction is a sensitive one but cadmium ions, which form a similar complex salt, interfere.

9. *Diphenylcarbazide* reacts with mercury(II) ions in a similar way to mercury(I). For details see Section **III.5**, reaction 14.

---

* 'en' is the usual short form of ethylenediamine $H_2N—CH_2—CH_2—NH_2$.

10. Cobalt(II) thyocyanate test To the test solution add an equal volume of the reagent (about 10%, freshly prepared), and stir the wall of the vessel with a glass rod. A deep-blue crystalline precipitate of cobalt tetrathiocyanatomercurate(II) is formed:

$$Hg^{2+} + Co^{2+} + 4SCN^- \rightarrow Co[Hg(SCN)_4]\downarrow$$

Drop test: Place a drop of the test solution on a spot plate, add a small crystal of ammonium thiocyanate followed by a little solid cobalt(II) acetate. A blue colour is produced in the presence of mercury(II) ions.

Sensitivity: 0·5 μg Hg^{2+}. *Concentration limit:* 1 in 10^5.

11. Dry test All mercury compounds, irrespective of their valency state, form mercury metal when heated with excess anhydrous sodium carbonate. For practical hints see Section **III.5**, reaction 15.

III.9 BISMUTH, Bi (A_r: 208·98) Bismuth is a brittle, crystalline, reddish-white metal. It melts at 271·5°C. It is insoluble in hydrochloric acid because of its standard potential (0·2V), but dissolves in oxidizing acids such as concentrated nitric acid (a), aqua regia (b), or hot, concentrated sulphuric acid (c).

$$2Bi + 8HNO_3 \rightarrow 2Bi^{3+} + 6NO_3^- + 2NO\uparrow + 4H_2O \quad (a)$$

$$Bi + 3HCl + HNO_3 \rightarrow Bi^{3+} + 3Cl^- + NO\uparrow + 2H_2O \quad (b)$$

$$2Bi + 6H_2SO_4 \rightarrow 2Bi^{3+} + 3SO_4^{2-} + 3SO_2\uparrow + 6H_2O \quad (c)$$

Bismuth forms tervalent and pentavalent ions. Tervalent bismuth ion Bi^{3+} is the most common. The hydroxide, $Bi(OH)_3$ is a weak base; bismuth salts therefore hydrolyse readily, when the following process occurs:

$$Bi^{3+} + H_2O \rightleftharpoons BiO^+ + 2H^+$$

The bismuthyl ion, BiO^+ forms insoluble salts, like bismuthyl chloride, BiOCl, with most ions. If we want to keep bismuth ions in solution, we must acidify the solution, when the above equilibrium shifts towards the left.

Pentavalent bismuth forms the bismuthate BiO_3^- ion. Most of its salts are insoluble in water.

Reactions of bismuth(III) ions These reactions can be studied with a 0·2M solution of bismuth(III) nitrate, which contains about 3–4 per cent nitric acid.

1. Hydrogen sulphide (gas or saturated aqueous solution): black precipitate of bismuth sulphide:

$$2Bi^{3+} + 3H_2S \rightarrow Bi_2S_3\downarrow + 6H^+$$

The precipitate is insoluble in cold, dilute acid and in ammonium sulphide.

Boiling concentrated hydrochloric acid dissolves the precipitate, when hydrogen sulphide gas is liberated.

$$Bi_2S_3\downarrow + 6HCl \rightarrow 2Bi^{3+} + 6Cl^- + 3H_2S\uparrow$$

Hot dilute nitric acid dissolves bismuth sulphide, leaving behind sulphur in the form of a white precipitate:

$$Bi_2S_3\downarrow + 8H^+ + 2NO_3^- \rightarrow 2Bi^{3+} + 3S\downarrow + 2NO\uparrow + 4H_2O$$

2. *Ammonia solution:* white basic salt of variable composition. The approximate chemical reaction is:

$$Bi^{3+} + NO_3^- + 2NH_3 + 2H_2O \rightarrow Bi(OH)_2NO_3\downarrow + 2NH_4^+$$

The precipitate is insoluble in excess reagent (distinction from copper or cadmium).

3. *Sodium hydroxide:* white precipitate of bismuth(III) hydroxide:

$$Bi^{3+} + 3OH^- \rightarrow Bi(OH)_3\downarrow$$

The precipitate is very slightly soluble in excess reagent in cold solution, 2–3 mg bismuth dissolved per 100 ml sodium hydroxide (2M). The precipitate is soluble in acids:

$$Bi(OH)_3\downarrow + 3H^+ \rightarrow Bi^{3+} + 3H_2O$$

When boiled, the precipitate loses water and turns yellowish-white:

$$Bi(OH)_3\downarrow \rightarrow BiO.OH\downarrow + H_2O$$

Both the hydrated and the dehydrated precipitate can be oxidized by 4–6 drops of concentrated hydrogen peroxide, when yellowish-brown bismuthate ions are formed:

$$BiO.OH\downarrow + H_2O_2 \rightarrow BiO_3^- + H^+ + H_2O$$

4. *Potassium iodide when added dropwise:* black precipitate of bismuth(III) iodide:

$$Bi^{3+} + 3I^- \rightarrow BiI_3\downarrow$$

The precipitate dissolves readily in excess reagent, when orange-coloured tetraiodo-bismuthate ions are formed:

$$BiI_3\downarrow + I^- \rightleftarrows [BiI_4]^-$$

When diluted with water, the above reaction is reversed and black bismuth iodide is reprecipitated. Heating the precipitate with water, it turns orange, owing to the formation of bismuthyl iodide:

$$BiI_3\downarrow + H_2O \rightarrow BiOI\downarrow + 2H^+ + 2I^-$$

5. *Potassium cyanide* (**POISON**): white precipitate of bismuth hydroxide. The reaction is a hydrolysis:

$$Bi^{3+} + 3H_2O + 3CN^- \rightarrow Bi(OH)_3\downarrow + 3HCN\uparrow$$

The precipitate is insoluble in excess reagent (distinction from cadmium ions).

6. *Sodium tetrahydroxostannate(II) (0·125M, freshly prepared):* in cold solution reduces bismuth(III) ions to bismuth metal which separates in the form of a black precipitate. First the sodium hydroxide present in the reagent reacts with bismuth(III) ions (a, cf. reaction 3), bismuth(III) hydroxide then is reduced by tetrahydroxostannate(II) ions, when bismuth metal and hexa-hydroxostannate(IV) ions are formed (b):

$$Bi^{3+} + 3OH^- \rightarrow Bi(OH)_3\downarrow \qquad (a)$$

$$2Bi(OH)_3\downarrow + 3[Sn(OH)_4]^{2-} \rightarrow 2Bi\downarrow + 3[Sn(OH)_6]^{2-} \qquad (b)$$

The reagent must be freshly prepared, and the test must be carried out in cold. The reagent slowly decomposes by disproportionation, when tin metal is formed as a black precipitate:

$$2[Sn(OH)_4]^{2-} \rightarrow Sn\downarrow + [Sn(OH)_6]^{2-} + 2OH^-$$

Note that tin is tetravalent in the hexahydroxostannate(IV) ion. Heating accelerates the decomposition.

Test by induced reaction: In the absence of bismuth(III) ions the reaction between tetrahydroxoplumbate(II) ions (cf. Section **III.4**, reaction 4) and tetrahydroxostannate(II) is slow:

$$[Pb(OH)_4]^{2-} + [Sn(OH)_4]^{2-} \xrightarrow{(slow)} Pb\downarrow + [Sn(OH)_6]^{2-} + 2OH^-$$

With dilute solutions (using 0·25M lead nitrate and 0·125M sodium tetrahydroxostannate(II) reagent) the formation of the black precipitate of lead metal is not observable within an hour. In the presence of bismuth the reaction is accelerated; at the same time bismuth is also precipitated:

$$[Pb(OH)_4]^{2-} + [Sn(OH)_4]^{2-} \xrightarrow{(fast)} Pb\downarrow + [Sn(OH)_6]^{2-} + 2OH^-$$

$$2Bi(OH)_3\downarrow + 3[Sn(OH)_4]^{2} \xrightarrow{(fast)} 2Bi\downarrow + 3[Sn(OH)_6]^{2-}$$

Such reactions are called **induced reactions** (that is, bismuth induces the reduction of lead). They have to be distinguished from catalytic processes by the fact, that the inductor is used up itself in the course of the reaction (and not regenerated, as in a catalytic process). Induced reactions are quite common among oxidation-reduction processes.

Silver, copper, and mercury interfere with the reaction. Copper can be made inactive by the addition of some potassium cyanide.

On a spot plate, mix a drop of the test solution, one drop lead nitrate, 1 drop potassium cyanide (**POISON**), and 2 drops freshly prepared sodium tetrahydroxostannate(II) reagent. A brown to black colouration (precipitate) is characteristic for bismuth.

7. Water When a solution of a bismuth salt is poured into a large volume of water, a white precipitate of the corresponding basic salt is produced, which is soluble in dilute mineral acids, but is insoluble in tartaric acid (distinction from antimony) and in alkali hydroxides (distinction from tin).

$$Bi^{3+} + NO_3^- + H_2O \rightarrow BiO(NO_3)\downarrow + 2H^+$$

$$Bi^{3+} + Cl^- + H_2O \rightarrow BiO.Cl\downarrow + 2H^+$$

8. Disodium hydrogen phosphate: white, crystalline precipitate of bismuth phosphate:

$$Bi^{3+} + HPO_4^- \rightarrow BiPO_4\downarrow + H^+$$

the precipitate is only sparingly soluble in dilute mineral acids (distinction from mercury(II), lead(II), copper(II), and cadmium ions).

Disodium hydrogen arsentate reacts in an analogous way; the product is the white bismuth arsenate precipitate:

$$Bi^{3+} + HAsO_4^- \rightarrow BiAsO_4\downarrow + H^+$$

9. Pyrogallol (10%, freshly prepared) reagent, when added in slight excess to a hot, faintly acid solution of bismuth ions: yellow precipitate of bismuth pyrogallate:

$$Bi + C_6H_3(OH)_3 \rightarrow Bi(C_6H_3O_3)\downarrow + 3H^+$$

It is best to neutralize the test solution first by ammonia against litmus paper, then add some drops of dilute nitric acid, and then the reagent. The test is a very sensitive one. Antimony interferes and should be absent.

10. Cinchonine–potassium iodide reagent (1%): orange-red colouration or precipitate in dilute acid solution.

Test on filter paper: Moisten a piece of drop-reaction paper with the reagent and place a drop of the slightly acid test solution upon it. An orange-red spot is obtained.

Sensitivity: 0·15 μg Bi. *Concentration limit:* 1 in 350,000.

The test may also be carried out on a spot plate.

Lead, copper, and mercury salts interfere because they react with the iodide. Nevertheless, bismuth may be detected in the presence of salts of these metals as they diffuse at different rates through the capillaries of the paper, and are fixed in distinct zones. When a drop of the test solution containing bismuth, lead, copper, and mercury ions is placed upon absorbent paper impregnated with the reagent, four zones can be observed: (i) a white central ring, containing the mercury; (ii) an orange ring, due to bismuth; (iii) a yellow ring of lead iodide; and (iv) a brown ring of iodine liberated by the reaction with copper. The thicknesses of the rings will depend upon the relative concentrations of the various metals.

11. Thiourea (10%): intense yellow complex with bismuth(III) ions in the presence of dilute nitric acid. The test may be carried out on drop-reaction paper, on a spot plate, or in a micro test-tube.

Sensitivity: 6 μg Bi^{3+}. *Concentration limit:* 1 in 3×10^4.

Mercury(I), silver(I), antimony(III), iron(III), and chromate ions interfere and should therefore be absent.

12. 8-Hydroxyquinoline (5%) and potassium iodide (6M) in acidic medium: red precipitate of 8-hydroxyquinoline-tetraiodobismuthate

$$Bi^{3+} + C_9H_7ON + H^+ + 4I^- \rightarrow C_9H_7ON.H\,BiI_4\downarrow$$

If other halide ions are absent, the reaction is characteristic for bismuth.

13. Dry test (blowpipe test). When a bismuth compound is heated on charcoal with sodium carbonate in the blowpipe flame, a brittle bead of metal, surrounded by a yellow incrustation of the oxide, is obtained.

III.10 COPPER, Cu (A_r: 63·54) Copper is a light-red metal, which is soft, malleable, and ductile. It melts at 1038°C. Because of its positive standard electrode potential (+0·34 V for the Cu/Cu^{2+} couple) it is insoluble in hydrochloric acid and in dilute sulphuric acid, although in the presence of oxygen some dissolution might take place. Medium-concentrated nitric acid (8M) dissolves copper readily:

$$3Cu + 8HNO_3 \rightarrow 3Cu^{2+} + 6NO_3^- + 2NO\uparrow + 4H_2O$$

Hot, concentrated sulphuric acid dissolves copper also:

$$Cu + 2H_2SO_4 \rightarrow Cu^{2+} + SO_4^{2-} + SO_2\uparrow + 2H_2O$$

Copper is readily dissolved in aqua regia as well:

$$3Cu + 6HCl + 2HNO_3 \rightarrow 3Cu^{2+} + 6Cl^- + 2NO\uparrow + 4H_2O$$

There are two series of copper compounds. Copper(I) compounds are derived from the red copper(I) oxide Cu_2O and contain the copper(I) ion Cu^+. These compounds are colourless, most of the copper(I) salts are insoluble in water their behaviour generally resembling that of the silver(I) compounds. They are readily oxidized to copper(II) compounds, which are derivable from the black copper(II) oxide, CuO. Copper(II) compounds contain the copper(II) ions Cu^{2+}. Copper(II) salts are generally blue both in solid, hydrated form and in dilute aqueous solution; the colour is characteristic really for the teatraquocuprate(II) ion $[Cu(H_2O)_4]^{2+}$ only. The limit of visibility of the colour of the tetraquocuprate(II) complex (i.e. the colour of copper(II) ions in aqueous solutions) is 500 μg in a concentration limit of 1 in 10^4. Anhydrous copper(II) salts, like anhydrous copper(II) sulphate $CuSO_4$, are white (or slightly yellow). In aqueous solutions we always have the tetraquo complex ion present; for the sake of simplicity they will be denoted in this text as the bare copper(II) ions Cu^{2+}.

In practice only the copper(II) ions is important, therefore only the reactions of the copper(II) ion are described.

Reactions of copper(II) ions These reactions can be studied with a 0·25M solution of copper(II) sulphate.

1. Hydrogen sulphide (gas or saturated aqueous solution): black precipitate of copper(II) sulphide:

$$Cu^{2+} + H_2S \rightarrow CuS\downarrow + 2H^+$$

K_s(CuS; 25°) = 10^{-44}. *Sensitivity:* 1 μg Cu^{2+}. *Concentration limit:* 1 in 5×10^6.

The solution must be acidic (M in hydrochloric acid) in order to obtain a crystalline, well-filterable precipitate. In the absence of acid, or in very slightly acid solutions a colloidal, brownish-black precipitate or colouration is obtained. By adding some acid and boiling coagulation can be achieved.

The precipitate is insoluble in boiling dilute (M) sulphuric acid (distinction from cadmium), in sodium hydroxide, sodium sulphide, ammonium sulphide, and only very slightly soluble in polysulphides.

Hot, concentrated nitric acid dissolves the copper(II) sulphide, leaving behind sulphur as a white precipitate:

$$3CuS\downarrow + 8HNO_3 \rightarrow 3Cu^{2+} + 6NO_3^- + 3S\downarrow + 2NO\uparrow + 2H_2O$$

When boiled for longer, sulphur is oxidized to sulphuric acid and a clear, blue solution is obtained:

$$S\downarrow + 2HNO_3 \rightarrow 2H^+ + SO_4^{2-} + 2NO\uparrow$$

Potassium cyanide (**POISON**) dissolves the precipitate, when colourless tetracyanocuprate(I) ions and disulphide ions are formed:

$$2CuS\downarrow + 8CN^- \rightarrow 2[Cu(CN)_4]^{3-} + S_2^{2-}$$

Note that this is an oxidation-reduction process (copper is reduced, sulphur is oxidized) coupled with a formation of a complex.

When exposed to air, in the moist state, copper(II) sulphide tends to oxidize to copper(II) sulphate:

$$CuS\downarrow + 2O_2 \rightarrow CuSO_4$$

and therefore becomes water soluble. A considerable amount of heat is liberated during this process. A filter paper with copper(II) sulphide precipitate on it should never be thrown into a waste container, with paper or other inflammable substances in it, but the precipitate should be washed away first with running water.

2. *Ammonia solution when added sparingly:* blue precipitate of a basic salt (basic copper sulphate):

$$2Cu^{2+} + SO_4^{2-} + 2NH_3 + 2H_2O \rightarrow Cu(OH)_2.CuSO_4\downarrow + 2NH_4^+$$

which is soluble in excess reagent, when a deep blue colouration is obtained, owing to the formation of tetramminocuprate(II) complex ions:

$$Cu(OH)_2.CuSO_4\downarrow + 8NH_3 \rightarrow 2[Cu(NH_3)_4]^{2+} + SO_4^{2-} + 2OH^-$$

If the solution contains ammonium salts (or it was highly acidic and larger amounts of ammonia were used up for its neutralization), precipitation does not occur at all, but the blue colour is formed right away.

The reaction is characteristic for copper(II) ions in the absence of nickel.

3. *Sodium hydroxide in cold solution:* blue precipitate of copper(II) hydroxide:

$$Cu^{2+} + 2OH^- \rightarrow Cu(OH)_2\downarrow$$

The precipitate is insoluble in excess reagent.

When heated, the precipitate is converted to black copper(II) oxide by dehydration:

$$Cu(OH)_2\downarrow \rightarrow CuO\downarrow + H_2O$$

In the presence of a solution of tartaric acid or of citric acid, copper(II) hydroxide is not precipitated by solutions of caustic alkalis, but the solution is coloured an intense blue. If the alkaline solution is treated with certain reducing agents, such as hydroxylamine, hydrazine, glucose, and acetaldehyde, yellow copper(I) hydroxide is precipitated from the warm solution, which is converted into red copper(I) oxide Cu_2O on boiling. The alkaline solution of copper(II) salt containing tartaric acid is usually known as Fehling's solution; it contains the complex ion $[Cu(COO.CHO)]^{2-}$.

4. *Potassium iodide:* precipitates copper(I) iodide, which is white, but the solution is intensely brown because of the formation of tri-iodide ions (iodine):

$$2Cu^{2+} + 5I^- \rightarrow 2CuI\downarrow + I_3^-$$

Adding an excess of sodium thiosulphate to the solution, tri-iodide ions are reduced to colourless iodide ions and the white colour of the precipitate becomes visible. The reduction with thiosulphate yields tetrathionate ions:

$$I_3^- + 2S_2O_3^{2-} \rightarrow 3I^- + S_4O_6^{2-}$$

These reactions are used in quantitative analysis for the iodometric determination of copper.

5. *Potassium cyanide* (**POISON**): when added sparingly forms first a yellow precipitate of copper(II) cyanide:

$$Cu^{2+} + 2CN^- \rightarrow Cu(CN)_2\downarrow$$

the precipitate quickly decomposes into white copper(I) cyanide and cyanogen (HIGHLY **POISONOUS** GAS):

$$2Cu(CN)_2\downarrow \rightarrow 2CuCN\downarrow + (CN)_2\uparrow$$

In excess reagent the precipitate is dissolved, and colourless tetracyanocuprate(I) complex is formed:

$$CuCN\downarrow + 3CN^- \rightarrow [Cu(CN)_4]^{3-}$$

The complex is so stable (i.e. the concentration of copper(I) ions is so low) that hydrogen sulphide cannot precipitate copper(I) sulphide from this solution (distinction from cadmium, cf. Section **III.11**, reactions 1 and 5).

6. *Potassium hexacyanoferrate(II):* reddish-brown precipitate of copper hexacyanoferrate(II) in neutral or acid medium.

$$2Cu^{2+} + [Fe(CN)_6]^{4-} \rightarrow Cu_2[Fe(CN)_6]\downarrow$$

The precipitate is soluble in ammonia solution, when dark-blue copper tetrammine ions are formed:

$$Cu_2[Fe(CN)_6]\downarrow + 8NH_3 \rightarrow 2[Cu(NH_3)_4]^{2+} + [Fe(CN)_6]^{4-}$$

Sodium hydroxide decomposes the precipitate, when blue copper(II) hydroxide precipitate is formed:

$$Cu_2[Fe(CN)_6]\downarrow + 4OH^- \rightarrow 2Cu(OH)_2\downarrow + [Fe(CN)_6]^{4-}$$

7. *Potassium thiocyanate:* black precipitate of copper(II) thiocyanate:

$$Cu^{2+} + 2SCN^- \rightarrow Cu(SCN)_2\downarrow$$

The precipitate decomposes slowly to form white copper(I) thiocyanate and thiocyanogen is formed:

$$2Cu(SCN)_2\downarrow \rightarrow 2CuSCN\downarrow + (SCN)_2\uparrow$$

thiocyanogen decomposes rapidly in aqueous solutions.

Copper(II) thiocyanate can be transformed to copper(I) thiocyanate immediately by adding a suitable reducing agent. A saturated solution of sulphur dioxide is the most suitable reagent:

$$2Cu(SCN)_2\downarrow + SO_2 + 2H_2O \rightarrow 2CuSCN\downarrow + 2SCN^- + SO_4^{2-} + 4H^+$$

8. *Iron* If a clean iron nail or a blade of a penknife is immersed in a solution of a copper salt, a red deposit of copper is obtained: (See Section **I.42**):

$$Cu^{2+} + Fe \rightarrow Fe^{2+} + Cu$$

and an equivalent amount of iron dissolves. The electrode potential of copper (more precisely of the copper–copper(II) system) is more positive than that of iron (or the iron–iron(II) system).

9. α-Benzoinoxime (or cupron) (5% in alcohol)

$$(C_6H_5.CHOH.C(=NOH).C_6H_5):$$

forms a green precipitate of copper(II) benzoinoxime $Cu(C_{14}H_{11}O_2N)$, insoluble in dilute ammonia. In the presence of metallic salts which are precipitated by ammonia, their precipitation can be prevented by the addition of sodium potassium tartrate (10%). The reagent is specific for copper in ammoniacal tartrate solution. Large amounts of ammonium salts interfere and should be removed by evaporation and heating to glowing: the residue is then dissolved in a little dilute hydrochloric acid.

Treat some drop-reaction paper with a drop of the weakly acid test solution and a drop of the reagent, and then hold it over ammonia vapour. A green colouration is obtained.

Sensitivity: 0·1 μg Cu. *Concentration limit:* 1 in 5×10^5.

If other ions, precipitable by ammonia solution, are present, a drop of Rochelle salt solution (10%) is placed upon the paper before the reagent is added.

10. Salicylaldoxime (1%)

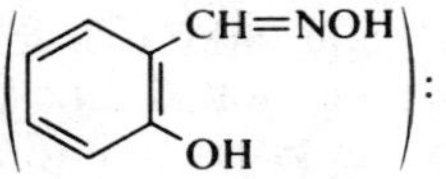

$$\left(C_6H_4(CH{=}NOH)(OH)\right):$$

forms a greenish-yellow precipitate of copper salicylaldoxime $Cu(C_7H_6O_2N)_2$ in acetic acid solution, soluble in mineral acids. Only palladium and gold interfere giving $Pd(C_7H_6O_2N)_2$ and metallic gold respectively in acetic acid solution; they should therefore be absent.

Place a drop of the test solution, which has been neutralized and then acidified with acetic acid in a micro test-tube and add a drop of the reagent. A yellow-green precipitate or opalescence (according to the amount of copper present) is obtained.

Sensitivity: 0·5 μg Cu. *Concentration limit:* 1 in 10^5.

11. Rubeanic acid (or dithio-oxamide) (0·5)

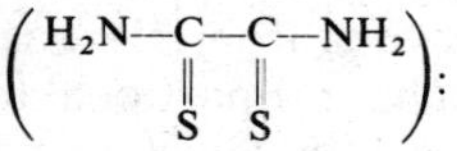

$$\left(H_2N{-}C(=S){-}C(=S){-}NH_2\right):$$

black precipitate of copper rubeanate $Cu[C(=NH)S]_2$ from ammoniacal or weakly acid solution. The precipitate is formed in the presence of alkali tartrates, but not in alkali-cyanide solutions. Only nickel and cobalt ions react under similar conditions yielding blue and brown precipitates respectively. Copper may however, be detected in the presence of these elements by utilizing the capillary separation method upon filter paper. Mercury(I) should be absent as it gives a black stain with ammonia.

Place a drop of the neutral test solution upon drop-reaction paper, expose it to ammonia vapour and add a drop of the reagent. A black or greenish-black spot is produced.

Sensitivity: 0·01 μg Cu. *Concentration limit:* 1 in 2×10^6.
Traces of copper in distilled water give a positive reaction, hence a blank test must be carried out with the distilled water.

In the presence of nickel, proceed as follows. Impregnate drop-reaction paper with the reagent and add a drop of the test solution acidified with acetic acid, (2M). Two zones or circles are formed: the central olive-green or black ring is due to copper and the outer blue-violet ring to nickel.

Sensitivity: 0·05 μg Cu in the presence of 20,000 times that amount of nickel. *Concentration limit:* 1 in 10^6.

In the presence of cobalt, the central green or black ring, due to copper, is surrounded by a yellow-brown ring of cobalt rubeanate.

Sensitivity: 0·25 μg Cu in the presence of 20,000 times that amount of cobalt. *Concentration limit:* 1 in 2×10^5.

12. Ammonium tetrathiocyanatomercurate(II):

$\{(NH_4)_2[Hg(SCN)_4]\}$:

deep-violet, crystalline precipitate in the presence of zinc or cadmium ions. Cobalt and nickel interfere since they yield green or blue precipitates of the corresponding tetrathiocyanatomercurates(II); the interference of iron(III) is avoided by carrying out the precipitation in the presence of alkali fluorides or oxalates.

Place a drop of the acid test solution upon a spot plate, add 1 drop zinc acetate (1%) solution and 1 drop of the reagent. The precipitated zinc tetrathiocyanatomercurate(II) is coloured violet owing to the coprecipitation of the copper complex; the composition of the precipitate is approximately $Zn[Hg(SCN)_4].Cu[Hg(SCN)_4]$. The addition of copper ions to a precipitate of zinc tetrathiocyanatomercurate(II), already formed, has no effect.

Sensitivity: 0·1 μg Cu^{2+}. *Concentration limit:* 1 in 5×10^5.

13. Catalytic test. Iron(III) salts react with thiosulphate according to the equations

$$Fe^{3+} + 2S_2O_3^{2-} \rightarrow [Fe(S_2O_3)_2]^- \quad \text{(a)}$$

$$[Fe(S_2O_3)_2]^- + Fe^{3+} \rightarrow 2Fe^{2+} + S_4O_6^{2-} \quad \text{(b)}$$

Reaction (a) is fairly rapid; reaction (b) is a slow one, but is enormously accelerated by traces of copper. If the reaction is carried out in the presence of a thiocyanate, which serves as an indicator for the presence of iron(III) ions and also retards reaction (b), then the reaction velocity, which is proportional to the time taken for complete decolourization, may be employed for detecting minute amounts of copper(II) ions. Tungsten and, to a lesser extent, selenium cause a catalytic acceleration similar to that of copper: they should therefore be absent.

Upon adjacent cavities of a spot plate place a drop of the test solution and a drop of distilled water. Add to each 1 drop iron(III) thiocyanate (0·05M) and 3 drops sodium thiosulphate (0·5M). The decolourization of the copper-free solution is complete in 1·5–2 minutes: if the test solution contains 1 μg copper, the decolourization is instantaneous. For smaller amounts of copper, the difference in times between the two tests is still appreciable.

Sensitivity: 0·2 μg Cu^{2+}. *Concentration limit:* 1 in 2×10^6.

14. Dry tests **a.** Blowpipe test When copper compounds are heated with alkali carbonate upon charcoal red metallic copper is obtained but no oxide is visible.

b. Borax bead Green while hot, and blue when cold after heating in the oxidizing flame; red in the reducing flame, best obtained by the addition of a trace of tin.

c. Flame test Green especially in the presence of halides, e.g. by moistening with concentrated hydrochloric acid before heating.

III.11 CADMIUM, Cd (*A*$_r$: 112·40) Cadmium is a silver-white, malleable and ductile metal. It melts at 321°C. It dissolves slowly in dilute acids with the evolution of hydrogen (owing to its negative electrode potential):

$$Cd + 2H^+ \rightarrow Cd^{2+} + H_2\uparrow$$

Cadmium forms bivalent ions which are colourless. Cadmium chloride, nitrate, and sulphate are soluble in water; the sulphide is insoluble with a characteristic yellow colour.

Reactions of the cadmium(II) ions These reactions can be studied most conveniently with a 0·25M solution of cadmium sulphate.

1. Hydrogen sulphide (gas or saturated aqueous solution): yellow precipitate of cadmium sulphide:

$$Cd^{2+} + H_2S \rightarrow CdS\downarrow + 2H^+$$

The reaction is reversible; if the concentration of strong acid in the solution is above 0·5M, precipitation is incomplete. Concentrated acids dissolve the precipitate for the same reason. The precipitate is insoluble in potassium cyanide (**POISON**) this distinguishes cadmium ions from copper.

2. Ammonia solution when added dropwise: white precipitate of cadmium(II) hydroxide:

$$Cd^{2+} + 2NH_3 + 2H_2O \rightleftharpoons Cd(OH)_2\downarrow + 2NH_4^+$$

The precipitate dissolves in acid, when the equilibrium shifts towards left.

An excess of the reagent dissolves the precipitate, when tetramminecadmiate(II) ions are formed:

$$Cd(OH)_2\downarrow + 4NH_3 \rightarrow [Cd(NH_3)_4]^{2+} + 2OH^-$$

the complex is colourless.

3. Sodium hydroxide: white precipitate of cadmium(II) hydroxide:

$$Cd^{2+} + 2OH^- \rightleftharpoons Cd(OH)_2\downarrow$$

The precipitate is insoluble in excess reagent; its colour and composition remains unchanged when boiled. Dilute acids dissolve the precipitate by shifting the equilibrium to the left.

4. *Potassium cyanide* (**POISON**): white precipitate of cadmium cyanide, when added slowly to the solution:

$$Cd^{2+} + 2CN^{-} \rightarrow Cd(CN)_2\downarrow$$

An excess of the reagent dissolves the precipitate, when tetracyanocadmiate(II) ions are formed:

$$Cd(CN)_2\downarrow + 2CN^{-} \rightarrow [Cd(CN)_4]^{2-}$$

The colourless complex is not too stable; when hydrogen sulohide gas is introduced, cadmium sulphide is precipitated:

$$[Cd(CN)_4]^{2-} + H_2S \rightarrow CdS\downarrow + 2H^{+} + 4CN^{-}$$

The marked difference in the stabilities of the copper and cadmium tetracyanato complexes serves as the basis for the separation of copper and cadmium ions (cf. Section **I.32**).

5. *Potassium thiocyanate:* forms no precipitate (distinction from copper).

6. *Potassium iodide:* forms no precipitate (distinction from copper).

7. *Dinitro-*p*-diphenyl carbazide (0·1%)*

$$\left(C{=}O(NH{-}NH{-}C_6H_4{-}NO_2)_2 \right):$$

forms a brown-coloured product with cadmium hydroxide, which turns greenish-blue with formaldehyde.

Place a drop of the acid, neutral or ammoniacal test solution on a spot plate and mix it with 1 drop sodium hydroxide (2M) solution and 1 drop potassium cyanide (10%) solution. Introduce 1 drop of the reagent and 2 drops formaldehyde solution (40%). A brown precipitate is formed, which very rapidly becomes greenish-blue. The reagent alone is red in alkaline solution and is coloured violet with formaldehyde, hence it is advisable to compare the colour produced in a blank test with pure water when searching for minute amounts of cadmium.

Sensitivity: 0·8 μg Cd. *Concentration limit:* 1 in 60,000.

In the presence of considerable amounts of copper, 3 drops each of the potassium cyanide and formaldehyde solution should be used; the sensitivity is 4 μg Cd in the presence of 400 times that amount of copper.

8. *4-Nitronaphthalene-diazoamino-azo-benzene ('Cadion 2B') (0·02%)*

$$\left(O_2N{-}C_{10}H_6{-}N{=}N{-}NH{-}C_6H_4{-}N{=}N{-}C_6H_5 \right):$$

Cadmium hydroxide forms a red-coloured lake with the reagent, which contrasts with the blue tint of the latter.

Place a drop of the reagent upon drop-reaction paper, add one drop of the test solution (which should be slightly acidified with acetic acid (2M) containing a little sodium potassium tartrate), and then one drop potassium hydroxide (2M). A bright-pink spot, surrounded by a blue circle, is produced.

Sensitivity: 0·025 μg Cd.

The interference of copper, nickel, cobalt, iron, chromium, and magnesium is prevented by adding sodium potassium tartrate to the test solution: only silver (removed as silver iodide by the addition of a little KI solution) and mercury then interfere. Mercury is best removed by adding a little sodium potassium tartrate, a few crystals of hydroxylamine hydrochloride, followed by sodium hydroxide solution until alkaline; the mercury is precipitated as metal. Tin(II) chloride is not suitable for this reduction since most of the cadmium is adsorbed on the mercury precipitate.

The reagent is prepared by dissolving 0·02 g 'cadion 2B' in 100 ml ethanol to which 1 ml 2M potassium hydroxide is added. The solution must not be warmed. It is destroyed by mineral acid.

9. Dry tests **a.** Blowpipe test All cadmium compounds when heated with alkali carbonate on charcoal give a brown incrustation of cadmium oxide CdO.

b. Ignition test Cadmium salts are reduced by sodium oxalate to elementary cadmium, which is usually obtained as a metallic mirror surrounded by a little brown cadmium oxide. Upon heating with sulphur, the metal is converted into yellow cadmium sulphide.

Place a little of the cadmium salt mixed with an equal weight of sodium oxalate in a small ignition tube, and heat. A mirror of metallic cadmium with brown edges is produced. Allow to cool, add a little flowers of sulphur and heat again. The metallic mirror is gradually converted into the orange-coloured sulphide, which becomes yellow after cooling. Do not confuse this with the yellow sublimate of sulphur.

III.12 ARSENIC, As (*A*$_r$: 74·92) – ARSENIC(III) Arsenic is a steel-grey, brittle solid with a metallic lustre. It sublimes on heating, and a characteristic, garlic-like odour is apparent; on heating in a free supply of air, arsenic burns with a blue flame yielding white fumes of arsenic(III) oxide As_4O_6. All arsenic compounds are poisonous. The element is insoluble in hydrochloric acid and in dilute sulphuric acid: it dissolves readily in dilute nitric acid yielding arsenite ions and concentrated nitric acid or in aqua regia or in sodium hypochlorite solution forming arsenate:

$$As + 4H^+ + NO_3^- \rightarrow As^{3+} + NO\uparrow + 2H_2O$$

$$3As + 5HNO_3(conc) + 2H_2O \rightarrow 3AsO_4^{3-} + 5NO\uparrow + 9H^+$$

$$2As + 5OCl^- + 3H_2O \rightarrow 2AsO_4^{3-} + 5Cl^- + 6H^+$$

Two series of compounds of arsenic are common: that of arsenic(III) and arsenic(V). Arsenic(III) compounds can be derived from the amphoteric arsenic trioxide As_2O_3, which yields salts both with strong acids (e.g. arsenic(III) chloride, $AsCl_3$), and with strong bases (e.g. sodium arsenite, Na_3AsO_3). In strongly acidic solutions therefore the arsenic(III) ion As^{3+} is stable. In strongly basic solutions the arsenite ion, AsO_3^{3-} is the stable one. Arsenic(V) compounds are derived from arsenic pentoxide, As_2O_5. This is the anhydride of arsenic acid, H_3AsO_4, which forms salts such as sodium arsenate Na_3AsO_4. Arsenic(V) therefore exists in solutions predominantly as the arsenate AsO_4^{3-} ion.

Reactions of arsenic(III) ions A 0·1M solution of arsenic(III) oxide, As_2O_3, or sodium arsenite, Na_3AsO_3, can be used for these experiments. Arsenic(III) oxide does not dissolve in cold water, but by boiling the mixture for 30 minutes, dissolution is complete. The mixture can be cooled without the danger of precipitating the oxide.

1. *Hydrogen sulphide*: yellow precipitate of arsenic(III) sulphide:

$$2As^{3+} + 3H_2S \rightarrow As_2S_3\downarrow + 6H^+$$

The solution must be strongly acidic; if there is not enough acid present a yellow colouration is visible only, owing to the formation of colloidal As_2S_3. The precipitate is insoluble in concentrated hydrochloric acid (dintinction and method of separation from Sb_2S_3 and SnS_2), but dissolves in hot concentrated nitric acid:

$$3As_2S_3 + 26HNO_3 + 8H_2O \rightarrow 6AsO_4^{3-} + 9SO_4^{2-} + 42H^+ + 26NO\uparrow$$

It is also readily soluble in solutions of alkali hydroxides, and ammonia:

$$As_2S_3 + 6OH^- \rightarrow AsO_3^{3-} + AsS_3^{3-} + 3H_2O$$

Ammonium sulphide also dissolves the precipitate:

$$As_2S_3 + 3S^{2-} \rightarrow 2AsS_3^{3-}$$

In both cases thioarsenite (AsS_3^{3-}) ions are formed. On reacidifying these both decompose, when arsenic(III) sulphide and hydrogen sulphide are formed:

$$2AsS_3^{3-} + 6H^+ \rightarrow As_2S_3\downarrow + 3H_2S\uparrow$$

Yellow ammonium sulphide (ammonium polysulphide), $(NH_4)_2S_2$ dissolves the precipitate, when thioarsenate AsS_4^{3-} ions are formed:

$$As_2S_3\downarrow + 4S_2^{2-} \rightarrow 2AsS_4^{3-} + S_3^{2-}$$

Upon acidifying this solution yellow arsenic(V) sulphide is precipitated, which is contaminated with sulphur because of the decomposition of the excess polysulphide reagent:

$$2AsS_4^{3-} + 6H^+ \rightarrow As_2S_5\downarrow + 3H_2S\uparrow$$
$$S_2^{2-} + 2H^+ \rightarrow H_2\uparrow + S\downarrow$$

2. *Silver nitrate:* yellow precipitate of silver arsenite in **neutral** solution (distinction from arsenates):

$$AsO_3^{3-} + 3Ag^+ \rightarrow Ag_3AsO_3\downarrow$$

The precipitate is soluble both in nitric acid (a) and ammonia (b):

$$Ag_3AsO_3\downarrow + 3H^+ \rightarrow H_3AsO_3 + 3Ag^+ \quad \text{(a)}$$
$$Ag_3AsO_3\downarrow + 6NH_3 \rightarrow 3[Ag(NH_3)_2]^+ + AsO_3^{3-} \quad \text{(b)}$$

3. Magnesia mixture (a solution containing $MgCl_2$, NH_4Cl, and a little NH_3): no precipitate (distinction from arsenate).

A similar result is obtained with the magnesium nitrate reagent (a solution containing $Mg(NO_3)_2$, NH_4NO_3, and a little NH_3).

4. Copper sulphate solution: green precipitate of copper arsenite (Scheele's green), variously formulated as $CuHAsO_3$ and $Cu_3(AsO_3)_2.xH_2O$, from neutral solutions, soluble in acids, and also in ammonia solution forming a blue solution. The precipitate also dissolves in sodium hydroxide solution; upon boiling, copper(I) oxide is precipitated.

5. Potassium tri-iodide (solution of iodine in potassium iodide): oxidizes arsenite ions while becoming decolourized:

$$AsO_3^{3-} + I_3^- + H_2O \rightarrow AsO_4^{3-} + 3I^- + 2H^+$$

The reaction is reversible, and an equilibrium is reached. If the hydrogen ions formed in this reaction, are removed by adding sodium hydrogen carbonate as a buffer, the reaction becomes complete.

6. Tin(II) chloride solution and concentrated hydrochloric acid (Bettendorff's Test) A few drops of the arsenite solution are added to 2 ml concentrated hydrochloric acid and 0·5 ml saturated tin(II) chloride solution, and the solution gently warmed; the solution becomes dark brown and finally black, due to the separation of elementary arsenic.

$$2As^{3+} + 3Sn^{2+} \rightarrow 2As\downarrow + 3Sn^{4+}$$

If the test is made with the sulphide precipitated in acid solution, only mercury will interfere; by converting the arsenic into magnesium ammonium arsenate and heating to redness, the pyroarsenate $Mg_2As_2O_7$ remains and any mercury salts present are volatilized. This forms the basis of a delicate test for arsenic.

Mix a drop of the test solution in a micro crucible with 1–2 drops concentrated ammonia solution, 2 drops '10-volume' hydrogen peroxide, and 2 drops M magnesium sulphate solution. Evaporate slowly and finally heat until fuming ceases. Treat the residue with 1–2 drops of a solution of tin(II) chloride in concentrated hydrochloric acid, and warm slightly. A brown or black precipitate or colouration is obtained.

Sensitivity: 1 μg As. *Concentration limit:* 1 in 50,000.

III.13 ARSENIC, As (A_r: 74·92) – ARSENIC(V) The properties of arsenic were summarized in Section **III.12**.

Reactions of arsenate ions A 0·1M solution of disodium hydrogen arsenate $Na_2HAsO_4.7H_2O$ can be used for the study of these reactions. The solution should contain some dilute hydrochloric acid.

1. Hydrogen sulphide: no immediate precipitate in the presence of dilute hydrochloric acid. If the passage of the gas is continued, a mixture of arsenic(III) sulphide, As_2S_3, and sulphur is slowly precipitated. Precipitation is more rapid in hot solution.

$$AsO_4^{3-} + H_2S \rightarrow AsO_3^{3-} + S\downarrow + H_2O$$
$$2AsO_3^{3-} + 3H_2S + 6H^+ \rightarrow As_2S_3\downarrow + 6H_2O$$

If a large excess of concentrated hydrochloric acid is present and hydrogen sulphide is passed rapidly into the cold solution, yellow arsenic pentasulphide As_2S_5 is precipitated; in the hot solution, the precipitate consists of a mixture of the tri- and penta-sulphides.

$$2AsO_4^{3-} + 5H_2S + 6H^+ \rightarrow As_2S_5\downarrow + 8H_2O$$

Arsenic pentasulphide, like the trisulphide, is readily soluble in alkali hydroxides or ammonia (a), ammonium sulphide (b), ammonium polysulphide (c), sodium or ammonium carbonate (d):

$$As_2S_5\downarrow + 6OH^- \rightarrow AsS_4^{3-} + AsO_3S^{3-} + 3H_2O \quad \text{(a)}$$
$$As_2S_5\downarrow + 3S^{2-} \rightarrow 2AsS_4^{3-} \quad \text{(b)}$$
$$As_2S_5\downarrow + 6S_2^{2-} \rightarrow 2AsS_4 + 3S_3^{2-} \quad \text{(c)}$$
$$As_2S_5\downarrow + 3CO_3^{2-} \rightarrow AsS_4^{3-} + AsO_3S^{3-} + 3CO_2 \quad \text{(d)}$$

Upon acidifying these solutions with hydrochloric acid, arsenic pentasulphide is reprecipitated:

$$2AsS_4^{3-} + 6H^+ \rightarrow As_2S_5\downarrow + 3H_2S\uparrow$$

For the rapid precipitation of arsenic from solutions of arsenates without using a large excess of hydrochloric acid, sulphur dioxide may be passed into the slightly acid solution in order to reduce the arsenic to the tervalent state and then the excess of sulphur dioxide is boiled off; on conducting hydrogen sulphide into the warm reduced solution immediate precipitation of arsenic trisulphide occurs.

The precipitation can be greatly accelerated by the addition of small amounts of an iodide, say, 1 ml of a 10 per cent solution, and a little concentrated hydrochloric acid. The iodide acts as a catalyst in that it reduces the arsenic acid thus:

$$AsO_4^{3-} + 2I^- + 2H^+ \rightarrow AsO_3^{3-} + I_2 + H_2O$$

The iodine liberated is reduced, in turn, to iodide:

$$I_2 + H_2S \rightarrow 2H^+ + 2I^- + S\downarrow$$

2. *Silver nitrate solution:* brownish-red precipitate of silver arsenate Ag_3AsO_4 from neutral solutions (distinction from arsenite and phosphate which yield yellow precipitates), soluble in acids and in ammonia solution but insoluble in acetic acid.

$$AsO_4^{3-} + 3Ag^+ \rightarrow Ag_3AsO_4\downarrow$$

This reaction may be adapted as a delicate test for arsenic in the following manner. The test is applicable only in the absence of chromates, hexacyano-ferrate(II) and (III) ions, which also give coloured silver salts insoluble in acetic acid.

Place a drop of the test solution in a micro crucible, add a few drops of concentrated ammonia solution and of '10-volume' hydrogen peroxide, and warm. Acidify with acetic acid and add 2 drops silver nitrate solution. A brownish-red precipitate or colouration appears.

Sensitivity: 6 μg As. *Concentration limit:* 1 in 8,000.

3. Magnesia mixture (see Section **III.12**, *reaction 3):* white, crystalline precipitate of magnesium ammonium arsenate $Mg(NH_4)AsO_4.6H_2O$ from neutral or ammoniacal solution (distinction from arsenite):

$$AsO_4^{3-} + Mg^{2+} + NH_4^+ \rightarrow MgNH_4AsO_4\downarrow$$

For some purposes (e.g. the detection of arsenate in the presence of phosphate), it is better to use the magnesium nitrate reagent (a solution containing $Mg(NO_3)_2$, NH_4Cl, and a little NH_3).

Upon treating the white precipitate with silver nitrate solution containing a few drops of acetic acid, red silver arsenate is formed (distinction from phosphate):

$$MgNH_4AsO_4\downarrow + 3Ag^+ \rightarrow Ag_3AsO_4\downarrow + Mg^{2+} + NH_4^+$$

4. Ammonium molybdate solution: when the reagent and nitric acid are added in considerable excess to a solution of an arsenate, a yellow crystalline precipitate of ammonium arsenomolybdate, $(NH_4)_3\ AsMo_{12}O_{40}$ is obtained on boiling (distinction from arsenites which give no precipitate, and from phosphates which yield a precipitate in the cold or upon gentle warming). The precipitate is insoluble in nitric acid, but dissolves in ammonia solution and in solutions of caustic alkalis.

$$AsO_4^{3-} + 12MoO_4^{2-} + 3NH_4^+ + 24H^+ \rightarrow (NH_4)_3\ AsMo_{12}O_{40}\downarrow + 12H_2O$$

The precipitate in fact contains trimolybdate ($Mo_3O_{10}^{2-}$) ions; each replacing one oxygen in AsO_4^{3-}. The composition of the precipitate should be written as $(NH_4)_3[As(Mo_3O_{10})_4]$.

5. Potassium iodide solution: in the presence of concentrated hydrochloric acid, iodine is precipitated; upon shaking the mixture with 1–2 ml of chloroform or of carbon tetrachloride, the latter is coloured blue by the iodine. The reaction may be used for the detection of arsenate in the presence of arsenite; oxidizing agents must be absent.

$$AsO_4^{3-} + 2H^+ + 2I^- \rightleftarrows AsO_3^{3-} + I_2\downarrow + H_2O$$

The reaction is reversible (cf. Section **III.12**, reaction 5); a large amount of acid must be present to complete this reaction.

6. Uranyl acetate solution: light yellow, gelatinous precipitate of uranyl ammonium arsenate $UO_2(NH_4)AsO_4$, xH_2O in the presence of excess of ammonium acetate, soluble in mineral acids but insoluble in acetic acid. If precipitation is carried out from a hot solution of an arsenate, the precipitate is obtained in granular form. This test provides an excellent method of distinction from arsenites, which do not give a precipitate with the reagent (an approximately 0·1M solution of uranyl acetate).

$$AsO_4^{3-} + UO_2^{2+} + NH_4^+ \rightarrow UO_2NH_4AsO_4\downarrow$$

III.14 SPECIAL TESTS FOR SMALL AMOUNTS OF ARSENIC These tests are applicable to all, arsenic compounds, and play an important role in forensic analysis.

(i) Marsh's test This test, which must be carried out in the fume chamber, is based upon the fact that all soluble compounds of arsenic are reduced by 'nascent' hydrogen in acid solution to arsine, AsH_3, a colourless, extremely poisonous gas with a garlic-like odour. If the gas, mixed with hydrogen, is conducted through a heated glass tube, it is decomposed into hydrogen and metallic arsenic, which is deposited as a brownish-black 'mirror' just beyond the heated part of the tube.

On igniting the mixed gases, composed of hydrogen and arsine (after all the air has been expelled from the apparatus), they burn with a livid blue flame and white fumes of arsenic(III) oxide are evolved; if the inside of a small porcelain dish is pressed down upon the flame, a black deposit of arsenic is obtained on the cool surface, and the deposit is readily soluble in sodium hypochlorite or bleaching powder solution (distinction from antimony).

The following reactions take place during these operations.

$$As^{3+} + 3Zn + 3H^+ \rightarrow AsH_3\uparrow + 3Zn^{2+}$$

$$AsO_4^{3-} + 4Zn + 11H^+ \rightarrow AsH_3\uparrow + 4Zn^{2+} + 4H_2O$$

$$4AsH_3\uparrow \rightarrow (\text{heat}) \rightarrow 4As\downarrow + 6H_2\uparrow$$

$$2AsH_3\uparrow + 3O_2 \rightarrow As_2O_3 + 3H_2O$$

$$2As\downarrow + 5OCl^- + 3H_2O \rightarrow 2AsO_4^{3-} + 5Cl^- + 6H^+$$

The Marsh test is best carried out as follows. The apparatus is fitted up as shown in Fig. III.1. A conical flask of about 125 ml capacity is fitted with a two-holed rubber stopper carrying a thistle funnel reaching nearly to the bottom of the flask and a 5–7 mm right-angle tube; the latter is attached by a short piece of 'pressure' tubing to a U-tube filled with glass wool moistened with lead acetate solution to absorb any hydrogen sulphide evolved (this may be dispensed with, if desired, as its efficacy has been questioned), then to a small tube containing anhydrous calcium chloride of about 8 mesh, then to a hard glass tube, c. 25 cm long and 7 mm diameter, constricted twice near the middle to about 2 mm diameter, the distance between the constrictions being 6–8 cm. The drying tubes and the tube ABC are securely supported by means of clamps.

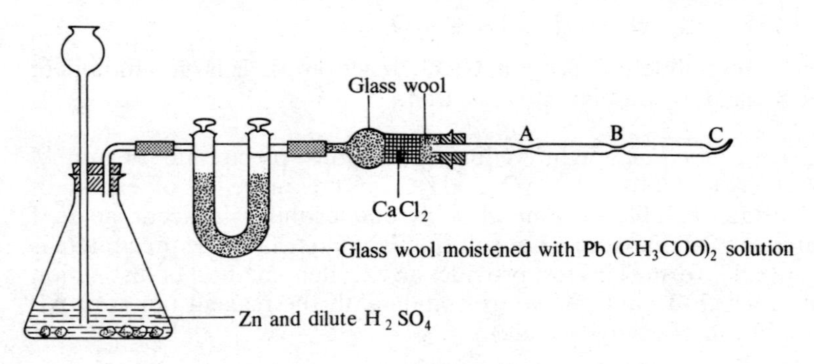

Fig. III.1

All reagents must be arsenic-free. Place 15–20 grams of arsenic-free zinc in the flask, add dilute sulphuric acid (1:3) until hydrogen is vigorously evolved. The purity of the reagents is tested by passing the gases, by means of a delivery tube attached by a short piece of rubber tubing to the end of C, through silver nitrate solution for several minutes; the absence of a black precipitate or suspension proves that appreciable quantities of arsenic are not present.

The black precipitate is silver:

$$AsH_3\uparrow + 6Ag^+ \rightarrow As^{3+} + 3H^+ + 6Ag\downarrow$$

The solution containing the arsenic compound is then added in small amounts at a time to the contents of the flask. If much arsenic is present, there will be an almost immediate blackening of the silver nitrate solution. Disconnect the rubber tube at C. Heat the tube at A to just below the softening point; a mirror of arsenic is deposited in the cooler, less constricted portion of the tube. A second flame may be applied at B to ensure complete decomposition (arsine is extremely poisonous). When a satisfactory mirror has been obtained, remove the flames at A and B and apply a light at C. Hold a cold porcelain dish in the flame, and test the solubility of the black or brownish deposit in sodium hypochlorite solution.

Smaller amounts of arsenic may be present in the silver nitrate solution as arsenious acid and can be detected by the usual tests, e.g. by hydrogen sulphide after removing the excess of silver nitrate with dilute hydrochloric acid, or by neutralizing and adding further silver nitrate solution, if necessary.

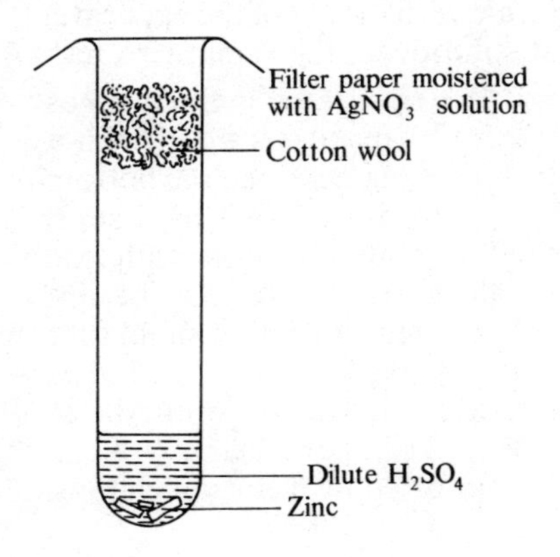

Fig. III.2

The original Marsh test involved burning and deposition of the arsenic upon a cold surface. Nowadays the mirror test is usually applied. The silver nitrate reaction (sometimes known as Hofmann's test) is very useful as a confirmatory test.

(ii) Gutzeit's test This is essentially a modification of Marsh's test, the chief difference being that only a test-tube is required and the arsine is detected by means of silver nitrate or mercury(II) chloride. Place 1–2 g arsenic-free zinc in a test-tube, add 5–7 ml dilute sulphuric acid, loosely plug the tube with purified cotton wool and then place a piece of filter paper moistened with 20 per cent silver nitrate solution on top of the tube (Fig. III.2). It may be necessary to warm the tube gently to produce a regular evolution of hydrogen. At the end of a definite period, say 2 minutes, remove the filter paper and examine the part that covered the test-tube; usually a light-brown spot is obtained owing to the traces of arsenic present in the reagents. Remove the cotton-wool plug, add 1 ml of the solution to be tested, replace the cotton wool and silver nitrate paper, the latter displaced so that a fresh portion is exposed. After 2 minutes, assuming that the rate of evolution of gas is approximately the same, remove the filter paper and examine the two spots. If much arsenic is present, the second spot (due to metallic silver), will appear black.

Hydrogen sulphide, phosphine PH_3, and stibine SbH_3, give a similar reaction. They may be removed by means of a purified cotton-wool plug impregnated with copper(I) chloride.

The use of mercury(II) chloride paper, prepared by immersing filter paper in a 5 per cent solution of mercuric chloride in alcohol and drying in the atmosphere out of contact with direct sunlight, constitutes an improvement. This is turned yellow by a little arsine and reddish-brown by larger quantities. Filter paper, impregnated with a 0·3M aqueous solution of 'gold chloride' $NaAuCl_4.2H_2O$, may also be employed when a dark-red to blue-red stain is produced. A blank test must be performed with the reagent in all cases. In this reaction gold is formed:

$$AsH_3 + 2Au^{3+} \rightarrow As^{3+} + 2Au\downarrow + 3H^+$$

The test may be performed on the semimicro scale with the aid of the apparatus shown in Fig. III.3. Place 10 drops of the test solution in the semimicro test-

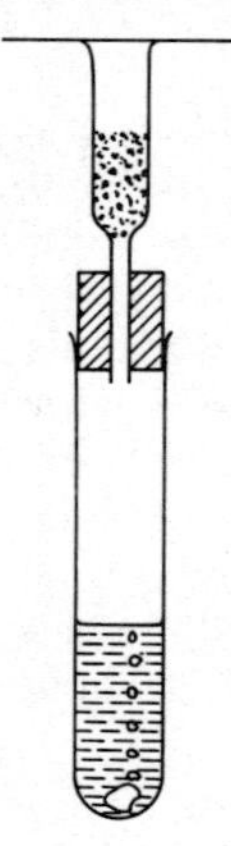

Fig. III.3

tube, add a few granules of arsenic-free zinc and 1 ml dilute sulphuric acid. Insert a loose wad of pure cotton wool moistened with lead nitrate solution in the funnel, and on top of this place a disc of drop-reaction paper impregnated with 20 per cent silver nitrate solution; the paper may be held in position by a watch glass or microscope slide. Warm the test-tube gently (if necessary) on a water bath to accelerate the reaction, and allow to stand. Examine the silver nitrate paper after about 5 minutes. A grey spot will be obtained; this is occasionally yellow, due to the complex $Ag_3As.3AgNO_3$.

For minute quantities of arsenic, it is convenient to use the apparatus depicted in Fig. II.56. Mix a drop of the test solution with a few grains of zinc and a few drops of dilute sulphuric acid in the micro test-tube. Insert the funnel with a flat rim, and place a small piece of drop-reaction paper moistened with 20 per cent silver nitrate solution on the flat surface. A grey stain will be obtained.

Sensitivity: 1 μg As. *Concentration limit:* 1 in 50,000.

A more sensitive test is provided by drop-reaction paper impregnated with gold chloride reagent (a 0·3M solution of $NaAuCl_4 . 2H_2O$). Perform the test as described in the previous paragraph: a blue to blue-red stain of metallic gold is obtained after standing for 10–15 minutes. It is essential to perform a blank test with the reagents to confirm that they are arsenic-free.

Sensitivity: 0·05 μg As. *Concentration limit:* 1 in 100,000.

(iii) Fleitmann's test This test depends upon the fact that nascent hydrogen generated in alkaline solution, e.g. from aluminium or zinc and sodium hydroxide solution, reduces arsenic(III) compounds to arsine, but does not affect antimony compounds. A method of distinguishing arsenic and antimony compounds is thus provided. Arsenates must first be reduced to the tervalent state before applying the test. The *modus operandi* is as for the Gutzeit test, except that zinc or aluminium and sodium hydroxide solution replace zinc and dilute sulphuric acid. It is necessary to warm the solutions. A black stain of silver is produced by the action of the arsine.

The apparatus of Fig. III.3 may be used on the semimicro scale. Place 1 ml test solution in the test-tube, add some pure aluminium turnings, and 1 ml 2M potassium hydroxide solution. Gentle warming is usually necessary. A yellow or grey stain is produced after several minutes.

(iv) Reinsch's test If a strip of bright copper foil is boiled with a solution of an arsenic(III) compound acidified with at least one-tenth of its bulk of concentrated hydrochloric acid, the arsenic is deposited upon the copper as a grey film of copper arsenide Cu_5As_2. Antimony, mercury, silver, and other metals are precipitated under similar conditions. It is therefore necessary to test for arsenic in the deposit in the dry way. The strip is washed with distilled water, dried between filter paper, and then gently heated in a test-tube; a white crystalline deposit of arsenic(III) oxide is obtained. The latter is identified by examination with a hand lens, when it will be seen to consist of colourless octahedral and tetrahedral crystals; it may also be dissolved in water and tested for arsenic by Fleitmann's or Bettendorff's test.

Arsenates are also reduced by copper, but only slowly even on boiling.

(v) Dry tests **a.** Blowpipe test Arsenic compounds when heated upon charcoal with sodium carbonate give a white incrustation of arsenic(III) oxide, and an odour of garlic is apparent while hot.

b. When heated with excess of potassium cyanide and of anhydrous sodium carbonate in a dry bulb tube, a black mirror of arsenic, soluble in sodium hypochlorite solution, is produced in the cooler part of the tube.

III.15 ANTIMONY, Sb (A_r: 121·75) – ANTIMONY(III) Antimony is a lustrous, silver-white metal, which melts at 630°C. It is insoluble in hydrochloric acid, and in dilute sulphuric acid. In hot, concentrated sulphuric acid it dissolves slowly forming antimony(III) ions:

$$2Sb + 3H_2SO_4 + 6H^+ \rightarrow 2Sb^{3+} + 3SO_2\uparrow + 6H_2O$$

Nitric acid oxidizes antimony to an insoluble product, which can be regarded as a mixture of Sb_2O_3 and Sb_2O_5. These anhydrides, in turn, can be dissolved in tartaric acid. A mixture of nitric acid and tartaric acid dissolves antimony easily.

Aqua regia dissolves antimony, when antimony(III) ions are formed:

$$Sb + HNO_3 + 3HCl \rightarrow Sb^{3+} + 3Cl^- + NO\uparrow + 2H_2O$$

Two series of salts are known, with antimony(III) and antimony(V) ions in them; these are derived from the oxides Sb_2O_3 and Sb_2O_5.

Antimony(III) compounds are easily dissolved in acids, when the ion Sb^{3+} is stable. If the solution is made alkaline, or the concentration of hydrogen ions is decreased by dilution, hydrolysis occurs when antimonyl, SbO^+, ions are formed:

$$Sb^{3+} + H_2O \rightleftarrows SbO^+ + 2H^+$$

Antimony(V) compounds contain the antimonate ion, SbO_4^{3-}. Their characteristics are similar to the corresponding arsenic compounds.

Reactions of antimony(III) ions A 0·2M solution of antimony(III) chloride, $SbCl_3$, can be used to study these reactions. This can be prepared either by dissolving the solid antimony(III) chloride or antimony(III) oxide, Sb_2O_3, in dilute hydrochloric acid.

1. Hydrogen sulphide: orange-red precipitate of antimony trisulphide, Sb_2S_3, from solutions which are not too acid. The precipitate is soluble in warm concentrated hydrochloric acid (distinction and method of separation from arsenic(III) sulphide and mercury(II) sulphide), in ammonium polysulphide (forming a thioantimonate), and in alkali hydroxide solutions (forming antimonite and thioantimonite).

$$2Sb^{3+} + 3H_2S \rightarrow Sb_2S_3\downarrow + 6H^+$$
$$Sb_2S_3\downarrow + 6HCl \rightarrow 2Sb^{3+} + 6Cl^- + 3H_2S\uparrow$$
$$Sb_2S_3\downarrow + 4S_2^{2-} \rightarrow 2SbS_4^{3-} + S_3^{2-}$$
$$2Sb_2S_3\downarrow + 4OH^- \rightarrow SbO_2^- + 3SbS_2^- + 2H_2O$$

Upon acidification of the thioantimonate solution with hydrochloric acid, antimony pentasulphide is precipitated initially but usually decomposes partially into the trisulphide and sulphur:

$$2SbS_4^{3-} + 6H^+ \rightarrow Sb_2S_5\downarrow + 3H_2S\uparrow$$
$$Sb_2S_5\downarrow \rightarrow Sb_2S_3\downarrow + 2S\downarrow$$

Acidification of the antimonite-thioantimonite mixture leads to the precipitation of the trisulphide:

$$SbO_2^- + 3SbS_2^- + 4H^+ \rightarrow 2Sb_2S_3\downarrow + 2H_2O$$

2. *Water:* when the solution is poured into water, a white precipitate of antimonyl chloride $SbO.Cl$ is formed, soluble in hydrochloric acid and in tartaric acid solution (difference from bismuth). With a large excess of water the hydrated oxide $Sb_2O_3.xH_2O$ is produced.

3. *Sodium hydroxide or ammonia solution:* white precipitate of the hydrated antimony(III) oxide $Sb_2O_3.xH_2O$ soluble in concentrated (5M) solutions of caustic alkalis forming antimonites.

$$2Sb^{3+} + 6OH^- \rightarrow Sb_2O_3\downarrow + 3H_2O$$
$$Sb_2O_3\downarrow + 2OH^- \rightarrow 2SbO_2^- + H_2O$$

4. *Zinc:* a black precipitate of antimony is produced. If a little of the antimony trichloride solution is poured upon platinum foil and a fragment of metallic zinc is placed on the foil, a black stain of antimony is formed upon the platinum; the stain (or deposit) should be dissolved in a little warm dilute nitric acid and hydrogen sulphide passed into the solution after dilution; an orange precipitate of antimony trisulphide will be obtained.

$$2Sb^{3+} + 3Zn\downarrow \rightarrow 2Sb\downarrow + 3Zn^{2+}$$

Some stibine SbH_3 may be evolved when zinc is used; it is preferable to employ tin.

$$2Sb^{3+} + 3Sn\downarrow \rightarrow 2Sb\downarrow + 3Sn^{2+}$$

A modification of the above test is to place a drop of the solution containing antimony upon a genuine silver coin and to touch the coin through the drop with a piece of tin or zinc; a black spot will form on the coin.

5. *Iron wire:* black precipitate of antimony. This may be confirmed as described in reaction 4.

$$2Sb^{3+} + 3Fe \rightarrow 2Sb\downarrow + 3Fe^{2+}$$

6. *Potassium iodide solution:* yellow colouration owing to the formation of a complex salt.

$$Sb^{3+} + 6I^- \rightarrow [SbI_6]^{3-}$$

7. *Rhodamine-B (or tetraethylrhodamine) reagent*

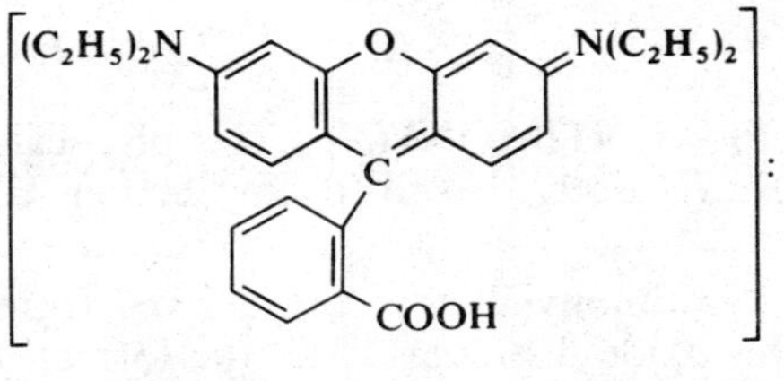

violet or blue colouration with quinquevalent antimony. Tervalent antimony does not respond to this test, hence it must be oxidized with potassium or sodium nitrite in the presence of strong hydrochloric acid. In Group IIB $SbCl_3$ is always formed together with $SnCl_4$ when the precipitate is treated with hydrochloric acid: by oxidizing Sb(III) to Sb(V) with a little solid nitrite, an excellent means of testing for Sb in the presence of a large excess of Sn is available. Mercury, gold, thallium, molybdates, vanadates, and tungstates in solution give similar colour reactions.

The test solution should be strongly acid with hydrochloric acid and the antimony(III) oxidized by the addition of a little solid sodium or potassium nitrite: a large excess of nitrite should be avoided. Place 1 ml reagent on a spot plate and add 1 drop test solution. The bright-red colour of the reagent changes to blue.

Sensitivity: 0·5 μg Sb and applicable in the presence of 12,500 times that amount of Sn. *Concentration limit:* 1 in 100,000.

The reagent is prepared by dissolving 0·01 g of rhodamine-B in 100 ml of water. A more concentrated reagent is obtained by dissolving 0·05 g of rhodamine-B in a 15 per cent solution of potassium chloride in 2M hydrochloric acid.

8. Phosphomolybdic acid reagent ($H_3[PMo_{12}O_{40}]$): 'molybdenum blue' is produced by antimony(III) salts. Of the ions in Group II, only tin(II) interferes with the test. The test solution may consist of the filtered solution obtained by treating the Group IIB precipitate with hydrochloric acid: the antimony is present as Sb^{3+} and the tin as Sn^{4+}, which has no effect upon the reagent.

Place a drop of the test solution upon drop-reaction paper which has been impregnated with the phosphomolybdic acid reagent and hold the paper in steam. A blue colouration appears within a few minutes.

Sensitivity: 0·2 μg Sb. *Concentration limit:* 1 in 250,000.

Alternatively, place 1 ml test solution in a semimicro test-tube, add 0·5–1 ml reagent, and heat for a short time. The reagent is reduced to a blue compound, which can be extracted with amyl alcohol.

Tin(II) chloride reduces not only the reactive phosphomolybdic acid but also its relatively unreactive (e.g. ammonium or potassium) salts to 'molybdenum blue'. However, antimony(III) salts do not reduce ammonium phosphomolybdate: Sn^{2+} may thus be detected in the presence of Sb^{3+}.

Impregnate drop-reaction paper with a solution of phosphomolybdic acid and then hold it for a short time over ammonia gas to form the yellow, sparingly soluble ammonium salt; dry. Place a drop of the test solution on this paper: a blue spot appears if tin(II) is present.

Sensitivity: 0·03 μg Sn. *Concentration limit:* 1 in 650,000.

The reagent consists of a 5 per cent aqueous solution of phosphomolybdic acid. It does not keep well.

III.16 ANTIMONY, Sb (*A*$_r$: 121·75) – ANTIMONY(V) The physical and chemical properties of antimony have been described in Section **III.15**.

Reactions of the antimony(V) ions Antimony(V) ions are derived from the amphoteric oxide Sb_2O_5. In **acids** this oxide dissolves under the formation of the **antimony(V) cation Sb^{5+}**:

$$Sb_2O_5 + 10H^+ \rightleftarrows 2Sb^{5+} + 5H_2O$$

In acid solutions therefore we have the Sb^{5+} ion present.

In **alkalis**, on the other hand, the antimonate SbO_4^{3-} ion is formed:

$$Sb_2O_5 + 3OH^- \rightleftarrows 2SbO_4^{3-} + 3H^+$$

In alkaline medium therefore we have the SbO_4^{3-} present in solutions. SbO_4^{3-} is a simplified expression of the composition of the antimonate ion; in fact it exists in the hydrated form, which may be termed hexahydroxoantimonate(V). Its formation from Sb_2O_5 with alkalis may be described by the reaction:

$$Sb_2O_5 + 2OH^- + 5H_2O \rightleftarrows 2[Sb(OH)_6]^-$$

For the study of these reactions, an acidified 0·2M solution of potassium hexahydroxoantimonate $K[Sb(OH)_6]$ can be used. Alternately, antimony pentoxide Sb_2O_5 may be dissolved in concentrated hydrochloric acid.

1. Hydrogen sulphide: orange-red precipitate of antimony pentasulphide, Sb_2S_5, in moderately acid solutions. The precipitate is soluble in ammonium sulphide solution (yielding a thioantimonate) in alkali hydroxide solutions, and is also dissolved by concentrated hydrochloric acid with the formation of antimony trichloride and the separation of sulphur. The thio-salt is decomposed by acids, the pentasulphide being precipitated.

$$2Sb^{5+} + 5H_2S \rightarrow Sb_2S_5\downarrow + 10H^+$$

$$Sb_2S_5\downarrow + 3S^{2-} \rightarrow 2SbS_4^{3-}$$

$$Sb_2S_5\downarrow + 6OH^- \rightarrow SbSO_3^{3-} + SbS_4^{3-} + 3H_2O$$

$$Sb_2S_5\downarrow + 6H^+ \rightarrow 2Sb^{3+} + 2S\downarrow + 3H_2S\uparrow$$

$$2SbS_4^{3-} + 6H^+ \rightarrow Sb_2S_5\downarrow + 3H_2S\uparrow$$

$$SbSO_3^{3-} + SbS_4^{3-} + 6H^+ \rightarrow Sb_2S_5\downarrow + 3H_2O$$

Note that the thioantimonate $SbSO_3^{3-}$ ion is the derivative of the antimonate ion SbO_4^{3-} (and has nothing whatsoever to do with the sulphite ion SO_3^{2-}).

2. Water: white precipitate of basic salts with various compositions; ultimately antimonic acid is formed:

$$Sb^{5+} + 4H_2O \rightleftarrows H_3SbO_4\downarrow + 5H^+$$

The precipitate dissolves both in acids and alkalis (but not in alkali carbonates):

$$H_3SbO_4\downarrow + 5H^+ \rightleftarrows Sb^{5+} + 4H_2O$$

$$H_3SbO_4\downarrow + 3OH^- \rightleftarrows SbO_4^{3-} + 3H_2O$$

3. Potassium iodide solution: in acidic solution iodine is separated:

$$Sb^{5+} + 2I^- \rightleftarrows Sb^{3+} + I_2\downarrow$$

If the Sb^{5+} ions are in excess, iodine crystals precipitate out and float on the surface of the solution. When heated the characteristic violet vapour of iodine appears. If the reagent is added in excess, brown tri-iodide ions are formed which screen the yellow colour of hexaiodoantimonate(III) ions:

$$Sb^{5+} + 9I^- \rightarrow [SbI_6]^{3-} + I_3^-$$

4. Zinc or tin: black precipitate of antimony in the presence of hydrochloric acid:

$$2Sb^{5+} + 5Zn\downarrow \rightarrow 2Sb\downarrow + 5Zn^{2+}$$
$$2Sb^{5+} + 5Sn\downarrow \rightarrow 2Sb\downarrow + 5Sn^{2+}$$

Some stibine (SbH_3) is produced with zinc.

5. Rhodamine-B reagent: see antimony(III) compounds (Section **III.15**, reaction 7). There is no need for preliminary oxidation in this case.

III.17 SPECIAL TESTS FOR SMALL AMOUNTS OF ANTIMONY

(i) Marsh's test This test is carried out exactly as described for arsenic. The stibine, SbH_3 (mixed with hydrogen), which is evolved burns with a faintly bluish-green flame and produces a dull black spot upon a cold porcelain dish held in the flame; this deposit in insoluble in sodium hypochlorite or bleaching powder solution, but is dissolved by a solution of tartaric acid (difference from arsenic).

The gas is also decomposed by passage through a tube heated to dull redness. A lustrous mirror of antimony is formed in a similar manner to the arsenic mirror, but it is deposited on both sides of the heated portion of the tube because of the greater instability of the stibine. This mirror may be converted into orange-red antimony trisulphide by dissolving it in a little boiling hydrochloric acid and passing hydrogen sulphide into the solution.

When the stibine-hydrogen mixture is passed into a solution of silver nitrate (Hofmann's test) a black precipitate of silver antimonide Ag_3Sb is obtained; this is decomposed by the excess of silver nitrate into silver and antimony(III) oxide:

$$SbH_3 + 3Ag^+ \rightarrow Ag_3Sb\downarrow + 3H^+$$
$$2Ag_3Sb\downarrow + 6Ag^+ + 3H_2O \rightarrow 12Ag\downarrow + Sb_2O_3\downarrow + 6H^+$$

It is best to dissolve the precipitate in a solution of tartaric acid and to test for antimony with hydrogen sulphide in the usual manner.

(ii) Gutzeit's test A brown stain is produced which is soluble in 80 per cent alcohol, provided the antimony concentration of the solution is not too high.

(iii) Fleitmann's or Bettendorff's tests Negative results are obtained when these are applied to antimony compounds (distinction from arsenic).

(iv) Reinsch's test A grey to black deposit is formed upon the copper. Ignition of this deposit in a dry test-tube gives a non-crystalline sublimate of antimony(III) oxide, soluble in potassium hydrogen tartrate solution; hydrogen sulphide precipitates orange-red antimony sulphide from this solution, acidified with hydrochloric acid.

(v) Dry test (blowpipe test). When antimony compounds are heated with sodium carbonate upon charcoal, a brittle metallic bead, surrounded by a white incrustation, is obtained.

III.18 TIN, Sn (A_r: 118·69) – Tin(II) Tin is a silver-white metal which is malleable and ductile at ordinary temperatures, but at low temperatures it becomes brittle due to transformation into a different allotropic modification. It melts at 231·8°C. The metal dissolves slowly in dilute hydrochloric and sulphuric acid with the formation of tin(II) (stannous) salts:

$$Sn + 2H^+ \rightarrow Sn^{2+} + H_2\uparrow$$

Dilute nitric acid dissolves tin slowly without the evolution of any gas, tin(II) and ammonium ions being formed:

$$4Sn + 10H^+ + NO_3^- \rightarrow 4Sn^{2+} + NH_4^+ + 3H_2O$$

With concentrated nitric acid a vigorous reaction occurs, a white solid, usually formulated as hydrated tin(IV) oxide $SnO_2.xH_2O$ and sometimes known as metastannic acid, being produced:

$$3Sn + 4HNO_3 + (x-2)H_2O \rightarrow 4NO\uparrow + 3SnO_2.xH_2O\downarrow$$

In the presence of antimony and tartaric acid tin dissolves readily in nitric acid (induced dissolution) because of complex formation. If larger amounts of iron are present, the formation of metastannic acid is again prevented.

In hot, concentrated sulphuric acid tin(IV) ions are formed at dissolution:

$$Sn + 4H_2SO_4 \rightarrow Sn^{4+} + 2SO_4^{2-} + 2SO_2\uparrow + 4H_2O$$

Aqua regia dissolves tin readily, when again tin(IV) (stannic) ions are formed:

$$3Sn + 4HNO_3 + 12HCl \rightarrow 3Sn^{4+} + 12Cl^- + 4NO\uparrow + 8H_2O$$

Tin can be bivalent and tetravalent in its compounds.

The tin(II) or stannous compounds are usually colourless. In acid solution the tin(II) ions Sn^{2+} are present, while in alkaline solutions tetrahydroxostannate(II) or stannite ions $[Sn(OH)_4]^{2-}$ are to be found. These two are readily transformed into each other:

$$Sn^{2+} + 4OH^- \rightleftarrows [Sn(OH)_4]^{2-}$$

Tin(II) ions are strong reducing agents.

Tin(IV) or stannic compounds are more stable. In their aqueous solution they can be present as the tin(IV) ions Sn^{4+} or as hexahydroxostannate(IV) (or simply stannate) ions $[Sn(OH)_6]^{2-}$. They again form an equilibrium system:

$$Sn^{4+} + 6OH^- \rightleftarrows [Sn(OH)_6]^{2-}$$

In acid solutions the equilibrium is shifted towards the left, while in alkaline medium it is shifted towards the right.

Reactions of tin(II) ions A 0·25M solution of tin(II) chloride, $SnCl_2.2H_2O$, can be used for studying these reactions. The solution should contain at least 4 per cent free hydrochloric acid (100 ml concentrated HCl per litre).

1. Hydrogen sulphide: brown precipitate of tin(II) sulphide, SnS, from not too acid solutions (say in the presence of 0·25–0·3M hydrochloric acid or *p*H c. 0·6). The precipitate is soluble in concentrated hydrochloric acid (distinction from arsenic(III) sulphide and mercury(II) sulphide); it is also soluble in yellow

[$(NH_4)_2S_x$], but not in colourless [$(NH_4)_2S$], ammonium sulphide solution to form a thiostannate. Treatment of the solution of ammonium thiostannate with an acid yields a yellow precipitate of tin(IV) sulphide, SnS_2.

$$Sn^{2+} + H_2S \rightarrow SnS\downarrow + 2H^+$$
$$SnS\downarrow + S_2^{2-} \rightarrow SnS_3^{2-}$$
$$SnS_3^{2-} + 2H^+ \rightarrow SnS_2\downarrow + H_2S\uparrow$$

Tin(II) sulphide is practically insoluble in solutions of caustic alkalis; hence, if potassium hydroxide solution is employed for separating Group IIA and Group IIB, the tin must be oxidized to the quadrivalent state with hydrogen peroxide before precipitation with hydrogen sulphide.

2. *Sodium hydroxide solution:* white precipitate of tin(II) hydroxide, which is soluble in excess alkali:

$$Sn^{2+} + 2OH^- \rightleftarrows Sn(OH)_2\downarrow$$
$$Sn(OH)_2\downarrow + 2OH^- \rightleftarrows [Sn(OH)_4]^{2-}$$

With ammonia solution, white tin(II) hydroxide is precipitated, which cannot be dissolved in excess ammonia.

3. *Mercury(II) chloride solution:* a white precipitate of mercury(I) chloride (calomel) is formed if a large amount of the reagent is added quickly:

$$Sn^{2+} + 2HgCl_2 \rightarrow Hg_2Cl_2\downarrow + Sn^{4+} + 2Cl^-$$

If however tin(II) ions are in excess, the precipitate turns grey, especially on warming, owing to further reduction to mercury metal:

$$Sn^{2+} + Hf_2Cl_2\downarrow \rightarrow 2Hg\downarrow + Sn^{4+} + 2Cl^-$$

4. *Bismuth nitrate and sodium hydroxide solutions:* black precipitate of bismuth metal (cf. Section **III.9**, reaction 6).

5. *Metallic zinc:* spongy tin is deposited which adheres to the zinc. If the zinc rests upon platinum foil, as described in Section **III.15**, reaction 4, and the solution is weakly acid, the tin is partially deposited upon the zinc in a spongy form but does not stain the platinum. The precipitate should be dissolved in concentrated hydrochloric acid and the mercury(II) chloride test applied.

6. *Dimethylglyoxime–iron(III) chloride test** No colouration is produced when iron(III) ions are mixed with the dimethylglyoxime reagent and a little ammonia solution, but if a trace of iron(II) is present (produced by reduction with tin(II) ions), a deep red colouration, due to the iron(II) dimethylglyoxime complex, is formed. If tin is present in the quadrivalent state, it is reduced to Sn(II) by treatment with aluminium or magnesium filings and hydrochloric acid and the solution is filtered. This procedure may be used with the Group IIB precipitate after treatment with hydrochloric acid.

Vanadium, uranyl, titanium, cobalt, and nickel ions give a similar reaction; the ions of copper, chromium, manganese, gold, palladium, platinum, selenium,

* For formulae, see Section **III.27**, reaction 8.

and tellurium interfere, as do also molybdates and tungstates. Reducing agents, which affect iron(III) chloride, must be absent.

Place 0·2 ml test solution containing tin(II) ions (this may consist of the solution obtained from the Group IIB precipitate, reduced with magnesium) in a micro test-tube, acidify (if necessary) with dilute hydrochloric acid, add 0·2 ml 0·1M iron(III) chloride solution, followed by 0·3 ml 5 per cent tartaric acid solution (to prevent the formation of iron(III) hydroxide), 3 drops dimethylglyoxime reagent, and about 0·5 ml 2M ammonia solution. A red colouration is produced.

Sensitivity: 0·05 μg Sn. *Concentration limit:* 1 in 1,250,000

The reagent consists of a 1 per cent solution of dimethylglyoxime in ethanol.

7. *Cacotheline reagent (a nitro-derivative of brucine, $C_{21}H_{21}O_7N_3$)*: violet colouration with stannous salts. The test solution should be acid (2M HCl), and if tin is in the quadrivalent state, it should be reduced previously with aluminium or magnesium, and the solution filtered.

The following interfere with the test: strong reducing agents (hydrogen sulphide, dithionites, sulphites and selenites); V, U, Te, Hg, Bi, Au, Pd, Se, Te, Sb, Mo, W, Co and Ni. The reaction is not selective, but is fairly sensitive: it can be used in the analysis of the Group IIB precipitate. Since iron(II) ions have no influence on the test, it may be applied to the tin solution which has been reduced with iron wire.

Impregnate some drop-reaction paper with the reagent and, before the paper is quite dry, add a drop of the test solution. A violet spot, surrounded by a less coloured zone, appears on the yellow paper.

Alternatively, treat a little of the test solution in a micro test-tube with a few drops of the reagent. A violet (purple) colouration is produced.

Sensitivity: 0·2 μg Sn. *Concentration limit:* 1 in 250,000.

The reagent consists of a 0·25 per cent aqueous solution of cacotheline.

8. *Diazine green reagent (dyestuff formed by coupling diazotized safranine with dimethylaniline):* tin(II) chloride reduces the blue diazine green to the red safranine, hence the colour change is blue → violet → red. Titanium(III) chloride reacts similarly, but iron(II) salts and similar reducing agents have no effect. The reagent is therefore useful in testing for tin in the mixed sulphides of antimony and tin obtained in routine qualitative analysis. The solution of the sulphides in hydrochloric acid is reduced with iron wire, aluminium, or magnesium powder and a drop of the reduced solution employed for the test.

Mix 1 drop of the test solution on a spot plate with 1 ml of the reagent. The colour changes from blue to violet or red. It is advisable to carry out a blank test.

Sensitivity: 2 μg Sn. *Concentration limit:* 1 in 25,000.

The reagent consists of a 0·01 per cent aqueous solution of diazine green.

9. *4-Methyl-1, 2-dimercapto-benzene (or 'dithiol') reagent*

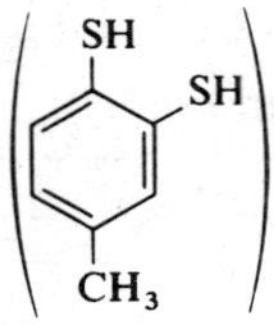

SH
SH
CH$_3$
):

red precipitate when warmed with tin(II) salts in acid solution. The following interfere: silver, lead, mercury, cadmium, arsenic, and antimony (yellow precipitates); copper, nickel, and cobalt (black precipitates); bismuth (red precipitate); colloidal organic substances (starch, etc.); phosphates and nitrites. The hydrochloric acid concentration of the solution should not exceed 15 per cent.

Place 2 drops of the test solution, acidified with dilute hydrochloric acid, in a micro crucible and add 3 drops of the reagent. Warm (not above 60°C): a red colour or precipitate develops.

Concentration limit: 1 in 500,000.

The reagent is prepared by dissolving 0·2 g 4-methyl-1, 2-dimercapto-benzene in 100 ml 1 per cent sodium hydroxide solution and adding 0·3–0·5 g thioglycollic acid. The use of the latter in the reagent is not imperative, but it serves to facilitate the reduction of any tin(IV) ions present. The reagent is discarded if a white precipitate of the disulphide forms.

III.19 TIN, Sn (A_r: 118·69) – TIN(IV) The properties of metallic tin were discussed at the beginning of Section **III.18**.

Reactions of tin(IV) ions To study these reactions use a 0·25M solution of ammonium hexachlorostannate(IV) by dissolving 92 g $(NH_4)_2[SnCl_6]$ in 250 ml concentrated hydrochloric acid and diluting the solution to 1 litre with water.

1. Hydrogen sulphide: yellow precipitate of tin(IV) sulphide SnS_2 from dilute acid solutions (0·3M). The precipitate is soluble in concentrated hydrochloric acid (distinction from arsenic(III) and mercury(II) sulphides), in solutions of alkali hydroxides, and also in ammonium sulphide and ammonium polysulphide. Yellow tin(IV) sulphide is precipitated upon acidification.

$$Sn^{4+} + 2H_2S \rightarrow SnS_2\downarrow + 4H^+$$
$$SnS_2\downarrow + S^{2-} \rightarrow SnS_3^{2-}$$
$$SnS_2\downarrow + 2S_2^{2-} \rightarrow SnS_3^{2-} + S_3^{2-}$$
$$SnS_3^{2-} + 2H^+ \rightarrow SnS_2\uparrow + H_2S^3\downarrow$$

No precipitation of tin(IV) sulphide occurs in the presence of oxalic acid, due to the formation of the stable complex ion of the type $[Sn(C_2O_4)_4(H_2O)_2]^{4-}$; this forms the basis of a method of separation of antimony and tin.

2. Sodium hydroxide solution: gelatinous white precipitate of tin(IV) hydroxide $Sn(OH)_4$, soluble in excess of the precipitant forming hexahydroxostannate(IV)

$$Sn^{4+} + 4OH^- \rightarrow Sn(OH)_4\downarrow$$
$$Sn(OH)_4\downarrow + 2OH^- \rightleftarrows [Sn(OH)_6]^{2-}$$

With ammonia and with sodium carbonate solutions, a similar precipitate is obtained which, however, is insoluble in excess reagent.

3. Mercury(II) chloride solution: no precipitate (difference from tin(II)).

4. Metallic iron: reduces tin(IV) ions to tin(II):

$$Sn^{4+} + Fe \rightarrow Fe^{2+} + Sn^{2+}$$

If pieces of iron are added to a solution, and the mixture is filtered, tin(II) ions can be detected with mercury(II) chloride reagent. A similar result is obtained on boiling the solution with copper or antimony.

5. Dry tests **a.** Blowpipe test All tin compounds when heated with sodium carbonate, preferably in the presence of potassium cyanide, on charcoal give white, malleable and metallic globules of tin which do not mark paper. Part of the metal is oxidized to tin(IV) oxide, especially on strong heating, which forms a white incrustation upon the charcoal.

b. Borax bead test A borax bead which has been coloured pale blue by a trace of copper salt becomes a clear ruby red in the reducing flame if a minute quantity of tin in added.

III.20 THIRD GROUP OF CATIONS: IRON(II) AND (III), ALUMINIUM, CHROMIUM(III) AND (VI), NICKEL, COBALT, MANGANESE(II) AND (VII), AND ZINC

Group reagent: hydrogen sulphide (gas or saturated aqueous solution) in the presence of ammonia and ammonium chloride, or ammonium sulphide solution.

Group reaction: precipitates of various colours: iron(II) sulphide (black), aluminium hydroxide (white), chromium(III) hydroxide (green), nickel sulphide (black), cobalt sulphide (black), manganese(II) sulphide (pink), and zinc sulphide (white).

The metals of this group are not precipitated by the group reagents for Groups I and II, but are all precipitated, in the presence of ammonium chloride, by hydrogen sulphide from their solutions made alkaline with ammonia solution. The metals, with the exception of aluminium and chromium which are precipitated as the hydroxides owing to the complete hydrolysis of the sulphides in aqueous solution, are precipitated as the sulphides. Iron, aluminium, and chromium (often accompanied by a little manganese) are also precipitated as the hydroxides by ammonia solution in the presence of ammonium chloride, whilst the other metals of the group remain in solution and may be precipitated as sulphides by hydrogen sulphide. It is therefore usual to subdivide the group into the iron group (iron, aluminium, and chromium) or Group IIIA and the zinc group (nickel, cobalt, manganese, and zinc) or Group IIIB.

III.21 IRON, Fe (A_r: 55·85) – IRON(II) Chemically pure iron is a silver-white, tenacious, and ductile metal. It melts at 1535°C. The commercial metal is rarely pure and usually contains small quantities of carbide, silicide, phosphide, and sulphide of iron, and some graphite. These contaminants play an important role in the strength of iron structures. Iron can be magnetized. Dilute or concentrated hydrochloric acid and dilute sulphuric acid dissolve iron, when iron(II) salts and hydrogen gas are produced.

$$Fe + 2H^+ \rightarrow Fe^{2+} + H_2\uparrow$$
$$Fe + 2HCl \rightarrow Fe^{2+} + 2Cl^- + H_2\uparrow$$

Hot, concentrated sulphuric acid yields iron(III) ions and sulphur dioxide:

$$2Fe + 3H_2SO_4 + 6H^+ \rightarrow 2Fe^{3+} + 3SO_2\uparrow + 6H_2O$$

With cold dilute nitric acid, iron(II) and ammonium ions are formed:

$$4Fe+10H^{+}+NO_3^{-} \rightarrow 4Fe^{2+}+NH_4^{+}+3H_2O$$

Cold, concentrated nitric acid renders iron *passive*; in this state it does not react with dilute nitric acid nor does it displace copper from an aqueous solution of a copper salt. 1+1 Nitric acid, or hot, concentrated nitric acid dissolves iron with the formation of nitrogen oxide gas and iron(III) ions:

$$Fe+HNO_3+3H^{+} \rightarrow Fe^{3+}+NO\uparrow+2H_2O$$

Iron forms two important series of salts.

The **iron(II)** (or ferrous) salts are derived from iron(II) oxide FeO. In solution they contain the cation Fe^{2+} and normally possess a slight-green colour. Intensively coloured ion-association and chelate complexes are also common. Iron(II) ions can be oxidized easily to iron(III), they are therefore strong reducing agents. The less acidic the solution the more pronounced this effect; in neutral or alkaline media even atmospheric oxygen will oxidize iron(II) ions. Iron(II) solutions must therefore be slightly acid when kept for a longer time.

Iron(III) (or ferric) salts are derived from iron(III) oxide Fe_2O_3. They are more stable than the iron(II) salts. In their solution the pale-yellow Fe^{3+} cations are present; if the solution contains chloride, the colour becomes stronger. Reducing agents convert iron(III) ions to iron(II).

Reactions of iron(II) ions Use a freshly prepared 0·5M solution of iron(II) sulphate $FeSO_4.7H_2O$ or iron(II) ammonium sulphate (Mohr's salt; $FeSO_4.(NH_4)_2SO_4.6H_2O$), acidified with 50 ml MH_2SO_4 per litre, for the study of these reactions.

1. Sodium hydroxide solution: white precipitate of iron(II) hydroxide, $Fe(OH)_2$, in the complete absence of air, insoluble in excess, but soluble in acids. Upon exposure to air, iron(II) hydroxide is rapidly oxidized, yielding ultimately reddish-brown iron(III) hydroxide. Under ordinary conditions it appears as a dirty-green precipitate; the addition of hydrogen peroxide immediately oxidizes it to iron(III) hydroxide.

$$Fe^{2+}+2OH^{-} \rightarrow Fe(OH)_2\downarrow$$

$$4Fe(OH)_2+2H_2O+O_2 \rightarrow 4Fe(OH)_3\downarrow$$

$$2Fe(OH)_2+H_2O_2 \rightarrow 2Fe(OH)_3\downarrow$$

2. Ammonia solution: precipitation of iron(II) hydroxide occurs (cf. reaction 1). If, however, larger amounts of ammonium ions are present, the dissociation of ammonium hydroxide is suppressed, (cf. Section **I.15**), and the concentration of hydroxyl ions is lowered to such an extent that the solubility product of iron(II) hydroxide, $Fe(OH)_2$, is not attained and precipitation does not occur. Similar remarks apply to the other divalent elements of Group III, nickel, cobalt, zinc and manganese and also to magnesium.

3. Hydrogen sulphide: no precipitation takes place in acid solution since the sulphide ion concentration, $[S^{2-}]$, is insufficient to exceed the solubility product of iron(II) sulphide. If the hydrogen-ion concentration is reduced, and the sulphide-ion concentration correspondingly increased, by the addition of sodium acetate solution, partial precipitation of black iron(II) sulphide, FeS, occurs.

4. *Ammonium sulphide solution:* black precipitate of iron(II) sulphide FeS, readily soluble in acids with evolution of hydrogen sulphide. The moist precipitate becomes brown upon exposure to air, due to its oxidation to basic iron(III) sulphate $Fe_2O(SO_4)_2$.

$$Fe^{2+} + S^{2-} \rightarrow FeS\downarrow$$
$$FeS\downarrow + 2H^+ \rightarrow Fe^{2+} + H_2S\uparrow$$
$$4FeSO\downarrow + 9O_2 \rightarrow 2Fe_2O(SO_4)_2\downarrow$$

5. *Potassium cyanide solution* (**POISON**): yellowish-brown precipitate of iron(II) cyanide, soluble in excess reagent when a pale-yellow solution of hexacyanoferrate(II) (ferrocyanide) $[Fe(CN)_6]^{4-}$ ions is obtained:

$$Fe^{2+} + 2CN^- \rightarrow Fe(CN)_2\downarrow$$
$$Fe(CN)_2\downarrow + 4CN^- \rightarrow [Fe(CN)_6]^{4-}$$

The hexacyanoferrate(II) ion being a complex ion does not give the typical reactions of iron(II) (cf. Sections **I.31** to **I.33**). The iron present in such solutions may be detected by decomposing the complex ion by boiling the solution with concentrated sulphuric acid *in a fume cupboard with good ventilation*, when carbon monoxide gas is formed (together with hydrogen cyanide, if potassium cyanide is present in excess):

$$[Fe(CN)^6]^{4-} + 6H_2SO_4 + 6H_2O \rightarrow Fe^{2+} + 6CO\uparrow + 6NH_4^+ + 6SO_4^{2-}$$

A dry sample which contains alkali hexacyanoferrate(II), decomposes on ignition to iron carbide, alkali cyanide, and nitrogen. By dissolving the residue in acid iron can be detected in the solution (all these operations must be carried out in a fume cupboard).

6. *Potassium hexacyanoferrate(II) solution:* in the complete absence of air a white precipitate of potassium iron(II) hexacyanoferrate(II) is formed:

$$Fe^{2+} + 2K^+ + [Fe(CN)_6]^{4-} \rightarrow K_2Fe[Fe(CN)_6]\downarrow$$

under ordinary atmospheric conditions a pale-blue precipitate is obtained (cf. reaction 7).

7. *Potassium hexacyanoferrate(III) solution:* a dark-blue precipitate is obtained. First hexacyanoferrate(III) ions oxidize iron(II) to iron(III), when hexacyanoferrate(II) is formed:

$$Fe^{2+} + [Fe(CN)_6]^{3-} \rightarrow Fe^{3+} + [Fe(CN)_6]^{4-}$$

and these ions combine to a precipitate called Turnbull's blue:

$$4Fe^{3+} + 3[Fe(CN)_6]^{4-} \rightarrow Fe_4[Fe(CN)_6]_3$$

Note, that the composition of this precipitate is identical to that of Prussian blue (cf. Section **III.22**, reaction 6). Earlier it was suggested that its composition was iron(II) hexacyanoferrate(III), $Fe_3[Fe(CN)_6]_2$, hence the different name. The identical composition and structure of Turnbull's blue and Prussian blue has recently been proved by Mössbauer spectroscopy. The precipitate is decomposed by sodium or potassium hydroxide solution, iron(III) hydroxide being precipitated.

8. Ammonium thiocyanate solution: no colouration is obtained with pure iron(II) salts (distinction from iron(III) ions).

9. α,α′-Dipyridyl reagent

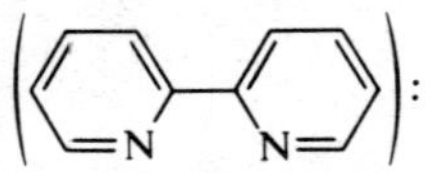

deep-red complex bivalent cation $[Fe(C_5H_4N)_2]^{2+}$ with iron(II) salts in mineral acid solution. Iron(III) ion does not react. Other metallic ions react with the reagent in acid solution, but the intensities of the resulting colours are so feeble that they do not interfere with the test for iron provided excess reagent is employed. Large amounts of halides and sulphates reduce the solubility of the iron(II) dipyridyl complex and a red precipitate may be formed.

Treat a drop of the faintly acidified test solution with 1 drop of the reagent on a spot plate: a red colouration is obtained. Alternatively, treat drop-reaction paper (Whatman No. 3 M.M., 1st quality) which has been impregnated with the reagent and dried, with a drop of the test solution: a red or pink spot is produced.

Sensitivity: 0·3 μg Fe^{2+}. *Concentration limit:* 1 in 1,600,000.

If appreciable quantities of iron(III) salts are present and traces of iron(II) salts are sought, it is best to carry out the reaction in a micro crucible lined with paraffin wax and to mask iron(III) ions as $[FeF_6]^{3-}$ by the addition of a few drops of potassium fluoride solution.

The reagent is prepared by dissolving 0·01 g α,α′-dipyridyl in 0·5 ml alcohol or in 0·5 ml 0·1M hydrochloric acid.

10. Dimethylglyoxime reagent: soluble red iron(II) dimethylglyoxime in ammoniacal solution. Iron(III) salts give no colouration, but nickel, cobalt, and large quantities of copper salts interfere and must be absent. The test may be carried out in the presence of potassium cyanide solution in which nickel dimethylglyoxime (cf. Section **III.27**, reaction 8) dissolves.

Mix a drop of the test solution with a small crystal of tartaric acid; introduce a drop of the reagent, followed by 2 drops ammonia solution. A red colouration appears.

Sensitivity: 0·04 μg Fe^{2+}. *Concentration limit:* 1 in 125,000.

The colouration fades on standing owing to the oxidation of the iron(II) complex.

If it is desired to detect iron(III) ions by this test, they must first be reduced by a little hydroxylamine hydrochloride.

The reagent consists of a 1 per cent solution of dimethylglyoxime in alcohol.

11. o-*Phenanthroline reagent*

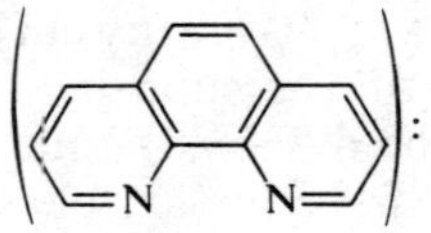

red colouration, due to the complex cation $[Fe(C_{18}H_8N_2)_3]^{2+}$ in faintly acid solution. Iron(III) has no effect and must first be reduced to the bivalent state with hydroxylamine hydrochloride if the reagent is to be used in testing for iron.

Place a drop of the faintly acid test solution on a spot plate and add 1 drop of the reagent. A red colour is obtained.

Concentration limit: 1 in 1,500,000.

The reagent consists of a 0·1 per cent solution of *o*-phenanthroline in water.

III.22 IRON, Fe (A_r: 55·85) – IRON(III) The most important characteristics of the metal have been described in Section III.21.

Reactions of iron(III) ions Use a 0·5M solution of iron(III) chloride $FeCl_3 . 6H_2O$. The solution should be clear yellow. If it turns brown, due to hydrolysis, a few drops of hydrochloric acid should be added.

1. Ammonia solution: reddish-brown, gelatinous precipitate of iron(III) hydroxide $Fe(OH)_3$, insoluble in excess of the reagent, but soluble in acids.

$$Fe^{3+} + 3NH_3 + 3H_2O \rightarrow Fe(OH)_3\downarrow + 3NH_4^+$$

The solubility product of iron(III) hydroxide is so small ($3{\cdot}8 \times 10^{-38}$) that complete precipitation takes place even in the presence of ammonium salts (distinction from iron(II), nickel, cobalt, manganese, zinc, and magnesium). Precipitation does not occur in the presence of certain organic acids (see reaction 8 below). Iron(III) hydroxide is converted on strong heating into iron(III) oxide; the ignited oxide is soluble with difficulty in dilute acids, but dissolves on vigorous boiling with concentrated hydrochloric acid.

$$2Fe(OH)_3\downarrow \rightarrow Fe_2O_3 + 3H_2O$$

$$Fe_2O_3 + 6H^+ \rightarrow 2Fe^{3+} + 3H_2O$$

2. Sodium hydroxide solution: reddish-brown precipitate of iron(III) hydroxide, insoluble in excess of the reagent (distinction from aluminium and chromium):

$$Fe^{3+} + 3OH^- \rightarrow Fe(OH)_3\downarrow$$

3. Hydrogen sulphide gas: in acidic solution reduces iron(III) ions to iron(II) and sulphur is formed as a milky-white precipitate:

$$2Fe^{3+} + H_2S \rightarrow 2Fe^{2+} + 2H^+ + S\downarrow$$

If a neutral solution of iron(III) chloride is added to a freshly prepared, saturated solution of hydrogen sulphide, a bluish colouration appears first, followed by the precipitation of sulphur. The blue colour is due to a colloid solution of sulphur of extremely small particle size. This reaction can be used to test the freshness of hydrogen sulphide solutions.

The finely distributed sulphur cannot be readily filtered with ordinary filter papers. By boiling the solution with a few torn pieces of filter paper the precipitate coagulates and can be filtered.

4. Ammonium sulphide solution: a black precipitate, consisting of iron(II) sulphide and sulphur is formed:

$$2Fe^{3+} + 3S^{2-} \rightarrow 2FeS\downarrow + S\downarrow$$

In hydrochloric acid the black iron(II) sulphide precipitate dissolves and the white colour of sulphur becomes visible:

$$FeS\downarrow + 2H^+ \rightarrow H_2S\uparrow + Fe^{2+}$$

From alkaline solutions black iron(III) sulphide is obtained:

$$2Fe^{3+} + 3S^{2-} \rightarrow Fe_2S_3\downarrow$$

On acidification with hydrochloric acid iron(III) ions are reduced to iron(II) and sulphur is formed:

$$Fe_2S_3\downarrow + 4H^+ \rightarrow 2Fe^{2+} + 2H_2S\uparrow + S\downarrow$$

The humid iron(II) sulphide precipitate, when exposed to air, is slowly oxidized to brown iron(III) hydroxide:

$$4FeS\downarrow + 6H_2O + 3O_2 \rightarrow 4Fe(OH)_3\downarrow + 4S\downarrow$$

The reaction is exothermal. Under certain conditions so much heat may be produced that the precipitate dries out, and the filter paper, with the finely distributed sulphur on it, catches fire. **Sulphide precipitates** therefore **should never be disposed of into a waste bin**, but should rather be washed away under running water; only the filter paper should be thrown away.

5. *Potassium cyanide* (**POISON**): when added slowly, produces a reddish-brown precipitate of iron(III) cyanide:

$$Fe^{3+} + 3CN^- \rightarrow Fe(CN)_3\downarrow$$

In excess reagent the precipitate dissolves giving a yellow solution, when hexacyanoferrate(III) ions are formed:

$$Fe(CN)_3\downarrow + 3CN^- \rightarrow [Fe(CN)_6]^{3-}$$

These reactions should be carried out in a fume cupboard, as the free acid present in the iron(III) chloride solution forms hydrogen cyanide gas with the reagent:

$$H^+ + CN^- \rightarrow HCN\uparrow$$

Iron(III) ions cannot be detected in a solution of hexacyanoferrate(III) with the usual reactions. The complex has to be decomposed first by evaporating with concentrated sulphuric acid or by igniting a solid sample, as described with hexacyanoferrate(II) (cf. Section **III.21**, reaction 5).

6. *Potassium hexacyanoferrate(II) solution:* intense blue precipitate of iron(III) hexacyanoferrate (Prussian blue):

$$4Fe^{3+} + 3[Fe(CN)_6]^{4-} \rightarrow Fe_4[Fe(CN)_6]_3$$

(cf. Section **III.21**, reaction 7).

The precipitate is insoluble in dilute acids, but decomposes in concentrated hydrocholoric acid. A large excess of the reagent dissolves it partly or entirely, when an intensive blue solution is obtained. Sodium hydroxide turns the precipitate to red, as iron(III) oxide and hexacyanoferrate(II) ions are formed:

$$Fe_4[Fe(CN)_6]_3\downarrow + 12OH^- \rightarrow 4Fe(OH)_3\downarrow + 3[Fe(CN)_6]^{4-}$$

Oxalic acid also dissolves Prussian Blue forming a blue solution; this process was once used to manufacture blue writing inks.

If iron(III) chloride is added to an excess of potassium hexacyanoferrate(II) a product with the composition of $KFe[Fe(CN)_6]$ is formed. This tends to form colloid solutions ('Soluble Prussian Blue') and cannot be filtered.

7. Potassium hexacyanoferrate(III): a brown colouration is produced, due to the formation of an undissociated complex, iron(III) hexacyanoferrate(III):

$$Fe^{3+} + [Fe(CN)_6]^{3-} \rightarrow Fe[Fe(CN)_6]$$

Upon adding hydrogen peroxide or some tin(II) chloride solution, the hexacyanoferrate(III) part of the compound is reduced and Prussian blue is precipitated.

8. Disodium hydrogen phosphate solution: a yellowish-white precipitate of iron(III) phosphate is formed:

$$Fe^{3+} + HPO_4^{2-} \rightarrow FePO_4\downarrow + H^+$$

The reaction is reversible, because a strong acid is formed which dissolves the precipitate. It is advisable to add small amounts of sodium acetate, which acts as a buffer against the strong acid:

$$CH_3COO^- + H^+ \rightleftarrows CH_3COOH$$

Acetic acid, formed in this reaction, does not dissolve the precipitate. The overall reaction, in the presence of sodium acetate, can be written

$$Fe^{3+} + HPO_4^{2-} + CH_3COO^- \rightarrow FePO_4\downarrow + CH_3COOH$$

9. Sodium acetate solution: a reddish-brown colouration is obtained, attributed to the formation of a complex ion with the composition $[Fe_3(OH)_2(CH_3COO)_6]^+$. The reaction

$$3Fe^{3+} + 6CH_3COO^- + 2H_2O \rightleftarrows [Fe_3(OH)_2(CH_3COO)_6]^+ + 2H^+$$

leads to an equilibrium, because a strong acid is formed, which decomposes the complex. If the reagent is added in excess, sodium acetate acts as a buffer and the reaction becomes complete.

If the solution is diluted and boiled, a reddish-brown precipitate of basic iron(III) acetate is formed:

$$[Fe_3(OH)_2(CH_3COO)_6]^+ + 4H_2O \rightarrow$$
$$\rightarrow 3Fe(OH)_2CH_3COO\downarrow + 3CH_3COOH + H^+$$

The excess of acetate ions acts again as a buffer and the reaction becomes complete.

Reactions 8 and 9 are combined for the removal of the phosphate ion from solutions in which it interferes with the normal course of analysis (cf. Section **V.13**).

10. Cupferron reagent, the ammonium salt of nitrosophenylhydroxylamine, $C_6H_5N(NO)ONH_4$*:* reddish-brown precipitate is formed in the presence of hydrochloric acid:

$$Fe^{3+} + 3C_6H_5N(NO)ONH_4 \rightarrow Fe[C_6H_5N(NO)O]_3\downarrow + 3NH_4^+$$

The precipitate is soluble in ether. It is insoluble in acids, but can be decomposed by ammonia or alkali hydroxides, when iron(III) hydroxide precipitate is formed.

The reagent is prepared by dissolving 2 g solid in 100 ml distilled water: it does not keep well. It is recommended that a piece of ammonium carbonate be placed in the stock bottle; this enhances the stability.

11. Ammonium thiocyanate solution: in slightly acidic solution a deep-red colouration is produced (difference from iron(II) ions), due to the formation of a non-dissociated iron(III) thiocyanate complex:

$$Fe^{3+} + 3SCN^- \rightarrow Fe(SCN)_3$$

This chargeless molecule can be extracted by ether or amyl alcohol. In addition to this a set of complex ions, such as $[Fe(SCN)]^{2+}$, $[Fe(SCN)_2]^+$, $[Fe(SCN)_4]^-$, $[Fe(SCN)_5]^{2-}$, and $[Fe(SCN)_6]^{3-}$ are also formed. The composition of the product in aqueous solution depends mainly on the relative amounts of iron and thiocyanate present. Phosphates, arsenates, borates, iodates, sulphates, acetates, oxalates, tartrates, citrates, and the corresponding free acid interfere due to the formation of stable complexes with iron(III) ions.

The dibasic organic acids form complex ions of the type:

$$Fe^{3+} + 3(COO)_2^{2-} \rightleftarrows \{Fe[(COO)_2]_3\}$$

Fluorides and mercury(II) ions bleach the colour because of the formation of the more stable hexafluoroferrate(III) $[FeF_6]^{3-}$ complex and the non-dissociated mercury(II) thiocyanate species:

$$Fe(SCN)_3 + 6F^- \rightarrow [FeF_6]^{3-} + 3SCN^-$$
$$2Fe(SCN)_3 + 3Hg^{2+} \rightarrow 2Fe^{3+} + 3Hg(SCN)_2$$

The presence of nitrites should be avoided for in acid solution they form nitrosyl thiocyanate NOSCN which yields a red colour, disappearing on heating, similar to that with iron (III).

The reaction is well adapted as a spot test and may be carried out as follows. Place a drop of the test solution on a spot place and add 1 drop of 1 per cent ammonium thiocyanate solution. A deep-red colouration appears.

Sensitivity: 0·25 g Fe^{3+}. *Concentration limit:* 1 in 200,000.
Coloured salts, e.g. those of copper, chromium, cobalt, and nickel, reduce the sensitivity of the test.

12. 7-Iodo-8-hydroxyquinoline-5-sulphonic acid (or ferron reagent)

(SO_3H, I, N, OH — structural formula):

green or greenish-blue colouration with iron(III) salts in faintly acid solution (*p*H 2·5–3·0). Iron(II) does not react: only copper interferes.

Place a few drops of the slightly acid test solution in a micro test-tube and add 1 drop of the reagent. A green colouration appears.

Sensitivity: 0·5 μg *Concentration limit:* 1 in 1,000,000.

The reagent consists of a 0·2 per cent aqueous solution of 7-iodo-8-hydroxyquinoline-5-sulphonic acid.

13. Reduction of iron(III) to iron(II) ions In acid solution this may be accomplished by various agents. Zinc or cadmium metal, or their amalgams (i.e. alloys with mercury) may be used:

$$2Fe^{3+} + Zn \rightarrow Zn^{2+} + 2Fe^{2+}$$
$$2Fe^{3+} + Cd \rightarrow Cd^{2+} + 2Fe^{2+}$$

The solution will contain zinc or cadmium ions respectively after the reduction. In acid solutions these metals will dissolve further with liberation of hydrogen; they should therefore be removed from the solution once the reduction has been accomplished.

Tin(II) chloride, potassium iodide, hydroxylamine hydrochloride, hydrazine sulphate, or ascorbic acid can also be used:

$$2Fe^{3+} + Sn^{2+} \rightarrow 2Fe^{2+} + Sn^{4+}$$
$$2Fe^{3+} + 2I^- \rightarrow 2Fe^{2+} + I_2$$
$$4Fe^{3+} + 2NH_2OH \rightarrow 4Fe^{2+} + N_2O + H_2O + 4H^+$$
$$4Fe^{3+} + N_2H_4 \rightarrow 4Fe^{2+} + N_2 + 4H^+$$
$$2Fe^{3+} + C_6H_8O_6 \rightarrow 2Fe^{2+} + C_6H_6O_6 + 2H^+$$

the product of the reduction with ascorbic acid being dehydroascorbic acid.

Hydrogen sulphide (cf. reaction 3) and sulphur dioxide gas reduce iron(III) ions also:

$$2Fe^{3+} + H_2S \rightarrow 2Fe^{2+} + S\downarrow + 2H^+$$
$$2Fe^{3+} + SO_2 + 2H_2O \rightarrow 2Fe^{2+} + SO_4^{2-} + 4H^+$$

14. Oxidation of iron(II) ions to iron(III): oxidation occurs slowly upon exposure to air. Rapid oxidation is effected by concentrated nitric acid, hydrogen peroxide, concentrated hydrochloric acid with potassium chlorate, aqua regia, potassium permanganate, potassium dichronate, and cerium(IV) sulphate in acid solution.

$$4Fe^{2+} + O_2 + 4H^+ \rightarrow 4Fe^{3+} + 2H_2O$$
$$3Fe^{2+} + HNO_3 + 3H^+ \rightarrow NO\uparrow + 3Fe^{3+} + 2H_2O$$
$$2Fe^{2+} + H_2O_2 + 2H^+ \rightarrow 2Fe^{3+} + 2H_2O$$
$$6Fe^{2+} + ClO_3^- + 6H^+ \rightarrow 6Fe^{3+} + Cl^- + 3H_2O$$
$$2Fe^{2+} + HNO_3 + 3HCl \rightarrow 2Fe^{3+} + NOCl\uparrow + 2Cl^- + 2H_2O$$
$$5Fe^{2+} + MnO_4^- + 8H^+ \rightarrow 5Fe^{3+} + Mn^{2+} + 4H_2O$$
$$Fe^{2+} + Ce^{4+} \rightarrow Fe^{3+} + Ce^{3+}$$

15. Distinctive tests for iron(II) and iron(III) ions. **Iron(II)** can be detected most reliably with α,α′-dipyridyl (cf. reaction 9, Section **III.21**); the test is conclusive also in the presence of iron(III). In turn, **iron(III) ions** can be detected with ammonium thiocyanate solution (cf. reaction 11). It must be remembered that even freshly prepared solutions of pure iron(II) salts contain some iron(III) and the thiocyanate test will be positive with these. If however iron(III) is reduced by one of the ways described in reaction 13, the thiocyanate test will become negative.

16. Dry tests **a.** Blowpipe test When iron compounds are heated on charcoal with sodium carbonate, grey metallic particles of iron are produced; these are ordinarily difficult to see, but can be separated from the charcoal by means of a magnet.

b. Borax bead test With a small quantity of iron, the bead is yellowish-brown while hot and yellow when cold in the oxidizing flame, and pale green in the reducing flame; with large quantities of iron the bead is reddish-brown in the oxidizing flame.

III.23 ALUMINIUM, Al (A_r: 26·98) Aluminium is a white, ductile and malleable metal; the powder is grey. It melts at 659°C. Exposed to air, aluminium objects are oxidized on the surface, but the oxide layer protects the object from further oxidation. Dilute hydrochloric acid dissolves the metal readily, dissolution is slower in dilute sulphuric or nitric acid:

$$2Al + 6H^+ \rightarrow 2Al^{3+} + 3H_2\uparrow$$

The dissolution process can be speeded up by the addition of some mercury(II) chloride to the mixture. Concentrated hydrochloric acid also dissolves aluminium:

$$2Al + 6HCl \rightarrow 2Al^{3+} + 3H_2\uparrow + 6Cl^-$$

Concentrated sulphuric acid dissolves aluminium with the liberation of sulphur dioxide:

$$2Al + 6H_2SO_4 \rightarrow 2Al^{3+} + 3SO_4^{2-} + 3SO_2\uparrow + 6H_2O$$

Concentrated nitric acid renders the metal passive. With alkali hydroxides a solution of tetrahydroxoaluminate is formed:

$$2Al + 2OH^- + 6H_2O \rightarrow 2[Al(OH)_4]^- + 3H_2\uparrow$$

Aluminium is tervalent in its compounds. Aluminium ions (Al^{3+}) form colourless salts with colourless anions. Its halides, nitrate, and sulphate are water-soluble; these solutions display acidic reactions owing to hydrolysis. Aluminium sulphide can be prepared in the dry state only, in aqueous solutions it hydrolyses and aluminium hydroxide, $Al(OH)_3$, is formed. Aluminium sulphate forms double salts with sulphates of monovalent cations with attractive crystal shapes, these are called alums.

Reactions of aluminium(III) ions Use a 0·33M solution of aluminium chloride, $AlCl_3$, or a 0·166M solution of aluminium sulphate, $Al_2(SO_4)_3 . 16H_2O$, or potash alum, $K_2SO_4 . Al_2(SO_4)_3 . 24H_2O$, to study these reactions.

1. Ammonium solution: white gelatinous precipitate of aluminium hydroxide, $Al(OH)_3$, slightly soluble in excess of the reagent. The solubility is decreased in the presence of ammonium salts owing to the common ion effect (Section **I.27**).

A small proportion of the precipitate passes into solution as colloidal aluminium hydroxide (aluminium hydroxide sol): the sol is coagulated on boiling the solution or upon the addition of soluble salts (e.g. ammonium chloride) yielding a precipitate of aluminium hydroxide, known as aluminium hydroxide gel. To ensure complete precipitation with ammonia solution, the aluminium solution is added in slight excess and the mixture boiled until the liquid has a slight odour of ammonia. When freshly precipitated, it dissolves readily in strong acids and bases, but after boiling it becomes sparingly soluble:

$$Al^{3+} + 3NH_3 + 3H_2O \rightarrow Al(OH)_3\downarrow + 3NH_4^+$$

2. *Sodium hydroxide solution:* white precipitate of aluminium hydroxide:

$$Al^{3+} + 3OH^- \rightarrow Al(OH)_3\downarrow$$

The precipitate dissolves in excess reagent, when tetrahydroxoaluminate ions are formed:

$$Al(OH)_3 + OH^- \rightarrow [Al(OH)_4]^-$$

The reaction is a reversible one (compare Section **I.28**), and any reagent which will reduce the hydroxyl-ion concentration sufficiently should cause the reaction to proceed from right to left with the consequent precipitation of aluminium hydroxide. This may be effected with a solution of ammonium chloride (the hydroxyl-ion concentration is reduced owing to the formation of the weak base ammonia, which can be readily removed as ammonia gas by heating) or by the addition of acid; in the latter case a large excess of acid causes the precipitated hydroxide to redissolve.

$$[Al(OH)_4]^- + NH_4^+ \rightarrow Al(OH)_3\downarrow + NH_3\uparrow + H_2O$$

$$[Al(OH)_4]^- + H^+ \rightleftarrows Al(OH)_3\downarrow + H_2O$$

$$Al(OH)_3 + 3H^+\downarrow \rightleftarrows Al^{3+} + 3H_2O$$

The precipitation of aluminium hydroxide by solutions of sodium hydroxide and ammonia does not take place in the presence of tartaric acid, citric acid, sulphosalicylic acid, malic acid, sugars, and other organic hydroxy compounds, because of the formation of soluble complex salts. These organic substances must therefore be decomposed by gentle ignition or by evaporating with concentrated sulphuric or nitric acid before aluminium can be precipitated in the ordinary course of qualitative analysis.

3. *Ammonium sulphide solution:* a white precipitate of aluminium hydroxide:

$$2Al^{3+} + 3S^{2-} + 6H_2O \rightarrow 2Al(OH)_3\downarrow + 3H_2S\uparrow$$

The characteristics of the precipitate are the same as mentioned under reaction 2.

4. *Sodium acetate solution:* no precipitate is obtained in cold, neutral solutions, but on boiling with excess reagent, a voluminous precipitate of basic aluminium acetate $Al(OH)_2CH_3COO$ is formed:

$$Al^{3+} + 3CH_3COO^- + 2H_2O \rightarrow Al(OH)_2CH_3COO\downarrow + 2CH_3COOH$$

5. *Sodium phosphate solution:* white gelatinous precipitate of aluminium phosphate:

$$Al^{3+} + HPO_4^{2-} \rightleftarrows AlPO_4\downarrow + H^+$$

The reaction is reversible; strong acids dissolve the precipitate. However, the precipitate is insoluble in acetic acid (difference from phosphates of alkaline earths, which are soluble). The precipitate can also be dissolved in sodium hydroxide:

$$AlPO_4\downarrow + 4OH^- \rightarrow [Al(OH)_4]^- + PO_4^{3-}$$

6. *Sodium carbonate solution:* white precipitate of aluminium hydroxide, because the sodium carbonate neutralizes the acid liberated at the hydrolysis of aluminium, when carbon dioxide gas is formed:

$$Al^{3+} + 3H_2O \rightleftarrows Al(OH)_3\downarrow + 3H^+$$

$$CO_3^{2-} + 2H^+ \rightarrow H_2CO_3 \rightarrow H_2O + CO\uparrow$$

The precipitate dissolves in excess reagent:

$$Al(OH)_3\uparrow + CO_3^{2-} + H_2O \rightarrow [Al(OH)_4]^- + HCO_3^-$$

7. *'Aluminon' reagent (a solution of the ammonium salt of aurine tricarboxylic acid)*

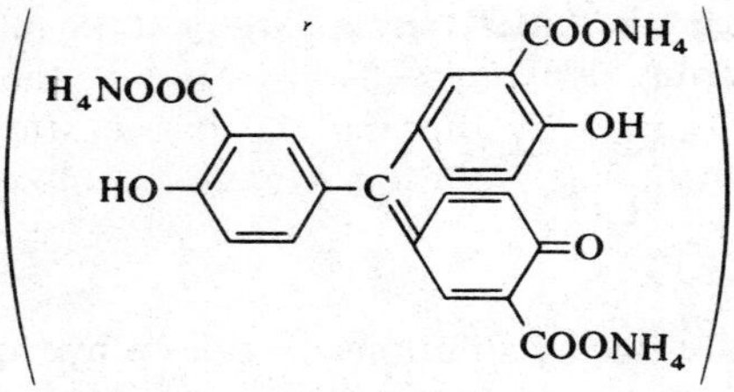

this dye is adsorbed by aluminium hydroxide giving a bright-red adsorption complex or 'lake'. The test is applied to the precipitate of aluminium hydroxide obtained in the usual course of analysis, since certain other elements interfere. Dissolve the aluminium hydroxide precipitate in 2 ml 2M hydrochloric acid, add 1 ml 10M ammonium acetate solution and 2 ml 0·1 per cent aqueous solution of the reagent. Shake, allow to stand for 5 minutes, and add excess of ammoniacal ammonium carbonate solution to decolourize excess dyestuff and lakes due to traces of chromium(III) hydroxide and silica. A bright-red precipitate (or colouration), persisting in the alkaline solution, is obtained.

Iron interferes with the test and should be absent. Chromium forms a similar lake in acetate solution, but this is rapidly decomposed by the addition of the ammoniacal ammonium carbonate solution. Beryllium gives a lake similar to that formed by aluminium. Phosphates, when present in considerable quantity, prevent the formation of the lake. It is then best to precipitate the aluminium phosphate by the addition of ammonia solution; the resultant precipitate is redissolved in dilute acid, and the test applied in the usual way.

The reagent is prepared by dissolving 0·25 g of 'aluminon' in 250 ml of water.

8. Alizarin reagent

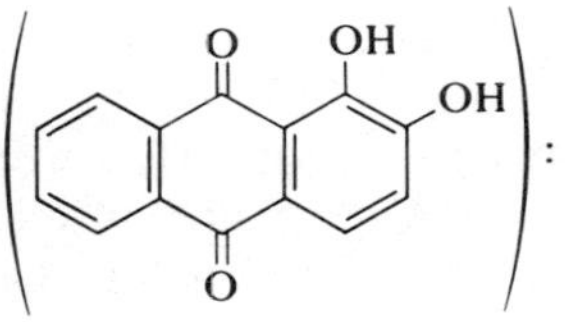

red lake with aluminium hydroxide.

Soak some quantitative filter (or drop-reaction) paper in a saturated alcoholic solution of alizarin and dry it. Place a drop of the acid test solution on the paper and hold it over ammonia fumes until a violet colour (due to ammonium alizarinate) appears. In the presence of large amounts of aluminium, the colour is visible almost immediately. It is best to dry the paper at 100°C when the violet colour due to ammonium alizarinate disappears owing to its conversion into ammonia and alizarin: the red colour of the alizarin lake is then clearly visible.

Sensitivity: 0·15 μg Al. *Concentration limit:* 1 in 333,000.

Iron, chromium, uranium, thorium, titanium, and manganese interfere, but this may be obviated by using paper previously treated with potassium hexacyanoferrate(II). The interfering ions are thus 'fixed' on the paper as insoluble hexacyanoferrate(II)s, and the aluminium solution diffuses beyond as a damp ring. Upon adding a drop of a saturated alcoholic solution of alizarin, exposing to ammonia vapour and drying, a red ring of the alizarin-aluminium lake forms round the precipitate. Uranium hexacyanoferrate(II), owing to its slimy nature, has a tendency to spread outwards from the spot and thus obscure the aluminium lake; this difficulty is surmounted by dipping the paper after the alizarin treatment in ammonium carbonate solution which dissolves the uranium hexacyanoferrate(II).

9. Alizarin-S (or sodium alizarin sulphonate) reagent

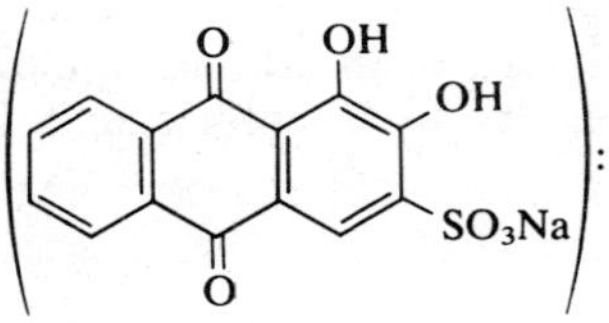

red precipitate or lake in ammoniacal solution, which is fairly stable to dilute acetic acid.

Place a drop of the test solution (which has been treated with just sufficient of M sodium hydroxide solution to give the tetrahydroxaluminate ion $[Al(OH)_4]^-$) on a spot plate, add 1 drop of the reagent, then drops of acetic acid until the violet colour just disappears and 1 drop in excess. A red precipitate or colouration appears.

Sensitivity: 0·7 μg Al. *Concentration limit:* 1 in 80,000.

A blank test should be made on the sodium hydroxide solution. Salts of Cu, Bi, Fe, Be, Zr, Ti, Co, Ce, rare earths, Zn. Th, U, Ca, Sr, and Ba interfere.

The reagent consists of a 0·1 per cent aqueous solution of alizarin-S.

10. Quinalizarin (or 1,2,5,8-tetrahydroxyanthraquinone) reagent

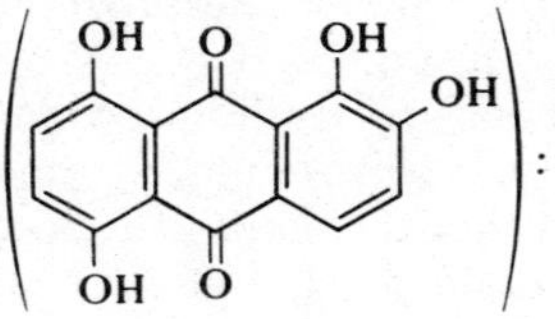

red precipitate or colouration under the conditions given below.

Place a drop of the test solution upon the reagent paper; hold it for a short time over a bottle containing concentrated ammonia solution and then over glacial acetic acid until the blue colour (ammonium quinalizarinate) first formed disappears and the unmoistened paper regains the brown colour of free quinalizarin. A red-violet or red spot is formed.

Sensitivity: 0·005 μg Al (drop of 0·1 ml). *Concentration limit:* 1 in 2,000,000.

The reagent paper is prepared by soaking quantitative filter paper in a solution obtained by dissolving 0·01 g quinalizarin in 2 ml pyridine and then diluting with 18 ml acetone.

11. Dry tests (blowpipe test) Aluminium compounds when heated with sodium carbonate upon charcoal in the blowpipe flame give a white infusible solid, which glows when hot. If the residue be moistened with one or two drops of cobalt nitrate solution and again heated, a blue infusible mass (Thenard's blue, or cobalt meta-aluminate) is obtained. It is important not to use excess cobalt nitrate solution since this yields black cobalt oxide Co_3O_4 upon ignition, which masks the colour of the Thenard's blue.

$$2Al_2O_3 + 2Co^{2+} + 4NO_3^- \rightarrow 2CoAl_2O_4 + 4NO_2\uparrow + O_2\uparrow$$

An alternative method for carrying out this test is to soak some ashless filter paper in aluminium salt solution, add a drop or two of cobalt nitrate solution and then to ignite the filter paper in a crucible; the residue is coloured blue.

III.24 CHROMIUM, Cr (A_r: 51·996) – CHROMIUM(III) Chromium is a white, crystalline metal and is not appreciably ductile or malleable. It melts at 1765°C. The metal is soluble in dilute or concentrated hydrochloric acid. If air is excluded, chromium(II) ions are formed:

$$Cr + 2H^+ \rightarrow Cr^{2+} + H_2\uparrow$$

$$Cr + 2HCl \rightarrow Cr^{2+} + 2Cl^- + H_2\uparrow$$

In the presence of atmospheric oxygen chromium gets partly or wholly oxidized to the tervalent state:

$$4Cr^{2+} + O_2 + 4H^+ \rightarrow 4Cr^{3+} + 2H_2O$$

Dilute sulphuric acid attacks chromium slowly, with the formation of hydrogen. In hot, concentrated sulphuric acid chromium dissolves readily, when chromium(III) ions and sulphur dioxide are formed:

$$2Cr + 6H_2SO_4 \rightarrow 2Cr^{3+} + 3SO_4^{2-} + 3SO_2\uparrow + 6H_2O$$

Both dilute and concentrated nitric acid render chromium passive, so does cold, concentrated sulphuric acid and aqua regia.

In aqueous solutions chromium forms three types of ions: the chromium(II) and chromium(III) cations and the chromate (and dichromate) anion, in which chromium has an oxidation state of +6.

The chromium(II) (or chromous) ion (Cr^{2+}) is derived from chromium(II) oxide CrO. These ions form blue coloured solutions. Chromium(II) ions are rather unstable, as they are strong reducing agents – they decompose even water slowly with the formation of hydrogen. Atmospheric oxygen oxidizes them readily to chromium(III) ions. As they are only rarely encountered in inorganic qualitative analysis, they will not be dealt with here.

Chromium(III) (or chromic) ions (Cr^{3+}) are stable and are derived from dichromium trioxide (or chromium trioxide, Cr_2O_3. In solution they are either green or violet. In the green solutions the pentaquomonochlorochromate(III) $[Cr(H_2O)_5Cl]^{2+}$ or the tetraquodichlorochromate $[Cr(H_2O)_4Cl_2]^+$ complex is present (chloride may be replaced by another monovalent cation), while in the violet solutions the hexaquochromate(III) ion $[Cr(H_2O)_6]^{3+}$ is present. Chromium(III) sulphide, like aluminium sulphide, can be prepared only in dry; it hydrolyses readily with water to form chromium(III) hydroxide and hydrogen sulphide.

In the chromate, CrO_4^{2-}, or dichromate, $Cr_2O_7^{2-}$, anions chromium is hexavalent with an oxidation state of +6. These ions are derived from chromium trioxide CrO_3. Chromate ions are yellow while dichromates have an orange colour. Chromates are readily transformed into dichromates on addition of acid:

$$2CrO_4^{2-} + 2H^+ \rightleftarrows Cr_2O_7^{2-} + H_2O$$

The reaction is reversible. In neutral (or alkaline) solutions the chromate ion is stable, while if acidified, dichromate ions will be predominant. Chromate and dichromate ions are strong oxidizing agents. Their reactions will be dealt with among the reactions of anions.

Reaction of chromium(III) ions To study these reactions use a 0·33M solution of chromium(III) chloride $CrCl_3 . 6H_2O$ or a 0·166M solution of chromium(III) sulphate $Cr_2(SO_4)_3 . 15H_2O$.

1. Ammonia Solution: grey-green to grey-blue gelatinous precipitate of chromium(III) hydroxide, $Cr(OH)_3$, slightly soluble in excess of the precipitant in the cold forming a violet or pink solution containing complex hexammine-chromate(III) ion; upon boiling the solution, chromium hydroxide is precipitated. Hence for complete precipitation of chromium as the hydroxide, it is essential that the solution be boiling and excess aqueous ammonia solution be avoided.

$$Cr^{3+} + 3NH_3 + 3H_2O \rightarrow Cr(OH)_3\downarrow + 3NH_4^+$$

$$Cr(OH)_3\downarrow + 6NH_3 \rightarrow [Cr(NH_3)_6]^{3+} + 3OH^-$$

In the presence of acetate ions and the absence of other tervalent metal ions, chromium(III) hydroxide is not precipitated. The precipitation of chromium(III) hydroxide is also prevented by tartrates and by citrates.

2. Sodium hydroxide solution: precipitate of chromium(III) hydroxide

$$Cr^{3+} + 3OH^- \rightarrow Cr(OH)_3\downarrow$$

The reaction is reversible; on the addition of acids the precipitate dissolves. In excess reagent the precipitate dissolves readily, tetrahydroxochromate(III) ions (or chromite ions) being formed:

$$Cr(OH)_3 + OH^- \rightleftarrows [Cr(OH)_4]^-$$

The solution is green. This reaction is reversible; on (slight) acidification and also on boiling chromium(III) hydroxide precipitates again.

On adding hydrogen peroxide to the alkaline solution of tetrahydroxochromate(III), a yellow solution is obtained, owing to the oxidation of chromium(III) to chromate:

$$2[Cr(OH)_4]^- + 3H_2O_2 + 2OH^- \rightarrow 2CrO_4^{2-} + 8H_2O$$

After decomposing the excess of hydrogen peroxide by boiling, chromate ions may be identified in the solution by one of its reactions (cf. Section **IV.33**).

3. *Sodium carbonate solution:* precipitate of chromium(III) hydroxide:

$$2Cr^{3+} + 3CO_3^{2-} + 3H_2O \rightarrow 2Cr(OH)_3\downarrow + 3CO_2\uparrow$$

4. *Ammonium sulphide solution:* precipitate of chromium(III) hydroxide:

$$2Cr^{3+} + 3S^{2-} + 6H_2O \rightarrow 2Cr(OH)_3\downarrow + 3H_2S\uparrow$$

5. *Sodium acetate solution:* no precipitate even on boiling the solution.

In the presence of considerable amounts of iron and aluminium salts, the precipitate of basic acetates obtained on boiling contains the basic acetate $Cr(OH)_2.CH_3COO$. If the chromium is present in excess, only a fraction of all the metals will be precipitated as basic acetates, whilst some aluminium, iron, and chromium will remain in the filtrate. The basic acetate method of removal of phosphates (see Section **V.13**) is therefore uncertain in the presence of a large concentration of chromium.

The precipitation of chromium(III) hydroxide is prevented by the presence of certain organic compounds in solution (see under aluminium, Section **V.13**).

6. *Sodium phosphate solution:* green precipitate of chromium(III) phosphate:

$$Cr^{3+} + HPO_4^{2-} \rightleftarrows CrPO_4\downarrow + H^+$$

The precipitate is soluble in mineral acids but is practically insoluble in cold dilute acetic acid.

7. *Chromate test* Chromium(III) ions can be oxidized to chromate in several ways.

(*a*) Adding an excess of sodium hydroxide to a chromium(III) salt followed by a few ml of 6 per cent ('20-volume') hydrogen peroxide (cf. reaction 2). The excess of hydrogen peroxide can be removed by boiling the mixture for a few minutes.

Note. Formerly sodium peroxide was employed to effect this oxidation. The sodium peroxide must be added to the cold solution; in hot solution it reacts almost explosively. The use of sodium peroxide is not recommended.

(*b*) Hydrogen peroxide can be replaced by a little solid sodium perborate

$NaBO_3.4H_2O$ in experiment (*a*). Perborate ions, when hydrolysed, form hydrogen peroxide. In alkaline medium this reaction can be written: as

$$BO_3^- + 2OH^- \rightarrow BO_3^{3-} + H_2O_2$$

(*c*) Oxidation can be carried out with bromine water in alkaline solution (i.e. by hypobromite):

$$2Cr^{3+} + 3OBr^- + 10OH^- \rightarrow 2CrO_4^{2-} + 3Br^- + 5H_2O$$

The excess of bromine can be removed by the addition of phenol, which gives tribromophenol.

(d) In acid solution chromium(III) ions can be oxidized by potassium (or ammonium) peroxodisulphate:

$$2Cr^{3+} + 3S_2O_8^{2-} + 8H_2O \rightarrow 2CrO_4^{2-} + 16H^+ + 6SO_4^{2-}$$

One drop of dilute silver nitrate has to be added to speed up the reaction. Silver ions act as catalysts; the catalytic action is due to the transitional formation of silver(III), Ag^{3+}. Halides must be absent; they can be removed easily by evaporating the solution with concentrated sulphuric acid until fumes of sulphur trioxide appear. After cooling, the solution can be diluted and the test carried out. The excess of peroxodisulphate can be decomposed by boiling:

$$2S_2O_8^{2-} + 2H_2O \rightarrow 4SO_4^{2-} + 4H^+ + O_2\uparrow$$

8. *Identification of chromium after oxidation to chromate.* Having carried out the oxidation with one of the methods described under reaction 7, chromate ions may be identified with one of the following tests:

a. Barium chloride test After acidifying the solution with acetic acid and adding barium chloride solution, a yellow precipitate of barium chromate is formed:

$$Ba^{2+} + CrO_4^{2-} \rightarrow BaCrO_4\downarrow$$

b. Chromium pentoxide (chromium peroxide, peroxochromic acid) test. Acidifying the solution with dilute sulphuric acid, adding 2–3 ml of ether or amyl alcohol to the mixture and finally adding some hydrogen peroxide, a blue colouration is formed, which can be extracted into the organic phase by gently shaking. During the reaction chromium pentoxide is formed:

$$CrO_4^{2-} + 2H^+ + 2H_2O_2 \rightarrow CrO_5 + 3H_2O$$

Chromium pentoxide has the following structure:

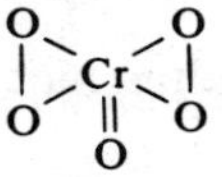

```
O       O
| \   / |
|  Cr   |
| /  ‖ \ |
O    ‖   O
     O
```

Because of the two peroxide groups the compound is often called chromium peroxide. The name peroxochromic acid is less appropriate, because the compound does not contain hydrogen at all. In aqueous solution the blue colour fades rapidly, because chromium pentoxide decomposes to chromium(III) and oxygen:

$$4CrO_5 + 12H^+ \rightarrow 4Cr^{3+} + 7O_2\uparrow + 6H_2O$$

c. Diphenylcarbazide test In dilute mineral acid solution diphenylcarbazide produces a soluble violet colour, which is a characteristic test for chromium. During the reaction chromate is reduced to chromium(II), and diphenylcarbazone is formed; these reaction products in turn produce a complex with the characteristic colour:

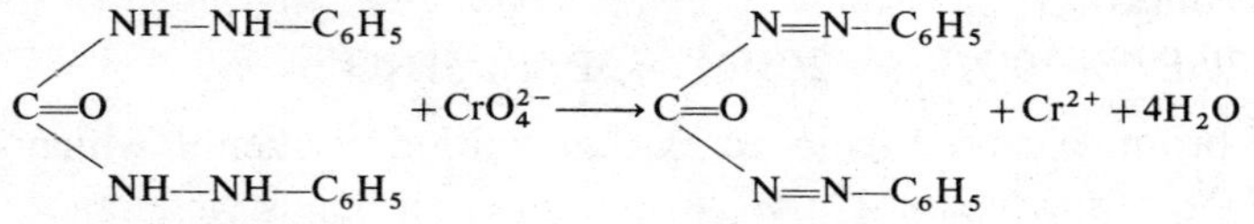

$$C{=}O(NH{-}NH{-}C_6H_5)_2 + CrO_4^{2-} \longrightarrow C{=}O(N{=}N{-}C_6H_5)_2 + Cr^{2+} + 4H_2O$$

diphenylcarbazide → diphenylcarbazone

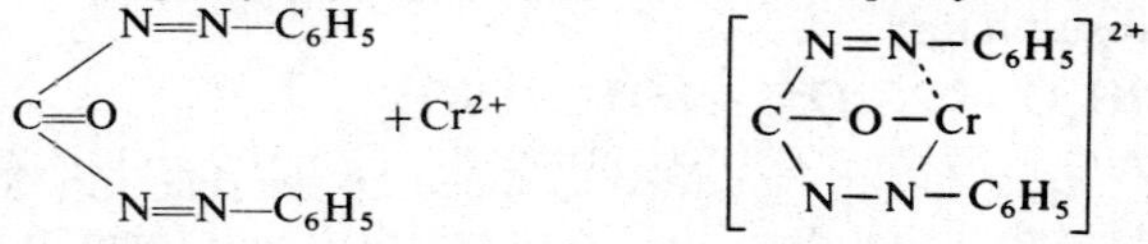

$$C{=}O(N{=}N{-}C_6H_5)_2 + Cr^{2+} \longrightarrow \left[\begin{array}{l} N{=}N{-}C_6H_5 \\ C{-}O{-}Cr \\ N{-}N{-}C_6H_5 \end{array}\right]^{2+}$$

diphenlycarbazone-chromium(II) complex

The reaction can be applied as a drop test for chromium; in this case the preliminary oxidation to chromate can also be made on the spot plate.

Place a drop of the test solution in mineral acid on a spot plate, introduce a drop of saturated bromine water followed by 2–3 drops 2M potassium hydroxide (the solution must be alkaline to litmus). Mix thoroughly, add a crystal of phenol, then a drop of the diphenylcarbazide reagent, and finally M sulphuric acid dropwise until the red colour (from the reaction between diphenylcarbazide and alkali) disappears. A blue-violet colouration is obtained.

Sensitivity: 0·25 μg Cr. *Concentration limit:* 1 in 2,000,000.

Copper, manganese, nickel, and cobalt salts interfere in this procedure because of their precipitation by the alkali.

Alternatively, mix a drop of the acidified test solution on a spot plate with 2 drops 0·1M potassium peroxodisulphate solution and 1 drop 0·1M silver nitrate solution, and allow to stand for 2–3 minutes. Add a drop of the diphenylcarbazide reagent. A violet or red colour is formed.

Sensitivity: 0·8 μg Cr. *Concentration limit:* 1 in 600,000.

Manganese ions interfere (oxidized to permanganate) as do also mercury(II) salts, molybdates, and vanadates, which give blue to violet compounds with the reagent in acid solution. The influence of molybdates can be eliminated by the addition of saturated oxalic acid solution thereby forming the complex $H_2[MoO_3(C_2O_4)]$.

The reagent consists of a 1 per cent solution of diphenylcarbazide in alcohol.

d. Chromotropic acid test. 1,8-dihydroxynaphthalene-3,6-disulphonic acid (or chromotropic acid)

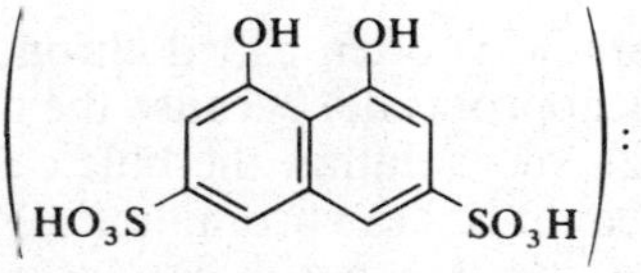

(structure: naphthalene with OH, OH at 1,8 and HO_3S, SO_3H at 3,6):

gives a red colouration (by transmitted light) with an alkali chromate in the presence of nitric acid. Chromium(III) salts may be oxidized with hydrogen peroxide and alkali to chromate, and then acidified with nitric acid before applying the test. The nitric acid serves to eliminate the influence of Fe, U, and Ti, which otherwise interfere.

Place a drop of the test solution in a semimicro test-tube, add a drop of the reagent, a drop of dilute nitric acid (1 : 1), and dilute to about 2 ml. Chromates give a red colouration: this is best observed with a white light behind the tube.

The reagent consists of a saturated solution of chromotropic acid in water.

Concentration limit: 1 in 5,000.

9. Dry tests **a.** Blowpipe test All chromium compounds when heated with sodium carbonate on charcoal yield a green infusible mass of chromium(III) oxide Cr_2O_3.

b. Borax bead test The bead is coloured green in both the oxidizing and reducing flames, but is not very characteristic.

c. Fusion with sodium carbonate and potassium nitrate in a loop of platinum wire or upon platinum foil or upon the lid of a nickel crucible results in the formation of a yellow mass of alkali chromate.

$$Cr_2(SO_4)_3 + 5Na_2CO_3 + 3KNO_3 = 2Na_2CrO_4 + 3KNO_2 + 3Na_2SO_4 + 5CO_2\uparrow$$

III.25 OXOANIONS OF GROUP III METALS: CHROMATE AND PERMANGANATE The oxoanions of certain Group III metals, like chromate CrO_4^{2-} (and dichromate, $Cr_2O_7^{2-}$) as well as permanganate MnO_4^-, are reduced by hydrogen sulphide in hydrochloric acid media to chromium(III) and manganese(II) ions respectively. In the course of the systematic analysis of an unknown sample (cf. Chapter V) these anions will already be converted into the corresponding Group III cations when the separating process reaches this stage. If therefore chromium(III) and/or manganese(II) are found, the initial oxidation states of these metals should be tested in the original sample.

The reactions of these ions are described in Chapter IV; those of chromates and dichromates in Section **IV.33**, and the reactions of permanganates in Section **IV.34**.

III.26 COBALT, Co (A_r: 58·93) Cobalt is a steel-grey, slightly magnetic metal. It melts at 1490°C. The metal dissolves readily in dilute mineral acids:

$$Co + 2H^+ \rightarrow Co^{2+} + H_2\uparrow$$

The dissolution in nitric acid is accompanied with the formation of nitrogen oxide:

$$3Co + 2HNO_3 + 6H^+ \rightarrow 3Co^{2+} + 2NO\uparrow + 4H_2O$$

In aqueous solutions cobalt is normally present as the cobalt(II) ion Co^{2+}; sometimes, especially in complexes, the cobalt(III) ion Co^{3+} is encountered. These two ions are derived from the oxides CoO and Co_2O_3 respectively. The cobalt(II)–cobalt(III) oxide Co_3O_4 is also known.

In the aqueous solutions of cobalt(II) compounds the red Co^{2+} ions are present. Anhydrous or undissociated cobalt(II) compounds are blue. If the dissociation of cobalt compounds is suppressed, the colour of the solution turns gradually to blue.

Cobalt(III) ions Co^{3+} are unstable, but their complexes are stable both in solution and in dry form. Cobalt(II) complexes can easily be oxidized to cobalt(III) complexes.

Reactions of the cobalt(II) ions The reactions of cobalt(II) ions can be studied with a 0·5M solution of cobalt(II) chloride $CoCl_2.6H_2O$ or cobalt(II) nitrate $Co(NO_3)_2.6H_2O$.

1. Sodium hydroxide solution: In cold a blue basic salt is precipitated:

$$Co^{2+} + OH^- + NO_3^- \rightarrow Co(OH)NO_3\downarrow$$

Upon warming with excess alkali (or sometimes merely upon addition of excess reagent) the basic salt is converted into pink cobalt(II) hydroxide precipitate:

$$Co(OH)NO_3\downarrow + OH^- \rightarrow Co(OH)_2\downarrow + NO_3^-$$

Some of the precipitate however passes into solution.

The hydroxide is slowly transformed into the brownish-black cobalt(III) hydroxide on exposure to the air:

$$4Co(OH)_2\downarrow + O_2 + 2H_2O \rightarrow 4Co(OH)_3\downarrow$$

The change takes place more rapidly if an oxidizing agent, like sodium hypochlorite or hydrogen peroxide is added:

$$2Co(OH)_2\downarrow + H_2O_2 \rightarrow 2Co(OH)_3\downarrow$$
$$2Co(OH)_2\downarrow + OCl^- + H_2O \rightarrow 2Co(OH)_3\downarrow + Cl^-$$

Cobalt(II) hydroxide precipitate is readily soluble in ammonia or concentrated solutions of ammonium salts, provided that the mother liquor is alkaline:

$$Co(OH)_2\downarrow + 6NH_3 \rightarrow [Co(NH_3)_6]^{2+} + 2OH^-$$
$$Co(OH)_2\downarrow + 6NH_4^+ + 4OH^- \rightarrow [Co(NH_3)_6]^{2+} + 6H_2O$$

The yellowish-brown solution of hexamminecobaltate(II) ions turns slowly to brownish-red if exposed to air; hydrogen peroxide oxidizes the complex ion more rapidly to hexamminecobaltate(III) ions:

$$4[Co(NH_3)_6]^{2+} + O_2 + 2H_2O \rightarrow 4[Co(NH_3)_6]^{3+} + 4OH^-$$
$$2[Co(NH_3)_6]^{2+} + H_2O_2 \rightarrow 2[Co(NH_3)_6]^{3+} + 2OH^-$$

In the presence of ammonium salts, alkali hydroxides do not precipitate cobalt(II) hydroxide at all. The same is true for solutions containing citrates or tartrates.

2. Ammonia solution: in the absence of ammonium salts small amounts of ammonia precipitate the basic salt as in reaction 1:

$$Co^{2+} + NH_3 + H_2O + NO_3^- \rightarrow Co(OH)NO_3\downarrow + NH_4^+$$

The excess of the reagent dissolves the precipitate, when hexamminecobaltate(II) ions are formed:

$$Co(OH)NO_3\downarrow + 6NH_3 \rightarrow [Co(NH_3)_6]^{2+} + NO_3^- + OH^-$$

The precipitation of the basic salt does not take place at all if larger amounts of ammonium ions are present, but the complex is formed in one step. Under such conditions the equilibrium

$$Co^{2+} + 6NH_4^+ \rightleftarrows [Co(NH_3)_6]^{2+} + 6H^+$$

gets shifted towards the right by the removal of hydrogen ions by ammonia:

$$H^+ + NH_3 \rightarrow NH_4^+$$

The characteristics of the precipitate and the complex are identical to those described under reaction 1.

3. *Ammonium sulphide solution:* black precipitate of cobalt(II) sulphide from neutral or alkaline solution:

$$Co^{2+} + S^{2-} \rightarrow CoS\downarrow$$

The precipitate is insoluble in dilute hydrochloric or acetic acids (though precipitation does not take place from such media). Hot, concentrated nitric acid or aqua regia dissolve the precipitate, when white sulphur remains:

$$3CoS\downarrow + 2NHO_3 + 6H^+ \rightarrow 3Co^{2+} + 3S\downarrow + 2NO\uparrow + 4H_2O$$
$$CoS\downarrow + HNO_3 + 3HCl \rightarrow Co^{2+} + S\downarrow + NOCl\uparrow + 2Cl^-$$

On longer heating the mixture becomes clear because sulphur gets oxidized to sulphate:

$$S\downarrow + 2HNO_3 \rightarrow SO_4^{2-} + 2H^+ \quad 2NO\uparrow$$
$$S\downarrow + 3HNO_3 + 9HCl \rightarrow SO_4^{2-} + 6Cl^- + 3NOCl\uparrow + 8H^+ + 2H_2O$$

4. *Potassium cyanide solution* (**POISON**): reddish-brown precipitate of cobalt(II) cyanide:

$$Co^{2+} + 2CN^- \rightarrow Co(CN)_2\downarrow$$

The precipitate dissolves in excess reagent; a brown solution of hexacyanocobaltate(II) is formed:

$$Co(CN)_2\downarrow + 4CN^- \rightarrow [Co(CN)_6]^{4-}$$

On acidification *in the cold* with dilute hydrochloric acid the precipitate reappears:

$$[Co(CN)_6]^{4-} + 4H^+ \rightarrow Co(CN)_2\downarrow + 4HCN\uparrow$$

The experiment must be carried out in a fume cupboard with good ventilation.

If the brown solution is boiled for a longer time in air, or if some hydrogen peroxide is added and the solution heated, it turns yellow because hexacyanocobaltate(III) ions are formed:

$$4[Co(CN)_6]^{4-} + O_2 + 2H_2O \rightarrow 4[Co(CN)_6]^{3-} + 4OH^-$$
$$2[Co(CN)_6]^{4-} + H_2O_2 \rightarrow 2[Co(CN)_6]^{3-} + 2OH^-$$

5. Potassium nitrite solution: yellow precipitate of potassium hexanitrito-cobaltate(III) $K_3[Co(NO_2)_6].3H_2O$:

$$Co^{2+} + 7NO_2^- + 2H^+ + 3K^+ \rightarrow K_3[Co(NO_2)_6]\downarrow + NO\uparrow + H_2O \quad \text{(a)}$$

This reaction takes place in two steps. First, nitrite oxidizes cobalt(II) to cobalt(III):

$$Co^{2+} + NO_2^- + 2H^+ \rightarrow Co^{3+} + NO\uparrow + H_2O \quad \text{(b)}$$

then cobalt(III) ions react with nitrite and potassium ions:

$$Co^{3+} + 6NO_2^- + 3K^+ \rightarrow K_3[Co(NO_2)_6] \quad \text{(c)}$$

Adding reactions (b) and (c) we obtain reaction (a).

The test can be carried out most conveniently as follows: to a neutral solution of cobalt(II) add acetic acid, then a freshly prepared saturated solution of potassium nitrite. If the concentration of cobalt(II) in the test solution is high enough, the precipitate appears immediately. Otherwise the mixture should either be slightly heated, or the wall of the vessel should be rubbed with a glass rod.

The reaction is also characteristic for potassium and for nitrite ions. Nickel ions do not react if acetic acid is present.

6. Ammonium thiocyanate test (Vogel reaction): adding a few crystals of ammonium thiocyanate to a neutral or acid solution of cobalt(II) a blue colour appears owing to the formation of tetrathiocyanatocobaltate(II) ions:

$$Co^{2+} + 4SCN^- \rightarrow [Co(SCN)_4]^{2-}$$

If amyl alcohol or ether is added the free acid $H_2[Co(SCN)_4]$ is formed and dissolved by the organic solvent (distinction from nickel). The test is rendered more sensitive if the solution is acidified with concentrated hydrochloric acid, when the equilibrium

$$2H^+ + [Co(SCN)_4]^{2-} \rightleftarrows H_2[Co(SCN)_4]$$

shifts towards the formation of the free acid, which then can be extracted with amyl alcohol or ether.

Interference of iron(III) ions can be eliminated by adding some potassium fluoride. The colourless hexafluoroferrate(III) $[FeF_6]^{3-}$ complex does not interfere. Alternately, iron(III) can be reduced with sodium thiosulphate or ascorbic acid (a few ml of a 0·1M solution of either), and iron(II) ions, formed in the reduction, do not interfere any more.

The reaction may be employed as a spot test as follows. Upon a spot plate, mix 1 drop test solution with 5 drops saturated solution of ammonium thiocyanate in acetone. A green to blue colouration appears.

Sensitivity: 0·5 μg Co. *Concentration limit:* 1 in 100,000.

If iron is present mix 1–2 drops of the slightly acid test solution with a few milligrams of ammonium or sodium fluoride on a spot plate and then add 5 drops saturated solution of ammonium thiocyanate in acetone. A blue colouration is produced.

Sensitivity: 1 μg Co in the presence of 100 times that amount of Fe. *Concentration limit:* 1 in 50,000.

7. *α-Nitroso-β-naphthol reagent*

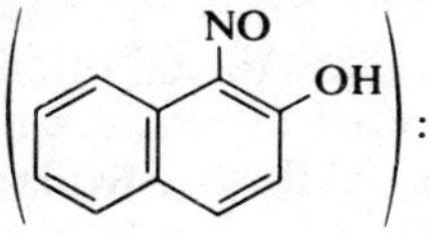

reddish-brown precipitate of slightly impure cobalt(III)-nitroso-β-naphthol $Co(C_{10}H_6O_2N)_3$ (a chelate complex) in solutions acidified with dilute hydrochloric acid or dilute acetic acid: the precipitate may be extracted by carbon tetrachloride to give a claret-coloured solution. Precipitates are also given by iron (Fe^{2+} and Fe^{3+}), nickel and uranyl salts in Group III. The nickel complex is soluble in dilute hydrochloric acid; iron(III) is rendered innocuous by the addition of sodium fluoride; the uranyl iron may be removed, as uranyl phosphate, by ammonium phosphate solution. Copper, mercury(II), palladium, and many other metals interfere. The reaction may, however, be used as a confirmatory test for cobalt among the Group(III) ions.

The reagent consists of a 1 per cent solution of α-nitroso-β-naphthol in 50 per cent acetic acid, or in ethanol, or in acetone.

The technique for using the reaction as a spot test is as follows. Place a drop of the faintly acid test solution on drop-reaction paper, and add a drop of the reagent. A brown stain is produced.

Sensitivity: 0·05 μg Co. *Concentration limit:* 1 in 1,000,000.

8. *Sodium 1-nitroso-2-hydroxynaphthalene-3:* 6-disulphonate (nitroso-R-salt) reagent

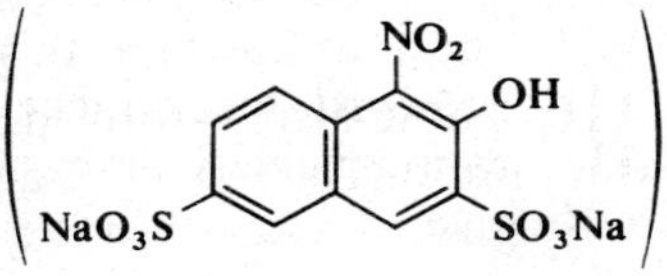

deep-red colouration. The test is applicable in the presence of nickel; tin and iron interfere and should be removed; the colouration produced by iron is prevented by the addition of an alkali fluoride.

Place a drop of the neutral test solution (buffered with sodium acetate) on a spot plate, and add 2–3 drops of the reagent. A red colouration is obtained.

Concentration limit: 1 in 500,000.

The reagent consists of a 1 per cent solution of nitroso-R-salt in water.

9. *Rubeanic acid (or dithio-oxamide) reagent*

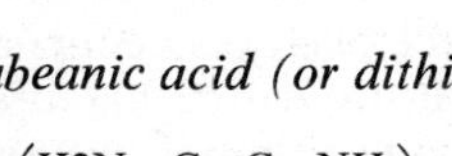

yellowish-brown precipitate. Under similar conditions nickel and copper salts give blue and black precipitates respectively (see Section **III.10**). Large quantities of ammonium salts reduce the sensitivity.

Place a drop of the test solution upon drop-reaction paper, hold it in ammonia vapour, and then add a drop of the reagent. A brown spot or ring is formed.

Sensitivity: 0·03 μg Co. *Concentration limit:* 1 in 660,000.

The reagent consists of an 0·5 per cent solution of rubeanic acid in alcohol.

10 Dry tests **a.** Blowpipe test All cobalt compounds when ignited with sodium carbonate on charcoal give grey, slightly metallic beads of cobalt. If these are removed, placed upon filter paper, and dissolved by the addition of a few drops of dilute nitric acid, a few drops of concentrated hydrochloric acid added and the filter paper dried, the latter is coloured blue by the cobalt chloride produced.
b. Borax bead test This gives a blue bead in both the oxidizing and reducing flames. Cobalt metaborate $Co(BO_2)_2$, or the complex salt $Na_2Co(BO_2)_4$, is formed. The presence of a large proportion of nickel does not interfere.

III.27 NICKEL, Ni (*A*$_r$: 58·71) Nickel is a hard, silver-white metal, it is ductile, malleable and very tenacious. It melts at 1455°C. It is slightly magnetic.

Hydrochloric acid (both dilute and concentrated) and dilute sulphuric acid dissolve nickel with the formation of hydrogen:

$$Ni+2H^+ \rightarrow Ni^{2+}+H_2\uparrow$$

$$Ni+2HCl \rightarrow Ni^{2+}+2Cl^-+H_2\uparrow$$

These reactions accelerate if the solution is heated. Concentrated, hot sulphuric acid dissolves nickel with the formation of sulphur dioxide:

$$Ni+H_2SO_4+2H^+ \rightarrow Ni^{2+}+SO_2\uparrow+2H_2O$$

Dilute and concentrated nitric acid dissolve nickel readily in cold:

$$3Ni+2HNO_3+6H^+ \rightarrow 3Ni^{2+}+2NO\uparrow+4H_2O$$

The stable nickel(II) salts are derived from nickel(II) oxide, NiO, which is a green substance. The dissolved nickel(II) salts are green, owing to the colour of the hexaquonickelate(II) complex $[Ni(H_2O)_6]^{2+}$; in short however, this will be regarded as the simple nickel(II) ion Ni^{2+}. A brownish-black nickel(III) oxide Ni_2O_3 also exists, but this dissolves in acids forming nickel(II) ions. With dilute hydrochloric acid this reaction yields chlorine gas:

$$Ni_2O_3+6H^++2Cl^- \rightarrow 2Ni^{2+}+Cl_2\uparrow+3H_2O$$

Reactions of the nickel(II) ions For the study of these reactions use a 0·5M solution of nickel sulphate $NiSO_4.7H_2O$ or nickel chloride $NiCl_2.6H_2O$.

1. Sodium hydroxide solution: green precipitate of nickel(II) hydroxide:

$$Ni^{2+}+2OH^- \rightarrow Ni(OH)_2\downarrow$$

The precipitate is insoluble in excess reagent. No precipitation occurs in the presence of tartrate or citrate, owing to complex formation. Ammonia dissolves the precipitate; in the presence of excess alkali hydroxide ammonium salts will also dissolve the precipitate:

$$Ni(OH)_2\downarrow+6NH_3 \rightarrow [Ni(NH_3)_6]^{2+}+2OH^-$$

$$Ni(OH)_2\downarrow+6NH_4^++4OH^- \rightarrow [Ni(NH_3)_6]^{2+}+6H_2O$$

The solution of hexamminenickelate(II) ions is deep blue; it can be easily mistaken for copper(II) ions, which form the blue tetramminecuprate(II) ions in an analogous reaction (cf. Section **III.10**). The solution does not oxidize on boiling with free exposure to air or upon the addition of hydrogen peroxide (difference from cobalt).

The green nickel(II) hydroxide precipitate can be oxidized to black nickel(III) hydroxide with sodium hypochlorite solution:

$$2Ni(OH)_2 + ClO^- + H_2O \rightarrow 2Ni(OH)_3\downarrow + Cl^-$$

Hydrogen peroxide solution however does not oxidize nickel(II) hydroxide, but the precipitate catalyses the decomposition of hydrogen peroxide to oxygen and water

$$2H_2O_2 \xrightarrow{Ni(OH)_2\downarrow} 2H_2O + O_2\uparrow$$

without any other visible change.

2. Ammonia solution: green precipitate of nickel(II) hydroxide:

$$Ni^{2+} + 2NH_3 + 2H_2O \rightarrow Ni(OH)_2\downarrow + 2NH_4^+$$

which dissolves in excess reagent:

$$Ni(OH)_2\downarrow + 6NH_3 \rightarrow [Ni(NH_3)_6]^{2+} + 2OH^-$$

the solution turns deep blue (cf. reaction 1). If ammonium salts are present, no precipitation occurs, but the complex is formed immediately.

3. Ammonium sulphide solution: black precipitate of nickel sulphide from neutral or slightly alkaline solutions:

$$Ni^{2+} + S^{2-} \rightarrow NiS\downarrow$$

If the reagent is added in excess, a dark-brown colloidal solution is formed which runs through the filter paper. If the colloidal solution is boiled or if it is rendered slightly acid with acetic acid and boiled, the colloidal solution (hydrosol) is coagulated and can then be filtered. The presence of large quantities of ammonium chloride usually prevents the formation of the sol. Nickel sulphide is practically insoluble in cold dilute hydrochloric acid (distinction from the sulphides of manganese and zinc) and in acetic acid, but dissolves in hot concentrated nitric acid and in aqua regia with the separation of sulphur:

$$3NiS\downarrow + 2HNO_3 + 6H^+ \rightarrow 3Ni^{2+} + 2NO\uparrow + 3S\downarrow + 4H_2O$$

$$NiS\downarrow + HNO_3 + 3HCl \rightarrow Ni^{2+} + S\downarrow + NOCl\uparrow + 2Cl^- + 2H_2O$$

On longer heating sulphur dissolves and the solution becomes clear:

$$S\downarrow + 2HNO_3 \rightarrow SO_4^{2-} + 2H^+ + 2NO\uparrow$$

$$S\downarrow + 3HNO_3 + 9HCl \rightarrow SO_4^{2-} + 6Cl^- + 3NOCl\uparrow + 8H^+ + 2H_2O$$

4. Hydrogen sulphide (gas or saturated aqueous solution): only part of the nickel is slowly precipitated as nickel sulphide from neutral solutions; no precipitation occurs from solutions containing mineral acid or much acetic acid. Complete precipitation occurs, however, from solutions made alkaline with ammonia solution, or from solutions containing excess of alkali acetate made slightly acid with acetic acid (compare Section **I.28**).

5. Potassium cyanide solution (**POISON**): green precipitate of nickel(II) cyanide

$$Ni^{2+} + 2CN^- \rightarrow Ni(CN)_2\downarrow$$

The precipitate is readily soluble in excess reagent, when a yellow solution appears owing to the formation of tetracyanonickelate(II) complex ions:

$$Ni(CN)_2\downarrow + 2CN^- \rightarrow [Ni(CN)_4]^{2-}$$

Dilute hydrochloric acid decomposes the complex, and the precipitate appears again. A fume cupboard with good ventilation must be used for this test:

$$[Ni(CN)_4]^{2-} + 2H^+ \rightarrow Ni(CN)_2\downarrow + 2HCN\uparrow$$

If the solution of tetracyanonickelate(II) is heated with sodium hypobromite solution (prepared *in situ* by adding bromine water to sodium hydroxide solution), the complex decomposes and a black nickel(III) hydroxide precipitate is formed (difference from cobalt ions):

$$2[Ni(CN)_4]^{2-} + OBr^- + 4OH^- + H_2O \rightarrow 2Ni(OH)_3\downarrow + 8CN^- + Br^-$$

Excess potassium cyanide and/or excess bromine water should be avoided, because these react with the formation of cyanogen bromide, which is poisonous and causes watering of the eyes:

$$CN^- + Br_2 \rightarrow BrCN\uparrow + Br^-$$

6. *Potassium nitrite solution:* no precipitate is produced in the presence of acetic acid (difference from cobalt).

7. *α-Nitroso-β-naphthol reagent (cf. Section* **III.26***, reaction 7):* brown precipitate of composition $Ni(C_{10}H_6O_2N)_2$, which is soluble in hydrochloric acid (different from cobalt, which produces a reddish-brown precipitate, insoluble in dilute hydrochloric acid).

8. *Dimethylglyoxime reagent* ($C_4H_8O_2N_2$)*:* red precipitate of nickel dimethylglyoxime from solutions just alkaline with ammonia or acid solutions buffered with sodium acetate:

$$Ni^{2+} + 2\begin{matrix} CH_3-C=N-OH \\ | \\ CH_3-C=N-OH \end{matrix} \longrightarrow \begin{matrix} & & & H & & & \\ & & O & & O & & \\ H_3C & & N & & N & & CH_3 \\ & C & & & & C & \\ & | & & Ni & & | & \\ & C & & & & C & \\ H_3C & & N & & N & & CH_3 \\ & & O & & O & & \\ & & & H & & & \end{matrix} + 2H^+$$

Iron(II) (red colouration), bismuth (yellow precipitate), and larger amounts of cobalt (brown colouration) interfere in ammoniacal solution. The influence of interfering elements (Fe^{2+} must be oxidized to Fe^{3+}, say, by hydrogen peroxide) can be eliminated by the addition of a tartrate. When large quantities of cobalt salts are present, they react with the dimethylglyoxime and a special procedure must be adopted (see below). Oxidizing agents must be absent. Palladium, platinum, and gold give precipitates in acid solution.

The reagent is prepared by dissolving 1 g dimethylglyoxime in 100 ml ethanol.

The spot test technique is as follows. Place a drop of the test solution on drop-reaction paper, add a drop of the above reagent and hold the paper over ammonia vapour. Alternatively, place a drop of the test solution and a drop of the reagent on a spot plate, and add a drop of dilute ammonia solution. A red spot or precipitate (or colouration) is produced.

Sensitivity: 0·16 μg Ni. *Concentration limit:* 1 in 300,000.

Detection of traces of nickel in cobalt salts. The solution containing the cobalt and nickel is treated with excess concentrated potassium cyanide solution, followed by 30 per cent hydrogen peroxide whereby the complex cyanides $[Co(CN)_6]^{3-}$ and $[Ni(CN)_4]^{4-}$ respectively are formed. Upon adding 40 per cent formaldehyde solution the hexacyanocobaltate(III) is unaffected (and hence remains inactive to dimethylglyoxime) whereas the tetracyanatonickelate(II) decomposes with the formation of nickel cyanide, which reacts immediately with the dimethylglyoxime.

$$[Ni(CN)_4]^{2-} + 2HCHO \rightarrow Ni(CN)_2\downarrow + 2CH_2(CN)O^-$$

$$Ni(CN)_2\downarrow + 2C_4H_8O_2N_2 \rightarrow Ni(C_4H_7O_2N_2)_3\downarrow + 2HCN\uparrow$$

9. α-Furil-dioxime reagent

$$\left(\begin{array}{l} C_4H_3O{-}C{=}NOH \\ \qquad\quad | \\ C_4H_3O{-}C{=}NOH \end{array}\right):$$

red precipitate in slightly ammoniacal solution.

Place a few drops of the slightly ammoniacal test solution in a micro test-tube and add a few drops of the reagent. A red precipitate or colouration is formed. Alternatively, the reaction may be carried out on a spot plate.

Sensitivity: 0·02 μg Ni. *Concentration limit:* 1 in 6,000,000.

The reaction is not disturbed by silver or copper, or by iron(III), chromium or aluminium in the presence of ammoniacal tartrate solution: if zinc is present, ammonium chloride should first be added; cobalt(III) ions represss the sensitivity and should be oxidized to the tervalent state with hydrogen peroxide; iron(II) interferes and should be oxidized and alkaline tartrate solution added before applying the test.

The reagent consists of a 10 per cent solution of α-furil-dioxime in alcohol.

10. Rubeanic acid reagent $(CS.NH_2)_2$*:* blue or violet precipitate or colouration in ammoniacal solution. Copper and cobalt, as well as iron salts interfere with the reaction and should be absent.

Place a drop of the test solution upon drop-reaction paper, hold it over ammonia vapour and add a drop of the reagent. A blue or blue-violet spot is obtained.

Sensitivity: 0·03 μg Ni. *Concentration limit:* 1 in 1,250,000.

The reagent consists of a 1 per cent solution of rubeanic acid in alcohol.

11. Dry tests **a.** Blowpipe test All nickel compounds when heated with sodium carbonate on charcoal yield grey, slightly magnetic scales of metallic nickel. If these are placed upon a strip of filter paper, dissolved by means of a few drops of nitric acid, a few drops of concentrated hydrochloric acid added and the filter paper dried by moving it back and forth in the flame or by placing it on the outside of a test-tube containing water which is heated to boiling point, the paper acquires a green colour owing to the formation of nickel(II) chloride. On moistening the filter paper with ammonia solution and adding a few drops dimethylglyoxime reagent, a red colour is produced.

b. Borax bead test This is coloured brown in the oxidizing flame, due to the formation of nickel metaborate, $Ni(BO_2)_2$, or of the complex metaborate $Na_2[Ni(BO_2)_4]$, and grey, due to metallic nickel, in the reducing flame.

III.28 MANGANESE, Mn (A_r: 54·938) – MANGANESE(II) Manganese is a greyish-white metal, similar in appearance to cast iron. It melts at about 1250°C. It reacts with warm water forming manganese(II) hydroxide and hydrogen:

$$Mn + 2H_2O \rightarrow Mn(OH)_2\downarrow + H_2\uparrow$$

Dilute mineral acids and also acetic acid dissolve it with the production of manganese(II) salts and hydrogen:

$$Mn + 2H^+ \rightarrow Mn^{2+} + H_2\uparrow$$

When it is attacked by hot, concentrated sulphuric acid, sulphur dioxide is evolved:

$$Mn + 2H_2SO_4 \rightarrow Mn^{2+} + SO_4^{2-} + SO_2\uparrow + 2H_2O$$

Six oxides of manganese are known: MnO, Mn_2O_3, MnO_2, MnO_3, Mn_2O_7, and Mn_3O_4. The first five of these correspond to oxidation states $+2$, $+3$, $+4$, $+6$ and $+7$ respectively, while the last, Mn_3O_4, is a manganese(II)–manganese(III) oxide, $(MnO.Mn_2O_3)$.

The manganese(II) cations are derived from manganese(II) oxide. They form colourless salts, though if the compound contains water of crystallization, and in solutions, they are slightly pink; this is due to the presence of the hexaquo-manganate(II) ion, $[Mn(H_2O)_6]^{2+}$.

Manganese(III) ions are unstable; some complexes, containing manganese in the $+3$ oxidation state, are however known. They are easily reduced to manganese(II) ions. Although they can be derived from the manganese(III) oxide, Mn_2O_3, the latter, when treated with mineral acids, produces manganese(II) ions. If hydrochloric acid is used, chlorine is the by-product:

$$Mn_2O_3\downarrow + 6HCl \rightarrow 2Mn^{2+} + Cl_2\uparrow + 4Cl^- + 3H_2O$$

With sulphuric acid, oxygen is formed:

$$2Mn_2O_3 + 4H_2SO_4 \rightarrow 4Mn^{2+} + O_2\uparrow + 4SO_4^{2-} + 4H_2O$$

Manganese(IV) compounds, with the exception of the manganese(IV) oxide (or manganese dioxide), MnO_2, are unstable, as both the manganese(IV) ion, Mn^{4+}, and the manganate(IV) (or manganite) ion, MnO_3^{2-}, are easily reduced to manganese(II). When dissolved in concentrated hydrochloric or sulphuric acid, manganese(IV) oxide produces manganese(II) ions as well as chlorine and oxygen gas respectively:

$$MnO_2\downarrow + 4HCl \rightarrow Mn^{2+} + Cl_2\uparrow + 2Cl^- + 2H_2O$$
$$2MnO_2\downarrow + 2H_2SO_4 \rightarrow 2Mn^{2+} + O_2\uparrow + 2SO_4^{2-} + 2H_2O$$

Manganese(VI) compounds contain the manganate(VI) MnO_4^{2-} anion. This is stable in alkaline solutions, and possesses a green colour. Upon neutralization a disproportionation reaction takes place; manganese dioxide precipitate and manganate(VII) (permanganate) ions are formed:

$$3MnO_4^{2-} + 2H_2O \rightarrow MnO_2\downarrow + 2MnO_4^- + 4OH^-$$

If manganese(VI) oxide is treated with acids, manganese(II) ions are produced. With hot, concentrated sulphuric acid, the reaction

$$2MnO_3 + 2H_2SO_4 \rightarrow 2Mn^{2+} + O_2\uparrow + 2SO_4^{2-} + 2H_2O$$

takes place.

Manganese(VII) compounds contain the manganate(VII) or permanganate ion, MnO_4^-. Alkali permanganates are stable compounds, producing violet-coloured solutions. They are all strong oxidizing agents.

In this section the reactions of manganese(II) ions will be dealt with, while reactions of permanganates will be described among the reactions of anions.

Reactions of manganese(II) ions For the study of these reactions a 0·25M solution of manganese(II) chloride $MnCl_2.4H_2O$ or manganese(II) sulphate $MnSO_4.4H_2O$ can be used.

1. Sodium hydroxide solution: an initially white precipitate of manganese(II) hydroxide:

$$Mn^{2+} + 2OH^- \rightarrow Mn(OH)_2\downarrow$$

The precipitate is insoluble in excess reagent. The precipitate rapidly oxidizes on exposure to air, becoming brown, when hydrated manganese dioxide, $MnO(OH)_2$, is formed:

$$Mn(OH)_2\downarrow + O_2 + H_2O \rightarrow MnO(OH)_2\downarrow + 2OH^-$$

Hydrogen peroxide converts manganese(II) hydroxide rapidly into hydrated manganese dioxide:

$$Mn(OH)_2\downarrow + H_2O_2 \rightarrow MnO(OH)_2\downarrow + H_2O$$

2. Ammonia solution: partial precipitation of (initially) white manganese(II) hydroxide:

$$Mn^{2+} + 2NH_3 + 2H_2O \rightleftharpoons Mn(OH)_2\downarrow + 2NH_4^+$$

The precipitate is soluble in ammonium salts, when the reaction proceeds towards the left.

No precipitation takes place in the presence of ammonium salts owing to the lowering of the hydroxyl-ion concentration and the consequent failure to attain $Mn(OH)_2$. On exposure to air, brown hydrated manganese dioxide is precipitated from the ammoniacal solution. This is important in connection with the separation from the Group IIIA metals. In precipitating iron, aluminium, and chromium, the solution should contain a large excess of ammonium chloride, and be boiled to expel most of the dissolved air, then a slight excess of ammonia solution should be added and the precipitate filtered as quickly as possible. Under these conditions very little manganese will be precipitated.

3. *Ammonium sulphide solution:* pink precipitate of manganese(II) sulphide:

$$Mn^{2+} + S^{2-} \rightarrow MnS\downarrow$$

The precipitate also contains loosely-bound water.

The precipitate is readily soluble in mineral acids (different from nickel and cobalt) and even in acetic acid (distinction from nickel, cobalt, and zinc):

$$MnS\downarrow + 2H^+ \rightarrow Mn^{2+} + H_2S\uparrow$$

$$MnS\downarrow + 2CH_3COOH \rightarrow Mn^{2+} + H_2S\uparrow + 2CH_3COO^-$$

The presence of ammonium chloride assists precipitation; it has a salting-out effect on the colloidal precipitate.

The precipitate turns slowly brown on exposure to air, owing to oxidation to manganese dioxide.

Boiling the pink precipitate with excess reagent, a yellowish-green, less hydrated sulphide ($3MnS + H_2O$) is formed.

4. *Sodium phosphate solution:* pink precipitate of manganese ammonium phosphate $Mn(NH_4)PO_4 . 7H_2O$, in the presence of ammonia (or ammonium ions):

$$Mn^{2+} + NH_3 + HPO_4^{2-} \rightarrow Mn(NH_4)PO_4\downarrow$$

If ammonium salts are absent, manganese(II) phosphate is formed:

$$3Mn^{2+} + 2HPO_4^{2-} \rightarrow Mn_3(PO_4)_2\downarrow + 2H^+$$

Both precipitates are soluble in acids.

5. *Lead dioxide and concentrated nitric acid* On boiling a dilute solution of manganese(II) ions, free from hydrochloric acid and chlorides, with lead dioxide (or red lead, which yields the dioxide in the presence of nitric acid) and a little concentrated nitric acid, diluting somewhat and allowing the suspended solid containing unattacked lead dioxide to settle, the supernatant liquid acquires a violet-red (or purple) colour due to permanganic acid. The latter is decomposed by hydrochloric acid and hence chlorides should be absent.

$$5PbO_2 + 2Mn^{2+} + 4H^+ \rightarrow 2MnO_4^- + 5Pb^{2+} + 2H_2O$$

6. *Ammonium or potassium peroxodisulphate* Solid $(NH_4)_2S_2O_8$ or $K_2S_2O_8$ is added to a dilute solution of manganese(II) ions, free of chloride. The solution is acidified with dilute sulphuric acid, and a few drops of dilute silver nitrate (which acts as a catalyst) are added; on boiling a reddish-violet solution is formed, owing to the presence of permanganate:

$$2Mn^{2+} + 5S_2O_8^{2-} + 8H_2O \rightarrow 2MnO_4^- + 10SO_4^{2-} + 16H^+$$

The catalytic action of silver is due to the transitional formation of silver(II), Ag^{2+}, and/or silver(III), Ag^{3+}, ions, which, as powerful oxidants, oxidize manganese(II) to permanganate.

The solution must be dilute (max 0·02M), otherwise manganese dioxide precipitate is formed.

Spot test. Place a drop of the test solution in a micro crucible, add 1 drop 0·1M silver nitrate solution, and stir. Introduce a few mg solid ammonium peroxodisulphate and heat gently. The characteristic colour of permanganate appears.

Sensitivity: 0·1 μg Mn. *Concentration limit:* 1 in 500,000.

7. Sodium bismuthate ($NaBiO_3$) When this solid is added to a cold solution of manganese(II) ions in dilute nitric acid or in dilute sulphuric acid, the mixture stirred and excess reagent filtered off (preferably through asbestos or glass wool or a sintered glass funnel), a solution of permanganate is produced.

$$2Mn^{2+} + 5NaBiO_3 + 14H^+ \rightarrow 2MnO_4^- + 5Bi^{3+} + 5Na^+ + 7H_2O$$

The spot-test technique is as follows. Place a drop of the test solution on a spot plate, add a drop of concentrated nitric acid and then a little sodium bismuthate. The purple colour of permanganic acid appears. If the solution is so dark that the colour cannot be detected, dilute the mixture with water until the colour appears.

Sensitivity: 25 μg Mn (in 5 ml). *Concentration limit:* 1 in 200,000.

8. Potassium periodate (KIO_4) The manganese(II) sulphate solution is rendered strongly acid with sulphuric or nitric acid or (best) phosphoric acid, 0·2–0·3 g potassium periodate added, and the solution boiled for 1 minute. A solution of permanganate is formed. Chlorides must be absent; if present, they must be removed by evaporation with sulphuric or nitric acid before applying the test.

$$2Mn^{2+} + 5IO_4^- + 3H_2O \rightarrow 2MnO_4^- + 5IO_3^- + 6H^+$$

9. Potassium periodate – 'tetrabase' test A solution of the 'tetrabase' (tetramethyl-diamono-diphenylmethane) in chloroform is employed as a sensitive test to identify very small amounts of manganese as permanganic acid. The latter oxidizes the 'tetrabase' to an intensely blue compound. Chromium salts should be absent for they are oxidized by periodates to chromates, which yield a similar colour with the 'tetrabase'.

Place a drop of the test solution on a spot plate, followed by a drop of saturated potassium periodate solution and 2 drops 1 per cent solution of the 'tetrabase' in chloroform. A deep-blue colour is formed.

10. Formaldoxime reagent ($HCH{=}NOH$): red colouration with alkaline solution of manganese(II) salts. Copper gives a blue-violet colouration, but the interference can be overcome by the use of alkali cyanide. Iron is best removed before applying the test (e.g. with a suspension of zinc hydroxide or as the basic acetate), although it may be rendered inactive by the addition of a tartrate. Chromium, cobalt, and nickel salts give colourations with the reagent and must therefore be absent.

Place 2 ml test solution, which has been rendered just alkaline with 2M sodium hydroxide, in a semimicro test-tube and add 1 drop of the reagent. A red colouration is obtained.

Sensitivity: 0·25 μg (in 5 ml). *Concentration limit:* 1 in 20,000,000.

The reagent consists of a 2·5 per cent solution of formaldoxime in water.

11. Dry tests **a.** Borax bead test The bead produced in the oxidizing flame by small amounts of manganese salts is violet whilst hot and amethyst-red when cold; with larger amounts of manganese the bead is almost brown and may be mistaken for that of nickel. In the reducing flame the manganese bead is colourless whilst that due to nickel is grey.

b. Fusion test Fusion of any manganese compound with sodium carbonate and an oxidizing agent (potassium chlorate or potassium nitrate) gives a green mass of alkali manganate. The test may be carried out either by heating upon a piece of platinum foil with potassium nitrate and sodium carbonate (if platinum foil is not available, a piece of broken porcelain may be employed), or by fusing a bead of sodium carbonate with a small quantity of the manganese compound in the oxidizing flame and dipping the fused mass whilst hot into a little powdered potassium chlorate or nitrate and reheating (cf. Section **II.1**).

$$MnSO_4 + 2KNO_3 + 2Na_2CO_3 = Na_2MnO_4 + 2KNO_2 + Na_2SO_4 + 2CO_2\uparrow$$
$$3MnSO_4 + 2KClO_3 + 6Na_2CO_3 = 3Na_2MnO_4 + 2KCl + 3Na_2SO_4 + 6CO_2\uparrow$$

III.29 ZINC, Zn (A_r: 65·38) Zinc is a bluish-white metal; it is fairly malleable and ductile at 110–150°C. It melts at 410°C and boils at 906°C.

The pure metal dissolves very slowly in acids and in alkalis; the presence of impurities or contact with platinum or copper, produced by the addition of a few drops of the solutions of the salts of these metals, accelerates the reaction. This explains the solubility of commercial zinc. The latter dissolves readily in dilute hydrochloric acid and in dilute sulphuric acid with the evolution of hydrogen:

$$Zn + 2H^+ \rightarrow Zn^{2+} + H_2\uparrow$$

Solution takes place in very dilute nitric acid, when no gas is evolved:

$$4Zn + 10H^+ + NO_3^- \rightarrow 4Zn^{2+} + NH_4^+ + 3H_2O$$

With increasing concentration of nitric acid, dinitrogen oxide (N_2O) nitrogen oxide (NO) are formed:

$$4Zn + 10H^+ + 2NO_3^- \rightarrow 4Zn^{2+} + N_2O\uparrow + 5H_2O$$
$$3Zn + 8HNO_3 \rightarrow 3Zn^{2+} + 2NO\uparrow + 6NO_3^- + 4H_2O$$

Concentrated nitric acid has little effect on zinc because of the low solubility of zinc nitrate in such a medium. With hot, concentrated sulphuric acid, sulphur dioxide is evolved:

$$Zn + 2H_2SO_4 \rightarrow Zn^{2+} + SO_2\uparrow + SO_4^{2-} + 2H_2O$$

Zinc also dissolves in alkali hydroxides, when tetrahydroxozincate(II) is formed:

$$Zn + 2OH^- + 2H_2O \rightarrow [Zn(OH)_4]^{2-} + H_2\uparrow$$

Zinc forms one series of salts only; these contain the zinc(II) cation, derived from zinc oxide, ZnO.

Reactions of zinc ions A 0·25M solution of zinc sulphate $ZnSO_4.7H_2O$ can be used to study these reactions.

1. Sodium hydroxide solution: white gelatinous precipitate of zinc hydroxide:

$$Zn^{2+} + 2OH^- \rightleftarrows Zn(OH)_2\downarrow$$

the precipitate is soluble in acids:

$$Zn(OH)_2\downarrow + 2H^+ \rightleftarrows Zn^{2+} + 2H_2O$$

and also in the excess of the reagent:

$$Zn(OH)_2\downarrow + 2OH^- \rightleftarrows [Zn(OH)_4]^{2-}$$

Zinc hydroxide is thus an amphoteric compound.

2. *Ammonia solution:* white precipitate of zinc hydroxide, readily soluble in excess reagent and in solutions of ammonium salts owing to the production of tetramminezincate(II). The non-precipitation of zinc hydroxide by ammonia solution in the presence of ammonium chloride is due to the lowering of the hydroxyl-ion concentration to such a value that the solubility product of $Zn(OH)_2$ is not attained.

$$Zn^{2+} + 2NH_3 + 2H_2O \rightleftarrows Zn(OH)_2\downarrow + 2NH_4$$

$$Zn(OH)_2\downarrow + 4NH_3 \rightleftarrows [Zn(NH_3)_4]^{2+} + 2OH^-$$

3. *Ammonium sulphide solution:* white precipitate of zinc sulphide, ZnS, from neutral or alkaline solutions; it is insoluble in excess reagent, in acetic acid, and in solutions of caustic alkalis, but dissolves in dilute mineral acids. The precipitate thus obtained is partially colloidal; it is difficult to wash and tends to run through the filter paper, particularly on washing. To obtain the zinc sulphide in a form which can be readily filtered, the precipitation is conveniently carried out in boiling solution in the presence of excess ammonium chloride, the precipitate is washed with dilute ammonium chloride solution containing a little ammonium sulphide.

$$Zn^{2+} + S^{2-} \rightarrow ZnS\downarrow$$

4. *Hydrogen sulphide:* partial precipitation of zinc sulphide in neutral solutions; when the concentration of acid produced is about 0·3M (*p*H about 0·6), the sulphide-ion concentration derived from the hydrogen sulphide is depressed so much by the hydrogen-ion concentration from the acid that it is too low to exceed the solubility product of ZnS, and consequently precipitation ceases.

$$Zn^{2+} + H_2S \rightleftarrows ZnS\downarrow + 2H^+$$

Upon the addition of alkali acetate to the solution, the hydrogen-ion concentration is reduced because of the formation of the feebly dissociated acetic acid, the sulphide-ion concentration is correspondingly increased, and precipitation is almost complete

$$Zn^{2+} + H_2S + 2CH_3COO^- \rightarrow ZnS\downarrow + 2CH_3COOH$$

Zinc sulphide is also precipitated from alkaline solutions of tetrahydroxozincate:

$$[Zn(OH)_4]^{2-} + H_2S \rightarrow ZnS\downarrow + 2OH^- + 2H_2O$$

5. *Disodium hydrogen phosphate solution:* white precipitate of zinc phosphate:

$$3Zn^{2+} + 2HPO_4^{2-} \rightleftarrows Zn_3(PO_4)_2\downarrow + 2H^+$$

In the presence of ammonium ions zinc ammonium phosphate is formed:

$$Zn^{2+} + NH_4^+ + HPO_4^{2-} \rightleftarrows Zn(NH_4)PO_4\downarrow + H^+$$

Both precipitates are soluble in dilute acids, when the reactions are reversed. Also, both precipitates are soluble in ammonia:

$$Zn_3(PO_4)_2 + 12NH_3 \rightarrow 3[Zn(NH_3)_4]^{2+} + 2PO_4^{3-}$$

$$Zn(NH_4)PO_4 + 3NH_3 \rightarrow [Zn(NH_3)_4]^{2+} + HPO_4^{2-}$$

6. *Potassium hexacyanoferrate(II) solution:* white precipitate of variable composition; if the reagent is added in some excess the composition of the precipitate is $K_2Zn_3[Fe(CN)_6]_2$:

$$3Zn^{2+} + 2K^+ + 2[Fe(CN)_6]^{4-} \rightarrow K_2Zn_3[Fe(CN)_6]_2$$

The precipitate is insoluble in dilute acids, but dissolves readily in sodium hydroxide:

$$K_2Zn_3[Fe(CN)_6]_2 + 12OH^- \rightarrow 2[Fe(CN)_6]^{4-} + 3[Zn(OH)_4]^{2-} + 2K^+$$

This reaction can be used to distinguish zinc from aluminium.

7. *Quinaldic acid reagent (quinoline-α-carboxylic acid, $C_9H_6N.CO_2H$)* Upon the addition of a few drops of the reagent to a solution of a zinc salt which is faintly acid with acetic acid, a white precipitate of the zinc complex $Zn(C_{10}H_6NO_2)_2.H_2O$ is obtained. The precipitate is soluble in ammonia solution and in mineral acids, but is reprecipitated on neutralization. Copper, cadmium, uranium, iron, and chromium ions give precipitates with the reagent and should be absent. Cobalt, nickel, and manganese ions have no effect. This is an extremely sensitive test for zinc ions, and is a useful confirmatory test for zinc isolated from the Group IIIB separation. The reagent is, however, expensive. The reaction is best carried out on the semimicro scale or as a spot test.

The reagent is prepared by neutralizing 1 g quinaldic acid with sodium hydroxide solution and diluting to 100 ml.

8. *Ammonium tetrathiocyanatomercurate(II)–copper sulphate test.* The faintly acid (sulphuric acid or acetic acid) solution is treated with 0·1 ml 0·25M copper sulphate solution, followed by 2 ml ammonium tetrathiocyanatomercurate(II) reagent. A violet precipitate is obtained. The test is rendered still more sensitive by boiling the mixture for 1 minute, cooling and shaking with a little amyl alcohol; the violet precipitate collects at the interface. Iron salts produce a red colouration; this disappears when a little alkali fluoride is added.

Copper salts alone do not form a precipitate with the ammonium tetrathiocyanatomercurate(II) reagent, whilst zinc ions, if present alone, form a white precipitate:

$$Zn^{2+} + [Hg(SCN)_4]^{2-} = Zn[Hg(SCN)_4]$$

In the presence of copper ions, the copper complex coprecipitates with that of zinc, and the violet (or blackish-purple) precipitate consists of mixed crystals of $Zn[Hg(SCN)_4] + Cu[Hg(SCN)_4]$.

The ammonium tetrathiocyanatomercurate(II) reagent is prepared by dissolving 8 g mercury(II) chloride and 9 g ammonium thiocyanate in 100 ml water.

Place a drop or two of the test solution, which is slightly acid (preferably with sulphuric acid), on a spot plate, add 1 drop 0·25M copper sulphate solution and 1 drop ammonium mercurithiocyanate reagent. A violet (or blackish-purple) precipitate appears.

The reaction may also be conducted in a semimicro test-tube; here 0·3–0·5 ml amyl alcohol is added. The violet precipitate collects at the interface.

Concentration limit: 1 in 10,000.

9. Ammonium tetrathiocyanatomercurate(II)–cobalt sulphate test This test is similar to that described under reaction 8, except that a minute amount of a dilute solution of a cobalt salt (nitrate, sulphate or acetate) is added. Coprecipitation of the cobalt tetrathiocyanatomercurate(II) yields a blue precipitate composed of mixed crystals of $Zn[Hg(SCN)_4]+Co[Hg(SCN)_4]$. Iron(II) salts give a red colouration but this can be eliminated by the addition of a little alkali fluoride (colourless $[FeF_6]^{3-}$ ions are formed). Copper salts should be absent.

Place a drop of the test solution (which should be slightly acid, preferably with dilute sulphuric acid), a drop of 0·5M cobalt nitrate solution, and a drop of the ammonium tetrathiocyanatomercurate(II) reagent on a spot plate or in a micro crucible. A blue precipitate is formed.

The reaction may also be carried out in a semimicro test-tube in the presence of 0·3–0·5 ml of amyl alcohol.

Sensitivity: 0·2–0·5 μg Zn. *Concentration limit:* 1 in 100,000.

*10. Potassium hexacyanoferrate(III)–*p*-phenetidine test* Potassium hexacyanoferrate(III) oxidizes *p*-phenetidine (4-ethoxyaniline) and other aromatic amines slowly with change of colour and the formation of potassium hexacyanoferrate(II). If the hexacyanoferrate(II) formed is removed by zinc ions as the sparingly soluble, white, zinc hexacyanoferrate(II), the oxidation proceeds rapidly: the white zinc hexacyanoferrate(II) is deeply coloured by the adsorption of the coloured oxidation products.

Prepare the reagent by mixing 10 drops 0·033M potassium hexacyanoferrate(III) solution, 2 drops M sulphuric acid, and 6 drops 2 per cent *p*-phenetidine hydrochloride solution.

To 0·1 ml freshly prepared reagent on a spot plate, add a drop of the test solution. A purple to blue colouration or precipitate appears in the presence of zinc. A blank test is desirable.

Sensitivity: 1μg Zn. *Concentration limit:* 1 in 100,000.

The test is especially useful in the presence of Cr, Al, and Mg. Cations (Cu^{2+}, Co^{2+}, Ni^{2+}, Fe^{2+}, etc.) which give coloured precipitates with potassium hexacyanoferrate(II) should be absent.

11. Rinmann's green test The test depends upon the production of Rinmann's green (largely cobalt zincate $CoZnO_2$) by heating the salts or oxides of zinc and cobalt. An excess of cobalt(II) oxide must be avoided for it leads to a red-brown colouration: oxidation to cobalt(III) oxide produces a darkening of colour. Optimum experimental conditions are obtained by converting the zinc into the hexacyanocobaltate(III).

$$3K^+ + [Co(CN)_6]^{3-} + Zn^{2+} = KZn[Co(CN)_6] + 2K^+$$

Ignition of the latter leads to zinc oxide and cobalt(II) oxide in the correct proportion for the formation of cobalt zincate, whilst carbon from the filter paper (see below) prevents the formation of cobalt(III) oxide. All other metals must be removed.

Place a few drops of the neutral (litmus) test solution upon potassium hexacyanocobaltate(III) (or cobalticyanide, Rinmann's green) test paper. Dry the paper over a flame and ignite in a small crucible. Observe the colour of the ash against a white background: part of it will be green.

Sensitivity: 0·6 μg Zn. *Concentration limit:* 1 in 3,000.

The potassium hexacyanocobaltate(III) (or cobalticyanide) test paper is prepared by soaking drop-reaction paper or quantitative filter paper in a solution containing 4 g potassium hexacyanocobaltate(III) and 1 g potassium chlorate in 100 ml water, and drying at room temperature or at 100°C. The paper is yellow and keeps well.

12. Dithizone test Dithizone (diphenyl thiocarbazone) forms complexes with a number of metal ions, which can be extracted with carbon tetrachloride. The zinc complex, formed in neutral, alkaline or acetic acid solutions, is red:

$$2S{=}C\begin{matrix}\diagup NH{-}NH{-}C_6H_5\\ \diagdown N{=}N{-}C_6H_5\end{matrix} \quad + Zn^{2+} \longrightarrow$$

$$\longrightarrow S{=}C\begin{matrix}\diagup NH{-}N(C_6H_5)\diagdown \\ \diagdown N{=}N(C_6H_5)\cdots\end{matrix} Zn \begin{matrix}\cdots N(C_6H_5){=}N\diagdown \\ \diagup N(C_6H_5){=}NH\diagup\end{matrix} C{=}S$$

Acidify the test solution with acetic acid, and add a few drops of the reagent. The organic phase turns red in the presence of zinc. Cu^{2+}, Hg^{2+}, Hg_2^{2+}, Ag^{+}, Au^{3+}, and Pd^{2+} ions interfere. Upon adding sodium hydroxide solution to the mixture and shaking, the aqueous phase turns to red also. This reaction is characteristic for zinc only. (In alkaline solutions the sensitivity of the test is less.)

Limit of detection: in neutral medium: 0·025 μg Zn^{2+} in acetic acid or sodium hydroxide medium: 0·9 μg Zn^{2+}.

The test can be applied in the presence of 2,000 times those amounts of Ni and Al.

The **reagent** is freshly prepared by dissolving 1–2 g of dithizone in 100 ml carbon tetrachloride.

13. Dry tests (blowpipe test) Compounds of zinc when heated upon charcoal with sodium carbonate yield an incrustation of the oxide, which is yellow when hot and white when cold. The metal cannot be isolated owing to its volatility and subsequent oxidation. If the incrustation is moistened with a drop of cobalt nitrate solution and again heated, a green mass (Rinmann's green), consisting largely of cobalt zincate $CoZnO_2$ is obtained.

An alternative method is to soak a piece of ashless filter paper in the zinc salt solution, add 1 drop cobalt nitrate solution and to ignite in a crucible or in a coil of platinum wire. The residue is coloured green.

III.30 FOURTH GROUP OF CATIONS: BARIUM, STRONTIUM, AND CALCIUM

Group reagent: 1M solution of ammonium carbonate.

The reagent is colourless, and displays an alkaline reaction because of hydrolysis:

$$CO_3^{2-} + H_2O \rightleftarrows HCO_3^- + OH^-$$

The reagent is decomposed by acids (even by acetic acid), when carbon dioxide gas is formed:

$$Co_3^{2-} + 2CH_3COOH \rightarrow CO_2\uparrow + H_2O + 2CH_3COO^-$$

The reagent has to be used in neutral or slightly alkaline media.

Commercial ammonium carbonate always contains ammonium hydrogen carbonate (NH_4HCO_3) and ammonium carbamate $NH_4O(NH_2)CO$. These compounds have to be removed before attempting the group reaction as the the alkaline earth salts of both are soluble in water. This can be done by boiling the reagent solution for a while; both ammonium hydrogen carbonate and ammonium carbamate are converted to ammonium carbonate in this way:

$$2HCO_3^- \rightarrow CO_3^{2-} + CO_2\uparrow + H_2O$$

$$(^{-}O)(H_2N)C{=}O + H_2O \rightarrow NH_4^+ + CO_3^{2-}$$

Group reaction: cations of the fourth group react neither with hydrochloric acid, hydrogen sulphide nor ammonium sulphide, but ammonium carbonate (in the presence of moderate amounts of ammonia or ammonium ions) forms white precipitates. The test has to be carried out in neutral or alkaline solutions. In the absence of ammonia or ammonium ions, magnesium will also be precipitated. The white precipitates formed with the group reagent are: barium carbonate $BaCO_3$, strontium carbonate $SrCO_3$, and calcium carbonate $CaCO_3$.

The three alkaline earth metals decompose water at different rates, forming hydroxides and hydrogen gas. Their hydroxides are strong bases, although with different solubilities: barium hydroxide is the most soluble, while calcium hydroxide is the least soluble among them. Alkaline earth chlorides and nitrates are very soluble; the carbonates, sulphates, phosphates, and oxalates are insoluble. The sulphides can be prepared only in the dry; they all hydrolyse in water, forming hydrogen sulphides and hydroxides, e.g.

$$2BaS + 2H_2O \rightarrow 2Ba^{2+} + 2SH^- + 2OH^-$$

Unless the anion is coloured, the salts form colourless solutions.

Because the alkaline earth ions behave very similarly to each other in aqueous solutions, it is very difficult to distinguish them and especially to separate them.

There are however differences in the solubilities of some of their salts in non-aqueous media. Thus, 100 g anhydrous ethanol dissolve 12·5 g calcium chloride, 0·91 g strontium chloride, and only 0·012 g barium chloride (all anhydrous salts). One hundred grams of a 1:1 mixture diethyl ether and anhydrous ethanol dissolve more than 40 g anhydrous calcium nitrate, the solubilities of anhydrous strontium and barium nitrates in this solution are negligible. These differences can be utilized for separations.

III.31 BARIUM, Ba (A_r: 137·34) Barium is a silver-white, malleable and ductile metal, which is stable in dry air. It reacts with the water in humid air forming oxide or hydroxide. It melts at 710°C. It reacts with water at room temperature forming barium hydroxide and hydrogen:

$$Ba+2H_2O \rightarrow Ba^{2+}+H_2\uparrow+2OH^-$$

Dilute acids dissolve barium readily with the evolution of hydrogen:

$$Ba+2H^+ \rightarrow Ba^{2+}+H_2\uparrow$$

Barium is bivalent in its salts forming the barium(II) cation Ba^{2+}. Its chloride and nitrate are soluble, but on adding concentrated hydrochloric or nitric acids to barium solutions, barium chloride or nitrate may precipitate out, as a consequence of the law of mass action.

Reactions of barium ions Use a 0·25M solution of barium chloride $BaCl_2.2H_2O$ or barium nitrate $Ba(NO_3)_2$ for the study of these reactions.

1. Ammonia solution: no precipitate of barium hydroxide because of its relatively high solubility. If the alkaline solution is exposed to the atmosphere, some carbon dioxide is absorbed and a turbidity, due to barium carbonate, is produced.

A slight turbidity may occur when adding the reagent; this is due to small amounts of ammonium carbonate, often present in an aged reagent.

2. Ammonium carbonate solution: white precipitate of barium carbonate, soluble in acetic acid and in dilute mineral acids.

$$Ba^{2+}+CO_3^{2-} \rightarrow BaCO_3\downarrow$$

The precipitate is slightly soluble in solutions of ammonium salts of strong acids; this is because the ammonium ion, being a strong acid, reacts with the base, the carbonate ion, CO_3^{2-}, leading to the formation of the hydrogen carbonate ion, HCO_3^-, and hence the carbonate-ion concentration of the solution is decreased.

$$NH_4^++CO_3^{2-} \rightarrow NH_3\uparrow+HCO_3^-$$

or $$NH_4^++BaCO_3\downarrow \rightarrow NH_3\uparrow+HCO_3^-+Ba^{2+}$$

If the amount of barium carbonate precipitate is very small, it may well dissolve in high concentrations of ammonium salts.

3. Ammonium oxalate solution: white precipitate of barium oxalate $Ba(COO)_2$, slightly soluble in water (0·09 g per litre; $K_s = 1{\cdot}7\times10^{-7}$), but readily dissolved by hot dilute acetic acid (distinction from calcium) and by mineral acids (cf. Section **I.28**).

$$Ba^{2+} + (COO)_2^{2-} \rightleftarrows Ba(COO)_2\downarrow$$

4. Dilute sulphuric acid: heavy, white, finely divided precipitate of barium sulphate $BaSO_4$, practically insoluble in water (2·5 mg ℓ^{-1}; $K_s = 9{\cdot}2 \times 10^{-11}$), almost insoluble in dilute acids and in ammonium sulphate solution, and appreciably soluble in boiling concentrated sulphuric acid. By precipitation in boiling solution or preferably in the presence of ammonium acetate, a more readily filterable form is obtained.

$$Ba^{2+} + SO_4^{2-} \rightarrow BaSO_4\downarrow$$

$$BaSO_4\downarrow + H_2SO_4(conc.) \rightarrow Ba^{2+} + 2HSO_4^-$$

If barium sulphate is boiled with a concentrated solution of sodium carbonate, partial transformation into the less soluble barium carbonate occurs in accordance with the equation:

$$BaSO_4\downarrow + CO_3^{2-} \rightleftarrows BaCO_3\downarrow + SO_4^{2-}$$

Owing to the reversibility of the reaction, the transformation is incomplete. If the mixture is filtered and washed (thus removing the sodium sulphate), and the residue boiled with a fresh volume of sodium carbonate solution, more of the barium sulphate will be converted into barium carbonate. By repetition of this process, practically all of the sulphate can be converted into the corresponding carbonate. The carbonate may be dissolved in acids; this process therefore provides a method for bringing insoluble sulphates into solution. A more expeditious method of obtaining the same result is to fuse the barium sulphate with 4 to 6 times its weight of anhydrous sodium carbonate; the maximum concentration of carbonate is thus obtained and the reaction proceeds almost to completion in one operation. The melt is allowed to cool, extracted with boiling water, and filtered; the residue of barium carbonate can then be dissolved in the appropriate acid. It has been stated that by boiling barium sulphate with a 15 fold excess of sodium carbonate solution of at least 1·5 mol ℓ^{-1} concentration, 99 per cent is converted into barium carbonate in 1 hour.

Barium sulphate precipitate may also be dissolved in a hot 3 to 5 per cent solution of disodium ethylenediamine tetraacetate (Na_2EDTA) in the presence of ammonia.

5. Saturated calcium sulphate solution: immediate white precipitate of barium sulphate. A similar phenomenon occurs if saturated strontium sulphate reagent is used.

The explanation of these reactions is as follows: of the three alkaline earth sulphates barium sulphate is the least soluble. In the solutions of saturated calcium or strontium sulphate the concentration of sulphate ions is high enough to cause precipitation with larger amounts of barium, because the product of ion concentrations exceeds the value of the solubility product:

$$SO_4^{2-} + Ba^{2+} \rightleftarrows BaSO_4\downarrow$$

6. *Potassium chromate solution:* a yellow precipitate of barium chromate, practically insoluble in water (3·2 mg ℓ^{-1}, $K_s = 1{\cdot}6 \times 10^{-10}$):

$$Ba^{2+} + CrO_4^{2-} \rightarrow BaCrO_4\downarrow$$

The precipitate is insoluble in dilute acetic acid (distinction from strontium and calcium), but readily soluble in mineral acids (cf. Section **I.28**).

The addition of acid to potassium chromate solution causes the yellow colour of the solution to change to reddish-orange, owing to the formation of dichromate:

$$2CrO_4^{2-} + 2H^+ \rightleftarrows Cr_2O_7^{2-} + H_2O$$

Upon adding a base (e.g. OH^- ions) to dichromate solutions, the reaction will proceed from right to left as the hydrogen ions are removed, and thus chromate will form. In the presence of a large concentration of hydrogen ions, the chromate-ion concentration will be reduced to such a value that the solubility product of $BaCrO_4$ will not be attained. Hence to precipitate Ba^{2+} ions as $BaCrO_4$, strong acids must be removed or neutralized. The addition of sodium acetate acts as a buffer by reducing the hydrogen-ion concentration and complete precipitation of $BaCrO_4$ will take place.

The solubility products for $SrCrO_4$ and $CaCrO_4$ are much larger than for $BaCrO_4$ and hence they require a larger CrO_4^{2-} ion concentration to precipitate them. The addition of acetic acid to the K_2CrO_4 solution lowers the CrO_4^{2-} ion concentration sufficiently to prevent the precipitation of $SrCrO_4$ and $CaCrO_4$ but it is maintained high enough to precipitate $BaCrO_4$.

7. *Sodium rhodizonate reagent*

$$\left(\begin{array}{l} CO\text{—}CO\text{—}C.ONa \\ | \qquad\qquad\quad \| \\ CO\text{—}CO\text{—}C.ONa \end{array}\right):$$

reddish-brown precipitate of the barium salt of rhodizonic acid in neutral solution. Calcium and magnesium salts do not interfere; strontium salts react like those of barium, but only the precipitate due to the former is completely soluble in dilute hydrochloric acid. Other elements, e.g. those precipitated by hydrogen sulphide and by ammonium sulphide, should be absent. The reagent should be confined to testing for elements in Group IV.

Place a drop of the neutral or faintly acid test solution upon drop-reaction paper and add a drop of the reagent. A brown or reddish-brown spot is obtained.

Sensitivity: 0·25 μg Ba. *Concentration limit:* 1 in 200,000.

The reagent consists of a 0·5 per cent aqueous solution of sodium rhodizonate. It does not keep well so only small quantities should be prepared at a time.

In the presence of strontium the reddish-brown stain of barium rhodizonate is treated with 0·5M hydrochloric acid; the strontium rhodizonate dissolves, whilst the barium derivative is converted into the brilliant-red acid salt. The reaction is best carried out on drop-reaction paper as above.

Treat the reddish-brown spot with a drop of 0·5M hydrochloric acid when a bright-red stain is formed if barium is present. If barium is absent, the spot disappears.

Sensitivity: 0·5 μg Ba in the presence of 50 times that amount of Sr. *Concentration limit:* 1 in 90,000.

8 Anhydrous ethanol and ether: a 1+1 mixture of these solvents does not dissolve anhydrous barium nitrate or barium chloride (distinct from strontium and calcium). The salts must be heated to 180°C before the test to remove all water of crystallization. This test can be applied for the separation of barium from strontium and/or calcium.

9. Dry test (flame colouration) Barium salts, when heated in the non-luminous Bunsen flame, impart a yellowish-green colour to the flame. Since most barium salts, with the exception of the chloride, are non-volatile, the platinum wire is moistened with concentrated hydrochloric acid before being dipped into the substance. The sulphate is first reduced to the sulphide in the reducing flame, then moistened with concentrated hydrochloric acid, and reintroduced into the flame.

III.32 STRONTIUM, Sr (A_r: 87·62) Strontium is a silver-white, malleable and ductile metal. It melts at 771°C. Its properties are similar to those of barium.

Reactions of strontium ions For the study of these reactions a 0·25M solution of strontium chloride $SrCl_2.6H_2O$ or strontium nitrate $Sr(NO_3)_2$ can be used.

1. Ammonia solution: no precipitate.

2. Ammonium carbonate solution: white precipitate of strontium carbonate:

$$Sr^{2+} + CO_3^{2-} \rightarrow SrCO_3\downarrow$$

Strontium carbonate is somewhat less soluble than barium carbonate; otherwise its characteristics (slight solubility in ammonium salts, decomposition with acids) are similar to those of the latter.

3. Dilute sulphuric acid: white precipitate of strontium sulphate:

$$Sr^{2+} + SO_4^{2-} \rightarrow SrSO_4\downarrow$$

The solubility of the precipitate is not negligible (0·097 g ℓ^{-1}, $K_s = 2{\cdot}8 \times 10^{-7}$). The precipitate is insoluble in ammonium sulphate solution even on boiling (distinction from calcium) and slightly soluble in boiling hydrochloric acid. It is almost completely converted into the corresponding carbonate by boiling with a concentrated solution of sodium carbonate:

$$SrSO_4 + CO_3^{2-} \rightleftarrows SrCO_3\downarrow + SO_4^{2-}$$

Strontium carbonate is less soluble than strontium sulphate (solubility: 5·9 mg $SrCO_3$ ℓ^{-1}, $K_s = 1{\cdot}6 \times 10^{-9}$ at room temperature).

After filtering the solution the precipitate can be dissolved in hydrochloric acid, thus strontium ions can be transferred into the solution.

4. Saturated calcium sulphate solution: white precipitate of strontium sulphate, formed slowly in the cold but more rapidly on boiling (distinction from barium).

5. Ammonium oxalate solution: white precipitate of strontium oxalate

$$Sr^{2+} + (COO)_2^{2-} \rightarrow Sr(COO)_2\downarrow$$

The precipitate is sparingly soluble in water (0·039 g ℓ$^{-1}$, $K_s = 5 \times 10^{-8}$). Acetic acid does not attack it; mineral acids however dissolve the precipitate.

6. *Potassium chromate solution:* yellow precipitate of strontium chromate:

$$Sr^{2+} + CrO_4^{2-} \rightarrow SrCrO_4\downarrow$$

The precipitate is appreciably soluble in water (1·2 g ℓ$^{-1}$, $K_s = 3{\cdot}5 \times 10^{-5}$), no precipitate occurs therefore in dilute solutions of strontium. The precipitate is soluble in acetic acid (distinction from barium) and in mineral acids, for the same reasons as described under barium (cf. Section **III.31**, reaction 6).

7. *Sodium rhodizonate reagent:* reddish-brown precipitate of strontium rhodizonate in neutral solution. The test is applied to the elements of Group IV. Barium reacts similarly and a method for the detection of barium in the presence of strontium has already been described (Section **III.31**, reaction 7). To detect strontium in the presence of barium, the latter is converted into the insoluble barium chromate. Barium chromate does not react with sodium rhodizonate, but the more soluble strontium chromate reacts normally.

If barium is absent, place a drop of the neutral test solution on drop-reaction paper or on a spot plate, and add a drop of the reagent. A brownish-red colouration or precipitate is produced.

Sensitivity: 4 μg Sr. *Concentration limit:* 1 in 13,000.

If barium is present, proceed as follows. Impregnate some quantitative filter paper or drop-reaction paper with a saturated solution of potassium chromate, and dry it. Place a drop of the test solution on this paper and, after a minute, place 1 drop of the reagent on the moistened spot. A brownish-red spot or ring is formed.

Sensitivity: 4 μg Sr in the presence of 80 times that amount of Ba. *Concentration limit:* 1 in 13,000.

For further details of the reagent, see Section **III.31**, reaction 7.

8. *Anhydrous ethanol and ether:* a 1+1 mixture of these solvents does *not* dissolve anhydrous strontium nitrate, but *does* dissolve anhydrous strontium chloride. The test can be utilized for the separation of calcium, strontium, and barium.

The test can be carried out as follows: precipitate strontium as the carbonate. Filter the precipitate, dissolve one part of it in hydrochloric acid, another in nitric acid. Evaporate the two solutions on separate watch-glasses to dryness, heat the residue to 180°C for 30 minutes, and try to dissolve the residues in a few millilitres of the solvent.

9. *Dry test* (flame colouration) Volatile strontium compounds, especially the chloride, impart a characteristic carmine-red colour to the non-luminous Bunsen flame (see remarks under Barium).

III.33 CALCIUM, Ca (A_r: 40·08) Calcium is a silver-white, rather soft metal. It melts at 845°C. It is attacked by atmospheric oxygen and humidity, when calcium oxide and/or calcium hydroxide is formed. Calcium decomposes water under the formation of calcium hydroxide and hydrogen.

Calcium forms the calcium(II) cation Ca^{2+} in aqueous solutions. Its salts are normally white powders and form colourless solutions, unless the anion is coloured. Solid calcium chloride is hygroscopic and is often used as a drying agent. Calcium chloride and calcium nitrate dissolve readily in ethanol or in a 1+1 mixture of anhydrous ethanol and diethyl ether.

Reactions of calcium ions To study these reactions a 0·5M solution of calcium chloride $CaCl_2 . 6H_2O$ can be used.

1. Ammonia solution: no precipitate, as calcium hydroxide is fairly soluble. With an aged precipitant a turbidity may occur owing to the formation of calcium carbonate (cf. Section **III.31**, reaction 1).

2. Ammonium carbonate solution: white amorphous precipitate of calcium carbonate:

$$Ca^{2+} + CO_3^{2-} \rightarrow CaCO_3\downarrow$$

On boiling the precipitate becomes crystalline. The precipitate is soluble in water which contains excess carbonic acid (e.g. freshly prepared soda water), because of the formation of soluble calcium hydrogen carbonate:

$$CaCO_3\downarrow + H_2O + CO_2 \rightleftarrows Ca^{2+} + 2HCO_3^-$$

On boiling the precipitate appears again, because carbon dioxide is removed during the process and the reaction proceeds towards left. Barium and strontium ions react in a similar way.

The precipitate is soluble in acids, even in acetic acid:

$$CaCO_3\downarrow + 2H^+ \rightarrow Ca^{2+} + H_2O + CO_2\uparrow$$

$$CaCO_3\downarrow + 2CH_3COOH \rightarrow Ca^{2+} + H_2O + CO_2\uparrow + 2CH_3COO^-$$

Calcium carbonate is slightly soluble in solutions of ammonium salts of strong acids (cf. Section **III.31**, reaction 2).

3. Dilute sulphuric acid: white precipitate of calcium sulphate

$$Ca^{2+} + SO_4^{2-} \rightarrow CaSO_4\downarrow$$

The precipitate is appreciably soluble in water (0·61 g Ca^{2+}, 2·06 g $CaSO_4$ or 2·61 g $CaSO_4 . 2H_2O$ ℓ^{-1}, $K_s = 2{\cdot}3 \times 10^{-4}$); that is more soluble than barium or strontium sulphate. In the presence of ethanol the solubility is much less.

The precipitate dissolves in hot, concentrated sulphuric acid:

$$CaSO_4 + H_2SO_4 \rightleftarrows 2H^+ + [Ca(SO_4)_2]^{2-}$$

The same complex is formed if a precipitate is heated with a 10 per cent solution of ammonium sulphate:

$$CaSO_4 + SO_4^{2-} \rightleftarrows [Ca(SO_4)_2]^{2-}$$

Although dissolution in ammonium sulphate may not be complete, calcium ions can be detected in the filtrate with oxalate, after neutralization with ammonia.

4. Saturated calcium sulphate: no precipitate is formed (difference from strontium and barium).

5. *Ammonium oxalate solution:* white precipitate of calcium oxalate, immediately from concentrated and slowly from dilute solutions:

$$Ca^{2+} + (COO)_2^{2-} \rightarrow Ca(COO)_2\downarrow$$

Precipitation is facilitated by making the solution alkaline with ammonia. The precipitate is practically insoluble in water (6·53 mg $Ca(COO)_2$ ℓ^{-1}, $K_s = 2{\cdot}6 \times 10^{-9}$), insoluble in acetic acid, but readily soluble in mineral acids (cf. Section **I.28**).

6. *Potassium chromate solution:* no precipitate from dilute solutions, nor from concentrated solutions in the presence of acetic acid (cf. Section **III.31**, reaction 6).

7. *Potassium hexacyanoferrate(II) solution:* white precipitate of a mixed salt:

$$Ca^{2+} + 2K^{+} + [Fe(CN)_6]^{4-} \rightarrow K_2Ca[Fe(CN)_6]\downarrow$$

In the presence of ammonium chloride the test is more sensitive. In this case potassium is replaced by ammonium ions in the precipitate. The test can be used to distinguish calcium from strontium; barium and magnesium ions however interfere.

8. *Sodium dihydroxytartrate osazone reagent*

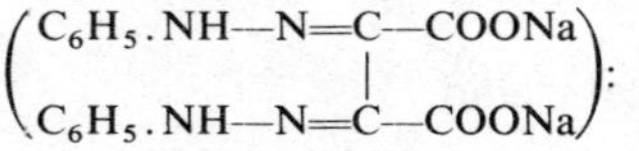

$$\left(\begin{array}{l} C_6H_5.NH{-}N{=}C{-}COONa \\ \qquad\qquad\quad | \\ C_6H_5.NH{-}N{=}C{-}COONa \end{array}\right):$$

yellow sparingly-soluble precipitate of the calcium salt. All other metals, with the exception of alkali and ammonium salts, must be absent. Magnesium does not interfere provided its concentration does not exceed 10 times that of the calcium.

Place a drop of the neutral test solution on a black spot plate or upon a black watch glass, and add a tiny fragment of the solid reagent. If calcium is absent, the reagent dissolves completely. The presence of calcium is revealed by the formation over the surface of the liquid of a white film which ultimately separates as a dense precipitate.

Sensitivity: 0·01 μg Ca. *Concentration limit:* 1 in 5,000,000.

This reagent is useful *inter alia* for the rapid differentiation between tap and distilled water: a positive result is obtained with a mixture of 1 part of tap water and 30 parts of distilled water.

9. *Picrolonic acid (or 1-p-nitrophenyl-3-methyl-4-nitro-5-pyrazolone) reagent*

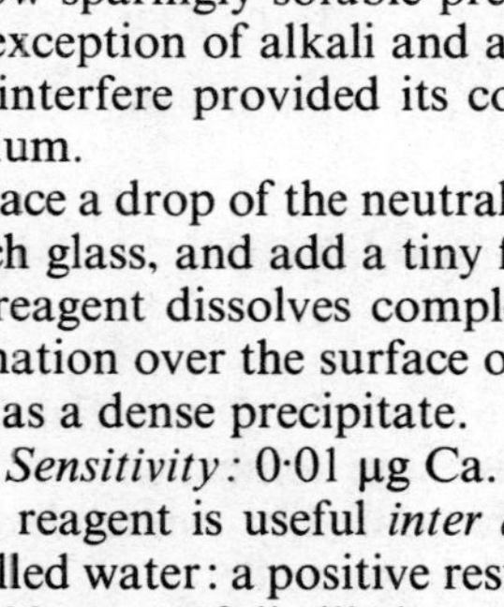

$$\left(\begin{array}{c} O_2N.CH{-}{-}C.CH_3 \\ |\qquad\quad \| \\ CO \qquad N \\ \diagdown\ \diagup \\ N \\ | \\ C_6H_4 \\ | \\ NO_2 \end{array}\right):$$

characteristic rectangular crystals of calcium picrolonate

$$Ca(C_{10}H_7O_5N_4)_2.8H_2O\downarrow$$

in neutral or acetic acid solutions. Strontium and barium give precipitates but of different crystalline form. Numerous elements, including copper, lead, thorium, iron, aluminium, cobalt, nickel, and barium, interfere.

Place a drop of the test solution (either neutral or acidified with acetic acid) in the depression of a *warm* spot plate and add 1 drop of a saturated solution of picrolonic acid. Characteristic rectangular crystals are produced.

Sensitivity: 100 μg (in 5 ml). *Concentration limit:* 1 in 50,000.
The sensitivity is 0·01 μg (in 0·01 ml) under the microscope.

10. Calcium sulphate dihydrate (microscope) test This is an excellent confirmatory test for calcium in Group IV; it involves the use of a microscope (magnification about 110×). The salts should preferably be present as nitrates.

Evaporate a few drops of the test solution on a watch glass to dryness on a water bath, dissolve the residue in a few drops of water, transfer to a microscope slide, and add a minute drop of dilute sulphuric acid. (It may be necessary to warm the slide gently on a water bath until crystallization just sets in at the edges.) Upon observation through a microscope, bundles of needles or elongated prisms will be visible if calcium is present.

Concentration limit: 1 in 6,000.

11. Anhydrous ethanol, or a 1+1 mixture of anhydrous ethanol and diethyl ether, dissolves both anhydrous calcium chloride and calcium nitrate. For practical details see Section **III.32**, reaction 8.

12. Dry test (flame colouration) Volatile calcium compounds impart a yellowish-red colour to the Bunsen flame (see remarks under Barium).

III.34 FIFTH GROUP OF CATIONS: MAGNESIUM, SODIUM, POTASSIUM, AND AMMONIUM.

Group reagent: there is no common reagent for the cations of this group.

Group reaction: cations of the fifth group do not react with hydrochloric acid, hydrogen sulphide, ammonium sulphide or (in the presence of ammonium salts) with ammonium carbonate. Special reactions or flame tests can be used for their identification.

Of the cations of this group, magnesium displays similar reactions to those of cations in the fourth group. However, magnesium carbonate, in the presence of ammonium salts, is soluble, and therefore during the course of systematic analysis (when considerable amounts of ammonium salts are building up in the solution) magnesium will not precipitate with the cations of the fourth group.

The reactions of ammonium ions are quite similar to those of the potassium ion, because the ionic radii of these two ions are almost identical.

III.35 MAGNESIUM, Mg (A_r: 24·305) Magnesium is a white, malleable and ductile metal. It melts at 650°C. It burns readily in air or oxygen with a brilliant white light, forming the oxide MgO and some nitride Mg_3N_2. The metal is slowly decomposed by water at ordinary temperature, but at the boiling point of water the reaction proceeds rapidly:

$$Mg+2H_2O \rightarrow Mg(OH)_2\downarrow+H_2\uparrow$$

Magnesium hydroxide, if ammonium salts are absent, is practically insoluble. Magnesium dissolves readily in acids:

$$Mg+2H^+ \rightarrow Mg^{2+}+H_2\uparrow$$

Magnesium forms the bivalent cation Mg^{2+}. Its oxide, hydroxide, carbonate, and phosphate are insoluble; the other salts are soluble. They taste bitter. Some of the salts are hygroscopic.

Reactions of magnesium ions To study these reactions a 0·5M solution of magnesium chloride $MgCl_6.6H_2O$ or magnesium sulphate $MgSO_4.7H_2O$ can be used.

1. Ammonia solution: partial precipitation of white, gelatinous magnesium hydroxide:

$$Mg^{2+}+2NH_3+2H_2O \rightarrow Mg(OH)_2\downarrow+2NH_4^+$$

The precipitate is very sparingly soluble in water (12 mg ℓ^{-1}, $K_s = 3{\cdot}4\times10^{-11}$), but readily soluble in ammonium salts.

As the reaction progresses, the concentration of the ammonium ions, due to the dissociation of the completely ionized ammonium salt, increases and consequently the concentration of the hydroxyl ions decreases owing to the common-ion effect (cf. Section **I.27**). The small hydroxyl-ion concentration, already low, is decreased still further so that much of the magnesium salt remains in solution. In the presence of a sufficient concentration of ammonium salts, the hydroxyl-ion concentration is reduced to such an extent that the solubility product of $Mg(OH)_2$ is not exceeded (cf. Section **I.28**); hence magnesium is not precipitated by ammonia solution in the presence of ammonium chloride or other ammonium salts.

2. Sodium hydroxide solution: white precipitate of magnesium hydroxide, insoluble in excess reagent, but readily soluble in solutions of ammonium salts:

$$Mg^{2+}+2OH^- \rightarrow Mg(OH)_2\downarrow$$

3. Ammonium carbonate solution: in the absence of ammonium salts a white precipitate of basic magnesium carbonate:

$$5Mg^{2+}+6CO_3^{2-}+7H_2O \rightarrow 4MgCO_3.Mg(OH)_2.5H_2O\downarrow+2HCO_3^-$$

In the presence of ammonium salts no precipitation occurs, because the equilibrium

$$NH_4^++CO_3^{2-} \rightleftarrows NH_3+HCO_3^-$$

is shifted towards the formation of hydrogen carbonate ions. The solubility product of the precipitate being high (K_s of pure $MgCO_3$ is 1×10^{-5}), the concentration of carbonate ions necessary to produce a precipitate is not attained.

4. Sodium carbonate solution: white, voluminous precipitate of basic carbonate (cf. reaction 3), insoluble in solutions of bases, but readily soluble in acids and in solutions of ammonium salts.

5. Disodium hydrogen phosphate solution: white crystalline precipitate of magnesium ammonium phosphate $Mg(NH_4)PO_4.6H_2O$ in the presence of ammonium chloride (to prevent precipitation of magnesium hydroxide) and ammonia solutions:

$$Mg^{2+} + NH_3 + HPO_4^{2-} \rightarrow Mg(NH_4)PO_4\downarrow$$

The precipitate is sparingly soluble in water, soluble in acetic acid and in mineral acids. The normal solubility of $Mg(NH_4)PO_4.6H_2O$ is increased by its hydrolysis in water:

$$Mg(NH_4)PO_4 + H_2O \rightleftarrows Mg^{2+} + HPO_4^{2-} + NH_3\uparrow + H_2O$$

This tendency is reduced by moderate amounts of ammonia (it is found that the compound is very sparingly soluble in 2·5 per cent ammonia solution). The precipitate separates slowly from dilute solutions because of its tendency to form supersaturated solutions; this may usually be overcome by cooling and by rubbing the test-tube or beaker beneath the surface of the liquid with a glass rod.

A white flocculant precipitate of magnesium hydrogen phosphate, $MgHPO_4$, is produced in neutral solutions.

$$Mg^{2+} + HPO_4^{2-} \rightarrow MgHPO_4\downarrow$$

6. Diphenylcarbazide reagent ($C_6H_5.NH.NH.CO.NH.NH.C_6H_5$) The magnesium salt solution is treated with sodium hydroxide solution – a precipitate of magnesium hydroxide will be formed – then with a few drops of the diphenylcarbazide reagent and the solution filtered. On washing the precipitate with hot water, it will be seen to have acquired a violet-red colour, due to the formation of a complex salt or an adsorption complex. Metals of Groups II and III interfere and should therefore be absent.

The reagent is prepared by dissolving 0·2 g diphenylcarbazide in 10 ml glacial acetic acid and diluting to 100 ml with ethanol.

7. 8-Hydroxyquinoline or 'oxine' reagent

(structure: N, OH):

when a solution of a magnesium salt, containing a little ammonium chloride, is treated with 1–2 ml reagent which has been rendered strongly ammoniacal by the addition of 3–4 ml dilute ammonia solution and the mixture heated to the boiling point, a yellow precipitate of the complex salt $Mg(C_9H_6N.O)_2,4H_2O$ is obtained. All other metals, except sodium and potassium, must be absent.

The reagent is prepared by dissolving 2 g oxine in 100 ml 2M acetic acid.

8. p-*Nitrobenzene-azo-resorcinol (or magneson I) reagent*

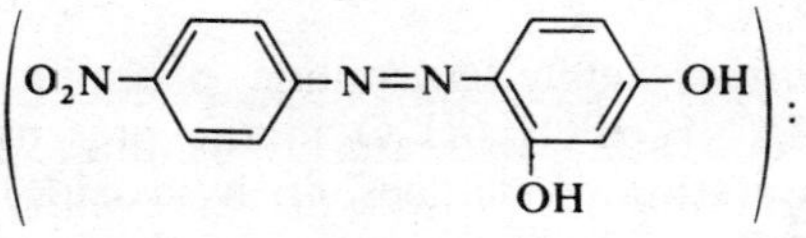

(O_2N–N=N–OH, OH):

this test depends upon the adsorption of the reagent, which is a dyestuff, upon $Mg(OH)_2$ in alkaline solution whereby a blue lake is produced. Two ml test solution, acidified slightly with hydrochloric acid, is treated with 1 drop of the reagent and sufficient 2M sodium hydroxide solution to render the solution strongly alkaline (say 2–3 ml). A blue precipitate appears. This is an excellent confirmatory test in macro analysis, but it is essential to perform a blank test with the reagents, which frequently yield a blue colouration. For this reason a blue precipitate should be looked for. All metals, except those of the alkalis, must be absent. Ammonium salts reduce the sensitivity of the test by preventing the precipitation of $Mg(OH)_2$, and should therefore be eliminated.

The reagent (for macro analysis) consists of a 0·5 per cent solution of *p*-nitrobenzene-azo-resorcinol in 0·2M sodium hydroxide.

The spot-test technique is as follows. Place a drop of the test solution on a spot plate and add 1–2 drops of the reagent. It is essential that the solution be strongly alkaline; the addition of 1 drop of 2M sodium hydroxide may be advisable. According to the concentration of magnesium a blue precipitate is formed or the reddish-violent reagent assumes a blue colour. A comparative test on distilled water should be carried out.

Sensitivity: 0·5 μg Mg. *Concentration limit:* 1 in 100,000.

Filter or drop-reaction paper should not be used.

An alternative agent is *p*-nitrobenzene-azo-α-naphthol or magneson II

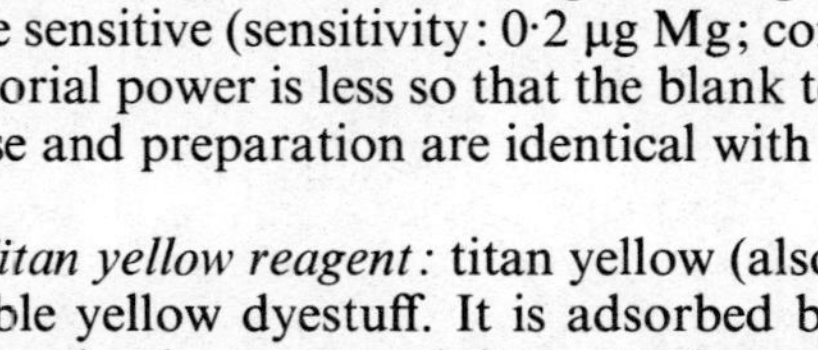

(O_2N–N=N–OH):

It yields the same colour changes as magneson I, but has the advantage that it is more sensitive (sensitivity: 0·2 μg Mg; concentration limit: 1 in 250,000) and its tinctorial power is less so that the blank test is not so deeply coloured. Its mode of use and preparation are identical with that described above for Magneson I.

9. Titan yellow reagent: titan yellow (also known as clayton yellow) is a water-soluble yellow dyestuff. It is adsorbed by magnesium hydroxide producing a deep-red colour or precipitate. Barium and calcium do not react but intensify the red colour. All elements of Groups I to III should be removed before applying the test.

Place a drop of the test solution on a spot plate, introduce a drop of reagent and a drop of 2M sodium hydroxide. A red colour or precipitate is produced.

An alternative technique is to treat 0·5 ml neutral or slightly acidic test solution with 0·2 ml 2M sodium hydroxide solution. A red precipitate or colouration is produced.

Sensitivity: 1·5 μg Mg. *Concentration limit:* 1 in 33,000.
The reagent consists of a 0·1 per cent aqueous solution of titan yellow.

10. Quinalizarin reagent: blue precipitate or cornflower-blue colouration with magnesium salts. The colouration can be readily distinguished from the blue-violet colour of the reagent. Upon the addition of a little bromine water, the colour disappears (difference from beryllium). The alkaline earth metals and aluminium do not interfere under the conditions of the test, but all elements of Groups I to III should be removed. Phosphates and large amounts of ammonium salts decrease the sensitivity of the reaction.

Place a drop of the test solution and a drop of distilled water in adjacent cavities of a spot plate and add 2 drops of the reagent to each. If the solution is acid, it will be coloured yellowish-red by the reagent. Add 2M sodium hydroxide until the colour changes to violet and a further excess to increase the volume by 25 to 50 per cent. A blue precipitate or colouration appears. The blank test has a blue-violet colour.

Sensitivity: 0·25 μg Mg. *Concentration limit:* 1 in 200,000.

The reagent is prepared by dissolving 0·01–0·02 g quinalizarin in 100 ml alcohol. Alternatively, a 0·05 per cent solution of 0·1M sodium hydroxide may be used.

11. Dry test (blowpipe test) All magnesium compounds when ignited on charcoal in the presence of sodium carbonate are converted into white magnesium oxide, which glows brightly when hot. Upon moistening with a drop or two of cobalt nitrate solution and reheating strongly, a pale-pink mass is obtained.

III.36 POTASSIUM, K (*A*$_r$: 39·098) Potassium is a soft, silver-white metal. Potassium melts at 63·5°C. It remains unchanged in dry air, but is rapidly oxidized in moist air, becoming covered with a blue film. The metal decomposes water violently, evolving hydrogen and burning with a violet flame:

$$2K^{+} + 2H_2O \rightarrow 2K^{+} + 2OH^{-} + H_2\uparrow$$

Potassium is usually kept under solvent naphtha.

Potassium salts contain the monovalent cation K^+. These salts are usually soluble and form colourless solutions, unless the anion is coloured.

Reactions of potassium ions A M solution of potassium chloride, KCl, can be used for these tests.

1. Sodium hexanitritocobaltate(III) solution $Na_3[Co(NO_2)_6]$: yellow precipitate of potassium hexanitritocobaltate(III):

$$3K^{+} + [Co(NO_2)_6]^{3-} \rightarrow K_3[Co(NO_2)_6]\downarrow$$

The precipitate is insoluble in dilute acetic acid. If larger amounts of sodium are present (or if the reagent is added in excess) a mixed salt, $K_2Na[Co(NO_2)_6]$, is formed. The precipitate forms immediately in concentrated solutions and slowly in diluted solutions; precipitation may be accelerated by warming. Ammonium salts give a similar precipitate and must be absent. In alkaline solutions a brown or black precipitate of cobalt(III) hydroxide $Co(OH)_3$ is obtained. Iodides and other reducing agents interfere and should be removed before applying the test.

2. Tartaric acid solution (or sodium hydrogen tartrate solution): white crystalline precipitate of potassium hydrogen tartrate:

$$K^+ + H_2.C_4H_4O_6 \rightleftarrows KHC_4H_4O_6\downarrow + H^+ \qquad (a)$$

and

$$K^+ + H.C_4H_4O_6^- \rightleftarrows KHC_4H_4O_6 \qquad (b)$$

If tartaric acid is used, the solution should be buffered with sodium acetate, because the strong acid, formed in reaction (a) dissolves the precipitate. Strong alkalis also dissolve the precipitate.

The precipitate is slightly soluble in water (3·26 g ℓ^{-1}, $K_s = 3 \times 10^{-4}$), but quite insoluble in 50 per cent ethanol. Precipitation is accelerated by vigorous agitation of the solution, by scratching the sides of the vessel with a glass rod, and by adding alcohol. Ammonium salts yield a similar precipitate and must be absent.

3. Perchloric acid solution ($HClO_4$): white crystalline precipitate of potassium perchlorate $KClO_4$ from not too dilute solutions.

$$K^+ + ClO_4^- \rightarrow KClO_4\downarrow$$

The precipitate is slightly soluble in water (3·2 g ℓ^{-1} and 198 g ℓ^{-1} at 0° and 100°C respectively), and practically insoluble in absolute alcohol. The alcoholic solution should not be heated as a dangerous explosion may result. This reaction is unaffected by the presence of ammonium salts.

4. Hexachloroplatinic(IV) acid ($H_2[PtCl_6]$) reagent: yellow precipitate of potassium hexachloroplatinate(IV):

$$2K^+ + [PtCl_6]^{2-} \rightarrow K_2[PtCl_6]\downarrow$$

Precipitation is instantaneous from concentrated solutions; in dilute solutions, precipitation takes place slowly on standing, but may be hastened by cooling and by rubbing the sides of the vessel with a glass rod. The precipitate is slightly soluble in water, but is almost insoluble in 75 per cent alcohol. Ammonium salts give a similar precipitate and must be absent.

The reagent is prepared by dissolving 2·6 g hydrated chloroplatinic acid $H_2[PtCl_6].6H_2O$ in 10 ml water. Owing to its expensive character, only small quantities should be employed and all precipitates placed in the platinum residues bottle.

5. Sodium hexanitritocobaltate(III)–silver nitrate test This is a modification of reaction 1 and is applicable to halogen-free solutions. Precipitation of potassium salts with sodium hexanitritocobaltate(III) and silver nitrate solution gives the compound $K_2Ag[Co(NH_2)_6]$, which is less soluble than the corresponding sodium compound $K_2Na[Co(NO_2)_6]$ and hence the test is more sensitive. Lithium, thallium, and ammonium salts must be absent for they give precipitates with sodium hexanitritocobaltate(III) solution.

Place a drop of the neutral or acetic acid test solution on a black spot plate, and add a drop of 0·1M silver nitrate solution and a small amount of finely powdered sodium hexanitritocobaltate(III). A yellow precipitate or turbidity appears.

Sensitivity: 1 μg K. *Concentration limit:* 1 in 50,000.
If silver nitrate solution is not added, the sensitivity is 4 μg K.

6. Dipicrylamine (or hexanitrodiphenylamine) reagent

$$\left(O_2N-C_6H_2(NO_2)_2-NH-C_6H_2(NO_2)_2-NO_2 \right):$$

the hydrogen atom of the NH group is replaceable by metals, the sodium salt is soluble in water to yield a yellow solution. With solutions of potassium salts, the latter gives a crystalline orange-red precipitate of the potassium derivative. The test is applicable in the presence of 80 times as much sodium and 130 times as much lithium. Ammonium salts should be removed before applying the test. Magnesium does not interfere.

Place a drop of the neutral test solution upon drop-reaction paper and immediately add a drop of the slightly alkaline reagent. An orange-red spot is obtained, which is unaffected by treatment with 1–2 drops of 2M hydrochloric acid.

Sensitivity: 3 μg K. *Concentration limit:* 1 in 10,000.

The reagent is prepared by dissolving 0·2 g dipicrylamine in 20 ml boiling 0·05M sodium carbonate and filtering the cooled liquid.

7. Sodium tetraphenylboron test Potassium forms a white precipitate in neutral solutions or in the presence of acetic acid:

$$K^+ + [B(C_6H_5)_4]^- \rightarrow K[B(C_6H_5)_4]\downarrow$$

The precipitate is almost insoluble in water (0·053 g ℓ^{-1}, $K_s = 2{\cdot}25 \times 10^{-8}$); potassium is precipitated quantitatively if a small excess of the reagent is applied (0·1–0·2 per cent). The precipitate is soluble in strong acids and alkalis, and also in acetone. Rubidium, caesium, thallium(I), and ammonium ions interfere.

The reagent is prepared by dissolving 3·42 g sodium tetraphenylboron $Na[B(C_6H_5)_4]$ (M_r: 342·2) in water and diluting it to 100 ml. This approximately 0·1M solution keeps for 2 weeks. If the solution is not clear, it must be filtered.

8. Dry test (flame colouration) Potassium compounds, preferably the chloride, colour the non-luminous Bunsen flame violet (lilac). The yellow flame produced by small quantities of sodium obscures the violet colour, but by viewing the flame through two thicknesses of cobalt blue glass, the yellow sodium rays are absorbed and the reddish-violet potassium flame becomes visible. A solution of chrome alum (310 g ℓ^{-1}), 3 cm thick, also makes a good filter.

III.37 SODIUM, Na (A_r: 22·99) Sodium is a silver-white, soft metal, melting at 97·5°C. It oxidizes rapidly in moist air and is therefore kept under solvent naphtha or xylene. The metal reacts violently with water forming sodium hydroxide and hydrogen:

$$2Na + 2H_2O \rightarrow 2Na^+ + 2OH^- + H_2\uparrow$$

In its salts sodium is present as the monovalent cation Na^+. These salts form colourless solutions unless the anion is coloured; almost all sodium salts are soluble in water.

Reactions of sodium ions To study these reactions a M solution of sodium chloride, NaCl, can be used.

1. Uranyl magnesium acetate solution: yellow, crystalline precipitate of sodium magnesium uranyl acetate $NaMg(UO_2)_3(CH_3COO)_9.9H_2O$ from concentrated solutions. The addition of about one-third volume of alcohol helps the precipitation.

$$Na^+ + Mg^{2+} + 3UO_2^{2+} + 9CH_3COO^- \rightarrow NaMg(UO_2)_3(CH_3COO)_9\downarrow$$

The reagent is prepared as follows. Dissolve 10 g uranyl acetate in 6 g glacial acetic acid and 100 ml water (solution *a*). Dissolve 33 g magnesium acetate in 10 g acetic acid and 100 ml water (solution *b*). Mix the two solutions *a* and *b*, allow to stand for 24 hours, and filter. Alternatively, a reagent of equivalent concentration may be prepared by dissolving uranyl magnesium acetate in the appropriate volume of water or of M acetic acid.

2. Chloroplatinic acid, tartaric acid or sodium hexanitritocobaltate(III) solution: no precipitate with solutions of sodium salts.

3. Uranyl zinc acetate reagent As a delicate test for sodium, the uranyl zinc acetate reagent is sometimes preferred to that employing uranyl magnesium acetate. The yellow crystalline sodium zinc uranyl acetate, $NaZn(UO_2)_3(CH_3COO)_9.9H_2O$, is obtained. The reaction is fairly selective for sodium. The sensitivity of the reaction is affected by copper, mercury, cadmium, aluminium, cobalt, nickel, manganese, zinc, calcium, strontium, barium, and ammonium when present in concentrations exceeding 5 g ℓ^{-1}; potassium and lithium salts are precipitated if their concentration in solution exceeds 5 g ℓ^{-1} and 1 g ℓ^{-1} respectively.

Place a drop of the neutral test solution on a black spot plate or upon a black watch glass, add 8 drops of the reagent, and stir with a glass rod. A yellow cloudiness or precipitate forms.

Sensitivity: 12·5 μg Na. *Concentration limit:* 1 in 4,000.

The reagent is prepared as follows. Dissolve 10 g uranyl acetate in 25 ml 30 per cent acetic acid, warming if necessary, and dilute with water to 50 ml (solution *a*). In a separate vessel stir 30 g zinc acetate with 25 ml 30 per cent acetic acid and dilute with water to 50 ml (solution *b*). Mix the two solutions *a* and *b*, and add a small quantity of sodium chloride. Allow to stand for 24 hours, and filter from the precipitated sodium zinc uranyl acetate.

Alternatively, a reagent of equivalent concentration may be prepared by dissolving uranyl zinc acetate in the appropriate volume of water or of M acetic acid.

4. Dry test (flame colouration) The non-luminous Bunsen flame is coloured an intense yellow by vapours of sodium salts. The colour is not visible when viewed through two thicknesses of cobalt blue glass. Minute quantities of sodium salts give this test, and it is only when the colour is intense and persistent that appreciable quantities of sodium are present.

III.38 AMMONIUM ION, NH_4^+ (M_r: 18·038) Ammonium ions are derived from ammonia, NH_3, and the hydrogen ion H^+. The characteristics of these ions are similar to those of alkali metal ions. By electrolysis with a mercury cathode, ammonium amalgam can be prepared, which has similar properties to the amalgams of sodium or potassium.

Ammonium salts are generally water-soluble compounds, forming colourless solutions (unless the anion is coloured). On heating, all ammonium salts decompose to ammonia and the appropriate acid. Unless the acid is non-volatile, ammonium salts can be quantitatively removed from dry mixtures by heating.

The reactions of ammonium ions are in general similar to those of potassium, because the sizes of the two ions are almost identical.

Reactions of ammonium ions To study these reactions a M solution of ammonium chloride NH_4Cl can be used.

1. Sodium hydroxide solution: ammonia gas is evolved on warming.

$$NH_4^+ + OH^- \rightarrow NH_3\uparrow + H_2O$$

This may be identified (*a*) by its odour (cautiously smell the vapour after removing the test-tube or small beaker from the flame); (*b*) by the formation of white fumes of ammonium chloride when a glass rod moistened with concentrated hydrochloric acid is held in the vapour; (*c*) by its turning moistened red litmus paper blue or turmeric paper brown; (*d*) by its ability to turn filter paper moistened with mercury(I) nitrate solution black (this is a very trustworthy test*); and (*e*) filter paper moistened with a solution of manganese(II) chloride and hydrogen peroxide gives a brown colour, due to the oxidation of manganese by the alkaline solution thus formed.

In test 1(*d*) a mixture of mercury(II) amidonitrate (white precipitate) and mercury (black precipitate) is formed:

$$2NH_3 + Hg_2^{2+} + NO_3^- \rightarrow Hg(NH_2)NO_3\downarrow + Hg\downarrow + NH_4^+$$

In test 1(*e*) hydrated manganese(IV) oxide is formed:

$$2NH_3 + Mn^{2+} + H_2O_2 + H_2O \rightarrow MnO(OH)_2\downarrow + 2NH_4^+$$

2. Nessler's reagent (alkaline solution of potassium tetraiodomercurate(II)): brown precipitate or brown or yellow colouration is produced according to the amount of ammonia or ammonium ions present. The precipitate is a basic mercury(II) amido-iodide:

$$NH_4^+ + 2[HgI_4]^{2-} + 4OH^- \rightarrow HgO.Hg(NH_2)I\downarrow + 7I^- + 3H_2O$$

The formula of the brown precipitate has also been given as $3HgO.Hg(NH_3)_2I_2$

---

* Arsine, however, blackens mercury(I) nitrate paper, and must therefore be absent.

(Britton and Wilson, 1933) and as $NH_2.Hg_2I_3$ (Nichols and Willits, 1934).

The test is an extremely delicate one and will detect traces of ammonia present in drinking water. All metals, except sodium or potassium, must be absent.

The reagent is prepared by dissolving 10 g potassium iodide in 10 ml ammonia-free water, adding saturated mercury(II) chloride solution (60 g ℓ^{-1}) in small quantities at a time, with shaking, until a slight, permanent precipitate is formed, then adding 80 ml 9M potassium hydroxide solution and diluting to 200 ml. Allow to stand overnight and decant the clear liquid. The reagent thus consists of an alkaline solution of potassium tetraiodomercurate(II) $K_2[HgI_4]$.

Nessler's original reagent has been described as a solution which is *c.* 0·09M in potassium tetraiodomercurate(II), $K_2[HgI_4]$, and 2·5M in potassium hydroxide.

An alternative method for the preparation of the reagent is as follows. Dissolve 23 g mercury(II) iodide and 16 g potassium iodide in ammonia-free water and make up the volume to 100 ml; add 100 ml 6M sodium hydroxide. Allow to stand for 24 hours, and decant the solution from any precipitate that may have formed. The solution should be kept in the dark.

The spot-test technique is as follows. Mix a drop of the test solution with a drop of concentrated sodium hydroxide solution on a watch glass. Transfer a micro drop of the resulting solution or suspension to drop-reaction paper and add a drop of Nessler's reagent. A yellow or orange-red stain or ring is produced.

Sensitivity: 0·3 μg NH_3 (in 0·002 ml).

A better procedure is to employ the technique described under the manganese(II) nitrate–silver nitrate reagent in reaction 9 below. A drop of Nessler's solution is placed on the glass knob of the apparatus. After the reaction is complete, the drop of the reagent is touched with a piece of drop-reaction or quantitative filter paper when a yellow colouration will be apparent.

Sensitivity: 0·25 μg NH_3.

3. *Sodium hexanitritocobaltate(III),* ($Na_3[Co(NO)_2)_6]$): yellow precipitate of ammonium hexanitritocobaltate(III), $(NH_4)_3[Co(NO_2)_6]$, similar to that produced by potassium ions:

$$3NH_4^+ + [Co(NO_2)_6]^{3-} \rightarrow (NH_4)_3[Co(NO_2)_6]\downarrow$$

4. *Hexachloroplatinic(IV) acid* ($H_2[PtCl_6]$): yellow precipitate of ammonium hexachloroplatinate(IV)

$$2NH_4^+ + [PtCl_6]^{2-} \rightarrow (NH_4)_2[PtCl_6]\downarrow$$

The characteristics of the precipitate are similar to that of the corresponding potassium salt, but differ from it in being decomposed by warming with sodium hydroxide solution with the evolution of ammonia gas.

5. *Saturated sodium hydrogen tartrate solution* ($NaH.C_4H_4O_6$):* white precipitate of ammonium acid tartrate $NH_4.H.C_4H_4O_6$, similar to but slightly more soluble than the corresponding potassium salt, from which it is distinguished by the evolution of ammonia gas on being heated with sodium hydroxide solution.

---

* Or, less effectively, tartaric acid solution.

$$NH_4^+ + HC_4H_4O_6^- \rightarrow NH_4HC_4H_4O_6\downarrow$$

6. *Perchloric acid or sodium perchlorate solution:* no precipitate (distinction from potassium).

7. *Tannic acid–silver nitrate test* The basis of this test is the reducing action of tannic acid (a glucoside of digallic acid) upon the silver ammine complex $[Ag(NH_3)_2]^+$ to yield black silver: it therefore precipitates silver in the presence of ammonia but not from a slightly acid silver nitrate solution.

Mix 2 drops 5 per cent tannic acid (tannin) solution with 2 drops 20 per cent silver nitrate solution, and place the mixture upon drop-reaction paper or upon a little cotton wool. Hold the paper in the vapour produced by heating an ammonium salt with sodium hydroxide solution. A black stain is formed on the paper or upon the cotton wool. The test is a sensitive one.

8. p-*Nitrobenzene-diazonium chloride reagent* The reagent (I) yields a red colouration (due to II) with an ammonium salt in the presence of sodium hydroxide solution.

$$O_2N-C_6H_4-N{=}N-Cl + NH_4^+ + 2OH^- \longrightarrow$$
(I)

$$\longrightarrow O_2N-C_6H_4-N{=}NONH_4 + Cl^- + H_2O$$
(II)

Place a drop of the neutral or slightly acid test solution on a spot plate, followed by a drop of the reagent and a fine granule of calcium oxide between the two drops. A red zone forms round the calcium oxide. A blank test should be carried out on a drop of water.

Sensitivity: 0·7 μg NH_3. *Concentration limit:* 1 in 75,000.

The reagent (sometimes known as Riegler's solution) is prepared as follows. Dissolve 1 g *p*-nitroaniline in 25 ml 2M hydrochloric acid (warming may be necessary) and dilute with 160 ml water. Cool, add 20 ml 2–5 per cent sodium nitrite solution with vigorous shaking. Continue the shaking until all dissolves. The reagent becomes turbid on keeping, but can be employed again after filtering.

9. *Ammonia-formation test* This is a modification of reaction 1 as adapted to delicate analysis. The apparatus is shown in Fig. II.55 and consists of a small glass tube of 1 ml capacity, which can be closed with a small ground-glass stopper carrying a small glass hook at the lower end.

Place a drop of the test solution or a little of the solid in the micro test-tube, and add a drop of 2M sodium hydroxide solution. Fix a small piece of red litmus paper on the glass hook and insert the stopper into position. Warm to 40°C for 5 minutes. The paper assumes a blue colour.

Sensitivity: 0·01 μg NH_3. *Concentration limit:* 1 in 5,000,000

Cyanides should be absent, for they give ammonia with alkalis:

$$CN^- + 2H_2O \rightarrow HCOO^- + NH_3\uparrow$$

If, however, a little mercury(II) oxide or a mercury(II) salt is added, the alkali-stable mercury(II) cyanide $Hg(CN)_2$ is formed and the interfering effect of cyanides is largely eliminated.

An alternative method for carrying out the test is to employ the manganese(II) nitrate–silver nitrate reagent. Upon treating a neutral solution of manganese(II) and silver salts with ammonia, a black precipitate is formed:

$$4NH_3 + Mn^{2+} + 2Ag^+ + 3H_2O \rightarrow MnO(OH)_2\downarrow + 2Ag\downarrow + 4NH_4^+$$

The sensitivity can be increased by treating the resultant precipitate with an acetic acid solution of benzidine whereupon the manganese dioxide oxidizes the benzidine to a blue oxidation product. (DANGER: THE REAGENT IS CARCINOGENIC).

Use the apparatus shown in Fig. II.53 or in Fig. II.54. Place a drop of the test solution and a drop of 2M sodium hydroxide in the micro test-tube; also place a drop of the reagent on the glass knob of the stopper and close the apparatus. Heat at 40°C for 5 minutes. Wash the drop of the reagent on to a piece of quantitative filter paper when a black or grey fleck will become transparent; this turns blue upon treatment with a solution of benzidine (the latter is prepared by dissolving 0·05 g benzidine or its hydrochloride in 10 ml glacial acetic acid, diluting to 100 ml with water, and filtering).

Sensitivity: 0·005 μg NH_3. *Concentration limit:* 1 in 10,000,000.

The manganese(II) nitrate–silver nitrate reagent is prepared thus: dissolve 2·87 g manganese(II) nitrate in 40 ml water, and filter. Add a solution of 3·55 g silver nitrate in 40 ml water, and dilute the mixture to 100 ml. Neutralize the acid formed by hydrolysis by adding dilute alkali dropwise until a black precipitate is formed and filter. Keep the reagent in a dark bottle.

10. Dry test All ammonium salts are either volatilized or decomposed when heated to just below red heat. In some cases, where the acid is volatile, the vapours recombine on cooling to form a sublimate of the salt, e.g. ammonium chloride.

CHAPTER IV **REACTIONS OF THE ANIONS**

IV.1 SCHEME OF CLASSIFICATION The methods available for the detection of anions are not as systematic as those which have been described in the previous chapter for cations. No really satisfactory scheme has yet been proposed which permits of the separation of the common anions into major groups, and the subsequent unequivocal separation of each group into its independent constituents. It must, however, be mentioned that it is possible to separate the anions into major groups dependent upon the solubilities of their silver salts, of their calcium or barium salts, and of their zinc salts; these however, can only be regarded as useful in giving an indication of the limitations of the method and for the confirmation of the results obtained by the simpler procedures to be described below.

The following scheme of classification has been found to work well in practice; it is not a rigid one since some of the anions belong to more than one of the subdivisions, and, furthermore, it has no theoretical basis. Essentially the processes employed may be divided into (A) those involving the identification by volatile products obtained on treatment with acids, and (B) those dependent upon reactions in solution. Class (A) is subdivided into (i) gases evolved with dilute hydrochloric acid or dilute sulphuric acid, and (ii) gases or vapours evolved with concentrated sulphuric acid. Class (B) is subdivided into (i) precipitation reactions, and (ii) oxidation and reduction in solution.

CLASS A

(i) Gases evolved with dilute hydrochloric acid or dilute sulphuric acid: Carbonate, hydrogen carbonate, sulphite, thiosulphate, sulphide, nitrite, hypochlorite, cyanide, and cyanate.

(ii) Gases or acid vapours evolved with concentrated sulphuric acid.

These include those of (i) with the addition of the following: fluoride, hexafluorosilicate,* chloride, bromide, iodide, nitrate, chlorate (**DANGER**), perchlorate, permanganate (**DANGER**), bromate, borate,* hexacyanoferrate(II), hexacyanoferrate(III), thiocyanate, formate, acetate, oxalate, tartrate, and citrate.

---

* This is often included in Class B(i).

CLASS B

(i) Precipitation reactions.
Sulphate, peroxodisulphate,* phosphate, phosphite, hypophosphite, arsenate, arsenite, chromate, dichromate, silicate, hexafluorosilicate, salicylate, benzoate, and succinate.

(ii) Oxidation and reduction in solution.
Manganate, permanganate, chromate and dichromate.

The reactions of all these anions will be systematically studied in the following pages. For convenience the reactions of certain organic acids are grouped together; these include acetates, formates, oxalates, tartrates, citrates, salicylates, benzoates, and succinates. It may be pointed out that acetates, formates, salicylates, benzoates, and succinates themselves form another group; all give a characteristic colouration or precipitate upon the addition of iron(III) chloride solution to a practically neutral solution.

IV.2 CARBONATES, CO_3^{2-} *Solubility* All normal carbonates, with the exception of those of the alkali metals and of ammonium, are insoluble in water. The hydrogen carbonates or bicarbonates of calcium, strontium, barium, magnesium, and possibly of iron exist in aqueous solution; they are formed by the action of excess carbonic acid upon the normal carbonates either in aqueous solution or suspension and are decomposed on boiling the solutions.

$$CaCO_3\downarrow + H_2O + CO_2 \rightarrow Ca^{2+} + 2HCO_3^-$$

The hydrogen carbonates of the alkali metals are soluble in water, but are less soluble than the corresponding normal carbonates.

To study these reactions a 0·5M solution of sodium carbonate $Na_2CO_3.10H_2O$ can be used.

1. Dilute hydrochloric acid: decomposition with effervescence, due to the evolution of carbon dioxide:

$$CO_3^{2-} + 2H^+ \rightarrow CO_2\uparrow + H_2O$$

the gas can be identified by its property of rendering lime water (or baryta water) turbid:

$$CO_2 + Ca^{2+} + 2OH^- \rightarrow CaCO_3\downarrow + H_2O$$

$$CO_2 + Ba^{2+} + 2OH^- \rightarrow BaCO_3\downarrow + H_2O$$

Some natural carbonates, such as magnesite, $MgCO_3$, siderite, $FeCO_3$, and dolomite, $(Ca,Mg)CO_3$, do not react appreciably in the cold; they must be finely powdered and the reaction mixture warmed.

The lime water or baryta water test is best carried out in the apparatus shown in Fig. IV.1. The solid substance is placed in the test-tube or small distilling flask (10–25 ml capacity), dilute hydrochloric acid added, and the cork immediately replaced. The gas which is evolved (warming may be necessary) is passed into lime water or baryta water contained in the test-tube; the production of a turbidity indicates the presence of a carbonate. It must be remembered that,

* Strictly speaking peroxodisulphates should be grouped with Class B(ii), but are best studied together with sulphates.

with prolonged passage of carbon dioxide, the turbidity slowly disappears as a result of the formation of a soluble hydrogen carbonate:

$$CaCO_3\downarrow + CO_2 + H_2O \rightarrow Ca^{2+} + 2HCO_3^-$$

Any acid, which is stronger than carbonic acid ($K_1 = 4{\cdot}31 \times 10^{-7}$) will displace it, especially on warming. Thus, even acetic acid ($K = 1{\cdot}76 \times 10^{-5}$) will decompose carbonates; the weak boric acid ($K_1 = 5{\cdot}8 \times 10^{-10}$) and hydrocyanic acid ($K = 4{\cdot}79 \times 10^{-10}$) will not.

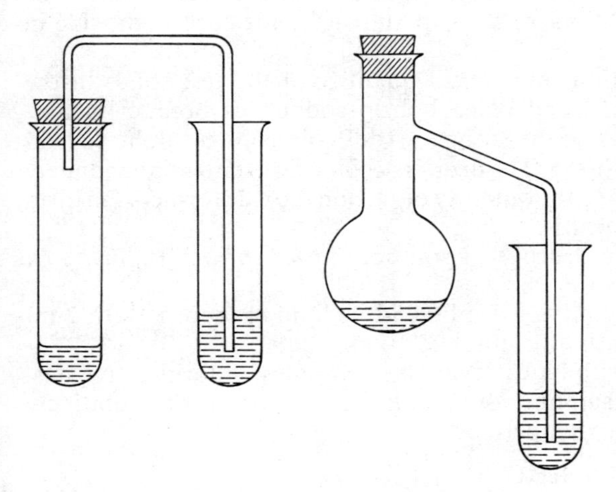

Fig. IV.1

2. *Barium chloride (or calcium chloride) solution:* white precipitate of barium (or calcium) carbonate:

$$CO_3^{2-} + Ba^{2+} \rightarrow BaCO_3\downarrow$$
$$CO_3^{2-} + Ca^{2+} \rightarrow CaCO_3\downarrow$$

Only normal carbonates react; hydrogen carbonates do not. The precipitate is soluble in mineral acids and carbonic acid:

$$BaCO_3 + 2H^+ \rightarrow Ba^{2+} + CO_2\uparrow + H_2O$$
$$BaCO_3 + CO_2 + H_2O \rightarrow Ba^{2+} + 2HCO_3^-$$

3. *Silver nitrate solution:* white precipitate of silver carbonate:

$$CO_3^{2-} + 2Ag^+ \rightarrow Ag_2CO_3\downarrow$$

The precipitate is soluble in nitric acid and in ammonia:

$$Ag_2CO_3 + 2H^+ \rightarrow 2Ag^+ + CO_2\uparrow + H_2O$$
$$Ag_2CO_3 + 4NH_3 \rightarrow 2[Ag(NH_3)_2]^+ + CO_3^{2-}$$

The precipitate becomes yellow or brown upon addition of excess reagent owing to the formation of silver oxide; the same happens if the mixture is boiled:

$$Ag_2CO_3\downarrow \rightarrow Ag_2O\downarrow + CO_2\uparrow$$

4. Sodium carbonate–phenolphthalein test This test depends upon the fact that phenolphthalein is turned pink by soluble carbonate and colourless by soluble bicarbonates. Hence if the carbon dioxide liberated by dilute acids from carbonates is allowed to come into contact with phenolphthalein solution coloured pink by sodium carbonate solution, it may be identified by the decolourization which takes place

$$CO_2 + CO_3^{2-} + H_2O \rightarrow 2HCO_3^-$$

The concentration of the sodium carbonate solution must be such as not to be decolourized under the conditions of the experiment by the carbon dioxide in the atmosphere.

Place 1–2 drops test solution (or a small quantity of the test solid) in the apparatus shown in Fig. II.53 and place 1 drop sodium carbonate–phenolphthalein reagent on the knob of the stopper. Add 3–4 drops M sulphuric acid and insert the stopper into position. The drop is decolourized either immediately or after a short time according to the quantity of carbon dioxide formed. Perform a blank test in a similar apparatus.

Sensitivity: 4 μg CO_2 (in 2 drops of solution). *Concentration limit:* 1 in 12,500.

The reagent is prepared by mixing 1 ml 0·05M sodium carbonate with 2 ml 0·5 per cent solution of phenolphthalein, and 10 ml water.

Sulphides, sulphites, thiosulphates, cyanides, cyanates, fluorides, nitrites, and acetates interfere. The sulphur-containing anions can be quantitatively oxidized to sulphates by hydrogen peroxide:

$$S^{2-} + 4H_2O_2 \rightarrow SO_4^{2-} + 4H_2O$$
$$SO_3^{2-} + H_2O_2 \rightarrow SO_4^{2-} + H_2O$$
$$S_2O_3^{2-} + 4H_2O_2 \rightarrow 2SO_4^{2-} + 3H_2O + 2H^+$$

The modified procedure in the presence of these anions is therefore to stir a drop of the test solution with 4 drops 3 per cent hydrogen peroxide, then to add 2 drops M sulphuric acid, and to continue as above. Cyanides are rendered innocuous by treating the test solution with 4 drops of a saturated solution of mercury(II) chloride, followed by 2 drops sulphuric acid, etc.; the slightly dissociated mercury(II) cyanide is formed. Nitrites can be removed by treatment with aniline hydrochloride.

IV.3 HYDROGEN CARBONATES, HCO_3^- Most of the reactions of hydrogen carbonates are similar to those of carbonates. The tests described here are suitable to distinguish hydrogen carbonates from carbonates.

A freshly prepared 0·5M solution of sodium hydrogen carbonate, $NaHCO_3$, or potassium hydrogen carbonate, $KHCO_3$, can be used to study these reactions.

1. Boiling When boiling, hydrogen carbonates decompose:

$$2HCO_3^- \rightarrow CO_3^{2-} + H_2O + CO_2\uparrow$$

carbon dioxide, formed in this way, can be identified with lime water or baryta water (cf. Section **IV.2**, reaction 1).

2. Magnesium sulphate Adding magnesium sulphate to a cold solution of hydrogen carbonate no precipitation occurs, while a white precipitate of magnesium carbonate, $MgCO_3$, is formed with normal carbonates.

Heating the mixture, a white precipitate of magnesium carbonate is formed:

$$Mg^{2+} + 2HCO_3^- \rightarrow MgCO_3 + H_2O + CO_2\uparrow$$

The carbon dioxide gas, formed in the reaction, can be detected with lime water or baryta water (cf. Section **IV.2**, reaction 1).

3. *Mercury(II) chloride* No precipitate is formed with hydrogen carbonate ions, while in a solution of normal carbonates a reddish-brown precipitate of basic mercury(II) carbonate ($3HgO.HgCO_3 = Hg_4O_3CO_3$) is formed:

$$CO_3^{2-} + 4Hg^{2+} + 3H_2O \rightarrow Hg_4O_3CO_3\downarrow + 6H^+$$

the excess of carbonate acts as a buffer, reacting with the hydrogen ions formed in the reaction:

$$CO_3^{2-} + 2H^+ \rightarrow CO_2\uparrow + H_2O$$

4. *Solid test* On heating some solid alkali hydrogen carbonate in a dry test-tube carbon dioxide is evolved:

$$2NaHCO_3 \rightarrow Na_2CO_3 + H_2O + CO_2\uparrow$$

The gas can be identified with lime water or baryta water (cf. Section **IV.2**, reaction 1). The residue evolves carbon dioxide if dilute hydrochloric acid is poured on it after cooling:

$$Na_2CO_3 + 2H^+ \rightarrow 2Na^+ + CO_2\uparrow + H_2O$$

5. *Test for hydrogen carbonate in the presence of normal carbonate* Adding an excess of calcium chloride to a mixture of carbonate and hydrogen carbonate the former is precipitated quantitatively:

$$CO_3^{2-} + Ca^{2+} \rightarrow CaCO_3\downarrow$$

On filtering the solution rapidly hydrogen carbonate ions pass into the filtrate. On adding ammonia to the filtrate, a white precipitate or cloudiness is obtained if hydrogen carbonates are present:

$$2HCO_3^- + 2Ca^{2+} + 2NH_3 \rightarrow 2CaCO_3\downarrow + 2NH_4^+$$

IV.4 SULPHITES, SO_3^{2-} *Solubility* Only the sulphites of the alkali metals and of ammonium are soluble in water; the sulphites of the other metals are either sparingly soluble or insoluble. The hydrogen sulphites of the alkali metals are soluble in water; the hydrogen sulphites of the alkaline earth metals are known only in solution.

A freshly prepared 0·5M solution of sodium sulphite $Na_2SO_3.7H_2O$ can be used to study these reactions.

1. *Dilute hydrochloric acid (or dilute sulphuric acid):* decomposition, more rapidly on warming, with the evolution of sulphur dioxide:

$$SO_3^{2-} + 2H^+ \rightarrow SO_2\uparrow + H_2O$$

The gas may be identified (i) by its suffocating odour of burning sulphur, (ii) by the green colouration, due to the formation of chromium(III) ions, produced when a filter paper, moistened with acidified potassium dichromate solution, is held over the mouth of the test-tube.

$$3SO_2 + Cr_2O_7^{2-} + 2H^+ \rightarrow 2Cr^{3+} + 3SO_4^{2-} + H_2O$$

Another method of identification of the gas is (iii) to hold a filter paper, moistened with potassium iodate and starch solution, in the vapour, when a blue colour, owing to the formation of iodine, is observable:

$$5SO_2 + 2IO_3^- + 4H_2O \rightarrow I_2 + 5SO_4^{2-} + 8H^+$$

2. *Barium chloride or strontium chloride solution:* white precipitate of barium or strontium) sulphite:

$$SO_3^{2-} + Ba^{2+} \rightarrow BaSO_3\downarrow$$

the precipitate dissolves in dilute hydrochloric acid, when sulphur dioxide evolves:

$$BaSO_3\downarrow + 2H^+ \rightarrow Ba^{2+} + SO_2\uparrow + H_2O$$

On standing, the precipitate is slowly oxidized to the sulphate and is then insoluble in dilute mineral acids; this change is rapidly effected by warming with bromine water or a little concentrated nitric acid or with hydrogen peroxide.

$$2BaSO_3\downarrow + O_2 \rightarrow 2BaSO_4\downarrow$$
$$BaSO_3\downarrow + Br_2 + H_2O \rightarrow BaSO_4\downarrow + 2Br^- + 2H^+$$
$$3BaSO_3\downarrow + 2HNO_3 \rightarrow 3BaSO_4\downarrow + 2NO\uparrow + H_2O$$
$$BaSO_3\downarrow + H_2O_2 \rightarrow BaSO_4\downarrow + H_2O$$

The solubilities at 18° of the sulphites of calcium, strontium, and barium are respectively 1·25 g ℓ^{-1}, 0·033 g ℓ^{-1}, 0·022 g ℓ^{-1}.

3. *Silver nitrate solution:* first, no visible change occurs because of the formation of sulphitoargentate ions:

$$SO_3^{2-} + Ag^+ \rightarrow [AgSO_3]^-$$

on the addition of more reagent, a white, crystalline precipitate of silver sulphite is formed:

$$[AgSO_3]^- + Ag^+ \rightarrow Ag_2SO_3\downarrow$$

The precipitate dissolves if sulphite ions are added in excess:

$$Ag_2SO_3\downarrow + SO_3^{2-} \rightarrow 2[AgSO_3]^-$$

On boiling the solution of the complex salt, or an aqueous suspension of the precipitate, grey metallic silver is precipitated:

$$2[AgSO_3]^- \rightarrow 2Ag\downarrow + SO_4^{2-} + SO_2\uparrow$$
$$Ag_2SO_3\downarrow + H_2O \rightarrow 2Ag\downarrow + SO_4^{2-} + 2H^+$$

The precipitate is soluble in dilute nitric acid, when sulphur dioxide gas is evolved:

$$Ag_2SO_3\downarrow + 2H^+ \rightarrow SO_2\uparrow + 2Ag^+ + H_2O$$

The precipitate also dissolves in ammonia, when the diammineargentate complex is formed:

$$Ag_2SO_3\downarrow + 4NH_3 \rightarrow 2[Ag(NH_3)_2]^+ + SO_3^{2-}$$

4. *Potassium permanganate solution, acidified with dilute sulphuric acid before the test:* decolourization owing to reduction to manganese(II) ions:

$$5SO_3^{2-} + 2MnO_4^- + 6H^+ \rightarrow 2Mn^{2+} + 5SO_4^{2-} + 3H_2O$$

5. *Potassium dichromate solution, acidified with dilute sulphuric acid before the test:* a green colouration, owing to the formation of chromium(III) ions, is produced:

$$3SO_3^{2-} + Cr_2O_7^{2-} + 8H^+ \rightarrow 2Cr^{3+} + 3SO_4^{2-} + 4H_2O$$

6. *Lead acetate or lead nitrate solution:* white precipitate of lead sulphite:

$$SO_3^{2-} + Pb^{2+} \rightarrow PbSO_3\downarrow$$

The precipitate dissolves in dilute nitric acid, when sulphur dioxide gas is formed:

$$PbSO_3\downarrow + 2H^+ \rightarrow SO_2\uparrow + Pb^{2+} + H_2O$$

The gas may be identified by one of the methods listed under reaction 1. On boiling, the precipitate is oxidized by atmospheric oxygen and lead sulphate is formed:

$$2PbSO_3\downarrow + O_2 \rightarrow 2PbSO_4\downarrow$$

This reaction can be used to distinguish sulphites and thiosulphates; the latter produce a black precipitate (cf. Section **IV.5**, reaction 5) on boiling.

7. *Zinc and sulphuric acid:* hydrogen sulphide gas is evolved, which may be detected by holding lead acetate paper to the mouth of the test-tube (cf. Section **IV.6**, reaction 1):

$$SO_3^{2-} + 3Zn + 8H^+ \rightarrow H_2S\uparrow + 3Zn^{2+} + 3H_2\uparrow + 3H_2O$$

8. *Lime water* This test is carried out by adding dilute hydrochloric acid to the solid sulphite, and bubbling the evolved sulphur dioxide through lime water (Fig. IV.1); a white precipitate of calcium sulphite $CaSO_3$ is formed.

$$SO_3^{2-} + Ca^{2+} \rightarrow CaSO_3\downarrow$$

The precipitate dissolves on prolonged passage of the gas, due to the formation of hydrogen sulphite ions:

$$CaSO_3\downarrow + SO_2 + H_2O \rightarrow Ca^{2+} + 2HSO_3^-$$

A turbidity is also produced by carbonates; sulphur dioxide must therefore be first removed when testing for the latter. This may be effected by adding potassium dichromate solution to the test-tube before acidifying. The dichromate oxidizes and destroys the sulphur dioxide without affecting the carbon dioxide (cf. Section **IV.2**).

9. Fuchsin test Dilute solutions of triphenylmethane dyestuffs, such as fuchsin (for formula, see Section **IV.15**, reaction 9) and malachite green, are immediately decolourized by neutral sulphites. Sulphur dioxide also decolourizes fuchsin solution, but the reaction is not quite complete: nevertheless it is a very useful test for sulphur and acid sulphites; carbon dioxide does not interfere, but nitrogen dioxide does. If the test solution is acid, it should preferably be just neutralized with sodium hydrogen carbonate. Thiosulphates do not interfere but sulphides, polysulphides, and free alkali do. Zinc, lead, and cadmium salts reduce the sensitivity of the test, hence the interference of sulphides cannot be obviated by the addition of these salts.

Place 1 drop of the fuchsin reagent on a spot plate and add 1 drop of the neutral test solution. The reagent is decolourized.

Sensitivity: 1 μg SO_2. *Concentration limit:* 1 in 50,000.

The fuchsin reagent is prepared by dissolving 0·015 g fuchsin in 100 ml water.

10. Nickel(II) hydroxide test The auto-oxidation of sulphur dioxide (or sulphurous acid) induces the oxidation of green nickel(II) hydroxide to the black nickel(III) hydroxide. The colour change is quite distinct, but for very small amounts of sulphur dioxide use may be made of the conversion of benzidine acetate to 'benzidine blue' by the nickel(III) hydroxide. (DANGER: THE REAGENT IS CARCINOGENIC.) Thiosulphates give a similar reaction and must therefore be absent; sulphides also interfere.

Place a drop of the test solution (or a little of the test solid) in the tube of the apparatus shown in Fig. II.53 and place a little washed nickel(II) hydroxide on the glass knob under the stopper. Add 1–2 drops 6M hydrochloric acid, close the apparatus and warm gently. The green hydroxide turns grey to black according to the amount of sulphite present. For small amounts of sulphites, transfer the nickel hydroxide to a quantitative filter paper and treat with a drop of the benzidine reagent: a blue colour is formed.

An alternative technique is to warm (water bath) the test solution in a semimicro test-tube with a little dilute hydrochloric acid, and expose the evolved gas to filter paper upon which a stain of nickel(II) hydroxide has been made. The stain acquires a black colour.

Sensitivity: 0·4 μg SO_2. *Concentration limit:* 1 in 125,000.

The nickel(II) hydroxide is prepared by precipitating nickel(II) chloride solution with sodium hydroxide solution and washing thoroughly until free from alkali. It should be freshly prepared.

The benzidine reagent is prepared by dissolving 0·05 g benzidine or its hydrochloride in 10 ml glacial acetic acid, diluting to 100 ml with water and filtering.

11. Sodium nitroprusside–zinc sulphate test Sodium nitroprusside solution reacts with a solution of a zinc salt to yield a salmon-coloured precipitate of zinc nitroprusside $Zn[Fe(CN)_5NO]$. The latter reacts with moist sulphur dioxide to give a red compound of unknown composition; the test is rendered more sensitive when the reaction product is held over ammonia vapour which decolourizes the unused zinc nitroprusside.

Place a drop of the test solution (or a grain of the solid test sample) in the tube of Fig. II.53 and coat the knob of the glass stopper with a thin layer of the zinc nitroprusside paste. Add a drop of 2M hydrochloric or sulphuric acid and close the apparatus. After the sulphur dioxide has been evolved, hold the stopper for a short time in ammonia vapour. The paste is coloured more or less deep red.

Sensitivity: 3·5 μg SO_2. *Concentration limit:* 1 in 14,000.

The zinc nitroprusside paste is prepared by precipitating sodium nitroprusside solution with an excess of zinc sulphate solution and boiling for a few minutes: the precipitate is filtered and washed, and kept in a dark glass bottle or tube.

The test is not applicable in the presence of sulphides and/or thiosulphates. These can be removed by the addition of mercury(II) chloride which reacts forming the acid-stable mercury(II) sulphide:

$$Hg^{2+} + S^{2-} \rightarrow HgS\downarrow$$

$$Hg^{2+} + S_2O_3^{2-} + H_2O \rightarrow HgS\downarrow + SO_4^{2-} + 2H^+$$

Place a drop of the test solution and 2 drops of saturated mercury(II) chloride solution in the same apparatus (Fig. II.53) and, after a minute, acidify with 2M hydrochloric or M sulphuric acid, and proceed as above. 20 μg Na_2SO_3 can be detected in the presence of 900 μg $Na_2S_2O_3$ and 1,500 μg of Na_2S.

12. Distinction between sulphites and hydrogen sulphites The solution of normal alkali sulphites shows an alkaline reaction against litmus paper, because of hydrolysis:

$$SO_3^{2-} + H_2O \rightleftarrows HSO_3^- + OH^-$$

while the solution of alkali hydrogen sulphites is neutral. Adding a neutral solution of dilute hydrogen peroxide to the solution of normal sulphites, sulphate ions are formed and the solution becomes neutral:

$$SO_3^{2-} + H_2O_2 \rightarrow SO_4^{2-} + H_2O$$

with hydrogen sulphites the same test yields hydrogen ions:

$$HSO_3^- + H_2O_2 \rightarrow SO_4^{2-} + H^+ + H_2O$$

and the solution shows a definite acid reaction. It must be emphasized that these tests alone are not specific for sulphites or hydrogen sulphites; their presence must be confirmed first by other reactions.

IV.5 THIOSULPHATES, $S_2O_3^{2-}$ *Solubility* Most of the thiosulphates that have been prepared are soluble in water; those of lead, silver, and barium are very sparingly soluble. Many of them dissolve in excess sodium thiosulphate solution forming complex salts.

To study these reactions use a 0·5M solution of sodium thiosulphate $Na_2S_2O_3.5H_2O$.

1. Dilute hydrochloric acid: no immediate change in the cold with a solution of a thiosulphate; the acidified liquid soon becomes turbid owing to the separation of sulphur, and sulphurous acid is present in solution. On warming the solution, sulphur dioxide is evolved which is recognized by its odour and its action upon filter paper moistened with acidified potassium dichromate solution. The sulphur first forms a colloidal solution, which is gradually coagulated by the free acid present. Side reactions also occur giving rise to thionic acids.

$$S_2O_3^{2-} + 2H^+ \rightarrow S\downarrow + SO_2\uparrow + H_2O$$

2. *Iodine solution:* decolourized when a colourless solution of tetrathionate ions is formed:

$$I_2 + 2S_2O_3^{2-} \rightarrow 2I^- + S_4O_6^{2-}$$

This reaction has important practical applications in the iodiometric and iodimetric methods of titrimetric analysis.*

3. *Barium chloride solution:* white precipitate of barium thiosulphate, BaS_2O_3, from moderately concentrated solutions.

$$S_2O_3^{2-} + Ba^{2+} \rightarrow BaS_2O_3\downarrow$$

Precipitation is accelerated by agitation and by rubbing the sides of the vessel with a glass rod. The solubility is 0·5 g ℓ^{-1} at 18°. No precipitate is obtained with calcium chloride solution since calcium thiosulphate is fairly soluble in water.

4. *Silver nitrate solution:* white precipitate of silver thiosulphate:

$$S_2O_3^{2-} + 2Ag^+ \rightarrow Ag_2S_2O_3\downarrow$$

At first no precipitation occurs because the soluble dithiosulphatoargentate(I) complex is formed:

$$2S_2O_3^{2-} + Ag^+ \rightarrow [Ag(S_2O_3)_2]^{3-}$$

The precipitate is unstable, turning dark on standing, when silver sulphide is formed:

$$Ag_2S_2O_3\downarrow + H_2O \rightarrow Ag_2S\downarrow + 2H^+ + SO_4^{2-}$$

This hydrolytic decomposition can be accelerated by warming.

5. *Lead acetate or lead nitrate solution:* first no change, but on further addition of the reagent a white precipitate of lead thiosulphate is formed:

$$S_2O_3^{2-} + Pb^{2+} \rightarrow PbS_2O_3\downarrow$$

The precipitate is soluble in excess thiosulphate; for this reason no precipitation occurs first. On boiling the suspension the precipitate darkens, forming finally a black precipitate of lead sulphide:

$$PbS_2O_3\downarrow + H_2O \rightarrow PbS\downarrow + 2H^+ + SO_4^{2-}$$

This reaction can be applied to distinguish sulphite and thiosulphate ions (cf. Section **IV.4**, reaction 6).

6. *Potassium cyanide solution* (**POISON**): after alkalizing the test solution first

---

* See A. I. Vogel's *A Textbook of Quantitative Inorganic Analysis, including Elementary Instrumental Analysis*, 4th ed., Longman 1978, p. 343 et f.

with sodium hydroxide and adding potassium cyanide, thiocyanate ions are formed on boiling:

$$S_2O_3^{2-} + CN^- \rightarrow SCN^- + SO_3^{2-}$$

On acidifying the cold solution, in a well-ventilated fume cupboard, with hydrochloric acid and adding iron(III) chloride, the red colour of iron(III) thiocyanate can be observed:

$$3SCN^- + Fe^{3+} \rightarrow Fe(SCN)_3$$

7. Blue ring test When a solution of thiosulphate mixed with ammonium molybdate solution is poured *slowly* down the side of a test tube which contains concentrated sulphuric acid, a blue ring is formed temporarily at the contact zone. The origin of this colour is as yet unexplained.

8. Iron(III) chloride solution: a dark-violet colouration appears, probably due to the formation of a dithiosulphatoiron(III) complex:

$$2S_2O_3^{2-} + Fe^{3+} \rightarrow [Fe(S_2O_3)_2]^-$$

on standing the colour disappears rapidly, while tetrathionate and iron(II) ions are formed:

$$[Fe(S_2O_3)_2]^- + Fe^{3+} \rightarrow 2Fe^{2+} + S_4O_6^{2-}$$

The overall reaction can be written as the reduction of iron(III) by thiosulphate:

$$2S_2O_3^{2-} + 2Fe^{3+} \rightarrow S_4O_6^{2-} + 2Fe^{2+}$$

9. Nickel ethylenediamine nitrate reagent $[Ni(NH_2.CH_2.CH_2.NH_2)_3](NO_3)_2$, abbreviated to $[Ni(en)_3](NO_3)_2$. When a neutral or slightly alkaline solution of a thiosulphate is treated with the reagent, a crystalline, violet precipitate of the complex thiosulphate is obtained:

$$[Ni(en)_3]^{2+} + S_2O_3^{2-} \rightarrow [Ni(en)_3]S_2O_3\downarrow$$

Sulphites, sulphates, tetrathionates, and thiocyanates do not interfere, but hydrogen sulphide and ammonium sulphide decompose the reagent with the precipitation of nickel sulphide.

The nickel ethylenediamine nitrate reagent is conveniently prepared when required by treating a little nickel chloride solution with ethylenediamine until a violet colour (due to the formation of the complex $[Ni(en)_3]^{2+}$ ion) appears.

Concentration limit: 1 in 25,000.

By obvious modification the reaction may be used for the detection of nickel; it is applicable in the presence of copper, cobalt, iron, and chromium.

10. Catalytic test Solutions of sodium azide, NaN_3, and iodine (as I_3^-) do not react, but on addition of a trace of thiosulphate, which acts as a catalyst, there is an immediate vigorous evolution of nitrogen:

$$2N_3^- + I_3^- \rightarrow 3I^- + 3N_2\uparrow$$

Sulphides and thiocyanates act similarly and must therefore be absent.

Mix a drop of the test solution and a drop of the iodine–azide reagent on a watch glass. A vigorous evolution of bubbles (nitrogen) ensues.

Sensitivity: 0·15 μg $Na_2S_2O_3$. *Concentration limit:* 1 in 330,000.

The sodium azide–iodine reagent consists of a solution of 3 g sodium azide in 100 ml 0·05M iodine.

IV.6 SULPHIDES, S^{2-} *Solubility* The acid, normal, and poly-sulphides of alkali metals are soluble in water; their aqueous solutions react alkaline because of hydrolysis.

$$S^{2-} + H_2O \rightleftarrows SH^- + OH^-$$
$$SH^- + H_2O \rightleftarrows H_2S + OH^-$$

The normal sulphides of most other metals are insoluble; those of the alkaline earths are sparingly soluble, but are gradually changed by contact with water into soluble hydrogen sulphides:

$$CaS + H_2O \rightarrow Ca^{2+} + SH^- + OH^-$$

The sulphides of aluminium, chromium, and magnesium can only be prepared in the dry, as they are completely hydrolysed by water:

$$Al_2S_3 + 6H_2O \rightarrow 2Al(OH)_3\downarrow + 3H_2S\uparrow$$

The characteristic colours and solubilities of many metallic sulphides have already been discussed in connection with the reactions of the cations in Chapter III. The sulphides of iron, manganese, zinc, and the alkali metals are decomposed by dilute hydrochloric acid with the evolution of hydrogen sulphide; those of lead, cadmium, nickel, cobalt, antimony, and tin(IV) require concentrated hydrochloric acid for decomposition; others, such as mercury(II) sulphide, are insoluble in concentrated hydrochloric acid, but dissolve in aqua regia with the separation of sulphur. The presence of sulphide in insoluble sulphides may be detected by reduction with nascent hydrogen (derived from zinc or tin and hydrochloric acid) to the metal and hydrogen sulphide, the latter being identified with lead acetate paper (see reaction 1 below). An alternative method is to fuse the sulphide with anhydrous sodium carbonate, extract the mass with water, and to treat the filtered solution with freshly prepared sodium nitroprusside solution, when a purple colour will be obtained; the sodium carbonate solution may also be treated with lead nitrate solution when black lead sulphide is precipitated.

For the study of these reactions a 2M solution of sodium sulphide $Na_2S.9H_2O$ can be used.

1. Dilute hydrochloric or sulphuric acid: hydrogen sulphide gas is evolved, which may be identified by its characteristic odour, and by the blackening of filter paper moistened with lead acetate solution:

$$S^{2-} + 2H^+ \rightarrow H_2S\uparrow$$
$$H_2S + Pb^{2+} \rightarrow PbS\downarrow$$

Alternatively, a filter paper moistened with cadmium acetate solution turns yellow:

$$H_2S + Cd^{2+} \rightarrow CdS\downarrow$$

A more sensitive test is attained by the use of sodium tetrahydroxoplumbate(II) solution, prepared by adding sodium hydroxide to lead acetate until the initial precipitate of lead hydroxide has just dissolved:

$$Pb^{2+} + 2OH^- \rightarrow Pb(OH)_2\downarrow$$
$$Pb(OH)_2\downarrow + 2OH^- \rightarrow [Pb(OH)_4]^{2-}$$
$$[Pb(OH)_4]^{2-} + H_2S \rightarrow PbS\downarrow + 2OH^- + 2H_2O$$

Hydrogen sulphide is a good reducing agent. It reduces (i) acidified potassium permanganate, (ii) acidified potassium dichromate, and (iii) potassium triiodide (iodine) solution:

$$2MnO_4^- + 5H_2S + 6H^+ \rightarrow 2Mn^{2+} + 5S\downarrow + 8H_2O \qquad \text{(i)}$$
$$Cr_2O_7^{2-} + 3H_2S + 8H^+ \rightarrow 2Cr^{3+} + 3S\downarrow + 7H_2O \qquad \text{(ii)}$$
$$I_3^- + H_2S \rightarrow 3I^- + 2H^+ + S\downarrow \qquad \text{(iii)}$$

in each case sulphur is precipitated. Small quantities of chlorine may be produced in (i) and (ii) if the hydrochloric acid is other than very dilute; this is avoided by using dilute sulphuric acid.

2. *Silver nitrate solution:* black precipitate of silver sulphide Ag_2S, insoluble in cold, but soluble in hot dilute nitric acid.

$$S^{2-} + 2Ag^+ \rightarrow Ag_2S\downarrow$$

3. *Lead acetate solution:* black precipitate of lead sulphide PbS (Section **III.4**, reaction 2).

4. *Barium chloride solution:* no precipitate.

5. *Silver* When a solution of a sulphide is brought into contact with a bright silver coin, a brown to black stain of silver sulphide is produced. The result is obtained more expeditiously by the addition of a few drops of dilute hydrochloric acid. The stain may be removed by rubbing the coin with moist lime.

6. *Sodium nitroprusside solution* ($Na_2[Fe(CN)_5NO]$)*:* transient purple colour in the presence of solutions of alkalis. No reaction occurs with solutions of hydrogen sulphide or with the free gas: if, however, filter paper is moistened with a solution of the reagent made alkaline with sodium hydroxide or ammonia solution, a purple colouration is produced with free hydrogen sulphide.

$$S^{2-} + [Fe(CN)_5NO]^{2-} \rightarrow [Fe(CN)_5NOS]^{4-}$$

The reagent must be freshly prepared by dissolving a crystal (about the size of a pea) of pure sodium nitroprusside in a little distilled water.

The spot-test technique is as follows. Mix on a spot plate a drop of the alkaline test solution with a drop of a 1 per cent solution of sodium nitroprusside. A violet colour appears. Alternatively, filter paper impregnated with an ammoniacal (2M) solution of sodium nitroprusside may be employed.

Sensitivity: 1 µg Na_2S. *Concentration limit:* 1 in 50,000.

7. *Methylene blue test* *p*-Aminodimethylaniline is converted by iron(III)

chloride and hydrogen sulphide in strongly acid solution into the water-soluble dyestuff, methylene blue:

$$(CH_3)_2N\text{-}C_6H_4\text{-}NH_2 + H_2S + H_2N\text{-}C_6H_4\text{-}N(CH_3)_2 + 6Fe^{3+} \longrightarrow$$

$$\longrightarrow 6Fe^{2+} + NH_4^+ + 4H^+ + \left[(CH_3)_2N\text{-}C_6H_3\text{<}^{S}_{N}\text{>}C_6H_3\text{=}\overset{+}{N}(CH_3)_2\right]^+$$

This is a sensitive test for soluble sulphides and hydrogen sulphide.

Place a drop of the test solution on a spot plate, add a drop of concentrated hydrochloric acid, mix, then dissolve a few grains of *p*-aminodimethylaniline in the mixture (or add 1 drop 1 per cent solution of the chloride or sulphate) and add a drop of 0·5M iron(III) chloride solution. A clear blue colouration appears after a short time (2–3 minutes).

Sensitivity: 1 μg H_2S. *Concentration limit:* 1 in 50,000.

8. Catalysis of iodine–azide reaction test Solutions of sodium azide, NaN_3, and of iodine (as I_3^-) do not react, but on the addition of a trace of a sulphide, which acts as a catalyst, there is an immediate evolution of nitrogen:

$$2N_3^- + I_3^- \rightarrow 3I^- + 3N_2\uparrow$$

Thiosulphates and thiocyanates act similarly and must therefore be absent. The sulphide can, however, be separated by precipitation with zinc or cadmium carbonate. The precipitated sulphide may then be introduced, say, at the end of a platinum wire into a semimicro test-tube or centrifuge tube containing the iodine–azide reagent, when the evolution of nitrogen will be seen.

Mix a drop of the test solution and a drop of the reagent on a watch glass. An immediate evolution of gas in the form of fine bubbles occurs.

Sensitivity: 0·3 μg Na_2S. *Concentration limit:* 1 in 166,000.

The sodium azide–iodine reagent is prepared by dissolving 3 g sodium azide in 100 ml 0·05M iodine. The solution is stable. [The test is rendered more sensitive (to 0·02 μg Na_2S) by employing a more concentrated reagent composed of 1 g sodium azide and a few crystals of iodine in 3 ml water.]

IV.7 NITRITES, NO_2^- *Solubility* Silver nitrite is sparingly soluble in water. All other nitrites are soluble in water.

Use freshly prepared 0·1M solution of potassium nitrite, KNO_2, to study these reactions.

1. Dilute hydrochloric acid Cautious addition of the acid to a solid nitrite in the cold yields a transient, pale-blue liquid (due to the presence of free nitrous acid, HNO_2, or its anhydride, N_2O_3) and the evolution of brown fumes of nitrogen dioxide, the latter being largely produced by combination of nitric oxide with the oxygen of the air. Similar results are obtained with the aqueous solution.

$$NO_2^- + H^+ \rightarrow HNO_2$$
$$(2HNO_2 \rightarrow H_2O + N_2O_3)$$
$$3HNO_2 \rightarrow HNO_3 + 2NO\uparrow + H_2O$$
$$2NO\uparrow + O_2\uparrow \rightarrow 2NO_2\uparrow$$

2. Iron(II) sulphate solution When the nitrite solution is added carefully to a concentrated (25%) solution of iron(II) sulphate acidified with dilute acetic acid or with dilute sulphuric acid, a brown ring, due to the compound $[Fe,NO]SO_4$, is formed at the junction of the two liquids. If the addition has not been made cautiously, a brown colouration results. This reaction is similar to the brown ring test for nitrates (see Section **IV.18**, reaction 3), for which a stronger acid (concentrated sulphuric acid) must be employed.

$$NO_2^- + CH_3COOH \rightarrow HNO_2 + CH_3COO^-$$
$$3HNO_2 \rightarrow H_2O + HNO_3 + 2NO\uparrow$$
$$Fe^{2+} + SO_4^{2-} + NO\uparrow \rightarrow [Fe,NO]SO_4$$

Iodides, bromides, coloured ions, and anions that give coloured compounds with iron(II) ions must be absent.

3. Barium chloride solution: no precipitate.

4. Silver nitrate solution: white crystalline precipitate of silver nitrite from concentrated solutions.

$$NO_2^- + Ag^+ \rightarrow AgNO_2\downarrow$$

5. Potassium iodide solution The addition of a nitrite solution to a solution of potassium iodide, followed by acidification with acetic acid or with dilute sulphuric acid, results in the liberation of iodine, which may be identified by the blue colour produced with starch paste. A similar result is obtained by dipping potassium iodide–starch paper moistened with a little dilute acid into the solution. An alternative method is to extract the liberated iodine with chloroform or carbon tetrachloride (see Section **IV.16**, reaction 4).

$$2NO_2^- + 2I^- + 2CH_3COOH \rightarrow I_2 + 2NO\uparrow + 2CH_3COO^- + 2H_2O$$

6. Acidified potassium permanganate solution: decolourized by a solution of a nitrite, but no gas is evolved.

$$5NO_2^- + 2MnO_4^- + 6H^+ \rightarrow 5NO_3^- + 2Mn^{2+} + 3H_2O$$

7. Ammonium chloride By boiling a solution of a nitrite with excess of the solid reagent, nitrogen is evolved and the nitrite is completely destroyed.

$$NO_2^- + NH_4^+ \rightarrow N_2\uparrow + 2H_2O$$

8. Urea $CO(NH_2)_2$ When a solution of a nitrite is treated with urea and the mixture acidified with dilute hydrochloric acid, the nitrite is decomposed, and nitrogen and carbon dioxide are evolved.

$$CO(NH_2)_2 + 2HNO_2 \rightarrow 2N_2\uparrow + CO_2\uparrow + 3H_2O$$

9. Thiourea $CS(NH_2)_2$ When a dilute acetic acid solution of a nitrite is treated with a little thiourea, nitrogen is evolved and thiocyanic acid is produced. The latter may be identified by the red colour produced with dilute HCl and $FeCl_3$ solution.

$$CS(NH_2)_2 + HNO_2 \rightarrow N_2\uparrow + H^+ + SCN^- + 2H_2O$$

Thiocyanates and iodides interfere and, if present, must be removed either with excess of solid Ag_2SO_4 or with dilute $AgNO_3$ solution before adding the acetic acid and thiourea.

10. Sulphamic acid ($HO.SO_2.NH_2$) When a solution of a nitrite is treated with sulphamic acid, it is completely decomposed:

$$HO.SO_2.NH_2 + HNO_2 \rightarrow N_2\uparrow + 2H^+ + SO_4^{2-} + H_2O$$

No nitrate is formed in this reaction, and it is therefore an excellent method for the complete removal of nitrite. Traces of nitrates are formed with ammonium chloride, urea, and thiourea (reaction 7, 8, and 9).

11. Sulphanilic acid-α-naphthylamine reagent. (Griess–Ilosvay test) This test depends upon the diazotization of sulphanilic acid by nitrous acid, followed by coupling with α-naphthylamine to form a red azo dye:

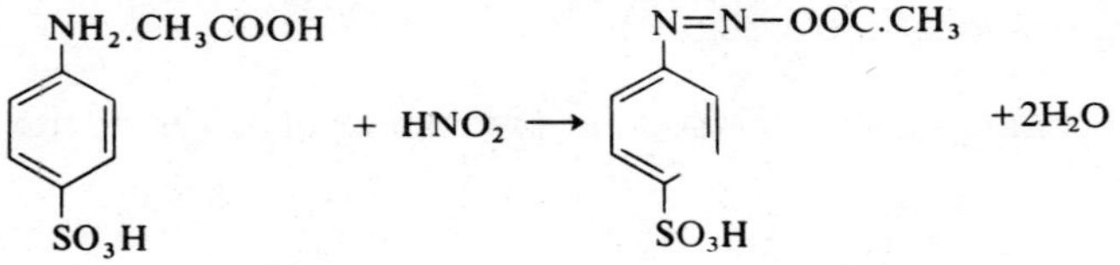

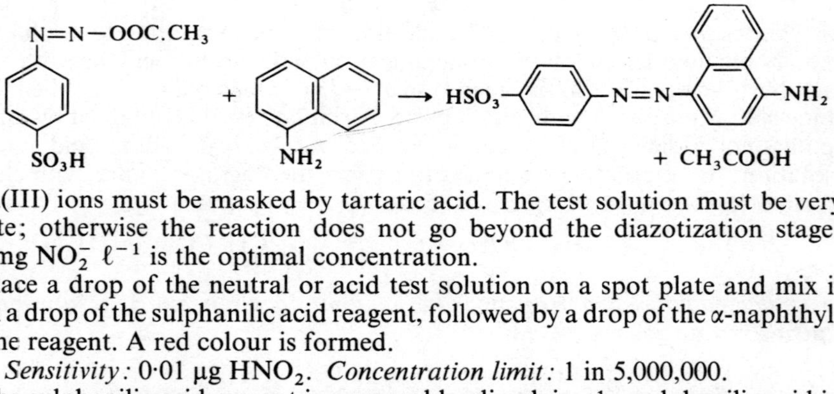

$NH_2.CH_3COOH$ / SO_3H + HNO_2 → $N{=}N{-}OOC.CH_3$ / SO_3H + $2H_2O$

$N{=}N{-}OOC.CH_3$ / SO_3H + NH_2 → HSO_3–N=N–NH_2 + CH_3COOH

Iron(III) ions must be masked by tartaric acid. The test solution must be very dilute; otherwise the reaction does not go beyond the diazotization stage: 0·2 mg NO_2^- ℓ^{-1} is the optimal concentration.

Place a drop of the neutral or acid test solution on a spot plate and mix it with a drop of the sulphanilic acid reagent, followed by a drop of the α-naphthylamine reagent. A red colour is formed.

Sensitivity: 0·01 μg HNO_2. *Concentration limit:* 1 in 5,000,000.

The sulphanilic acid reagent is prepared by dissolving 1 g sulphanilic acid in 100 ml warm 30 per cent acetic acid. The α-naphthylamine reagent is prepared by boiling 0·3 g α-naphthylamine with 70 ml water, filtering or decanting from the small residue, and mixing with 30 ml glacial acetic acid.

12. Indole reagent

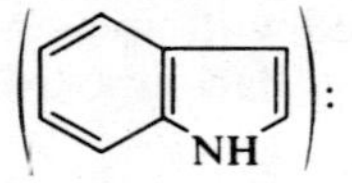

(NH):

a red-coloured nitroso-indole is formed.

Place a drop of the test solution in a semimicro test-tube, add 10 drops of the reagent and 5 drops of 8M sulphuric acid. A purplish-red colouration appears.

Concentration limit: 1 in 1,000,000.

The reagent consists of a 0·015 per cent solution of indole in 96 per cent alcohol.

IV.8 CYANIDES, CN^- *Solubility* Only the cyanides of the alkali and alkaline earth metals are soluble in water; the solutions react alkaline owing to hydrolysis.

$$CN^- + H_2O \rightarrow HCN + OH^-$$

Mercury(II) cyanide, $Hg(CN)_2$, is also soluble in water, but is practically a non-electrolyte and therefore does not exhibit the ionic reactions of cyanides. Many of the metallic cyanides dissolve in solutions of potassium cyanide to yield complex salts (cf. Section **I.34,c**).

Use a freshly prepared solution of 0·1M potassium cyanide, KCN, to study these reactions.

Note: All cyanides are highly poisonous. The free acid, HCN, is volatile and is particularly dangerous so that all experiments in which the gas is likely to be evolved, or those in which cyanides are heated, should be carried out in the fume cupboard.

1. Dilute hydrochloric acid: hydrocyanic acid, HCN, with an odour reminiscent of bitter almonds, is evolved in the cold. It should be smelled with great caution. A more satisfactory method for identifying hydrocyanic acid consists in converting it into ammonium thiocyanate by allowing the vapour to come into contact with a little ammonium polysulphide on filter paper. The paper may be conveniently placed over the test-tube or dish in which the substance is being treated with the dilute acid. Upon adding a drop of iron(III) chloride solution and a drop of dilute hydrochloric acid to the filter paper, the characteristic red colouration, due to the iron(III) thiocyanate complex, $Fe(SCN)_3$, is obtained (see reaction 6 below). Mercury(II) cyanide is not decomposed by dilute acids.

$$CN^- + H^+ \rightarrow HCN\uparrow$$

2. Silver nitrate solution: white precipitate of silver cyanide, AgCN, readily soluble in excess of the cyanide solution forming the complex ion, dicyanoargentate(I) $[Ag(CN)_2]^-$ (cf. Section **III.6**, reaction 7):

$$CN^- + Ag^+ \rightarrow AgCN\downarrow$$

$$AgCN\downarrow + CN^- \rightarrow [Ag(CN)_2]^-$$

Silver cyanide is soluble in ammonia solution and in sodium thiosulphate solution, but is insoluble in dilute nitric acid.

3. Concentrated sulphuric acid Heat a little of the solid salt with concentrated sulphuric acid; carbon monoxide is evolved which may be ignited and burns with a blue flame. All cyanides, complex and simple, are decomposed by this treatment.

$$2KCN + 2H_2SO_4 + 2H_2O \rightarrow 2CO\uparrow + K_2SO_4 + (NH_4)_2SO_4$$

4. *Prussian blue test* This is a delicate test and is carried out in the following manner. The solution of the cyanide is rendered strongly alkaline with sodium hydroxide solution, a few millilitres of a freshly prepared solution of iron(II) sulphate added (if only traces of cyanide are present, it is best to use a saturated (25 %) solution of iron(II) sulphate) and the mixture boiled. Hexacyanoferrate(II) ions are thus formed. Upon acidifying with hydrochloric acid (in order to neutralize any free alkali which may be present), a clear solution is obtained, which gives a precipitate of Prussian blue upon the addition of a little iron(III) chloride solution. If only a little cyanide was used, or is present, in the solution to be tested, a green solution is obtained at first; this deposits Prussian blue on standing.

$$6CN^- + Fe^{2+} \rightarrow [Fe(CN)_6]^{4-}$$
$$3[Fe(CN)_6]^{4-} + 4Fe^{3+} \rightarrow Fe_4[Fe(CN)_6]_3\downarrow$$

5. *Mercury(I) nitrate solution:* grey precipitate of metallic mercury (difference from chloride, bromide, and iodide):

$$2CN^- + Hg_2^{2+} \rightarrow Hg\downarrow + Hg(CN)_2$$

Mercury(II) cyanide is very little ionized in solution. To detect cyanide in the presence of mercury, an excess of potassium iodide should be added to the sample, when cyanide ions are liberated:

$$Hg(CN)_2 + 4I^- \rightarrow [HgI_4]^{2-} + 2CN^-$$

Mercury(II) cyanide is decomposed by hydrogen sulphide, when mercury(II) sulphide is precipitated ($K_s = 4 \times 10^{-53}$). If the precipitate is filtered off, cyanide ions can be tested for in the solution:

$$Hg(CN)_2 + H_2S \rightarrow HgS\downarrow + 2HCN$$

6. *Iron(III) thiocyanate test* This is another excellent test for cyanides and depends upon the direct combination of alkali cyanides with sulphur (best derived from an alkali or ammonium polysulphide). A little ammonium polysulphide solution is added to the potassium cyanide solution contained in a porcelain dish, and the whole evaporated to dryness on the water bath in the fume cupboard. The residue contains alkali and ammonium thiocyanates together with any residual polysulphide. The latter is destroyed by the addition of a few drops of hydrochloric acid. One or two drops of iron(III) chloride solution are then added. A blood-red colouration, due to the iron(III) thiocyanate complex, is produced immediately:

$$CN^- + S_2^{2-} \rightarrow SCN^- + S^{2-};$$
$$3SCN^- + Fe^{3+} \rightarrow Fe(SCN)_3$$ (cf. Section **IV.10**, reaction 6).

The spot-test technique is as follows. Stir a drop of the test solution with a drop of yellow ammonium sulphide on a watch glass and warm until a rim of sulphur is formed round the liquid (evaporation to dryness, other than on a water bath, should be avoided). Add 1–2 drops dilute hydrochloric acid, allow to cool, and add 1–2 drops 3 per cent iron(III) chloride solution. A red colouration is obtained.

Sensitivity: 1 μg CN^-. *Concentration limit:* 1 in 50,000.

The test is applicable in the presence of sulphide or sulphite; if thiocyanate is originally present, the cyanide must be isolated first by precipitation, e.g. as zinc cyanide.

7. Copper sulphide test Solutions of cyanides readily dissolve copper(II) sulphide forming the colourless tetracyanocuprate(I) ions:

$$2CuS\downarrow + 10CN^- \rightarrow 2[Cu(CN)_4]^{3-} + 2S^{2-} + (CN)_2\uparrow$$

Note that the oxidation number of copper in the solution is +1. The test is best carried out on filter paper or drop-reaction paper and is applicable in the presence of chlorides, bromides, iodides, hexacyanoferrate(II) and (III) ions.

Place a drop of a freshly prepared copper sulphide suspension on a filter paper (or on a spot plate) and add a drop of the test solution. The brown colour of copper sulphide disappears at once.

Sensitivity: 2·5 μg CN^-. *Concentration limit:* 1 in 20,000.

The copper sulphide suspension is prepared by dissolving 0·12 g crystallized copper sulphate in 100 ml water, adding a few drops of ammonia solution and rendering the solution cloudy with a little hydrogen sulphide.

An alternative procedure is to employ quantitative filter paper or drop-reaction paper which has been impregnated with an ammoniacal solution containing 0·1 g of copper sulphate per 100 ml and dried. Immediately before the test a little hydrogen sulphide is blown on to the paper so that it acquires a uniform brown colour. Place a drop of the test solution upon this paper when a white ring will be obtained.

Sensitivity: 2·5 μg CN^-. *Concentration limit:* 1 in 20,000.

8. Copper acetate–benzidine acetate test (DANGER: THE REAGENT IS CARCINOGENIC.) This reaction takes place because the oxidation-reduction potential of the copper(II)–copper(I) couple is increased if copper(I) ions are removed by cyanide ions.

Mix a few drops of the test solution with a little dilute sulphuric acid in a micro test-tube and tie (or otherwise fix) a piece of drop-reaction paper which has been moistened with a mixture of equal parts of the copper acetate and benzidine reagents to the top of the tube. A blue colouration is produced.

Sensitivity: 0·25 μg CN^-. *Concentration limit:* 1 in 200,000.

Oxidizing and reducing gases interfere: it has been recommended that the hydrocyanic acid be liberated by heating with sodium hydrogen carbonate.

Alternatively, place a drop of the test solution (or a few milligrams of the test solid) in the reaction bulb of Fig. II.56, add 2 thin pieces zinc foil and 2–3 drops dilute sulphuric acid. Place a small circle of quantitative filter paper (or drop-reaction paper) moistened with acetate–benzidine acetate reagent across the funnel. The paper is coloured blue by the hydrocyanic acid carried over with the hydrogen.

Sensitivity: 1 μg CN^-. *Concentration limit:* 1 in 50,000.

The copper(II) acetate reagent (solution I) is a 3 per cent solution of copper(II) acetate in water.

The benzidine reagent (solution II) is a 1 per cent solution of benzidine in 2M acetic acid.

These solutions are best kept apart in black well-stoppered bottles. Equal volumes of solutions I and II are mixed immediately before required.

IV.9 CYANATES, OCN^- *Solubility* The cyanates of the alkalis and of the alkaline earths are soluble in water. Those of silver, mercury(I), lead, and copper are insoluble. The free acid is a colourless liquid with an unpleasant odour; it is very unstable.

To study these reactions use a 0·2M solution of potassium cyanate, KOCN.

1. Dilute sulphuric acid: vigorous effervescence, due largely to the evolution of carbon dioxide. The free cyanic acid, HOCN, which is liberated initially, is decomposed into carbon dioxide and ammonia, the latter combining with the sulphuric acid present to form ammonium sulphate. A little cyanic acid, however, escapes decomposition and may be recognized in the evolved gas by its penetrating odour. If the resulting solution is warmed with sodium hydroxide, ammonia is evolved (test with mercury(I) nitrate paper).

$$OCN^- + H^+ \rightarrow HOCN$$

$$HOCN + H^+ + H_2O \rightarrow CO_2\uparrow + NH_4^+$$

2. Concentrated sulphuric acid The reaction is similar to that with the dilute acid, but is somewhat more vigorous.

3. Silver nitrate solution: white, curdy precipitate of silver cyanate, AgOCN, soluble in ammonia solution and in dilute nitric acid. The precipitate appears instantaneously, without complex formation (difference from cyanide):

$$OCN^- + Ag^+ \rightarrow AgOCN\downarrow$$

4. Barium chloride solution: no precipitate.

5. Cobalt acetate solution When the reagent is added to a concentrated solution of potassium cyanate, a blue colouration, due to tetracyanatocobaltate(II) ions, $[Co(OCN)_4]^{2-}$ is produced. The colour is stabilized and intensified somewhat by the addition of ethanol.

$$4OCN^- + Co^{2+} \rightarrow [Co(OCN)_4]^{2-}$$

6. Copper sulphate–pyridine test When a cyanate is added to a dilute solution of a copper salt to which a few drops of pyridine have been previously added, a lilac-blue precipitate is formed of the compound $[Cu(C_5H_5N)_2](OCN)_2$; this is soluble in chloroform with the production of a sapphire-blue solution. Thiocyanates interfere; excess of copper solution should be avoided.

$$2OCN^- + Cu^{2+} + 2C_5H_5N \rightarrow [Cu(C_5H_5N)_2](OCN)_2$$

Add a few drops of pyridine to 2–3 drops of a 0·25M solution of copper sulphate then introduce about 2 ml chloroform followed by a few drops of the neutral cyanate solution. Shake the mixture briskly; the chloroform will acquire a blue colour.

Concentration limit: 1 part cyanate in 20,000.

The blue complex is stable in the presence of a moderate excess of acetic acid; the reaction can therefore be applied to the detection of cyanates in alkaline solution. The solution to be tested is added to the copper-pyridine-chloroform mixture, acetic acid added slowly and the solution shaken vigorously after each addition. As soon as the solution is neutral, the chloroform will assume a blue colour.

IV.10 THIOCYANATES, SCN^-* *Solubility* Silver and copper(I) thiocyanates are practically insoluble in water, mercury(II) and lead thiocyanates are sparingly soluble; the solubilities, in g ℓ^{-1} at 20°, are 0·0003, 0·0005, 0·7, and 0·45 respectively. The thiocyanates of most other metals are soluble.

To study these reactions use a 0·1M solution of potassium thiocyanate KSCN.

1. Sulphuric acid With the concentrated acid a yellow colouration is produced in the cold: upon warming a violent reaction occurs, carbonyl sulphide (burns with a blue flame) is formed.

$$SCN^- + H_2SO_4 + H_2O \rightarrow COS\uparrow + NH_4^+ + SO_4^{2-}$$

The reaction however is more complex than this, because sulphur dioxide (fuchsin solution decolourized) and carbon dioxide may be detected also in the gaseous decomposition products. In the solution some sulphur is precipitated, and formic acid can also be detected.

With the 2·5M acid no reaction occurs in the cold, but on boiling a yellow solution is formed, sulphur dioxide and a little carbonyl sulphide are evolved. A similar but slower reaction takes place with M sulphuric acid.

2. Silver nitrate solution: white, curdy precipitate of silver thiocyanate, AgSCN, soluble in ammonia solution but insoluble in dilute nitric acid.

$$SCN^- + Ag^+ \rightarrow AgSCN\downarrow$$

$$AgSCN\downarrow + 2NH_3 \rightarrow [Ag(NH_3)_2]^+ + SCN^-$$

Upon boiling with M sodium chloride solution, the precipitate is converted into silver chloride:

$$AgSCN\downarrow + Cl^- \rightarrow AgCl\downarrow + SCN^-$$

(distinction and method of separation from silver halides). After acidification with hydrochloric acid, thiocyanate ions can be detected with iron(III) chloride (cf. reaction 6).

Silver thiocyanate also decomposes upon ignition or upon fusion with sodium carbonate.

3. Copper sulphate solution: first a green colouration, then a black precipitate of copper(II) thiocyanate is observed:

$$2SCN^- + Cu^{2+} \rightarrow Cu(SCN)_2\downarrow$$

Adding sulphurous acid (a saturated solution of sulphur dioxide) the precipitate turns into white copper(I) thiocyanate:

$$2Cu(SCN)_2 + SO_2 + 2H_2O \rightarrow 2CuSCN\downarrow + 2SCN^- + SO_4^{2-} + 4H^+$$

4. Mercury(II) nitrate solution: white precipitate of mercury(II) thiocyanate $Hg(SCN)_2$, readily soluble in excess of the thiocyanate solution. If the precipi-

*Raman spectra of thiocyanates appear to indicate that the ion is —S—C≡N rather than S=C=N—, since a line assignable to the triple link and none for the double link was observed. Salts will be written as, e.g. KSCN and the ion as SCN^-.

tate is heated, it swells up enormously forming 'Pharaoh's serpents', a polymerized cyanogen product.

$$2SCN^- + Hg^{2+} \rightarrow Hg(SCN)_2\downarrow$$
$$Hg(SCN)_2\downarrow + 2SCN^- \rightarrow [Hg(SCN)_4]^{2-}$$

5. *Zinc and dilute hydrochloric acid:* hydrogen sulphide and hydrogen cyanide (**POISONOUS**) are evolved:

$$SCN^- + Zn + 3H^+ \rightarrow H_2S\uparrow + HCN\uparrow + Zn^{2+}$$

6. *Iron(III) chloride solution:* blood-red colouration, due to the formation of a complex:

$$3SCN^- + Fe^{3+} \rightleftarrows Fe(SCN)_3$$

In fact there are a series of complex cations and anions formed (see Section **III.22**, reaction 11). The (uncharged) complex can be extracted by shaking with ether. The red colour is discharged by fluorides, mercury(II) ions, and oxalates; when colourless, more stable complexes are formed:

$$Fe(SCN)_3 + 6F^- \rightarrow [FeF_6]^{3-} + 3SCN^-$$
$$4Fe(SCN)_3 + 3Hg^{2+} \rightarrow 3[Hg(SCN)_4]^{2-} + 4Fe^{3+}$$
$$Fe(SCN)_3 + 3(COO)_2^{2-} \rightarrow [Fe\{(COO)_2\}_3]^{3-} + 3SCN^-$$

7. *Dilute nitric acid:* decomposition upon warming, a red colouration is produced, and nitrogen oxide and hydrogen cyanide (**POISONOUS**) are evolved:

$$SCN^- + H^+ + 2NO_3^- \rightarrow 2NO\uparrow + HCN\uparrow + SO_4^{2-}$$

With concentrated nitric acid a more vigorous reaction takes place, with the formation of nitrogen oxide and carbon dioxide.

8. *Distillation test* Free thiocyanic acid, HSCN, can be liberated by hydrochloric acid, distilled into ammonia solution, where it can be identified with iron(III) chloride (cf. reaction 6). This test may be applied to separate thiocyanate from mixtures with ions which would interfere with reaction 6.

Place a few drops of the test solution in a semimicro test-tube, acidify with dilute hydrochloric acid, add a small fragment of broken porcelain and attach a gas absorption pipette (Fig. II.29*c*) charged with a drop or two of ammonia solution. Boil the solution in the test-tube gently so as to distil any HSCN present into the ammonia solution. Rinse the ammonia solution into a clean semimicro test-tube, acidify slightly with dilute hydrochloric acid and add a drop of iron(III) chloride solution. A red colouration is obtained.

9. *Cobalt nitrate solution:* blue colouration, due to the formation of $[Co(SCN)_4]^{2-}$ (Section **III.26**, reaction 6), but no precipitate [distinction from cyanide, hexacyanoferrate(II) and (III)]:

$$4SCN^- + Co^{2+} \rightarrow [Co(SCN)_4]^{2-}$$

The spot-test technique is as follows. Mix a drop of the test solution in a micro crucible with a very small drop (0·02 ml) of 0·5M solution of cobalt nitrate and evaporate to dryness. The residue, whether thiocyanate is present or not, is coloured violet and the colour slowly fades. Add a few drops of acetone. A blue-green or green colouration is obtained.

Sensitivity: 1 μg SCN^-. *Concentration limit:* 1 in 50,000.

Nitrites yield a red colour due to nitrosyl thiocyanate and therefore interfere with the test.

10. Catalysis of iodine–azide reaction test Traces of thiocyanates act as powerful catalysts in the otherwise extremely slow reaction between triiodide (iodine) and sodium azide:

$$I_3^- + 2N_3^- \rightarrow 3I^- + 3N_2\uparrow$$

Sulphides (see Section **IV.6**, reaction 8) and thiosulphates (see Section **IV.5**, reaction 9) have a similar catalytic effect; these may be removed by precipitation with mercury(II) nitrate solution:

$$Hg^{2+} + S^{2-} \rightarrow HgS\downarrow$$

$$Hg^{2+} + S_2O_3^{2-} + H_2O \rightarrow HgS\downarrow + SO_4^{2-} + 2H^+$$

Considerable amounts of iodine retard the reaction; iodine is therefore best largely removed by the addition of an excess of mercury(II) nitrate solution whereupon the complex $[HgI_4]^{2-}$ ion, which does not affect the catalysis, is formed.

Mix a drop of the test solution with 1 drop of the iodine–azide reagent on a spot plate. Bubbles of gas (nitrogen) are evolved.

Sensitivity: 1·5 μg KSCN. *Concentration limit:* 1 in 30,000.

The reagent is prepared by dissolving 3 g sodium azide in 100 ml 0·05M iodine.

11. Copper(II) sulphate–pyridine reagent When a neutral solution of a thiocyanate is added to a dilute solution of a copper salt containing a few drops of pyridine, a yellowish-green precipitate of the composition $[Cu(C_5H_5N)_2](SCN)_2$ is formed; the compound is soluble in chloroform to which it imparts an emerald-green colouration. Cyanates interfere with this reaction; excess of pyridine should be avoided.

$$2SCN^- + Cu^{2+} + 2C_5H_5N \rightarrow [Cu(C_5H_5N)_2](SCN)_2\downarrow$$

Add a few drops of pyridine to 3–4 drops of a 0·25M solution of copper sulphate, then introduce about 2 ml chloroform, followed by a few drops of the neutral thiocyanate solution. Shake the mixture vigorously. The chloroform will acquire a green colour.

Concentration limit: 1 in 50,000.

IV.11 HEXACYANOFERRATE(II) IONS, $[Fe(CN)_6]^{4-}$ *Solubility* The alkali and alkaline earth hexacyanoferrate(II)s are soluble in water; those of the other metals are insoluble in water and in cold dilute acids, but are decomposed by alkalis.

Use a 0·025M solution of potassium hexacyanoferrate(II), (often called potassium ferrocyanide), $K_4[Fe(CN)_6].3H_2O$, to study these reactions.

1. Concentrated sulphuric acid: complete decomposition occurs on prolonged boiling with the evolution of carbon monoxide, which burns with a blue flame:

$$[Fe(CN)_6]^{4-} + 6H_2SO_4 + 6H_2O \rightarrow Fe^{2+} + 6NH_4^+ + 6CO\uparrow + 6SO_4^{2-}$$

a little sulphur dioxide may also be produced, due to the oxidation of iron(II) with sulphuric acid:

$$2Fe^{2+} + 2H_2SO_4 \rightarrow 2Fe^{3+} + SO_2\uparrow + SO_4^{2-} + 2H_2O$$

With dilute sulphuric acid, little reaction occurs in the cold, but on boiling, a partial decomposition of hexacyanoferrate(II) occurs with the evolution of hydrogen cyanide (**POISON**):

$$[Fe(CN)_6]^{4-} + 6H^+ \rightarrow 6HCN\uparrow + Fe^{2+}$$

Iron(II) ions, formed in this reaction, react with some of the undecomposed hexacyanoferrate, yielding initially a white precipitate of potassium iron(II) hexacyanoferrate(II):

$$[Fe(CN)_6]^{4-} + Fe^{2+} + 2K^+ \rightarrow K_2Fe[Fe(CN)_6]\downarrow$$

This precipitate is gradually oxidized to Prussian blue by atmospheric oxygen (cf. Section **III.21**, reaction 6).

2. *Silver nitrate solution:* white precipitate of silver hexacyanoferrate(II):

$$[Fe(CN)_6]^{4-} + 4Ag^+ \rightarrow Ag_4[Fe(CN)_6]\downarrow$$

The precipitate is insoluble in ammonia [distinction from hexacyanoferrate(III)] and nitric acid, but soluble in potassium cyanide and sodium thiosulphate:

$$Ag_4[Fe(CN)_6]\downarrow + 8CN^- \rightarrow 4[Ag(CN)_2]^- + [Fe(CN)_6]^{4-}$$
$$Ag_4[Fe(CN)_6]\downarrow + 8S_2O_3^{2-} \rightarrow 4[Ag(S_2O_3)_2]^{3-} + [Fe(CN)_6]^{4-}$$

Upon warming with concentrated nitric acid, the precipitate is converted to orange-red silver hexacyanoferrate(III), when it becomes soluble in ammonia:

$$3Ag_4[Fe(CN)_6]\downarrow + HNO_3 + 3H^+ \rightarrow$$
$$\rightarrow 3Ag_3[Fe(CN)_6]\downarrow + 3Ag^+ + NO\uparrow + 2H_2O$$
$$Ag_3[Fe(CN)_6]\downarrow + 6NH_3 \rightarrow 3[Ag(NH_3)_2]^+ + [Fe(CN)_6]^{3-}$$

3. *Iron(III) chloride solution:* precipitate of Prussian blue in neutral or acid solutions:

$$3[Fe(CN)_6]^{4-} + Fe^{3+} \rightarrow Fe_4[Fe(CN)_6]_3\downarrow$$

the precipitate is decomposed by solutions of alkali hydroxides, brown iron(III) hydroxide being formed (cf. Section **III.22**, reaction 6).

The spot-test technique is as follows. Mix a drop of the test solution on a spot plate with a drop of iron(III) chloride solution. A blue precipitate or stain is formed.

Sensitivity: 1 μg $[Fe(CN)_6]^{4-}$. *Concentration limit:* 1 in 400,000.

4. *Iron(II) sulphate solution:* white precipitate of potassium iron(II) hexacyanoferrate(II), $K_2Fe[Fe(CN)_6]$, which turns rapidly blue by oxidation (see reaction 1 and also Section **III.21**, reaction 6).

5. *Copper sulphate solution:* brown precipitate of copper hexacyanoferrate(II):

$$[Fe(CN)_6]^{4-} + 2Cu^{2+} \rightarrow Cu_2[Fe(CN)_6]\downarrow$$

The precipitate is insoluble in dilute acetic acid, but decomposes in solutions of alkali hydroxides.

6. Thorium nitrate solution: white precipitate of thorium hexacyanoferrate:

$$[Fe(CN)_6]^{4-} + Th^{4+} \rightarrow Th[Fe(CN)_6]\downarrow$$

It is difficult to filter this precipitate, as it tends to form a colloid. The reaction can be used to distinguish hexacyanoferrate(II) ions from hexacyanoferrate(III) and thiocyanate, which do not react.

7. Hydrochloric acid If a concentrated solution of potassium hexacyanoferrate(II) is mixed with 1:1 hydrochloric acid, hydrogen hexacyanoferrate(II) is formed, which can be extracted by ether:

$$[Fe(CN)_6]^{4-} + 4H^+ \rightleftarrows H_4[Fe(CN)_6]$$

By evaporating the ether, the substance is obtained as a white crystalline solid.

8. Ammonium molybdate solution: from a solution of potassium hexacyanoferrate(II), acidified with dilute hydrochloric acid, a brown precipitate of molybdenyl hexacyanoferrate(II) is formed. The exact composition of the precipitate is not known. The precipitate is insoluble in dilute acids, but soluble in solutions of alkali hydroxides. The test can be applied to differentiate hexacyanoferrate(II) ions from hexacyanoferrate(III) and thiocyanate, which do not react.

9. Titanium(IV) chloride solution: reddish-brown flocculent precipitate of titanium(IV) hexacyanoferrate(II):

$$[Fe(CN)_6]^{4-} + Ti^{4+} \rightarrow Ti[Fe(CN)_6]\downarrow$$

The precipitate is insoluble in 6M hydrochloric acid. Hexacyanoferrate(III) ions do not react. Oxidizing anions (like chromate, arsenate or nitrite) should be absent, as these oxidize hexacyanoferrate(II) ions in acid medium.

The drop test can be carried out as follows: place a drop of the test solution on a spot plate, just acidify with hydrochloric acid and introduce 1 drop of the reagent. A reddish-brown precipitate is obtained in the presence of hexacyanoferrate(II) ions.

The reagent is prepared by adding 10 ml liquid titanium(IV) chloride to 90 ml of 1:1 hydrochloric acid.

10. Cobalt nitrate solution: greyish-green precipitate of cobalt hexacyanoferrate, which is insoluble in dilute hydrochloric or dilute acetic acid:

$$[Fe(CN)_6]^{4-} + 2Co^{2+} \rightarrow Co_2[Fe(CN)_6]\downarrow$$

11. Uranyl acetate solution: brown precipitate of uranyl hexacyanoferrate(II):

$$[Fe(CN)_6]^{4-} + 2UO_2^{2+} \rightarrow (UO_2)_2[Fe(CN)_6]\downarrow$$

Dilute acetic acid should be present as well. Hexacyanoferrate(III) ions react only in concentrated solutions, after long standing or heating, to give a dirty-yellow precipitate. Filter paper may reduce some hexacyanoferrate(III) to hexacyanoferrate(II), hence the test should not be carried out on filter paper.

The spot test is carried out as follows: place a drop of the test solution on a spot plate, and add a drop of 0·1M uranyl acetate solution. A brown precipitate or spot is obtained within 2 minutes.

Sensitivity: 1 μg $[Fe(CN)_6]^{4-}$. *Concentration limit:* 1 in 50,000.

12. Dry test All hexacyanoferrate(II) compounds are decomposed on heating, nitrogen and cyanogen being evolved:

$$3K_4[Fe(CN)_6] \rightarrow N_2\uparrow + 2(CN)_2\uparrow + 12KCN + Fe_3C + C$$

IV.12 HEXACYANOFERRATE(III) IONS $[Fe(CN)_6]^{3-}$ *Solubility* Alkali and alkaline earth hexacyanoferrate(III)s are soluble in water, so is iron(III) hexacyanoferrate(III). Those of most other metals are insoluble or sparingly soluble. Metal hexacyanoferrate(III)s are in general more soluble than metal hexacyanoferrate(II)s.

To study these reactions use a 0·033M solution of potassium hexacyanoferrate(III) $K_3[Fe(CN)_6]$.

1. Concentrated sulphuric acid On warming a solid hexacyanoferrate(III) with this acid, it is decomposed completely, carbon monoxide gas being evolved:

$$K_3[Fe(CN)_6] + 6H_2SO_4 + 6H_2O \rightarrow 6CO\uparrow + Fe^{3+} + 3K^+ + 6NH_4^+ + 6SO_4^{2-}$$

With dilute sulphuric acid, no reaction occurs in the cold, but on boiling hydrocyanic acid (**POISON**) is evolved:

$$[Fe(CN)_6]^{3-} + 6H^+ \rightarrow Fe^{3+} + 6HCN\uparrow$$

This test must be carried out in a fume cupboard with good ventilation.

2. Silver nitrate solution: orange-red precipitate of silver hexacyanoferrate(III):

$$[Fe(CN)_6]^{3-} + 3Ag^+ \rightarrow Ag_3[Fe(CN)_6]\downarrow$$

The precipitate is soluble in ammonia [distinction from hexacyanoferrate(II)], but insoluble in nitric acid.

3. Iron(II) sulphate solution: dark-blue precipitate of Prussian blue (earlier termed Turnbull's blue), in neutral or acid solution. (Section **III.21**, reaction 7).

4. Iron(III) chloride solution: brown colouration, owing to the formation of undissociated iron(III) hexacyanoferrate(III):

$$[Fe(CN)_6]^{3-} + Fe^{3+} \rightarrow Fe[Fe(CN)_6]$$

(cf. Section **III.22**, reaction 7).

5. Copper sulphate solution: green precipitate of copper(II) hexacyanoferrate(III):

$$2[Fe(CN)_6]^{3-} + 3Cu^{2+} \rightarrow Cu_3[Fe(CN)_6]_2\downarrow$$

6. Concentrated hydrochloric acid: adding concentrated hydrochloric acid to a saturated solution of potassium hexacyanoferrate(III) in cold, a brown precipitate of free hydrogen hexacyanoferrate(III) (hexacyanoferric acid) is obtained:

$$[Fe(CN)_6]^{3-} + 3HCl \rightarrow H_3[Fe(CN)_6]\downarrow + 3Cl^-$$

7. Potassium iodide solution: iodine is liberated in the presence of dilute hydrochloric acid, which may be identified by the blue colour produced with starch solution:

$$2[Fe(CN)_6]^{3-} + 2I^- \rightarrow 2[Fe(CN)_6]^{4-} + I_2$$

The reaction is reversible; in neutral solution iodine oxidizes hexacyanoferrate(II) ions.

8. Cobalt nitrate solution: red precipitate of cobalt hexacyanoferrate(III), which is insoluble in hydrochloric acid, but soluble in ammonia:

$$2[Fe(CN)_6]^{3-} + 3Co^{2+} \rightarrow Co_3[Fe(CN)_6]_2\downarrow$$

9. Benzidine reagent: (DANGER: THE REAGENT IS CARCINOGENIC) blue precipitate of oxidation product. Other oxidizing agents (molybdates, chromates, etc.) must be absent. The test is applicable in the presence of hexacyanoferrate(II) for benzidine hexacyanoferrate(II) is white: the sensitivity is, however, reduced and it is best to precipitate the hexacyanoferrate(II) first as the white, soluble lead hexacyanoferrate(II) by the addition of lead acetate, the hexacyanoferrate(III) remaining in solution.

Mix a drop of the test solution on a spot plate with a drop of the benzidine reagent. If hexacyanoferrate(II) is present, a drop of 1 per cent lead nitrate solution should be added before the reagent. A blue precipitate or colouration appears.

Sensitivity: 1 μg $[Fe(CN)_6]^{3-}$. *Concentration limit:* 1 in 50,000.

10. Action of heat Alkali hexacyanoferrate(III)s decompose similarly to hexacyanoferrate(II)s (cf. Section **IV.11**, reaction 12):

$$6K_3[Fe(CN)_6] \rightarrow 6N_2\uparrow + 3(CN)_2\uparrow + 18KCN + 2Fe_3C + 10C$$

IV.13 HYPOCHLORITES, OCl^- *Solubility* All hypochlorites are soluble in water. They react alkaline because of hydrolysis:

$$OCl^- + H_2O \rightleftarrows HOCl + OH^-$$

In solution hypochlorites disproportionate, slowly in cold but fast in hot solution, when chlorate and chloride ions are formed:

$$3OCl^- \rightarrow ClO_3^- + 2Cl^-$$

Thus, if these reactions are to be studied, the solution should be freshly prepared. By saturating 2M sodium hydroxide with chlorine gas, a M solution of sodium hypochlorite is obtained:

$$Cl_2 + 2OH^- \rightarrow ClO^- + Cl^- + H_2O$$

Chloride ions are invariably present and these will interfere with some ionic reactions.

1. Dilute hydrochloric acid: the solution first turns to yellow, later effervescence occurs and chlorine is evolved:

$$OCl^- + H^+ \rightarrow HOCl$$
$$HOCl + H^+ + Cl^- \rightarrow Cl_2\uparrow + H_2O$$

The gas may be identified (*a*) by its yellowish-green colour and irritating odour, (*b*) by its bleaching a wet litmus paper, and (*c*) by its action upon potassium iodide-starch paper, which it turns blue-black:

$$Cl_2\uparrow + 2I^- \rightarrow 2Cl^- + I_2$$

2. Potassium iodide–starch paper: a bluish-black colour is formed in neutral or weakly alkaline solution as the result of separation of iodine:

$$OCl^- + 2I^- + H_2O \rightarrow I_2 + 2OH^- + Cl^-$$

If the solution is too alkaline the colour disappears because hypoiodite and iodide ions are formed:

$$I_2 + 2OH^- \rightarrow OI^- + I^- + H_2O$$

3. Lead acetate or lead nitrate solution: brown lead dioxide is produced on boiling:

$$OCl^- + Pb^{2+} + H_2O \rightarrow PbO_2\downarrow + 2H^+ + Cl^-$$

4. Cobalt nitrate solution: adding a few drops of the reagent to a solution of the hypochlorite, a black precipitate of cobalt(III) hydroxide is obtained:

$$2Co^{2+} + OCl^- + 5H_2O \rightarrow 2Co(OH)_3\downarrow + Cl^- + 4H^+$$

Hydrogen ions, formed in the reaction are neutralized by the excess alkali present. On warming, oxygen is liberated (identified by the rekindling of a glowing splint), the cobalt acting as a catalyst:

$$2OCl^- \rightarrow 2Cl^- + O_2\uparrow$$

5. Mercury On shaking a slightly acidified (use sulphuric acid) solution of a hypochlorite with mercury, a brown precipitate of basic mercury(II) chloride $(HgCl)_2O$ is formed:

$$2Hg + 2H^+ + 2OCl^- \rightarrow O\begin{matrix} \diagup HgCl \\ \diagdown HgCl\downarrow \end{matrix} + H_2O$$

The precipitate is soluble in dilute hydrochloric acid, when undissociated mercury(II) chloride is formed:

$$O\begin{matrix} \diagup HgCl \\ \diagdown HgCl \end{matrix} + 2H^+ + 2Cl^- \rightarrow 2HgCl_2 + H_2O$$

If the precipitate is separated from the excess of mercury, washed and dissolved in hydrochloric acid, mercury can be detected in the solution by passing hydrogen sulphide into it (cf. Section **III.8**).

Chlorine water, under similar conditions, produces a white precipitate of mercury(II) chloride Hg_2Cl_2.

IV.14 CHLORIDES, Cl^- *Solubility* Most chlorides are soluble in water. Mercury(I) chloride, Hg_2Cl_2, silver chloride, AgCl, lead chloride, $PbCl_2$ (this is sparingly soluble in cold but readily soluble in boiling water), copper(I) chloride, CuCl, bismuth oxychloride, BiOCl, antimony oxychloride, SbOCl, and mercury(II) oxychloride, Hg_2OCl_2, are insoluble in water.

To study these reactions use a 0·1M solution of sodium chloride, NaCl.

1. Concentrated sulphuric acid: considerable decomposition of the chloride occurs in the cold, completely on warming, with the evolution of hydrogen chloride,

$$Cl^- + H_2SO_4 \rightarrow HCl\uparrow + HSO_4^-$$

The product is recognized (*a*) by its pungent odour and the production of white fumes, consisting of fine drops of hydrochloric acid, on blowing across the mouth of the tube, (*b*) by the formation of white clouds of ammonium chloride when a glass rod moistened with ammonia solution is held near the mouth of the vessel, and (*c*) by its turning blue litmus paper red.

2. Manganese dioxide and concentrated sulphuric acid If the solid chloride is mixed with an equal quantity of precipitated manganese dioxide,* concentrated sulphuric acid added and the mixture gently warmed, chlorine is evolved which is identified by its suffocating odour, yellowish-green colour, its bleaching of moistened litmus paper, and turning of potassium iodide–starch paper blue. The hydrogen chloride first formed is oxidized to chlorine.

$$MnO_2 + 2H_2SO_4 + 2Cl^- \rightarrow Mn^{2+} + Cl_2\uparrow + 2SO_4^{2-} + 2H_2O$$

3. Silver nitrate solution: white, curdy precipitate of silver chloride, AgCl, insoluble in water and in dilute nitric acid, but soluble in dilute ammonia solution and in potassium cyanide and sodium thiosulphate solutions (see under Silver, Section **III.6**, reaction 1, also under Complex Ions, Section **I.33**):

$$Cl^- + Ag^+ \rightarrow AgCl\downarrow$$

$$AgCl\downarrow + 2NH_3 \rightarrow [Ag(NH_3)_2]^+ + Cl^-$$

$$[Ag(NH_3)_2]^+ + Cl^- + 2H^+ \rightarrow AgCl\downarrow + 2NH_4^+$$

If the silver chloride precipitate is filtered off, washed with distilled water, and then shaken with sodium arsenite solution it is converted into yellow silver arsenite (distinction from silver bromide and silver iodide, which are unaffected by this treatment). This may be used as a confirmatory test for a chloride.

$$3AgCl + AsO_3^{3-} \rightarrow Ag_3AsO_3\downarrow + 3Cl^-$$

---

* The commercial substance (pyrolusite) usually contains considerable quantities of chlorides.

4. Lead acetate solution: white precipitate of lead chloride, $PbCl_2$, from concentrated solutions (see under Lead, Section **III.4**, reaction 1):

$$2Cl^- + Pb^{2+} \rightarrow PbCl_2\downarrow$$

5. Potassium dichromate and sulphuric acid (chromyl chloride test) The solid chloride is intimately mixed with three times its weight of powdered potassium dichromate in a small distilling flask (Fig. IV.1), an equal bulk of concentrated sulphuric acid is added and the mixture gently warmed.* The deep-red vapours of chromyl chloride, CrO_2Cl_2, which are evolved are passed into sodium hydroxide solution contained in a test-tube. The resulting yellow solution in the test-tube contains sodium chromate; this is confirmed by acidifying with dilute sulphuric acid, adding 1–2 ml amyl alcohol† followed by a little hydrogen peroxide solution. The organic layer is coloured blue. Alternatively, the diphenylcarbazide reagent test (Section **IV.33**, reaction 10) may be applied. The formation of a chromate in the distillate indicates that a chloride was present in the solid substance, since chromyl chloride is a readily volatile liquid (b.p. 116·5°C).

$$4Cl^- + Cr_2O_7^{2-} + 6H^+ \rightarrow 2CrO_2Cl_2\uparrow + 3H_2O$$

$$CrO_2Cl_2\uparrow + 4OH^- \rightarrow CrO_4^{2-} + 2Cl^- + 2H_2O$$

Some chlorine may also be liberated, owing to the reaction:

$$6Cl^- + Cr_2O_7^{2-} + 14H^+ \rightarrow 3Cl_2\uparrow + 2Cr^{3+} + 7H_2O$$

and this decreases the sensitivity of the test.

Bromides and iodides give rise to the free halogens, which yield colourless solutions with sodium hydroxide: if the ratio of iodide to chloride exceeds 1:15, the chromyl chloride formation is largely prevented and chlorine is evolved.‡ Fluorides give rise to the volatile chromyl fluoride, CrO_2F_2, which is decomposed by water, and hence should be absent or removed. Nitrites and nitrates interfere, as nitrosyl chloride may be formed. Chlorates must, of course, be absent.

The chlorides of mercury, owing to their slight ionization, do not respond to this test. Only partial conversion to CrO_2Cl_2 occurs with the chlorides of lead, silver, antimony, and tin.

The spot-test technique is as follows. Into the tube of Fig. II.57 place a few milligrams of the solid sample (or evaporate a drop or two of the test solution in it), add a small quantity of powdered potassium dichromate and a drop of concentrated sulphuric acid. Place a column about 1 mm long of a 1 per cent solution of diphenylcarbazide in alcohol into the capillary of the stopper and heat the apparatus for a few minutes. The chromyl chloride evolved causes the reagent to assume a violet colour.

Sensitivity: 1·5 μg Cl^-. *Concentration limit:* 1 in 30,000.

---

* This test must not be carried out in the presence of chlorates because of the danger of forming explosive chlorine dioxide (Section **IV.19**, reaction 1).

† Diethyl ether may also be used, but owing to its highly inflammable character and the possible presence of peroxides (unless previously removed by special treatment), it is preferable to employ amyl alcohol ir, less efficiently amyl acetate.

‡ The iodine reacts with the chromic acid yielding iodic acid: the latter, in the presence of concentrated sulphuric acid and especially on warming, liberates chlorine from chlorides, regenerating iodide. This explains the failure to form chromyl chloride.

Alternatively, employ the same quantities of materials in the apparatus of Fig. II.53 and replace the diphenylcarbazide solution by a drop of dilute alkali on the glass knob. Warm for a few minutes and, after cooling, dip the glass knob into a few drops of the alcoholic diphenylcarbazide solution which has been treated with a little dilute sulphuric acid and is contained on a spot plate. A violet colouration is obtained.

Sensitivity: 0·3 μg Cl^-. *Concentration limit:* 1 in 150,000.
Small amounts of bromides (5 per cent) do not interfere, but large amounts of bromides give rise to sufficient bromine to oxidize the reagent. It is therefore best to add a little phenol to the reagent solution, whereupon the bromine is removed as tribromophenol. Nitrates interfere since nitrosyl chloride is formed, but they may be reduced to ammonium salts. The interference of iodides is discussed above.

IV.15 BROMIDES, Br^- *Solubility* Silver, mercury(I), and copper(I) are insoluble in water. Lead bromide is sparingly soluble in cold, but more soluble in boiling water. All other bromides are soluble.

To study these reactions use a 0·1M solution of potassium bromide KBr.

1. Concentrated sulphuric acid If concentrated sulphuric acid is poured on some solid potassium bromide, first a reddish-brown solution is formed, later reddish-brown vapours of bromine accompany the hydrogen bromide (fuming in moist air) which is evolved:

$$KBr+H_2SO_4 \rightarrow HBr\uparrow+HSO_4^-+K^+$$

$$2KBr+2H_2SO_4 \rightarrow Br_2\uparrow+SO_2\uparrow+SO_4^{2-}+2K^++2H_2O$$

These reactions are accelerated by warming. If concentrated phosphoric acid is substituted for the sulphuric acid and the mixture is warmed, only hydrogen bromide is formed:

$$KBr+H_3PO_4 \rightarrow HBr\uparrow+H_2PO_4^-+K^+$$

The properties of hydrogen bromide are similar to those of hydrogen chloride.

2. Manganese dioxide and concentrated sulphuric acid When a mixture of a solid bromide, precipitated manganese dioxide, and concentrated sulphuric acid is warmed, reddish-brown vapours of bromine are evolved, bromine is recognized (*a*) by its powerful irritating odour, (*b*) by its bleaching of litmus paper, (*c*) by its staining of starch paper orange-red and, (*d*) by the red colouration produced upon filter paper impregnated with fluorescein (see reaction 8 below):

$$2KBr+MnO_2+2H_2SO_4 \rightarrow Br_2\uparrow+2K^++Mn^{2+}+2SO_4^{2-}+2H_2O$$

3. Silver nitrate solution: curdy, pale-yellow precipitate of silver bromide, AgBr, sparingly soluble in dilute, but readily soluble in concentrated ammonia solution. The precipitate is also soluble in potassium cyanide and sodium thiosulphate solutions, but insoluble in dilute nitric acid.

$$Br^-+Ag^+ \rightarrow AgBr\downarrow$$

$$AgBr\downarrow+2NH_3 \rightarrow [Ag(NH_3)_2]^++Br^-$$

$$AgBr\downarrow+2CN^- \rightarrow [Ag(CN)_2]^-+Br^-$$

$$AgBr\downarrow+2S_2O_3^{2-} \rightarrow [Ag(S_2O_3)_2]^{3-}+Br^-$$

4. *Lead acetate solution:* white crystalline precipitate of lead bromide:

$$2Br^- + Pb^{2+} \rightarrow PbBr_2\downarrow$$

The precipitate is soluble in boiling water.

5. *Chlorine water** The addition of this reagent dropwise to a solution of a bromide liberates free bromine, which colours the solution orange-red; if carbon disulphide, chloroform or carbon tetrachloride (2 ml) is added and the liquid shaken, the bromine dissolves in the solvent (see The Distribution Law, Section **I.45**) and, after allowing to stand, forms a reddish-brown solution below the colourless aqueous layer. With excess chlorine water, the bromine is converted into yellow bromine monochloride or into colourless hypobromous or bromic acid, and a pale-yellow or colourless solution results (difference from iodide).

$$2Br^- + Cl_2\uparrow \rightarrow Br_2\uparrow + 2Cl^-$$
$$Br_2\uparrow + Cl_2\uparrow \rightleftarrows 2BrCl$$
$$Br_2\uparrow + Cl_2\uparrow + 2H_2O \rightarrow 2OBr^- + 2Cl^- + 4H^+$$
$$Br_2\uparrow + 5Cl_2\uparrow + 6H_2O \rightarrow 2BrO_3^- + 10Cl^- + 12H^+$$

6. *Potassium dichromate and concentrated sulphuric acid* On gently warming a mixture of a solid bromide, concentrated sulphuric acid, and potassium dichromate (see Chlorides, Section **IV.14**, reaction 5) and passing the evolved vapours into water, a yellowish-brown solution, containing free bromine but no chromium, is produced. A colourless (or sometimes a pale-yellow) solution is obtained on treatment with sodium hydroxide solution this does not give the chromate reaction with dilute sulphuric acid, hydrogen peroxide and amyl alcohol, or with the diphenyl carbazide reagent (distinction from chloride).

$$6KBr + K_2Cr_2O_7 + 7H_2SO_4 \rightarrow 3Br_2\uparrow + 2Cr^{3+} + 4SO_4^{2-} + 7H_2O$$

7. *Nitric acid* Hot, fairly concentrated (1:1) nitric acid oxidizes bromides to bromine:

$$6Br^- + 8HNO_3 \rightarrow 3Br_2\uparrow + 2NO\uparrow + 6NO_3^- + 4H_2O$$

8. *Fluorescein test* Free bromine converts the yellow dyestuff fluorescein(I) into the red tetrabromofluorescein or eosin(II). Filter paper impregnated with fluorescein solution is therefore a valuable reagent for bromine vapour since the paper acquires a red colour.

$$\text{(I)} + 4Br_2 \rightarrow \text{(II)} + 4Br^- + 4H^+$$

---

* In practice, it is more convenient to use dilute sodium hypochlorite solution, acidified with dilute hydrochloric acid.

Chlorine tends to bleach the reagent. Iodine forms the red-violet coloured iodo-eosin and hence must be absent. If the bromide is oxidized to free bromine by heating with lead dioxide and acetic acid, practically no chlorine is simultaneously evolved from chlorides, and hence the test may be conducted in the presence of chlorides.

$$2Br^- + PbO_2 + 4CH_3COOH \rightarrow Br_2\uparrow + Pb^{2+} + 4CH_3COO^- + 2H_2O$$

Place a drop of the test solution together with a few milligrams of lead dioxide and acetic acid in the apparatus of Fig. II.56 and close the tube with the funnel stopper carrying a piece of filter paper which has been impregnated with the reagent and dried. Warm the apparatus gently. A circular red spot is formed on the yellow test paper.

Alternatively, the apparatus of Fig. II.57 may be used; a column, about 1 mm long of the reagent, is employed.

Sensitivity: 2 µg Br_2. *Concentration limit:* 1 in 25,000.

The fluorescein reagent consists of a saturated solution of fluorescein in 50 per cent alcohol.

9. Fuchsin (or Magenta) test The dyestuff fuchsin(I) forms a colourless

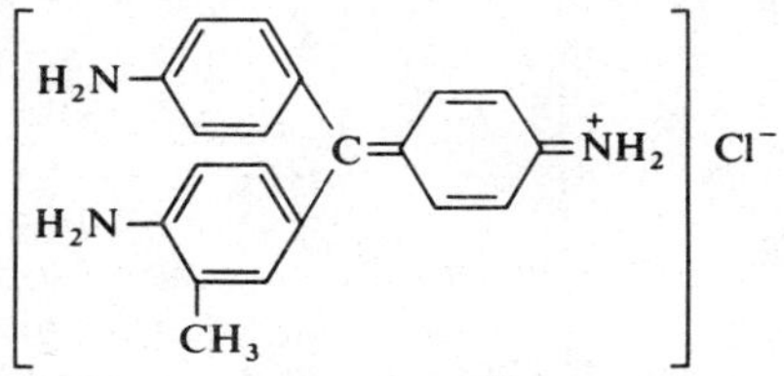

(I)

addition compound with hydrogen sulphite. Free bromine converts the thus decolourized fuchsin into a blue or violet brominated dyestuff. Neither free chlorine nor free iodine affect the colourless fuchsin hydrogen sulphite compound, hence the reaction may be employed for the detection of bromides in the presence of chlorides and iodides.

Place a drop of the test solution (or a few milligrams of the test solid) in the tube of the apparatus shown in Fig. II.57, add 2–4 drops 25 per cent chromic acid solution and close the apparatus with the 'head' which contains 1–2 drops of the reagent solution in the capillary. Warm the apparatus gently (do not allow it to boil). In a short time the liquid in the capillary assumes a violet colour.

Sensitivity: 3 µg Br^-. *Concentration limit:* 1 in 15,000.

The reagent consists of a 0·1 per cent fuchsin solution just decolourized by sodium hydrogen sulphite.

IV.16 IODIDES, I^- *Solubility* The solubilities of the iodides are similar to the chloride and bromides. Silver, mercury(I), mercury(II), copper(I), and lead iodides are the least soluble salts. These reactions can be studied with a 0·1M solution of potassium iodide, KI.

1. Concentrated sulphuric acid With a solid iodide, iodine is liberated; on warming, violet vapours are evolved, which turn starch paper blue. Some hydrogen iodide is formed – this can be seen by blowing across the mouth of the vessel, when white fumes are produced – but most of it reduces the sulphuric acid to sulphur dioxide, hydrogen sulphide, and sulphur, the relative proportions of which depend upon the concentrations of the reagents.

$$2I^- + 2H_2SO_4 \rightarrow I_2\uparrow + SO_4^{2-} + 2H_2O$$
$$I^- + H_2SO_4 \rightarrow HI\uparrow + HSO_4^-$$
$$6I^- + 4H_2SO_4 \rightarrow 3I_2\uparrow + S\downarrow + 3SO_4^{2-} + 4H_2O$$
$$8I^- + 5H_2SO_4 \rightarrow 4I_2\uparrow + H_2S\uparrow + 4SO_4^{2-} + 4H_2O$$

Pure hydrogen iodide is formed on warming with concentrated phosphoric acid:

$$I^- + H_3PO_4 \rightarrow HI\uparrow + H_2PO_4^-$$

If manganese dioxide is added to the mixture, only iodine is formed and the sulphuric acid does not get reduced:

$$2I^- + MnO_2 + 2H_2SO_4 \rightarrow I_2\uparrow + Mn^{2+} + 2SO_4^{2-} + 2H_2O$$

2. *Silver nitrate solution:* yellow, curdy precipitate of silver iodide AgI, readily soluble in potassium cyanide and in sodium thiosulphate solutions, very slightly soluble in concentrated ammonia solution, and insoluble in dilute nitric acid.

$$I^- + Ag^+ \rightarrow AgI$$
$$AgI\downarrow + 2CN^- \rightarrow [Ag(CN)_2]^- + I^-$$
$$AgI\downarrow + 2S_2O_3^{2-} \rightarrow [Ag(S_2O_3)_2]^{3-} + I^-$$

3. *Lead acetate solution:* yellow precipitate of lead iodide, PbI_2, soluble in much hot water forming a colourless solution, and yielding golden-yellow plates ('spangles') on cooling.

$$2I^- + Pb^{2+} \rightarrow PbI_2\downarrow$$

4. *Chlorine water** When this reagent is added dropwise to a solution of an iodide, iodine is liberated, which colours the solution brown; on shaking with 1–2 ml carbon disulphide, chloroform or carbon tetrachloride (see Section **I.45**), it dissolves forming a violet solution, which settles out below the aqueous layer. The free iodine may also be identified by the characteristic blue colour it forms with starch solution. If excess chlorine water is added, the iodine is oxidized to colourless iodic acid.

$$2I^- + Cl_2\uparrow \rightarrow I_2 + 2Cl^-$$
$$I_2 + 5Cl_2\uparrow + 6H_2O \rightarrow 2IO_3^- + 10Cl^- + 12H^+$$

5. *Potassium dichromate and concentrated sulphuric acid:* only iodine is liberated, and no chromate is present in the distillate (see Chlorides, Section **IV.14**, reaction 5) (difference from chloride).

$$6I^- + Cr_2O_7^{2-} + 7H_2SO_4 \rightarrow 3I_2\uparrow + 2Cr^{3+} + 7SO_4^{2-} + 7H_2O$$

---

* In practice it is more convenient to use dilute sodium hypochlorite solution acidified with dilute hydrochloric acid.

6. Sodium nitrite solution Iodine is liberated when this reagent is added to an iodide solution acidified with dilute acetic or sulphuric acid (difference from bromide and chloride). The iodine may be identified by colouring starch paste blue, or carbon tetrachloride violet.

$$2I^- + 2NO_2^- + 4H^+ \rightarrow I_2 + 2NO\uparrow + 2H_2O$$

7. Copper sulphate solution: brown precipitate consisting of a mixture of copper(I) iodide, CuI, and iodine. The iodine may be removed by the addition of sodium thiosulphate solution or sulphurous acid, and a nearly white precipitate of copper(I) iodide obtained.

$$4I^- + 2Cu^{2+} \rightarrow 2CuI\downarrow + I_2$$
$$I_2 + 2S_2O_3^{2-} \rightarrow 2I^- + S_4O_6^{2-}$$

8. Mercury(II) chloride solution: scarlet precipitate of mercury(II) iodide:

$$2I^- + HgCl_2 \rightarrow HgI_2\downarrow + 2Cl^-$$

(note that mercury(II) chloride is practically undissociated in solution). The precipitate dissolves in excess potassium iodide, forming a tetraiodomercurate(II) complex:

$$HgI_2\downarrow + 2I^- \rightarrow [HgI_4]^{2-}$$

9. Starch test Iodides are readily oxidized in acid solution to free iodine by a number of oxidizing agents; the free iodine may then be identified by the deep-blue colouration produced with starch solution. The best oxidizing agent to employ in the spot test reaction is acidified potassium nitrite solution (cf. reaction 6):

$$2I^- + 2NO_2^- + 4H^+ \rightarrow I_2 + 2NO\uparrow + 2H_2O$$

Cyanides interfere because of the formation of cyanogen iodide: they are therefore removed before the test either by heating with sodium hydrogen carbonate solution or by acidifying and heating:

$$I_2 + CN^- \rightarrow ICN\uparrow + I^-$$

Mix a drop of the acid test solution on a spot plate with a drop of the reagent and add a drop of 50 per cent potassium nitrite solution. A blue colouration is obtained.

Sensitivity: 2·5 μg I_2. *Concentration limit:* 1 in 20,000.

10. Catalytic reduction of cerium(IV) salts test The reduction of cerium(IV) salts in acid solution by arsenites takes place very slowly:

$$2Ce^{4+} + AsO_3^{3-} + H_2O \rightarrow 2Ce^{3+} + AsO_4^{3-} + 2H^+$$

Iodides accelerate this change, possibly owing to iodine liberated in the instantaneous reaction (i):

$$2Ce^{4+} + 2I^- \rightarrow 2Ce^{3+} + I_2 \qquad \text{(i)}$$

reacting further according to (ii):

$$AsO_3^{3-} + I_2 + H_2O \rightarrow AsO_4^{3-} + 2I^- + 2H^+ \qquad \text{(ii)}$$

the iodide ion reacting again as in (i). The completion of the reduction is indicated by the disappearance of the yellow colour of the cerium(IV) solution. Osmium and ruthenium salts have a similar catalytic effect. Moderate amounts of chlorides, bromides, sulphates, and nitrates have no influence, but cyanides and also mercury(II), silver, and manganese salts interfere.

Place a drop of the test solution together with a drop each of neutral or slightly acid 0·1M sodium arsenite solution and 0·1M cerium(IV) sulphate solution on a spot plate. The yellow colour soon disappears.

Sensitivity: 0·03 μg I^-. *Concentration limit:* 1 in 1,000,000.

11. Palladium(II) chloride test Solutions of iodides react with palladium(II) chloride solution to yield a brownish-red precipitate of palladium(II) iodide, PdI_2, insoluble in mineral acids.

$$2I^- + Pd^{2+} \rightarrow PdI_2\downarrow$$

Mix a drop of the test solution on drop-reaction paper with a drop of a 1 per cent aqueous solution of palladium chloride. A brownish-black precipitate forms.

Sensitivity: 1 μg I^-. *Concentration limit:* 1 in 50,000.

IV.17 FLUORIDES, F^- *Solubility* The fluorides of the common alkali metals and of silver, mercury, aluminium, and nickel are readily soluble in water, those of lead, copper, iron(III), barium, and lithium are slightly soluble, and those of the other alkaline earth metals are insoluble in water.

To study these reactions use a 0·1M solution of sodium fluoride, NaF.

1. Concentrated sulphuric acid With the solid fluoride, a colourless, corrosive gas, hydrogen fluoride, H_2F_2, is evolved on warming; the gas fumes in moist air, and the test-tube acquires a greasy appearance as a result of the corrosive action of the vapour on the silica in the glass, which liberates the gas, silicon tetrafluoride, SiF_4. By holding a moistened glass rod in the vapour, gelatinous silicic acid H_2SiO_3 is deposited on the rod; this is a product of the decomposition of the silicon tetrafluoride.

$$2F^- + H_2SO_4 \rightarrow H_2F_2\uparrow + SO_4^{2-}$$
$$SiO_2 + 2H_2F_2 \rightarrow SiF_4\uparrow + 2H_2O$$
$$3SiF_4\uparrow + 3H_2O \rightarrow 2[SiF_6]^{2-} + H_2SiO_3\downarrow + 4H^+$$

Note that at room temperature hydrogen fluoride gas is almost completely dimerized, therefore its formula has been written as H_2F_2. At elevated temperatures (say 90°C) it dissociates completely to monomer hydrogen fluoride:

$$H_2F_2 \rightleftarrows 2HF$$

The same result is more readily attained by mixing the solid fluoride with an equal bulk of silica, making into a paste with concentrated sulphuric acid and warming gently; silicon tetrafluoride is readily evolved.

The spot-test technique of the reaction utilizes the conversion of the silicic and fluosilicic acids by means of ammonium molybdate into silicomolybdic acid $H_4[SiMo_{12}O_{40}]$. The latter, unlike free molybdic acid, oxidizes benzidine in acetic acid solution to a blue dyestuff and 'molybdenum blue' is simultaneously produced. (DANGER: THE REAGENT IS CARCINOGENIC.)

Mix the solid test sample with a little pure silica powder in the tube of the apparatus shown in Fig. II.53 and moisten the silica with 1–2 drops concentrated sulphuric acid. Place a drop of water on the glass knob of the stopper, insert it in position and heat the apparatus gently for about 1 minute. Remove the source of heat and allow to stand for 5 minutes. Wash the drop of water into a micro crucible, add 1–2 drops ammonium molybdate reagent and warm the mixture until bubbling just commences. Allow to cool, introduce a drop of a 0·05 per cent solution of benzidine in 2M acetic acid and a few drops of saturated sodium acetate solution. A blue colour is obtained.

Sensitivity: 1 μg F.

2. The etching test A clean watch glass is coated on the convex side with paraffin wax, and part of the glass is exposed by scratching a design on the wax with a nail or wire. A mixture of about 0·3 g fluoride and 1 ml concentrated sulphuric acid is placed in a small lead or platinum crucible, and the latter immediately covered with watch glass, convex side down. A little water should be poured in the upper (concave) side of the watch glass to prevent the wax from melting. The crucible is very gently warmed (best on a boiling water bath). After 5–10 minutes, the hydrogen fluoride will have etched the glass. This is readily seen after removing the paraffin wax by holding above a flame or with hot water, and then breathing upon the surface of the glass.

The test may also be conducted in a small lead capsule, provided with a close-fitting lid made from lead foil. A small hole of about 3 mm diameter is pierced in the lid. About 0·1 g suspected fluoride and a few drops concentrated sulphuric acid are placed in the clean capsule, and a small piece of glass (e.g. a microscope slide) is placed over the hole in the lid. Upon warming very gently (best on a water bath) it will be found that an etched spot appears on the glass where it covers the hole.

Chlorates, silicates, and borates interfere and should therefore be absent.

3. Silver nitrate solution: no precipitate, since silver fluoride is soluble in water.

4. Calcium chloride solution: white, slimy precipitate of calcium fluoride, CaF_2, sparingly soluble in acetic acid, but slightly more soluble in dilute hydrochloric acid.

$$2F^- + Ca^{2+} \rightarrow CaF_2\downarrow$$

5. Iron(III) chloride solution: white crystalline precipitate of sodium hexafluoroferrate(III) from concentrated solutions of fluorides, sparingly soluble in water. The precipitate does not give the reactions of iron (e.g. with ammonium thiocyanate), except upon acidification.

$$6F^- + Fe^{3+} + 3Na^+ \rightarrow Na_3[FeF_6]\downarrow$$

6. Zirconium–alizarin lake test Hydrochloric acid solutions of zirconium salts are coloured reddish-violet by alizarin-S or by alizarin (see under Aluminium, Section **III.23**, reactions 8 and 9 and under Zirconium, Section **VII.18**, reaction 8); upon adding a solution of a fluoride the colour of such solutions changes immediately to a pale yellow (that of the liberated alizarin sulphonic acid or alizarin) because of the formation of the colourless hexafluorozirconate(IV) ion

$[ZrF_6]^{2-}$. The test may be performed on a spot plate.

Mix together on a spot plate 2 drops each (equal volumes) of a 0·1 per cent aqueous solution of alizarin-S (sodium alizarin sulphonate) and zirconyl chloride solution (0·1 g solid zirconyl chloride dissolved in 20 ml concentrated hydrochloric acid and diluted to 100 ml with water): upon the addition of a drop or two of the fluoride solution the zirconium lake is decolourized to a clear yellow solution.

Alternatively, mix 2 drops each of the alizarin-S and zirconyl chloride solutions in a semimicro test-tube, add a drop of dilute hydrochloric acid or of 1:1 (v/v) acetic acid, followed by 2 drops of the test solution. The pink colour will change to yellow.

The most sensitive method of carrying out the spot test is as follows. Impregnate some quantitative filter paper or drop-reaction paper with the zirconium–alizarin-S reagent, dry it and moisten with a drop of 1:1 (v/v) acetic acid. Place a drop of the neutral test solution upon the moist red spot; the spot will turn yellow.

Sensitivity: 1 μg F^-. *Concentration limit:* 1 in 50,000.

Large amounts of sulphates, thiosulphates, nitrites, arsenates, phosphates, and oxalates interfere with the test.

The zirconium–alizarin-S paper is prepared as follows. Immerse quantitative filter paper (or drop-reaction paper) in a 10 per cent solution of zirconium nitrate in 2M hydrochloric acid, drain and place in a 2 per cent aqueous solution of alizarin-S. Wash the paper, which is coloured red-violet by the zirconium lake, until the washings are nearly colourless and then dry in air.

IV.18 NITRATES, NO_3^- *Solubility* All nitrates are soluble in water. The nitrates of mercury and bismuth yield basic salts on treatment with water; these are soluble in dilute nitric acid.

These reactions can be studied with a 0·1M solution of potassium nitrate KNO_3.

1. Concentrated sulphuric acid: reddish-brown vapours of nitrogen dioxide, accompanied by pungent acid vapours of nitric acid which fume in the air, are formed on heating the solid nitrate with the reagent. Dilute sulphuric acid has no action (difference from nitrite):

$$4NO_3^- + 2H_2SO_4 \rightarrow 4NO_2\uparrow + O_2\uparrow + 2SO_4^{2-} + 2H_2O$$

2. Concentrated sulphuric acid and bright copper turnings On heating these with the solid nitrate, reddish-brown fumes of nitrogen dioxide are evolved, and the solution acquires a blue colour owing to the formation of copper(II) ions. A solution of the nitrate may also be used; the sulphuric acid is then added very cautiously.

$$2NO_3^- + 4H_2SO_4 + 3Cu \rightarrow 3Cu^{2+} + 2NO\uparrow + 4SO_4^{2-} + 4H_2O$$

$$2NO\uparrow + O_2\uparrow \rightarrow 2NO_2\uparrow$$

3. Iron(II) sulphate solution and concentrated sulphuric acid (brown ring test) This test is carried out in either of two ways: (*a*) Add 3 ml freshly prepared saturated solution of iron(II) sulphate to 2 ml nitrate solution, and pour 3–5 ml concentrated sulphuric acid slowly down the side of the test-tube so that the acid forms a layer beneath the mixture. A brown ring will form where the liquids meet. (*b*) Add 4 ml concentrated sulphuric acid slowly to 2 ml nitrate solution, mix the liquids thoroughly and cool the mixture under a stream of cold water from the tap. Pour a saturated solution iron(II) sulphate slowly down the side of the tube so that it forms a layer on top of the liquid. A brown ring will form at the zone of contact of the two liquids.

The brown ring is due to the formation of the $[Fe(NO)]^{2+}$. On shaking and warming the mixture the brown colour disappears, nitric oxide is evolved, and a yellow solution of iron(III) ions remains. The test is unreliable in the presence of bromide, iodide, nitrite, chlorate, and chromate (see Section **IV.45**, reactions 3 and 4).

$$2NO_3^- + 4H_2SO_4 + 6Fe^{2+} \rightarrow 6Fe^{3+} + 2NO\uparrow + 4SO_4^{2-} + 4H_2O$$
$$Fe^{2+} + NO\uparrow \rightarrow [Fe(NO)]^{2+}$$

Bromides and iodides interfere because of the liberated halogen; the test is not trustworthy in the presence of chromates, sulphites, thiosulphates, iodates, cyanides, thiocyanates, hexacyanoferrate(II) and (III) ions. All of these anions may be removed by adding excess of nitrate-free Ag_2SO_4 to an aqueous solution (or sodium carbonate extract), shaking vigorously for 3–4 minutes, and filtering the insoluble silver salts, etc.

Nitrites react similarly to nitrates. They are best removed by adding a little sulphamic acid (compare Section **IV.7**, reaction 10). The following reaction takes place in the cold:

$$H_2N\text{–}HSO_3 + NO_2^- \rightarrow N_2\uparrow + SO_4^{2-} + H^+ + H_2O$$

The spot-test technique is as follows. Place a crystal of iron(II) sulphate about as large as a pin head on a spot plate. Add a drop of the test solution and allow a drop of concentrated sulphuric acid to run in at the side of the drop. A brown ring forms round the crystal of iron(II) sulphate.

Sensitivity: 2·5 μg NO_3^-. *Concentration limit:* 1 in 25,000.

4. *Reduction of nitrates in alkaline medium* Ammonia is evolved (detected by its odour; by its action upon red litmus paper and upon mercury(I) nitrate paper; or by the tannic acid–silver nitrate test, Section **III.38**, reaction 7) when a solution of a nitrate is boiled with zinc dust or gently warmed with aluminium powder and sodium hydroxide solution. Excellent results are obtained by the use of Devarda's alloy (45 per cent Al, 5 per cent Zn, and 50 per cent Cu). Ammonium ions must, of course, be removed by boiling the solution with sodium hydroxide solution (and, preferably, evaporating almost to dryness) before the addition of the metal.

$$NO_3^- + 4Zn + 7OH^- + 6H_2O \rightarrow NH_3\uparrow + 4[Zn(OH)_4]^{2-}$$
$$3NO_3^- + 8Al + 5OH^- + 18H_2O \rightarrow 3NH_3\uparrow + 8[Al(OH)_4]^-$$

Nitrites give a similar reaction and may be removed most simply with the aid of sulphamic acid (see reaction 3). Another, but more expensive, procedure involves the addition of sodium azide to the acid solution; the solution is allowed to stand for a short time and then boiled in order to complete the reaction and to expel the readily volatile hydrogen azide:

$$NO_2^- + N_3^- + 2H^+ \rightarrow N_2\uparrow + N_2O\uparrow + H_2O$$
$$N_3^- + H^+ \rightarrow HN_3\uparrow$$

Other nitrogen compounds which evolve ammonia under the above conditions are cyanides, thiocyanates, hexacyanoferrate(II) and (III) ions. These may be removed by treating the aqueous solution (or a sodium carbonate extract) with excess of nitrate-free silver sulphate (cf. Section **IV.45**, 3), warming the mixture to about 60°, shaking vigorously for 3–4 minutes and filtering from the silver salts of the interfering anions and excess of precipitant. The excess silver ions are removed from the filtrate by adding an excess of sodium hydroxide solution, and filtering from the precipitated silver oxide. The filtrate is concentrated and tested with zinc, aluminium, or Devarda's alloy.

Attention is directed to the fact that arsenites are reduced in alkaline solution by aluminium, Devarda's alloy, etc., to arsine, which blackens mercury(I) nitrate paper and also gives a positive tannic acid–silver nitrate test. Hence neither the mercury(I) nitrate test nor the tannic acid–silver nitrate test for ammonia is applicable if arsenites are present.

The spot-test technique is carried out as follows. Place a drop of the test solution in the tube of Fig. II.56, add 1–2 drops 20 per cent sodium hydroxide solution and a few milligrams of Devarda's alloy. Place a watch glass with a drop of *p*-nitrobenzene-diazonium chloride reagent (for preparation, see under Ammonium, Section **III.38**, reaction 8) on its underside upon the funnel stopper. Heat the apparatus gently for a short time. Add a tiny fragment of calcium oxide to the drop of the reagent: a red ring forms within 10–15 seconds.

Sensitivity: 10 μg NO_3^-. *Concentration limit:* 1 in 5,000.

5. Diphenylamine reagent ($C_6H_5.NH.C_6H_5$) Pour the nitrate solution carefully down the side of the test-tube so that it forms a layer above the solution of the reagent; a blue ring is formed at the zone of contact of the two liquids. The test is a very sensitive one, but unfortunately is also given by a number of oxidizing agents, such as nitrites, chlorates, bromates, iodates, permanganates, chromates, vanadates, molybdates, and iron(III) salts.

The reagent is prepared by dissolving 0·5 g diphenylamine in 85 ml concentrated sulphuric acid diluted to 100 ml with water.

6. Nitron reagent (diphenyl-endo-anilo-dihydrotriazole $C_{20}H_{16}N_4$*)*

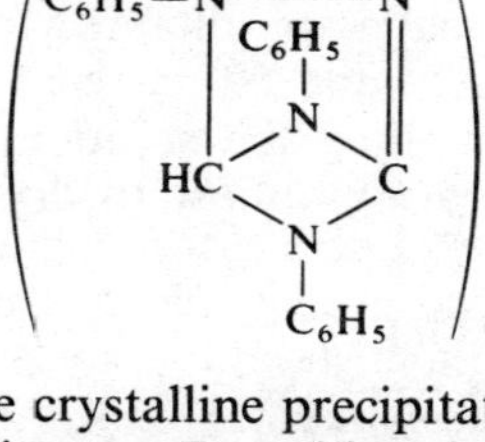

```
 / C6H5—N————N \
 |      | C6H5 ‖ |
 |      |  |   ‖ |
 |      |  N   ‖ |
 |      | / \  ‖ |  :
 |     HC     C  |
 |       \   /   |
 |         N     |
 |         |     |
 \        C6H5   /
```

white crystalline precipitate of nitron nitrate $C_{20}H_{16}N_4,HNO_3$ with solutions of nitrates. Bromides, iodides, nitrites, chromates, chlorates, perchlorates, thiocyanates, oxalates, and picrates also yield insoluble compounds, and hence the reaction is not very characteristic.

The reagent is prepared by dissolving 5 g nitron in 100 ml 2M acetic acid.

7. Action of heat The result varies with the metal. The nitrates of sodium and potassium evolve oxygen (test with glowing splint) and leave solid nitrites (brown fumes with dilute acid); ammonium nitrate yields dinitrogen oxide and steam; the nitrates of the noble metals leave a residue of the metal, and a mixture of nitrogen dioxide and oxygen is evolved; the nitrates of the other metals, such as those of lead and copper, evolve oxygen and nitrogen dioxide, and leave a residue of the oxide.

$$2NaNO_3 \rightarrow 2NaNO_2 + O_2\uparrow$$
$$NH_4NO_3 \rightarrow N_2O\uparrow + 2H_2O$$
$$2AgNO_3 \rightarrow 2Ag + 2NO_2\uparrow + O_2\uparrow$$
$$2Pb(NO_3)_2 \rightarrow 2PbO + 4NO_2\uparrow + O_2\uparrow$$

8. Reduction to nitrite test Nitrates are reduced to nitrites by metallic zinc in acetic acid solution; the nitrite can be readily detected by means of the sulphanilic acid-α-naphthylamine reagent (see under Nitrites, Section **IV.7**, reaction 11). Nitrites, of course, interfere and are best removed with sulphamic acid (see reaction 3 above).

Mix on a spot plate a drop of the neutral or acetic acid test solution with a drop of the sulphanilic acid reagent and a drop of the α-naphthylamine reagent, and add a few milligrams of zinc dust. A red colouration develops.

Sensitivity: 0·05 μg NO_3^-. *Concentration limit:* 1 in 1,000,000.

IV.19 CHLORATES, ClO_3^- *Solubility* All chlorates are soluble in water; potassium chlorate is one of the least soluble (66 g ℓ^{-1} at 18°) and lithium chlorate one of the most soluble (3150 g ℓ^{-1} at 18°).

To study these reactions use a 0·1M solution of potassium chlorate.

1. Concentrated sulphuric acid (**DANGER**) All chlorates are decomposed with the formation of the greenish-yellow gas, chlorine dioxide, ClO_2, which dissolves in the sulphuric acid to give an orange-yellow solution. On warming gently (**DANGER**) an explosive crackling occurs, which may develop into a violent explosion. In carrying out this test one or two small crystals of potassium chlorate (weighing not more than 0·1 g) are treated with 1 ml concentrated sulphuric acid in the cold; the yellow explosive chlorine dioxide can be seen on shaking the solution. The test-tube should not be warmed, and its mouth should be directed away from the student.

$$3KClO_3 + 3H_2SO_4 \rightarrow 2ClO_2\uparrow + ClO_4^- + 3SO_4^{2-} + 4H^+ + 3K^+ + H_2O$$

2. Concentrated hydrochloric acid All chlorates are decomposed by this acid, and chlorine, together with varying quantities of the explosive chlorine dioxide, is evolved; chlorine dioxide imparts a yellow colour to the acid. The mixture of gases is sometimes known as 'euchlorine'. The experiment should be conducted on a very small scale, and not more than 0·1 g potassium chlorate should be used. The following two chemical reactions probably occur simultaneously:

$$2KClO_3 + 4HCl \rightarrow 2ClO_2 + Cl_2\uparrow + 2K^+ + 2Cl^- + 2H_2O$$
$$KClO_3 + 6HCl \rightarrow 3Cl_2\uparrow + K^+ + Cl^- + 3H_2O$$

3. Sodium nitrite solution On warming this reagent with a solution of the chlorate, the latter is reduced to a chloride, which may be identified by adding silver nitrate solution after acidification with dilute nitric acid. The nitrite must, of course, be free from chloride. A solution of sulphurous acid or of formaldehyde (10 per cent; 1 part of formaline to 3 parts of water) acts similarly. Excellent results are obtained with zinc, aluminium or Devarda's alloy and sodium hydroxide solution (see under Nitrates, Section **IV.18**, reaction 4); the solution is acidified with dilute nitric acid after several minutes boiling* and silver nitrate solution added.

$$ClO_3^- + 3NO_2^- \rightarrow Cl^- + 3NO_3^-$$
$$ClO_3^- + 3H_2SO_3 \rightarrow Cl^- + 3SO_4^{2-} + 6H^+$$
$$ClO_3^- + 3HCHO \rightarrow Cl^- + 3HCOOH$$
$$ClO_3^- + 3Zn + 6OH^- + 3H_2O \rightarrow Cl^- + 3[Zn(OH)_4]^{4-}$$
$$ClO_3^- + 2Al + 2OH^- + 3H_2O \rightarrow Cl^- + 2[Al(OH)_4]^-$$

4. *Silver nitrate solution:* no precipitate in neutral solution or in the presence of dilute nitric acid. Upon the addition of a little pure (chloride-free) sodium nitrite to the dilute nitric acid solution, a white precipitate of silver chloride is obtained because of the reduction of the chlorate to chloride (see reaction 3 above).

No precipitate is obtained with barium chloride solution.

5. *Potassium iodide solution:* iodine is liberated if a mineral acid is present. If acetic acid is used, no iodine separates even on long standing (difference from iodate).

$$ClO_3^- + 6I^- + 6H^+ \rightarrow 3I_2 + Cl^- + 3H_2O$$

6. *Iron(II) sulphate solution:* reduction to chloride upon boiling in the presence of mineral acid (difference from perchlorate).

$$ClO_3^- + 6Fe^{2+} + 6H^+ \rightarrow Cl^- + 6Fe^{3+} + 3H_2O$$

7. *Indigo test* A dilute solution of indigo in concentrated sulphuric acid is added to the chlorate solution until the latter has a pale-blue colour. Dilute sulphurous acid or sodium sulphite solution is then added drop by drop; the blue colour is discharged. The chlorate is reduced by the sulphurous acid to chlorine or to hypochlorite, and the latter bleaches the indigo.

8. *Aniline sulphate test,* $(C_6H_5.NH_2)_2H_2SO_4$ A small quantity of the solid chlorate (say, 0·05 g) (**DANGER**) is mixed with 1 ml concentrated sulphuric acid, and 2–3 ml aqueous aniline sulphate solution added; a deep-blue colour is obtained (distinction from nitrate).

9. *Manganese(II) sulphate–phosphoric acid test:* manganese(II) sulphate in concentrated phosphoric acid solution reacts with chlorates to form the violet-coloured diphosphatomanganate(III) ion:

$$ClO_3^- + 6Mn^{2+} + 12PO_4^{3-} + 6H^+ \rightarrow 6[Mn(PO_4)_2]^{3-} + Cl^- + 3H_2O$$

Peroxodisulphates, nitrites, bromates, iodates, and also periodates react simi-

---

* It is best to filter off the excess metal before adding the silver nitrate solution.

larly. The first-named may be decomposed by evaporating the sulphuric acid solution with a little silver nitrate as catalyst.

Place a drop of the test solution in a micro crucible and add a drop of the reagent. Warm rapidly over a micro burner and allow to cool. A violet colouration appears. Very pale colourations may be intensified by adding a drop of 1 per cent alcoholic diphenylcarbazide solution when a deep-violet colour, due to an oxidation product of the diphenylcarbazide, is obtained.

Sensitivity: 0·05 μg ClO_3^-. *Concentration limit:* 1 in 1,000,000.

The reagent is prepared by mixing equal volumes of saturated manganese(II) sulphate solution and concentrated phosphoric acid.

10. Action of heat All chlorates are decomposed by heat into chlorides and oxygen. Some perchlorate is usually formed as an intermediate product. The chloride is identified in the residue by extracting with water and adding dilute nitric acid and silver nitrate solution. An insoluble chlorate should be mixed with sodium carbonate before ignition.

$$2KClO_3 \rightarrow 2KCl + 3O_2\uparrow$$
$$2KClO_3 \rightarrow KClO_4 + KCl + O_2\uparrow$$

IV.20 BROMATES, BrO_3^- *Solubility* Silver, barium, and lead bromates are slightly soluble in water, the solubilities being respectively 2·0 g ℓ$^{-1}$, 7·0 g ℓ$^{-1}$ and 13·5 g ℓ$^{-1}$ at 20°; mercury(I) bromate is also sparingly soluble. Most of the other metallic bromates are readily soluble in water.

To study these reactions use a 0·1M solution of potassium bromate.

1. Concentrated sulphuric acid Add 2 ml acid to 0·5 g solid bromate; bromine and oxygen are evolved in the cold.

$$4KBrO_3 + 2H_2SO_4 \rightarrow 2Br_2\uparrow + 5O_2\uparrow + 4K^+ + 2SO_4^{2-} + 2H_2O$$

2. Silver nitrate solution A white crystalline precipitate of silver bromate, $AgBrO_3$, is produced with a concentrated solution of a bromate. The precipitate is soluble in hot water, readily soluble in dilute ammonia solution forming a complex salt, and sparingly soluble in dilute nitric acid.

$$BrO_3^- + Ag^+ \rightarrow AgBrO_3\downarrow$$
$$AgBrO_3\downarrow + 2NH_3 \rightarrow [Ag(NH_3)_2]^+ + BrO_3^-$$

Precipitates of the corresponding bromates are also produced by the addition of solutions of barium chloride, lead acetate or mercury(I) nitrate to a concentrated solution of a bromate.

$$2BrO_3^- + Ba^{2+} \rightarrow Ba(BrO_3)_2\downarrow$$
$$2BrO_3^- + Pb^{2+} \rightarrow Pb(BrO_3)_2\downarrow$$
$$2BrO_3^- + Hg_2^{2+} \rightarrow Hg_2(BrO_3)_2\downarrow$$

If the solution of silver bromate in dilute ammonia solution is treated dropwise with sulphurous acid solution, silver bromide separates: the latter dissolves in concentrated ammonia solution (difference from iodate).

3. Sulphur dioxide If the gas is bubbled through a solution of a bromate, the latter is reduced to a bromide (see Bromides, Section **IV.15**). A similar result is obtained with hydrogen sulphide and with sodium nitrite solution (see under Chlorates, Section **IV.19**, reaction 3).

$$BrO_3^- + 3H_2SO_3 \rightarrow Br^- + 3SO_4^{2-} + 6H^+$$
$$BrO_3^- + 3H_2S \rightarrow Br^- + 3S\downarrow + 3H_2O$$
$$BrO_3^- + 3NO_2^- \rightarrow Br^- + 3NO_3^-$$

4. Hydrobromic acid Mix together solutions of potassium bromate and bromide, and acidify with dilute sulphuric acid; bromine is liberated as a result of interaction between the bromic and hydrobromic acids set free. The bromine may be extracted by adding a little chloroform or carbon tetrachloride.

$$BrO_3^- + 5Br^- + 6H^+ \rightarrow 3Br_2\uparrow + 3H_2O$$

5. Action of heat Potassium bromate on heating evolves oxygen and a bromide remains. Sodium and calcium bromates behave similarly, but cobalt, zinc, and other similar metallic bromates evolve oxygen and bromine, and leave an oxide.

$$2KBrO_3 \rightarrow 2KBr + 3O_2\uparrow$$

6. Manganese(II) sulphate test If a bromate solution is treated with a little of a 1:1 mixture of saturated manganese(II) sulphate solution and M sulphuric acid, a transient red colouration (due to manganese(III) ions) is observed. Upon concentrating the solution rapidly, brown, hydrated manganese dioxide separates. The latter is insoluble in dilute sulphuric acid, but dissolves in a mixture of dilute sulphuric and oxalic acids (difference from chlorates and iodates, which neither give the colouration nor yield the brown precipitate).

$$BrO_3^- + 6Mn^{2+} + 6H^+ \rightarrow 6Mn^{3+} + Br^- + 3H_2O$$
$$BrO_3^- + 6Mn^{3+} + 9H_2O \rightarrow 6MnO_2\downarrow + Br^- + 18H^+$$
$$MnO\downarrow + (COO)_2^{2-} + 4H^+ \rightarrow Mn^{2+} + 2CO_2\uparrow + 2H_2O$$

In the spot-test technique, the reaction is combined with a sensitive test for manganese (oxidation of benzidine by manganese dioxide to 'benzidine blue'). (DANGER: THE REAGENT IS CARCINOGENIC.) Place a drop of the test solution in a semimicro centrifuge tube, add a drop or two of 0·25M manganese(II) sulphate solution acidified with dilute sulphuric acid and warm for 2–3 minutes in a boiling water bath. Cool, add a few drops of the benzidine reagent and a few small crystals of sodium acetate. A blue colouration results.
Sensitivity: 30 μg BrO_3^-. *Concentration limit:* 1 in 2,500.

IV.21 IODATES, IO_3^- *Solubility* The iodates of the alkali metals are soluble in water; those of the other metals are sparingly soluble and, in general, less soluble than the corresponding chlorates and bromates. Some solubilities in g ℓ^{-1} at 20° are: lead iodate 0·03 (25°), silver iodate 0·06, barium iodate 0·22, calcium iodate 3·7, potassium iodate 81·3, and sodium iodate 90·0. Iodic acid is a crystalline solid, and has a solubility of 2,330 g ℓ^{-1} at 20°.

To study these reactions use a 0·1M solution of potassium iodate, KIO_3.

1. Concentrated sulphuric acid: no action in the absence of reducing agents; readily converted to iodide in the presence of iron(II) sulphate:

$$IO_3^- + 6Fe^{2+} + 6H^+ \rightarrow I^- + 6Fe^{3+} + 3H_2O$$

If the iodate is in excess, iodine is formed because of the interaction of iodate and iodide (cf. reaction 6):

$$IO_3^- + 5I^- + 6H^+ \rightarrow 3I_2\downarrow + 3H_2O$$

2. Silver nitrate solution: white, curdy precipitate of silver iodate, readily soluble in dilute ammonia solution, but sparingly soluble in dilute nitric acid.

$$IO_3^- + Ag^+ \rightarrow AgIO_3\downarrow$$
$$AgIO_3\downarrow + 2NH_3 \rightarrow [Ag(NH_3)_2]^+ + IO_3^-$$

If the ammoniacal solution of the precipitate is treated dropwise with sulphurous acid solution, silver iodide is precipitated; the latter is not dissolved by concentrated ammonia solution (difference from bromate).

$$[Ag(NH_3)_2]^+ + IO_3^- + 3H_2SO_3 \rightarrow AgI\downarrow + 3SO_4^{2-} + 2NH_4^+ + 4H^+$$

3. Barium chloride solution: white precipitate of barium iodate (difference from chlorate), sparingly soluble in hot water and in dilute nitric acid, but insoluble in alcohol (difference from iodide). If the precipitate of barium iodate is well washed, treated with a little sulphurous acid solution and 1–2 ml carbon tetrachloride, the latter is coloured violet by the liberated iodine, and barium sulphate is precipitated:

$$2IO_3^- + Ba^{2+} \rightarrow Ba(IO_3)_2\downarrow$$
$$Ba(IO_3)_2\downarrow + 5H_2SO_3 \rightarrow I_2 + BaSO_4\downarrow + 4SO_4^{2-} + 8H^+ + H_2O$$

4. Mercury(II) nitrate solution: white precipitate of mercury(II) iodate (difference from chlorate and bromate). Lead acetate solution similarly gives a precipitate of lead iodate. Mercury(II) chloride solution, which is practically un-ionized (as mercuric chloride is covalent) gives no precipitate.

$$2IO_3^- + Hg^{2+} \rightarrow Hg(IO_3)_2\downarrow$$
$$2IO_3^- + Pb^{2+} \rightarrow Pb(IO_3)_2\downarrow$$

5. Sulphur dioxide or hydrogen sulphide Passage of sulphur dioxide or of hydrogen sulphide into a solution of an iodate, acidified with dilute hydrochloric acid, liberates iodine, which may be recognized by the addition of starch solution or chloroform or carbon tetrachloride. With an excess of either reagent, the iodine is further reduced to iodide.

$$2IO_3^- + 5H_2SO_3 \rightarrow I_2 + 5SO_4^{2-} + 8H^+ + H_2O$$
$$I_2 + H_2SO_3 + H_2O \rightarrow 2I^- + SO_4^{2-} + 4H^+$$
$$2IO_3^- + 5H_2S + 2H^+ \rightarrow I_2 + 5S\downarrow + 6H_2O$$
$$I_2 + H_2S \rightarrow 2I^- + 2H^+ + S\downarrow$$

6. Potassium iodide solution Mix together solutions of potassium iodide and potassium iodate, and acidify with hydrochloric acid, acetic acid or with tartaric acid solution; iodine is immediately liberated (use the chloroform or carbon tetrachloride test).

$$IO_3^- + 5I^- + 6H^+ \rightarrow 3I_2 + 3H_2O$$

7. *Action of heat* The alkali iodates decomposed into oxygen and an iodide. Most iodates of the bivalent metals yield iodine and oxygen and leave a residue of oxide; barium iodate, exceptionally, gives the periodate (more precisely, hexoxoperiodate):

$$2KIO_3 \rightarrow 2KI + 3O_2\uparrow$$
$$2Pb(IO_3)_2 \rightarrow 2I_2\uparrow + 5O_2\uparrow + 2PbO$$
$$5Ba(IO_3)_2 \rightarrow Ba_5(IO_6)_2 + 4I_2\uparrow + 9O_2\uparrow$$

8. *Hypophosphorous acid–starch solution test* Iodates are reduced by hypophosphorous acid eventually to iodides. The reaction takes place in three stages (with the transitional formation of phosphorous acid):

$$IO_3^- + 3H_3PO_2 \rightarrow I^- + 3H_3PO_3 \quad \text{(i)}$$
$$5I^- + IO_3^- + 6H^+ \rightarrow 3I_2 + 3H_2O \quad \text{(ii)}$$
$$I_2 + H_3PO_3 + H_2O \rightarrow 2I^- + H_3PO_4 + 2H^+ \quad \text{(iii)}$$

The first two stages are rapid and the third stage is a slow reaction. The iodine can be readily identified by the starch reaction. Chlorates and bromates do not react under these conditions.

Place a drop of the neutral test solution on a spot plate and mix it with a drop of starch solution (for preparation, see under Iodides, Section **IV.16**, reaction 9) and a drop of a dilute solution of hypophosphorous acid. A transitory blue colouration is produced.

Sensitivity: 1 µg IO_3^-. *Concentration limit:* 1 in 50,000.

9. *Potassium thiocyanate test* Iodates react with thiocyanates in acid solution with the liberation of iodine:

$$6IO_3^- + 5SCN^- + H^+ + 2H_2O \rightarrow 3I_2 + 5HCN\uparrow + 5SO_4^{2-}$$

Treat a piece of starch paper successively with a drop of 0·1M potassium thiocyanate solution and a drop of the acid test solution. A blue spot is obtained.

Sensitivity: 3 µg IO_3^-. *Concentration limit:* 1 in 12,000.

IV.22 PERCHLORATES, ClO_4^- *Solubility* The perchlorates are generally soluble in water. Potassium perchlorate is one of the least soluble (7·5 g ℓ$^{-1}$ and 218 g ℓ$^{-1}$ at 0° and 100° respectively), and sodium perchlorate is one of the most soluble (2,096 g ℓ$^{-1}$ at 25°).

To study these reactions use a 2M solution of perchloric acid or solid sodium perchlorate, $NaClO_4$.

1. *Concentrated sulphuric acid:* no visible action with the solid salt in cold, on strong heating white fumes of perchloric acid monohydrate are produced:

$$NaClO_4 + H_2SO_4 + H_2O \rightarrow HClO_4.H_2O\uparrow + HSO_4^- + Na^+$$

2. *Potassium chloride solution:* white precipitate of $KClO_4$, insoluble in alcohol (see Section **III.36**, reaction 3). Ammonium chloride solution gives a similar reaction:

$$ClO_4^- + K^+ \rightarrow KClO_4\downarrow$$
$$ClO_4^- + NH_4^+ \rightarrow NH_4ClO_4\downarrow$$

3. *Barium chloride solution:* no precipitate. A similar result is obtained with silver nitrate solution.

4. *Indigo test:* no decolourization even in the presence of acid (difference from hypochlorite and chlorate).

5. *Sulphur dioxide, hydrogen sulphide, or iron(II) salts:* no reduction (difference from chlorate).

6. *Titanium(III) sulphate solution:* in the presence of sulphuric acid, perchlorates are reduced to chlorides:

$$ClO_4^- + 8Ti^{3+} + 8H^+ \rightarrow Cl^- + 8Ti^{4+} + 4H_2O$$

Chloride ions can be identified in the solution with the usual tests (see Section **IV.14**).

7. *Tetramminecadmium perchlorate test* When a neutral solution of perchlorate is treated with a saturated solution of cadmium sulphate in concentrated ammonia solution, a white crystalline precipitate of tetramminecadmium perchlorate is obtained:

$$2ClO_4^- + [Cd(NH_3)_4]^{2+} \rightarrow [Cd(NH_3)_4](ClO_4)_2\downarrow$$

Sulphides interfere by precipitating cadmium as cadmium sulphide, and should therefore be absent.

8. *Action of heat:* oxygen is evolved from a solid perchlorate, and in the residue chloride ions can be tested (see Section **IV.14**):

$$NaClO_4 \rightarrow NaCl + 2O_2\uparrow$$

IV.23 BORATES, BO_3^{3-}, $B_4O_7^{2-}$, BO_2^- The borates are derived from the three boric acids: orthoboric acid, H_3BO_3, pyroboric acid, $H_2B_4O_7$, and metaboric acid, HBO_2. Orthoboric acid is a white, crystalline solid, sparingly soluble in cold but more soluble in hot water; very few salts of this acid are definitely known. On heating orthoboric acid at 100°, it is converted into metaboric acid; at 140° pyroboric acid is produced. Most of the salts are derived from the meta- and pyro-acids. Owing to the weakness of boric acid, the soluble salts are hydrolysed in solution and therefore react alkaline.

$$BO_3^{3-} + 3H_2O \rightleftarrows H_3BO_3 + 3OH^-$$
$$B_4O_7^{2-} + 7H_2O \rightleftarrows 4H_3BO_3 + 2OH^-$$
$$BO_2^- + 2H_2O \rightleftarrows H_3BO_3 + OH^-$$

Solubility The borates of the alkali metals are readily soluble in water.

The borates of the other metals are, in general, sparingly soluble in water, but fairly soluble in acids and in ammonium chloride solution.

To study these reactions use a 0·1M solution of sodium tetraborate (sodium pyroborate, borax) $Na_2B_4O_7.10H_2O$.

1. Concentrated sulphuric acid: no visible action in the cold, although orthoboric acid, H_3BO_3, is set free. On heating, however, white fumes of boric acid are evolved. If concentrated hydrochloric acid is added to a concentrated solution of borax, boric acid is precipitated:

$$Na_2B_4O_7 + H_2SO_4 + 5H_2O \rightarrow 4H_3BO_3\uparrow + 2Na^+ + SO_4^{2-}$$

$$Na_2B_4O_7 + 2HCl + 5H_2O \rightarrow 4H_3BO_3\downarrow + 2Na^+ + 2Cl^-$$

2. Concentrated sulphuric acid and alcohol (flame test) If a little borax is mixed with 1 ml concentrated sulphuric acid and 5 ml methanol or ethanol (the former is to be preferred owing to its greater volatility) in a small porcelain basin, and the alcohol ignited, the latter will burn with a green-edge flame, due to the formation of methyl borate $B(OCH_3)_3$ or of ethyl borate $B(OC_2H_5)_3$. Both these esters are poisonous. Copper and barium salts may give a similar green flame. The following modification of the test, which depends upon the greater volatility of boron trifluoride, BF_3, can be used in the presence of copper and barium compounds; these do not form volatile compounds under the experimental conditions given below. Thoroughly mix the borate with powdered calcium fluoride and a little concentrated sulphuric acid, and bring a little of the paste thus formed on the loop of a platinum wire, or upon the end of a glass rod, very close to the edge of the base of a Bunsen flame without actually touching it; volatile boron trifluoride is formed and colours the flame green.

$$H_3BO_3 + 3CH_3OH \rightarrow B(OCH_3)_3\uparrow + 3H_2O$$

$$Na_2B_4O_7 + 6CaF_2 + 7H_2SO_4 \rightarrow 4BF_3\uparrow + 6CaSO_4 + 2Na^+ + SO_4^{2-} + 7H_2O$$

The reaction may be adapted as a spot test in the following manner. The methyl borate is distilled off and passed into an aqueous solution containing potassium fluoride, manganese(II) nitrate, and silver nitrate. The ester is hydrolysed by the water to boric acid:

$$B(OCH_3)_3\uparrow + 3H_2O \rightarrow H_3BO_3 + 3CH_3OH$$

The boric acid reacts with the alkali fluoride forming a borofluoride and liberating free caustic alkali:

$$H_3BO_3 + 4F^- \rightarrow [BF_4]^- + 3OH^-;$$

the free caustic alkali is identified by the formation of a black precipitate with manganese(II) nitrate–silver nitrate solution (see under Ammonium, Section **III.38**, reaction 9):

$$Mn2^+ + 2Ag^+ + 4OH^- \rightarrow MnO_2\downarrow + 2Ag\downarrow + 2H_2O$$

Place a drop of the alkaline test solution in the distillation apparatus of Fig. II.59 and evaporate to dryness. Add 5 drops concentrated sulphuric acid and 5 drops pure methanol, stopper the apparatus, and heat to 80°C in a water bath. Collect the methyl borate which distils over in a micro porcelain crucible, waxed on the inside, and containing about 1 ml reagent. A black precipitate forms. For very small amounts of borate it is best to add a few drops of benzidine acetate solution and thus to detect the traces of manganese dioxide by the resulting blue colour. (DANGER: THE REAGENT IS CARCINOGENIC.)

Sensitivity: 0·01 μg B. *Concentration limit:* 1 in 5,000,000.

The reagent (a manganese(II) nitrate–silver nitrate solution containing potassium fluoride) is prepared as follows. Dissolve 2·87 g manganese(II) nitrate and 1·69 g silver nitrate in 100 ml water, add a drop of dilute alkali and filter the solution from the black precipitate. Treat the filtrate with a solution of 3·5 g potassium fluoride in 50 ml water; a white precipitate will form which on heating becomes grey and black. Boil, filter, and use the clear solution as the reagent.

3. *Turmeric paper test* If a piece of turmeric paper is dipped into a solution of a borate acidified with dilute hydrochloric acid and then dried at 100°, it becomes reddish-brown. The drying of the paper is most simply carried out by winding it on the outside near the rim of a test-tube containing water, and boiling the water for 2–3 minutes. On moistening the paper with dilute sodium hydroxide solution, it becomes bluish-black or greenish-black. Chromates, chlorates, nitrites, iodides, and other oxidizing agents interfere because of their bleaching action on the turmeric.

4. *Silver nitrate solution:* white precipitate of silver metaborate, $AgBO_2$, from fairly concentrated borax solution, soluble in both dilute ammonia solution and in acetic acid. On boiling the precipitate with water, it is completely hydrolysed and a brown precipitate of silver oxide is obtained. A brown precipitate of silver oxide is produced directly in very dilute solutions.

$$B_4O_7^{2-} + 4Ag^+ + H_2O \rightarrow 4AgBO_2\downarrow + 2H^+$$

$$2AgBO_2\downarrow + 3H_2O \rightarrow Ag_2O\downarrow + 2H_3BO_3$$

The boric acid, formed in this reaction, is practically undissociated.

5. *Barium chloride solution:* white precipitate of barium metaborate, $Ba(BO_2)_2$, from fairly concentrated solutions; the precipitate is soluble in excess reagent, in dilute acids, and in solutions of ammonium salts. Solutions of calcium and strontium chloride behave similarly.

$$B_4O_7^{2-} + 2Ba^{2+} + H_2O \rightarrow 2Ba(BO_2)_2\downarrow + 2H^+$$

6. *Action of heat* Powdered borax when heated in an ignition tube, or upon a platinum wire, swells up considerably, and then subsides, leaving a colourless glass of the anhydrous salt. The glass possesses the property of dissolving many oxides on heating, forming metaborates, which often have characteristic colours. This is the basis of the borax bead test for various metals (see Section **II.2**, reaction 5).

7. p-*Nitrobenzene-azo-chromotropic acid* reagent*

OH OH
O_2N– –N=N–
HO_3S– –SO_3H
:

* Alternative names are: *p*-nitrobenzene-azo-1 : 8-dihydroxynaphthalene-3 : 6-disulphonic acid and 'Chromotrope 2B' (the latter is the sodium salt).

Borates cause the blue-violet reagent to assume a greenish-blue colour.

Evaporate a drop of the slightly alkaline solution to dryness in a semimicro crucible. Stir the warm residue with 2–3 drops of the reagent. A greenish-blue colouration is obtained on cooling. A blank test should be performed simultaneously.

Sensitivity: 0·1 μg B. *Concentration limit:* 1 in 500,000.

Oxidizing agents and fluorides interfere, the latter because of the formation of borofluorides. Oxidizing agents, including nitrates and chlorates, are rendered innocuous by evaporating with solid hydrazine sulphate, whilst fluorides may be removed as silicon tetrafluoride by evaporation with silicic acid and sulphuric acid.

The experimental details are as follows. Treat 2 drops of the test solution in a small porcelain crucible either with a little solid hydrazine sulphate or with a few specks of precipitated silica and 1–2 drops concentrated sulphuric acid, and heat cautiously until fumes of sulphuric acid appear. Add 3–4 drops of the reagent whilst the residue is still warm and observe the colour on cooling.

Sensitivity: 0·25 μg B in the presence of 12,000 times that amount of $KClO_3$ or KNO_3; 0·5 μg B in the presence of 2,500 times that amount of NaF.

The reagent consists of a 0·005 per cent solution of Chromotrope 2B in concentrated sulphuric acid.

8. Mannitol–bromothymol blue test Boric acid acts as a very weak monobasic aid ($K_a = 5{\cdot}8 \times 10^{-10}$), but on the addition of certain organic polyhydroxy compounds, such as mannitol (mannite), glycerol, dextrose or invert sugar, it is transformed into a relatively strong acid, probably of the type:

$$\begin{array}{l} -\overset{|}{C}-OH \\ \quad | \\ -\underset{|}{C}-OH \end{array} + \begin{array}{l} HO \diagdown \\ \qquad B-OH \\ HO \diagup \end{array} = \begin{array}{l} -C-O \diagdown \\ \qquad\qquad B-O^- \\ -C-O \diagup \end{array} + H^+ + 2H_2O$$

The *p*H of the solution therefore decreases. Hence if the solution is initially almost neutral to, say, bromothymol blue (green), then upon the addition of mannitol the colour becomes yellow. It is advisable when testing for minute quantities of borates to recrystallize the mannitol from a solution neutralized to bromothymol blue, wash with pure acetone and dry at 100°. The reagent (a 10 per cent aqueous solution of mannitol) may also be neutralized with 0·01M potassium hydroxide solution, using bromothymol blue (a 0·04 per cent solution in 96 per cent ethanol) as indicator. Only periodate interferes with the test: it can be decomposed by heating on charcoal.

Render the test solution almost neutral to bromothymol blue by treating it with dilute acid or alkali (as necessary) until the indicator turns green. Place a few drops of the test solution in a micro test-tube, and add a few drops of the reagent solution. A yellow colouration is obtained in the presence of a borate. It is advisable to carry out a blank test with distilled water simultaneously.

Sensitivity: 0·001 μg B. *Concentration limit:* 1 in 30,000,000.

IV.24 SULPHATES, SO_4^{2-} *Solubility* The sulphates of barium, strontium, and lead are practically insoluble in water,* those of calcium and mercury(II) are slightly soluble, and most of the remaining metallic sulphates are soluble. Some basic sulphates such as those of mercury, bismuth, and chromium, are also insoluble in water, but these dissolve in dilute hydrochloric or nitric acid.

Sulphuric acid is a colourless, oily, and hygroscopic liquid, of specific gravity 1·838. The pure, commercial, concentrated acid is a constant boiling point mixture, boiling point 338° and containing *c.* 98 per cent of acid. It is miscible with water in all proportions with the evolution of considerable heat; on mixing the two, the acid should always be poured in a thin stream into the water (if the water is poured into the heavier acid, steam may be suddenly generated which will carry with it some of the acid and may therefore cause considerable damage).

To study these reactions use a 0·1M solution of sodium sulphate, $Na_2SO_4.10H_2O$.

1. Barium chloride solution: white precipitate of barium sulphate, $BaSO_4$ (see under Barium, Section **III.31**), insoluble in warm dilute hydrochloric acid and in dilute nitric acid, but moderately soluble in boiling, concentrated hydrochloric acid.

$$SO_4^{2-} + Ba^{2+} \rightarrow BaSO_4\downarrow$$

The test is usually carried out by adding the reagent to the solution acidified with dilute hydrochloric acid; carbonates, sulphites, and phosphates are not precipitated under these conditions. Concentrated hydrochloric acid or concentrated nitric acid should not be used, as a precipitate of barium chloride or of barium nitrate may form; these dissolve, however, upon dilution with water. The barium sulphate precipitate may be filtered from the hot solution and fused on charcoal with sodium carbonate, when sodium sulphide will be formed. The latter may be extracted with water, and the extract filtered into a freshly prepared solution of sodium nitroprusside, when a transient, purple colouration is obtained (see under Sulphides, Section **IV.6**, reaction 5). An alternative method is to add a few drops of very dilute hydrochloric acid to the fused mass, and to cover the latter with lead acetate paper; a black stain of lead sulphide is produced on the paper. The so-called Hepar reaction, which is less sensitive than the above two tests, consists of placing the fusion product on a silver coin and moistening with a little water; a brownish-black stain of silver sulphide results.

$$BaSO_4 + 4C + Na_2CO_3 \rightarrow Na_2S + BaCO_3 + 4CO\uparrow$$

$$Na_2S \rightarrow 2Na^+ + S^{2-}$$

$$2S^{2-} + 4Ag + O_2 + 2H_2O \rightarrow 2Ag_2S\downarrow + 4OH^-$$

A more efficient method for decomposing most sulphur compounds is to heat them with sodium or potassium, and then test the solution of the product for sulphide. The test is rendered sensitive by heating the substance with potassium in an ignition tube, dissolving the melt in water, and testing for sulphide by the nitroprusside or methylene blue reactions (see under Sulphides, Section **IV.6**, reactions 6 and 7).

The reader is warned that the above tests (depending upon the formation of a sulphide) are not exclusive to sulphates but are given by most sulphur com-

* Of these three sulphates, that of strontium is the most soluble.

pounds. If, however, the barium sulphate precipitated in the presence of hydrochloric acid is employed, then the reaction may be employed as a confirmatory test for sulphates.

2. Lead acetate solution: white precipitate of lead sulphate, $PbSO_4$, soluble in hot concentrated sulphuric acid, in solutions of ammonium acetate and of ammonium tartrate (see under Lead, Section **III.4**, reaction 5), and in sodium hydroxide solution. In the last case sodium tetrahydroxoplumbate(II) is formed, and on acidification with hydrochloric acid, the lead crystallizes out as the chloride. If any of the aqueous solutions of the precipitate are acidified with acetic acid and potassium chromate solution added, yellow lead chromate is precipitated (see under Lead, Section **III.4**, reaction 6).

$$SO_4^{2-} + Pb^{2+} \rightarrow PbSO_4\downarrow$$

3. Silver nitrate solution: white, crystalline precipitate of silver sulphate, Ag_2SO_4 (solubility 5·8 g ℓ^{-1} at 18°), from concentrated solutions.

$$SO_4^{2-} + 2Ag^+ \rightarrow Ag_2SO_4\downarrow$$

4. Sodium rhodizonate test Barium salts yield a reddish-brown precipitate with sodium rhodizonate (see under Barium, Section **III.31**, reaction 7). Sulphates and sulphuric acid cause immediate decolourization because of the formation of insoluble barium sulphate. This test is specific for sulphates.

Place a drop of barium chloride solution upon filter or drop-reaction paper, followed by a drop of a freshly prepared 0·5 per cent aqueous solution of sodium rhodizonate. Treat the reddish-brown spot with a drop of the acid or alkaline test solution. The coloured spot disappears.

Sensitivity: 4 μg SO_4^{2-}. *Concentration limit:* 1 in 10,000.

5. Potassium permanganate–barium sulphate test If barium sulphate is precipitated in a solution containing potassium permanganate, it is coloured pink (violet) by adsorption of some of the permanganate. The permanganate which has been adsorbed on the precipitate cannot be reduced by the common reducing agents (including hydrogen peroxide); the excess of potassium permanganate in the mother liquor reacts readily with reducing agents, thus rendering the pink barium sulphate clearly visible in the colourless solution.

Place 3 drops test solution in a semimicro centrifuge tube, add 2 drops 0·02M potassium permanganate solution and 1 drop barium chloride solution. A pink precipitate is obtained. Add a few drops 3 per cent hydrogen peroxide solution or 0·5M oxalic acid solution (in the latter case it will be necessary to warm on a water bath until decolourization is complete). Centrifuge. The coloured precipitate is clearly visible.

Sensitivity: 2·5 μg SO_4^{2-}. *Concentration limit:* 1 in 20,000.

6. Mercury(II) nitrate solution: yellow precipitate of basic mercury(II) sulphate:

$$SO_4^{2-} + 3Hg^{2+} + 2H_2O \rightarrow HgSO_4.2HgO\downarrow + 4H^+$$

This is a sensitive test, given even by suspensions of barium or lead sulphates.

7. *Benzidine hydrochloride solution:* white precipitate of benzidine sulphate. (DANGER: THE REAGENT IS CARCINOGENIC.)

$$SO_4^{2-} + H_2N-C_6H_4-C_6H_4-NH_2.2HCl \rightarrow$$

$$\rightarrow H_2N-C_6H_4-C_6H_4-NH_2.H_2SO_4 + 2Cl^-$$

Chromates, oxalates, hexacyanoferrate(II) and (III) ions interfere with the reaction.

IV.25 PEROXODISULPHATES. $S_2O_8^{2-}$

Solubility The best known peroxodisulphates, those of sodium, potassium, ammonium, and barium are soluble in water, the potassium salt being the least soluble, (17·7 g ℓ^{-1} at 0°C).

To study these reactions use a freshly prepared 0·1M solution of ammonium peroxodisulphate $(NH_4)_2S_2O_8$.

1. *Water* All peroxodisulphates are decomposed on boiling with water into the sulphate, free sulphuric acid, and oxygen. The oxygen contains appreciable quantities of ozone, which may be detected by its odour or by its property of turning starch-iodide paper blue. A similar result is obtained with dilute sulphuric or nitric acid. With dilute hydrochloric acid, chlorine is evolved (see reaction 4 below). By dissolving the solid peroxodisulphate in concentrated sulphuric acid at 0°, peroxomonosulphuric acid (Caro's acid), H_2SO_5, is formed in solution; this possesses strong oxidizing properties.

$$2S_2O_8^{2-} + 2H_2O \rightarrow 4SO_4^{2-} + 4H^+ + O_2\uparrow$$
$$2S_2O_8^{2-} + 3H_2O \rightarrow 4SO_4^{2-} + 6H^+ + O_3\uparrow$$
$$O_3\uparrow + 2I^- + 2H^+ \rightarrow I_2 + O_2\uparrow + H_2O$$

2. *Silver nitrate solution:* black precipitate of silver peroxide, Ag_2O_2, from concentrated solutions. If only a little silver nitrate solution is added followed by dilute ammonia solution, the silver peroxide, or the silver ion, acts catalytically leading to the evolution of nitrogen and the liberation of considerable heat.

$$S_2O_8^{2-} + 2Ag^+ + 2H_2O \longrightarrow Ag_2O_2\downarrow + 2SO_4^{2-} + 4H^+$$

$$3S_2O_8^{2-} + 8NH_3 \xrightarrow{(Ag^+)} N_2\uparrow + 6SO_4^{2-} + 6NH_4^+$$

3. *Barium chloride solution:* no immediate precipitate in the cold with a solution of a pure peroxodisulphate; on standing for some time or on boiling, a precipitate of barium sulphate is obtained, due to the decomposition of the peroxodisulphate.

4. *Potassium iodide solution:* iodine is slowly liberated in the cold and rapidly on warming (test with starch solution) (distinction from perborate and percarbonate, which liberate iodine immediately). Traces of copper catalyse the reaction.

$$S_2O_8^{2-} + 2I^- \rightarrow I_2 + 2SO_4^{2-}$$

5. Manganese(II) sulphate solution: brown precipitate in neutral or preferably alkaline solution. The precipitate is manganese(IV) dioxide hydrate with a composition nearest to $MnO_2.H_2O$.

$$S_2O_8^{2-} + Mn^{2+} + 4OH^- \rightarrow MnO_2.H_2O\downarrow + 2SO_4^{2-} + H_2O$$

In the presence of nitric or sulphuric acid and small amounts of silver nitrate, permanganate ions are formed on warming, and the solution turns to violet:

$$5S_2O_8^{2-} + 2Mn^{2+} + 8H_2O \rightarrow 2MnO_4^- + 10SO_4^{2-} + 16H^+$$

Silver ions act as catalysts (cf. Section **III.28**, reaction 6).

6. Potassium permanganate solution: unaffected (distinction from hydrogen peroxide). Peroxodisulphates are also unaffected by a solution of titanium(IV) sulphate.

7. Benzidine acetate test (DANGER: THE REAGENT IS CARCINOGENIC) A neutral or weakly acetic acid solution of a peroxodisulphate converts benzidine into a blue oxidation product. Perborates, percarbonates, and hydrogen peroxide do not react. Chromates, hexacyanoferrate(III) ions, permanganates, and hypohalites react similarly to peroxodisulphates.

Mix 1 drop of the test solution (neutral or faintly acid with acetic acid) with 1 drop of the benzidine acetate reagent. A blue colouration is produced.

Sensitivity: 1 μg $S_2O_8^{2-}$. *Concentration limit:* 1 in 100,000.

The reagent consists of a 0·05 per cent solution of benzidine in dilute acetic acid.

IV.26 SILICATES, SiO_3^{2-} The silicic acids may be represented by the general formula $xSiO_2.yH_2O$. Salts corresponding to orthosilicic acid, H_4SiO_4, ($SiO_2.2H_2O$) metasilicic acid, H_2SiO_3 ($SiO_2.H_2O$), and disilicic acid $H_2Si_2O_5$ ($2SiO_2.H_2O$) are definitely known. The metasilicates are sometimes designated simply as silicates.

Solubility Only the silicates of the alkali metals are soluble in water; they are hydrolysed in aqueous solution and therefore react alkaline.

$$SiO_3^{2-} + 2H_2O \rightleftharpoons H_2SiO_3 + 2OH^-$$

To study these reactions use a M solution of sodium silicate, Na_2SiO_3. Commercial (30%) water-glass solutions should be diluted with a five-fold amount of water.

1. Dilute hydrochloric acid Add dilute hydrochloric acid to the solution of the silicate; a gelatinous precipitate of metasilicic acid is obtained, particularly on boiling. The precipitate is insoluble in concentrated acids. The freshly precipitated substance is appreciably soluble in water and in dilute acids. It is converted by repeated evaporation with concentrated hydrochloric acid on the water bath into a white insoluble powder (silica, SiO_2).

If a dilute solution (say, 1–10 per cent) of water glass is quickly added to moderately concentrated hydrochloric acid, no precipitation of silicic acid takes place; it remains in colloidal solution (sol).

$$SiO_3^{2-} + 2H^+ \rightarrow H_2SiO_3\downarrow$$

2. *Ammonium chloride or ammonium carbonate solution:* gelatinous precipitate of silicic acid. This reaction is important in routine qualitative analysis since silicates, unless previously removed, will be precipitated by ammonium chloride solution in Group IIIA.

$$SiO_3^{2-} + 2NH_4^+ \rightarrow H_2SiO_3\downarrow + 2NH_3$$
$$\text{Base}_1 + \text{Acid}_2 \rightarrow \text{Acid}_1 + \text{Base}_2$$

This reaction is essentially an acid–base process, which is most conveniently explained in terms of the Brønsted–Lowry theory of acids and bases (see Section **I.23**).

3. *Silver nitrate solution:* yellow precipitate of silver silicate, soluble in dilute acids and in ammonia solution:

$$SiO_3^{2-} + 2Ag^+ \rightarrow Ag_2SiO_3\downarrow$$

4. *Barium chloride solution:* white precipitate of barium silicate, soluble in dilute nitric acid.

$$SiO_3^{2-} + Ba^{2+} \rightarrow BaSiO_3\downarrow$$

Calcium chloride solution gives a similar precipitate of calcium silicate.

5. *Microcosmic salt bead test* Most silicates, and also silica, when fused in a bead of microcosmic salt, $Na(NH_4)HPO_4.4H_2O$, in a loop of platinum wire give this test. The microcosmic salt first fuses to a transparent bead consisting largely of sodium metaphosphate (see Section **II.2**, 6); when a minute quantity of the solid silicate or even of the solution is introduced into the bead (best by dipping the hot bead into the substance) and the whole again heated, the silica produced will not dissolve in the bead, but will swim about in the fused mass, and is visible as white opaque masses or 'skeletons' in both the fused and the cold bead.

$$CaSiO_3 + NaPO_3 \rightarrow CaNaPO_4 + SiO_2$$

Insoluble silicates are best brought into solution by fusing the powdered solid, mixed with six times its weight of fusion mixture, in a platinum crucible* or upon platinum foil; the alkali carbonates react with the silicate yielding an alkali silicate. The cold mass is then evaporated to dryness on the water bath with excess dilute hydrochloric acid; the alkali silicate is thereby decomposed yielding first gelatinous silicic acid and, ultimately, white, amorphous silica, whilst the metallic oxides derived from the insoluble silicate are converted into chlorides. The residue is extracted with boiling dilute hydrochloric acid; this removes the metals as chlorides and insoluble silica remains behind. A simpler, but not quantitative, method is to extract the fusion mixture melt with boiling water: sufficient sodium or potassium silicate passes into solution to give any of the reactions referred to above.

$$SiO_2 + Na_2CO_3 \rightarrow CO_2\uparrow + Na_2SiO_3$$
$$SiO_3^{2-} + 2H^+ \rightarrow SiO_2\downarrow + H_2O$$

---

* A nickel or iron crucible should be used if metals of Group I or II are likely to be present.

6. *Silicon tetrafluoride test* This test depends upon the fact that when silica (isolated from a silicate by treatment with ammonium chloride solution or with hydrochloric acid, etc.) is heated with less than an equivalent amount of calcium fluoride and some concentrated sulphuric acid, silicon tetrafluoride is evolved. The latter is identified by its action upon a drop of water held in a loop of platinum wire, when a turbidity (due to silica) is produced.

$$SiO_2+2CaF_2+2H_2SO_4 \rightarrow 2CaSO_4+SiF_4\uparrow+2H_2O;$$
$$3SiF_4\uparrow+2H_2O \rightarrow SiO_2\downarrow+2[SiF_6]^{2-}+4H^+$$

Excess of calcium fluoride should be avoided since a mixture of H_2F_2 and SiF_4 will be formed and interfere with the test.

Mix the solid substance (or, preferably, silicic acid isolated by treatment of the silicate with ammonium chloride solution) with one-third of its weight of calcium fluoride in a small lead (or platinum) capsule, and add sufficient concentrated sulphuric acid to form a thin paste: mix the contents of the capsule with a stout platinum wire. Warm gently (FUME CUPBOARD) and hold close above the mixture a loop of platinum wire supporting a drop of water. The drop of water will become turbid, due to the hydrolysis of the silicon tetrafluoride absorbed.

7. *Ammonium molybdate–benzidine test* (DANGER: THE REAGENT IS CARCINOGENIC.) Silicates react with molybdates in acid solution to form the complex silicomolybdic acid $H_4[SiMo_{12}O_{40}]$. The ammonium salt, unlike the analogous phosphoric acid and arsenic acid compounds, is soluble in water and acids to give a yellow solution. The test depends upon the reaction between silicomolybdic acid and benzidine in acetic acid solution whereby 'molybdenum blue' and a blue quinonoid oxidation compound of benzidine are produced.

Place a drop of the test solution and of the molybdate reagent upon drop-reaction paper, and warm gently over a wire gauze. Add a drop of 0·05 per cent benzidine acetate and hold the paper over ammonia vapour. A blue colouration results.

Sensitivity: 1 μg SiO_2. *Concentration limit:* 1 in 50,000.

A better method for conducting the test is the following. Place a drop of the slightly acid test solution (the acidity should not exceed 0·5M) in a small porcelain crucible of good quality, and add a drop of the molybdate reagent. Warm cautiously over a wire gauze (or upon a sheet of asbestos resting upon a hot plate) until bubbles escape. Cool, add a drop of the benzidine reagent followed by a drop of saturated sodium acetate solution. A blue colour is obtained. It is essential to carry out a blank test with a drop of water and a drop of the molybdate reagent in another crucible of similar quality.

Sensitivity: 0·1 μg SiO_2. *Concentration limit:* 1 in 500,000.

Phosphoric and arsenic acids form compounds analogous to silicomolybdic acid which also react with benzidine with colour formation hence these acids should be removed before applying the test. In the presence of phosphoric acid, the test is carried out as follows. Mix a drop of the test solution with 2 drops of the molybdate reagent in a micro centrifuge tube and centrifuge the mixture. Transfer the supernatant liquid to a micro crucible by means of a capillary tube, warm gently, cool and add 2 drops 0·5M oxalic acid solution (the latter decomposes the small quantity of residual phosphomolybdate $(NH_4)_3[PMo_{12}O_{40}]$ but has little action on the silicomolybdic acid complex), then introduce a drop of the benzidine reagent and 2–3 drops of saturated sodium acetate solution. A blue colour forms.

Sensitivity: 6 μg SiO_2 in the presence of 250 times that amount of P_2O_5. *Concentration limit:* 1 in 8,000.

8. Ammonium molybdate–tin(II) chloride test The silicate is separated by volatilization as silicon tetrafluoride and the latter collected in a little sodium hydroxide solution. The resulting silicate is treated with ammonium molybdate solution and the ammonium salt of silicomolybdic acid, $H_4[SiMo_{12}O_{40}]$, is reduced by tin(II) chloride solution to 'molybdenum blue'. Tin(II) chloride does not reduce ammonium molybdate solution.

Phosphates and arsenates give the same reaction, but do not interfere under the conditions of the test: large amounts of borates should be absent, but may be removed by warming with methanol and sulphuric acid.

Place a little of the solid silicate in a small lead or platinum crucible, add a little sodium fluoride and a few drops of concentrated sulphuric acid. Cover the crucible with a small sheet of cellophane from which is suspended a drop of 2M sodium hydroxide solution (freshly prepared from the A.R. solid). Warm gently for 3–5 minutes over a micro burner with the crucible about 8 cm from the flame. Transfer the drop of sodium hydroxide solution to a porcelain micro crucible, add 2 drops ammonium molybdate reagent and then 2M acetic acid until feebly acidic. Then add a few drops 0·25M tin(II) chloride followed by sufficient sodium hydroxide solution to dissolve the tin(II) hydroxide. A blue colouration is obtained.

Concentration limit: 1 in 10,000.

The reaction may be applied to aqueous solutions of silicates, but phosphates and arsenates must be absent.

IV.27 HEXAFLUOROSILICATES (SILICOFLUORIDES), $[SiF_6]^{2-}$

Solubility Most metallic hexafluorosilicates (with the exception of the barium and potassium salts which are sparingly soluble), are soluble in water. A solution of the acid (hydrofluosilicic acid $H_2[SiF_6]$) is one of the products of the action of water upon silicon tetrafluoride, and is also formed by dissolving silica in hydrofluoric acid.

$$SiO_2 + 6HF \rightarrow H_2[SiF_6] + 2H_2O$$

To study these reactions use a freshly prepared 0·1M solution of sodium hexafluorosilicate.

1. Concentrated sulphuric acid: silicon tetrafluoride and hydrogen fluoride are evolved on warming the reagent with the solid salt. If the reaction is carried out in a platinum or lead capsule or crucible, the escaping gas will etch glass and will cause a drop of water to become turbid (see under Silicates, Section **IV.26**, reaction 6).

$$Na_2[SiF_6] + H_2SO_4 \rightarrow SiF_4\uparrow + H_2F_2\uparrow + 2Na^+ + SO_4^{2-}$$

2. Barium chloride solution: white, crystalline precipitate of barium hexafluorosilicate, $Ba[SiF_6]$, sparingly soluble in water (0·25 g ℓ^{-1} at 25°) and insoluble in dilute hydrochloric acid. The precipitate is distinguished from barium sulphate by the evolution of hydrogen fluoride and silicon fluoride, which etch glass, on heating with concentrated sulphuric acid in a lead crucible.

$$[SiF_6]^{2-} + Ba^{2+} \rightarrow Ba[SiF_6]\downarrow$$

3. *Potassium chloride solution:* white, gelatinous precipitate of potassium hexafluorosilicate, $K_2[SiF_6]$, from concentrated solutions. The precipitate is slightly soluble in water (1·77 g ℓ^{-1} at 25°), less soluble in excess of the reagent and in 50 per cent alcohol).

$$[SiF_6]^{2-} + 2K^+ \rightarrow K_2[SiF_6]\downarrow$$

4. *Ammonia solution:* decomposition occurs with the separation of gelatinous silicic acid.

$$[SiF_6]^{2-} + 4NH_3 + 3H_2O \rightarrow H_2SiO_3\downarrow + 6F^- + 4NH_4^+$$

5. *Action of heat:* decomposition occurs into silicon tetrafluoride, which renders a drop of water turbid, and the metallic fluoride, which can be tested for in the usual manner (see under Fluorides, Section **IV.17**).

$$Na_2[SiF_6] \rightarrow SiF_4\uparrow + 2Na^+ + 2F^-$$

IV.28 ORTHOPHOSPHATES, PO_4^{3-} Three phosphoric acids are known: orthophosphoric, H_3PO_4; pyrophosphoric, $H_4P_2O_7$; and metaphosphoric acid, HPO_3. Salts of the three acids exist: the orthophosphates are the most stable and the most important;* solutions of pyro- and metaphosphates pass into orthophosphates slowly at the ordinary temperature, and more rapidly on boiling. Metaphosphates, unless prepared by special methods, are usually polymeric, i.e. are derived from $(HPO_3)_n$.

Orthophosphoric acid is a tribasic acid giving rise to three series of salts: primary orthophosphates, e.g. NaH_2PO_4; secondary orthophosphates, e.g. Na_2HPO_4; and tertiary orthophosphates, e.g. Na_3PO_4. If a solution of orthophosphoric acid is neutralized with sodium hydroxide solution using methyl orange as indicator, the neutral point is reached when the acid is converted into the primary phosphate; with phenolphthalein as indicator, the solution will react neutral when the secondary phosphate is formed; with 3 moles of alkali, the tertiary or normal phosphate is formed. NaH_2PO_4 is neutral to methyl orange and acid to phenolphthalein, Na_2HPO_4 is neutral to phenolphthalein and alkaline to methyl orange, Na_3PO_4 is alkaline to most indicators because of its extended hydrolysis. Ordinary 'sodium phosphate' is disodium hydrogen phosphate, $Na_2HPO_4.12H_2O$.

Solubility The phosphates of the alkali metals, with the exception of lithium, and of ammonium are soluble in water; the primary phosphates of the alkaline earth metals are also soluble. All the phosphates of the other metals, and also the secondary and tertiary phosphates of the alkaline earth metals are sparingly soluble or insoluble in water.

To study these reactions use a 0·033M solution of disodium hydrogen phosphate $Na_2HPO_4.12H_2O$.

---

* They are often referred to simply as phosphates.

1. Silver nitrate solution: yellow precipitate of normal silver orthophosphate, Ag_3PO_4 (distinction from meta- and pyrophosphate), soluble in dilute ammonia solution and in dilute nitric acid.

$$HPO_4^{2-} + 3Ag^+ \rightarrow Ag_3PO_4\downarrow + H^+$$
$$Ag_3PO_4\downarrow + 2H^+ \rightarrow H_2PO_4^- + 3Ag^+$$
$$Ag_3PO_4\downarrow + 6NH_3 \rightarrow 3[Ag(NH_3)_2]^+ + PO_4^{3-}$$

2. Barium chloride solution: white, amorphous precipitate of secondary barium phosphate, $BaHPO_4$, from neutral solutions, soluble in dilute mineral acids and in acetic acid. In the presence of dilute ammonia solution, the less soluble tertiary phosphate, $Ba_3(PO_4)_2$, is precipitated.

$$HPO_4^{2-} + Ba^{2+} \rightarrow BaHPO_4\downarrow$$
$$2HPO_4^{2-} + 3Ba^{2+} + 2NH_3 \rightarrow Ba_3(PO_4)_2\downarrow + 2NH_4^+$$

3. Magnesium nitrate reagent or magnesia mixture The former is a solution containing $Mg(NO_3)_2$, NH_4NO_3, and a little aqueous NH_3, and the latter is a solution containing $MgCl_2$,NH_4Cl, and a little aqueous NH_3. The magnesium nitrate reagent is generally preferred since it may be employed in any subsequent test with silver nitrate solution. With either reagent a white, crystalline precipitate of magnesium ammonium phosphate, $Mg(NH_4)PO_4.6H_2O$, is produced: this precipitate is soluble in acetic acid and in mineral acids, but practically insoluble in 2·5 per cent ammonia solution (see under Magnesium, Section **III.35**, reaction 5; also under Arsenic(V), Section **III.13**, reaction 3).

$$HPO_4^{2-} + Mg^{2+} + NH_3 \rightarrow MgNH_4PO_4\downarrow$$

Arsenates give a similar precipitate $Mg(NH_4)AsO_4.6H_2O$ with either reagent. They are most simply distinguished from one another by treating the washed precipitate with silver nitrate solution containing a few drops of dilute acetic acid: the phosphate turns yellow (Ag_3PO_4) whilst the arsenate assumes a brownish-red colour (Ag_3AsO_4).

4. Ammonium molybdate reagent The addition of a large excess (2–3 ml) of this reagent to a small volume (0·5 ml) of a phosphate solution produces a yellow, crystalline precipitate of ammonium phosphomolybdate, to which the formula $(NH_4)_3PO_4.12MoO_3$ was formerly assigned. The correct formula is $(NH_4)_3[PMo_{12}O_{40}]$ or $(NH_4)_3[P(Mo_3O_{10})_4]$, the Mo_3O_{10} group replacing each oxygen atom in phosphate. The resulting solution should be strongly acid with nitric acid; the latter is usually present in the reagent and addition is therefore unnecessary. Precipitation is accelerated by warming to a temperature not exceeding 40°, and by the addition of ammonium nitrate solution.

The precipitate is soluble in ammonia solution and in solutions of caustic alkalis. Large quantities of hydrochloric acid interfere with the test and should preferably be removed by evaporation to a small volume with excess concentrated nitric acid. Reducing agents, such as sulphides, sulphites, hexacyanoferrate(II)s, and tartrates, seriously affect the reaction, and should be destroyed before carrying out the test.

Arsenates give a similar reaction on boiling (see under Arsenic, Section **III,13**, reaction 4). Both ammonium phosphomolybdate and ammonium arsenomolybdate dissolve on boiling with ammonium acetate solution, but only the latter yields a white precipitate on cooling.

$$HPO_4^{2-} + 3NH_4^+ + 12MoO_4^{2-} + 23H^+ \rightarrow (NH_4)_3[P(Mo_3O_{10})_4]\downarrow + 12H_2O$$

Note: Commercial ammonium molybdate has the formula $(NH_4)_6Mo_7O_{24}.4H_2O$ (a paramolybdate) and not $(NH_4)_2MoO_4$; but the ion MoO_4^{2-} is employed in the equations for purposes of simplicity. These ions may exist under the experimental conditions of the reaction.

5. Iron(III) chloride solution: yellowish-white precipitate of iron(III) phosphate, $FePO_4$, soluble in dilute mineral acids, but insoluble in dilute acetic acid.

$$HPO_4^{2-} + Fe^{3+} \rightarrow FePO_4\downarrow + H^+$$

Precipitation is incomplete owing to the free mineral acid produced. If the hydrogen ions, arising from the complete ionization of the mineral acid, are removed by the addition of the salt of a weak acid, such as ammonium or sodium acetate, precipitation is almost complete. This is the basis of one of the methods for the removal of phosphates, which interfere with the precipitation of Group IIIA metals, in qualitative analysis.

6. Zirconium nitrate reagent When the reagent is added to a solution of a phosphate containing hydrochloric acid not exceeding M in concentration, a white gelatinous precipitate of zirconyl phosphate $ZrO(HPO_4)$ is obtained. This reaction forms the basis of a simple method for the removal of phosphate prior to the precipitation of Group IIIA (cf. Section **V.13**, 4).

$$HPO_4^{2-} + ZrO^{2+} \rightarrow ZrO(HPO_4)\downarrow$$

In fact the precipitate is of variable composition, depending on the concentrations of zirconium, phosphate, and hydrogen ions. Species like $ZrO(H_2PO_4)_2$ and $ZrPO_4$ may also be formed.

7. Cobalt nitrate test Phosphates when heated on charcoal and then moistened with a few drops of cobalt nitrate solution, give a blue mass of the phosphate, $NaCoPO_4$. This must not be confused with the blue mass produced with aluminium compounds (see Section **III.23**).

$$Na(NH)_4\ HPO_4 \rightarrow NaPO_3 + NH_3\uparrow + H_2O;$$
$$NaPO_3 + CoO \rightarrow NaCoPO_4$$

8. Magnesium The only simple method for reducing the stable phosphates consists in heating with magnesium powder, whereby a phosphide is produced. The latter is readily identified by the odour and the inflammability of the phosphine formed on the addition of water. Intimately mix a small quantity of sodium phosphate with magnesium powder and heat in an ignition tube. Moisten the cold mass with water and observe the unpleasant odour of phosphine.

$$Na_3PO_4 + 4Mg \rightarrow 4MgO + Na_3P;$$
$$Na_3P + 3H_2O \rightarrow PH_3\uparrow + 3Na^+ + 3OH^-$$

9. Ammonium molybdate–benzidine test (DANGER: THE REAGENT IS

CARCINOGENIC.) In this test use is made of the fact that benzidine, which is unaffected by normal molybdates and by free molybdic acid, is oxidized in acetic acid solution by phosphomolybdic acid or by its insoluble ammonium salt (see reaction 4 above). This reaction is extremely sensitive; two coloured products are formed, viz. the blue reduction product of molybdenum compounds ('molybdenum blue') and the blue oxidation product of benzidine ('benzidine blue'). Moreover, solutions of phosphates which are too dilute to show a visible precipitate with the ammonium molybdate reagent will react with the molybdate reagent and benzidine to give a blue colouration.

Arsenates and silicates with ammonium molybdate yield the ammonium salts of arsenomolybdic, $H_3[AsMo_{12}O_{40}]$ or $H_3[As(Mo_3O_{10})_4]$, and silicomolybdic, $H_4[SiMo_{12}O_{40}]$ or $H_4[Si(Mo_3O_{10})_4]$, acids respectively; these complex acids and their salts react similarly with benzidine. However, phosphates may be detected in the presence of arsenates and silicates by preventing the formation of the corresponding molybdo-acids by the use of a tartaric acid–ammonium molybdate reagent which does not react with arsenic and silicic acids but does react with phosphoric acid when the reaction is carried out on filter paper.

Hydrogen peroxide, oxalates, and fluorides interfere with the precipitation of the phosphomolybdate and should therefore be absent.

Place a drop of the acid solution under test upon quantitative filter paper, add a drop of the ammonium molybdate reagent, followed by a drop of 0·05 per cent benzidine acetate. Hold the paper over ammonia vapour. A blue stain is formed when most of the mineral acid has been neutralized.

Sensitivity: 1·25 μg P_2O_5. *Concentration limit:* 1 in 40,000.

In the presence of silicates and/or arsenates, proceed as follows. Place a drop of the test solution upon quantitative filter paper, followed by a drop of the tartaric acid–ammonium molybdate reagent. Hold the paper over a hot, wire gauze (or over a sheet of asbestos heated on a hot plate) to accelerate the reaction. Then add a drop of the benzidine reagent and develop over ammonia vapour. A blue colouration results.

Sensitivity: 1·5 μg P_2O_5 in the presence of 500 times that amount of SiO_2. *Concentration limit:* 1 in 50,000.

The tartaric acid–ammonium molybdate reagent is prepared by dissolving 15 g crystallized tartaric acid in 100 ml ammonium molybdate reagent.

10. Ammonium molybdate–quinine sulphate reagent Phosphates give a yellow precipitate with this reagent: the exact composition appears to be unknown.

Reducing agents (sulphides, thiosulphates, etc.) interfere since they yield 'molybdenum blue'; hexacyanoferrate(II) ions give a red colouration. Arsenates (warming is usually required), arsenites, chromates, oxalates, tartrates, and silicates give a similar reaction with some variation in the colour of the precipitate. All should be removed before applying the test.

Place 1 ml test solution in a semimicro test-tube and add 1 ml reagent. A yellow precipitate is produced within a few minutes; gentle warming (water bath) is sometimes necessary.

Concentration limit: 1 in 20,000.

The reagent is prepared by dissolving 4·0 g finely powdered ammonium molybdate, $(NH_4)_6Mo_7O_{24}.4H_2O$ in 20 ml water and adding, with stirring, a solution of 0·1 g quinine sulphate in 80 ml concentrated nitric acid.

IV.29 PYROPHOSPHATES, $P_2O_7^{4-}$, AND METAPHOSPHATES, PO_3^-

Sodium pyrophosphate is prepared by heating disodium hydrogen phosphate:

$$2Na_2HPO_4 \rightarrow Na_4P_2O_7 + H_2O\uparrow$$

This is the normal salt. Acid salts, e.g. $Na_2H_2P_2O_7$, are known.

Sodium metaphosphate (polymeric) may be prepared by heating microcosmic salt or sodium dihydrogen phosphate:

$$Na(NH_4)HPO_4 \rightarrow NaPO_3 + NH_3\uparrow + H_2O\uparrow$$
$$NaH_2PO_4 \rightarrow NaPO_3 + H_2O\uparrow$$

A number of metaphosphates are known and these may be regarded as derived from the polymeric acid, $(HPO_3)_n$, i.e. polymetaphosphoric acid. *Calgon*, used for water softening, is probably $(NaPO_3)_6$ or $Na_2[Na_4(PO_3)_6]$.

Pyro- and metaphosphates give the ammonium molybdate test on warming for some time; this is doubtless due to their initial conversion in solution into orthophosphates. The chief differences between ortho-, pyro- and meta-phosphates are incorporated in the following table. The student should carry out all the tests. To study these reactions use freshly prepared solutions of sodium pyrophosphate and sodium metaphosphate.

Table IV.1 Reactions of ortho-, pyro- and metaphosphates

Reagent	Orthophosphate	Pyrophosphate	Metaphosphate
1. Silver nitrate solution.	Yellow ppt., soluble in dilute HNO_3 and in dilute NH_3 solution.	White ppt., soluble in dilute HNO_3 and in dilute NH_3 solution: sparingly soluble in dilute acetic acid.	White ppt. (separates slowly), soluble in dilute HNO_3, in dilute NH_3 solution and in dilute acetic acid.
2. Albumin and dilute acetic acid.	No coagulation.	No coagulation.	Coagulation.
3. Copper sulphate solution.	Pale blue ppt.	Very pale blue ppt.	No ppt.
4. Magnesia mixture or $Mg(NO_3)_2$ reagent.	White ppt., insoluble in excess of the reagent.	White ppt., soluble in excess of reagent, but reprecipitated on boiling.	No ppt., even on boiling.
5. Cadmium acetate solution and dilute acetic acid.	No ppt.	White ppt.	No ppt.
6. Zinc sulphate solution.	White ppt., soluble in dilute acetic acid.	White ppt., insoluble in dilute acetic acid; soluble in dilute NH_3 solution, yielding a white ppt. on boiling.	White ppt., on warming; soluble in dilute acetic acid.

IV.30 PHOSPHITES, HPO_3^{2-} *Solubility* The phosphites of the alkali metals are soluble in water; all other metallic phosphites are insoluble in water.

To study these reactions use a freshly prepared 0·1M solution of sodium phosphite $Na_2HPO_3.5H_2O$.

1. Silver nitrate solution: white precipitate of silver phosphite, Ag_2HPO_3, which soon passes in the cold into black metallic silver. Warming is necessary with dilute solutions. Upon adding the reagent to a warm solution of a phosphite, a black precipitate of metallic silver is obtained immediately.

$$HPO_3^{2-} + 2Ag^+ \rightarrow Ag_2HPO_3\downarrow$$
$$Ag_2HPO_3 + H_2O \rightarrow 2Ag\downarrow + H_3PO_4$$
$$HPO_3^{2-} + 2Ag^+ + H_2O \rightarrow 2Ag\downarrow + H_3PO_4$$

2. Barium chloride solution: white precipitate of barium phosphite, $BaHPO_3$, soluble in dilute acids.

$$HPO_3^{2-} + Ba2^+ \rightarrow BaHPO_3\downarrow$$

3. Mercury(II) chloride solution: white precipitate of calomel in the cold; on warming with excess of the phosphite solution, grey, metallic mercury is produced.

$$HPO_3^{2-} + 2HgCl_2 + H_2O \rightarrow Hg_2Cl_2\downarrow + H_3PO_4 + 2Cl^-$$
$$Hg_2Cl_2\downarrow + HPO_3^{2-} + H_2O \rightarrow 2Hg\downarrow + H_3PO_4 + 2Cl^-$$

Note that $HgCl_2$ and H_3PO_4 are practically undissociated.

4. Potassium permanganate solution: no action in the cold with a solution acidified with acetic acid, but decolourized on warming.

5. Concentrated sulphuric acid: no reaction in the cold with the solid salt, but on warming sulphur dioxide is evolved.

$$Na_2HPO_3 + H_2SO_4 \rightarrow SO_2\uparrow + 2Na^+ + HPO_4^{2-} + H_2O$$
$$H_3PO_3 + H_2SO_4 \rightarrow SO_2\uparrow + H_3PO_4 + H_2O$$

6. Zinc and dilute sulphuric acid Phosphites are reduced by the nascent hydrogen to phosphine, PH_3, which may be identified as described in the Gutzeit test under Arsenic (Section **III.14**); the silver nitrate paper is stained first yellow and then black.

$$HPO_3^{2-} + 3Zn + 8H^+ \rightarrow PH_3\uparrow + 3Zn^{2+} + 3H_2O$$
$$PH_3\uparrow + 6Ag^+ + 3NO_3^- \rightarrow Ag_3P.3AgNO_3\downarrow(\text{yellow}) + 3H^+$$
$$Ag_3P.3AgNO_3\downarrow + 3H_2O \rightarrow 6Ag\downarrow(\text{black}) + H_3PO_3 + 3H^+ + 3NO_3^-$$

7. Copper sulphate solution: light-blue precipitate of copper phosphite, $CuHPO_3$; the precipitate merely dissolves when it is boiled with acetic acid (cf. Hypophosphites, Section **IV.31**).

$$HPO_3^{2-} + Cu^{2+} \rightarrow CuHPO_3\downarrow$$

8. Lead acetate solution: white precipitate, insoluble in acetic acid.

$$HPO_3^{2-} + Pb^{2+} \rightarrow PbHPO_3\downarrow$$

9. Action of heat: inflammable phosphine is evolved, and a mixture of phosphates is produced.

$$8Na_2HPO_3 \rightarrow 2PH_3\uparrow + 4Na_3PO_4 + Na_4P_2O_7 + H_2O$$

IV.31 HYPOPHOSPHITES, $H_2PO_2^-$ *Solubility* All hypophosphites are soluble in water.

To study these reactions use a freshly prepared 0·1M solution of sodium hypophosphite $NaH_2PO_2.H_2O$.

1. Silver nitrate solution: white precipitate of silver hypophosphite, AgH_2PO_2, which is slowly reduced to silver at the ordinary temperature, but more rapidly on warming, hydrogen being simultaneously evolved.

$$H_2PO_2^- + Ag^+ \rightarrow AgH_2PO_2\downarrow$$
$$2AgH_2PO_2\downarrow + 4H_2O \rightarrow 2Ag\downarrow + 2H_3PO_4 + 3H_2\uparrow$$

2. Barium chloride solution: no precipitate.

3. Mercury(II) chloride solution: white precipitate of calomel in the cold, converted by warming into grey, metallic mercury.

$$H_2PO_2^- + 4HgCl_2 + 2H_2O \rightarrow 2Hg_2Cl_2\downarrow + H_3PO_4 + 3H^+ + 4Cl^-$$
$$H_2PO_2^- + 2Hg_2Cl_2\downarrow + 2H_2O \rightarrow 4Hg\downarrow + H_3PO_4 + 3H^+ + 4Cl^-$$

4. Copper sulphate solution: no precipitate in the cold, but on warming red, copper(I) hydride is precipitated.

$$3H_2PO_2^- + 4Cu^{2+} + 6H_2O \rightarrow 4CuH\downarrow + 3H_3PO_4 + 5H^+$$

When heated with concentrated hydrochloric acid, the precipitate decomposes to white copper(I) chloride precipitate and hydrogen gas.

$$CuH\downarrow + HCl \rightarrow CuCl\downarrow + H_2$$

5. Potassium permanganate solution: reduced to colourless immediately in the cold.

$$5H_2PO_2^- + 4MnO_4^- + 17H^+ \rightarrow 5H_3PO_4 + 4Mn^{2+} + 6H_2O$$

6. Concentrated sulphuric acid: reduced to sulphur dioxide by the solid salt on warming.

$$NaH_2PO_2 + 4H_2SO_4 \rightarrow 2SO_2\uparrow + H_3PO_4 + 3H^+ + Na^+ + 2SO_4^{2-} + 2H_2O$$

Some elementary sulphur may also be formed during the process.

7. Concentrated sodium hydroxide solution: hydrogen gas is evolved and phosphate ions formed on warming.

$$H_2PO_2^- + 2OH^- \rightarrow 2H_2\uparrow + PO_4^{3-}$$

8. Zinc and dilute sulphuric acid: inflammable phosphine is evolved (see under Phosphites, Section **IV.30**, reaction 6).

9. Ammonium molybdate solution: reduced to 'molybdenum blue' in solution acidified with dilute sulphuric acid (difference from phosphite).

10. Action of heat: phosphine is evolved, and a pyrophosphate is produced.

$$4NaH_2PO_2 \rightarrow Na_4P_2O_7 + 2PH_3\uparrow + H_2O$$

IV.32 ARSENITES, AsO_3^{3-}, AND ARSENATES, AsO_4^{3-} See under Arsenic, Section **III.12** and **III.13**.

IV.33 CHROMATES, CrO_4^{2-}, AND DICHROMATES, $Cr_2O_7^{2-}$ The metallic chromates are usually coloured solids, yielding yellow solutions when soluble in water. In the presence of dilute mineral acids, i.e. of hydrogen ions, chromates are converted into dichromates; the latter yield orange-red aqueous solutions. The change is reversed by alkalis, i.e. by hydroxyl ions.

$$2CrO_4^{2-} + 2H^+ \rightleftarrows Cr_2O_7^{2-} + H_2O$$

or

$$Cr_2O_7^{2-} + 2OH^- \rightleftarrows 2CrO_4^{2-} + H_2O$$

The reactions may also be expressed as:

$$2CrO_4^{2-} + 2H^+ \rightleftarrows 2HCrO_4^- \rightleftarrows Cr_2O_7^{2-} + H_2O$$

Solubility The chromates of the alkali metals and of calcium and magnesium are soluble in water; strontium chromate is sparingly soluble. Most other metallic chromates are insoluble in water. Sodium, potassium, and ammonium dichromates are soluble in water.

To study these reactions use a 0·1M solution of potassium chromate, K_2CrO_4, or potassium dichromate, $K_2Cr_2O_7$.

1. Barium chloride solution: pale-yellow precipitate of barium chromate, $BaCrO_4$, insoluble in water and in acetic acid, but soluble in dilute mineral acids (for explanation, see Section **I.28**).

$$CrO_4^{2-} + Ba^{2+} \rightarrow BaCrO_4\downarrow$$

Dichromate ions produce the same precipitate, but as a strong acid is formed, precipitation is only partial:

$$Cr_2O_7^{2-} + 2Ba^{2+} + H_2O \rightleftarrows 2BaCrO_4\downarrow + 2H^+$$

If sodium hydroxide or sodium acetate is added, precipitation becomes quantitative.

2. Silver nitrate solution: brownish-red precipitate of silver chromate, Ag_2CrO_4, with a solution of a chromate. The precipitate is soluble in dilute nitric acid and in ammonia solution, but is insoluble in acetic acid. Hydrochloric acid converts the precipitate into silver chloride (white).

$$CrO_4^{2-} + 2Ag^+ \rightarrow Ag_2CrO_4\downarrow$$
$$2Ag_2CrO_4\downarrow + 2H^+ \rightarrow 4Ag^+ + Cr_2O_7^{2-} + H_2O$$
$$Ag_2CrO_4\downarrow + 4NH_3 \rightarrow 2[Ag(NH_3)_2]^+ + CrO_4^{2-}$$
$$Ag_2CrO_4\downarrow + 2Cl^- \rightarrow 2AgCl\downarrow + CrO_4^{2-}$$

A reddish-brown precipitate of silver dichromate, $Ag_2Cr_2O_7$, is formed with a concentrated solution of a dichromate; this passes, on boiling with water, into the less soluble silver chromate.

$$Cr_2O_7^{2-} + 2Ag^+ \rightarrow Ag_2Cr_2O_7\downarrow$$
$$Ag_2Cr_2O_7^{2-} + H_2O \rightarrow Ag_2CrO_4\downarrow + CrO_4^{2-} + 2H^+$$

3. Lead acetate solution: yellow precipitate of lead chromate, $PbCrO_4$, insoluble in acetic acid, but soluble in dilute nitric acid.

$$CrO_4^{2-} + Pb^{2+} \rightarrow PbCrO_4\downarrow$$
$$2PbCrO_4\downarrow + 2H^+ \rightleftarrows 2Pb^{2+} + Cr_2O_7^{2-} + H_2O$$

The precipitate is soluble in sodium hydroxide solution; acetic acid reprecipitates lead chromate from such solutions.

The solubility in sodium hydroxide solution is due to the formation of the soluble complex tetrahydroxoplumbate(II) ion, which suppresses the Pb^{2+} ion concentration to such an extent that the solubility product of lead chromate is no longer exceeded, and consequently the latter dissolves (cf. Section **I.28**).

$$PbCrO_4\downarrow + 4OH^- \rightleftarrows [Pb(OH)_4]^{2-} + CrO_4^{2-}$$

4. Hydrogen peroxide If an acid solution of a chromate is treated with hydrogen peroxide, a deep-blue solution of chromium pentoxide is obtained (cf. Chromium, Section **III.24**, reaction 8*b*). The blue solution is very unstable and soon decomposes, yielding oxygen and a green solution of a chromium(III) salt. The blue compound is soluble in amyl alcohol and also in amyl acetate and in diethyl ether, and can be extracted from aqueous solutions by these solvents to yield somewhat more stable solutions. Amyl alcohol is recommended. Diethyl ether is not recommended for general student use owing to its highly inflammable character and also because it frequently contains peroxides after storage for comparatively short periods: which may interfere. Peroxides may be removed from diethyl ether by shaking with a concentrated solution of iron(II) salt or with sodium sulphite.

The blue colouration is attributed to the presence of chromium pentoxide, CrO_5:

$$CrO_4^{2-} + 2H^+ + 2H_2O_2 \rightarrow CrO_5 + 3H_2O$$

The enhanced stability in solutions of amyl alcohol, ether, etc., is due to the formation of complexes with these oxygen-containing compounds.

Just acidify a cold solution of a chromate with dilute sulphuric acid or dilute nitric acid, add 1–2 ml amyl alcohol, then 1 ml of 10-volume (3 per cent) hydrogen peroxide solution dropwise and with shaking after each addition: the organic layer is coloured blue. The chromium pentoxide is more stable below 0°C than at laboratory temperature.

5. Hydrogen sulphide An acid solution of a chromate is reduced by this reagent to a green solution of chromium(III) ions, accompanied by the separation of sulphur.*

$$2CrO_4^{2-} + 3H_2S + 10H^+ \rightarrow 2Cr^{3+} + 3S\downarrow + 8H_2O$$

---

* In qualitative analysis the production of sulphur in the reduction of chromates may be troublesome, because the precipitate is almost colloidal and can hardly be filtered. On adding a few strips of filter paper to the mixture and boiling, the precipitate coagulates and can be filtered easily. Alternatively, before the use of hydrogen sulphide, chromates can be reduced by heating the solid substance with concentrated hydrochloric acid, evaporating most of the acid and then diluting with water. With potassium dichromate the reaction is as follows:

$$K_2Cr_2O_7 + 14HCl \rightarrow 2Cr^{3+} + 3Cl_2\uparrow + 2K^+ + 8Cl^- + 7H_2O$$

Chlorine gas must be removed from the mixture; this is done during evaporation.

6. Reduction of chromates and dichromates (*a*) Sulphur dioxide, in the presence of dilute mineral acid, reduces chromates or dichromates:

$$2CrO_4^{2-} + 3SO_2 + 4H^+ \rightarrow 2Cr^{3+} + 3SO_4^{2-} + 2H_2O$$

The solution becomes green owing to the formation of chromium(III) ions.

(*b*) Potassium iodide, in the presence of dilute mineral acid, can be used as a reductant. The colour of the solution becomes brown or bluish, according to the amounts of iodine and chromium(III) ions, which are formed in the reaction:

$$2CrO_4^{2-} + 6I^- + 16H^+ \rightarrow 2Cr^{3+} + 3I_2 + 8H_2O$$

(*c*) Iron(II) sulphate, in the presence of mineral acid, reduces chromates or dichromates smoothly:

$$CrO_4^{2-} + 3Fe^{2+} + 8H^+ \rightarrow Cr^{3+} + 3Fe^{3+} + 4H_2O$$

(*d*) Ethanol, again in the presence of mineral acid, reduces chromates or dichromates slowly in cold, but rapidly if the solution is heated:

$$2CrO_4^{2-} + 3C_2H_5OH + 10H^+ \rightarrow 2Cr^{3+} + 3CH_3CHO\uparrow + 8H_2O$$

The smell of acetaldehyde (CH_3CHO) is easily observed if the mixture is heated.

7. Reduction of solid chromates or dichromates with concentrated hydrochloric acid On heating a solid chromate or dichromate with concentrated hydrochloric acid, chlorine is evolved, and a solution containing chromium(III) ions is produced:

$$2K_2CrO_4 + 16HCl \rightarrow 2Cr^{3+} + 3Cl_2\uparrow + 4K^+ + 10Cl^- + 8H_2O$$
$$K_2Cr_2O_7 + 14HCl \rightarrow 2Cr^{3+} + 3Cl_2\uparrow + 2K^+ + 8Cl^- + 7H_2O$$

If the solution is evaporated almost to dryness (to remove chlorine), after dilution chromium(III) ions can be tested in the solution (cf. Section **III.24**).

8. Concentrated sulphuric acid and a chloride See chromyl chloride test under Chlorides, Section **IV.14**, reaction 5; also Section **IV.45**, 5.

9. Diphenylcarbazide reagent The solution is acidified with dilute sulphuric acid or with dilute acetic acid, and 1–2 ml of the reagent added. A deep-red colouration is produced. With small quantities of chromates, the solution is coloured violet.

Full details of the use of the reagent as a spot test are given under Chromium, Section **III.24**, reaction 8**c**.

Easier ways to reduce chromates are also available. Thus, when warming with hydrochloric acid and alcohol (a) or hydrochloric acid and 10 per cent formaldehyde solution (b) the reduction is achieved:

$$2CrO_4^{2-} + 3C_2H_5OH + 10H^+ \rightarrow 2Cr^{3+} + 3CH_3CHO + 8H_2O \quad \text{(a)}$$
$$2CrO_4^{2-} + 3HCHO + 10H^+ \rightarrow 2Cr^{3+} + 3HCOOH + 5H_2O \quad \text{(b)}$$

In reaction (a) the acetaldehyde, which is formed, may reduce some chromate and form acetic acid. The reaction is analogous to (b).

The use of sulphur dioxide is not recommended, as sulphuric acid is formed, and this will precipitate lead, strontium, and barium if these metals are present.

10. Chromotropic acid test A red colouration, best seen by transmitted light, is given by chromates. For details, see under Chromium, Section **III.24**, reaction 8**d**.

IV.34 PERMANGANATES, MnO_4^- *Solubility* All permanganates are soluble in water, forming purple (reddish-violet) solutions.

To study these reactions use a 0·02M solution of potassium permanganate, $KMnO_4$.

1. Hydrogen peroxide The addition of this reagent to a solution of potassium permanganate, acidified with dilute sulphuric acid, results in decolourization and the evolution of pure but moist oxygen.

$$2MnO_4^- + 5H_2O_2 + 6H^+ \rightarrow 5O_2\uparrow + 2Mn^{2+} + 8H_2O$$

2. Reduction of permanganates In acid solutions the reduction proceeds to the formation of colourless manganese(II) ions. The following reducing agents may be used:

(*a*) Hydrogen sulphide: in the presence of dilute sulphuric acid the solution decolourizes and sulphur is precipitated.*

$$2MnO_4^- + 5H_2S + 6H^+ \rightarrow 5S\downarrow + 2Mn^{2+} + 8H_2O$$

(*b*) Sulphur dioxide, in the presence of sulphuric acid, reduces permanganate instantaneously:

$$2MnO_4^- + 5SO_2 + 2H_2O \rightarrow 5SO_4^{2-} + 2Mn^{2+} + 4H^+$$

Some dithionate ($S_2O_6^{2-}$) may also be formed in this reaction, the quantity being dependent upon the experimental conditions:

$$2MnO_4^- + 6SO_2 + 2H_2O \rightarrow 2Mn^{2+} + S_2O_6^{2-} + 4SO_4^{2-} + 4H^+$$

(*c*) Iron(II) sulphate, in the presence of sulphuric acid, reduces permanganate to manganese(II). The solution becomes yellow because of the formation of iron(III) ions.

$$MnO_4^- + 5Fe^{2+} + 8H^+ \rightarrow 5Fe^{3+} + Mn^{2+} + 4H_2O$$

The yellow colour disappears if phosphoric acid or potassium fluoride is added; they form colourless complexes with iron(III).

(*d*) Potassium iodide, in the presence of sulphuric acid, reduces permanganate with the formation of iodine:

$$2MnO_4^- + 10I^- + 16H^+ \rightarrow 5I_2 + 2Mn^{2+} + 8H_2O$$

(*e*) Sodium nitrite, in the presence of sulphuric acid, reduces permanganate with formation of nitrate ions:

$$2MnO_4^- + 5NO_2^- + 6H^+ \rightarrow 2Mn^{2+} + 5NO_3^- + 3H_2O$$

---

* To avoid the production of sulphur by the reduction of potassium permanganate by hydrogen sulphide in systematic qualitative analysis, the solution may be reduced with formaldehyde solution, or the solid substance may be boiled with concentrated hydrochloric acid. The use of sulphur dioxide is not recommended (see footnote in connection with Chromates, Section **IV.33**, reaction 5).

$$2MnO_4^- + 5HCHO + 6H^+ \rightarrow 2Mn^{2+} + 5HCOOH + 3H_2O$$

Some side reactions also take place, and the mixture has the smell of nitrogen oxide gas.

(*f*) Oxalic acid, in the presence of sulphuric acid, produces carbon dioxide gas:

$$2MnO_4^- + 5(COO)_2^{2-} + 16H^+ \rightarrow 10CO_2\uparrow + 2Mn^{2+} + 8H_2O$$

This reaction is slow at room temperature, but becomes fast at 60°C. Manganese(II) ions catalyse the reaction; thus, the reaction is **autocatalytic**; once manganese(II) ions are formed, it becomes faster and faster.

In alkaline solution, the permanganate is decolourized, but manganese dioxide is precipitated. In the presence of sodium hydroxide solution, potassium iodide is converted into potassium iodate, and sodium sulphite solution into sodium sulphate on boiling.

$$2MnO_4^- + I^- + H_2O \rightarrow 2MnO_2\downarrow + IO_3^- + 2OH^-$$
$$2MnO_4^- + 3SO_3^{2-} + H_2O \rightarrow 2MnO_2\downarrow + 3SO_4^{2-} + 2OH^-$$

These reactions indicate that permanganate oxidizes more strongly in alkaline medium than in acidic solutions (cf. Table I.18).

3. Concentrated hydrochloric acid All permanganates on boiling with concentrated hydrochloric acid evolve chlorine.

$$2MnO_4^- + 16HCl \rightarrow 5Cl_2\uparrow + 2Mn^{2+} + 6Cl^- + 8H_2O$$

4. Concentrated sulphuric acid (**GREAT DANGER**) Permanganates dissolve in this reagent to yield a green solution, which contains manganese heptoxide (permanganic anhydride), Mn_2O_7; the solution is liable to explode spontaneously at the ordinary temperature, and a very vigorous explosion may result on warming. The student is therefore warned not to carry out this experiment except with minute quantities of materials (not more than 0·05 g) and then only under the direct supervision of the teacher. The experiment is best omitted.

$$2KMnO_4 + H_2SO_4 \rightarrow Mn_2O_7 + 2K^+ + SO_4^{2-} + H_2O$$

5. Sodium hydroxide solution Upon warming a concentrated solution of potassium permanganate with concentrated sodium hydroxide solution, a green solution of potassium manganate is produced and oxygen is evolved. When the manganate solution is poured into a large volume of water or is acidified with dilute sulphuric acid, the purple colour of the potassium permanganate is restored, and manganese dioxide is precipitated.

$$4MnO_4^- + 4OH^- \rightarrow 4MnO_4^{2-} + O_2\uparrow + 2H_2O$$
$$3MnO_4^{2-} + 2H_2O \rightarrow 2MnO_4^- + MnO_2\downarrow + 4OH^-$$

The latter reaction is in fact a **disproportionation**, when manganese(VI) is partly oxidized to manganese(VII) and partly reduced to manganese(IV).

A manganate is produced when a manganese compound is fused with potassium nitrate and sodium carbonate (see Section **III.28**, Dry tests).

6. Action of heat When potassium permanganate is heated in a test-tube, pure oxygen is evolved, and a black residue of potassium manganate K_2MnO_4 and manganese dioxide remains behind. Upon extracting with a little water and filtering, a green solution of potassium manganate is obtained.

$$2KMnO_4 \rightarrow K_2MnO_4 + MnO_2 + O_2\uparrow$$

IV.35 ACETATES, CH_3COO^- *Solubility* All normal acetates, with the exception of silver and mercury(I) acetates which are sparingly soluble, are readily soluble in water. Some basic acetates, e.g. those of iron, aluminium, and chromium, are insoluble in water. The free acid, CH_2COOH, is a colourless liquid with a pungent odour, boiling point 117°, melting point 17° and is miscible with water in all proportions; it has a corrosive action on the skin.

To study these reactions use a 2M solution of sodium acetate $CH_3COONa.3H_2O$.

1. Dilute sulphuric acid: acetic acid, easily recognized by its vinegar-like odour, is evolved on warming.

$$CH_3COO^- + H^+ \rightarrow CH_3COOH\uparrow$$

2. Concentrated sulphuric acid: acetic acid is evolved on heating, together with sulphur dioxide, the latter tending to mask the penetrating odour of the concentrated acetic acid vapour. The test with dilute sulphuric acid, in which the acetic acid vapour is diluted with steam, is therefore to be preferred as a test for an acetate.

3. Ethanol and concentrated sulphuric acid 1 g of the solid acetate is treated with 1 ml concentrated sulphuric acid and 2–3 ml rectified spirit in a test-tube, and the whole gently warmed for several minutes; ethyl acetate $CH_3COO.C_2H_5$ is formed, which is recognized by its pleasant, fruity odour. On cooling and dilution with water on a clock glass, the fragrant odour will be more readily detected.

$$CH_3COONa + H_2SO_4 \rightarrow CH_3COOH + Na^+ + HSO_4^-$$
$$CH_3COOH + C_2H_5OH \rightarrow CH_3COO.C_2H_5\uparrow + H_2O$$

(In the second reaction the sulphuric acid acts as a dehydrating agent.)

It is preferable to use iso-amyl alcohol because the odour of the resulting iso-amyl acetate is more readily distinguished from the alcohol itself than is the case with ethanol. It is as well to run a parallel test with a known acetate and to compare the odours of the two products.

4. Silver nitrate solution: a white, crystalline precipitate of silver acetate is produced in concentrated solutions in the cold. The precipitate is more soluble in boiling water (10·4 g ℓ^{-1} at 30° and 25·2 g ℓ^{-1} at 80°) and readily soluble in dilute ammonia solution.

$$CH_3COO^- + Ag^+ \rightleftarrows CH_3COOAg\downarrow$$

When heating the mixture, the precipitate redissolves without the formation of black precipitate of silver metal (distinction from formate ions).

5. Barium chloride, calcium chloride or mercury(II) chloride solution: no change in the presence of acetates (distinction from oxalates and formates).

6. *Iron(III) chloride solution:* deep-red colouration, owing to the formation of a complex ion, $[Fe_3(OH)_2(CH_3COO)_6]^+$. On boiling the red solution it decomposes and a brownish-red precipitate of basic iron(III) is formed:

$$6CH_3COO^- + 3Fe^{3+} + 2H_2O \rightarrow [Fe_3(OH_2(CH_3COO)_6]^+ + 2H^+$$

$$[Fe_3(OH)_2(CH_3COO)_6]^+ + 4H_2O \rightarrow$$
$$\rightarrow 3Fe(OH)_2CH_3COO\downarrow + 3CH_3COOH + H^+$$

7. *Cacodyl oxide reaction* If a dry acetate, preferably that of sodium or potassium, is heated in an ignition tube or test-tube with a small quantity of arsenic(III) oxide, an extremely nauseating odour of cacodyl oxide is produced. All cacodyl compounds are extremely **POISONOUS**; the experiment must therefore be performed on a very small scale, and preferably in the fume chamber. Mix not more than 0·2 g sodium acetate with 0·2 g arsenic(III) oxide in an ignition tube and warm; observe the extremely unpleasant odour that is produced.

$$4CH_3COONa + As_2O_3 \rightarrow \begin{matrix} CH_3 & & CH_3 \\ & As{-}O{-}As & \\ CH_3 & & CH_3 \end{matrix} \quad 2Na_2CO_3 + 2CO_2\uparrow$$

8. *Lanthanum nitrate test* Treat 0·5 ml of the acetate solution with 0·5 ml 0·1M lanthanum nitrate solution, add 0·5 ml iodine solution and a few drops of dilute ammonia solution, and heat slowly to the boiling point. A blue colour is produced; this is probably due to the adsorption of the iodine by the basic lanthanum acetate. This reaction provides an extremely sensitive test for an acetate.

Sulphates and phosphates interfere, but can be removed by precipitation with barium nitrate solution before applying the test. Propionates give a similar reaction.

The spot-test technique is as follows. Mix a drop of the test solution on a spot plate with a drop of 0·1M lanthanum nitrate solution and a drop of 0·005M iodine. Add a drop of M ammonia solution. Within a few minutes a blue to blue-brown ring will develop round the drop of ammonia solution.

Sensitivity: 50 μg CH_3COO^-. *Concentration limit:* 1 in 2000.

9. *Formation of indigo test* The test depends upon the conversion of acetone, formed by the dry distillation of acetates (see reaction 10), into indigo. No other fatty acids give this test, but the sensitivity is reduced in their presence.

Mix the solid test sample with calcium carbonate or, alternatively, evaporate a drop of the test solution to dryness with calcium carbonate; both operations may be carried out in the hard glass tube of Fig. II.58. Cover the open end of the tube with a strip of quantitative filter paper moistened with a freshly prepared solution of *o*-nitrobenzaldehyde in 2M sodium hydroxide, and hold the paper in position with a small glass cap or a small watch glass. Insert the tube into a hole in an asbestos or 'uralite' sheet and heat the tube gently. Acetone is evolved which colours the paper blue or bluish-green. For minute amounts of acetates, it is best to remove the filter paper after the reaction and treat it with a drop of dilute hydrochloric acid; the original yellow colour of the paper is thus bleached and the blue colour of the indigo is more readily apparent.

Sensitivity: 60 μg CH_3COO^-.

10. Action of heat All acetates decompose upon strong ignition, yielding the highly inflammable acetone, $CH_3.CO.CH_3$, and a residue which consists of the carbonates for the alkali acetates, of the oxides for the acetates of the alkaline earth and heavy metals, and of the metal for the acetates of silver and the noble metals. Carry out the experiment in an ignition tube with sodium acetate and lead acetate.

$$2CH_3COONa \rightarrow CH_3.CO.CH_3\uparrow + Na_2CO_3$$

IV.36 FORMATES, HCOO⁻ *Solubility* With the exception of the lead, silver, and mercury(I) salts which are sparingly soluble, most formates are soluble in water. The free acid HCOOH, is a pungent smelling liquid (boiling point 100·5°, melting point 8°), miscible with water in all proportions, and producing blisters when allowed to come into contact with the skin.

To study these reactions use a M solution of sodium formate, HCOONa.

1. Dilute sulphuric acid: formic acid is liberated, the pungent odour of which can be detected on warming the mixture.

$$HCOO^- + H^+ \rightarrow HCOOH\uparrow$$

2. Concentrated sulphuric acid: carbon monoxide (**HIGHLY POISONOUS**) is evolved on warming; the gas should be ignited and the characteristic blue flame obtained. The test can be successfully carried out with solid sodium formate.

$$HCOONa + H_2SO_4 \rightarrow CO\uparrow + Na^+ + HSO_4^- + H_2O$$

3. Ethanol and concentrated sulphuric acid: a pleasant odour, owing to ethyl formate, is apparent on warming (for details, see under Acetates, Section **IV.35**, reaction 3).

$$HCOONa + H_2SO_4 \rightarrow HCOOH + Na^+ + HSO_4^-$$
$$HCOOH + C_2H_5OH \rightarrow HCOO.C_2H_5\uparrow + H_2O$$

4. Silver nitrate solution: white precipitate of silver formate in neutral solutions, slowly reduced at room temperature and more rapidly on warming, a black precipitate of silver being formed (distinction from acetate). With very dilute solutions, the silver may be deposited in the form of a mirror on the walls of the tube.

$$HCOO^- + Ag^+ \rightarrow HCOOAg\downarrow$$
$$2HCOOAg\downarrow \rightarrow 2Ag\downarrow + HCOOH + CO_2\uparrow$$

5. Barium chloride or calcium chloride solution: no change (distinction from oxalates).

6. Iron(III) chloride: a red colouration, due to the complex $[Fe_3(HCOO)_6]^{3+}$, is produced in practically neutral solutions; the colour is discharged by hydrochloric acid. If the red solution is diluted and boiled, a brown precipitate of basic iron(III) formate is formed.

$$6HCOO^- + 3Fe^{3+} \rightarrow [Fe_3(HCOO)_6]^{3+}$$
$$[Fe_3(HCOO)_6]^{3+} + 4H_2O \rightarrow 2Fe(OH)_2HCOO\downarrow + 4HCOOH\uparrow + Fe^{3+}$$

7. Mercury(II) chloride solution: white precipitate of mercury(I) chloride (calomel) is produced on warming; this passes into grey, metallic mercury, in the presence of excess formate solution (distinction from acetate).

$$2HCOO^- + 2HgCl_2 \rightarrow Hg_2Cl_2\downarrow + 2Cl^- + CO\uparrow + CO_2\uparrow + H_2O$$
$$2HCOO^- + Hg_2Cl_2\downarrow \rightarrow 2Hg\downarrow + 2Cl^- + CO\uparrow + CO_2\uparrow + H_2O$$

8. Mercury(II) formate test Free formic acid is necessary for this test. The solution of sodium formate must be acidified with dilute sulphuric acid and shaken vigorously with a little mercury(II) oxide. The solution must be then filtered. The filtrate, when boiled, gives momentarily a white precipitate of mercury(I) formate, which rapidly changes to a grey precipitate of metallic mercury.

$$2HCOOH + HgO \rightarrow (HCOO)_2Hg + H_2O$$
$$2(HCOO)_2Hg \rightarrow (HCOO)_2Hg_2\downarrow + HCOOH + CO_2\uparrow$$
$$(HCOO)_2Hg_2\downarrow \rightarrow 2Hg\downarrow + HCOOH + CO_2\uparrow$$

Formic acid and mercury(II) formate are almost completely undissociated under these circumstances.

9. Formaldehyde–chromotropic acid test Formic acid, H.COOH, is reduced to formaldehyde H.CHO by magnesium and hydrochloric acid. The formaldehyde is identified by its reaction with chromotropic acid (see Section **III.24**, reaction 9**d**) in strong sulphuric acid when a violet-pink colouration appears. Other aliphatic aldehydes do not give the violet colouration.

Place a drop or two of the test solution in a semimicro test-tube, add a drop or two of dilute hydrochloric acid, followed by magnesium powder until the evolution of gas ceases. Introduce 3 ml sulphuric acid (3 vol. acid + 2 vol. water) and a little solid chromotropic acid, and warm to 60°C. A violet-pink colouration appears within a few minutes.

Sensitivity: 1·5 μg H.COOH. *Concentration limit:* 1 in 20,000.

10. Action of heat Cautious ignition of the formates of the alkali metals yields the corresponding oxalates (for tests, see Section **IV.37**) and hydrogen.

$$2HCOONa \rightarrow (COO)_2Na_2 + H_2\uparrow$$

IV.37 OXALATES, $(COO)_2^{2-}$ *Solubility* The oxalates of the alkali metals and of iron(II) are soluble in water; all other oxalates are either insoluble or sparingly soluble in water. They are all soluble in dilute acids. Some of the oxalates dissolve in a concentrated solution of oxalic acid by virtue of the formation of soluble acid or complex oxalates. Oxalic acid (a dibasic acid) is a colourless, crystalline solid, $(COOH)_2.2H_2O$, and becomes anhydrous on heating to 110°; it is readily soluble in water (111 g ℓ^{-1} at 20°).

To study these reactions a 0·1M solution of sodium oxalate, $(COONa)_2$, or ammonium oxalate, $(COONH_4)_2.H_2O$, should be used.

1. Concentrated sulphuric acid: decomposition of all solid oxalates occurs with the evolution of carbon monoxide and carbon dioxide; the latter can be detected by passing the escaping gases through lime water (distinction from formate) and the former by burning it at the mouth of the tube. With dilute sulphuric acid, there is no visible action; in the presence of manganese dioxide, however, carbon dioxide is evolved.

$$(COOH)_2 \rightarrow H_2O + CO\uparrow + CO_2\uparrow$$
$$(COO)_2^{2-} + MnO_2 + 4H^+ \rightarrow Mn^{2+} + 2CO_2\uparrow + 2H_2O$$

(the concentrated sulphuric acid acts as a dehydrating agent).

2. *Silver nitrate solution:* white, curdy precipitate of silver oxalate sparingly soluble in water, soluble in ammonia solution and in dilute nitric acid.

$$(COO)_2^{2-} + 2Ag^+ \rightarrow (COOAg)_2\downarrow$$
$$(COOAg)_2\downarrow + 4NH_3 \rightarrow 2[Ag(NH_3)_2]^+ + (COO)_2^{2-}$$
$$(COOAg)_2\downarrow + 2H^+ \rightarrow Ag^+ + (COO)_2^{2-} + 2H^+$$

3. *Calcium chloride solution:* white, crystalline precipitate of calcium oxalate from neutral solutions, insoluble in dilute acetic acid, oxalic acid, and in ammonium oxalate solution, but soluble in dilute hydrochloric acid and in dilute nitric acid. It is the most insoluble of all oxalates (0·0067 g ℓ^{-1} at 13°) and is even precipitated by calcium sulphate solution and acetic acid. Barium chloride solution similarly gives a white precipitate of barium oxalate sparingly soluble in water (0·016 g ℓ^{-1} at 8°), but soluble in solutions of acetic and of oxalic acids.

$$(COO)_2^{2-} + Ca^{2+} \rightarrow (COO)_2Ca\downarrow$$
$$(COO)_2^{2-} + Ba^{2+} \rightarrow (COO)_2Ba\downarrow$$

4. *Potassium permanganate solution:* decolourized when warmed in acid solution to 60°–70°. The bleaching of permanganate solution is also effected by many other organic compounds but if the evolved carbon dioxide is tested for by the lime water reaction (Section **IV.2**, reaction 1), the test becomes specific for oxalates.

$$5(COO)_2^{2-} + 2MnO_4^- + 16H^+ \rightarrow 10CO_2\uparrow + 2Mn^{2+} + 8H_2O$$

5. *Resorcinol test* Place a few drops of the test solution in a test-tube: add several drops of dilute sulphuric acid and a speck or two of magnesium powder. When the metal has dissolved, add about 0·1 g resorcinol, and shake until dissolved. Cool. Carefully pour down the side of the tube 3–4 ml concentrated sulphuric acid. A blue ring will form at the junction of the two liquids. Upon warming the sulphuric acid layer at the bottom of the tube very gently (**CAUTION**), the blue colour spreads downwards from the interface and eventually colours the whole of the sulphuric acid layer.

Citrates do not interfere. In the presence of tartrates, a blue ring is obtained in the cold or upon very gently warming (compare similar test under Tartrates in the following section).

6. *Manganese(II) sulphate test* A solution of manganese(II) sulphate is treated with sodium hydroxide and the resulting mixture warmed gently to oxidize the precipitate by atmospheric oxygen to hydrated manganese dioxide (cf. Section **III.28**, reaction 1). After cooling, the solution of the oxalate, made

acid with dilute sulphuric acid, is added. The precipitate dissolves and a red colour is produced, which is probably due to the formation of trioxalato-manganate(III) complex ion:

$$7(COO)_2^{2-} + 2MnO(OH)_2\downarrow + 8H^+ \rightarrow 2\{Mn[(COO)_2]_3\}^{3-} + 2CO_2\uparrow + 6H_2O$$

7. Action of heat All oxalates decompose upon ignition. Those of the alkali metals and of the alkaline earths yield chiefly the carbonates and carbon monoxide; a little carbon is also formed. The oxalates of the metals whose carbonates are easily decomposed into stable oxides, are converted into carbon monoxide, carbon dioxide, and the oxide, e.g. magnesium and zinc oxalates. Silver oxalate yields silver and carbon dioxide; silver oxide decomposes on heating. Oxalic acid decomposes into carbon dioxide and formic acid, the latter being further partially decomposed into carbon monoxide and water.

$$7(COONa)_2 \rightarrow 7Na_2CO_3 + 3CO\uparrow + 2CO_2\uparrow + 2C$$

$$(COO)_2Ba \rightarrow BaCO_3 + CO\uparrow$$

$$(COO)_2Mg \rightarrow MgO + CO\uparrow + CO_2\uparrow$$

$$(COOAg)_2 \rightarrow 2Ag + 2CO_2\uparrow$$

$$(COOH)_2 \rightarrow HCOOH + CO_2\uparrow$$

$$HCOOH \rightarrow CO\uparrow + H_2O\uparrow$$

8. Formation of aniline blue test Upon heating insoluble oxalates with concentrated phosphoric acid and diphenylamine or upon heating together oxalic acid and diphenylamine, the dyestuff aniline blue (or diphenylamine blue) is formed. Formates, acetates, tartrates, citrates, succinates, benzoates, and salts of other organic acids do not react under these experimental conditions. In the presence of other anions which are precipitated by calcium chloride solution, e.g. tartrate, sulphate, sulphite, phosphate, and fluoride, it is best to heat the precipitate formed by calcium chloride with phosphoric acid as detailed below.

Place a few milligrams of the test sample (or, alternatively, use the residue obtained by evaporating 2 drops of the test solution to dryness) in a micro test-tube, add a little pure diphenylamine and melt over a free flame. When cold, take up the melt in a drop or two of alcohol when the latter will be coloured blue.

Sensitivity: 5 μg $(COOH)_2$. *Concentration limit:* 1 in 10,000.

In the presence of anions which are precipitated by calcium chloride solution, proceed as follows. Precipitate the acetic acid test solution with calcium chloride solution, and collect the precipitate on a filter or in a centrifuge tube. Remove the water from the precipitate either by drying or by washing with alcohol and ether. Mix a small amount of the precipitate with diphenylamine in a dry micro test-tube, add a little concentrated phosphoric acid, and heat gently over a free flame. Calcium phosphate and free oxalic acid are formed, and the latter condenses with the diphenylamine to aniline blue and colours the hot phosphoric acid blue. The colour disappears on cooling. Dissolve the melt in alcohol, when a blue colouration appears. Pour the alcoholic solution into water thus precipitating the excess of diphenylamine, which is coloured light blue by the adsorption of the dyestuff. The dye may be extracted from aqueous solution by ether.

IV.38 TARTRATES, $C_4H_4O_6^{2-}$ *Solubility* Tartaric acid, HOOC.[CH(OH)]$_2$COOH or $H_2.C_4H_4O_6$, is a crystalline solid, which is extremely soluble in water; it is a dibasic acid. Potassium and ammonium hydrogen tartrates are sparingly soluble in water; those of the other alkali metals are readily soluble. The normal tartrates of the alkali metals are easily soluble, those of the other metals being sparingly soluble in water, but dissolve in solutions of alkali tartrates forming complex salts, which often do not give the typical reactions of the metals present.

The most important commercial salts are 'tartar emetic', $K(SbO).C_4H_4O_6.0{\cdot}5H_2O$; 'Rochelle salt', $KNa.C_4H_4O_6.4H_2O$; and 'cream of tartar', $KH.C_4H_4O_6$.

To study these reactions use a 0·1M solution of tartaric acid, $H_2.C_4H_4O_6$, or a 0·1M solution of Rochelle salt, $KNa.C_4H_4O_6.4H_2O$. The latter solution is neutral.

1. Concentrated sulphuric acid When a solid tartrate is heated with concentrated sulphuric acid, it is decomposed in a complex manner. Charring occurs almost immediately (owing to the separation of carbon), carbon dioxide and carbon monoxide are evolved, together with some sulphur dioxide; the last-named probably arises from the reduction of the sulphuric acid by the carbon. An empyreumatic odour, reminiscent of burnt sugar, can be detected in the evolved gases. Dilute sulphuric acid has no visible action.

$$H_2.C_4H_4O_6 \rightarrow CO\uparrow + CO_2\uparrow 2C + 3H_2O\uparrow$$

$$C + 2H_2SO_4 \rightarrow 2SO_2\uparrow + CO_2\uparrow + 2H_2O\uparrow$$

2. Silver nitrate solution: white, curdy precipitate of silver tartrate, $Ag_2.C_4H_4O_6$, from neutral solutions of tartrates, soluble in excess of the tartrate solution, in dilute ammonia solution and in dilute nitric acid. On warming the ammoniacal solution, metallic silver is precipitated; this can be deposited in the form of a mirror under suitable conditions.

$$C_4H_4O_6^{2-} + 2Ag^+ \rightarrow Ag_2.C_4H_4O_6\downarrow$$

$$Ag_2.C_4H_4O_6\downarrow + 4NH_3 \rightarrow 2[Ag(NH_3)_2]^+ + C_4H_4O_6^{2-}$$

The silver mirror test is best performed as follows. The solution of the tartrate is acidified with dilute nitric acid, excess of silver nitrate solution added and any precipitate present filtered off. Very dilute ammonia solution (approximately 0·02M) is then added to the solution until the precipitate at first formed is nearly redissolved, the solution is filtered, and the filtrate collected in a clean test-tube; the latter is then placed in a beaker of boiling water. A brilliant mirror is formed on the sides of the tube after a few minutes. The test-tube may be cleaned either by boiling with chromic acid mixture or by boiling it with a little sodium hydroxide solution, and then rinsing it well with distilled water.

An alternative method for carrying out the silver mirror test is the following. Prepare ammoniacal silver nitrate solution by placing 5 ml silver nitrate solution in a thoroughly clean test-tube and add 2–3 drops of dilute sodium hydroxide solution; add dilute ammonia solution dropwise until the precipitated silver oxide is almost redissolved (this procedure reduces the danger of the formation of the explosive silver azide, AgN_3, to a minimum). Introduce about 0·5 ml neutral tartrate solution. Place the tube in warm water. A silver mirror is formed in a few minutes.

The reaction is interfered with by other acids. It is best to isolate the potassium hydrogen tartrate, as in reaction 4 below, before carrying out the test.

Immediately discard all the solution which remains after the test.

3. Calcium chloride solution: white, crystalline precipitate of calcium tartrate, $Ca.C_4H_4O_6$, with a concentrated neutral solution. The precipitate is soluble in dilute acetic acid (difference from oxalate), in dilute mineral acids, and in cold alkali solutions. An excess of the reagent must be added because calcium tartrate dissolves in an excess of an alkali tartrate solution forming a complex tartrate ion, $[Ca(C_4H_4O_6)_2]^{2-}$. Precipitation is slow in dilute solutions, but may be accelerated by vigorous shaking or by rubbing the walls of the test-tube with a glass rod.

$$C_4H_4O_6^{2-} + Ca^{2+} \rightarrow Ca.C_4H_4O_6\downarrow$$

4. Potassium chloride solution When a concentrated neutral solution of a tartrate is treated with a solution of a potassium salt (e.g. potassium chloride or potassium acetate) and then acidified with acetic acid, a colourless crystalline precipitate of potassium hydrogen tartrate, $KH.C_4H_4O_6$, is obtained. The precipitate forms slowly in dilute solutions; crystallization is induced by vigorous shaking or by rubbing the walls of the vessel with a glass rod.

$$C_4H_4O_6^{2-} + K^+ + CH_3COOH \rightarrow KH.C_4H_4O_6\downarrow + CH_3COO^-$$

5. Fenton's test Add 1 drop of a 25 per cent solution of iron(II) sulphate to about 5 ml of the neutral or acid solution of the tartrate, and then 2–3 drops of hydrogen peroxide solution (10 volume). A deep-violet or blue colouration is developed on adding excess sodium hydroxide solution. The colour becomes more intense upon the addition of a drop of iron(III) chloride solution.

The violet colour is due to the formation of a salt or dihydroxymaleic acid, $CO_2H.C(OH){=}C(OH).CO_2H$.

The test is not given by citrates, malates or succinates.

6. Resorcinol test Place a few drops of the test solution in a test-tube; add several drops of dilute sulphuric acid and a speck or two of magnesium powder. When the metal has dissolved, add about 0·1 g resorcinol, and shake until dissolved. Cool. Carefully pour down the side of the tube 3–4 ml concentrated sulphuric acid. Upon warming the sulphuric acid layer at the bottom of the tube very gently (**CAUTION**), a red layer (or ring) forms at the junction of the two liquids. With continued gentle heating, the red colour spreads downwards from the interface and eventually colours the whole of the sulphuric acid layer.

Citrates do not interfere. In the presence of oxalate, a blue ring is formed in the cold and on gentle warming of the sulphuric acid layer at the bottom of the tube the blue colouration spreads downwards into the concentrated acid layer and a red ring forms at the interface.

The colour is due to the formation of a condensation product of resorcinol, $C_6H_4(OH)_2$, and glycollic aldehyde, $CH_2OH.CHO$, the latter arising from the action of the sulphuric acid upon the tartaric acid. The formula of the condensation product is $CH_2OH.CH[C_6H_3(OH)_2]_2$.

7. Copper hydroxide test Tartrates dissolve copper hydroxide in the presence of excess alkali hydroxide solution to form the dark-blue ditartratocuprate(II) ion, $[Cu(C_4H_4O_6)_2]^{2-}$, which is best detected by filtering the solution. If only small quantities of tartrate are present, the filtrate should be acidified with acetic acid and tested for copper by the potassium hexacyanoferrate(II) test.

Arsenites, ammonium salts, and organic compounds convertible into tartaric acid also give a blue colouration, and should therefore be absent or be removed.

The experimental details are as follows. Treat the test solution with an equal volume of 2M sodium or potassium hydroxide (or treat the test solid directly with a few ml of the alkali hydroxide solution, warm for a few minutes with stirring and then cool), add a few drops of 0·25M copper sulphate solution (i.e. just sufficient to yield a visible precipitate of copper(II) hydroxide), shake the mixture vigorously for 5 minutes, and filter. If the filtrate is not clear, warm to coagulate the colloidal copper hydroxide and filter again. A distinct blue colouration indicates the presence of a tartrate. If a pale colouration is obtained, it is advisable to add concentrated ammonia solution dropwise when the blue colour intensifies; it is perhaps better to acidify with 2M acetic acid and add potassium hexacyanoferrate(II) solution to the clear solution whereupon a reddish-brown precipitate or (with a trace of a tartrate) a red colouration is obtained.

8. ββ′-Dinaphthol test

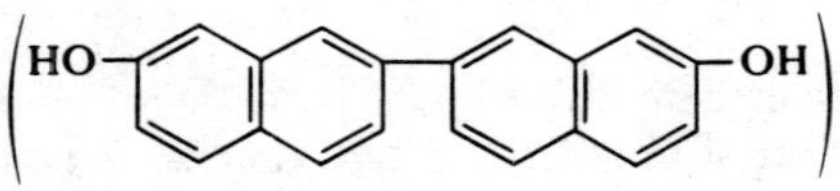

HO OH :

when a solution of ββ′-dinaphthol in concentrated sulphuric acid is heated with tartaric acid, a green colouration is obtained. Oxalic, citric, succinic, and cinnamic acids do not interfere.

Heat a few milligrams of the solid test sample or a drop of the test solution with 1–2 ml of the dinaphthol reagent in a water bath at 85°C for 30 minutes. A luminous green fluorescence appears during the heating; it intensifies on cooling, and at the same time the violet fluorescence of the reagent disappears.

Sensitivity: 10 μg tartaric acid. *Concentration limit:* 1 in 5,000.

The reagent consists of a solution of 0·05 g ββ′-dinaphthol in 100 ml concentrated sulphuric acid.

9. Action of heat The tartrates and also tartaric acid decompose in a complex manner on heating; charring takes place, a smell of burnt sugar is apparent and inflammable vapours are evolved.

IV.39 CITRATES, $C_6H_5O_7^{3-}$ *Solubility* Citric acid, $HOOC.CH_2.C(OH)\text{-}CO_2H.CH_2.COOH.H_2O$ or $H_3.C_6H_5O_7.H_2O$ is a crystalline solid which is very soluble in water; it becomes anhydrous at 55° and melts at 160°. It is a tribasic acid, and therefore gives rise to three series of salts. The normal citrates of the alkali metals dissolve readily in water; other metallic citrates are sparingly soluble. The acid citrates are more soluble than the acid tartrates.

To study these reactions use a 0·5M solution of sodium citrate, $Na_3.C_6H_5O_7.2H_2O$.

1. Concentrated sulphuric acid When a solid citrate is heated with concentrated sulphuric acid, carbon monoxide and carbon dioxide are evolved; the solution slowly darkens owing to the separation of carbon, and sulphur dioxide is evolved (compare Tartrates, Section **IV.38**, which char almost immediately).

The first products of the reaction are carbon monoxide and acetone dicarboxylic acid(I); the latter undergoes further partial decomposition into acetone(II) and carbon dioxide.

$$HOOC.CH_2.C(OH)CO_2H.CH_2.COOH \rightarrow$$
$$\rightarrow CO\uparrow + H_2O\uparrow + HOOC.CH_2.CO.CH_2.COOH(I);$$
$$HOOC.CH_2CO.CH_2.COOH \rightarrow$$
$$\rightarrow CH_3.CO.CH_3.CO.CH_3\uparrow(II) + 2CO_2\uparrow$$

Use may be made of the intermediate formation of acetone dicarboxylic acid and of the interaction of the latter with sodium nitroprusside solution to yield a red colouration as a test for citrates. When about 0·5 g of a citrate or of citric acid is treated with 1 ml concentrated sulphuric acid for 1 minute, the mixture cooled, cautiously diluted with water, rendered alkaline with sodium hydroxide solution and then a few millilitres of a freshly prepared solution of sodium nitroprusside added, an intense red colouration results.

2. *Silver nitrate solution:* white, curdy precipitate of silver citrate, $Ag_3.C_6H_5O_7$, from neutral solutions. The precipitate is soluble in dilute ammonia solution, and this solution undergoes only very slight reduction to silver on boiling (distinction from tartrate).

$$C_6H_5O_7^{3-} + 3Ag^+ \rightarrow Ag_3.C_6H_5O_7\downarrow$$
$$Ag_3.C_6H_5O_7\downarrow + 6NH_3 \rightarrow 3[Ag(NH_3)_2]^+ + C_6H_5O_7^{3-}$$

3. *Calcium chloride solution:* no precipitate with neutral solutions in the cold (difference from tartrate), but on boiling for several minutes a crystalline precipitate of calcium citrate, $Ca_3(C_6H_5O_7)_2.4H_2O$, is produced. If sodium hydroxide solution is added to the cold solution containing excess calcium chloride, there results an immediate precipitation of amorphous calcium citrate, insoluble in solutions of caustic alkalis, but soluble in ammonium chloride solution; on boiling the ammonium chloride solution, crystalline calcium citrate is precipitated, which is now insoluble in ammonium chloride.

$$2C_6H_5O_7^{3-} + 3Ca^{2+} \rightarrow Ca_3(C_6H_5O_7)_2\downarrow$$

4. *Cadmium acetate solution:* white, gelatinous precipitate of cadmium citrate $Cd_3(C_6H_5O_7)_2$, practically insoluble in boiling water, but readily soluble in warm acetic acid (tartrate gives no precipitate).

$$2C_6H_5O_7^{3-} + 3Cd^{2+} \rightarrow Cd_3(C_6H_5O_7)_2\downarrow$$

5. *Deniges's test* Add 0·5 ml acid mercury(II) sulphate solution to 3 ml citrate solution, heat to boiling and then add a few drops of a 0·02M solution of potassium permanganate. Decolourization of the permanganate will take place rapidly and then, somewhat suddenly, a heavy white precipitate, consisting of the double salt of basic mercury(II) sulphate with mercury(II) acetone dicarboxylate

$$HgSO_4.2HgO.2[CO(CH_2.CO_2)_2Hg]\downarrow$$

is formed. Salts of the halogen acids interfere and must therefore be removed before carrying out the test.

The reagent (an acid solution of mercury(II) sulphate) is prepared by adding 10 ml concentrated sulphuric acid slowly to 50 ml water, and dissolving 2·2 g mercury(II) oxide in the hot solution.

6. *Action of heat* Citrates and citric acid char on heating; carbon monoxide, carbon dioxide, and acrid-smelling vapours are evolved.

IV.40 SALICYLATES, $C_6H_4(OH)COO^-$ OR $C_7H_5O_3^-$ *Solubility* Salicylic acid, $C_6H_4(OH)COOH$ (*o*-hydroxybenzoic acid), forms colourless needles, which melt at 155°. The acid is sparingly soluble in cold water, but more soluble in hot water, from which it can be recrystallized. It is readily soluble in alcohol and ether. With the exception of the lead, mercury, silver, and barium salts, the monobasic salts, $C_6H_4(OH)COOM$ – these are the most commonly occurring salts – are readily soluble in cold water.

To study these reactions use a 0·5M solution of sodium salicylate, $C_6H_4(OH)COONa$ or $Na.C_7H_5O_3$.

1. *Concentrated sulphuric acid:* solid sodium salicylate dissolves on warming; charring occurs slowly, accompanied by the evolution of carbon monoxide and sulphur dioxide.

2. *Concentrated sulphuric acid and methanol* ('oil of wintergreen' test) When 0·5 g of a salicylate or of salicylic acid is treated with a mixture of 1·5 ml concentrated sulphuric acid and 3 ml methanol and the whole gently warmed, the characteristic, fragrant odour of the ester, methyl salicylate(I) ('oil of wintergreen'), is obtained. The odour is readily detected by pouring the mixture into dilute sodium carbonate solution contained in a porcelain dish.

$$C_6H_4(OH)COOH + CH_3OH \rightarrow C_6H_4(OH)COO.CH_3\uparrow(I) + H_2O$$

(The sulphuric acid acts as a dehydrating agent.)

3. *Dilute hydrochloric acid:** crystalline precipitate of salicylic acid from solutions of the salts. The precipitate is moderately soluble in hot water, from which it crystallizes on cooling.

4. *Silver nitrate solution:* heavy, crystalline precipitate of silver salicylate, $Ag.C_7H_5O_3$, from neutral solutions; it is soluble in boiling water and crystallizes from the solution on cooling.

$$C_6H_4(OH)COO^- + Ag^+ \rightarrow C_6H_4(OH)COOAg\downarrow$$

5. *Iron(III) chloride solution:* intense violet-red colouration with neutral solutions of salicylates or with free salicylic acid; the colour disappears upon the addition of dilute mineral acids, but not of a little acetic acid. The presence of a large excess of many organic acids (acetic, tartaric, and citric) prevents the development of the colour; but the addition of a few drops of dilute ammonia solution will cause it to appear.

---

*Or any other mineral acid.

6. Dilute nitric acid When a salicylate or the free acid is boiled with dilute nitric acid (2M) and the mixture poured into 4 times its volume of cold water, a crystalline precipitate of 5-nitrosalicylic acid is obtained. The precipitate is filtered off and recrystallized from boiling water; the acid, after drying, has a melting point of 226°.

$$C_6H_4(OH)COOH + HNO_3 \rightarrow C_6H_3(NO_2)(OH)COOH\downarrow + H_2O$$

Polynitro salicylic acids are formed when the concentration of nitric acid exceeds 2–3M. It must be emphasized that nitrosalicylic acids may explode on strong heating, and this must be borne in mind when removing salicylates by oxidation with nitric acid and evaporating to dryness prior to the precipitation of Group IIIA. If salicylate is suspected, it is probably best to remove it by acidifying the concentrated solution with dilute hydrochloric acid: most of the salicylic acid will separate out upon cooling.

7. Soda lime When salicylic acid or one of its salts is heated with excess of soda lime in an ignition tube, phenol, $C_6H_5(OH)$, is evolved, which may be recognized by its characteristic odour.

$$C_6H_4(OH)COONa + NaOH \rightarrow C_6H_5OH\uparrow + Na_2CO_3$$

8. Action of heat Salicylic acid, when gradually heated above its melting point, sublimes. If it is rapidly heated, it is decomposed into carbon dioxide and phenol. Salicylates char on heating and phenol is evolved.

$$C_6H_4(OH)COOH \rightarrow C_6H_5OH\uparrow + CO_2\uparrow$$

IV.41 BENZOATES, $C_6H_5COO^-$ OR $C_7H_5O_2^-$ *Solubility* Benzoic acid, C_6H_5COOH or $H.C_7H_5O_2$, is a white crystalline solid; it has a melting point of 121°. The acid is sparingly soluble in cold water, but more soluble in hot water, from which it crystallizes out on cooling; it is soluble in alcohol and in ether. All the benzoates, with the exception of the silver and basic iron(III) salts, are readily soluble in cold water.

To study these reactions use a 0·5M solution of potassium benzoate, C_6H_5COOK or $K.C_7H_5O_2$.

1. Concentrated sulphuric acid: no charring occurs on heating solid benzoic acid with this reagent; the acid forms a sublimate on the sides of the test-tube and irritating fumes are evolved.

*2. Dilute sulphuric acid:** white, crystalline precipitate of benzoic acid from cold solutions of benzoates. The acid may be filtered off, dried between filter paper or upon a porous tile and identified by means of its melting point (121°). If the latter is somewhat low, it may be recrystallized from hot water.

$$C_6H_5COO^- + H^+ \rightarrow C_6H_5COOH\downarrow$$

3. Concentrated sulphuric acid and ethanol Heat 0·5 g of a benzoate or of benzoic acid with a mixture of 1·5 ml concentrated sulphuric acid and 3 ml

* Or any other mineral acid.

ethanol for a few minutes. Allow the mixture to cool, and note the pleasant and characteristic aromatic odour of the ester, ethyl benzoate. The odour is more apparent if the mixture is poured into dilute sodium carbonate solution contained in an evaporating basin; oily drops of ethyl benzoate will separate out.

$$C_6H_5.COOH + C_2H_5OH \rightarrow C_6H_5.COOC_2H_5\uparrow + H_2O$$

4. Silver nitrate solution: white precipitate of silver benzoate, $Ag.C_7H_5O_2$, from neutral solutions. The precipitate is soluble in hot water and crystallizes from the solution on cooling; it is also soluble in dilute ammonia solution.

$$C_6H_5.COO^- + Ag^+ \rightarrow C_6H_5.COOAg\downarrow$$

$$C_6H_5.COOAg\downarrow + 2NH_3 \rightarrow [Ag(NH_3)_2]^+ + C_6H_5.COO^-$$

5. Iron(III) chloride solution: buff-coloured precipitate of basic iron(III) benzoate from neutral solutions. The precipitate is soluble in hydrochloric acid, with the simultaneous separation of benzoic acid (see reaction 2).

$$3C_6H_5COO^- + 2Fe^{3+} + 3H_2O \rightarrow (C_6H_5COO)_3Fe.Fe(OH)_3\downarrow + 3H^+$$

6. Soda lime Benzoic acid and benzoates, when heated in an ignition tube with excess soda lime, are decomposed into benzene, C_6H_6, which burns with a smoky flame, and carbon dioxide, the latter combining with the alkali present.

$$C_6H_5.COOH + 2NaOH \rightarrow C_6H_6\uparrow + Na_2CO_3 + H_2O\uparrow$$

7. Action of heat When benzoic acid is heated in an ignition tube, it melts, evaporates, and condenses in the cool parts of the tube. An irritating vapour is simultaneously evolved, but no charring occurs. If a little of the acid or of one of its salts is heated upon platinum foil or broken porcelain it burns with a blue, smoky flame (distinction from succinate).

IV.42 SUCCINATES, $C_4H_4O_4^{2-}$ *Solubility* Succinic acid, $HOOC.CH_2.CH_2.COOH$ or $H_2.C_4H_4O_4$ (a dibasic acid), is a white, crystalline solid with a melting point of 182°; it boils at 235° with the formation of succinic anhydride, $C_4H_4O_3$, by the loss of one molecule of water. The acid is fairly soluble in water (68·4 g ℓ^{-1} at 20°), but more soluble in hot water; it is moderately soluble in alcohol and in acetone, slightly soluble in ether, and sparingly soluble in chloroform. Most normal succinates are soluble in cold water; the silver, calcium, barium, and basic iron(III) salts are sparingly soluble.

To study these reactions use a 0·5M solution of sodium succinate, $C_2H_4(COONa)_2.6H_2O$ or $Na_2.C_4H_4O_4.6H_2O$.

1. Concentrated sulphuric acid The acid or its salts dissolve in warm, concentrated sulphuric acid without charring; if the solution is strongly heated, slight charring takes place and sulphur dioxide is evolved.

Dilute sulphuric acid has no visible action.

2. Silver nitrate solution: white precipitate of silver succinate, $Ag_2.C_4H_4O_4$, from neutral solutions, readily soluble in dilute ammonia solution.

$$C_2H_4(COO)_2^{2-} + 2Ag^+ \rightarrow C_2H_4(COOAg)_2\downarrow$$

$$C_2H_4(COOAg)_2\downarrow + 4NH_3 \rightarrow 2[Ag(NH_3)_2]^+ + C_2H_4(COO)_2^{2-}$$

3. *Iron(III) chloride solution:* light-brown precipitate of basic iron(III) succinate with a neutral solution; some free succinic acid is simultaneously produced, and the solution becomes slightly acidic.

$$3C_2H_4(COO)_2^{2-} + 2Fe^{3+} + 2H_2O \rightarrow$$
$$\rightarrow 2C_2H_4(COO)_2Fe(OH)\downarrow + C_2H_4(COOH)_2$$

4. *Barium chloride solution:* white precipitate of barium succinate from neutral or slightly ammoniacal solutions (distinction from benzoate). Precipitation is slow from dilute solutions, but may be accelerated by vigorous shaking or by rubbing the inner wall of the test-tube with a glass rod.

$$C_2H_4(COO)_2^{2-} + Ba^{2+} \rightarrow C_2H_4(COO)_2Ba\downarrow$$

5. *Calcium chloride solution:* a precipitate of calcium succinate is very slowly produced from concentrated neutral solutions.

$$C_2H_4(COO)_2^{2-} + Ca^{2+} \rightarrow C_2H_4(COO)_2Ca\downarrow$$

6. *Fluorescein test* About 0·5 g of the acid or one of its salts is mixed with 1 g resorcinol, a few drops of concentrated sulphuric acid added, and the mixture gently heated; a deep-red solution is formed. On pouring the latter into a large volume of water, an orange-yellow solution is obtained, which exhibits an intense green fluorescence. The addition of excess sodium hydroxide solution intensifies the fluorescence, and the colour of the solution changes to bright red.

Under the influence of concentrated sulphuric acid, succinic anhydride(I) is first formed, and this condenses with the resorcinol(II) to yield succinylfluorescein(III).

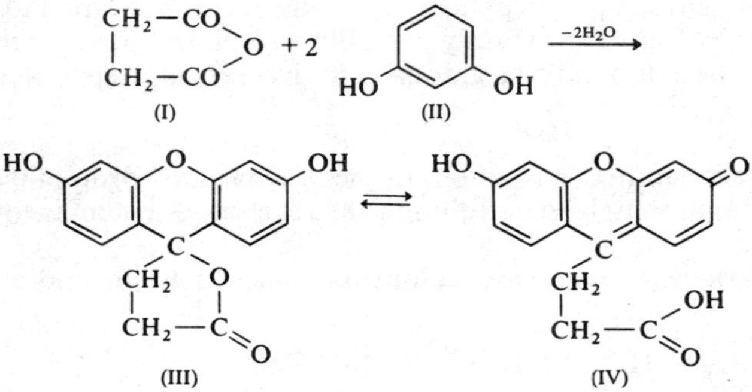

(I) + 2 (II) $\xrightarrow{-2H_2O}$ (III) ⇌ (IV)

The quinoid structure (IV) is predominant in alkaline solutions.

7. *Action of heat* When succinic acid or its salts are strongly heated in an ignition tube, a white sublimate of succinic anhydride, $C_4H_4O_3$, is formed and irritating vapours are evolved. If the ignition is carried out on platinum foil or upon broken porcelain, the vapour burns with a non-luminous, blue flame, leaving a residue of carbon (distinction from benzoate).

IV.43 HYDROGEN PEROXIDE, H_2O_2 This substance is marketed in the form of '10-, 20-, 40-, and 100-volume' solutions. It is formed upon adding sodium peroxide in small portions to ice water:

$$Na_2O_2+2H_2O = H_2O_2+2Na^++2OH^-$$

Owing to the heat liberated in the reaction, part of the hydrogen peroxide is decomposed:

$$2H_2O_2 \rightarrow 2H_2O+O_2\uparrow$$

Hydrogen peroxide is prepared in industrial scale by an electrochemical method.* A convenient source is the inexpensive sodium perborate, $NaBO^3.4H_3O$; this yields hydrogen peroxide when its aqueous solution is heated:

$$BO_3^-+H_2O \rightarrow H_2O_2+BO_2^-$$

1. Chromium pentoxide test If an acidified solution of hydrogen peroxide is mixed with a little diethyl ether (highly inflammable), amyl alcohol or amyl acetate and a few drops of potassium dichromate solution added and the mixture gently shaken, the upper organic layer is coloured a beautiful blue. For further details, see Section **IV.33**, reaction 4, and Section **III.24**, reaction 8**b**. The test will detect 0·1 mg of hydrogen perodide.

If ether is employed, a blank test must always be carried out with the ether alone because after standing it may contain organic peroxides which give the test. The peroxides may be removed by shaking 1 ℓ of ether with a solution containing 60 g $FeSO_4.7H_2O$, 6 ml concentrated H_2SO_4, and 110 ml water, then with chromic acid solution (to oxidize any acetaldehyde produced), followed by washing with alkali and redistilling. The peroxides may also be removed by treatment with aqueous sodium sulphite. The use of amyl alcohol is, however, to be preferred.

2. Potassium iodide and starch If potassium iodide and starch are added to hydrogen peroxide, acidified previously by dilute sulphuric acid, iodine is formed slowly and the solution turns gradually to deeper and deeper blue:

$$H_2O_2+2H^++2I^- \rightarrow I_2+2H_2O$$

Molybdate ions accelerate this reaction. In the presence of 1 drop of 0·025M (0·4 per cent) ammonium molybdate solution the reaction is instantaneous.

3. Potassium permanganate solution: decolourized in acid solution and oxygen is evolved:

$$2MnO_4^-+5H_2O_2+6H^+ \rightarrow 2Mn^{2+}+5O_2\uparrow+8H_2O$$

4. Titanium(IV) chloride solution: orange-red colouration (see under Titanium, Section **VII.17**, reaction 6). This is a very delicate test.

5. Potassium hexacyanoferrate(III) and iron(III) chloride To a nearly neutral solution of iron(III) chloride add some potassium hexacyanoferrate(III) solution. A yellow solution is obtained. To this, add a nearly neutral solution of hydrogen peroxide. The solution turns to green and Prussian blue separates slowly.

* Cf. Mellor's Modern Inorganic Chemistry, 6th edn., Longman 1967, p. 323.

Hydrogen peroxide reduces hexacyanoferrate(III) ions to hexacyanoferrate(II):

$$2[Fe(CN)_6]^{3-} + H_2O_2 \rightarrow 2[Fe(CN)_6]^{4-} + 2H^+ + O_2\uparrow$$

Prussian blue is then produced from iron(III) and hexacyanoferrate(II) ions:

$$4Fe^{3+} + 3[Fe(CN)_6]^{4-} \rightarrow Fe_4[Fe(CN)_6]_3\downarrow$$

(cf. Section **III.22**, reaction 6 and Section **III.21**, reaction 5).

Other reducing agents, e.g. tin(II) chloride, sodium sulphite, sodium thiosulphate, etc. will reduce hexacyanoferrate(III) ions to hexacyanoferrate(II), so that this reaction is not always a reliable test.

The spot-test technique is as follows. In adjacent depressions of a spot plate place a drop of distilled water and a drop of the test solution. Add a drop of the reagent (equal volumes of 0·5M iron(III) chloride solution and 0·033M potassium hexacyanoferrate(III) solution) to each. A blue colouration or precipitate is formed.

Sensitivity: 0·1 μg H_2O_2. *Concentration limit:* 1 in 600,000.

6. *Gold chloride solution:* reduced to finely divided metallic gold, which appears greenish-blue by transmitted light and brown by reflected light.

$$2Au^{3+} + 3H_2O_2 \rightarrow 2Au\downarrow + 3O_2\uparrow + 6H^+$$

The spot-test technique is as follows. Mix a drop of the neutral test solution with a drop of 0·33M. gold chloride solution in a micro crucible and warm. After a short time, the solution is coloured red or blue with colloidal gold.

Sensitivity: 0·07 μg H_2O_2. *Concentration limit:* 1 in 700,000.

7. *Lead sulphate test* Hydrogen peroxide converts black lead sulphide into white lead sulphate:

$$PbS\downarrow + 4H_2O_2 \rightarrow PbSO_4\downarrow + 4H_2O$$

Place a drop of the neutral or slightly acid test solution upon drop-reaction paper impregnated with lead sulphide. A white spot is formed on the brown paper.

Sensitivity: 0·04 μg H_2O_2. *Concentration limit:* 1 in 1,250,000.

The lead sulphide paper is prepared by soaking drop-reaction paper in a 0·0025M solution of lead acetate, exposing it to a little hydrogen sulphide gas and then drying in a vacuum desiccator. The paper will keep in a stoppered bottle.

8. *Chemiluminescence test* This test has to be carried out in a dark room.

To 20 ml 2M sodium hydroxide add 5 ml 1 per cent aqueous solution of lucigenine (dimethyl diacridylium dinitrate). Switch off the light. Add 5 ml 3 per cent hydrogen peroxide solution and mix. A steady greenish-yellow light is emitted; the phenomenon lasts for about 15 minutes.

To one-half of the solution add 1 drop of 1 per cent osmium tetroxide solution and mix. The light flares up, turns to bluish-violet, then quickly fades away.

For the explanation of these phenomena consult the literature.*

---

* See L. Erdey in E. Bishop (ed.), *Indicators*, Pergamon Press 1972, p. 713 et f.

IV.44 DITHIONITES, $S_2O_4^{2-}$ Dithionites are obtained by the action of reducing agents, such as zinc, upon hydrogen sulphites

$$4HSO_3^- + Zn \rightarrow S_2O_4^{2-} + 2SO_3^{2-} + Zn^{2-} + 2H_2O$$

Sulphur dioxide may also be passed into a cooled suspension of zinc dust in water:

$$Zn + 2SO_2 \rightarrow Zn^{2+} + S_2O_4^{2-}$$

Zinc ions, which are produced in both processes can be removed from the solution with sodium carbonate, when zinc carbonate is precipitated. By saturating the solution with sodium chloride, sodium dithionite, $Na_2S_2O_4.2H_2O$, is precipitated.

Sodium dithionite is a powerful reducing agent. A solution of the salt, containing an excess of sodium hydroxide, is used as an absorbent for oxygen in gas analysis.

To study these reactions use a freshly prepared 0·5M solution of sodium dithionite, $Na_2S_2O_4$.

1. Dilute sulphuric acid: orange colouration, which soon disappears, accompanied by the evolution of sulphur dioxide and the precipitation of sulphur:

$$2S_2O_4^{2-} + 4H^+ \rightarrow 3SO_2\uparrow + S\downarrow + 2H_2O$$

2. Concentrated sulphuric acid: immediate evolution of sulphur dioxide and the precipitation of pale-yellow sulphur.

$$2S_2O_4^{2-} + 2H_2SO_4 \rightarrow 3SO_2\uparrow + S\downarrow + 2SO_4^{2-} + 2H_2O$$

3. Silver nitrate solution: black precipitate of metallic silver.

$$S_2O_4^{2-} + 2Ag^+ + 2H_2O \rightarrow 2Ag\downarrow + 2SO_3^{2-} + 4H^+$$

If the reagent is added in excess, white silver sulphite is also precipitated (cf. Section **IV.4**, reaction 3).

4. Copper sulphate solution: red precipitate of metallic copper.

$$S_2O_4^{2-} + Cu^{2+} + 2H_2O \rightarrow Cu\downarrow + 2SO_3^{2-} + 4H^+$$

5. Mercury(II) chloride solution: grey precipitate of metallic mercury.

$$S_2O_4^{2-} + HgCl_2 + 2H_2O \rightarrow Hg\downarrow + 2SO_3^{2-} + 2Cl^- + 4H^+$$

6. Methylene blue solution: immediate decolourization in the cold. The methylene blue is reduced to the colourless leuco compound.

7. Indigo solution: reduced to the colourless leuco compound, known as indigo white.

$$S_2O_4^{2-} + \underset{\text{(indigo)}}{C_{16}H_{10}O_2N_2} + 2H_2O \rightarrow \underset{\text{(indigo white)}}{C_{16}H_{12}O_2N_2} + 2SO_3^{2-} + 2H^+$$

8. Potassium hexacyanoferrate(II) and iron(II) sulphate If potassium hexacyanoferrate(II) and iron(II) sulphate are mixed, white precipitate of dipotassium iron(II) hexacyanoferrate(II) turns quite rapidly into Prussian blue (cf. Section **III.21**, reaction 6), because of oxidation caused by dissolved and atmospheric oxygen. In the presence of dithionite the precipitate remains durably white.

9. Action of heat Upon heating solid sodium dithionite it suddenly evolves sulphur dioxide at about 190° (exothermic reaction). The residue has lost its reducing power to indigo and methylene blue solution, being a mixture of sodium sulphite and sodium thiosulphate.

$$2Na_2S_2O_4 \rightarrow Na_2SO_3 + Na_2S_2O_3 + SO_2\uparrow$$

IV.45 SPECIAL TESTS FOR MIXTURES OF ANIONS When the reactions of individual anions have been studied, the student should try his skill in carrying out these special tests. All of these are of great practical importance, as samples often contain mixtures of ions which might interfere with each other's tests.

Full experimental details of the various tests referred to in this section will be found under the reactions of the anions. Reference will be made to the 'soda extract' and to the 'neutralized soda extract' (see 20 below) as these are normally used in systematic analysis. A solution of the sodium salts of the various anions may, of course, be used when trying out these procedures.

1. Carbonate in the presence of sulphite Sulphites, on treatment with dilute sulphuric acid, liberate sulphur dioxide which, like carbon dioxide, produces a turbidity with lime or baryta water. The dichromate test for sulphites is, however, not influenced by the presence of carbonates. To detect carbonates in the presence of sulphites, treat the solid mixture with dilute sulphuric acid and pass the evolved gases through a small wash-bottle or boiling tube containing potassium dichromate solution and dilute sulphuric acid. The solution will be turned green and the sulphur dioxide will, at the same time, be completely removed; the residual gas is then tested with lime or baryta water in the usual manner.

An alternative procedure is to add a little solid potassium dichromate or a small volume of 3 per cent hydrogen peroxide solution to the mixture and then to warm with dilute sulphuric acid; the evolved gas is then passed through lime or baryta water.

2. Nitrate in the presence of nitrite The nitrite is readily identified in the presence of a nitrate by treatment with dilute mineral acid, potassium iodide, and starch paste (or potassium iodide–starch paper). The nitrate cannot, however, be detected in the presence of a nitrite since the latter gives the brown ring test with iron(II) sulphate solution and dilute sulphuric acid. The nitrite is therefore completely decomposed first by one of the following methods:

(i) Boil with ammonium chloride solution until effervescence ceases.*
(ii) Warm with urea and dilute sulphuric acid until evolution of gas ceases.*
(iii) Add the solution to solid hydrazine sulphate; reaction takes place in

* Traces of nitrate are always formed in this reaction.

the cold, and the danger of slight oxidation of nitrite to nitrate is thereby avoided.

$$2NO_2^- + N_2H_4 + 2H^+ \rightarrow N_2\uparrow + N_2O\uparrow + 3H_2O$$
$$NO_2^- + N_2H_4 + 2H^+ \rightarrow NH_4^+ + N_2O\uparrow + H_2O$$

(iv) Add some sulphamic acid to the solution:

$$NO_2^- + HO.SO_2.NH_2 \rightarrow N_2\uparrow + SO_4^{2-} + H^+ + H_2O$$

The sulphamic acid procedure is probably the simplest and most efficient method for the removal of nitrite in aqueous solution.

The nitrate can then be tested for by the brown ring test.

3. Nitrate in the presence of bromide and iodide The brown ring test for nitrates cannot be applied in the presence of bromides and iodides since the liberation of the free halogen with concentrated sulphuric acid will obscure the brown ring due to the nitrate. The solution is therefore boiled with sodium hydroxide solution until ammonium salts, if present, are completely decomposed; powdered Devarda's alloy or aluminium powder (or wire) is then added and the solution gently warmed. The evolution of ammonia, detected by its smell and its action upon mercury(I) nitrate paper (see Section **III.38**, reaction 1) and upon red litmus paper, indicates the presence of a nitrate.

An alternative method is to remove the halides by adding saturated silver sulphate solution. Since silver sulphate is only slightly soluble in water, it is better to use ammoniacal silver sulphate solution. This reagent contains $[Ag(NH_3)_2]_2SO_4$ and provides a high concentration of silver ions: it is prepared by dissolving 7·8 g pure silver sulphate in 50 ml of 2M ammonia solution and diluting to 100 ml with water. The test solution is treated with sufficient of the reagent to precipitate all the halides; the silver halides are filtered off, a portion of the filtrate is carefully mixed with a five-fold excess of concentrated sulphuric acid, and the iron(II) sulphate solution cautiously introduced down the side of the tube. (Discard the ammoniacal silver reagent immediately after use. cf. Section **III.6**, reaction 3).

4. Nitrate in the presence of chlorate The chlorate obscures the brown ring test. The nitrate is reduced to ammonia as described under 3; the chlorate is at the same time reduced to chloride, which may be tested for with silver nitrate solution and dilute nitric acid.

If a chloride is originally present, it may be removed first by the addition of silver sulphate solution or ammoniacal silver sulphate solution.

5. Chloride in the presence of bromide and/or iodide

Method A This procedure involves the removal of the bromide and iodide (but not chloride) by oxidation with potassium or ammonium peroxodisulphate in the presence of dilute sulphuric acid. The free halogens are thus liberated, and may be removed either by simple evaporation (addition of water may be necessary to maintain the original volume) or by evaporation at about 80° in a stream of air.

$$2I^- + S_2O_8^{2-} \rightarrow I_2\uparrow + 2SO_4^{2-}$$
$$2Br^- + S_2O_8^{2-} \rightarrow Br_2\uparrow + 2SO_4^{2-}$$

It must be emphasized that these reactions take place only in acid media. Add solid potassium peroxodisulphate and dilute sulphuric acid to the 'neutralized soda extract' of the mixed halides contained in a conical flask; heat the flask to about 80°, and aspirate a current of air through the solution with the aid of a filter pump until the solution is colourless (Fig. IV.2. T is a drawn out capillary

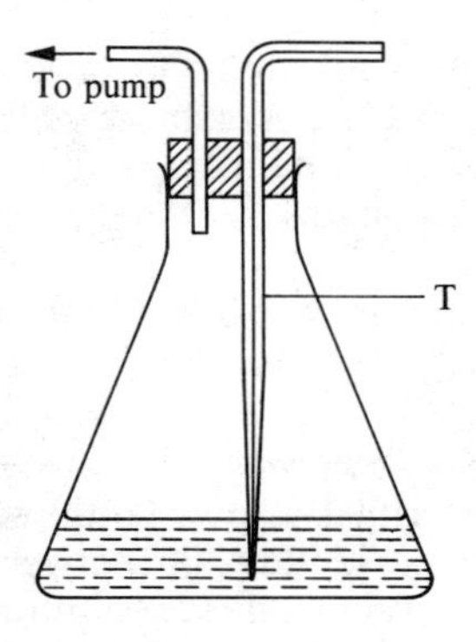

Fig. IV.2

tube). Add more potassium peroxodisulphate or water as may be found necessary. Test the residual colourless liquid for chloride with silver nitrate solution and dilute nitric acid.

Do not mistake precipitated silver sulphate for silver chloride. In a modification of this method, lead dioxide is substituted for potassium peroxodisulphate. Acidify the solution with acetic acid, add lead dioxide, and boil the mixture until bromine and iodine are no longer evolved. Filter, and test the filtrate, which should be colourless, with silver nitrate solution and dilute nitric acid.

Method B Acidify the neutralized soda extract (see 20 below) of the halides with dilute nitric acid, add excess silver nitrate solution, filter, and wash until the washings no longer give the reactions of silver ions. Shake the precipitate with ammonium carbonate solution, filter, and add a few drops of potassium bromide solution to the filtrate; a yellowish-white precipitate of silver bromide is obtained if a chloride is present.

The hydrolysis of the ammonium carbonate in aqueous solution gives rise to free dilute ammonia solution in which silver chloride, but not silver bromide or silver iodide, is appreciably soluble. The addition of bromide ions to the solution of silver chloride in ammonia results in the solubility product of silver bromide being exceeded, and precipitation occurs.

Alternatively, wash the precipitated halides until free from excess silver nitrate solution, and then treat the precipitate with a cold, freshly prepared solution containing 4 per cent formaldehyde in 0·05M sodium carbonate solution: pour a little of the latter reagent through the filter several times. Test the filtrate for chloride with silver nitrate solution and dilute nitric acid. Silver bromide reacts slightly with the reagent yielding a faint opalescence; silver iodide is unaffected.

6. Chloride in the presence of bromide Warm the solid mixture* with a little potassium dichromate and concentrated sulphuric acid in a small distilling flask (Fig. IV.1), and pass the vapours, which contain chromyl chloride, into sodium hydroxide solution. Test for chromate, which proves the presence of a chloride, with hydrogen peroxide and amyl alcohol or with the diphenylcarbazide reagent.

7. Chloride in the presence of iodide Add excess silver nitrate solution to the neutralized soda extract (see 20) and filter; reject the filtrate. Wash the precipitate with dilute ammonia solution and filter again. Add dilute nitric acid to the washings; a white precipitate of silver chloride indicates the presence of chloride.

The separation is based upon the solubility of silver chloride in dilute ammonia solution and the practical insolubility of silver iodide in this reagent.

8. Bromide and iodide in the presence of each other and of chloride The presence of a chloride does not interfere with the reactions described below. To the soda extract, strongly acidified with dilute hydrochloric acid add 1–2 drops of chlorine water (the solution obtained by carefully acidifying a dilute solution of sodium hypochlorite with dilute hydrochloric acid may also be used) and 2–3 ml chloroform or carbon tetrachloride; shake; a violet colour indicates iodide. Continue the addition of chlorine water or of acidified sodium hypochlorite solution drop by drop to oxidize the iodine to iodate and shake after each addition. The violet colour will disappear, and a reddish-brown colouration of the chloroform or carbon tetrachloride, due to dissolved bromine (and/or bromine monochloride, BrCl), will be obtained if a bromide is present. If iodide alone is present, the solution will be colourless after the violet colour has disappeared.

9. Chloride, chlorate, and perchlorate in the presence of each other Each anion must be tested for separately. Divide the soda extract into 3 parts.

a. Chloride Acidify with dilute nitric acid and boil to expel carbon dioxide. Add silver nitrate solution; a white precipitate of silver chloride indicates the presence of chloride. Silver chlorate and perchlorate are soluble in water.

b. Chlorate Any of the following methods may be used.

(*a*) Acidify with dilute nitric acid and add excess silver nitrate solution; filter off the precipitated silver chloride. Now introduce a little chloride-free sodium nitrite (which reduces the chlorate to chloride) and more silver nitrate solution into the filtrate; a white, curdy precipitate of silver chloride indicates the presence of chlorate.

Alternatively, the reduction may be effected by heating the filtrate with a few ml of 10 per cent formaldehyde solution: a white precipitate of silver chloride forms if a chlorate is present.

$$ClO_3^- + 3H.CHO \rightarrow Cl^- + 3H.COOH$$

* If a solution is being analysed, it should be evaporated to dryness before applying the test. Chlorates must, of course, be absent.

(*b*) Acidify with dilute sulphuric acid, add a little indigo solution followed by sulphurous acid solution: the indigo is bleached if a chlorate is present.

c. Perchlorate Pass excess sulphur dioxide into the solution to reduce chlorate to chloride, boil off the excess sulphur dioxide, and precipitate the chloride with silver sulphate solution, afterwards removing the excess silver with a solution of sodium carbonate. Evaporate the resulting solution to dryness and heat to dull redness (better in the presence of a little halide-free lime) to convert the perchlorate into chloride. Extract the residue with water, and test for chloride with silver nitrate solution and dilute nitric acid.

10. Iodate and iodide in the presence of each other The addition of dilute acid to a mixture of iodate and iodide results in the separation of free iodine, due to the interaction between these ions:

$$IO_3^- + 5I^- + 6H^+ \rightarrow 3I_2 + 3H_2O$$

Neither iodates nor iodides alone do this when acidified with dilute hydrochloric, sulphuric or acetic acids in the cold.

Test for iodide in the neutralized soda extract or in a solution of the sodium salts by the addition of a few drops of chlorine water (or acidified sodium hypochlorite solution) and 2–3 ml chloroform or carbon tetrachloride; the latter is coloured violet. Add excess silver sulphate solution to another portion of the neutral solution and filter off the silver iodide; remove the excess silver sulphate with sodium carbonate solution. Pass sulphur dioxide into the filtrate to reduce iodate to iodide, boil off the excess sulphur dioxide, and add silver nitrate solution and dilute nitric acid. A yellow precipitate of silver iodide confirms the presence of iodate in the original substance.

11. Phosphate in the presence of arsenate Both arsenate and phosphate give a yellow precipitate on warming with ammonium molybdate solution and nitric acid, the latter on gently warming and the former on boiling.

Acidify the soda extract with dilute hydrochloric acid, pass in sulphur dioxide to reduce the arsenate to arsenite, boil off the excess sulphur dioxide (test with potassium dichromate paper), and pass hydrogen sulphide into the solution to precipitate the arsenic as arsenic(III) sulphide; continue the passage of hydrogen sulphide until no more precipitate forms. Filter, boil off the hydrogen sulphide, and test the filtrate for phosphate by the ammonium molybdate test or with the magnesium nitrate reagent.

An alternative method for the elimination of arsenate is the following. Acidify the soda extract with dilute hydrochloric acid and then add about one-quarter of the volume of concentrated hydrochloric acid (the total volume should be about 10 ml). Add 0·5 ml of 10 per cent ammonium iodide solution, heat to boiling, and pass hydrogen sulphide into the boiling solution until precipitation is complete (5–10 minutes). Filter off the arsenic(III) sulphide, and boil off the hydrogen sulphide from the filtrate. Add ammonia solution until alkaline, and excess of the magnesium nitrate reagent. A white precipitate indicates the presence of a phosphate.

$$AsO_4^{3-} + 2I^- + 8H^+ \rightarrow As^{3+} + I_2 + 4H_2O$$

$$2As^{3+} + 3H_2S \rightarrow As_2S_3\downarrow + 6H^+$$

$$I_2 + H_2S \rightarrow S\downarrow + 2I^- + 2H^+$$

12. Phosphate, arsenate, and arsenite in the presence of each other Treat the soda extract with dilute sulphuric acid until acidic, warm to expel carbon dioxide, render just alkaline with dilute ammonia solution and filter, if necessary. Add a few millilitres of the magnesium nitrate reagent and allow to stand for 5–10 minutes; shake or stir from time to time. Filter off the white, crystalline precipitate of magnesium ammonium phosphate and/or magnesium ammonium arsenate(*A*), and keep the filtrate (*B*). Test for arsenite in the filtrate (*B*) by acidifying with dilute hydrochloric acid and passing hydrogen sulphide, when a yellow precipitate of arsenic(III) sulphide is immediately produced if arsenite is present.

Wash the precipitate (*A*) with 2M ammonia solution, and remove half of it to a small beaker with the aid of a clean spatula. Treat the residue on the filter with a little silver nitrate solution containing a few drops of dilute acetic acid: a brownish-red residue (due to Ag_3AsO_4) indicates arsenate present. If the residue is yellow (largely Ag_3PO_4), pour 6M hydrochloric acid through the filter a number of times, and add a little potassium iodide solution and 1–2 ml chloroform or carbon tetrachloride to the extract and shake; if the organic layer acquires a purple colour, an arsenate is present.

Dissolve the white precipitate in the beaker in dilute hydrochloric acid, reduce the arsenate (if present) with sulphur dioxide or, better, with ammonium iodide solution (as described under 11), precipitate the arsenic as arsenic(III) sulphide with hydrogen sulphide and boil off the hydrogen sulphide in the filtrate. Render the filtrate (*C*) slightly ammoniacal and add a little magnesium nitrate reagent: white, crystalline magnesium ammonium phosphate will be precipitated if a phosphate is present. Filter off the precipitate, wash with a little water, and pour a little silver nitrate solution (containing a few drops of dilute acetic acid) over it: yellow silver phosphate will be formed. Alternatively, the ammonium molybdate test may be applied to the filtrate (*C*) after evaporating to a small volume.

13. Sulphide, sulphite, thiosulphate, and sulphate in the presence of each other Upon the addition of dilute acid to the mixture, the hydrogen sulphide, liberated from the sulphide, and the sulphur dioxide, liberated from the sulphite and thiosulphate, react and sulphur is precipitated; this complication necessitates the use of a special procedure for their separation (see Table IV.2).

A mixture of the sodium salts or soda extract is employed for this separation.

An alternative procedure for the removal of the sulphide is the following. Treat the solution of the sodium salts with an alkaline solution of hexamminezincate hydroxide $[Zn(NH_3)_6](OH)_2$ prepared by adding ammonia solution to zinc nitrate solution until the initial precipitate of zinc hydroxide has redissolved, and then adding a further excess:

$$Zn^{2+} + 8NH_3 + 2H_2O \rightarrow [Zn(NH_3)_6]^{2+} + 2OH^- + 2NH_4^+$$

when zinc sulphide will be precipitated and is filtered off. The precipitate may be treated with dilute hydrochloric acid and a few drops of copper sulphate solution, when a black precipitate of copper(II) sulphide will indicate sulphide, or it may be tested for sulphide in the usual manner.

A simple method for separating sulphite and sulphate consists in acidifying the neutral solution of the alkali salts with dilute hydrochloric acid and adding excess barium chloride solution; the sulphate is precipitated as barium sulphate

Table IV.2 Separation of sulphide, sulphite, thiosulphate, and sulphate Shake the solution with excess of freshly precipitated $CdCO_3$, and filter.

<table>
<tr><th>Residue</th><th colspan="3">Filtrate</th></tr>
<tr><td rowspan="5">CdS and excess of $CdCO_3$. Wash and reject washings. Digest residue with dilute acetic acid to remove excess carbonate. A yellow residue indicates sulphide. Confirm by warming with dilute HCl and test the evolved H_2S with lead acetate paper.</td><td colspan="3">Add $Sr(NO_3)_2$ solution* in slight excess, shake, allow to stand overnight and filter.</td></tr>
<tr><td colspan="2">Residue</td><td>Filtrate</td></tr>
<tr><td colspan="2">$SrSO_3$ and $SrSO_4$. Wash, treat with dilute HCl and filter.</td><td rowspan="3">Contains SrS_2O_3. Acidify with dilute HCl and boil; SO_2 is evolved and S is slowly precipitated.
Thiosulphate present.</td></tr>
<tr><td>Residue</td><td>Filtrate</td></tr>
<tr><td>$SrSO_4$.
White.
Sulphate present.
Confirm by fusion with Na_2CO_3 and apply sodium nitroprusside test.</td><td>Contains sulphurous acid. Add a few drops of a dilute solution of iodine; the latter is decolourized.†
Sulphite present.</td></tr>
</table>

* Solubility in g ℓ^{-1}: SrS_2O_3, 250 at 13°; $SrSO_3$, 0·033 at 17°.
† Alternatively, add a few drops of fuchsin reagent; if it is decolourized, sulphite is present.

and is removed by filtration. The production of a further precipitate of barium sulphate upon the addition to the filtrate of a little concentrated nitric acid or bromine water, and warming, proves the presence of sulphite.

14. Sulphide, sulphite, and thiosulphate in the presence of each other It is assumed that the solution is slightly alkaline and contains the sodium salts of the anions, e.g. the soda extract is used.

(i) Test a portion of the solution for sulphide with sodium nitroprusside solution. The formation of a purple colouration indicates the presence of a sulphide.

(ii) If sulphide is present, remove it by shaking with freshly precipitated cadmium carbonate and filter. Test a portion of the filtrate with sodium nitroprusside solution to see that all the sulphide has been removed. When no sulphide is present, treat the remainder of the filtrate with a drop of phenolphthalein solution and pass in carbon dioxide until the solution is decolourized by it. Test 2–3 ml of the colourless solution with 0·5–1 ml of the fuchsin reagent. If the reagent is decolourized, the presence of sulphite is indicated.

(iii) Treat the remainder of the colourless solution with dilute hydrochloric acid and boil for a few minutes. If sulphur separates, a thiosulphate is indicated. Alternatively, the nickel ethylenediamine nitrate test (Section **IV.5**, reaction 9) may be applied.

15. Borate in the presence of copper and barium salts When carrying out the ethyl borate test for borates (Section **IV.23**, reaction 2), it must be remembered that copper and barium salts may also impart a green colour to the alcohol

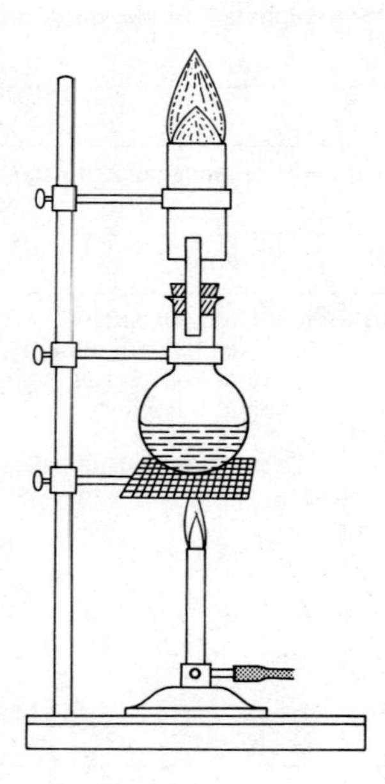

Fig. IV.3

flame. The test may be carried out in the following way when salts of either or both of these metals are present. The mixture of the borate, concentrated sulphuric acid, and ethanol is placed in a small round-bottomed flask, fitted with a glass jet and surmounted by a wide glass tube, which acts as a 'chimney' (Fig. IV.3). The mixture is gently warmed, and the vapours ignited at the top of the wide glass tube. A green flame confirms the presence of a borate.

An alternative procedure is given in Section **V.2**, test 10.

16. Fluoride, hexafluorosilicate, and sulphate in the presence of each other The following differences in solubilities of the lead salts are utilized in this separation: lead hexafluorosilicate is soluble in water; lead fluoride is insoluble in water but soluble in dilute acetic acid; lead sulphate is insoluble in water and in boiling dilute acetic acid, but soluble in concentrated ammonium acetate solution.

Table IV.3 Detection of fluoride, hexafluorosilicate, and sulphate in the presence of each other Add excess lead acetate solution to the solution of the alkali metal salts, and filter cold.

Residue	**Filtrate**
Wash well with cold water. Divide into two parts. (i) Smaller portion. Add excess dilute acetic acid and boil (this will dissolve any lead fluoride). White residue indicates sulphate. The residue is soluble in ammonium acetate solution. (ii) Larger portion. Treat cautiously with concentrated sulphuric acid and test with a moist glass rod. Milkiness of the water and etching of the tube indicate fluoride.	Add $Ba(NO_2)_2$ solution and warm gently. White crystalline precipitate indicates hexafluorosilicate.

17. Oxalate in the presence of fluoride Both calcium fluoride and calcium oxalate are precipitated by ammonium oxalate solution in the presence of dilute acetic acid. The fluoride may be identified in the usual manner with concentrated sulphuric acid or as described below. The oxalate is most simply detected by dissolving a portion of the precipitate in hot dilute sulphuric acid and then adding a few drops of a very dilute solution of potassium permanganate. The latter will be decolourized if an oxalate is present.

Use may be made of the fact that even solid calcium fluoride reacts with the zirconium–alizarin-S reagent (compare Section **IV.17**, reaction 6) and, in consequence, the fluoride test may be carried out in the presence of oxalate and phosphate, which interfere in aqueous solution. The calcium salts are precipitated in neutral or faintly basic solution. The precipitate is ignited and digested with dilute acid. The residue is then tested for fluoride by the zirconium–alizarin-S test; the red hue of the reagent disappears and a yellow colouration results (see Section **IV.17**, reaction 6).

18 Chloride and cyanide in the presence of each other Both silver chloride and silver cyanide are insoluble in water, but soluble in dilute ammonia solution. Three methods are available for the detection of cyanide in the presence of chloride.

(i) The concentrated solution, or preferably the solid mixture of sodium salts is treated with about 5 times its weight of '100 volume' hydrogen peroxide and the mixture gently warmed; ammonia is evolved, which is recognized by its action upon mercury(I) nitrate paper. The solution is then boiled to decompose all the hydrogen peroxide, and then tested for chloride with silver nitrate solution and dilute nitric acid.

$$CN^- + H_2O_2 \rightarrow OCN^- + H_2O;$$
$$OCN^- + 2H_2O \rightarrow HCO_3^- + NH_3$$

(ii) The cyanide is precipitated as the pale-green nickel cyanide by the addition of excess of a solution of nickel sulphate to the neutral solution; the whole is filtered and the excess nickel sulphate removed by boiling with sodium hydroxide solution and filtering off the precipitate of nickel hydroxide. The filtrate is acidified with dilute nitric acid and excess silver nitrate solution added. A white, curdy precipitate of silver chloride confirms the presence of chloride.

(iii) The solution of the mixed sodium salts is treated with excess solid sodium hydrogen carbonate (see Fig. IV.1), the mixture warmed, and the evolved hydrogen cyanide passed into silver nitrate solution acidified with dilute nitric acid; a white precipitate of silver cyanide is obtained. This method can be employed in the presence of hexacyanoferrate(II), hexacyanoferrate(III), and thiocyanate ions.

For the detection of cyanide, chloride, bromide, and iodide in the presence of one another the cyanide is first completely removed, and the separation of chloride, bromide, and iodide carried out as detailed under 8.

The cyanide can also be detected by the usual Prussian blue test or 'iron(III) thiocyanate' test (Section **IV.8**, reactions 4 and 6).

19. Hexacyanoferrate(II), hexacyanoferrate(III), and thiocyanate in the presence of each other

Table IV.4 Separation of hexacyanoferrate(II), hexacyanoferrate(III), and thiocyanate
Acidify the neutralized soda extract of the sodium salts with acetic acid, add excess of a solution of thorium nitrate and a little Gooch asbestos and shake well; filter.

<table>
<tr><th>Residue</th><th colspan="2">Filtrate</th></tr>
<tr><td rowspan="3">$Th[Fe(CN)_6]$ and asbestos (the latter is added to facilitate filtration of the gelatinous precipitate).* Wash with a little cold water. Treat the ppt. on the filter paper with dilute NaOH solution; acidify the alkaline extract with dilute HCl and add a few drops of $FeCl_3$ solution.
A precipitate of Prussian blue indicates hexacyanoferrate(II).</td><td colspan="2">Add $CdSO_4$ solution and a little Gooch asbestos, shake, filter.</td></tr>
<tr><th>Residue</th><th>Filtrate</th></tr>
<tr><td>Orange. $Cd_3[Fe(CN)_6]_2$ and asbestos. Extract with NaOH solution, acidify extract wit dilute HCl, add freshly prepared $FeSO_4$ solution. Precipitate of Prussian blue indicates hexacyanoferrate(III).</td><td>Add $FeCl_3$ solution and ether. Red colouration of ether proves presence of thiocyanate.</td></tr>
</table>

* Filter paper pulp (e.g. in the form of a Whatman filtration accelerator) cannot be used as the organic matter may cause appreciable reduction of the hexacyanoferrate(III).

20. Detection of organic acids (oxalate, tartrate, citrate, benzoate, succinate, acetate, formate, and salicylate) in the presence of each other Intimately mix the mixture of acids and/or their salts with 3–4 times the bulk of anhydrous sodium carbonate, add just sufficient water to dissolve the sodium carbonate, heat under reflux for 15 minutes, and filter. Wash and neglect the residue. The filtrate will be termed the soda extract. Treat the combined filtrate and washings with dilute nitric acid a little at a time with stirring until acid (use litmus paper test), boil gently to expel carbon dioxide and allow to cool. Add dilute ammonia solution portionwise and with stirring until just alkaline. Filter, if not clear. It is advisable to boil gently for a minute or two to remove any appreciable excess of ammonia. The resulting solution will be free from heavy metals and almost neutral. It will be referred to as the neutralized soda extract (see Table IV.5).

Note: With Fenton's reagent and a neutral solution of a citrate, a yellow or brown colour is obtained, but there is no precipitation of iron; with an oxalate, the iron is precipitated as the hydroxide.

Procedure A Separation of benzoate and succinate Boil the precipitated iron(III) salts with a little dilute ammonia solution, filter off the precipitated iron(III) hydroxide and discard the precipitate; boil off the excess ammonia from the filtrate. Divide the filtrate into two parts.

Treat one portion with barium chloride solution and filter off the precipitate formed. The residue consists of barium succinate. Confirm the succinate by the fluorescein test. Acidify the second portion of the filtrate with dilute hydrochloric acid, when benzoic acid will separate out on cooling. It can be identified by its melting point.

Table IV.5 Separation and detection of organic acids* Add $CaCl_2$ solution to the cold neutralized soda extract, rub sides of vessel with a glass rod, allow to stand for 10 minutes with occasional shaking; filter.

<table>
<tr><th colspan="2">Residue</th><th colspan="4">Filtrate</th></tr>
<tr><td colspan="2" rowspan="2">May be Ca oxalate (formed immediately) or Ca tartrate (formed on standing).†
Wash. Boil with dilute acetic acid and filter.</td><td colspan="4">Add more $CaCl_2$ solution, boil under reflux for at least 5 minutes. A ppt. may form gradually; filter.</td></tr>
<tr><th>Residue</th><th colspan="3">Filtrate</th></tr>
<tr><th>Residue</th><th>Filtrate</th><td rowspan="5">Ca citrate. White.
Citrate present.
Divide into 2 parts.
(i) Confirm by Deniges's test.
(ii) Dissolve in dilute HCl, convert into neutral solution with NH_3 solution, etc., and add $CdCl_2$ solution. White gelatinous precipitate.</td><td colspan="3">If citrate present, evaporate to dryness on water bath, add a little cold water, and filter from any residue of Ca citrate.
If citrate absent, proceed with solution.
Add $FeCl_3$ solution; filter (if necessary).</td></tr>
<tr><td rowspan="4">CaC_2O_4.
Dissolve in a little dilute HCl, add NH_3 solution in slight excess White ppt. insoluble in dilute acetic acid.
Oxalate present.
Confirm by resorcinol test, or by decolourization of a dilute solution of acidified $KMnO_4$ at 60–70°.</td><td rowspan="4">May contain Ca tartrate. Test neutral solution with Fenton's reagent, violet colouration (see Note). or by silver mirror test for tartrate. Confirm by copper hydroxide test.</td><th>Residue</th><th colspan="2">Filtrate</th></tr>
<tr><td rowspan="3">Iron (III) benzoate and/or iron(III) succinate. Separate by procedure A.</td><td colspan="2">Coloured. Dilute, boil and filter.</td></tr>
<tr><th>Residue</th><th>Filtrate</th></tr>
<tr><td>Basic iron(III) formate and basic iron(III) acetate. Separate by procedure B.</td><td>Coloured.
(a) Violet – salicylate. Confirm by 'oil of wintergreen' test.
(b) Deep blue or greenish black – gallate or tannate.‡</td></tr>
</table>

* See Section **V.18**, Table V.29 for a slightly modified procedure.

† It must be remembered that phosphates, fluorides, sulphates, etc., are precipitated by calcium chloride solution.

‡ Gallates and tannates are of rare occurrence in general qualitative analysis.

Procedure B Formates and acetates in the presence of each other **Formate.**
(i) In the absence of tartrates, citrates, and reducing agents. Add silver nitrate solution to the neutralized soda extract; a white precipitate, which is converted into a black deposit of silver on boiling, indicates formate. A little of the solid substance may be treated with concentrated sulphuric acid, when carbon monoxide (burns with blue flame) will be evolved in the cold.

(ii) In the presence of tartrates, citrates, and reducing agents. Acidify the mixture with dilute sulphuric acid and distil. A solution of formic and acetic acids will pass over. Neutralize the distillate with dilute ammonia solution, boil off excess ammonia (if necessary) and test as in (i).

Acetate If other organic acids are present, obtain the solution of mixed acids as in (ii); otherwise, use the neutralized soda extract. Boil under reflux with an equal volume of potassium dichromate solution and dilute sulphuric acid; this will decompose the formic acid. Distil off the acetic acid, neutralize and test with iron(III) chloride solution.

If the solid is available, the cacodyl test may be employed.

For other methods of separation and detection of selected anions, see Section **VI.7.**

CHAPTER V SYSTEMATIC QUALITATIVE INORGANIC ANALYSIS

V.1 INTRODUCTION In the scheme of analysis to be described in the following pages, it is assumed that the student is already familiar with the tests and operations described in the preceding chapters. It will be shown how these isolated facts are incorporated in the systematic methods of qualitative analysis, applicable not only to simple solid substances, but also to mixtures of solid substances, to liquids, to alloys, and to 'insoluble' substances, i.e. substances which are insoluble in aqua regia and acid solvents.

It must be emphasized that the object of qualitative analysis is not simply to detect the constituents of a given mixture; an equally important aim is to ascertain the approximate relative amounts of each component. For this purpose 0·5–1 g of the substance is usually employed for the analysis; the relative magnitudes of the various precipitates will provide a rough guide as to the proportions of the constituents present.

Every analysis is divided into three parts:

1. **The preliminary examination.** This includes preliminary examination by dry tests, examination of the volatile products with sodium hydroxide solution (for ammonium), and with dilute and concentrated sulphuric acid (for acid radicals or anions).
2. **The examination for metal ions (cations) in solution.**
3. **The examination for anions in solution.**

The substance to be analysed may be: (*A*) solid and non-metallic, (*B*) a liquid (solution), (*C*) a metal or an alloy, and (*D*) an 'insoluble' substance. Each of these will be discussed separately.

V.2 PRELIMINARY TESTS ON NON-METALLIC SOLID SAMPLES

1. Appearance The appearance of the substance should be carefully noted; a lens or microscope should be used if necessary. Observe whether it is crystalline or amorphous, whether it is magnetic and whether it possesses any characteristic odour or colour.

Some of the commonly occurring coloured compounds are listed below:

Red: Pb_3O_4, As_2S_2, HgO, HgI_2, HgS, Sb_2S_3, CrO_3, Cu_2O, $K_3[Fe(CN)_6]$; dichromates are orange-red; permanganates and chrome alum are reddish-purple.

Pink: hydrated salts of manganese and of cobalt.

Yellow: CdS, As_2S_3, SnS_2, PbI_2, HgO (precipitated), $K_4[Fe(CN)_6].3H_2O$; chromates; iron(III) chloride and nitrate.

Green: Cr_2O_3, Hg_2I_2, $Cr(OH)_3$; iron(II) salts, e.g. $FeSO_4.7H_2O$,

$FeSO_4.(NH_4)_2SO_4.6H_2O$, $FeCl_2.4H_2O$; nickel salts; $CrCl_3.6H_2O$, $CuCl_2.2H_2O$, $CuCO_3$, K_2MnO_4.

Blue: anhydrous cobalt salts; hydrated copper(II) salts; Prussian blue.

Brown: PbO_2, $CdO.Fe_3O_4$, Ag_3AsO_4, SnS, Fe_2O_3 and $Fe(OH)_3$ (reddish-brown).

Black: PbS, CuS, CuO, HgS, FeS, MnO_2, Co_3O_4, CoS, NiS, Ni_2O_3, Ag_2S, C.

The colour of the solution obtained when the substance is dissolved in water or in dilute acids should be noted, as this may often give valuable information. The following colours are shown by the ions (the cations are usually hydrated) present in the dilute solution:

Blue: copper(II); green: nickel, iron(II), chromium(III), manganates; yellow: chromates, hexacyanoferrate(II), iron(II); orange-red: dichromates; purple: permanganates; pink: cobalt, manganese(II).

The substance should be reduced to a fine powder in a suitable mortar before proceeding with the following tests. These tests usually give a great deal of useful information; they are quickly performed (10–15 minutes) and should never be omitted.

2. Heating in a test tube Place a small quantity (4–5 mg) of the substance in a dry test-tube so that none of it remains adhering to the sides, and heat cautiously; the tube should be held in an almost horizontal position. The temperature is gradually raised, and any changes which take place carefully noted.

Table V.1 Heating test

Observation	Inference
(*a*) The substance changes colour.	
1. Blackening from separation of carbon, often accompanied by burning.	Organic substances, e.g. tartrates and citrates.
2. Blackening, not accompanied by burning or odour.	Cu, Mn, and Ni salts at a very high temperature.
3. Yellow when hot, white when cold.	ZnO and many Zn salts.
4. Yellowish-brown when hot, yellow when cold.	SnO_2 or Bi_2O_2.
5. Yellow when hot, yellow when cold.	PbO and some Pb salts.
6. Brown when hot, brown when cold.	CdO and many Cd salts.
7. Red to black when hot, brown when cold.	Fe_2O_3.
(*b*) A sublimate is formed.	
1. White sublimate.*	$HgCl_2$, $HgBr_2$, Hg_2Cl_2, ammonium halides, As_2O_2, Sb_2O_3, certain volatile organic compounds (oxalic acid, benzoic acid).
2. Grey sublimates, easily rubbed to globules.	Hg.
3. Steel-grey sublimate; garlic odour.	As.
4. Yellow sublimate.	S (melts on heating), As_2S_3, HgI_2 (red when rubbed with a glass rod).
5. Blue-black sublimate; violet vapour.	I.
6. Black; red on trituration.	HgS.

* If a white sublimate forms, heat with 4 times the bulk of anhydrous sodium carbonate and a little potassium cyanide in an ignition tube. A grey mirror, convertible into globules on rubbing with a glass rod, indicates Hg; a brownish-black mirror, yielding a white sublimate and an odour of garlic when heated in a wide tube, indicates As; ammonia evolved (test with mercury(I) nitrate paper) indicates ammonium salts.

Table V.1 Heating test (contd.)

Observation	Inference
(*c*) A gas or vapour is evolved.	
1. Water is evolved; test with litmus paper.	Compounds with water of crystallization (often accompanied by change of colour), ammonium salts, acid salts, oxy-acids, hydroxides.
The water is alkaline.	Ammonium salts.
The water is acid	Readily decomposed salts of strong acids, also acids.
2. Oxygen is evolved (rekindles glowing splint).	Nitrates, chlorates, perchlorates, bromates, iodates, peroxides, per-salts, and permanganates.
3. Dinitrogen oxide (rekindles glowing splint) and steam are evolved.	Ammonium nitrate or nitrate mixed with an ammonium salt.
4. Dark-brown or reddish fumes (oxides of nitrogen); acidic in reaction.	Nitrates or nitrites of heavy metals.
5. Carbon dioxide is evolved (lime water rendered turbid).	Carbonates, hydrogen carbonates, oxalates, and organic compounds.
6. Carbon monoxide is evolved (burns with a blue flame forming carbon dioxide); poisonous gas.	Oxalates.
7. Cyanogen is evolved (burns with violet flame and characteristic odour); very poisonous gas.	Cyanides of heavy metals, e.g. of Hg and Ag; $K_3[Fe(CN)_6]$.
8. Acetone is evolved (burns with luminous flame).	Acetates.
9. Ammonia is evolved (odour; turns red litmus paper blue; turns mercury(I) nitrate paper black).	Ammonium salts; certain complex ammines.
10. Phosphine is evolved (odour of fish; inflammable); very poisonous.	Phosphites and hypophosphites.
11. Sulphur dioxide is evolved (odour of burning sulphur; turns potassium dichromate paper green; decolourizes fuchsin solution).	Normal and acid sulphites; thiosulphates; certain sulphates.
12. Hydrogen sulphide is evolved (odour of rotten eggs; turns lead acetate paper black or cadmium acetate paper yellow).	Acid sulphides; hydrated sulphides.
13. Chlorine is evolved (yellowish-green gas; bleaches litmus paper; turns potassium iodide–starch paper blue); very poisonous.	Unstable chlorides, e.g. of Cu, Au and Pt; chlorides in presence of oxidizing agents.
14. Bromine is evolved (reddish-brown vapour; choking odour; turns fluorescein paper red).	Sources similar to chlorine.
15. Iodine is evolved (violet vapours condensing to black crystals).	Free iodine and certain iodides.

3. Flame colourations Place a small quantity (3–4 mg) of the substance on a watch glass, moisten with a little concentrated hydrochloric acid, and introduce a little of the substance on a clean platinum wire into the base of the non-luminous Bunsen flame (Table V.2). An alternative method is to dip the platinum wire into concentrated hydrochloric acid contained in a watch glass and then into the substance; sufficient will adhere to the platinum wire for the test to be carried out.

The sodium flame masks that of other elements, e.g. that of potassium. Mixtures can be readily detected with the direct vision spectroscope (see Fig. II.4). A less delicate method is to view the flame through two thicknesses of cobalt blue glass, whereby the yellow colour due to sodium is masked or absorbed, and the other colours are modified as listed in Table V.3.

Table V.2 Flame test

Observation	Inference
Persistent golden-yellow flame.	Sodium.
Violet (lilac) flame.	Potassium.
Carmine-red flame.	Lithium.
Brick-red (yellowish-red) flame.	Calcium.
Crimson flame.	Strontium.
Yellowish-green flame.	Barium (molybdenum).
Green flame.	Borates, copper (thallium).
Livid blue flame (wire slowly corroded).	Lead, arsenic, antimony, bismuth, copper.

Table V.3 Flame test with cobalt glass

Flame colouration	Flame colouration through cobalt glass	Inferences
Golden-yellow	Nil.	Sodium.
Violet.	Crimson.	Potassium.
Brick-red.	Light-green.	Calcium.
Crimson.	Purple.	Strontium.
Yellowish-green.	Bluish-green.	Barium.

Table V.4 Charcoal block reduction test

Observation	Inference
1. The substance decrepitates.	Crystalline salts, e.g. NaCl, KCl.
2. The substance deflagrates.	Nitrates, nitrites, chlorates, perchlorates, iodates, permanganates.
3. The substance fuses and is absorbed by the charcoal, or forms a liquid bead.	Salts of the alkalis and some salts of the alkaline earths.
4. The substance is infusible and incandescent, or forms an incrustation upon the charcoal.	Apply test (*b*) below.

Table V.5 Ignition with Na_2CO_3 on charcoal

Observation	Inference
1. White, infusible and incandescent when hot.	BaO, SrO, CaO, MgO (residue alkaline to litmus paper). Al_2O_3, ZnO, SiO_2 (residue not alkaline to litmus paper).
2. Incrustation without metal:	
White, yellow when hot.	ZnO.
White, garlic odour.	As_2O_3.
Brown.	CdO.
3. Incrustation with metal:	
White incrustation; brittle metal.	Sb.
Yellow incrustation; brittle metal.	Bi.
Yellow incrustation; malleable metal, marks paper.	Pb.
4. Metal without incrustation:	
Grey metallic particles, attracted by magnet.	Fe, Ni, Co.
Malleable beads.	Ag and Sn (white), Cu (red flakes) Au.

4. Charcoal block reductions (*a*) Heat a little of the substance (3–4 mg) in a small cavity scooped in a charcoal block in a blowpipe flame (Table V.4).

(*b*) Mix the substance (3–4 mg) with twice its bulk of anhydrous sodium carbonate, place the mixture in a cavity of a piece of charcoal and heat in the reducing flame of the blowpipe.

The sodium carbonate converts a metallic salt into a carbonate or oxide on heating, and thus reduction occurs more rapidly than with the charcoal alone, as in (*a*). Further, the sodium carbonate acts as a flux and, in the fused state, protects any metallic globules, which may have formed beneath it, from oxidation.

Sulphur compounds are reduced to sulphide by this treatment; the residue may be moistened with water and placed in contact with a silver coin when a brown to black stain of silver sulphide is obtained (Hepar reaction), or it may be extracted with a little water and filtered into a freshly prepared sodium nitroprusside solution, when an unstable purple colouration will indicate the presence of sulphur (see Section **IV.6**, reaction 5) (Table V.5).

(*c*) Moisten the substance or the infusible residue of test (*b*) with one or two drops of cobalt nitrate solution and ignite strongly (Table V.6).

Table V.6 Ignition with Co-salts

Observation	Inference
1. Blue residue.	Al_2O_3, phosphates, arsenates, silicates, borates.
2. Green residue.	ZnO.
3. Pink residue.	MgO.

5. Borax bead reactions (Table V.7.) Prepare a borax bead in a loop of platinum wire by dipping the hot wire into borax and heating until colourless and transparent. Bring a minute quantity of the substance into contact with the hot bead and heat in the outer or oxidizing flame. Observe the colour when the bead is hot and also when it is cold. Heat the bead in the inner or reducing flame and observe the colours in the hot and cold states. Coloured beads are obtained with compounds of copper, iron, chromium, manganese, cobalt, and nickel (see however, Section **VII.25**, Table VII.8).

Table V.7 Borax bead test

Oxidizing flame	Reducing flame	Metal
1. Green when hot; blue when cold.	Colourless when hot, opaque-red when cold.	Copper.
2. Yellowish-brown or red when hot; yellow when cold.	Green, hot and cold.	Iron.
3. Dark yellow when hot, green when cold.	Green, hot and cold.	Chromium.
4. Violet (amethyst), hot and cold.	Colourless, hot and cold.	Manganese.
5. Blue, hot and cold.	Blue, hot and cold.	Cobalt.
6. Reddish-brown when cold.	Grey or black and opaque when cold.	Nickel.

The presence of manganese and of chromium is confirmed by fusing the substance with sodium carbonate and potassium nitrate on platinum foil or broken porcelain. A green mass on cooling indicates manganese; a yellow mass, chromium. The sodium carbonate bead test (Section **II.2**, test 7) may also be employed.

A useful reaction which may be carried out at this stage is the microcosmic bead test (Section **II.2**, test 6). This test is carried out in a loop of platinum wire exactly as for the borax bead test. The presence of a white skeleton (of silica) in the coloured glass indicates silicate. Tin(IV) oxide, SnO_2, dissolves slowly in the bead may be mistaken for silica.

6. Test for ammonium ions Boil a little (*c.* 0·1 g) of the substance with sodium hydroxide solution. The evolution of ammonia, detected by its odour and its action upon red litmus paper and upon filter paper soaked in mercury(I) nitrate solution,* indicates the presence of an ammonium salt.

Note. Great care must be exercised when heating mixtures containing solutions of alkali hydroxides because of their destructive effects upon the eyes. These mixtures tend to bump and should preferably be heated in the fume cupboard. Under no circumstances should one attempt to smell the vapour whilst heating the mixture.

The following experimental details in testing for ammonia are of value for other gases (with suitable modification of reagents). In order to avoid holding the test paper (litmus, etc.) in the vapour, the apparatus shown in Fig. V.1 may be employed; the test paper is supported on the upper end of the wide glass tube. If ammonia is present, the litmus paper should show a gradual development of colour from the bottom upwards and should eventually become uniformly blue; scattered blue spots indicate that droplets of the alkaline solution have come into contact with the paper. The spray may be trapped, if desired, by a loosely fitting plug of cotton wool inserted in the upper part of the test-tube.

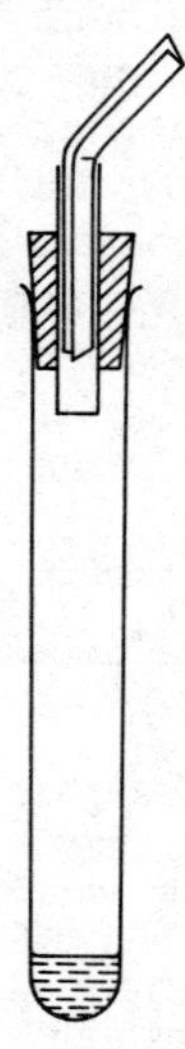

Fig. V.1

* Drop-reaction paper, treated with tannic acid and silver nitrate solution (see Section **III.38**, reaction 7), may also be used.

If the evolved gas is soluble in water and a solution is required for further testing, the apparatus shown in Fig. IV.1 (with the delivery tube of fairly wide tubing and about twice the length of the test-tube or distilling flask in order to avoid the danger of 'sucking back') may be employed. In the present case the ammonia may be absorbed in 3–4 ml of distilled water. Upon adding Nessler's reagent (Section **III.38**, reaction 2 – the second method of preparation is recommended), an orange or brown precipitate confirms ammonia. This test is an extremely sensitive one and in order to establish the presence of ammonia evolved in the reaction a precipitate and not a colouration must be obtained.

Table V.8 Test with dilute H_2SO_4

Observation	Inference
1. Colourless gas is evolved with effervescence; gas is odourless and produces turbidity when passed into lime water.* (*C*)	CO_2 from carbonate or hydrogen carbonate.
2. Nitrous fumes evolved; recognized by reddish-brown colour, and turning starch–potassium iodide paper bluish-black. (*C*)	NO_2 from nitrite.
3. Yellowish-green gas evolved; suffocating odour, reddens and then bleaches litmus paper; turns starch–KI paper blue; very poisonous.	Cl_2 from hypochlorite.
4. Odour of acetylene; burns with luminous, smoky flame. (*C*)	C_2H_2 from carbide.
5. Colourless gas is evolved with suffocating odour; turns filter paper moistened with acidified potassium dichromate solution green; decolourizes fuschsin soln.	SO_2 from sulphite.
6. Colourless gas is evolved; gives above tests for SO_2; sulphur is deposited in the solution.	SO_2 and S from thiosulphate.
7. Colourless gas is evolved; odour of rotten eggs; blackens filter paper moistened with lead acetate solution; cadmium acetate paper turned yellow.	H_2S from sulphide.†
8. Colourless gas is evolved; gives above tests for H_2S; sulphur is deposited.	H_2S and S from polysulphide.
9. Odour of vinegar.	CH_3COOH from acetate.
10. Colourless gas is evolved; odour of bitter almond;‡ highly poisonous.	HCN from cyanide or from soluble hexacyanoferrate(III)s and hexacyanoferrate(II)s.
11. Colourless gas is evolved; rekindles glowing splint.	O_2 from peroxides and peroxo-salts of alkali and alkaline earth metals.
12. Colourless gas is evolved; pungent odour, reminiscent of SO_2; produces turbidity when passed into lime water.	CO_2 and a little HCNO from cyanate.
13. Upon boiling, yellow solution formed and SO_2 (fuchsin solution decolourized, etc.) evolved.	SO_2, etc., from thiocyanate.

* Magnesite, dolomite, and a few other native carbonates give little or no CO_2 in the cold; CO_2 is readily evolved upon warming.

† Many sulphides, especially native ones, are not affected by dilute H_2SO_4; some H_2S is evolved upon warming with concentrated HCl alone or with a little tin.

‡ If cyanide is suspected, gently warm (water bath) 5 mg of the mixture in a small test-tube (e.g. 75×10 mm) with 5–10 drops of M H_2SO_4; place a filter paper moistened at the centre with 1 drop of NaOH solution over the mouth of the tube. After 2 minutes, treat the drop of the paper with 1 drop of $FeSO_4$ solution, warm, and add a drop or two of 6M HCl. A blue colour indicates the presence of a cyanide. For another test, see Section **IV.8**, reaction 1. Mercury(II) cyanide is attacked slowly.

7. ***Action of dilute sulphuric acid*** Treat 0·1 g of the substance in a small test tube with 2 ml M sulphuric acid, and note whether any reaction takes place in the cold (indicated by *C*). Warm gently and observe the effect produced (Table V.8).

8. ***Action of concentrated sulphuric acid*** Treat a small quantity (say, 0·1 g) of the substance with 1–2 ml concentrated sulphuric acid, and warm the mixture gently; incline the mouth of the test tube away from the observer. (If chlorates or permanganates are suspected from the preliminary charcoal reduction tests to be present, very small quantities must be used (about 0·02 g) as a dangerous explosion may occur on warming.)

If the substance reacted with dilute sulphuric acid, the addition of the concentrated acid may result in vigorous reaction and rapid evolution of gas, which may be accompanied by a very fine spray of the acid. In such a case, it is best to add dilute sulphuric acid dropwise to another portion of the substance until reaction ceases, and then to add 2–3 ml concentrated sulphuric acid.

The following results may be obtained (Table V.9).

Table V.9 Test with concentrated H_2SO_4

Observation	Inference
1. Colourless gas evolved with pungent odour and which fumes in the air; white fumes of NH_4Cl in contact with glass rod moistened with concentrated NH_2 solution; Cl_2 evolved on addition of precipitated MnO_2 (bleaches litmus paper; turns KI–starch paper blue).	HCl from chloride.
2. Gas evolved with pungent odour, reddish colour, and fumes in moist air; on addition of precipitated MnO_2, increased amount of red fumes with odour of bromine (fumes colour moist starch paper orange-red or fluorescein paper red).	HBr and Br_2 from bromide.
3. Violet vapours evolved, accompanied by pungent acid fumes, and often SO_2 and even H_2S.	HI and I_2 from iodide.
4. Reddish-brown fumes evolved (similar in colour to bromine); on passing into water, obtain chromic and hydrochloric acids, both readily identified (yellow ppt. of $PbCrO_4$ with excess NH_3 solution, lead acetate solution, and acetic acid; or by 'chromium pentoxide' test).	CrO_2Cl_2 from chloride in presence of chromate.
5. Pungent acid fumes evolved, often coloured brown by NO_2; colour depends upon addition of copper turnings (if nitrites absent).	HNO_2 and NO_2 from nitrate.
6. Yellow gas evolved in the cold with characteristic odour; explosion or crackling noise on warming (DANGER).	ClO_2 from chlorate.
7. Yellowish-green gas evolved; irritating odour; bleaches litmus paper; turns KI–starch paper blue; very poisonous.	Cl_2 from chloride in presence of oxidizing agent.
8. 'Oily' appearance of tube in cold; on warming, pungent gas evolved which corrodes the glass; if moistened glass rod introduced into the vapour, gelatinous ppt. of silicic acid is deposited upon it.	HF from fluoride or silicofluoride.
9. Purple fumes evolved with explosion (GREAT DANGER).	Mn_2O_7 from permanganate.
10. Colourless gas evolved; burns with blue flame; no charring.	CO from formate, oxalate, cyanide, hexacyanoferrate(III) or (II).*

Table V.9 Test with concentrated H_2SO_4 (contd.)

Observation	Inference
11. Colourless gas evolved; renders lime water turbid and also burns with a blue flame; no blackening.	CO and CO_2 from oxalate.
12. Colourless gas evolved; burns with a blue flame and/or renders lime water turbid; as heating is continued, SO_2 is evolved and residue in tube (*a*) chars rapidly (odour of burnt sugar), (*b*) chars slowly, accompanied by irritant vapours.	CO, CO_2 and SO_2 from (*a*) tartrate, (*b*) citrate.
13. Irritating fumes evolved.	Benzoate.
14. Pungent odour of vinegar.	CH_3COOH from acetate.
15. Dark-crimson colouration of acid.	Gallate.
16. Brownish-purple colouration of acid.	Tannate.
17. Colourless gas evolved; rekindles glowing splint.	O_2 from peroxides, some peroxo salts or chromate.
18. Colourless gas evolved, burns with blue flame, deep blue solution produced.	CO and anhydrous $CoSO_4$ from hexacyanocobaltate(III).†
19. Yellow colouration in cold: upon warming, vigorous reaction, COS (burns with blue flame), SO_2 (decolourizes fuchsin solution, etc.) and free S produced.	COS, SO_2, and S from thiocyanate.
20. Red fumes of Br_2 (turn fluorescein paper red) and also O_2 evolved.	Br_2 and O_2 from bromate.

* If hexacyanoferrate(II) or (III) are present, they must be destroyed before proceeding with the analysis for cations because they would yield precipitates when the solution is acidified and boiled, and would also introduce other disturbing effects. This may be effected by heating about 1 g of the mixture with 3–4 ml concentrated sulphuric acid in a porcelain crucible placed in an inclined position over the flame and directing the flame against the upper part of the crucible. The heating is continued until fumes of sulphur trioxide cease to be evolved. The residue is then treated with a little concentrated sulphuric acid, warmed gently and water added portionwise. The whole is then boiled for about 5 minutes, and filtered when cold. The filtrate may be analysed for all metals except lead, strontium, and barium which, if present, will be found in the residue.

Alternatively, the complex cyanides may be eliminated from the mixture by fusing with an equal weight of sodium or potassium carbonate in a porcelain crucible. Soluble cyanates and cyanides are formed, which may be extracted with water; the residual metallic iron may be dissolved in dilute hydrochloric acid.

$$K_4[Fe(CN)_6]+K_2CO_3 \rightarrow 5KCN+KCNO+CO_2\uparrow+Fe$$

† Rarely encountered in routine qualitative analysis; a characteristic test is the white precipitate, insoluble in nitric acid, produced with iron(II) sulphate solution (see under Cobalt, Section **III.26**, reaction 4).

9. Test for nitrate (or nitrite) If ammonium has been found by test 6, boiling is continued until ammonia can no longer be detected by its action upon mercury(I) nitrate paper or upon red litmus paper. Add a little (say, 0·1 g) aluminium powder or zinc dust or finely powdered Devarda's alloy to the cooled solution and warm the mixture gently. Remove the flame as soon as evolution of hydrogen commences (with aluminium powder the reaction may become vigorous; cooling with tap water may be necessary to moderate the vigour of the reaction).

If ammonia is evolved, as detected by its odour, its action upon litmus paper and upon filter paper soaked in mercury(I) nitrate solution,* then the presence

* It must be emphasized that the mercury(I) nitrate paper test for ammonia is not applicable in the presence of arsenite. Arsenite is reduced by alkaline reducing agents to arsine, which blackens mercury(I) nitrate paper. The tannic acid–silver nitrate test (Section **III.38**, reaction 7) may also be used; this test is likewise not applicable in the presence of arsenite.

of a nitrate or nitrite is indicated (compare Section **IV. 18**, reaction 4). The presence of a nitrite will generally also be detected in the reaction with dilute sulphuric acid see test 7; if nitrite be absent, then the presence of nitrate alone is indicated.

Cyanides, thiocyanates, hexacyanoferrate(II)s, and hexacyanoferrate(III)s also yield ammonia under these experimental conditions. The reaction is somewhat slower for these anions; up to 5 minutes may elapse before ammonia can be detected from hexacyanoferrate(II)s and hexacyanoferrate(III)s. If these are present, or are suspected as a result of the preliminary tests, particularly that with concentrated sulphuric acid, they must first be removed as follows. Treat the soda extract with excess of nitrate-free silver sulphate, warm the mixture to about 60°, shake vigorously for 3–4 minutes, and filter from the silver salts of the interfering anions and excess of precipitant. Remove the excess silver ions from the filtrate by adding excess sodium hydroxide solution and filter off the precipitated silver oxide. Evaporate the filtrate to about half bulk and test with zinc, aluminium or Devarda's alloy. If cyanides alone are present, they may be rendered innocuous by the addition of a little mercury(II) chloride solution.

10. Test for borate Make a paste of the original substance with calcium fluoride and concentrated sulphuric acid. Hold some of this in a platinum loop just outside the base of the Bunsen flame. A green flame, due to boron trifluoride, indicates borates. Barium and copper do not interfere when the test is carried out in the above manner.

Alternatively fit up the apparatus shown in Fig. V.2, using rubber stoppers; the end of the right-angled bend should be drawn out into a capillary of not more than 0·5 mm bore and 3–4 cm long. The empty test-tube acts as a trap between the mouth and the test tube containing the solution under test to prevent the solution from reaching the mouth should the capillary become blocked.

Place about 0·1 g of the substance in the test-tube, add 1–2 ml concentrated sulphuric acid with the aid of a dropper or small pipette, followed by 5–6 ml methanol in 1 ml portions (**CAUTION**). Introduce the modified 'wash-bottle'

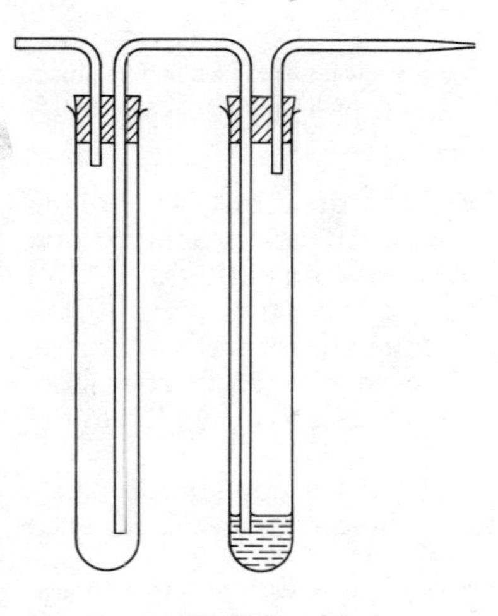

Fig. V.2

tubes and connect by a short length of rubber tubing to the trap. Blow gently through the liquid and direct the vapours issuing from the capillary into a colourless Bunsen flame. If a borate is present, the flame will acquire a characteristic green colour due to the volatile methyl borate, $B(OCH_3)_3$. Under these experimental conditions copper and barium, which colour the flame green, do not interfere.*

V.3 PRELIMINARY TESTS ON METAL SAMPLES The analysis of a metal or of an alloy is simplified by the fact that no anions need be looked for. Many alloys contain small amounts of P, Si, C, and S; phosphorus is converted by the usual solution process into phosphate, and may be identified as such (Section **IV.28**).

The alloy or metal should be in the form of borings, turnings or filings. About 0·5 g is treated with 10 ml 1:1 nitric acid in a porcelain basin in the fume chamber, warmed gently until the evolution of red fumes ceases, and evaporated almost to dryness.† About 10 ml water are then added, the mixture heated for a few minutes and filtered, if necessary.

These cases may arise:

1. The metal or alloy dissolves completely. Examine solution according to Sections **V.8** to **V.19** remembering that phosphate may be present and should be tested for.

2. The metal or alloy does not dissolve completely. The solution (*A*) is examined as in case 1. If the residue is black, it may be either carbon or gold and/or platinum. Test for carbon by igniting on a crucible lid; carbon glows and burns. Gold and platinum dissolve in aqua regia (compare Sections **VII.7** and **VII.8**).

If the residue is white, it may contain, among other things, stannic acid or antimony pentoxide, bismuth oxide, together with traces of copper, lead, and iron, and is best analysed as outlined in Table V.10.

3. The metal or alloy is unattacked. If the alloy is not attacked by nitric acid (1:1), treat a separate 0·5 g sample with 20 ml aqua regia (15 ml concentrated hydrochloric acid and 5 ml concentrated nitric acid) in a porcelain dish. Cover the latter with a clock glass and heat gently until the alloy has disintegrated completely. Raise the clock glass, boil down to about 5 ml, and finally evaporate to dryness on a water bath. Add 5 ml concentrated hydrochloric acid, heat gently, dilute with 15 ml water, stir and heat to boiling. Cool to room temperature and filter. The residue may consist of AgCl, $PbCl_2$, and SiO_2. The filtrate may contain the metals of the remaining groups together with arsenate, orthophosphate, and sulphate. Examine by Table V.12 (Section **V.8**).

If the alloy resists the action of aqua regia, fuse it with sodium hydroxide pellets in a silver dish or crucible **(CAUTION)**. When decomposition is complete, allow to cool, transfer the silver vessel to a beaker and extract the melt with water; remove the silver vessel from the beaker. Strongly acidify the contents of the beaker with nitric acid, evaporate to dryness on a water bath, and proceed as above.

The alkali fusion is sometimes replaced by warming on a water bath with

* This procedure is simpler than that of Section **IV.45**, 15; the latter, however, is foolproof and perfectly unambiguous.

† The nitric acid will oxidize the P to H_3PO_4, S to H_2SO_4, As to H_3AsO_4, Sb to $Sb_2O_3.xH_2O$ (converted by gently heating into Sb_2O_4), Sn to $SnO_2.xH_2O$ and Si to gelatinous silicic acid.

Table V.10 Analysis of portion of alloy insoluble in concentrated nitric acid Wash residue with water and dry by heating in a crucible. Add 6 times its weight of an intimate mixture of equal parts of anhydrous Na_2CO_3 and sulphur, mix well, cover the crucible and heat over a small flame until the excess sulphur has burned off. This operation usually occupies about 20 minutes. Allow to cool and extract contents of the crucible with hot water and filter.

Residue	**Filtrate**	
May contain PbS, Bi_2S_3, CuS, and FeS. Dissolve in hot dilute HNO_3, filter off any S and evaporate almost to dryness. Dissolve in water and add to original solution *A*.	May contain SnS_3^{2-}, SbS_4^{3-},* AsS_4^{3-}, PO_4^{3-} and, possibly, S^{2-}. Acidify with dilute HCl (test with litmus paper), and filter.	
	Residue	**Filtrate**
	May contain SnS_3, Sb_2S_5, As_2S_5 and S. Examine by Tables V.18 and V.19 (in Section **V.12**).	Test for phosphate. Boil off H_2S, add excess NH_3 solution and then magnesium nitrate reagent. A white crystalline ppt. indicates phosphate present. Confirm.

* These are formed in accordance with the following equations;

$$2SnO_2 + 2CO_3^{2-} + 9S \rightarrow 2SnS_3^{2-} + 3SO_2\uparrow + 2CO_2\uparrow$$

$$2Sb_2O_4 + 6CO_3^{2-} + 23S \rightarrow 4SbS_4^{3-} + 7SO_2\uparrow + 6CO_2\uparrow$$

$$Sb_2O_5 + 3CO_3^{2-} + 12S \rightarrow 2SbS_4^{3-} + 4SO_2\uparrow + 3CO_2\uparrow$$

concentrated hydrochloric acid and 10–20 per cent of its volume of liquid bromine.

The compositions of some of the common alloys are given below; the chief constituents are listed in the order of their predominance:

Brasses	Cu, Zn, Sn, Pb
Bronzes	Cu, Sn, Zn, Pb
Phosphor bronzes	Cu, Sn, Pb, P
Solders	Sn, Pb, Bi
Pewter	Sn, Sb, Pb, Cu
Type metals	Pb, Sb, Sn
German silver	Cu, Ni, Zn
Monel metal	Ni, Cu
Constantan	Cu, Ni
Nichrome	Ni, Fe, Cr
Manganin	Cu, Mn, Ni
Wood's alloy	Bi, Pb, Sn, Cd
Rose's alloy	Bi, Pb, Sn

V.4 PRELIMINARY TESTS ON LIQUID SAMPLES (SAMPLES IN SOLUTION) 1. Observe the colour, odour, and any special physical properties.

2. Test its reaction towards litmus paper or a suitable narrow range indicator paper.

(*a*) The solution is neutral: free acids, free bases, acid salts, and salts which give an acid or alkaline reaction owing to hydrolysis, are absent.

(*b*) The solution reacts alkaline: this may be due to the hydroxides of the alkali and alkaline earth metals, to the carbonates, borates, sulphides, cyanides, hypochlorites, silicates, peroxy-salts, and peroxides of the alkali metals, etc.

(*c*) The solution reacts acid: this may be due to free acids, acid salts, salts which yield an acid reaction because of hydrolysis, or to solutions of salts in acids.

3. Evaporate a known volume of the liquid to dryness on the water bath; carefully smell vapours evolved from time to time. If a solid residue remains, examine as described under *A* for solid and non-metallic substances (Section **V.2**). If a liquid remains, evaporate cautiously on a wire gauze in the fume cupboard; a solid residue should be examined as already stated. If charring occurs, organic matter is present, and must be removed before testing for Group IIIA in the subsequent systematic analysis. If no residue remains, then the liquid consists of some volatile substance which may be water or water containing certain gases or volatile substances, such as CO_2, NH_3, SO_2, H_2S, HCl, HBr, HI, H_2O_2, $(NH_4)_2CO_3$, etc., all of which can be readily detected by special tests. It is best to neutralize the solution with sodium carbonate and test for acid radicals (anions).

4. If the solution reacts alkaline, the following tests should be performed:

(*a*) Peroxides and peroxo-salts (e.g. H_2O_2 and $NaBO_3$).

(i) Heat a little of the solution with a few drops of cobalt nitrate solution; a black precipitate of a higher oxide of cobalt is obtained (sulphides and hypochlorites interfere and must be absent).

(ii) Add a little titanium(III) sulphate or chloride solution, and carefully acidify with cold dilute sulphuric acid; a yellow colouration is obtained in the presence of hydrogen peroxide.

(iii) Add a little iron(III) chloride and potassium hexacyanoferrate(III) solutions; Prussian blue precipitate.

(*b*) Hydroxides and carbonates. Boil to decompose hydrogen peroxide, if present. Add a slight excess of barium chloride solution; if the solution now reacts alkaline, hydroxyl ions are present. Filter off the precipitate, and examine for carbonate with dilute acid (Section **IV.2**, reaction 1).

5. If the original solution is acidic, render a known volume (say, 5 ml) alkaline with aqueous ammonia before evaporating it on a water bath. This will prevent the loss of volatile acids, such as hydrochloric and boric acids. Examine the residue.

V.5 PRELIMINARY TESTS ON INSOLUBLE SUBSTANCES A substance which cannot be dissolved by concentrated acids (hydrochloric or nitric) or by aqua regia is described as 'insoluble'. Special methods for solution must therefore be employed, the actual process chosen depending largely upon the nature of the insoluble substance.

The most common insoluble substances encountered in qualitative analysis are:

AgCl, AgBr, AgI, AgCN;

$SrSO_4$, $BaSO_4$, $PbSO_4$;

the strongly ignited oxides Al_2O_3, Cr_2O_3, Fe_2O_3, SnO_2, Sb_2O_4, TiO_2, ThO_2, $WO_3.xH_2O$;

fused $PbCrO_4$, and certain minerals, e.g. CaF_2 (fluorspar), $FeCr_2O_4$ (chrome ironstone);

$Cu_2[Fe(CN)_6]$, $Zn_2[Fe(CN)_6]$, Prussian blue;
SiO_2, and various silicates; SnS_2 (mosaic gold); C and S; metallic silicides; carborundum.

The insoluble substance should be subjected to the tests enumerated below in the order given until it is brought into solution. The substance should be in the form of a fine powder; use an agate mortar, if necessary.

1. Note colour and appearance The following substances are coloured: Cr_2O_3 (green), Fe_2O_3 (dark red), SnS_2 (bronze), $PbCrO_4$ (brown), $Fe_4[Fe(CN)_6]_3$ (Prussian blue), $Cu_2[Fe(CN)_6]$ (dark brown), $FeCr_2O_4$ (dark grey), AgBr* (very pale yellow), AgI* (light yellow), C and S.

The remaining substances listed above are white or nearly so, but may be slightly coloured by traces of impurities; the effect of the latter is less marked when the substance is in the form of a fine powder.

2. Examine the effect of heat Heat a small quantity in a small crucible or upon platinum foil. Sulphur will melt to a yellow liquid, and burn with a blue flame with the production of sulphur dioxide (test with potassium dichromate paper or with fuchsin solution).

Carbon will glow and burn away almost completely; a light-coloured ash may remain. If the black substance is dropped in very small quantities into a little fused potassium nitrate contained in a hard glass tube, the oxidation will be more vigorous and some potassium carbonate will be formed. The residue will evolve carbon dioxide on treatment with dilute acids. Additional confirmation of carbon is obtained by heating an intimate mixture of the substance with dry copper(II) oxide in a hard glass tube; carbon dioxide will be evolved (test with lime water), and red metallic copper will remain.

$$2CuO + C \rightarrow CO_2\uparrow + 2Cu$$

The silver salts, AgCl, AgBr, and AgI, will melt without further change; AgCN decomposes to give a residue of silver, and cyanogen gas is evolved.

Antimony oxide, Sb_2O_4, melts to a yellow liquid.

3. Heat with sodium carbonate upon charcoal The following observations may be made:

a. No metallic button is produced This indicates the absence of Ag, Sn, and Pb. Either moisten residue with a few drops of dilute hydrochloric acid and place in contact with lead acetate paper or extract with a little water and filter into freshly prepared sodium nitroprusside solution; a black stain or a transient violet colouration respectively indicates the presence of sulphide and therefore of sulphate in the original substance.

The test should be repeated with another portion of the original substance if a white residue is obtained. Add one or two drops of cobalt nitrate solution and heat again. A blue mass indicates aluminium.

b. A metallic button or bead is obtained The solubility of the bead in nitric acid and in hydrochloric acid is tested.

---

* The silver halides become violet on exposure to light.

(i) The bead dissolves in nitric acid forming a clear solution, and yielding a curdy, white precipitate of silver chloride, soluble in dilute ammonia solution, upon the addition of a little hydrochloric acid. The presence of silver and the absence of tin is indicated.

(ii) The bead gives a clear solution with hydrochloric acid, and the addition of mercury(II) chloride solution produces a white precipitate of mercury(I) chloride. With nitric acid, a white insoluble powder ('metastannic acid') is produced. This test indicates the presence of tin and the absence of silver.

*4. **Heat with concentrated sulphuric acid*** (i) Escaping gas renders a drop of water upon a glass rod turbid. Fluoride is indicated.

(ii) Carbon monoxide evolved, which burns with a blue flame. Hexacyanoferrate(II) is indicated.

*5. **Heat upon a platinum wire in the reducing zone of the Bunsen flame*** This will reduce any sulphate present to sulphide (as already indicated in test 3). Upon moistening with dilute hydrochloric acid, the sulphide will be converted into the comparatively volatile chloride, and the usual flame test is applied. The presence of barium or of a mixture of strontium and barium will be indicated.

*6. **Apply the microcosmic bead test*** If a skeleton bead is obtained, silica or a silicate is indicated. A negative result does not definitely prove that silica or a silicate is absent, as a skeleton is not always formed. The silicon tetrafluoride test should then be employed (Section **IV.26**, reaction 6).

Heat the bead in the reducing flame in order to test for titanium. If the bead is violet when cold (the colour is produced more readily by the addition of a minute speck of tin or of tin(II) chloride), the presence of titanium is indicated. If iron is also present, the bead will be coloured brownish-red in the reducing flame.

When titanium is found, it is best to fuse with potassium, pyrosulphate in a silica or platinum crucible, and to extract the residue with cold water whereby a solution of titanium(IV) sulphate is obtained.

*7. **Heat with sodium carbonate and potassium nitrate*** This test may be carried out in a loop of platinum wire or upon platinum foil or upon a piece of broken porcelain. If chromium is present, a yellow melt is produced. This should be dissolved in water, acidified with dilute acetic acid, and (*a*) silver nitrate solution added, when brownish-red silver chromate is precipitated, (*b*) lead acetate solution added, when yellow lead chromate is precipitated, or (*c*) 1–2 ml diphenylcarbazide reagent added, when a deep-red colouration is produced.

*8. **Boil with sodium hydroxide solution*** (i) Lead chromate dissolves:

$$PbCrO_4\downarrow + 4OH^- \rightarrow [Pb(OH)_4]^{2-} + CrO_4^{2-}$$

(ii) Prussian blue yields iron(III) hydroxide and sodium hexacyanoferrate(II) (see under Iron, Section **III.22**, reaction 6).

(iii) Copper hexacyanoferrate(II) yields copper oxide and sodium hexacyanoferrate(II).

(iv) Zinc hexacyanoferrate(II) yields sodium tetrahydroxozincate(II) and sodium hexacyanoferrate(II), i.e. it dissolves completely. The zinc is readily identified by passing hydrogen sulphide into the solution; the hexacyanoferrate(II) (see Section **IV.11**) is detected in the filtrate after acidifying and boiling off the hydrogen sulphide.

(v) Alumina and silica may dissolve, forming solutions of sodium tetrahydroxoaluminate and sodium silicate respectively.

9. Heat with concentrated hydroidic acid The powdered substance (0·5 g) should be heated to just below the boiling point with hydriodic acid, sp.gr. 1·7* (2·5 ml).

(i) Tin(IV) oxide dissolves. A pink to red colouration is produced when the solution cools; the colouration disappears on warming to 90–100°C. A yellow to orange sublimate of tin(IV) iodide is frequently observed.

$$SnO_2\downarrow + 4HI \rightarrow SnI_4\uparrow + 2H_2O$$

Upon filtration and dilution, the tin may be precipitated with hydrogen sulphide.

(ii) The sulphates of lead, strontium, and barium are gradually decomposed and hydrogen sulphide is evolved, which is identified by ammoniacal sodium nitroprusside paper (see under Sulphides, Section **IV.6**, reaction 6).

$$PbSO_4\downarrow + 11HI \rightarrow [PbI_3]^- + 4I_2\uparrow + H_2S\uparrow + H^+ + 4H_2O$$
$$BaSO_4\downarrow + 10HI \rightarrow Ba^{2+} + 2I^- + 4I_2\uparrow + H_2S + 4H_2O$$

Upon filtration through a sintered glass crucible and dilution, golden yellow lead iodide is precipitated from lead sulphate. Barium may be detected by the addition of dilute sulphuric acid, and strontium by sulphuric acid and ethanol.

(iii) The silver halides dissolve readily in the cold, owing to the formation of a complex:

$$AgI\downarrow + HI \rightarrow [AgI_2]^- + H^+$$
$$AgX\downarrow + 2HI \rightarrow [AgI_2]^- + 2H^+ + X^- \quad (X = Cl \text{ or } Br)$$

Upon warming the solution, the hydrogen halides are expelled from solution with effervescence. When the solution is diluted, the complex di-iodoargentate is decomposed and silver iodide is precipitated; it is best, however, to expel the excess of hydriodic acid by evaporation before diluting with water. If lead sulphate was originally present, lead iodide will be precipitated on dilution; this can be separated from silver iodide by extraction with ammonium acetate solution.

(iv) Calcium fluoride is attacked by the hot acid, hydrogen fluoride being evolved, which will etch glass:

$$CaF_2\downarrow + 2HI \rightarrow H_2F_2\uparrow + Ca^{2+} + O2I^-$$

Upon dilution, neutralization with ammonia solution and addition of ammonium oxalate solution, the calcium is precipitated as calcium oxalate.

10. Treat with ammonium sulphide solution If the insoluble unknown or the washed residue from the aqua regia extraction is white or light-coloured, treat it in a porcelain dish or crucible with a few drops of ammonium sulphide

* Where colour reactions are to be observed, it is recommended that the hydriodic acid be decolourized by the addition of 1–2 per cent by volume of 50 per cent hypophosphorous acid or by warming with a little potassium hypophosphite.

solution and stir. Lead and silver compounds are probably absent if the colour is unchanged. A blackening of the solid indicates that lead and silver compounds may be present.

V.6 DISSOLUTION OF THE SAMPLE The preliminary tests, described in the previous sections, have already revealed whether the substance is soluble in water or in acids. If such information is not available, the following procedure should be adapted.

Use small quantities (5–10 mg) of the powdered solid and examine the solubility in the following solvents in the order given: (1) water, (2) dilute hydrochloric acid, (3) concentrated hydrochloric acid, (4) dilute nitric acid, (5) concentrated nitric acid, (6) and aqua regia. Try the solubility first in the cold and then on warming: if in doubt whether the substance or a portion of the substance has dissolved, evaporate a little of the clear solution on a watch glass. If the substance dissolves in water, proceed immediately to test for the metal ions. If the use of dilute hydrochloric acid results in the formation of a precipitate, this may consist of metals of Group I; the precipitate may either be filtered off and examined for this group, or else the original substance may be dissolved in dilute nitric acid. If concentrated hydrochloric acid is employed for solution, it will be necessary to evaporate off most of the acid since certain metals of Group II (e.g. cadmium and lead) are not completely precipitated in the presence of large concentrations of acid. Where nitric acid has been used for the process of solution, all of the acid must be removed by evaporating nearly to dryness, adding a little hydrochloric acid, evaporating again to a small bulk and then diluting with water; these remarks also apply to aqua regia. For this reason nitric acid is often omitted from the solvents, and the tests confined to solvents 1, 2, 3 and 6.

When a suitable solvent has been found, the solution for analysis is prepared with 0·5–1 g of the solid; the volume of the final solution should be 15–20 ml.

If the substance is insoluble in aqua regia (and in concentrated acids), it is regarded as insoluble and is treated by the special methods detailed in Section **V.7** below.

If the sample is a metal, the procedure described in Section **V.3** should be adopted. The insoluble part, if any, should be treated according to Section **V.7**.

Unless the sample is readily soluble in water, the solution is unsuitable for testing for anions, as during dissolution in acids some of these might decompose. For the test for anions we can use either the aqueous solution, or, if the sample is not soluble in water, sodium carbonate extract should be prepared either from the whole of the sample, or the sample should first be extracted with hot water and the residue treated with sodium carbonate. This procedure is described in detail in Section **V.18**.

V.7 EXAMINATION OF THE INSOLUBLE RESIDUE It may well happen that after treating the sample with the various solvents, as described in Section **V.6**, some insoluble residue remains or the whole sample may prove to be insoluble. With this insoluble residue the preliminary tests, described in Section **V.5**, should be carried out; these tests may supply valuable information as to the composition of the insoluble substance. In every case, the following methods of bringing the substance into solution for systematic analysis should be used.

1. Removal of lead salts Treat about 1 g of the insoluble unknown, or the residue from the aqua regia extraction, with 3 ml 10M ammonium acetate solution slightly acidified with acetic acid. (Excess of ammonia solution must be absent to avoid the solvent action upon any silver chloride which may be present.) Heat the mixture, with stirring, to about 70°C and filter; wash with about 5 ml water. Test separate portions of the combined filtrate and washings for Pb^{2+}, SO_4^{2-}, and Cl^-. The silver nitrate test for chloride must be conducted in the presence of about 10 per cent of the volume of concentrated nitric acid and the mixture heated to boiling; silver acetate will dissolve under these conditions. If lead salts are found, repeat the extraction with ammonium acetate solution and wash the residue with hot water until the washings give no colouration with dilute ammonium sulphide solution.

Lead silicate is insoluble in ammonium acetate solution; it will be detected in 3.

2. Removal of silver salts Warm the insoluble unknown or the residue from 1 (if lead salts are present) with a concentrated solution of potassium cyanide. (If it dissolves completely, only AgCl, AgBr, AgI, and AgCN are present.) Filter and reserve the residue, *R*, for subsequent treatment. Dilute the filtrate considerably and treat with hydrogen sulphide. Filter off any black precipitate (Ag_2S), wash, dissolve in hot dilute nitric acid and add dilute hydrochloric acid. A white precipitate of silver chloride indicates the presence of silver.

If silver is found, the halogen with which the metal was originally combined is identified by melting another portion of the insoluble substance, immersing it in dilute sulphuric acid, placing a piece of zinc in contact with the acid and the fused mass, warming and allowing to stand for a few minutes. The silver salt is reduced to metallic silver, whilst the anions are present in solution in the presence of zinc ions, i.e. as zinc salts. Filter. The filtrate is tested for chloride, bromide, and iodide in the usual manner; the tests for mixtures of these anions are described in Section **IV.45**, and in Section **V.18**, Table V.30.

3. Sodium carbonate fusion The residue left undissolved after operations 1 and 2 have been carried out, should be treated according to the scheme outlined in Table V.11.

4. Fusion of silicides Metallic silicides and carborundum are rarely encountered in routine qualitative analysis. They are best brought into solution by fusion with sodium or potassium hydroxide in a silver crucible (**CAUTION**):

$$Cu_2Si + 2KOH + H_2O \rightarrow K_2SiO_3 + 2Cu + 2H_2\uparrow$$
$$SiC + 4KOH + 2H_2O \rightarrow K_2SiO_3 + K_2CO_3 + 4H_2\uparrow$$

During the fusion the liberated hydrogen catches fire forming water by combination with the oxygen of the air. Upon treating the melt with water, the soluble alkali silicate is extracted.

Carborundum, when in the form of a fine powder, is readily decomposed by fusion with potassium carbonate in a platinum crucible. Upon removing the cover of the crucible the blue flame of burning carbon monoxide may be seen.

$$SiC + 3K_2CO_3 \rightarrow K_2SiO_3 + 2K_2O + 4CO\uparrow$$

Table V.11 Sodium carbonate fusion The residue, *R*, free from lead and silver salts, or the original substance, if lead and silver salts are absent, is mixed with 5–6 times its weight of pure, sulphate-free, anhydrous Na_2CO_3 or with a mixture of equal parts of Na_2CO_3 and K_2CO_3 (fusion mixture). The mixture is heated upon Pt foil or in a Pt crucible until a tranquil melt is obtained. (It may be necessary to heat over the blowpipe flame.) Allow to cool, extract the melt thoroughly by boiling it with water. Filter.

Residue		**Filtrate**	
Wash well, first with 0·05M Na_2CO_3 solution, and then with hot water. May contain *inter alia* $BaCO_3$, $SrCO_3$, $CaCO_3$, other insoluble carbonates and unattacked CaF_2, SnO_2, Sb_2O_4, Al_2O_3, Fe_2O_3, etc. Extract with dilute HNO_3 and filter.		May contain CrO_4^{2-}, $[Al(OH)_4]^-$, $[Sn(OH)_6]^{2-}$, AsO_4^{3-}, F^-, and SO_4^{2-}. Acidify with concentrated HCl and evaporate to dryness in the fume cupboard. Triturate the dry mass with concentrated HCl, add water, warm and filter.	
Residue	**Filtrate**	**Residue**	**Filtrate**
If white, may contain CaF_2, SnO_2, Sb_2O_4, Al_2O_2, SiO_2, etc., which are incompletely attacked by Na_2CO_3. Fuse with NaOH in a Ni crucible.* Allow to cool, extract with water and filter. The filtrate may contain $[Sn(OH)_6]^{2-}$, SbO_4^{3-}, $[Al(OH)_4]^{2-}$, and SiO_3^{2-}. Test for Sn, Sb, and Al.	Evaporate almost to dryness to remove HNO_3, add dilute HCl and examine for metal ions by Table V.12 (Section **V.8**).	May contain SiO_2.† Confirm by microcosmic bead test or by the SiF_4 test in lead capsule.	Test a portion for sulphate. Examine for metals of Groups II and III.

* An alternative procedure is the Na_2CO_3–S fusion method already described under the analysis of metals and alloys (Table V.10, Section **V.3**).

The original insoluble substance may also be subjected to the Na_2CO_3–S fusion, and the extract, after acidifying with 1:1 HCl, filtered. Any precipitate is washed, dissolved in concentrated HCl, and H_2S removed by boiling: Sb may be detected with Rhodamine-B and Sn with $HgCl_2$ solution.

† This precipitate may also contain $WO_3.xH_2O$ from mineral tungstates insoluble in aqua regia. Confirm W by digesting with dilute NH_3 solution to dissolve $WO_3.xH_2O$ and then apply the $SnCl_2$–HCl test.

V.8 SEPARATION OF CATIONS INTO GROUPS Once the sample is dissolved, the separation of cations into groups can be attempted. This can be done according to the scheme outlined in Table V.12.

Before describing the general scheme for the separation of the metal ions into groups, the following facts are brought to the notice of the student as it is believed that by their proper understanding and appreciation, many of the usual pitfalls will be avoided.

1. The analysis should not be conducted with large quantities of the substance because much time will be spent in filtering the precipitates and difficulty will be experienced in washing and dissolving them. It is therefore recommended that 0·5–1 g should be employed for the analysis. After a little experience the student will be able to judge from the relative sizes of the precipitates, the relative quantities of the various components present in the mixture.

2. The tests must, in the first place, be carried out in the order given. A group reagent will separate its particular group only from those which follow it and not from those which precede it. Thus hydrogen sulphide in the presence of 0·4M hydrochloric acid will separate Group II from Groups IIIA, IIIB, IV, and V, but does not separate Group II from Group I. It is most important therefore that one group should be completely precipitated before precipitation of the next group is attempted, otherwise the group precipitates will be contaminated by metals from the preceding groups, and misleading results will be obtained.

3. The conditions for precipitation and for solution must be rigidly followed.

4. All precipitates must be washed to remove adhering solution in order to avoid contamination by the metals remaining in the filtrate. The first washings should be added to the solution from which the precipitate has been filtered; later washings may be discarded.

5. If the volume of the solution at any stage of the analysis becomes too large, it should be reduced by evaporation.

6. All the apparatus employed in the analysis must be scrupulously clean. The use of dirty apparatus may be sufficient to introduce impurities into the substance under examination.

Notes and explanations to Table V.12: 1. If the original substance was completely soluble in dilute HCl it is evident that no silver or mercury(I) salt is

Table V.12 Separation of cations into groups (Anions of organic acids, borates, fluorides, silicates, and phosphates being present). Add a few drops of dilute HCl to the cold solution. If a ppt. forms, continue adding dilute HCl until no further precipitation takes place. Filter. (1)

<table>
<tr><th>Residue</th><th colspan="2">Filtrate</th></tr>
<tr><td rowspan="3">White.
Group I present.
Examine by Group Separation Table V.13 (Section V.9).</td><td colspan="2">Add 1 ml of 3% H_2O_2 solution (2). Adjust the HCl concentration to 0·3M (3). Heat nearly to boiling, and saturate with H_2S under 'pressure' (4). Filter.</td></tr>
<tr><th>Residue</th><th>Filtrate</th></tr>
<tr><td>Coloured.
Group II present.
Examine by Group Separation Tables V.14–V.19 (Sections V.10 to V.12.</td><td>Boil down in a porcelain dish to about 10 ml and thus ensure that all H_2S has been removed (test with lead acetate paper). Add 3–4 ml concentrated HNO_3 to oxidize any Fe^{2+} to Fe^{3+} etc. (5), and evaporate cautiously to dryness; moisten with 2–3 ml concentrated HNO_3 and heat gently: this will remove organic acids.
If borate and fluoride are present (6), evaporate the residue repeatedly with 5–10 ml concentrated HCl.
If borate is present and fluoride is absent, treat the residue with 5 ml methanol and 10 ml concentrated HCl and evaporate on a water bath (7).
Add about 15 ml dilute HCl, warm and filter from any residue, originating from any silicate present (8).
Test 0·5 ml of the filtrate (or solution, if silicate is absent) for phosphate with 3 ml ammonium molybdate reagent and a few drops of concentrated HNO_3, and warm to about 40°C: a yellow ppt. indicates phosphate is present. If phosphate is present (9), remove all phosphate ions as detailed in Tables V.20–V.22 (Section V.13).
If phosphate is absent, add 1–2 g solid NH_4Cl, heat to boiling, add dilute NH_3 solution until mixture is alkaline and then 1 ml in excess, boil for 1 minute and filter immediately.</td></tr>
</table>

Table V.12 Separation of cations into groups (contd.)

<table>
<tr><th>Residue</th><th colspan="3">Filtrate</th></tr>
<tr><td rowspan="5">Group IIIA present. Examine by Group Separation Table V.23 (Section V.14).</td><td colspan="3">Add 2–3 ml of dilute aqueous NH_3, heat, pass H_2S (under 'pressure') for 0·5–1 minute (10), filter and wash (11).</td></tr>
<tr><th>Residue</th><th colspan="2">Filtrate (12)</th></tr>
<tr><td rowspan="3">Group IIIB present. Examine by Group Separation Table V.24 or V.25 (Section V.15).</td><td colspan="2">Transfer to a porcelain dish and acidify with dilute acetic acid (13). Evaporate to a pasty mass (FUME CUPBOARD), allow to cool, add 3–4 ml concentrated HNO_3 so as to wash the solid round the walls to the centre of the dish, and heat cautiously until the mixture is dry. Then heat more strongly until all ammonium salts are volatilized (14). Cool. Add 3 ml dilute HCl and 10 ml water: warm and stir to dissolve the salts. Filter, if necessary. Add 0·25 g solid NH_4Cl (or 2·5 ml of 20% NH_4Cl solution), render alkaline with concentrated NH_3 solution and then add, with stirring, $(NH_4)_2CO_3$ solution in slight excess. Keep and stir the solution in a water bath at 50–60°C for 3–5 minutes (15). Filter and wash with a little hot water.</td></tr>
<tr><th>Residue</th><th>Filtrate (16)</th></tr>
<tr><td>White. Group IV present. Examine by Group Separation Table V.26 or V.27 (Section V.16).</td><td>Evaporate to a pasty mass in a porcelain dish (FUME CUPBOARD), add 3 ml concentrated HNO_3 so as to wash solid from walls to centre of dish, evaporate cautiously to dryness and then heat until white fumes of ammonium salts cease to be evolved. White residue. Group V present. Examine by Group Separation Table V.28 (Section V.17).</td></tr>
</table>

For notes 1 to 16 see the text.

present. When lead is present, the solution may be clear whilst hot, but $PbCl_2$ is deposited on cooling the solution. Any lead missed in this group will be precipitated with H_2S in Group II.

A precipitate may form upon the addition of HCl to certain neutral or slightly acid solutions even when none of the Group I metals is present. This may occur under the following conditions:

(*a*) Aqueous solutions of Sb, Bi, and Sn, not containing free HCl, precipitate as the oxychlorides upon the addition of this acid. The precipitate dissolves, however, upon the addition of excess acid.

(*b*) Concentrated solutions of certain chlorides, e.g. NaCl and $BaCl_2$, may form precipitates upon the addition of HCl; these dissolve on diluting with water.

(*c*) Borates may yield a white crystalline precipitate of boric acid, particularly if the acid is concentrated; only partial precipitation may occur here.

(*d*) Silicates may yield a gelatinous precipitate of silicic acid; only partial precipitation may occur here.

(*e*) The thio-salts of arsenic, antimony, and tin will give the corresponding sulphides.

2. The H_2O_2 solution is added to oxidize Sn^{2+} to Sn^{4+}, thus leading ultimately to the precipitation of SnS_2 in place of the somewhat gelatinous SnS. The excess H_2O_2 should preferably be decomposed by boiling before passing H_2S, otherwise some S may be precipitated. The subsequent separation of Groups IIA and IIB by means of aqueous KOH is thus rendered more complete since SnS_2 dissolves entirely and SnS dissolves only partially in 2M KOH.

If it is intended to use ammonium polysulphide in the separation of Groups IIA and IIB, the addition of H_2O_2 is not essential since $(NH_4)_2S_z$ will oxidize SnS to SnS_2 and the latter dissolves as the thiostannate SnS_3^{2-}.

3. It is important that the concentration of HCl be approximately correct, i.e. 0·3M, before passing H_2S: with higher concentrations of acid, lead, cadmium, and tin(II) will be incompletely precipitated, if the acidity is too low, sulphides of Group IIIB (NiS, CoS, and ZnS) may be precipitated.

Either of two methods may be employed to adjust the acid concentration.

(*a*) Concentrate the solution until it has a volume of 10–15 ml. Cool. Add concentrated NH_3 solution dropwise from a dropper pipette, with constant stirring, until the mixture is alkaline. (Ignore any precipitate which may form: this either dissolves when the HCl is added or else is converted into the sulphide by the H_2S treatment.) Introduce dilute HCl dropwise until the mixture is just acid (use litmus paper). Then add 3·0 ml of 2M HCl (measured from a graduated pipette or a calibrated dropper) and dilute the solution to a volume of 20 ml with distilled water.

(*b*) An alternative procedure is to make use of the indicator methyl violet (0·1 per cent aqueous solution or, better, the purchased or prepared indicator paper). The following table gives the colour of the indicator at various concentrations of acid.

Acid concentration	*p*H	Methyl violet indicator
Neutral or alkaline	7+	Violet
0·1M HCl	1·0	Blue
0·25M HCl	0·6	Blue-green
0·33M HCl	0·5	Yellow-green
0·50M HCl	0·3	Yellow

Add 1 drop of the methyl violet indicator solution and introduce dilute HCl or dilute aqueous NH_3 (as necessary) dropwise and with constant stirring until the colour of the solution is yellow-green: a blue-green colour is almost but not quite acid enough, yet is acceptable for most analyses. (If the indicator paper is available, the thoroughly stirred solution should be spotted on fresh portions of the paper.) It is recommended that a comparison solution containing, say, 10 ml of 0·3M HCl and 1 drop of indicator be freshly prepared: this will facilitate the correct adjustment of the acidity. A more satisfactory standard is a buffer solution prepared by mixing 5 ml of 2M sodium acetate and 10 ml of 2M HCl: this solution has a pH of 0·5.

4. For the passage of H_2S into the solution, the latter is placed in a small conical flask (one of 50 ml capacity is suitable) and the 'pressure' method used as detailed in Section **II.3**, 7). Heat the solution almost to boiling and pass in H_2S, whilst slowly shaking the flask with a swirling motion, until precipitation is complete: the latter will be apparent when bubbling either stops altogether or is reduced to a very slow rate of 1–2 bubbles per minute. Saturation is normally reached in 2–5 minutes. The best method of determining whether precipitation is complete is to filter off a portion of the solution and to test the filtrate with H_2S. If only a white precipitate or suspension of sulphur is obtained, the presence of an oxidizing agent is indicated.

If an oxidizing agent is present e.g. a permanganate, dichromate or iron(III) salt, as is shown by the gradual separation of a fine white precipitate of sulphur and/or change in colour of the solution, it is usual to pass SO_2 into the hot solution until reduction is complete, then to boil off the excess of SO_2 (test with $K_2Cr_2O_7$ paper) and finally to pass H_2S. Arsenates, in particular, are slowly precipitated by H_2S: they are therefore generally reduced by SO_2 to arsenites and then precipitated as As_2S_3 with H_2S, after prior removal of the excess SO_2 in order to avoid interaction of the latter with H_2S and the consequent separation of S. Tin(IV) compounds may be very slightly reduced to the divalent state by this treatment; the amount of reduction is, however, so small that it may be neglected. The original solution or substance must be tested for the valence state of the arsenic.

The objection to the use of SO_2 is that some sulphuric acid may be formed, especially upon boiling, and this may partially precipitate Pb, Sr, and Ba as sulphates. Any precipitate formed in this process should accordingly be examined for these cations: $PbSO_4$ is soluble in ammonium acetate solution.

An alternative procedure to be borne in mind when arsenate is present, which does not possess the disadvantages associated with SO_2 and is perhaps more expeditious, is to add 2–3 ml concentrated HCl and 0·5 ml of 10 per cent ammonium iodide solution. The arsenate is thereby reduced to arsenite, and upon saturation of the hot solution with H_2S under 'pressure', the arsenic is completely precipitated as As_2S_3. This reduction can be carried out after the sulphides of the other elements have been precipitated in the presence of 0·3M HCl.

The precipitated sulphides may be washed with a wash liquid prepared by dissolving 0·25 g NH_4NO_3 in 5 ml water and treating this solution with H_2S. The H_2S must be present in the wash liquid to prevent oxidation of some of the moist sulphides to sulphates.

5. Nitric acid is added to oxidize iron(II) to iron(III); if iron(III) was originally present it will have been reduced by the H_2S. Alternatively, bromine water may be used for the oxidation; the excess of bromine must be removed by boiling. Iron(II) is incompletely precipitated by NH_3 solution in the presence of NH_4Cl.

Organic acids (and their anions) interfere with the normal course of analysis. Thus in the presence of oxalic, tartaric or citric acids, the addition of the group reagent NH_4Cl and NH_3 solution might fail to cause the precipitation of the hydroxides of Fe, Al, and Cr. Furthermore, if oxalic acid is present, the oxalates of some of the metals of Groups IIIB, IV, and of Mg, which are insoluble in ammoniacal solution, might be precipitated at this stage. It is therefore essential to destroy the organic acids, the presence of which has been indicated in the preliminary tests, before precipitation of Group IIIA. This is best effected by evaporation with concentrated nitric acid until the black residue is completely oxidized. It is important not to heat the residue too strongly as it may convert any Fe, Al or Cr present into the form of an oxide which is sparingly soluble in concentrated HCl.

If benzoic and/or salicylic acid are present, the free acids will separate upon the addition of HCl and they should therefore be looked for in the Group I precipitate. If any salicylic acid should pass through to Group III, great care must be taken when evaporating to dryness with concentrated HNO_3 since the nitrosalicylic acids explode on strong heating.

6. Borates and fluorides of the metals of Groups IIIB, IV, and of Mg are insoluble (or sparingly soluble) in ammoniacal solution, and are therefore liable to be precipitated at this stage. They may be removed by repeated evaporation with concentrated HCl; the boric acid will slowly volatilize in the steam and the hydrogen fluoride with the excess HCl.

7. Boric acid alone is more rapidly eliminated as the volatile methyl borate, $B(OCH_3)_3$, (highly poisonous). If much boric acid is present, two treatments with CH_3OH and HCl may be required.

8. Unless silicates are removed here, they are likely to be confused with $Al(OH)_3$ in the group separation. Repeated evaporation with concentrated HCl converts silicates into a granular form of hydrated silica, which is readily filtered, particularly after a final digestion with dilute HCl. The precipitate should be subjected to the microcosmic salt bead test or to the silicon tetrafluoride test (Section **IV.26**).

Solutions of silicates are decomposed by dilute HCl into silicic acid, which may partially separate in Group I in the gelatinous form. That not precipitated in Group I will be precipitated by NH_4Cl solution in Group IIIA (see under Silicates, Section **IV.26**, reaction 2).

9. The phosphates of the metals of Groups IIIA, IIIB, IV, and of Mg are insoluble in water and in ammoniacal solution, and may be precipitated at this stage. An excellent method for the removal of phosphate is given in Table V.20 (Section **V.13**).

10. It is recommended that a small portion of the filtrate from Group IIIA be tested first with a little aqueous NH_3 and H_2S. If a precipitate is obtained, the main solution should be treated in this manner. If there is no precipitate, metals of Group IIIB are absent and the main filtrate from Group IIIA may then be employed in testing for Group IV, etc.

The student should remember that the conditions for the precipitation of Group IIIB differ from those in Group II. In the latter group, H_2S is passed into an acid solution in which the gas is only slightly soluble and hence much of it escapes unabsorbed unless the 'pressure' technique is employed. In the former group, the solution is alkaline and therefore the H_2S is readily absorbed. Moreover, if too much H_2S is employed, NiS may partially form a colloidal solution. This is largely avoided by passing H_2S for 30–60 seconds, and testing for completeness of precipitation. Some authors recommend that H_2S be passed into a solution acidified with acetic acid and thus avoid the complication due to colloidal NiS.

11. The wash liquid for the Group IIIB precipitate may consist of 1 per cent NH_4Cl to which 1 per cent by volume of ammonium sulphide solution is added. Oxidation of the moist sulphides to the soluble sulphates is thus considerably reduced.
12. If the solution or the filtrate from Group IIIB is brown or dark in colour, Ni may be suspected. The dark-coloured solution contains colloidal NiS, which runs through the filter paper. It may be acidified with acetic acid, and then boiled until the NiS has coagulated: this may either be added to the Group IIIB precipitate or tested separately for Ni. As a general rule, the addition of macerated filter paper (e.g. in the form of a portion of a Whatman filtration accelerator or ashless tablet) to the suspension before filtration will lead to a clear or colourless solution.
13. The filtrate must be acidified immediately and concentrated to remove H_2S. Ammonium sulphide solution on exposure to air slowly oxidizes to ammonium sulphate and would then precipitate any barium or strontium present as $BaSO_4$ or $SrSO_4$. Another reason for the immediate acidification of the filtrate from Group IIIB is to prevent the absorption of CO_2 from the air with the attendant formation of carbonate ions: the latter would also precipitate the ions of Group IV.
14. The initial filtrate from Group IIIB will contain a very high concentration of ammonium salts. The concentration of ammonium ions is much greater than is necessary to prevent the precipitation of $Mg(OH)_2$ and it may also lead to incomplete precipitation of the carbonates of Group IV metals. The latter effect is due to the equilibrium.

$$NH_4^+ + CO_3^{2-} \rightleftarrows NH_3 + HCO_3^-$$

which reduces the concentration of free CO_3^{2-} ions in solutions which contain NH_4^+ ions. For these reasons most of the ammonium salts should be eliminated first. This can either be done by simple volatilization, or by adding concentrated HNO_3 as recommended in Table V.12, when ammonium salts are decomposed at a lower temperature than required for volatilization:

$$NH_4^+ + HNO_3 \rightarrow N_2O\uparrow + H^+ + 2H_2O$$

Loss by decrepitation and spurting during this operation must be avoided.
15. The 'ammonium carbonate solution' contains much ammonium hydrogen carbonate, NH_4HCO_3, and will accordingly form soluble hydrogen carbonates with the alkaline earth metals unless excess ammonia solution is present; in the latter case the amount of normal carbonate, $(NH_4)_2CO_3$, present will be increased. When precipitating Group IV, the solution should be warm (*c.* 50°C) to decompose any hydrogen carbonates formed. The solution must not, however, be boiled because the equilibrium

$$MCO_3\downarrow + 2NH_4^+ \rightleftarrows M^{2+} + 2NH_3\uparrow + CO_2\uparrow + H_2O$$

is shifted towards the right as NH_3 and CO_2 are removed. The excess of ammonium carbonate decomposes also, especially above 60°C:

$$2NH_4^+ + CO_3^{2-} \rightarrow 2NH_3\uparrow + CO_2\uparrow + H_2O$$

The digestion also improves the filtering properties of the precipitate.
16. Owing to the slight solubility of $CaCO_3$, $SrCO_3$, and $BaCO_3$ in solutions of ammonium salts, the filtrate from Group IV will, when these metals are

present, contain minute amounts of the ions of the alkaline earth metals. Since the Group IV metals may interfere to a limited extent with the flame tests for Na and K and also with the Na_2HPO_4 test for Mg (if employed), it has been recommended that the filtrate from Group IV be heated with a little (say, 1 ml) of $(NH_4)_2SO_4$ solution and $(COONH_4)_2$ solution and filtered from any precipitate which forms. Owing to the comparatively small concentration of ammonium salts, this is generally unnecessary if the procedure described in Table V.28 is adopted.

V.9 SEPARATION AND IDENTIFICATION OF THE GROUP I CATIONS (SILVER GROUP) The residue after filtering off the precipitate obtained with dilute hydrochloric acid (cf. Table V.12 in Section **V.8**) may contain lead, silver, and mercury(I) ions. Their separation and identification can be carried out according to the scheme shown in Table V.13.

Table V.13 Separation and identification of Group I cations (silver group) The ppt. may contain $PbCl_2$, AgCl, and Hg_2Cl_3. Wash the ppt. on the filter first with 2 ml of 2M HCl, then 2–3 times with 1 ml portions of cold water and reject the washings with water. Transfer the ppt. to a small beaker or to a boiling tube, and boil with 5–10 ml water. Filter hot. (1)

<table>
<tr><th colspan="2">Residue</th><th>Filtrate</th></tr>
<tr><td colspan="2">May contain Hg_2Cl_2 and AgCl. Wash the ppt. several times with hot water until the washings give no ppt. with K_2CrO_4 solution: this ensures the complete removal of the Pb.
Pour 3–4 ml warm dilute NH_3 solution over the ppt. and collect the filtrate (2).</td><td rowspan="3">May contain $PbCl_2$.
Cool a portion of the solution: a white crystalline ppt. of $PbCl_2$ is obtained if Pb is present in any quantity.
Divide the filtrate into 3 parts:
(i) Add K_2CrO_4 solution. Yellow ppt. of $PbCrO_4$, insoluble in dilute acetic acid.
(ii) Add KI solution. Yellow ppt. of PbI_2, soluble in boiling water to a colourless solution, which deposits brilliant yellow crystals upon cooling.
(iii) Add dilute H_2SO_4. White ppt. of $PbSO_4$, soluble in ammonium acetate solution.
Pb present.</td></tr>
<tr><th>Residue</th><th>Filtrate (5)</th></tr>
<tr><td>If black, consists of $Hg(NH_2)Cl + Hg$ (3).
Hg_2^{2+} present (4).</td><td>May contain $[Ag(NH_3)_2]^+$.
Divide into 3 parts:
(i) Acidify with dilute HNO_3. White ppt. of AgCl.
(ii) Add a few drops of KI solution. Pale-yellow ppt. of AgI.
(iii) Add a few drops of the rhodanine* reagent.
Ag present.</td></tr>
</table>

* See Section **III.6**, reaction 11.

Notes and explanations to Table V.13: 1. Lead, silver, and mercury(I) ions are precipitated by dilute HCl as the insoluble chlorides $PbCl_2$, AgCl, and Hg_2Cl_2 respectively; the chlorides of all the other common metals are soluble. Among these $PbCl_2$ is soluble in boiling water, whilst Hg_2Cl_2 and AgCl are insoluble. Extraction of the precipitate with hot water therefore removes the $PbCl_2$. Confirmatory tests for lead ions are applied to the solution so obtained, e.g. with potassium dichromate

$$Pb^{2+} + CrO_4^{2-} \rightarrow PbCrO_4\downarrow$$

2. AgCl is soluble in dilute NH_3 solution, yielding the soluble complex ion $[Ag(NH_3)_2]^+$. This complex is decomposed by both dilute HNO_3 and by KI solution with the precipitation of the insoluble salts AgCl and AgI respectively.

$$AgCl + 2NH_3 \rightarrow [Ag(NH_3)_2]^+ + Cl^-$$
$$[Ag(NH_3)_2]^+ + Cl^- + 2H^+ \rightleftarrows AgCl\downarrow + 2NH_4^+$$
$$[Ag(NH_3)_2]^+ + I^- \rightarrow AgI\downarrow + 2NH_3$$

The rhodanine reagent (cf. Section **III.6**, reaction 11) gives a precipitate with a solution of $[Ag(NH_3)_2]^+$; any interference by mercury and copper(I) is thus eliminated.

3. The conversion of Hg_2Cl_2 by dilute NH_4OH solution into the insoluble black mixture of mercury amido chloride, $Hg(NH_2)Cl$, and finely divided mercury.

$$Hg_2Cl_2\downarrow + 2NH_3 \rightarrow Hg(NH_2)Cl\downarrow + Hg\downarrow + NH_4^+ + Cl^-$$

Aqua regia converts the black mixture into $HgCl_2$ (see under Mercury, Section **III.5**, reaction 1). Mercury(II) is then detected with $SnCl_2$ solution.

$$2HgCl_2 + Sn^{2+} \rightarrow Hg_2Cl_2\downarrow + Sn^{4+} + 2Cl^-$$
$$Hg_2Cl_2\downarrow + Sn^{2+} \rightarrow 2Hg\downarrow + Sn^{4+} + 2Cl^-$$

4. This is a conclusive test for mercury(I). It may be further confirmed by dissolving the precipitate in 3–4 ml of boiling aqua regia, diluting, filtering if necessary, and adding $SnCl_2$ solution (cf. note 3).

5. If Hg is present in reasonably large quantity and Ag has not been detected, carry out the above confirmatory test for Hg, and treat the thoroughly washed residue, insoluble in aqua regia, with dilute NH_3 solution. Filter, if necessary, and add dilute HNO_3 to the clear solution. A white precipitate of AgCl will form if small amounts of Ag are present. This is an alternative procedure for separating Hg and Ag in Hg_2Cl_3–AgCl mixture.

When the amount of $Hg(NH_2)Cl + Hg$ is large and that of AgCl is small, the latter may be reduced to metallic silver according to the reaction,

$$2Hg\downarrow + 2AgCl\downarrow \rightarrow Hg_2Cl_2\downarrow + 2Ag\downarrow$$

and thus escape detection.

V.10 SEPARATION OF GROUP II CATIONS INTO GROUPS IIA AND IIIB Having precipitated the sulphides of the Group II cations (cf. Section **V.8**, Table V.12) the next task is to separate these into Groups IIA (Hg^{2+}, Bi^{3+}, Pb^{2+}, Cu^{2+} and Cd^{2+}) and IIB (As^{3+}, As^{5+}, Sb^{3+}, Sb^{5+}, Sn^{2+}, Sn^{4+}). There are two methods recommended for the purpose, one making use of (yellow) ammonium polysulphide, the other making use of potassium hydroxide.

(a) The ammonium polysulphide method The separation can be carried out according to the scheme shown in Table V.14.

Notes and explanations to Table V.14: 1. The sulphides of mercury, lead, bismuth, copper, and cadmium and of arsenic, antimony, and tin are precipitated by H_2S in the presence of dilute (0·3M) HCl. The sulphides of arsenic (As_2S_3), antimony (Sb_2S_3 and Sb_2S_5), and of tin (SbS_2) (Group IIB) are soluble in (yellow) ammonium polysulphide solution forming thio-salts, whilst those of the other metals (Group IIA) are insoluble. Tin(II) sulphide, SnS, is insoluble

Table V.14 Separation of Group II into Group IIA and Group IIB with the ammonium polysulphide method The ppt. may consist of the sulphides of the Group IIA metals (HgS, PbS, Bi_2S_3, CuS, CdS) and those of Group IIB (As_2S_3, Sb_2S_3, Sb_2S_5, SnS, SnS_2). Wash the precipitated sulphides with a little M NH_4Cl solution that has been saturated with H_2S (the latter to prevent conversion of CuS into $CuSO_4$ by atmospheric oxidation), transfer to a porcelain dish, add about 5 ml yellow ammonium polysulphide (1) solution, heat to 50–60°C, (2) and maintain at this temperature for 3–4 minutes with constant stirring. Filter (3).

Residue	Filtrate
May contain HgS, PbS, Bi_2S_3, CuS, and CdS. Wash once or twice with small volumes of dilute (1:100) ammonium sulphide solution, then with 2% NH_4NO_3 solution and reject all washings. Group IIA present.	May contain solutions of the thio-salts $(NH_4)_3AsS_4$, $(NH_4)_2SbS_4$, and $(NH_4)_2SnS_3$. Just acidify by adding concentrated HCl dropwise (test with litmus paper), and warm gently. A yellow or orange ppt., (4) which may contain As_2S_5, Sb_2S_5, and SnS_2 (5) indicates Group IIB present (6).

and antimony trisulphide, Sb_2S_3, is sparingly soluble in colourless ammonium sulphide solution, but both of these dissolve readily in the yellow ammonium polysulphide solution; the SnS is converted into the thiostannate. The yellow ammonium polysulphide solution contains an excess of sulphur and may be formulated $(NH_4)_2S_2$. This is often used for the last-named in writing equations, but there is some evidence that the solution contains a number of sulphides up to the pentasulphide, $(NH_4)_2S_5$. Yellow ammonium polysulphide and not colourless ammonium sulphide is therefore generally used in one method for the separation of Groups IIA and IIB. The separation of sulphur upon acidification is a disadvantage.

The following processes take place during dissolution:

$$As_2S_3\downarrow + 4S_2^{2-} \rightarrow S_3^{2-} + 2AsS_4^{3-} \quad \text{(thioarsenate)}$$
$$Sb_2S_3\downarrow + 4S_2^{2-} \rightarrow S_3^{2-} + 2SnS_4^{3-} \quad \text{(thioantimonate)}$$
$$Sb_2S_5\downarrow + 6S_2^{2-} \rightarrow 3S_3^{2-} + 2SbS_4^{3-} \quad \text{(thioantimonate)}$$
$$SnS\downarrow + S_2^{2-} \rightarrow SnS_3^{2-} \quad \text{(thiostannate)}$$
$$SnS_2\downarrow + 2S_2^{2-} \rightarrow S_3^{2-} + SnS_3^{2-} \quad \text{(thiostannate)}$$

2. If the ammonium polysulphide extract is boiled for some time in air, a red antimony oxide sulphide Sb_2OS_2 may be precipitated.
3. If the precipitate is completely soluble in ammonium polysulphide, Group IIA is absent.
4. Upon acidification the sulphides of the higher oxidation states are precipitated, mixed with sulphur which originates from the reagent. Hydrogen sulphide gas is also liberated:

$$2AsS_4^{3-} + 6H^+ \rightarrow As_2S_5\downarrow + 3H_2S\uparrow$$
$$2SbS_4^{3-} + 6H^+ \rightarrow Sb_2S_5\downarrow + 3H_2S\uparrow$$
$$SnS_3^{2-} + 2H^+ \rightarrow SnS_2\downarrow + H_2S\uparrow$$
$$S_2^{2-} + 2H^+ \rightarrow S\downarrow + H_2S\uparrow$$

5. If much S and little sulphide is suspected, shake with a little benzene: the sulphides collect on the boundary surface.
6. The disadvantages related to the use of ammonium polysulphide in the separation of Groups IIA and IIB are:

(i) Some CuS and HgS are dissolved.

(ii) In the presence of large quantities of the IIA Group, small amounts of tin may escape detection.

(iii) Acidification of the ammonium polysulphide extract leads to the precipitation of sulphur, which may obscure the presence of sulphides of Group IIB.

(iv) Ammonium polysulphide solution, unless freshly prepared, may contain sulphate and this will lead to the precipitation of barium as $BaSO_4$, etc.

(b) The potassium hydroxide method The method is outlined in Table V.15.

Table V.15 Separation of Group II into Group IIA and Group IIB with the potassium hydroxide (1) method The ppt. may consist of the sulphides of the Group IIA metals (HgS, PbS, Bi_2S_3, CuS, and CdS) and those of Group IIB (As_2S_3, Sb_2S_3, Sb_2S_5, and SnS_2 (2). Wash the precipitated sulphides with a small volume of M NH_4Cl solution that has been saturated with H_2S. Transfer the ppt. to a 100 ml beaker or porcelain basin, add 10 ml 2M KOH solution and boil with constant stirring (**CAUTION**) (3) for 2–3 minutes. Add 3 ml freshly prepared, saturated H_2S water, stir and filter (preferably through a double filter). Wash the residue with a little water and collect the washings with the filtrate (4).

Residue	**Filtrate**
May contain HgS, PbS, Bi_2S_3, CuS, and CdS. Group IIA present.	May contain AsO_3^{3-}, AsS_3^{3-}, SbO_2^-, SbS_2^-, $SbSO_3^{3-}$, SbS_4^{3-}, SnO_3^{2-}, SnS_3^{2-}, and $[HgS_2]^{2-}$. Transfer to a small conical flask or beaker, add concentrated HCl dropwise and with stirring until the mixture is distinctly acid to litmus paper (5). Treat with H_2S for 2 minutes to ensure complete precipitation of the sulphides. The formation of a ppt. indicates the possible presence of Hg, As, Sb or Sn. Filter and wash the ppt. with a little water. Group IIB present.

Notes and explanations to Table V.15: 1. Potassium hydroxide is employed in preference to sodium hydroxide since sodium antimonate is sparingly soluble.
2. Tin(II) sulphide is not completely soluble in KOH. The oxidation with H_2O_2 solution will have converted Sn^{2+} to Sn^{4+} (cf. Table V.12).
3. Potassium hydroxide is an extremely dangerous substance because of its destructive effect upon the eyes. Precipitates, when heated with KOH solution, tend to bump. Suitable precautions should therefore be taken: the process should be carried out in the fume cupboard, and protective eye-glasses should be worn.
4. By digesting the Group II precipitate with 2M KOH solution, the sulphides of As, Sb, and Sn(IV) dissolve whilst those of Hg, Pb, Bi, Cu, and Cd are largely unattacked. The very slight solubility of PbS is reduced by adding a little H_2S water though some HgS will dissolve in the presence of H_2S in consequence, provision is made for the detection of Hg in both Groups IIA and IIB.

The following processes take place during dissolution (cf. also note 2):

$$As_2S_3\downarrow + 6OH^- \rightarrow AsO_3^{3-} + AsS_3^{3-} \text{ (thioarsenite)} + 3H_2O$$
$$2Sb_2S_3\downarrow + 4OH^- \rightarrow SbO_2^- + 3SbS_2^- \text{ (thioantimonite)} + 2H_2O$$
$$Sb_2S_5\downarrow + 6OH^- \rightarrow SbSO_3^{3-} \text{ (monothioantimonate)} + SbS_4^{3-} \text{ (thioantimonate)} + 3H_2O$$
$$3SnS_2\downarrow + 6OH^- \rightarrow SnO_3^{2-} + 2SnS_3^{2-} \text{ (thiostannate)} + 3H_2O$$

In the presence of H_2S some of the HgS dissolves:

$$HgS\downarrow + H_2S + 2OH^- \rightarrow [HgS_2]^{2-} + 2H_2O$$

5. Upon acidification the sulphides are reprecipitated:

$$AsO_3^{3-} + AsS_3^{3-} + 6H^+ \rightarrow As_2S_3\downarrow + 3H_2O$$
$$SbO_2^- + 3SbS_2^- + 4H^+ \rightarrow 2Sb_2S_3\downarrow + 2H_2O$$
$$SbSO_3^{3-} + SbS_4^{3-} + 6H^+ \rightarrow Sb_2S_5\downarrow + 3H_2O$$
$$SnO_3^{2-} + 2SnS_3^{2-} + 6H^+ \rightarrow 3SnS_2\downarrow + 3H_2O$$
$$[HgS_2]^{2-} + 2H^+ \rightarrow HgS\downarrow + H_2S\uparrow$$

V.11 SEPARATION AND IDENTIFICATION OF GROUP IIA CATIONS

For the separation and identification of Group IIA cations (Hg^{2+}, Pb^{2+}, Bi^{3+}, Cu^{2+}, and Cd^{2+}) two, somewhat different methods are recommended. From the reagents used, these are called (*a*) sulphuric acid method and (*b*) sodium hydroxide method.

(a) The sulphuric acid method The method is described in Table V.16.

Table V.16 Separation of Group IIA cations with the sulphuric acid method The ppt. may contain HgS, PbS, Bi_2S_3, CuS, and CdS. Transfer to a beaker or porcelain basin, add 5–10 ml dilute HNO_3, boil gently for 2–3 minutes, filter and wash with a little water. (1).

<table>
<tr><th>Residue</th><th colspan="3">Filtrate</th></tr>
<tr><td rowspan="6">Black. HgS.
Dissolve in a mixture of of M NaOCl solution and 0·5 ml of dilute HCl. Add 1 ml dilute HCl, boil off excess Cl_2 and cool. Add $SnCl_2$ solution.
White ppt. turning grey or black.
Hg^{2+} present (2).</td><td colspan="3">May contain nitrates of Pb, Bi, Cu, ond Cd. Test a small portion for Pb by adding dilute H_2SO_4 and alcohol. A white ppt. of $PbSO_4$ indicates Pb present. If Pb present, add dilute H_2SO_4 to the remainder of the solution, concentrate in the fume cupboard until white fumes (from the dissociation of the H_2SO_4) appear. Cool, add 10 ml of water, stir, allow to stand 2–3 minutes, filter and wash with a little water.</td></tr>
<tr><th>Residue</th><th colspan="2">Filtrate</th></tr>
<tr><td rowspan="4">White: $PbSO_4$. Pour 2 ml of 10% ammonium acetate through the filter several times, add to the filtrate a few drops of dilute acetic acid and then K_2CrO_4 solution.</td><td colspan="2">May contain nitrates and sulphates of Bi, Cu and Cd. Add concentrated NH_3 solution until solution is distinctly alkaline. Filter (4).</td></tr>
<tr><th>Residue</th><th>Filtrate</th></tr>
<tr><td>White: may be $Bi(OH)_3$. Wash. Dissolve in the minimum volume of dilute HCl and pour into cold sodium tetrahydroxo-stannate(II).</td><td>May contain $[Cu(NH_3)_4]^{2+}$ and $[Cd(NH_3)_4]^{2+}$.
If deep blue in colour, Cu is present in quantity. Confirm Cu by acidifying a portion of the filtrate with dilute acetic acid and add $K_4[Fe(CN)_6]$ solution (6).
Reddish-brown ppt.
Cu present (7).</td></tr>
</table>

Table V.16 Separation of Group IIA cations with the sulphuric acid method (contd.)

	Yellow ppt. of $PbCrO_4$. Pb present (3).	solution (5). Bi present. Alternatively, dissolve a little of ppt. in 2–3 drops dilute HNO_3. Place 1 drop of this solution upon filter paper moistened with cinchonine–KI reagent. Orange-red spot. Bi present.	To the remainder of the filtrate, add KCN solution dropwise until colour is discharged, and add a further ml in excess. Pass H_2S for 20–30 seconds (8). Yellow ppt., sometimes discoloured, of CdS. Cd present. Filter off ppt. and dissolve a portion of it in 1 ml dilute HCl: boil to expel H_2S and most of the acid and apply the 'cadion-2B' test on 1 drop of the solution. A pink spot confirms Cd (9).

Notes and explanations to Table V.16: 1. HgS is insoluble in dilute HNO_3, while PbS, Bi_2S_3, CuS, and CdS all dissolve;

$$3PbS\downarrow + 8H^+ + 2NO_3^- \rightarrow 3Pb^{2+} + 3S\downarrow + 2NO\uparrow + 4H_2O$$
$$Bi_2S_3\downarrow + 8H^+ + 2NO_3^- \rightarrow 2Bi^{3+} + 3S\downarrow + 2NO\uparrow + 4H_2O$$
$$3CuS\downarrow + 8H^+ + 2NO_3^- \rightarrow 3Cu^{2+} + 3S\downarrow + 2NO\uparrow + 4H_2O$$
$$3CdS\downarrow + 8H^+ + 2NO_3^- \rightarrow 3Cd^{2+} + 3S\downarrow + 2NO\uparrow + 4H_2O$$

2. HgS dissolves in NaOCl–HCl mixture by forming (undissociated) $HgCl_2$:

$$HgS\downarrow + OCl^- + 2H^+ + Cl^- \rightarrow HgCl_2 + S\downarrow + H_2O$$

Alternatively, the precipitate can be dissolved in aqua regia (cf. Section **III.8**, reaction 1).

The identification of mercury with tin(II) chloride is based on the reduction first to (white) calomel, then to (grey) mercury:

$$2HgCl_2 + Sn^{2+} \rightarrow Hg_2Cl_2\downarrow + Sn^{4+} + 2Cl^-$$
$$Hg_2Cl_2\downarrow + Sn^{2+} \rightarrow 2Hg\downarrow + Sn^{4+} + 2Cl^-$$

3. The filtrate from the nitric acid treatment contains the nitrates of Pb, Bi, Cu, and Cd. Dilute H_2SO_4 precipitates Pb as $PbSO_4$, leaving the other metals (which form soluble sulphates) in solution.

$$Pb^{2+} + SO_4^{2-} \rightarrow PbSO_4\downarrow$$

The object of the evaporation with H_2SO_4 until white fumes appear is to eliminate the HCl and HNO_3, which have a slight solvent action upon the $PbSO_4$. The $PbSO_4$ is dissolved in ammonium acetate forming tetra-acetato-plumbate(II) complex:

$$PbSO_4\downarrow + 4CH_3COO^- \rightarrow [Pb(CH_3COO)_4]^{2-} + SO_4^{2-}$$

from which K_2CrO_4 (or $K_2Cr_2O_7$) precipitates yellow $PbCrO_4$:

$$[Pb(CH_3COO)_4]^{2-} + CrO_4^{2-} \rightarrow PbCrO_4\downarrow + 4CH_3COO^-$$

4. The addition of excess NH_3 results in the precipitation of $Bi(OH)_3$ and the formation of soluble tetrammine complexes of Cu^{2+} and Cd^{2+}

$$Bi^{3+} + 3NH_3 + 3H_2O \rightarrow Bi(OH)_3\downarrow + 3NH_4^+$$

$$Cu^{2+} + 4NH_3 \rightarrow [Cu(NH_3)_4]^{2+} \text{ (blue)}$$
$$Cd^{2+} + 4NH_3 \rightarrow [Cd(NH_3)_4]^{2+} \text{ (colourless)}$$

If the solution turns blue, copper is present.

5. Sodium tetrahydroxostannate(II) (or sodium stannite) solution is prepared by adding NaOH to $SnCl_2$ solution until the white precipitate of $Sn(OH)_2$ dissolves completely.

The test is based on the formation of bismuth metal:

$$2Bi^{3+} + 3[Sn(OH)_4]^{2-} + 6OH^- \rightarrow 2Bi\downarrow + 3[Sn(OH)_6]^{2-}$$

6. Acetic acid decomposes the blue $[Cu(NH_3)_4]^{2+}$ complex to Cu^{2+}, which in turn is precipitated by $K_4[Fe(CN)_6]$:

$$[Cu(NH_3)_4]^{2+} + 4CH_3COOH \rightarrow Cu^{2+} + 4NH_4^+ + 4CH_3COO^-$$
$$2Cu^{2+} + [Fe(CN)_6]^{4-} \rightarrow Cu_2[Fe(CN)_6]\downarrow$$

7. An alternative procedure is as follows: just acidify a portion with H_2SO_4. Add 1 drop of this solution to a few drops of $ZnSO_4$ solution, and introduce a little ammonium tetrathiocyanatomercurate(II) reagent. A violet precipitate confirms the presence of Cu^{2+} (cf. Section **III.10**, reaction 12). The precipitate can be rendered clearly visible by adding a little amyl alcohol and shaking: it collects in and colours the organic layer.

8. If the solution is colourless, it indicates that Cu^{2+} is absent. In this case there is no need for the addition of KCN, but H_2S can be introduced immediately. A yellow precipitate indicates that Cd^{2+} is present:

$$[Cd(NH_3)_4]^{2+} + H_2S \rightarrow CdS\downarrow + 2NH_4^+ + 2NH_3$$

If Cu^{2+} is also present, KCN has to be added, which forms the colourless complex ions:

$$[Cd(NH_3)_4]^{2+} + 4CN^- \rightarrow [Cd(CN)_4]^{2-} + 4NH_3$$
$$2[Cu(NH_3)_4]^{2+} + 10CN^- \rightarrow 2[Cu(CN)_4]^{3-} + (CN)_2\uparrow + 8NH_3$$

Note that copper is univalent in the complex. The cyanogen, which is formed, partly reacts to form cyanate and cyanide ions:

$$(CN)_2 + 2NH_3 + H_2O \rightarrow CNO^- + CN^- + 2NH_4^+$$

The tetracyanocuprate(I) complex is so stable that H_2S does not produce a precipitate (cf. Section **I.31**). From the tetracyanocadmiate(II) complex yellow CdS is precipitated with H_2S:

$$[Cd(CN)_4]^{2-} + H_2S + 2NH_3 \rightarrow CdS\downarrow + 2NH_4^+ + 4CN^-$$

9. Cd may be confirmed in the precipitate by the ignition test with $(COONa)_2$ (cf. Section **III.11**, reaction 9).

(b) The sodium hydroxide method This method is described in Table V.17.

Table V.17 Separation of Group IIA cations with the sodium hydroxide method The precipitate may contain HgS, PbS, Bi_2S_3, CdS, and CuS. Transfer the ppt. to a small beaker or porcelain basin, add *c.* 5 ml dilute HNO_3, boil gently for a few minutes, filter, and wash with a little water. (1).

<table>
<tr><th>Residue</th><th colspan="3">Filtrate</th></tr>
<tr><td rowspan="6">Black: HgS. Dissolve in a mixture of 2·5 ml of M NaOCl solution and 0·5 ml dilute HCl. Add 1 ml dilute HCl, boil off excess Cl_2 and cool. Add $SnCl_2$ solution. White ppt. turning grey or black.
Hg^{2+} present (2).</td><td colspan="3">May contain Pb^{2+}, Bi^{3+}, Cu^{2+}, and Cd^{2+}. Add excess concentrated NH_3 solution until precipitation is complete. Filter (3).</td></tr>
<tr><th colspan="2">Residue</th><th>Filtrate</th></tr>
<tr><td colspan="2">May contain $Bi(OH)_3$ and $Pb(OH)_2$. Warm with 5 ml of NaOH solution and filter (4).</td><td rowspan="3">May contain $[Cu(NH_3)_4]^{2+}$ and $[Cd(NH_3)_4]^{2+}$. If colourless, Cu is absent; test then directly for Cd by passing H_2S for 10–20 seconds into the ammoniacal solution.
Yellow ppt. of CdS.
Cd present.
If blue, Cu present. Divide the solution into two unequal parts.
Smaller portion. Acidify with dilute acetic acid and add $K_4[Fe(CN)_6]$ solution (7).
Reddish-brown ppt.
Cu present (8).
Larger portion. Add KCN solution dropwise until solution is decolourized. Pass H_2S for 10–20 seconds.
Immediate yellow ppt. of CdS (9).
Cd present (10).</td></tr>
<tr><th>Residue</th><th>Filtrate</th></tr>
<tr><td>May be $Bi(OH)_3$. Wash with a little water. Pour sodium tetrahydroxo-stannate(II) solution over filter.
Blackening of ppt.
Bi present (5).
Alternatively, dissolve a little of the ppt. in 2–3 drops of dilute HNO_3; place 1 drop of this solution upon filter paper moistened with cinchonine–potassium iodide reagent.
Orange-red spot.
Bi present.</td><td>May contain tetrahydroxo-plumbate, $[Pb(OH)_4]^{2-}$ Acidify with dilute acetic acid and add K_2CrO_4 solution
Yellow ppt. of $PbCrO_4$.
Pb present (6).</td></tr>
</table>

Notes and explanations to Table V.17: 1. See note 1 to Table V.16.

2. See note 2 to Table V.16.
3. If an excess of NH_3 is added, Cu^{2+} and Cd^{2+} are complexed and remain in solution:

$$Cu^{2+} + 4NH_3 \rightarrow [Cu(NH_3)_4]^{2+}$$
$$Cd^{2+} + 4NH_3 \rightarrow [Cd(NH_3)_4]^{2+}$$

Bi^{3+} and Pb^{2+} however are precipitated as hydroxides:

$$Bi^{3+} + 3NH_3 + 3H_2O \rightarrow Bi(OH)_3\downarrow + 3NH_4^+$$
$$Pb^{2+} + 2NH_3 + 2H_2O \rightarrow Pb(OH)_2\downarrow + 2NH_4^+$$

4. $Bi(OH)_3$ precipitate is unaffected if NaOH is added, while $Pb(OH)_2$ dissolves:

$$Pb(OH)_2 + 2OH^- \rightarrow [Pb(OH)_4]^{2-}$$

5. See note 5 to Table V.16.
6. See note 3 to Table V.16.

7. See note 6 to Table V.16.
8. See note 7 to Table V.16.
9. See note 8 to Table V.16.
10. See note 9 to Table V.16.

V.12 SEPARATION AND IDENTIFICATION OF GROUP IIB CATIONS

For the separation of Group II cations into Groups IIA and IIB two methods were suggested (cf. Section **V.10**, Tables V.14 and V.15). To separate and identify the Group IIB cations, one of the two recommended procedures must be followed, according to the method chosen in the previous step. If the ammonium polysulphide method (method *a*) has been followed in Section **V.10**, then method (*a*) has to be applied here. Method (*b*), on the other hand, has to be used if the potassium hydroxide method was followed to separate Groups IIA and IIB.

(a) The ammonium polysulphide method Separation and identification of As, Sb, and Sn can be carried out according to the scheme shown in Table V.18.

Table V.18 Separation of Group IIB cations with the ammonium polysulphide method

Treat the yellow ammonium polysulphide extract of the Group II ppt. (see Table V.14) with dilute HCl, with constant stirring, until it is slightly acid (test with litmus paper), warm and shake or stir for 1–2 minutes. A fine white or yellow ppt. is sulphur only. A yellow or orange flocculant ppt. indicates As, Sb, and/or Sn present. (1). Filter and wash the ppt., which may contain As_2S_5, As_2S_3, Sb_2S_5 and S, with a little H_2S water; reject the washings. (2).

Transfer the ppt. to a small conical flask, add 5–10 ml concentrated HCl and boil gently for 5 minutes (with funnel in mouth of flask). Dilute with 2–3 ml water, pass H_2S for 1 minute to reprecipitate small amounts of arsenic that may have dissolved, and filter (3).

Residue	Filtrate
May contain As_2S_5, As_2S_3, and S (yellow). Dissolve the ppt. in 3–4 ml warm dilute NH_3 solution, filter (if necessary), add 3–4 ml 3% H_2O_2 solution and warm for a few minutes to oxidize arsenite to arsenate. Add a few ml of the $Mg(NO_3)_2$ reagent. Stir and allow to stand. White, crystalline ppt. of $Mg(NH)_4)AsO_4$, $6H_2O$. As present. Filter off ppt., and pour 1 ml of $AgNO_3$ solution containing 6–7 drops of 2M acetic acid on to residue on filter. Brownish-red residue of Ag_3AsO_4 (4).	May contain Sb^{3+} and Sn^{4+}. Boil to expel H_2S, and divide the cold solution into four parts. (i) Render just alkaline with dilute NH_3 solution, disregard any slight ppt., add 1–2 g solid oxalic acid, boil, and pass H_2S for *c.* 1 minute into the hot filtrate. Orange ppt. of Sb_2S_3. Sb present. (ii) To 2 drops of the solution on a spot plate, add a minute crystal of $NaNO_2$ and then 2 drops of Rhodamine-B reagent. Violet solution or ppt. Sb present (5). (iii) Partially neutralize the liquid, add 10 cm clean iron wire to 1 ml of the solution. (If much Sb is present, it is better to reduce with Mg powder.) Warm gently to reduce the tin to the divalent state, and filter into a solution of $HgCl_2$. White ppt. of Hg_2Cl_2 or grey ppt. of Hg. Sn present. (iv) Treat 0·2–0·3 ml of the solutions with 5–10 mg of Mg powder, add 2 drops of $FeCl_2$ solution, 2–3 drops of M tartaric acid solution, 1–2 drops of dimethylglyoxime reagent and then dilute NH_3 solution until basic. Red colouration. Sn present (6).

Notes and explanations to Table V.18: 1. If Cu is present in the original Group II precipitate, a small amount may be dissolved by the ammonium polysulphide reagent, and coprecipitate here (brownish-red).
2. It is impossible to decide at this stage whether the ions were originally present in the lower (As^{3+}, Sb^{2+}, Sn^{2+}) or higher (As^{5+}, Sb^{5+}, Sn^{4+}) oxidation state. This can be tested from the original sample (for specific reactions see the appropriate sections in which the reactions of these ions are described).
3. The precipitate, obtained after acidification, contains As_2S_5, some As_2S_3, Sb_2S_5, SnS_2, and some S (from the decomposition of the reagent). For the chemical reactions see note 4 of Table V.14, Section **V.10**.

The separation with concentrated HCl is based on the fact that As_2S_5 and also As_2S_3 are insoluble in concentrated HCl, while Sb_2S_5 and SnS_2 dissolve:

$$Sb_2S_5\downarrow + 6HCl \rightarrow 2Sb^{3+} + 3H_2S\uparrow + 2S\downarrow + 6Cl^-$$
$$SnS_2\downarrow + 4HCl \rightarrow Sn^{4+} + 2H_2S\uparrow + 4Cl^-$$

4. The As_2S_3 and As_2S_5 precipitates are converted by NH_3 and H_2O_2 to soluble arsenate:

$$As_2S_5\downarrow + 16NH_3 + 20H_2O_2 \rightarrow 2AsO_4^{3-} + 16NH_4^+ + 5SO_4^{2-} + 12H_2O$$
$$As_2S_3\downarrow + 12NH_3 + 14H_2O_2 \rightarrow AsO_4^{3-} + 12NH_4^+ + 3SO_4^{2-} + 8H_2O$$

The identification reactions of As are as follows:

$$AsO_4^{3-} + NH_4^+ + Mg^{2+} \rightarrow MgNH_4AsO_4\downarrow \quad \text{(white)}$$
$$AsO_4^{3-} + 3Ag^+ \rightarrow Ag_3AsO_4\downarrow \quad \text{(brown)}$$

5. The filtrate from the HCl treatment contains Sb^{3+} and Sn^{4+} ions. When oxalic acid (with NH_3) is added, a stable trioxalatostannate(IV) complex is formed:

$$Sn^{4+} + 3(COO)_2^{2-} \rightarrow [Sn\{(COO)_2\}_3]^{2-}$$

from which H_2S does not produce a precipitate. Sb_2S_3 is precipitated under these conditions:

$$2Sb^{3+} + 3H_2S + 6NH_3 \rightarrow Sb_2S_3\downarrow + 6NH_4^+$$

Antimony is identified with the rhodamine-B reaction (see Section **III.15**, reaction 7).

Continued treatment with H_2S (beyond 5–6 minutes) may lead to the partial decomposition of the trioxalatostannate(IV) complex; this will be indicated by the darkening of the orange Sb_2S_3 precipitate.
6. The addition of metallic iron to the dilute solution results in the reduction of Sn^{4+} to Sn^{2+} and Sb^{3+} to Sb metal:

$$Sn^{4+} + Fe \rightarrow Sn^{2+} + Fe^{2+}$$
$$2Sb^{3+} + 3Fe \rightarrow 2Sb\downarrow + 3Fe^{2+}$$

The test for Sn^{2+} with $HgCl_2$ involves first the formation of white Hg_2Cl_2:

$$Sn^{2+} + 2HgCl_2 \rightarrow Hg_2Cl_2\downarrow + Sn^{4+} + 2Cl^-$$

then, if Sn^{2+} is present in excess, the white precipitate turns grey when Hg is formed:

$$Sn^{2+} + Hg_2Cl_2\downarrow \rightarrow 2Hg\downarrow + Sn^{4+} + 2Cl^-$$

For the $FeCl_3$–dimethylglyoxime test see Section **III.18**, reaction 6.

(b) The potassium hydroxide method Separation and identification of As, Sb, and S from the potassium hydroxide extract can be carried out according to the scheme given in Table V.19.

Table V.19 Separation of Group IIB cations with the potassium hydroxide method The filtrate from the Copper Group (KOH extract) may contain AsO_3^{3-}, AsS_3^{3-}, SbO_2^-, SbS_2^-, $SbSO_3^{3-}$, SbS_4^{3-}, SnO_3^{2-}, SnS_3^{2-}, and $[HgS_2]^{2-}$ (1). Add concentrated HCl cautiously (dropwise and with cautious stirring) until the solution is distinctly acid to litmus paper. Treat with H_2S for 2–3 minutes to ensure complete precipitation of the sulphides. The formation of a ppt. indicates the possible presence of HgS, As_2S_3, As_2S_5, Sb_2S_3, Sb_2S_5, and SnS_2 (2). Filter; wash the ppt. with a little water and discard the washings. Transfer the ppt. to a small conical flask, add 5–10 ml concentrated HCl, heat to boiling (FUME CUPBOARD) and maintain the mixture near the boiling point over a free flame for 3–5 minutes; stir constantly. Dilute with a little water and filter (3).

Residue		**Filtrate**
May contain HgS, As_2S_3, As_2S_5, and some S. If yellow, only As is present. Wash with water. Pour 5 ml dilute NH_3 solution through the filter 2 or 3 times.		May contain Sb^{3+} and Sn^{4+}. Divide into four parts. (i) Render just alkaline with dilute NH_3 solution, disregard any slight ppt., add 1–2 g solid oxalic acid, boil and pass H_2S for *c.* 1 minute into the hot solution. Orange ppt. of Sb_2S_2. Sb present. (ii) To 2 drops of the solution on a spot plate, add a minute crystal of $NaNO_2$ and then 2 drops of Rhodamine-B reagent. Violet ppt. or solution. Sb present (5). (iii) Partially neutralize the liquid, add 10 cm clean iron wire to 1 ml of the solution. (If much Sb is present, it is better to reduce with Mg powder.) Warm gently to reduce the tin to the bivalent state, and filter into a solution of $HgCl_2$. White ppt. of Hg_2Cl_2 or grey ppt. of Hg. Sn present. (iv) Treat 0·2–0·3 ml of the solution with 5–10 mg of mg powder, add 2 drops of $FeCl_2$ solution, 2–3 drops of M tartaric acid solution, 1–2 drops of dimethylglyoxime reagent, and then dilute NH_3 solution until basic. Red colouration. Sn present (6).
Residue	**Filtrate**	
If dark-coloured (HgS). Hg present. Confirm as in Table V.16, if Hg not found in Group IIA.	Add dilute HNO_3 until distinctly acid. Yellow ppt. of As_2S_3. As present. Confirm by redissolving ppt. in 3–4 ml warm, dilute NH_3 solution, add 3–4 ml of 3% H_2O_2 solution and warm for a few minutes to oxidize arsenite to arsenate. Add a few ml of $Mg(NO_3)_2$ reagent, stir and allow to stand. White ppt. of $Mg(NH_4)AsO_4.6H_2O$. Pour on the filter 1 ml of $AgNO_3$ solution containing 6–7 drops of 2M acetic acid. Brownish-red residue of Ag_3AsO_4 (4).	

Notes and explanations to Table V.19: 1. See note 2 to Table V.18.
2. See note 5 to Table V.15.
3. See second part of note 3 to Table V.18.
4. See note 4 to Table V.18.
5. See note 5 to Table V.18.
6. See note 6 to Table V.18.

V.13 REMOVAL OF INTERFERING IONS BEFORE THE PRECIPITATION OF THE GROUP III CATIONS When discussing the general scheme of separation of cations into groups (Section **V.18**, Table V.12), references have been made to the fact that certain ions might interfere with the separation and identification of cations of Groups III to V. These ions are phosphate, silicate, borate, fluoride, and anions of organic acids. From the point of view of their interfering action, they can be divided into two main groups.

(i) Those which, when present in solution, combine with various metals to form stable complex ions (compare Section **I.31** and **I.32**); this may result in the failure of these metals to precipitate with the usual group reagent. To this class belong such organic acids as oxalic, citric, and tartaric acid, and also hydroxy compounds, such as sugar and starch. In their presence, iron, chromium, and aluminium are either incompletely precipitated or not precipitated at all by ammonium chloride and ammonia solution.

(ii) Those which, under certain conditions, form insoluble compounds with some of the metals of the later groups. It is conceivable therefore, that under conditions which result in the precipitation of the Group IIIA metals, metals of the subsequent groups will also be precipitated in consequence of the formation of compounds which are insoluble or sparingly soluble in the presence of ammonium chloride and ammonia solution. The most important of these anions are oxalates, tartrates, citrates, borates, fluorides, and phosphates. Silicates are included with these owing to the precipitation of gelatinous silicic acid by ammonium chloride and ammonia solution (see under Silicates, Section **IV.26**, reaction 2).

Reference to a table of solubilities will show:

(*a*) That the oxalates, tartrates and citrates of most of the metals of Groups IIIA, IIIB, IV, and of magnesium are either insoluble or sparingly soluble in water. (It should be noted that most oxalates and tartrates, with the exception of those of calcium, strontium, and barium, form soluble complex salts with an excess of alkali oxalate and alkali tartrate respectively; compare (i) above.)

(*b*) That the borates of the metals of Groups IIIA, IIIB, and IV are insoluble in water; magnesium borate is sparingly soluble. (Some of these are, however, soluble in ammonium chloride solution.)

(*c*) That the fluorides of calcium and magnesium are insoluble, and those of nickel, cobalt, strontium, and barium are sparingly soluble in water.

(*d*) That the phosphates of the metals of Groups IIIA, IIIB, IV, and of magnesium are insoluble in water.

Furthermore, the borates, fluorides, phosphates, oxalates, tartrates, and citrates of the Group IIIA, IIIB, IV metals and of magnesium are insoluble in alkaline solution, but dissolve in acid solution.

It is evident that in the presence of these acids, Group IIIA metals cannot be separated from those of the remaining groups by the addition of the customary group reagent, NH_4Cl and NH_3 solution. The usual scheme of systematic analysis is therefore modified when any or all of these anions are present before proceeding to the precipitation of Group IIIA.

The presence of some of these anions will have been indicated in the preliminary tests. Organic acids are revealed in the dry tests, and particularly by the action of concentrated sulphuric acid. In view of the common occurrence of oxalates and the somewhat indecisive indications of the preliminary tests, it is recommended that this ion be tested for in a portion of the filtrate from Group II, from which all the hydrogen sulphide has been expelled by boiling. This is most simply carried out by adding 1 ml $CaCl_2$ solution and 1 ml CH_3COONa solution to 1 ml of the solution; if a crystalline precipitate forms (the precipitate should not be confused with gelatinous CaF_2), filter, wash, dissolve in a little hot dilute H_2SO_4, filter and add 2–3 drops of dilute $KMnO_4$ solution to the filtrate. If the colour of the $KMnO_4$ solution is discharged, an oxalate is present.

The presence of fluoride is indicated in the preliminary treatment with concentrated sulphuric acid; a moistened glass rod introduced into the tube becomes covered with a film of gelatinous silicic acid.

The detection of borate is provided for in the preliminary test, 10, of Section **V.2**. Alternatively, it may be tested for after the precipitation of Group II.

The presence of phosphate is always tested for in the Group II filtrate from which all the H_2S has been expelled. To 1 ml of the solution, add 1 ml HNO_3 and 3 ml ammonium molybdate solution, and warm to a temperature not exceeding 40°. The production of a bright-yellow precipitate proves the presence of a phosphate (Section **IV.28**, reaction 4).

The anions are removed in the following order:

1. Organic acids. The filtrate from Group II is evaporated to dryness, when some carbon may be liberated and the organic acids decomposed. By repeated evaporation with concentrated HNO_3, the black residue is completely oxidized. The residue must not be heated too strongly as Fe_2O_3, Cr_2O_3, and Al_2O_3 may be rendered very sparingly soluble in HCl.

2. Borates and fluorides. The residue from 1 is repeatedly evaporated almost to dryness with concentrated HCl; the HF volatilizes with the HCl and the H_3BO_3 volatilizes in the steam.

If borate is present and fluoride is absent, the former may be removed by repeated evaporation to dryness on a water bath with a mixture of 5 ml methanol (inflammable!) and 10 ml concentrated HCl. The borate slowly volatilizes as methyl borate, $(CH_3)_3BO_3$ (**POISONOUS**).

3. Silicates. The evaporation with concentrated HCl converts silicates into an insoluble form of SiO_2. Hence by complete evaporation to dryness with a further quantity of concentrated HCl, extraction with water or dilute HCl, and filtration, the silicate is completely eliminated.

4. Phosphates. There are several methods for the removal of phosphate ions. Of these, three methods will be outlined here. The first one, the **zirconium nitrate method** is by far the most convenient; if correctly performed, all phosphate ions are quantitatively and rapidly removed as the highly insoluble zirconium phosphate. The **iron(III)–acetate method** has been quite extensively used in the past; it can be adapted if zirconium nitrate is not available in the laboratory; its theoretical background is quite interesting too. Finally, the **tin(IV) chloride method** is also described because this method can be applied in the presence of the 'rarer' elements (cf. Section **VII.23**).

(a) Removal of phosphate by the zirconium method The scheme outlined in Table V.20 should be followed when removing phosphate by the zirconium nitrate method.

Table V.20 Removal of phosphate by the zirconium nitrate method Reduce the volume of the solution (filtrate of Group II) to about 10 ml; by evaporation, if necessary. Adjust the HCl concentration so that it does not exceed M (1). Add 0·5–1 g solid NH_4Cl, stir until dissolved, and then add zirconium(IV) nitrate reagent (2) slowly and with stirring until precipitation is complete (3); a large excess of the reagent must be avoided. Heat the contents of the test-tube or small conical flask just to boiling and stir with a glass rod to prevent bumping. Filter through a Whatman No. 32 filter paper (4). Wash the ppt. with a little hot water, and combine the washings with the filtrate.

Residue	**Filtrate**	
Zirconium phosphate. Reject.	Test if all phosphate has been precipitated by the addition of a drop of the zirconium(IV) nitrate reagent. Add about 0·5 g solid NH_4Cl, heat to boiling, add a slight excess of dilute NH_3 solution (i.e. until the odour of NH_3 is permanent in the boiling solution), boil for 2–3 minutes and filter.	
	Residue	**Filtrate**
	Examine for Group IIIA by Group Separation Table V.23 (Section **V.14**). The excess of Zr will be found in the residue after treatment with H_2O_2 and NaOH solution (or with sodium peroxoborate, $NaBO_3.4H_2O$, and boiling), and will accompany any $Fe(OH)_3$, if Fe is present.	Examine for Groups IIIB, IV, and V as detailed in General Table V.12 (Section **V.8**).

Notes and explanations to Table V.20: 1. The HCl concentration should not exceed M, otherwise a turbid supernatant liquid is obtained and the precipitation of phosphate is not quantitative.

2. The reagent is prepared by heating 10 g of commercially pure zinconyl nitrate $ZrO(NO_3)_2.2H_2O$ and 100 ml 2M HNO_3 to the boiling point with constant stirring. Allow any undissolved solid to settle for 24 hours, and then decant the clear solution.

3. It is important that the excess of the zirconium nitrate reagent should not exceed 25 per cent, otherwise a turbid supernatant liquid will be obtained: this turbidity cannot be removed by filtration or centrifugation. It is best, therefore, to add the zirconium nitrate solution slowly and with stirring until precipitation appears complete, heat just to boiling, filter and test the filtrate with the reagent etc.

4. The addition of half a Whatman filtration accelerator (or a little filter paper pulp) assists filtration; the precipitate must be washed thoroughly with hot water. The filtration accelerator consists of compressed filter paper which disintegrates on boiling; it increases the speed of filtration by retaining part of the precipitate and thus preventing the clogging of the pores of the filter paper.

(b) Removal of phosphates by the iron(III)–acetate method The theoretical basis of the iron(III)–acetate method may be summarized as follows:

(i) $FePO_4$, $AlPO_4$ and, to a lesser extent, $CrPO_4$ are insoluble in warm, dilute acetic acid buffered by a solution of ammonium acetate; the phosphates of the metals of Group IIIB and IV and of Mg are soluble.

(ii) The precipitation of phosphate ions will be complete only if the trivalent ions are present in excess. The procedure depends upon having excess iron(III) ions present. This ion is chosen because (*a*) it is easy to test for Fe^{3+} in the filtrate from Group II, (*b*) iron(III) phosphate is the least soluble in the acetic acid medium, and (*c*) it can most easily be detected by its colour in aqueous solution. Furthermore, the excess of iron(III) ions can be readily removed since iron(III) acetate is considerably hydrolysed upon dilution and boiling, and the excess iron is precipitated as basic iron(III) acetate (or as iron(III) hydroxide occluding large amounts of acetate ions).

$$Fe^{3+} + 3CH_3COO^- + 2H_2O \rightleftarrows Fe(OH)_2CH_3COO\downarrow + 2CH_3COOH$$
$$Fe^{3+} + 3CH_3COO^- + 3H_2O \rightleftarrows Fe(OH)_3\downarrow + 3CH_3COOH$$

The reaction tends to reverse upon cooling, hence filtration of the hot solution is desirable.

(iii) The addition of ammonium acetate solution to a solution containing iron, aluminium, and chromium ions yields the acetates of these metals; upon boiling the highly diluted solutions, the basic acetate of iron is almost completely and that of aluminium is largely precipitated. Chromium acetate does not appear to form a basic acetate under these conditions, but in the presence of considerable quantities of iron and aluminium, chromium is coprecipitated to a considerable extent with these two elements either as the basic acetate or as an adsorption complex.

The separation can be carried out according to the scheme outlined in Table V.21.

Table V.21 Removal of phosphate using the iron(III)–acetate method If a phosphate has been found (separation of cations into groups, Section **V.8**), proceed as follows. Dissolve the ppt. produced by the action of NH_4Cl and a slight excess of NH_3 solution in the minimum volume of dilute HCl. [The ppt. may contain $Fe(OH)_3$, $Al(OH)_3$, $Cr(OH)_3$, $MnO_2.xH_2O$, traces of CaF_2, and the phosphates of Mg and of the Group IIIA, IIIB, and IV metals.] Test about 0·5 ml for Fe by the addition of $K_4[Fe(CN)_6]$ or NH_4SCN solution. To the main volume of the cold solution, add dilute NH_3 solution dropwise, with stirring, until either a faint permanent precipitate is just obtained or until the solution is just alkaline (test with litmus paper). Then add 2–3 ml dilute acetic acid (1 : 1) and 5 ml 10M ammonium acetate solution. Disregard any ppt. which may form at this stage. If the solution is red or brownish-red, sufficient iron(III) is present (1) in the solution to combine with all the phosphate ions. If the solution is not red in colour, add 'neutral' $FeCl_3$ solution (2), drop by drop and with stirring, until the solution acquires a deep brownish-red colour (3). Dilute the solution to about 150 ml with hot water, boil gently for 1–2 minutes, filter hot (4) and wash the residue with a little boiling water.

<table>
<tr><th colspan="2">Residue</th><th colspan="2">Filtrate</th></tr>
<tr><td colspan="2">May contain the phosphates (and, possibly, the basic acetates) of Fe, Al, and Cr and also $Fe(OH)_3$.
Group IIIA present.
Rinse the ppt. into a porcelain dish by means of 10 ml cold water, add 1–1·5 g sodium peroxoborate, $NaBO_3.4H_2O$ (or add 5 ml NaOH solution, followed by 5 ml 3% H_2O_2 solution), and boil gently until the evolution of O_2 ceases (2–3 minutes). Filter and wash with a little hot water.</td><td colspan="2">Boil down in an evaporating dish to 20–25 ml. Add 0·5 g NH_4Cl and then dilute NH_3 solution in slight excess. Filter, if necessary.</td></tr>
<tr><th>Residue</th><th>Filtrate</th><th>Residue</th><th>Filtrate</th></tr>
<tr><td>$FePO_4^+$
$Fe(OH)_3$.
Reject.</td><td>May contain $[Al(OH)_4]^-$ and CrO_4^{2-}.
Examine for Al and Cr as described in Group Separation Table V.23 (Section V.14).</td><td>Examine for Al and Cr, if not previously tested for.
In general, no ppt. will be obtained here.</td><td>Examine for Groups IIIB, IV and for Mg as detailed in Table V.12 (Section V.8).</td></tr>
</table>

Notes to Table V.21: 1. If the reddish-brown colour cannot be seen owing to the formation of a precipitate or the presence of coloured ions, filter a small portion of the mixture and render the filtrate alkaline with dilute ammonia solution. If a light-coloured precipitate is obtained, more iron(III) chloride solution should be added to the main mixture with stirring until a similar test gives a reddish-brown precipitate, indicating that Fe^{3+} is present in excess.
2. 'Neutral' $FeCl_3$ solution is prepared by adding dilute ammonia solution dropwise to the side-shelf $FeCl_3$ solution until a slight permanent precipitate forms; this is filtered off. The side-shelf reagent usually contains excess free hydrochloric acid added during its preparation to produce a clear solution; the free acid leads to incomplete precipitation.
3. If the colour cannot be seen because of the presence of a precipitate or of coloured ions, filter a small portion of the mixture and test the filtrate with dilute ammonia solution as in note 1. A large excess of $FeCl_3$ solution must be avoided since it exerts a solvent action on iron(III) phosphate and the precipitate of the latter will be incomplete.
4. The addition of filter paper pulp or a Whatman filtration accelerator will facilitate filtration.

(c) Removal of phosphate by the tin(IV) chloride method It has been found that a freshly prepared solution of tin(IV) chloride readily removes phosphates from solution. It is very probable that the phosphate is removed largely as the insoluble tin(IV) phosphate, $Sn_3(PO_4)_4$; an alternative view is that an adsorption complex of phosphoric acid and hydrated tin(IV) oxide is formed, but it may well be that both compound formation and adsorption play a part. When small amounts of phosphates are present the precipitate filters with difficulty and the filtrate may be turbid; in such cases the addition of more phosphate, say as diammonium hydrogen phosphate, $(NH_4)_2HPO_4$, to the solution before precipitation with the reagent is recommended.

The experimental procedure is outlined in Table V.22.

Table V.22 Removal of phosphate by the tin(IV) chloride method Boil the filtrate from Group II to remove hydrogen sulphide, dilute to 50 ml, add 3 ml 0·5M $(NH_4)_2HPO_4$ solution (if necessary), followed by dilute NH_3 solution until the precipitate which forms initially does not redissolve on shaking. (The *p*H should be 3·5–4; the use of congo red or bromophenol blue indicator paper is advantageous.) Add 1 ml dilute hydrochloric acid, heat to boiling, and introduce 2 ml of the $SnCl_4$ reagent (prepared by dissolving 5 g of the crystalline salt in 5 ml water). Filter 1–2 ml of the suspension and test the filtrate for phosphate with ammonium molybdate and nitric acid. If the test is positive, add a further 0·5 ml of the $SnCl_4$ reagent to the boiling suspension; repeat the process until the phosphate test is negative. Introduce a Whatman filtration accelerator and filter the boiling suspension, preferably with the aid of suction (compare Section **II.2**, 10); wash well with hot water.

<table>
<tr><th>Residue</th><th colspan="2">Filtrate</th></tr>
<tr><td rowspan="3">Test for Cr^{3+}, and reject (1).</td><td colspan="2">Treat with H_2S until SnS_2 is removed. Filter.</td></tr>
<tr><th>Residue</th><th>Filtrate</th></tr>
<tr><td>Reject.</td><td>Concentrate by boiling with HCl, examine for Groups IIIA, IIIB, IV and V, (see Table V.12, Section V.8).</td></tr>
</table>

Note to Table V.22: 1. If the $Sn_3(PO_4)_4$ precipitate is green in colour, Cr might be present. To test for this boil a small portion of the precipitate with 2M NaOH, allow to cool, filter, add a little 3 per cent H_2O_2 and 1–2 ml amyl alcohol: a blue colouration of the amyl alcohol layer confirms Cr.

V.14 SEPARATION AND IDENTIFICATION OF GROUP IIIA CATIONS

In the filtrate, obtained after the separation of the Group II cations, the presence of phosphates, silicates, borates, fluorides, and anions of organic acids must be tested. If any of these are present, they must be removed, according to the methods described in Section **V.13**. The solution, from which the Group IIIA cations are to be separated, must therefore be free of the ions mentioned above, and also free of hydrogen sulphide (cf. Table V.12 in Section **V.8**). The separation and identification of Group IIIA cations can then be carried out according to the procedure outlined in Table V.23.

Table V.23 Separation of Group IIIA cations The ppt. produced by adding NH_4Cl and NH_3 solution and boiling may contain $Fe(OH)_3$, $Cr(OH)_3$, $Al(OH)_3$, and a little $MnO_2.xH_2O$ (1). Wash with a little hot 1 per cent NH_4Cl solution. Transfer the ppt. with the aid of 5–10 ml water to a small evaporating basin or a small beaker, add 1–1·5 g sodium peroxoborate, $NaBO_3.4H_2O$ (or add 5 ml NaOH solution, followed by 5 ml 3% H_2O_2 solution). Boil gently until the evolution of O_2 ceases (2–3 minutes). Filter.

Residue	**Filtrate**
May contain $Fe(OH)_3$ and $MnO_2.xH_2O$. Wash with a little hot water. Dissolve a small portion of the ppt. in 1 ml of 1:1 HNO_3 with the aid (if necessary) of 3–4 drops of 3% H_2O_2 solution or 1 drop of saturated H_2SO_3 solution. Boil (to decompose H_2O_2), cool thoroughly, add 0·05–0·1 g sodium bismuthate, shake, and allow the solid to settle. Violet solution of MnO_4^-. Mn present. Dissolve another portion of the ppt. in dilute HCl (filter, if necessary). Either – add a few drops of KSCN solution. Deep red colouration. Fe present. Or – add $K_4[Fe(CN)_4]$ solution. Blue ppt. Fe present. The original solution or substance should be tested with $K_4[Fe(CN)_6]$ and with KSCN to determine whether Fe^{2+} or Fe^{3+} (5).	May contain CrO_4^{2-} (yellow) and $[Al(OH)_4]^-$ (colourless). If colourless, Cr is absent and need not be considered further. If the solution is yellow, Cr is indicated. Divide the liquid into three portions. (i) Acidify with acetic acid and add lead acetate solution. Yellow ppt. of $PbCrO_4$. Cr present. (ii) Acidify 2 ml with dilute HNO_3, cool thoroughly, add 1 ml amyl alcohol, and 4 drops of 3% H_2O_2 solution. Shake well and allow the two layers to separate. Blue upper layer (containing chromium pentoxide; it does not keep well). Cr present (3). (iii) Acidify with dilute HCl (test with litmus paper), then add dilute NH_3 solution until just alkaline. Heat to boiling. Filter. White gelatinous ppt. of $Al(OH)_3$. Al present. Dissolve a small portion of the ppt. in 1 ml hot dilute HCl. Cool, add 1 ml 10M ammonium acetate solution and 0·5 ml of the 'aluminon' reagent. Stir the solution and render basic with ammonium carbonate solution. A red ppt. confirms the presence of Al (4).

Notes and explanations to Table V.23: 1. The precipitate may contain $Fe(OH)_3$, $Cr(OH)_3$, $Al(OH)_3$, and a little $MnO_2.xH_2O$: Co, Ni, Zn and some Mn remain in solution as the complex ammine ions. The $MnO_2.xH_2O$ owes its

formation to oxidation of some $Mn(OH)_2$ (which is held in solution by NH_4Cl and aqueous NH_3) by air and also partially to the HNO_3 or Br_2 water treatment made primarily to oxidize Fe^{2+} to Fe^{3+}: some Mn^{2+} is simultaneously oxidized to Mn^{4+} and the latter is precipitated as $MnO_2.xH_2O$ by the group reagent.
2. When the precipitate is boiled with NaOH and H_2O_2 solution (or with $NaBO_3.4H_2O$ solution), the $Al(OH)_3$ is converted into the soluble tetrahydroxo-aluminate $[Al(OH)_4]^-$ and the $Cr(OH)_3$ is oxidized to the yellow, soluble chromate CrO_4^{2-}. The $Fe(OH)_3$ and $MnO_2.xH_2O$ remain undissolved. The excess H_2O_2 is decomposed upon boiling: any reduction of chromate upon acidification is thus avoided.

$$Al(OH)_3\downarrow + OH^- \rightarrow [Al(OH)_4]^-$$
$$2Cr(OH)_3\downarrow + 3H_2O_2 + 4OH^- \rightarrow 2CrO_4^{2-} + 8H_2O$$
$$BO_3^- + H_2O \rightarrow BO_2^- + H_2O_2$$

3. The tests for Cr in the yellow solution include the precipitation of $PbCrO_4$ with acetic acid and lead acetate solution (Section **IV.33**, reaction 3) and the formation of chromium pentoxide – best in the presence of a little amyl alcohol (Section **IV.33**, reaction 4).
4. Aluminium is identified by reprecipitation as $Al(OH)_3$ by boiling in the presence of NH_4Cl and by the 'aluminon' test (Section **III.23**, reaction 7).

$$[Al(OH)_4]^- + NH_4^+ \rightarrow Al(OH)_3\downarrow + NH_3 + H_2O$$

5. The precipitate ($Fe(OH)_3 + MnO_2.xH_2O$), remaining after boiling with $NaOH + H_2O_2$ or with $NaBO_3.4H_2O$, is tested portionwise for Fe and Mn. Extraction with dilute HCl dissolves the $Fe(OH)_3$ as $FeCl_3$: the solution may be treated with $K_4[Fe(CN)_6]$ or with KSCN solution (Section **III.22**, reactions 6 and 11) and the presence of Fe established.

Another portion of the precipitate is dissolved in dilute HNO_3 and a few drops of H_2O_2 solution:

$$MnO_2\downarrow + H_2O_2 + 2H^+ \rightarrow Mn^{2+} + O_2\uparrow + 2H_2O$$

and the Mn^{2+} is oxidized to MnO_4^- with sodium bismuthate, $NaBiO_3$ (Section **III.28**, reaction 7) or with PbO_2 and HNO_3 (Section **III.28**, reaction 5). HCl should not be used for acidification, as Cl_2 gas is formed with MnO_2.

V.15 SEPARATION AND IDENTIFICATION OF GROUP IIIB CATIONS

For the separation of Group IIIB cations two methods are described. The first (*a*) is used most generally and is recommended for the beginner. The second one (*b*) needs somewhat more experience, but is equally efficient. It can be recommended especially if manganese is absent; in this case it is much faster than method (*a*).

(a) The hydrochloric acid–hydrogen peroxide method The separation can be carried out by following the scheme described in Table V.24.

Table V.24 Separation of Group IIIB cations with the hydrochloric acid–hydrogen peroxide method The ppt. may contain CoS, NiS, MnS, and ZnS. Wash well with 1per cent NH_4Cl solution to which 1 per cent by volume of $(NH_4)_2S$ has been added; reject the washings. Transfer the ppt. to a small beaker. Add 5 ml water and 5 ml 2M HCl, stir well, allow to stand for 2–3 minutes and filter (1).

Table V.24 (contd.)

<table>
<tr><th rowspan="3">Residue</th><th colspan="2">Filtrate</th></tr>
<tr><td colspan="2">May contain Mn^{2+} and Zn^{2+} and, possibly, traces of Co^{2+} and Ni^{2+}. Boil until H_2S removed (test with lead acetate paper), cool, add excess NaOH solution, followed by 1 ml 3% H_2O_2 solution. Boil for 3 minutes. Filter (3).</td></tr>
<tr><th>Residue</th><th>Filtrate</th></tr>
<tr><td>If black, may contain CoS and NiS. Test residue with borax bead. If blue, Co is indicated.
Dissolve the ppt. in a mixture of 1·5 ml M NaOCl solution and 0·5 ml dilute HCl. Add 1 ml dilute HCl, and boil until all Cl_2 is expelled. Cool and dilute to about 4 ml.
Divide the solution into two equal parts (2).
(i) Add 1 ml amyl alcohol, 2 g solid NH_4SCN and shake well. Amyl alcohol layer coloured blue.
Co present.
(ii) Add 2 ml NH_4Cl solution, NH_3 solution until alkaline, and then excess of dimethylglyoxime reagent.
Red ppt.
Ni present.</td><td>Largely $MnO_2.xH_2O$ and perhaps traces of $Ni(OH)_2$ and $Co(OH)_3$ (4). Dissolve the ppt. in 5 ml of 1:1 HNO_3. with the addition of a few drops of 3% H_2O_2 solution, if necessary (5). Boil to decompose excess H_2O_2 and cool. Add 0·05 g $NaBiO_3$, stir and allow to settle. Purple solution of MnO_4^-.
Mn present.</td><td>May contain $[Zn(OH)_4]^{2-}$.
Divide into two parts.
(i) Acidify with acetic acid and pass H_2S. White ppt. of ZnS.
Zn present.
(ii) Just acidify with dilute H_2SO_4, add 0·5 ml of 0·1M cobalt acetate solution and 0·5 ml of the ammonium tetrathiocyanatomercurate(II) reagent: stir.
Pale-blue ppt.
Zn present.</td></tr>
</table>

Notes and explanations to Table V.24: 1. MnS and ZnS dissolve readily in cold, very dilute (~ M) HCl, whilst NiS and CoS dissolve only slightly during the short period (2–3 minutes) that the sulphides are in contact with the acid.

$$MnS\downarrow + 2H^+ \rightarrow Mn^{2+} + H_2S\uparrow$$
$$ZnS\downarrow + 2H^+ \rightarrow Zn^{2+} + H_2S\uparrow$$

The presence of small amounts of Ni^{2+} and Co^{2+} ions in the filtrate (containing Mn^{2+} and Zn^{2+} ions) causes no serious intererence in the subsequent tests.
2. The detection of Co and of Ni in a mixture of CoS and NiS is carried out upon separate portions of the solutions prepared by dissolving the sulphides in aqua regia or in a mixture of NaOCl solution and dilute HCl:

$$3CoS\downarrow + 2HNO_3 + 6HCl \rightarrow 3Co^{2+} + 2NO\uparrow + 3S\downarrow + 6Cl^- + 4H_2O$$
$$CoS\downarrow + OCl^- + 2H^+ \rightarrow Co^{2+} + S\downarrow + Cl^- + H_2O$$

The sulphur, which precipitates, dissolves on prolonged heating with aqua regia, the nitric acid being active in this case:

$$S\downarrow + 2HNO_3 \rightarrow SO_4^{2-} + 2H^+ + 2NO\uparrow$$

The tests employed are highly sensitive; moderate quantities of other elements do not interfere and separation is therefore unnecessary.

The NH_4SCN test for Co is based on the formation of the blue tetrathiocyanato cobaltate(II) ion, $[Cu(SCN)_4]^{2-}$, which, paired with H^+ or NH_4^+ dissolves in amyl alcohol. The extracted compound is stable. The interference of iron (due to the red $Fe(SCN)_3$), may be eliminated by the addition of a soluble fluoride when the complex and highly stable hexafluoroferrate(III) ion, $[FeF_6]^{3-}$, is formed (Section **III.26**, reaction 6).

The dimethylglyoxime test for Ni (Section **III.27**, reaction 8) is applicable in the presence of Co provided excess reagent is added.

3. The solution containing Mn^{2+} and Zn^{2+} and possibly traces of Co^{2+} and Ni^{2+} is boiled to remove H_2S, and then warmed with a little H_2O_2 solution to oxidize the Co^{2+} to Co^{3+} and the Mn^{2+} to Mn^{4+}. $Co(OH)_2$ is slightly soluble but $Co(OH)_3$ is insoluble in excess NaOH solution. Upon adding excess NaOH solution, the $Zn(OH)_2$ precipitated initially will dissolve to form $[Zn(OH)_4]^{2-}$ whilst $MnO_2.xH_2O$ and traces of $Ni(OH)_2$ and $Co(OH)_3$ will be precipitated.

The Mn is readily identified in the precipitate by dissolving it in dilute HNO_3 and a little H_2O_2 and applying the sodium bismuthate test.

The Zn may be identified as white ZnS by passing H_2S into the NaOH extract as such or acidified with acetic acid:

$$[Zn(OH)_4]^{2-} + H_2S \rightarrow ZnS\downarrow + 2OH^- + 2H_2O$$

A characteristic test for Zn is the reaction with ammonium tetrathiocyanatomercurate(II) and cobalt nitrate. A pale-blue precipitate is formed, which is the mixture of zinc and cobalt tetrathiocyanatomercurate(II)s:

$$Zn^{2+} + Co^{2+} + 2[Hg(SCN)_4]^{2-} \rightarrow Zn[Hg(SCN)_4]\downarrow + Co[Hg(SCN)_4]\downarrow$$

If zinc ions are present *alone*, a white precipitate is formed. In the presence of traces of copper(II) ions, a blue-violet precipitate forms.

4. The Mn may be separated, if desired, from any Co or Ni by reprecipitating as MnO_2 in ammoniacal solution. To the nitric acid solution add 5 ml NH_4Cl solution and about 5 ml excess of dilute aqueous NH_3. Then add 1 g solid $K_2S_2O_8$, boil for 30 seconds, filter, and wash. The Co and Ni remain in solution as $[Co(NH_3)_6]^{3+}$ and $[Ni(NH_3)_6]^{2+}$. The MnO_2 may be filtered off, dissolved in $HNO_3 + H_2O_2$, and identified. This separation is not essential.

5. Any $Ni(OH)_2$ and $Co(OH)_3$ present will dissolve in the $HNO_3 + H_2O_2$ to form Ni^{2+} and Co^{2+} ions respectively. They dissolve less readily than MnO_2, so that if any difficulty is experienced in effecting complete solution of the precipitate, the undissolved solid may be discarded and the clear solution used to test for Mn.

(b) The hydrochloric acid–potassium chlorate–hydrogen peroxide method This method is in many ways a modification of the previous method, and yields quick results in skilled hands. The separation scheme is described in Table V.25.

Table V.25 Separation of Group IIIB cations with the hydrochloric acid–potassium chlorate–hydrogen peroxide method The ppt. may contain CoS, NiS, MnS, and ZnS. Wash well with 1 per cent NH_4Cl solution to which 1 per cent by volume of yellow ammonium sulphide solution has been added; reject the washings. Transfer the ppt. to a porcelain basin, add 5 ml water and 5 ml concentrated HCl and stir for 2–3 minutes. If a black residue is obtained, the presence of NiS and CoS is indicated: if complete dissolution takes place, only small amounts of Ni and Co are likely to be present. Evaporate the mixture to 2–3 ml in the FUME CUPBOARD, add 4 ml concentrated HNO_3 (1) and concentrate to 2–3 ml. If the solution is not clear, dilute with 8–10 ml water, filter off the sulphur and return the filtrate to the porcelain basin in the fume cupboard. Boil down to 1–2 ml, taking great care not to evaporate to dryness. Add 5 ml concentrated HNO_3 and evaporate again to 1–2 ml (2).

Remove about 0·2 ml of the solution with the aid of a dropper to a small test-tube, add 5 ml dilute HNO_3 and then 0·1 g sodium bismuthate: stir and allow to stand. A purple colouration, due to MnO_4^-, indicates Mn present. If Mn is present, treat the remainder of the solution in the porcelain basin in accordance with the procedure commencing at (*A*): if Mn is absent, continue the analysis from (*B*) onwards.

(*A*) Add 10–15 ml concentrated HNO_3, and heat just to boiling in the fume cupboard. By means of a spatula (preferably of glass), add finely-powdered $KClO_3$ in 0·1 g portions [GREAT CARE: DANGER OF EXPLOSION (3)]: after each addition boil gently until the greenish-yellow vapours of chlorine dioxide are expelled. Continue the addition of 0·1 g portions, boiling gently after each addition, until 1·5 g have been added. A brown or black ppt. of hydrated MnO_2 will form. Dilute with 3 ml water. Filter the ppt. through a small sintered glass funnel or crucible, and wash the ppt. with 2–3 ml water.

Residue	Filtrate	
$MnO_2.xH_2O$.	Evaporate in a porcelain basin (FUME CUPBOARD) to 2–3 ml. Allow to cool. (*B*) Dilute with 5 ml water. Add excess NaOH solution and 1 ml 3% H_2O_2 solution (4). Boil for 3 minutes and filter.	
	Residue	**Filtrate**
	May contain $Ni(OH)_2$ and $Co(OH)_2$. Dissolve the ppt. in dilute HCl, and test for Co and Ni as detailed in Table V.24.	May contain $[Zn(OH)_4]^{2-}$. Test for Zn as detailed in Table V.24.

Notes and explanations to Table V.25: 1. The precipitate is first treated with dilute HCl alone, partly to indicate the presence of Ni and Co in quantity but also because much free sulphur and sulphate would be formed if an oxidizing agent were used with the HCl at this stage.

2. Boiling of the undissolved sulphides with HCl in the presence of concentrated HNO_3, or with concentrated HNO_3 alone, results in the solution of the sulphides and the separation of sulphur:

$$3CoS\downarrow + 2HNO_3 + 6HCl \rightarrow 3Co^{2+} + 2NO\uparrow + 3S\downarrow + 6Cl^- + 4H_2O$$

or

$$3CoS\downarrow + 8HNO_3 \rightarrow 3Co^{2+} + 2NO\uparrow + 3S\downarrow + 6NO_3^- + 4H_2O$$

If the action of concentrated HNO_3 is continued, the S will gradually pass into solution as SO_4^{2-}:

$$S\downarrow + 2HNO_3 \rightarrow SO_4^{2-} + 2H^+ + 2NO\uparrow$$

The repeated evaporation with concentrated HNO_3 completely removes chloride ion, which interferes with the subsequent precipitation of MnO_2.

3. The Mn is precipitated in a chloride-free medium in the presence of concentrated HNO_3 with $KClO_3$:

$$Mn^{2+} + 2ClO_3^- \rightarrow MnO_2\downarrow + 2ClO_2\uparrow$$

The action of $KClO_3$ in concentrated HNO_3 is rapid and vigorous, and one of the decomposition products is the explosive, greenish-yellow gas, ClO_2. If the chlorate is added in small quantities as instructed and the solution is boiled gently between each addition to prevent the accumulation of ClO_2, the procedure is reasonably safe and the only evidence of explosion will be a gentle puff of gas. Under no circumstances may more than 0·1 g $KClO_3$ be added at one time.

4. In the subsequent separation of Co, Ni, and Zn, the addition of excess NaOH solution will yield a precipitate of $Co(OH)_2$ and $Ni(OH)_2$, and a solution of tetrahydroxozincate:

$$Co^{2+} + 2OH^- \rightarrow Co(OH)_2\downarrow$$
$$Ni^{2+} + 2OH^- \rightarrow Ni(OH)_2\downarrow$$
$$Zn^{2+} + 4OH^- \rightarrow [Zn(OH)_4]^{2-}$$

$Co(OH)_2$ is slightly soluble in alkali hydroxide solution. The addition of H_2O_2 (or Br_2 water) oxidizes it to $Co(OH)_3$, which is precipitated quantitatively:

$$2Co(OH)_2\downarrow + H_2O_2 \rightarrow 2Co(OH)_3\downarrow$$
$$2Co(OH)_2\downarrow + Br_2 + 2OH^- \rightarrow 2Co(OH)_3\downarrow + 2Br^-$$

V.16 SEPARATION AND IDENTIFICATION OF GROUP IV CATIONS

The Group IV cations cannot be precipitated directly from the filtrate obtained after the precipitation of Group IIIB, because it contains too high a concentration of ammonium salts. This was necessary to maintain when the Group IIIA cations were precipitated, to prevent the precipitation of magnesium hydroxide with this group (cf. Section **I.27** and Section **III.35**, reaction 1). This high concentration of ammonium ions however prevents the quantitative precipitation of the alkaline earth carbonates, because they decrease the concentration of carbonate ions in the solution through the equilibrium:

$$CO_3^{2-} + NH_4^+ \rightleftarrows HCO_3^- + NH_3$$

(cf. Section **III.31**, reaction 2).

Some textbooks recommend the removal of ammonium salts simply by sublimation. This can be done by first evaporating the acidified filtrate on a water bath, and then heating the dry residue at a higher temperature, until the white fumes of ammonium salts cease to appear. It is much easier however to remove ammonium salts with concentrated nitric acid, as described in Table V.12, when a chemical decomposition takes place:

$$NH_4^+ + HNO_3 \rightarrow N_2O\uparrow + 2H_2O + H^+$$

This reaction proceeds at a much lower temperature than is needed for the volatilization of ammonium salts.

A small amount of ammonium ions must however be present to prevent the precipitation of magnesium again. Thus a small amount of ammonium chloride is added after the removal of ammonium salts.

These operations are all incorporated into Table V.12, describing the general separation scheme of cations into groups.

It may be noted that the precipitation with ammonium carbonate takes place in hot solution, due to the fact that the carbonates are thus obtained in a more granular form and are therefore more easily filtered and washed. The solution must not be boiled as this decomposes the reagent and may result in some redissolution of the carbonates:

$$(NH_4)_2CO_3 \rightarrow 2NH_3\uparrow + CO_2\uparrow + H_2O$$

When precipitating $SrCO_3$ and/or $CaCO_3$ in the subsequent separation, it is probably better to use a little Na_2CO_3 than $(NH_4)_2CO_3$.

Once the carbonates are precipitated and filtered, the three cations can be separated and identified. For this, two methods are described. The 'sulphate' method (*a*) can be utilized in all cases, while the 'nitrate' method (*b*) is based on an unusual phenomenon, namely the 'salting out' of strontium nitrate with nitric acid.

(a) The sulphate method Separation and identification of Group IV cations can be carried out according to the scheme outlined in Table V.26.

Table V.26 Separation of Group IV cations with the sulphate method The ppt. may contain $BaCO_3$, $SrCO_3$, and $CaCO_3$. Wash with a little hot water and reject the washings. Dissolve the ppt. in 5 ml hot 2M acetic acid by pouring the acid repeatedly through the filter paper. Test 1 ml for barium by adding K_2CrO_4 solution dropwise to the nearly boiling solution. A yellow ppt. indicates Ba.

Ba present. Heat the remainder of the solution almost to boiling and add a slight excess of K_2CrO_4 solution (i.e. until the solution assumes a yellow colour and precipitation is complete). Filter and wash the ppt. (*C*) with a little hot water. Render the hot filtrate and washings basic with NH_3 solution and add excess $(NH_4)_2CO_3$ solution or, better, a little solid Na_3CO_2. A white ppt. indicates the presence of $SrCO_3$ and/or $CaCO_3$. Wash the ppt. with hot water, and dissolve it in 4 ml warm 2M acetic acid: boil to remove excess CO_3 (solution *A*).

Ba absent. Discard the portion used in testing for barium, and employ the remainder of the solution (*B*), after boiling for 1 minute to expel CO_2, to test for strontium and calcium (1).

<table>
<tr><th>Residue (C)</th><th colspan="2">Solution A or Solution B</th></tr>
<tr><td rowspan="3">Yellow: $BaCrO_4$.
Wash well with hot water. Dissolve the ppt. in a little concentrated HCl, evaporate almost to dryness and apply to the flame test.
Green (or yellowish-green) flame.
Ba present.
(Use spectroscope, if available.)</td><td colspan="2">The volume should be about 4 ml.
Either – To 2 ml of the cold solution, add 2 ml saturated $(NH_4)_2SO_4$ solution, followed by 0·2 g sodium thiosulphate, heat in a beaker of boiling water for 5 minutes and allow to stand for 1–2 minutes. Filter (2).
Or – To 2 ml of the solution add 2 ml triethanolamine, 2 ml saturated $(NH_4)_2SO_4$ solution, heat on a boiling water bath with continuous stirring for 5 minutes and allow to stand for 1–2 minutes. Dilute with an equal volume of water and filter (3).</td></tr>
<tr><th>Residue</th><th>Filtrate</th></tr>
<tr><td>Largely $SrSO_4$. Wash with a little water. Transfer ppt. and filter paper to a small crucible, heat until ppt. has charred (or burn filter paper and ppt., held in a Pt wire, over a crucible), moisten ash with a few drops concentrated HCl and apply the flame test (4).
Crimson flame.
Sr present.
(Use spectroscope, if available.)</td><td>May contain Ca complex.
(If Sr is absent, use 2 ml of solution A or B.) Add a little $(NH_4)_2C_2O_4$ solution, 2 ml 2M CH_3COOH and warm on a water bath.
White ppt. of CaC_2O_4.
Ca present (5).
Confirm by flame test on ppt. – brick-red flame.
(Use spectroscope, if available.)</td></tr>
</table>

Notes and explanations to Table V.26: 1. $BaCrO_4$ ($K_s = 1{\cdot}6\times10^{-10}$) is almost insoluble in dilute acetic acid, whilst $SrCrO_4$ ($K_s = 3{\cdot}6\times10^{-5}$) and $CaCrO_4$ ($K_s = 2{\cdot}3\times10^{-2}$) are soluble and are therefore not precipitated in a dilute acetic acid medium.

$$Ba^{2+} + CrO_4^{2-} \rightarrow BaCrO_4\downarrow$$

The function of the acetic acid is to convert some of the CrO_4^{2-} ions into $Cr_2O_7^{2-}$ ions, thus lowering the CrO_4^{2-} ion concentration so that the solubility products of $SrCrO_4$ and $CaCrO_4$ are not attained and in consequence they remain in solution:

$$2CrO_4^{2-} + 2CH_3COOH \rightleftharpoons Cr_2O_7^{2-} + 2CH_3COO^- + H_2O$$

It will be realized that in the presence of a large H^+ ion concentration (as distinct from the limited one due to the weak acetic acid), the CrO_4^{2-} may be such that the solubility product of $BaCrO_4$ is not reached and under such conditions $BaCrO_4$ will not be precipitated: this accounts for the solubility of this salt in dilute mineral acids.

2. Upon adding saturated $(NH_4)_2SO_4$ solution and warming, $SrSO_4$ is largely precipitated upon standing: the addition of some $Na_2S_2O_3$ increases the solubility of $CaSO_4$ and this reduces the amount of coprecipitation with $SrSO_4$. The formation of a soluble complex salt, such as $(NH_4)_2\,Ca(SO_4)_2$, has been suggested to account for the failure of $CaSO_4$ to precipitate in appreciable quantity with $(NH_4)_2SO_4$ solution.

3. In the alternative procedure, using triethanolamine, $N(C_2H_4OH)_3$, and saturated $(NH_4)_2SO_4$ solution, it is probable that the hexa-triaethanolamino-calciumate(II) $[Ca\{N(C_6H_4OH)_3\}_6]^{2+}$ complex ion is formed, and, in consequence, very little $CaSO_4$ is precipitated: however, the concentration of Ca^{2+} ions due to the dissociation of the complex ion is such that the solubility product of CaC_2O_4 ($2{\cdot}6 \times 10^{-9}$) is exceeded upon adding $(NH_4)_2C_2O_4$ solution.

4. $SrSO_4$ does not give a flame colouration easily; it is therefore partially reduced to SrS by the carbon of the filter paper. Treatment of the residue containing SrS with HCl gives the relatively volatile $SrCl_2$.

5. Calcium is readily identified by precipitation as CaC_2O_4, followed by the flame test (best observed through a spectroscope). It may also be converted into $CaSO_4.2H_2O$, and the characteristic crystals examined in a microscope – magnification of about 100 diameters.

The following is an alternative test for Ca. Place a drop or two or the filtrate upon a glass slide, add a drop or two of dilute H_2SO_4, and concentrate by placing the slide on a small crucible and warming until crystallization just commences. Examine the crystals in a microscope (magnification: *c.* 100×). Bundles of needles or elongated prisms indicate the presence of calcium.

(b) The nitrate method The separation and identification of Group IV cations can be carried out according to the scheme shown in Table V.27.

Table V.27 Separation of Group IV cations using the nitrate method The ppt. may contain $BaCO_3$, $SrCO_3$, and $CaCO_3$. Wash with a little hot water and reject the washings. Dissolve the ppt. in 5 ml hot 2M acetic acid by pouring the acid repeatedly through the filter paper. Test 1 ml for barium by adding K_2CrO_4 solution dropwise to the nearly boiling solution. A yellow ppt. indicates the presence of Ba.

Ba present. Heat the remainder of the solution almost to boiling and add a slight excess of K_2CrO_4 solution (i.e. until precipitation is complete and the solution assumes a yellow colour). Filter and wash the ppt. (*C*) with a little hot water. Combine the filtrate and washings (solution *A*) (1).

Ba absent. Discard the portion used in testing for barium, and employ the remainder of the solution (*B*), after boiling for 1 minute to expel CO_2 to test for strontium and barium.

Table V.27 (contd.)

Residue (*C*)	**Solution *A* or Solution *B***	
Yellow: $BaCrO_4$. Wash well with hot water. Dissolve the ppt. in a little concentrated HCl, evaporate almost to dryness and apply the flame test. Green (or yellowish-green) flame. Ba present. (Use spectroscope, if available.)	Render alkaline with NH_3 solution, and add a slight excess of $(NH_4)_2CO_3$ solution, or better, a little solid Na_2CO_3. Place the test-tube in a boiling water bath for 5 minutes. A white ppt. indicates the presence of $SrCO_3$ and/or $CaCO_3$. Filter and wash the precipitated carbonates with hot water until the washings are colourless; discard the washings. Drain the liquid from ppt. as completely as possible. Transfer the ppt. to a small beaker, add cautiously 15–20 ml 83% HNO_3 stir for 3–4 minutes with a glass rod and filter through a sintered glass crucible (2).	
	Residue	**Filtrate**
	White: $Sr(NO_3)_3$. Sr present. Confirm by crimson flame. (Use spectroscope, if available.)	May contain Ca^{2+}. Transfer most of the liquid to a porcelain basin and evaporate almost to dryness (FUME CUPBOARD). Dissolve the residue in 2 ml water (solution *D*), render alkaline with NH_3 solution, then acid with dilute acetic acid, add excess $(NH_4)_2(COO)_2$ solution and warm on a water bath. White ppt. of $Ca(COO)_2$. Ca present. Filter off the ppt. and apply the flame test: brick-red flame. (Use spectroscope, if available.) Alternatively – Place a drop or two of solution *D* on a glass slide, and add a drop of two of dilute H_2SO_4. Concentrate by placing the slide on a small crucible and warm until crystallization just commences. Examine the crystals in a microscope (magnification, *c*. 100 ×). Bundles of needles or elongated prisms. Ca present (3).

Notes and explanations to Table V.27: 1. The barium is separated as $BaCrO_4$ as in Table V.26 and the strontium and calcium are precipitated as carbonates by $(NH_4)_2CO_3$ solution in the presence of a little aqueous ammonia. It is better to employ Na_2CO_3 to precipitate the carbonates; the influence of ammonium salts in tending to reduce the carbonate-ion concentration is thus eliminated.

2. The mixture of $SrCO_3$ and $CaCO_3$ is treated cautiously with 83 per cent HNO_3; the carbonates are thus converted into the nitrates. $Sr(NO_3)_2$ is insoluble in the medium whilst the $Ca(NO_3)_2$ dissolves. The $Sr(NO_3)_2$ is collected by filtration through a sintered glass crucible or funnel: the presence of Sr is confirmed by the flame test, preferably with the aid of a hand spectroscope.

3. Upon gentle evaporation of the filtrate to remove HNO_3, $Ca(NO_3)_2$ is obtained. The latter is dissolved in water, rendered ammoniacal and then faintly acid with dilute acetic acid, and $(NH_4)_2C_2O_4$ solution added whereupon CaC_3O_4 is precipitated; the flame test may be applied to the CaC_2O_4 precipitate. Alternatively, the calcium may be converted into the sulphate $CaSO_4.2H_2O$, and the latter identified under the microscope.

V.17 IDENTIFICATION OF GROUP V CATIONS The Group V cations (Mg^{2+}, Na^+, K^+, and NH_4^+) can be identified one by one without preliminary separation. In the scheme described in Table V.28 a partial separation of Mg is followed by tests for K and Na.

Table V.28 Identification of Group V cations Treat the dry residue from Group IV with 4 ml water, stir, warm for 1 minute and filter (1).

If the residue dissolves completely (or almost completely) in water, dilute the resulting solution (after filtration, if necessary) to about 6 ml and divide it into three approximately equal parts: (i) Use the major portion to test for Mg with the prepared 'oxine' solution: confirm Mg by applying the 'mageson' test to 3–4 drops of the solution; (ii) and (iii) Test for Na and K, respectively, as described below.

Residue	**Filtrate**
Dissolve in a few drops of dilute HCl and add 2–3 ml water. Divide the solution into two unequal parts. (i) Larger portion: Treat 1 ml 2 per cent oxine solution in 2M acetic acid with 5 ml 2M ammonia solution and, if necessary, warm to dissolve any precipitated oxine. Add a little NH_4Cl solution to the test solution, followed by the ammoniacal oxine reagent, and heat to boiling point for 1–2 minutes (the odour of NH_3 should be discernible). Pale yellow ppt. of Mg 'oxinate'. Mg present (2). (ii) Smaller portion: To 3–4 drops, add 2 drops of the 'magneson' reagent, followed by several drops of NaOH solution until alkaline. A blue ppt. confirms Mg.	Divide into two parts (*a*) and (*b*). (*a*) Add a little uranyl magnesium acetate reagent, shake and allow to stand for a few minutes. Yellow crystalline ppt. Na present (3). Confirm by flame test: persistent yellow flame. (*b*) Add a little sodium hexanitrito-cobaltate(III) solution (or *c.* 4 mg of the solid) and a few drops of dilute acetic acid. Stir and, if necessary, allow to stand for 1–2 minutes. Yellow ppt. of $K_3[Co(NO_2)_6]$. K present (4). Confirm by flame test and view through two thicknesses of cobalt glass: red colouration (usually transient).

Examination for ammonium. This has already been carried out with the original substance in the preliminary tests [Section **V.2** test 6] (5).

Notes and explanations to Table V.28 1. The filtrate from Group IV may contain Mg, Na, K, and ammonium salts. It is evaporated to dryness and heated until all the ammonium salts have been volatilized; a residue is indicative of the presence of one or more of these metals. The dry residue is extracted with water to separate the soluble Na and K salts and filtered: the residue is tested for Mg and the filtrate is examined for Na and K.

2. The residue is dissolved in dilute HCl and the larger portion of the resulting solution is tested for Mg with the oxine reagent (Section **III.35**, reaction 7): the smaller portion is subjected to the 'magneson' test (Section **III.35**, reaction 8) when a blue precipitate should be obtained.

The precipitation of Mg as $Mg(NH_4)PO_4.6H_2O$ by the addition of a little NH_4Cl and excess Na_2HPO_4 to the ammoniacal solution (Section **III.35**, reaction 5) is sometimes rather slow; also traces of Group IV metals tend to interfere. For these reasons the oxine and 'magneson' tests are preferred.

If it is desired however, to carry out the less satisfactory Na_2HPO_4 test for comparison with the oxine test for Mg, treat the acid solution with a little NH_4Cl, followed by dilute NH_3 solution until basic, and add Na_2HPO_4 solution. Shake and stir vigorously. A white crystalline precipitate of $Mg(NH_4)PO_4.6H_2O$ indicates Mg. This precipitate sometimes separates slowly.

3. The most satisfactory precipitation for sodium ions is that with uranyl magnesium or zinc acetate (Section **III.37**, reactions 1 and 3). The flame test, in which an intense persistent yellow colouration is produced, is characteristic. Traces of sodium may be introduced from the reagents during the analysis, and hence it is important to look for a strong persistent yellow colouration; a feeble yellow colouration may be ignored.
4. The precipitation of the yellow potassium hexanitritocobaltate(III) with $Na_3[Co(NO_2)_6]$ (Section **III.36**, reaction 1) and the flame test (best observed through a spectroscope) are characteristic.

It is advisable to test the cobalt glass with a potassium salt to be certain that the glass is satisfactory: some samples of cobalt glass completely absorb the red lines from potassium. It is recommended that a hand spectroscope be used, if available.
5. Ammonium was tested for in preliminary test 6 (Section **V.2**). By heating the original substance with NaOH solution, NH_3 will be evolved from ammonium salts. The NH_3 gas may be identified by its odour, by its action upon red litmus paper or upon filter paper moistened with mercury(I) nitrate solution, or by the tannic acid–silver nitrate test (Section **III.38**, reaction 7). The insertion of a loose plug of cotton wool in the upper part of the tube will eliminate the danger of NaOH solution spray affecting the reagent paper.

V.18 PRELIMINARY TESTS FOR AND SEPARATION OF CERTAIN ANIONS After the systematic separation and detection of cations, the search for anions should be started. At this stage already considerable information is available about the presence or absence of certain anions. Not only the preliminary wet and dry tests supplied this information, but during the separation of cations a number of facts became available; thus the presence of phosphate, borate, fluoride, citrate, tartrate, and oxalate has been detected, before the precipitation of Group III cations. Similarly, the presence of chromate (or dichromate), permanganate and arsenate has been established at various stages of these separations. When the systematic examination for anions is carried out, such findings should be kept in mind.

Before describing these preliminary tests, a summary of the solubilities of the salts of the more common anions in water may be found useful in the subsequent deductions.

Nitrates, chlorates, acetates, manganates, and **permanganates** are all soluble; exceptions are a few basic nitrates (e.g. Bi and Sb) and basic acetates (e.g. Fe); silver and mercury(I) acetates are sparingly soluble.

Nitrites are all soluble; silver nitrite is sparingly soluble.

Chlorides are generally soluble; exceptions are AgCl, Hg_2Cl_2, TlCl, CuCl, SbOCl, BiOCl, which are insoluble; $PbCl_2$ is sparingly soluble.

Bromides have similar solubilities to the chlorides.

Iodides are generally soluble; exceptions are AgI, Hg_2I_2, HgI_2, CuI, BiOI, SbOI, which are insoluble; PbI_2, BiI_3, and SnI_2 are slightly soluble.

Carbonates are generally insoluble; those of Na, K, and NH_4 are soluble. The **hydrogen carbonates** of the alkali metals and of Ca, Sr, Ba, Mg, Fe, and Mn are also soluble.

Sulphides are generally insoluble; those of Ba, Sr, and Ca are slightly soluble; those of Na, K, and NH_4 are readily soluble.

Sulphites are generally insoluble; exceptions are those of the alkali metals, and the hydrogen sulphites of the alkaline earth group.

Sulphates are generally soluble; those of Pb, Hg(I), Sr, and Ba are insoluble; those of Ag, Hg(II), Ca, and a few basic sulphates (e.g. Bi and Hg) are slightly soluble.

Phosphates, arsenates, and **arsenites** are generally insoluble; those of Na, K, and NH_4 are soluble.

Fluorides are generally insoluble; those of Na, K, NH_4, Ag, and Hg(I) are soluble.

Borates, with the exception of those of the alkali metals, are insoluble.

Silicates possess solubilities similar to those of the borates.

Chromates are generally insoluble or sparingly soluble; exceptions are those of the alkali metals and Ca, Sr, Mg, Mn, Zn, Fe, and Cu.

Thiocyanates of Hg(II), Cu(II), Fe, Ca, Sr, Ba, Mg, Na, K, and NH_4 are insoluble.

Thiosulphates are generally soluble; the Ag and Ba salts are sparingly soluble.

Oxalates, formates, tartrates, and **citrates** are generally insoluble; those of Na, K, and NH_4 are soluble.

For the examination of anions, it is necessary to obtain a solution containing all (or most) of the anions free, as far as possible, from heavy metals. This is best prepared by boiling the substance with concentrated sodium carbonate solution; double decomposition occurs (either partially or completely) with the production of the insoluble carbonates (in some instances basic carbonates of hydroxides) of the metals (other than alkali metals) and the soluble sodium salts of the anions, which pass into solution.

The preparation of this 'soda extract' can be carried out according to the instructions in Table V.29.

Table V.29 Preparation of the soda extract for testing for anions

Boil 1·0 g of the finely divided substance or mixture (1) with a saturated solution of pure sodium carbonate (~1·5M) (prepared from 4 g anhydrous sodium carbonate and 25 ml distilled water) for 10 minutes, or until no further action appears to be taking place (3), in a small conical flask with a small funnel in the mouth to reduce the loss by evaporation; alternatively a reflux condenser may be employed. Filter (4), wash the residue with hot distilled water and collect the washings together with the main filtrate; the total volume should be 30–35 ml. Keep the residue (5).

Notes to Table V.29: 1. If a solution is supplied for analysis, use sufficient to contain 1·0 g of solid material, render it strongly alkaline with saturated sodium carbonate solution and evaporate it down to 10–15 ml.

2. It is essential to use pure sodium carbonate; only the analytical grade reagent is satisfactory. Some 'pure' samples may contain traces of sulphate or chloride: the absence of these impurities should be confirmed by a blank experiment.

3. If ammonium is present, ammonia will be evolved (indicated by the smell above the solution); continue boiling until ammonia is completely expelled, otherwise those metals which form ammonia complexes may pass into the sodium carbonate extract.

4. If no precipitate is obtained, the substance is largely free from heavy metals, and the sodium carbonate treatment may be omitted if more of the solution is required. Certain amphoteric elements may, however, be present in the sodium carbonate solution, e.g. Cu, Sn, Sb, As, Al, Cr, and Mn, dissolved in appreciable quantities; manganates, permanganates, and chromates may also be present

so that the colour of the solution should be noted. If cyanide is present, certain cations, such as Ag, Hg, Fe, Ni, and Co, may pass into the alkaline extract by virtue of the formation of complex cyanides (anions).

5. The residue from the Na_2CO_3 treatment may contain, in addition to the carbonates, basic carbonates and hydroxides of the heavy metals, certain phosphates (e.g. those of Pb, Ag, Cd, Cr, Mo, Mn, Zn, Ca, and Mg, which are less than 50 per cent transposed), arsenates, sulphides, fluorides, silicates, and complex cyanides, and the halides of silver. This may be due, in the case of products produced by high-temperature processes or of certain natural products, to the slow rate of reaction with the alkali carbonate or it may be due to the fact that the solubilities are such that very little metathesis can take place. Hence the residue should be kept until the tests for anions have been completed, and then tested for the above anions if they are not found in the alkaline extract or in the preliminary tests: the original substance can of course be employed for these tests. After extraction of the sodium carbonate residue with dilute hydrochloric acid, the undissolved portion may be treated as an insoluble substance and investigated according to Section **V.7**. It is generally more convenient, however, to test the residue for the various anions as follows:

Phosphate and arsenate. If the soda extract, when acidified with dilute nitric acid, boiled for 1 minute, and rendered alkaline with ammonia solution, does not give a precipitate with the magnesium nitrate reagent or with magnesia mixture, phosphate and arsenate are absent. Heat a small portion of the residue with concentrated nitric acid; if brown fumes are evolved, indicating the presence of a reducing agent, add more concentrated nitric acid and continue the heating until all the reducing agent is oxidized. Dilute the solution with water, heat to boiling, and filter. Render the filtrate alkaline with concentrated ammonia solution, and add the magnesium nitrate reagent or magnesia mixture. A white precipitate indicates arsenate and/or phosphate. Separate as described in Section **IV.45**, 11 and 12.

Sulphide. Reduce a portion of the residue with zinc and dilute sulphuric acid. If hydrogen sulphide is evolved (lead acetate paper test), sulphide is present. It is preferable to employ the sodium carbonate residue rather than the original mixture in testing for sulphide, for the latter may contain sulphite and thiocyanate, both of which give hydrogen sulphide when reduced (cf. Section **IV.4**, reaction 8 and Section **IV.10**, reaction 5).

Cyanide. If this is suspected, hydrogen cyanide will be evolved in the reduction with zinc and sulphuric acid, particularly in warming. Identify the hydrogen cyanide by the ammonium sulphide test (Section **IV.8**, reaction 1). It will also be identified in the preliminary test with dilute sulphuric acid (Section **V.2**, test 7).

Halides of silver. Treat a portion of the residue with zinc and dilute sulphuric acid (the solution remaining after testing for sulphide may be used):

$$2AgCl + Zn \rightarrow 2Ag\downarrow + Zn^{2+} + 2Cl^-$$

Filter to remove the excess zinc and the precipitated silver, boil the solution to remove all the hydrogen sulphide (if sulphide is present), and test the solution for Cl^-, Br^-, and I^- as detailed in Section **IV.45**.

Fluoride. A portion of the dried residue (or of the original mixture) may be decomposed with concentrated sulphuric acid. Heat in a lead capsule or crucible with concentrated sulphuric acid and apply the etching test (Section **IV.17**, reaction 2). Alternatively, the 'water' test (Section **V.18**, reaction 8) may be used.

With this soda extract the following tests should then be carried out:

1. Sulphate test To 2 ml of the soda extract add dilute hydrochloric acid until acid (test with litmus paper) and then add 1–2 ml in excess. Boil for 1–2 minutes to expel carbon dioxide completely, and then add about 1 ml barium chloride solution. A white precipitate indicates sulphate. Confirm by the charcoal test (Section **IV.24**, reaction 1).

Hexafluorosilicates also give a white precipitate under the above conditions, but are of comparatively rare occurrence. They can be readily distinguished from sulphates by the action of concentrated sulphuric acid (Section **IV.27**, reaction 1) and, of course, by the charcoal test.

2. Test for reducing agents Potassium permanganate test. Acidify 2 ml of the soda extract with dilute sulphuric acid and add 1 ml dilute sulphuric acid in excess. Add 0·5 ml of 0·004M potassium permanganate solution (prepared by diluting 1 ml of 0·02M $KMnO_4$ to 5 ml) from a dropper. Bleaching of the permanganate solution indicates the presence of one or more of the following reducing anions: sulphite, thiosulphate, sulphide, nitrite, cyanide, thiocyanate, bromide, iodide, arsenite, and hexacyanoferrate(II). If the permanganate is not decolourized, heat and observe the result. If the reagent is bleached only on heating, the presence of oxalate, formate or tartrate is indicated. A negative test points to the absence of the above anions with the exception of cyanide, which, if present in low concentration, may not act upon the permanganate solution.

3. Test for oxidizing agents Manganese(II) chloride reagent test. This test depends upon the fact that a saturated solution of manganese(II) chloride ($MnCl_2.4H_2O$) in concentrated hydrochloric acid is converted by even mild oxidizing agents to a dark-brown-coloured manganese(III) salt, probably containing the complex $[McCl_4]^-$ ions.

To 2 ml of the soda extract add 1 ml concentrated hydrochloric acid and 2 ml of the manganese(II) chloride reagent. A brown (or black) colouration indicates the presence of nitrate, nitrite, hexacyanoferrate(III), chlorate, bromate, iodate, chromate or permanganate. A negative test indicates the absence of the above oxidizing agents except small amounts of nitrates and nitrites and of arsenate; if reducing anions have been found, this test is inconclusive.

4. Tests with silver nitrate solution The separation of a comparatively large number of anions from the soda extract into several groups is possible with this reagent.

The presence of thiosulphate, sulphide, cyanide, sulphite, hexacyanoferrate(II), and (III), however, introduces difficulties in the subsequent separations, hence these anions must be tested for first and, if present, removed.

Thiosulphate will be detected in the preliminary test with dilute sulphuric acid: if it is found, it should be eliminated by heating the original mixture with dilute sulphuric acid until no more sulphur dioxide is evolved, evaporating the residual mixture just to dryness and then heating with 1·5M sodium carbonate solution, etc. The occurrence of thiosulphates in mixtures is comparatively rare and hence this special treatment prior to making the soda extract is rarely necessary. The interference of the $S_2O_3^{2-}$ ion with the silver nitrate reaction may arise from (i) the formation of a precipitate of sulphur upon acidification, (ii) the formation of white silver thiosulphate, $Ag_2S_2O_3$, which rapidly passes into the black silver sulphide, Ag_2S, thus giving a false test for S^{2-} ion, and (iii) under certain circumstances it will convert CN^- into SCN^- and Fe^{3+} into Fe^{2+}.

Sulphide. This may have been detected in the preliminary test with dilute sulphuric acid. Sulphide can be readily found by adding a little lead nitrate solution to 0·5 ml of the soda extract when a black precipitate of lead sulphide is produced.

Cyanide. This should have been detected and confirmed in the preliminary test with dilute sulphuric acid (Prussian blue test or as Section **IV.8**, reaction 1).

Sulphite. This anion will have been detected in the preliminary test with dilute sulphuric acid (potassium dichromate paper or fuchsin solution test).

Hexacyanoferrate(II) (and Thiocyanate). Acidify 1 ml of the soda extract with dilute hydrochloric acid and add a few drops of iron(III) chloride solution. A deep-blue precipitate indicates hexacyanoferrate(II) present. Now add 0·5–1 ml iron(III) chloride solution, 0·2 g sodium chloride and half a Whatman filtration accelerator, shake the mixture vigorously and filter. A deep-red filtrate indicates thiocyanate present.

Hexacyanoferrate(III). Acidify 1 ml of the soda extract with dilute hydrochloric acid and add a few drops of freshly prepared iron(II) sulphate solution. A deep-blue precipitate indicates hexacyanoferrate(III) present. This precipitate should not be confused with that of light-grey-blue colour produced by a hexacyano-ferrate(II).

Use is made of the following facts in the preliminary and partial separation of anions with silver nitrate solution:

1. In dilute nitric acid solution ($\sim$ 1–2M), chloride, bromide, iodide, iodate, and thiocyanate are precipitated.

It will be appreciated that iodate is incompatible with both iodide (cf. Section **IV.21**, reaction 6) and with thiocyanate (Section **IV.21**, reaction 9) since iodine is liberated in acid solution. Also sulphide is incompatible with both bromate and iodate (oxidation to sulphate occurs), and an arsenite is oxidized by iodate in acid solution. These facts should therefore be borne in mind when interpreting Table V.30. An independent test for iodate (test 11) is provided below: this can be performed before the silver nitrate tests.

2. Upon treatment of the filtrate from 1 with sodium nitrite solution, chlorate and bromate are reduced to the simple halides, the presence of which is revealed by the separation of silver chloride and silver bromide respectively. Chromates (which, of course, yield a coloured solution) are simultaneously reduced to chromium(III) salts.

3. If the acidity of the filtrate from 2 is reduced (to about *p*H 5) by just neutralizing with sodium hydroxide solution and adding dilute acetic acid, silver nitrate solution will precipitate phosphate, arsenate, arsenite,* oxalate, and possibly other organic acids.

It cannot be too strongly emphasized that Table V.30 is intended to act merely as a guide and to indicate the presence of groups of anions; it should therefore be considered carefully in conjunction with the various observations which have

* It must be remembered that arsenite may be partially oxidized by the dilute nitric acid treatment and also by other oxidizing anions which may be present. Arsenite is, however, readily detected in the analysis for cations (cf. Section **III.12**, reaction 1).

been made in the preliminary tests and particular note be taken of possible interferences. Table V.30 contains the various steps of the separation.

Table V.30 Separation of anions with silver nitrate It is essential to remove interfering anions first in the following order. These anions, if present, will have been detected in the preliminary tests.

1. Sulphide, cyanide, and **sulphite.** Acidify 10 ml of the soda extract with dilute acetic acid and boil gently for 3–4 minutes: make certain that the solution remains acid (e.g. to litmus) throughout. If sulphite is present, it is advisable to heat the solution for 10–15 minutes (maintaining the volume, if necessary) whilst a stream of air is drawn through it (cf. Fig. IV.2): test for complete removal of sulphur dioxide with fuchsin solution.

2. Hexacyanoferrate(II) and **(III).** Employ the original solution from **1** or, if sulphide, cyanide, and sulphite are absent, use 10 ml of the soda extract made acid (to Congo red) with dilute acetic acid. Add 0·5M zinc nitrate solution until precipitation ceases, introduce a Whatman filtration accelerator, stir, and filter the zinc and hexacyanoferrate(II) and (III) in the cold; wash with a little 0·1M zinc nitrate solution.

Use the filtrate from **2** or, if hexacyanoferrate(II) and/or (III) and other interfering anions are absent, acidify 10 ml of the soda extract cautiously with dilute HNO_3 (to litmus or other indicator paper). Add one-tenth of the volume of concentrated HNO_3, stir for 30 seconds and then add $AgNO_3$ solution with stirring until precipitation is complete. Heat to the boiling point, allow the ppt. to settle, cool, and filter. Wash the ppt. with 2–3 ml 2M nitric acid.

<table>
<tr><th>Residue (A)</th><th colspan="3">Filtrate</th></tr>
<tr><td rowspan="5">Yellow or white. May contain: AgCl, AgBr, AgI (or $AgIO_3$), AgSCN. Follow Procedure A.</td><td colspan="3">Add 1 ml $AgNO_3$ solution, then 5 per cent $NaNO_2$ solution (prepared from the A.R. solid) dropwise and with stirring until precipitation is complete. (If no ppt. forms, do not add more than 1 ml $NaNO_2$ solution.) Filter and wash with 2–3 ml 2M nitric acid.</td></tr>
<tr><td>Residue (B)</td><td colspan="2">Filtrate</td></tr>
<tr><td rowspan="3">May contain AgCl and AgBr derived from $AgClO_3$ and $AgBrO_3$. Follow Procedure B.</td><td colspan="2">Add NaOH solution (use a dropper) with vigorous stirring until neutral to litmus (1), then 0·5 ml dilute acetic acid, followed by 1 ml $AgNO_3$ solution. Heat the mixture to about 80°C (2). If a permanent precipitate forms, add more $AgNO_3$ solution until precipitation is complete. Filter and wash with hot water.</td></tr>
<tr><td>Residue (C) (3)</td><td>Filtrate</td></tr>
<tr><td>Ag_3PO_4 – yellow.
Ag_2AsO_4 – brownish-red.
Ag_3AsO_3 – yellow.
$Ag_2(COO)_2$ – white (4).
Follow Procedure C.</td><td>Discard.</td></tr>
</table>

Notes to Table V.30: 1. The sodium hydroxide solution should be added until the first permanent precipitate of C or of brown silver oxide appears. The solution must not be allowed to become alkaline to litmus for this will produce a large precipitate of silver oxide which redissolves only slowly.

The *p*H required for the precipitation of C is about 5·5; this can be more conveniently achieved by the use of either nitrazine (sodium dinitrophenyl-azo-naphthol sulphonate) or bromocresol purple (dibromo-*o*-cresol-sulphone-phthalein) test-papers. All that is necessary is to add the sodium hydroxide solution until the appropriate colour change is produced; it is best to use a

standard for comparison (nitrazine, *p*H range 5·5–7·2, yellow to blue; bromocresol purple, *p*H range 5·2–6·8, yellow to purple).
2. A crystalline precipitate of silver acetate may separate here; this dissolves when the solution is heated.
3. Chromate, if present, would normally be precipitated at a *p*H of about 5·5, but the sodium nitrite treatment in *B* reduces it to the tervalent state. Chromium(III) hydroxide is not precipitated in the acetic acid–acetate solution unless present in very large amounts.
4. Other organic acids may also separate here. If only a white precipitate is obtained, use the procedure described in Section **IV.45**, 20. The presence of organic acids will also be indicated by some of the preliminary tests.

Procedure A *(examination of residue A)* *Thiocyanate test* Test one-quarter of the precipitate *A* for thiocyanate by heating for 3–4 minutes with 5 ml of M NaCl solution (this dissolves part of the AgSCN) cool and allow the precipitate to settle. Treat the supernatant liquid with 1 ml dilute HCl and a few drops of $FeCl_3$ solution. Red colouration. Thiocyanate present.

If thiocyanate is present, it must be destroyed since it interferes with the tests for the halides. Heat the remainder of the residue *A* (previously dried in the steam oven or in an air bath) in a porcelain crucible, gently at first and then gradually to dull redness until all the thiocyanate is decomposed, e.g. until blackening of the precipitate and/or burning of S ceases.

To the residue in the crucible (if thiocyanate is present) or to the remainder of the residue *A* transferred to a small beaker (if thiocyanate is absent), add 1–2 g granulated zinc (but preferably of 20 mesh) and 5–10 ml dilute H_2SO_4. Allow the reduction to proceed for 10 minutes with frequent stirring; gentle warming may be necessary to start the reaction. Filter and wash the precipitate with a little dilute H_2SO_4. Divide the filtrate into three equal parts.*

Iodide test. Add 1–2 ml of CCl_4 and 3 ml of 10-volume H_2O_2 or 3 ml of 25 per cent $Fe_2(SO_4)_3$ solution to one-third of the filtrate. Shake vigorously and allow to settle. Purple to violet colouration of CCl_4 layer. Iodide present.

Bromide. (*a*) If iodide is present, this must be removed by treating one-third of the solution with 5 ml dilute H_2SO_4 and 1 ml of 50 per cent KNO_2 solution (chloride-free). Boil the solution with stirring, concentrate to 3 ml and allow to cool. Test for bromide as under (*b*).

(*b*) If iodide is absent, use one-third of the solution directly. Add an equal volume of concentrated HNO_3, immerse the test-tube in a beaker of boiling water for 1 minute and cool to room temperature with cold water. Add 1–2 ml of CCl_4 and stir vigorously with a glass rod (if a glass-stoppered tube or conical flask is available, this will facilitate shaking). Yellow or brown colour of the CCl_4 layer. Bromide present.

Chloride test. (*a*) If iodide and/or bromide present, dilute the remaining third of the solution to 15 ml, add 8 ml of concentrated HNO_3 and boil (glass rod in beaker) for 5 minutes or until no more Br_2 is given off. Cool and test for chloride as in (*b*) below with $AgNO_3$ solution only.

---

* The preliminary test with concentrated H_2SO_4 will generally indicate the presence of any of the three halides or mixtures of them. For individual halides, the confirmatory tests of Section **V.19** may be used. If mixtures are present, the methods described in Section **IV.45**, 5–8 may be adopted. For those who prefer systematic testing for any or all of the halides, the procedures described below are recommended.

(*b*) If bromide and iodide are absent, add 3–4 ml dilute HNO_3 and 3 ml $AgNO_3$ solution; stir and heat to boiling. White precipitate. Chloride present.

Procedure B *(examination of residue B)* Suspend the residue *B* in 5–10 ml dilute sulphuric acid in a beaker, and add 1–2 g granulated zinc (but preferably of 20 mesh). Allow the reduction to proceed for 10 minutes with frequent stirring: gentle warming may be necessary to start the reaction. Filter and wash the precipitate with a little dilute sulphuric acid. Examine the solution for chloride and bromide as in procedure *A*.

Procedure C *(examination of residue C)* If the precipitate is white, only organic acids may be present and the other anions need not be tested for. Furthermore, the preliminary tests of heating alone and heating with concentrated sulphuric acid will have indicated the presence of organic acids. If organic acids are indicated or suspected, follow the procedure given in Section **IV.45**, 20.

If the precipitate is yellow phosphate and/or arsenite may be present and arsenate is absent. The scheme described in Table **V.31** provides for the separation of phosphate, arsenate, and arsenite.

Table V.31 Separation of phosphate, arsenite, and arsenate Dissolve the residue *C* (1), which may contain Ag_3PO_4, Ag_3AsO_4 and Ag_3AsO_3, by pouring 10–15 ml 2M HCl repeatedly through it. Filter and wash with 5 ml of 0·1M HCl.

Residue	Filtrate
AgCl. Reject.	Render alkaline with dilute aqueous NH_3 and add 5 ml in excess (2). Add 10 ml of the $Mg(NO_3)_2$ reagent, allow to stand for 10 minutes, stirring frequently, and filter. Wash the ppt. with 5 ml 0·1M NH_3 solution.

Residue	Filtrate
May contain $Mg(NH_4)PO_4.6H_2O$ and $Mg(NH_4)AsO_4.6H_2O$ (3). Dissolve the ppt. by pouring 10 ml dilute HCl repeatedly through the filter. Add to the cold solution 0·5 g solid $NaHCO_3$ and 0·5 g solid KI. An immediate yellow to brown iodine colour indicates arsenate. Saturate the solution with H_2S (under 'pressure'). Filter and wash with 0·5M HCl.	Add 4 ml of 3% H_2O_2 solution, heat nearly to boiling (to oxidize arsenite to arsenate), cool and allow to stand for 10 minutes; shake frequently. Filter.

Residue	Filtrate
Yellow: As_2S_3. Arsenate present.	Boil until H_2S is expelled and the volume is ~10 ml. Filter and discard any ppt. of S. Render alkaline with concentrated NH_3 solution and add 3 ml in excess. Add 5 ml of the $Mg(NO_3)_2$ reagent and allow to stand, with frequent stirring, for 10 minutes. White ppt. of $Mg(NH_4)PO_4.6H_2O$. Phosphate present.

Residue	Filtrate
$Mg(NH_4)AsO_4.6H_2O$. Arsenite present.	Reject.

Notes to Table V.31: 1. The soda extract may also be used for the detection of arsenate, phosphate, and arsenite. Acidify 10 ml of the soda extract with dilute HCl and then render alkaline with dilute aqueous NH_3. Filter, if necessary, and reject any precipitate. Treat the clear solution with the $Mg(NO_3)_2$ reagent, etc., and proceed as in Table V.31.
2. If chromate is present, it will have been reduced to chromium(III) ion by the KNO_2 solution and may be partly precipitated as $Cr(OH)_3$. This will be dissolved by HCl and reprecipitated by the aqueous NH_3. Hence any precipitate formed at this point should be filtered off before the $Mg(NO_3)_2$ reagent is added.
3. If phosphate has been detected previously (e.g. in the cation analysis), a qualitative test for arsenate may be made as follows. Pour 1 ml $AgNO_3$ solution, to which 2 drops of dilute acetic acid have been added, over the white precipitate. A brownish-red colouration of the precipitate confirms the presence of arsenate. The acetic acid is added to increase the solubility of the magnesium salt and thus facilitate the conversion of magnesium ammonium arsenate to the characteristic silver arsenate.

5. ***Test with calcium chloride solution*** For tests 5–7 a neutralized soda extract is required. This is prepared as follows. Take 10 ml soda extract in a porcelain dish and render it faintly acid with dilute nitric acid (use litmus paper or other equivalent test-paper). Boil for 1–2 minutes to expel carbon dioxide, allow to cool, then add dilute ammonia solution until just alkaline* and boil for 1 minute to expel the slight excess of ammonia. Divide the solution into three equal parts for tests 5, 6, and 7.

Add $CaCl_2$ solution (equal in volume to that of the solution) and allow to stand for several minutes. A white precipitate indicates fluoride, oxalate, phosphate, arsenate, and tartrate;† a precipitate which separates on boiling for 1–2 minutes is citrate. Of these only oxalate and fluoride are insoluble in dilute acetic acid. Hence extract the white precipitate with dilute acetic acid and filter. A residue (*R*) insoluble in dilute acetic acid, indicates oxalate and/or fluoride. Exactly neutralize the acetic acid solution by adding sodium hydroxide solution from a dropper and testing with an indicator paper or solution (bromothymol blue or nitrazine yellow is suitable); a white precipitate indicates the presence of phosphate, arsenate, and/or tartrate. The precipitate often separates slowly. Add a little silver nitrate solution to the suspension or solution: a yellow precipitate indicates the presence of phosphate; a brownish-red precipitate indicates arsenate or arsenate plus phosphate.

It is convenient to test the residue *R* for oxalate here. Dissolve it by pouring hot dilute sulphuric acid into the filter. Treat the hot filtrate with a few drops of 0·004M potassium permanganate solution. If the permanganate solution is reduced, oxalate is present. If no reduction occurs, the presence of fluoride is indicated.

6. ***Test with iron(III) chloride solution*** Treat the second third of the neutralized soda extract drop by drop with iron(III) chloride solution until no further change occurs. The results of this test are summarized in Table V.32.

---

* If a precipitate forms on neutralizing the solution, the presence of arsenic, antimony and tin sulphides and possibly salts of amphoteric elements (lead, tin, aluminium, and zinc) is indicated. The precipitate should be filtered off and rejected.

† $CaSO_4$ and $Ca(BO_2)_2$ may separate from sufficiently concentrated solutions.

(The bench reagent contains free hydrochloric acid; add dilute ammonia solution until a precipitate just forms, filter and use the filtrate for the test. The filtrate is sometimes termed 'neutral' $FeCl_3$ solution.)

Table V. 32 Test with iron(III) chloride

Yellow to brown precipitate.	Benzoate, succinate (also chromate, phosphate, arsenate and borate, all of which have been or will be found by other tests).
Blue precipitate.	Hexacyanoferrate(II).
Reddish-brown colouration ppt. after dilution and boiling.	Acetate, formate.
Blood-red colouration discharged by $HgCl_2$ solution.	Thiocyanate.
Reddish-purple colouration; colour vanishes on warming.	Thiosulphate.*
Brown colouration; blue ppt. with $FeSO_4$ solution.	Hexacyanoferrate(III).
Violet colouration.	Salicylate.
Greenish-black colouration.	Gallate.
Bluish-black colouration.	Tannate.

* As a general rule, thiosulphate will not be found here as it should have been more or less completely decomposed in the preparation of the neutralized soda extract.

*7. **Test for silicate*** To the remaining third of the neutralized soda extract add ammonium chloride and ammonium carbonate solution. A gelatinous precipitate indicates silicate.

A more satisfactory reagent for the precipitation of silicates is hexammine zinc hydroxide, $[Zn(NH_3)_6](OH)_2$. This precipitates zinc silicate, $ZnSiO_3$, which is much more sparingly soluble in dilute alkaline solution than is the free silicic acid.

$$SiO_3^{2-} + [Zn(NH_3)_6]^{2+} \rightarrow ZnSiO_3\downarrow + 6NH_3$$

The reagent is added in slight excess and the solution boiled until all the ammonia is expelled.

The reagent is prepared by treating 0·5M zinc nitrate solution with potassium hydroxide solution, filtering off the precipitated zinc hydroxide, washing well, and dissolving the precipitate in dilute ammonia solution.

*8. **Test for fluoride*** The presence of fluoride will have been indicated in the preliminary test with concentrated sulphuric acid by the 'oily' or 'greasy' appearance of the tube and also by the calcium chloride solution test. It may be confirmed by the following test.

Fit a small test-tube with a cork carrying a piece of glass tubing open at both ends; cut a V-shaped groove in the side of the cork to allow for the expansion of the air in the tube when heated. Mix a small quantity of the original substance in a crucible with about three times its bulk of ignited silica (quartz powder is preferable, but results are obtained with precipitated silica), transfer it carefully to the tube, and add by means of a dropper or small pipette about twice as much (by volume) of concentrated sulphuric acid as there is solid. Introduce the glass tube, wet on the inside with water thus forming a ring of water at the lower end (Fig. V.3), into the test-tube and adjust its height so that the bottom is at a distance of approximately one and a half times the diameter of the tube from the

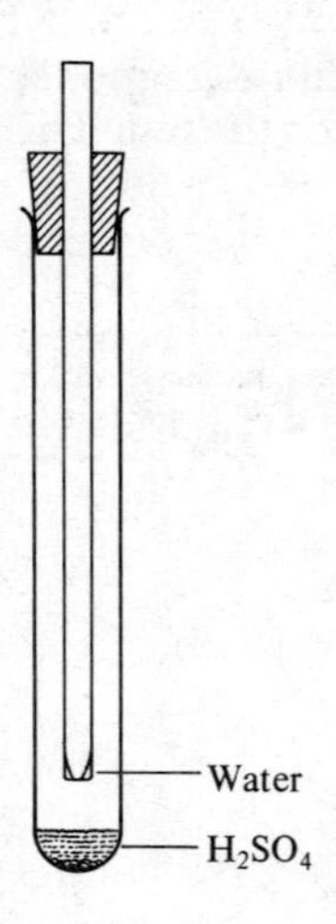

Fig. V. 3

paste in the test-tube. Heat the mixture gently over a small flame for 2–3 minutes. A white film of silicic acid in the water confirms the presence of fluoride.

The test is based on the following reactions (M indicates a bivalent metal):

$$MF_2+H_2SO_4 \rightarrow M^{2+}+SO_4^{2-}+H_2F_2$$
$$SiO_2+2H_2F_2 \rightarrow SiF_4\uparrow+2H_2O$$
$$3SiF_4\uparrow+2H_2O \rightarrow SiO_2\downarrow+2[SiF_6]^{2-}+4H^+$$

9. Test for cyanide This is sometimes missed in the preliminary test with dilute sulphuric acid and in the potassium permanganate test for reducing anions. The following test is conclusive. Use the apparatus described in the test for carbonates (Fig. IV.1). Place 0·2 g of the substance in the test-tube together with three or four fragments of marble. Introduce 5 ml of 2M hydrochloric acid, replace the cork immediately and allow the evolved gas to bubble through 5 ml of 2M sodium hydroxide solution. After 5–10 minutes add 0·5 ml saturated iron(II) sulphate solution to the alkaline solution, heat to boiling point, cool thoroughly, acidify with concentrated hydrochloric acid and add a few drops of iron(III) chloride solution. A blue precipitate (Prussian blue) is obtained; with small amounts of cyanide the solution acquires a blue or blue-green colour.

The test may also be conducted with 2 ml of the soda extract. Carbonate, sulphite, and thiosulphate have no influence upon the reaction. Nitrite interferes, presumably owing to the oxidation of the hydrogen cyanide. In the presence of sulphide, the test is complicated by the precipitation of black iron(II) sulphide when sulphate is added to the alkaline solution. It is best to boil the solution containing the suspended iron(II) sulphide, acidify with hydrochloric acid, and boil again to expel most of the dissolved hydrogen sulphide; upon adding a drop of iron(III) chloride solution, a blue precipitate is produced if cyanide is present.

Complex cyanides, such as hexacyanoferrate(II) and (III), do not interfere when the test is conducted in the cold.

10. Test for chromate If the soda extract is colourless, chromate is absent. If yellow, chromate may be present; hexacyanoferrate(II) and (III) also impart a yellow colour to the solution. The presence of chromate will have been indicated by the precipitation of green chromium(III) hydroxide in the silver nitrate solution tests (Table V.31) and also in the analysis for cations.

To confirm chromate, acidify 2 ml of the soda extract with dilute sulphuric acid, boil for one minute to expel CO_2, etc., filter if necessary, add 1–2 ml amyl alcohol, followed by 1–2 ml of 10-volume hydrogen peroxide and shake gently (compare Section **IV.33**, reaction 4). A blue colouration of the amyl alcohol layer confirms chromate.

11. Test for iodate This anion will give a positive test for oxidizing agents, but will not normally be detected in the systematic analysis.

The presence of iodate in the soda extract can be readily detected as follows. Treat 2 ml of the solution with silver nitrate solution until precipitation ceases, heat to boiling for 2–3 minutes and filter. Render the filtrate strongly acid with hydrochloric acid, add 2 ml of 0·5M iron(II) sulphate solution (or 0·5M sodium sulphite solution) and shake it with 2 ml carbon tetrachloride. A purple colouration of the organic layer indicates iodate.

This test utilizes the fact that silver iodide, but not silver iodate, is precipitated upon the addition of silver nitrate to the soda extract.

An iodate does not react with concentrated sulphuric acid in the cold or upon gentle heating. A solution of an iodate gives with silver nitrate solution a white precipitate of silver iodate, insoluble in M nitric acid but soluble in dilute ammonia solution, thus simulating the behaviour of a chloride towards these reagents. Iodates, however, give a white precipitate of barium iodate, $Ba(IO_3)_2$, with barium chloride solution; the precipitate is sparingly soluble in dilute nitric acid.

12. Test for periodate (see note 1 below) This anion will give a positive test for oxidizing agents, but will not be detected in the systematic analysis. It will be necessary to remove first iodide or iodate by precipitation with silver nitrate in acid solution, and the excess silver ions with sodium chloride solution; the resulting solution is strongly acidified with hydrochloric acid and an iron(II) salt is added. If a periodate is present, it will be reduced to iodine, which can be identified with carbon tetrachloride.

Acidify 3 ml of the soda extract with concentrated perchloric acid and add 1 ml in excess (note 2 below); add silver nitrate solution, slowly and with stirring, until precipitation is complete. Filter and collect the filtrate in a ground-glass stoppered conical flask or test-tube. Add M sodium chloride solution, 0·5 ml at a time, to the filtrate until no more precipitate forms. Stopper the flask or test-tube and shake the mixture vigorously after each addition. Filter off the silver chloride (note 3 below), transfer the filtrate to the stoppered vessel, add an equal volume of concentrated hydrochloric acid (if a precipitate of sodium chloride forms, filter), cool, then add 1–2 g solid iron(II) ammonium sulphate and 2 ml carbon tetrachloride. Shake the mixture intermittently for 5 minutes. A purple colour in the carbon tetrachloride indicates periodate present.

Notes: 1. This anion is rarely encountered and need not be tested for in general qualitative analysis. It is given here for the sake of completeness as no reactions for periodates are included in Chapter IV. Four tests which distinguish periodates from iodates are:

(*a*) With mercury(II) nitrate solution an orange-red precipitate of $I_2O_7.5HgO$ is obtained [cf. iodates which give white $Hg(IO_3)_2$].
(*b*) Upon boiling with 1:1 nitric acid and a little manganese(II) sulphate solution, a purple solution of permanganate is produced (cf. iodates, which do not yield permanganate).
(*c*) With silver nitrate solution a chocolate-brown precipitate (Ag_5IO_6) is obtained; this darkens on boiling and is soluble in dilute nitric acid (cf. silver iodate, $AgIO_3$, which is white and sparingly soluble in dilute nitric acid).
(*d*) With barium chloride solution a white precipitate is produced, which is soluble in dilute nitric acid (cf. barium iodate, which is sparingly soluble in dilute nitric acid).

2. Nitric acid tends to oxidize iron(II) ions hence it is advisable to maintain the nitrate-ion concentration as low as possible and to keep the solution cold.
3. The excess silver ions are removed as silver chloride because they are reduced by iron(II) ion salts to silver.

The preliminary dry tests and the reactions in solution described above will give a general (and, in some cases, a particular) indication of the nature of the anions present. For a number of anions (e.g. thiocyanate, chloride, bromide, iodide, iodate, bromate, chlorate, phosphate, silicate, fluoride, cyanide, and chromate), the tests are more or less conclusive. However, in the presence of mixtures of anions, it will be necessary to distinguish between those which give analogous reactions, for example, (i) sulphite, thiosulphate, and sulphate; (ii) chloride, chlorate, and perchlorate; (iii) arsenite, arsenate, and phosphate; (iv) hexacyanoferrate(II) and (III); (v) acetate and formate; (vi) succinate and benzoate; and (vii) oxalate, tartrate, and citrate. In some cases one anion interferes with the reaction of the other, e.g. (viii) carbonate and sulphite; (ix) nitrite and nitrate; (x) nitrate in the presence of bromide, iodide, chlorate, and perchlorate; (xi) iodate and iodide; (xii) oxalate and fluoride; and (xiii) chloride and cyanide. This subject is fully discussed in Section **IV.45**. Particular attention is directed to the procedure for separation of organic acids (Section **IV.45**, 20): this must be used with due consideration as to the influence of interfering organic acids (cf. tests 5 and 6 above).

V.19 CONFIRMATORY TESTS FOR ANIONS In every case where the presence of any anion has been indicated, it must be confirmed by at least one distinctive confirmatory test. Conclusive tests for anions (**halides**, **sulphate**, **oxy-halides**, **thiocyanate**, **phosphate**, **silicate**, **fluoride**, **cyanide** and **chromate**) have already been given in Section **V.18**, it will, of course, be unnecessary to confirm these further. The following list, which, for the sake of completeness includes those anions already referred to, will assist the student in the choice of suitable tests. Full experimental details will be found in Chapter IV under the reactions of the anions; the reference to these will be abbreviated as follows: thus (**IV.2**, 3) is to be interpreted as Section **IV.2**, reaction 3. It is assumed, of course, that interfering acids are absent or have been removed.

Chloride. (i) Heat solid with concentrated H_2SO_4 and MnO_2; Cl_2 evolved (bleaches litmus paper and also turns potassium iodide–starch paper blue) (**IV.14**, 2). (ii) Chromyl chloride test (**IV.14**, 5). (iii) Silver chloride–sodium arsenite test (**IV.14**, 3).

Bromide. (i) Heat solid with concentrated H_2SO_4 and MnO_2; Br_2 evolved

(**IV.15**, 2). (ii) NaOCl–HCl and $CHCl_3$ or CCl_4 test; yellowish-brown to yellow colouration (**IV.15**, 5).

Iodide. (i) NaOCl–HCl (or chlorine water) and $CHCl_3$ or CCl_4 test; violet colouration (**IV.16**, 4). (ii) NaOCl–HCl and starch paste test; blue colouration (**IV.16**, 4).

Fluoride. (i) Etching test (**IV.17**, 2). (ii) Silicon tetrafluoride test heat with concentrated sulphuric acid in a test-tube (**IV.17**, 1); better, test 8 in **VII.16**. (iii) Zirconium–alizarin test (**IV.17**, 6).

Cyanide. (i) Prussian blue test (**IV.8**, 4); better, test 9 in **V.18**. (ii) Ammonium sulphide test (**IV.8**, 1, 6).

Nitrite. (i) Brown ring test with dilute acetic acid or with dilute sulphuric acid (**IV.7**, 2). (ii) Thiourea test (**IV.7**, 9). (iii) Sulphanilic acid-α-naphthylamine reagent test (**IV.7**, 11).

Nitrate. (i) Brown ring test with concentrated H_2SO_4 (**IV.18**, 3), if bromide, iodide, and nitrite absent. (ii) Ammonia test with Devarda's alloy (**IV.18**, 4).

Sulphide. (i) Dilute H_2SO_4 on solid, and action of H_2S on lead or cadmium acetate paper (**IV.6**, 1). (ii) Sodium nitroprusside test (**IV.6**, 6).

Sulphite. (i) Dilute H_2SO_4 on solid, and action of SO_2 upon potassium dichromate paper (**IV.4**, 1). (ii) $BaCl_2$–Br_2 water test (**IV.4**, 2). (iii) Fuchsin solution test (**IV.4**, 9).

Thiosulphate. (i) Action of dilute H_2SO_4 upon solid, and liberation of SO_2 (dichromate paper test or fuchsin solution test) and sulphur (**IV.5**, 1). (ii) Potassium cyanide test (**IV.5**, 6). (iii) Nickel ethylenediamine test (**IV.5**, 9).

Sulphate. (i) $BaCl_2$ solution and dilute HCl test, and reduction to sulphide test for latter with sodium nitroprusside or lead acetate solution (**IV.24**, 1). (ii) Lead acetate test and solubility of $PbSO_4$ in ammonium acetate solution (**IV.24**, 2).

Carbonate. (i) Action of dilute H_2SO_4 upon solid, and lime water or baryta water test (**IV.2**, 1).

Hypochlorite. (i) Action of dilute HCl, and test for Cl_2 evolved (**IV.13**, 1). (ii) Cobalt nitrate solution test (**IV.13**, 4).

Chlorate. (i) Sodium nitrite test (**IV.19**, 4; also test 4 in **VII.16**). (ii) Iron(II) sulphate test (**IV.19**, 6).

Bromate. (i) Action of concentrated H_2SO_4; Br_2 and O_2 evolved (**IV.20**, 1). (ii) Hydrobromic acid test (**IV.20**, 4).

Iodate. (i) Potassium iodide test (**IV.21**, 6). (ii) Sulphur dioxide or hydrogen sulphide test (**IV.21**, 5).

Perchlorate. (i) Action of heat and test for chloride (**IV.22**, 7), and non-reduction with $NaNO_2$ or $FeSO_4$ in acid solution (**IV.22**, 5).

Borate. (i) Flame test (**IV.23**, 2; better, test (10) in Section (**V.2**). (ii) Turmeric paper test (**IV.23**, 3).

Silicate. (i) Microcosmic bead test (**IV.26**, 5) and silicon tetrafluoride test (**IV.26**, 6). (ii) Ammonium molybdate solution and $SnCl_2$ solution test (**IV.26**, 8).

Silicofluoride. (i) Action of concentrated H_2SO_4 and $BaCl_2$ solution test (**IV.27**, 1, 2).

Peroxodisulphate. (i) Action of boiling water (**IV.25**, 1). (ii) Manganese(II) sulphate solution test (**IV.25**, 5).

Chromate. (i) Hydrogen peroxide test (**IV.33**, 4). (ii) Lead acetate solution test (**IV.33**, 3). (iii) Action of hydrogen sulphide or sulphur dioxide (**IV.33**, 5, 6). (iv) Solid when warmed with a solid chloride and concentrated H_2SO_4 evolves chromyl chloride (**IV.14**, 5). (v) Diphenylcarbazide reagent test (**IV.33**, 9).

Permanganate. (i) Hydrogen peroxide test, and identification of Mn^{2+} ion (**IV.34**, 1). (ii) Action of oxalic acid, and identification of Mn^{2+} ion (**IV.37**, 4). (iii) Action of hydrogen sulphide or sulphur dioxide (**IV.34**, 2).

Arsenite. (i) Action of H_2S upon acid solution (**III.12**, 1). (ii) Silver nitrate solution test (**III.12**, 2), and absence of precipitate with magnesium nitrate reagent (**III.12**, 3) or on boiling with ammonium molybdate solution and nitric acid. (iii) Bettendorff's test (**III.12**, 6).

Arsenate. (i) Action of H_2S on acid solution (**III.13**, 1), and silver nitrate solution test upon neutral solution (**III.13**, 2). (ii) Magnesium nitrate reagent test (**III.13**, 3). (iii) Ammonium molybdate test (**III.13**, 4).

Orthophosphate. (i) Ammonium molybdate test (temperature not above 40°) (**IV.28**, 4). (ii) Magnesium nitrate reagent test (**IV.28**, 3).

Phosphite. (i) Silver nitrate solution test (**IV.30**, 1) and Zn + dilute H_2SO_4 test (**IV.30**, 6).

Hypophosphite. (i) Silver nitrate solution test (**IV.31**, 1), copper sulphate solution test (**IV.31**, 4), and ammonium molybdate test (**IV.31**, 9).

Cyanate. (i) Dilute H_2SO_4 test (**IV.9**, 1). (ii) Cobalt acetate solution test (**IV.9**, 5). (iii) Copper sulphate–pyridine test (**IV.9**, 6).

Hexacyanoferrate(II). (i) Iron(III) chloride solution test (**IV.11**, 3). (ii) Iron(II) sulphate solution test (**IV.11**, 4). (iii) Uranyl acetate solution test (**IV.11**, 11). (iv) Titanium(IV) chloride test (**IV.11**, 9).

Hexacyanoferrate(III). (i) Iron(II) sulphate solution test (**IV.12**, 3). (ii) Iron(III) chloride solution test (**IV.12**, 4).

Thiocyanate. (i) Iron(III) chloride solution test; colour discharged by $HgCl_2$ solution or by NaF solution, but not by dilute HCl (**IV.10**, 6). (ii) Cobalt nitrate solution test (**IV.10**, 9). (iii) Copper sulphate–pyridine test (**IV.10**, 11).

Acetate. (i) Action of ethanol or of iso-amyl alcohol and concentrated H_2SO_4 (**IV.35**, 3), $AgNO_3$ solution test (**IV.35**, 4), and $FeCl_3$ solution test (**IV.35**, 6). (ii) Cacodyl oxide test (**IV.35**, 7). (iii) Indigo test (**IV.35**, 9).

Formate. (i) Mercury(II) chloride solution test (**IV.36**, 7) or $AgNO_3$ solution test (**IV.36**, 4) and $FeCl_3$ solution test (**IV.36**, 6). (iii) Mercury(II) formate test (**IV.36**, 8).

Oxalate. (i) Immediate precipitation with $CaCl_2$ solution in neutral solution; precipitate decolourizes a dilute solution of $KMnO_4$ (**IV,37**, 3, 4). (ii) Resorcinol test (**IV.37**, 5).

Tartrate. (i) Resorcinol test (**IV.38**, 6). (ii) Copper hydroxide test (**IV.38**, 7).

Citrate. (i) Deniges's test (**IV.39**, 5). (ii) Cadmium acetate solution test (**IV.39**, 4), and negative results with resorcinol and copper hydroxide tests (**IV.38**, 6, 7).

Salicylate. (i) Violet colouration with $FeCl_3$ solution, discharged by mineral acids (**IV.40**, 5), and soda lime test (**IV.40**, 7). (ii) 'Oil of winter-green' test (**IV.40**, 2).

Benzoate. (i) Buff-coloured precipitate with $FeCl_3$ solution, soluble in dilute HCl with precipitation of benzoic acid (**IV.41**, 5). (ii) Dilute H_2SO_4 test (**IV.41**, 2), and no precipitate with $BaCl_2$ solution in neutral solution.

Succinate. (i) Light-brown precipitate with $FeCl_3$ solution, soluble in dilute HCl, but no precipitate of acid occurs (**IV.42**, 3). (ii) Fluorescein test (**IV.42**, 6).

CHAPTER VI SEMIMICRO QUALITATIVE INORGANIC ANALYSIS

VI.1 INTRODUCTION It is assumed that the student is already familiar with the operations (ignition tests, blowpipe tests, flame tests, borax bead tests, precipitation, filtration, evaporation, etc.) described in Sections **II.1** to **II.3**. A detailed description of semimicro apparatus and of semimicro analytical operations is given in Section **II.4**. The student should read this carefully and thus acquire a general knowledge of semimicro technique. To secure the introductory practical experience the following course of instruction may be followed. Reactions are normally carried out in 4 ml test-tubes or in 3 ml centrifuge tubes, unless otherwise stated. The solutions employed in testing for cations or anions contain 10 mg of cation or anion per ml. For purposes of calculation a drop of solution may be assumed to have a volume of 0·05 ml; this will help in rough computations as to volumes required in adding excess of reagents, etc.

For the cations, a limited number of preliminary reactions should be studied first, followed by an analysis of a mixture or solution containing a member or members of each group. This will give practical experience in the routine operations of semimicro analysis and also provide practice from the very outset in the use of Group Separation Tables. Particular attention should be paid to:

(i) the exact experimental conditions of the reaction;

(ii) the colour and physical characteristics (e.g. whether crystalline, amorphous or gelatinous) of each precipitate; and

(iii) the solubility of each precipitate in excess of precipitant, or in solutions of other reagents.

VI.2 THE STUDY OF REACTIONS OF CATIONS AND ANIONS ON THE SEMIMICRO SCALE Before attempting to analyse unknown samples with semimicro techniques, it is worth while to study, in semimicro scale, some of the reactions of cations and anions described in Chapters III and IV. To illustrate the manner in which the simple reactions should be carried out, a few selected examples will be given.

Reactions of lead ions, Pb^{2+}

1. Dilute hydrochloric acid Place 2 drops of the test solution in a 3 ml centrifuge tube, and add 2–3 drops of dilute HCl. Note the colour and characteristics of the precipitate. Centrifuge the mixture, balancing the tube with another similar tube containing an approximately equal volume of water. Remove the supernatant liquid, termed the centrifugate, by means of a capillary dropper to another tube, and test with a further drop of dilute HCl to ensure that precipitation is complete: if this is so, discard the solution. Treat the precipitate, usually termed the residue, with 2 drops of water, stir, and centrifuge. Reject the solution, called the washings. Add 5–6 drops of water to the washed residue, stir, and place the tube containing it in the rack in the boiling water bath. Stir frequently whilst heating. Observe that the precipitate dissolves completely. Cool the clear solution in a stream of water from the cold water tap; observe that colourless crystals of $PbCl_2$ separate.

2. Dilute sulphuric acid To 2 drops of the test solution in a 3 ml centrifuge tube or 4 ml test-tube, add 2–3 drops dilute H_2SO_4. Centrifuge; remove the centrifugate and test for complete precipitation with 1 drop dilute H_2SO_4. Wash the precipitate with 2–3 drops water, centrifuge, and discard the washings. Now introduce 2–3 drops ammonium acetate solution and stir: the preparation will dissolve.

3. Potassium chromate solution Treat 2 drops of the test solution with 1–2 drops K_2CrO_4 solution. Note the formation of a yellow precipitate. Centrifuge and wash. Add 2 drops dilute acetic acid to the precipitate and stir; note that the precipitate does not dissolve.

The above examples should suffice to give a general conception as to how the various reactions should be performed. In the reactions involving the passage of hydrogen sulphide into a solution, the apparatus depicted in Fig. II.27 should be used.

The following reactions should be examined:

CATIONS

Pb^{2+}	Section **III.4**, reactions 1, 5, 6, 7, 16a.
Hg_2^{2+}	Section **III.5**, reactions 1, 3, 10.
Ag^{+}	Section **III.6**, reactions 1, 5, 6, 12.
Hg^{2+}	Section **III.8**, reactions 1, 3, 6, 11.
Bi^{3+}	Section **III.9**, reactions 1, 2, 6, 7, 11.
Cu^{2+}	Section **III.10**, reactions 1, 2, 3, 4, 5, 6, 8, 14.
Cd^{2+}	Section **III.11**, reactions 1, 2, 4, 9.
As^{3+}	Section **III.12**, reactions 1, 2, 3.
As^{5+}	Section **III.13**, reactions 1, 2, 3, 4.
Sb^{3+}	Section **III.15**, reactions 1, 2, 8.
Sb^{5+}	Section **III.16**, reactions 1, 2, 4.
Sn^{2+}	Section **III.18**, reactions 1, 2, 3.
Sn^{4+}	Section **III.19**, reactions 1, 3, 4.
Fe^{2+}	Section **III.21**, reactions 1, 2, 3, 4, 6, 7, 8, 9.
Fe^{3+}	Section **III.22**, reactions 1, 2, 3, 4, 6, 7, 11.
Al^{3+}	Section **III.23**, reactions 1, 2, 3, 7, 11.
Cr^{3+}	Section **III.24**, reactions 1, 2, 4, 7a, 8b, 9.
Co^{2+}	Section **III.26**, reactions 1, 2, 3, 5, 6, 7, 10b.
Ni^{2+}	Section **III.27**, reactions 1, 2, 3, 8, 11b.

Mn^{2+}	Section **III.28**, reactions 1, 2, 3, 6, 7, 11b.
Zn^{2+}	Section **III.29**, reactions 1, 2, 3, 6, 8, 9, 11, 13.
Ba^{2+}	Section **III.31**, reactions 1, 2, 3, 4, 6, 10.
Sr^{2+}	Section **III.32**, reactions 1, 2, 3, 5, 6, 9.
Ca^{2+}	Section **III.33**, reactions 1, 2, 3, 5, 6, 7, 12.
Mg^{2+}	Section **III.35**, reactions 1, 2, 3, 5, 7, 8, 11.
K^{+}	Section **III.36**, reactions 1, 2, 7.
Na^{+}	Section **III.37**, reactions 1 (or 3), 4.
NH_4^{+}	Section **III.38**, reactions 1, 2, 3, 7, 10.

ANIONS

CO_3^{2-}	Section **IV.2**, reactions 1, 2.
HCO_3^{-}	Section **IV.3**, reactions 1, 2, 3.
SO_3^{2-}	Section **IV.4**, reactions 1, 2, 4, 5, 8, 9.
$S_2O_3^{2-}$	Section **IV.5**, reactions 1, 2, 4, 6, 8.
S^{2-}	Section **IV.6**, reactions 1, 2, 3, 5, 6.
NO_2^{-}	Section **IV.7**, reactions 1, 2, 7, 10.
CN^{-}	Section **IV.8**, reactions 1, 2, 3, 4, 6.
SCN^{-}	Section **IV.10**, reactions 1, 2, 3, 6, 8.
Cl^{-}	Section **IV.14**, reactions 1, 2, 3.
Br^{-}	Section **IV.15**, reactions 1, 2, 5, 8.
I^{-}	Section **IV.16**, reactions 1, 2, 4, 7, 9.
F^{-}	Section **IV.17**, reactions 1, 2, 6.
NO_3^{-}	Section **IV.18**, reactions 1, 2, 3, 4.
ClO_3^{-}	Section **IV.19**, reactions 1, 2, 3, 6, 7, 10.
$B_4O_7^{2-}$	Section **IV.23**, reactions 1, 2, 3, 8.
SO_4^{2-}	Section **IV.24**, reactions 1, 2, 5, 6.
PO_4^{3-}	Section **IV.28**, reactions 1, 3, 4, 5, 6.
CrO_4^{2-}	Section **IV.33**, reactions 1, 2, 3, 5, 6.
CH_3COO^{-}	Section **IV.35**, reactions 1, 2, 3, 4, 6.
$(COO)_2^{2-}$	Section **IV.37**, reactions 1, 2, 3, 4, 5.

Note that not all the anions, described in Chapter IV, are included into this scheme.

VI.3 SYSTEMATIC ANALYSIS ON THE SEMIMICRO SCALE. GENERAL CONSIDERATIONS The object of systematic qualitative analysis on the semimicro scale is not only to detect the constituents of a given mixture, an equally important aim is to ascertain the approximate relative amounts of each component. For this purpose about 0·2 g of material is usually employed for the analysis: the relative magnitudes of the various precipitates will provide a rough guide as to the proportions of the constituents present.

Systematic semimicro analysis is carried out in the following steps:

1. The preliminary examination of the solid sample This includes preliminary examination by dry tests, examination of the volatile products with sodium hydroxide solution (for ammonium), and with dilute and concentrated sulphuric acid (certain anions). A special preliminary test for nitrate and/or nitrite is also made here.

*2. **Testing for anions in solution*** This includes the preparation of a 'soda extract', followed by systematic tests. These, in turn, can be made conclusive by confirmatory tests.

*3. **Testing for cations in solution*** The sample must be dissolved first, then subjected to preliminary tests and separations. Finally, the presence of each cation must be confirmed by suitable reactions.

Note that the order here is reversed, i.e. anions are tested for first, followed by tests for cations. Experience has shown that once the preliminary tests are carried out, considerable information is collected about the presence or absence of certain anions, and it is worthwhile to carry on with anion testing at this stage, always keeping in mind the results obtained by the preliminary tests. The systematic analysis for cations follows this, based again on the separation of each single cation as in macro analysis, and on specific tests carried out after the separation of the cations.

VI.4 PRELIMINARY TESTS ON THE SEMIMICRO SCALE In most cases samples submitted for analysis are solid. The preliminary tests 1–8 described in this section are to be carried out on solid samples. If the sample is liquid (e.g. a solution), these tests can be omitted, though tests 3–8, with little modifications, might be useful even for such samples. Preliminary tests to be carried out with liquid samples are described under test 9.

Testing for *appearance* [Section **V.2**, test 1] will not be described again: it is useful however to carry out such an examination, using a magnifying glass or a microscope, as described there.

*1. **Heating in a test tube*** Place about 5 mg of the substance in a dry test-tube (70×4–5 mm) so that none of it remains adhering to the sides and heat cautiously with a semimicro burner; the tube should be held in an almost horizontal position. Raise the temperature gradually and carefully note any changes which take place.

The results of this test are summarized in Table VI.1.

Table VI.1 Heating test

Observation	Inference
(*a*) A gas or vapour is evolved.	
1. Vapour is evolved; test with litmus paper.	Compounds with water of crystallization (often accompanied by change of colour), ammonium salts, acid salts, and hydroxides.
The vapour is alkaline.	Ammonium salts.
The vapour is acid.	Readily decomposable salts of strong acids.
2. Oxygen is evolved (rekindles a glowing splint).	Nitrates, chlorates, and certain oxides.
3. Dinitrogen oxide (rekindles a glowing splint) and steam are evolved.	Ammonium nitrate or nitrate mixed with an ammonium salt.
4. Dark-brown or reddish fumes (oxides of nitrogen); acidic in reaction.	Nitrates or nitrites of heavy metals.

Table VI.1 Heating test (contd.)

Observation	Inference
5. Carbon dioxide is evolved (lime water is rendered turbid).	Carbonates or hydrogen carbonates.
6. Ammonia is evolved (turns red litmus paper blue).	Ammonium salts.
7. Sulphur dioxide is evolved (odour of burning sulphur; turns potassium dichromate paper green; decolourizes fuchsin solution).	Sulphites and hydrogen sulphites; thiosulphates; certain sulphates.
8. Hydrogen sulphide is evolved (odour of rotten eggs; turns lead acetate paper black or cadmium acetate paper yellow).	Hydrated sulphides or sulphides in the presence of water.
9. Chlorine is evolved (yellowish-green gas; bleaches litmus paper; turns potassium iodide–starch paper blue); very poisonous.	Unstable chlorides, e.g. of copper; chlorides in the presence of oxidizing agents.
10. Bromine is evolved (reddish-brown vapour; choking odour; turns fluorescein paper red).	Unstable bromides, bromides in the presence of oxidizing agents.
11. Iodine is evolved (violet vapours condensing to black crystals).	Free iodine and certain iodides.
(*b*) A sublimate is formed.	
1. White sublimate.*	Ammonium and mercury salts; As_2O_3; Sb_2O_2.
2. Grey sublimate, easily rubbed to globules.	Hg.
3. Steel-grey sublimate; garlic odour.	As.
4. Yellow sublimate.	S (melts on heating), As_2S_3, HgI_2 (red when rubbed with a glass rod).

* If a white sublimate forms, heat with four times the bulk of anhydrous Na_2CO_3 and a little KCN in an ignition tube. A grey mirror, convertible into globules on rubbing with a glass rod, indicates Hg (**Note**: Hg vapour is very poisonous); a brownish-black mirror, yielding a white sublimate and an odour of garlic when heated in a wide tube, indicates As; ammonia evolved (test with mercury(I) nitrate paper) indicates ammonium salts.

*2. **Charcoal block reductions*** (*a*) Heat a little of the substance (say, 2–3 mg) in a small cavity in a charcoal block in a blowpipe flame. Evaluate results with the aid of Table VI.2.

Table VI.2 Charcoal block reductions

Observation	Inference
1. The substance decrepitates.	Crystalline salts, e.g. NaCl, KCl.
2. The substance deflagrates.	Oxidizing agents, e.g. nitrates, nitrites, and chlorates.
3. The substance is infusible or incandescent, or forms an incrustation upon the charcoal.	Apply test (*b*) below.

(*b*) Mix the substance (3–5 mg) with twice its bulk of anhydrous sodium carbonate, place it in a cavity in a charcoal block, and heat in the reducing flame of the blowpipe. Evaluate results with the help of Table VI.3.

*3. **Flame test*** Place 2–3 mg of the substance in a depression of a spot-plate, moisten with a few drops of concentrated hydrochloric acid, and introduce a little of the substance on a clean platinum wire into the base of a non-luminous

Table VI.3 Charcoal block reductions in the presence of soda

Observation	Inference
1. Incrustation without metal:	
White, yellow when hot.	ZnO.
White, garlic odour.	As_2O_3.
Brown.	CdO.
2. Incrustation with metal:	
White incrustation; brittle metal.	Sb.
Yellow incrustation; brittle metal.	Bi.
Yellow incrustation; grey and soft metal, marks paper.	Pb.
3. Metal without incrustation:	
Grey metallic particles, attracted by magnet.	Fe, NI, Co.
Malleable beads.	Ag and Sn (white); Cu (red flakes).

flame of a semimicro burner. Alternatively, dip the platinum wire into concentrated hydrochloric acid contained in a depression of a spot plate and then into the substance; sufficient will adhere to the platinum wire for the test to be carried out. Possible results are shown in Table VI.4.

Table VI.4 Flame test

Observation	Inference
Persistent golden-yellow flame.	Na.
Violet (lilac) flame (crimson through cobalt-blue glass).	K.
Brick-red (yellowish-red) flame.	Ca.
Crimson flame.	Sr.
Yellowish-green flame.	Ba.
Livid-blue flame (wire slowly corroded).	Pb, As, Sb, Bi, Cu.

The sodium flame masks that of other elements, e.g. that of potassium. Mixtures can be readily detected with a direct-vision spectroscope (see Section **II.2**, 4). A less delicate method is to view the flame through two thicknesses of cobalt-blue glass, whereby the yellow colour due to sodium is masked or absorbed; potassium then appears crimson.

*4. **Borax bead reactions*** Make a borax bead in a loop of platinum wire by dipping the hot wire into borax and heating until colourless and transparent. Bring a minute quantity of the substance into contact with the hot bead and heat in the outer or oxidizing flame. Observe the colour when the bead is hot and also when it is cold. Heat the bead in the inner or reducing flame and observe the colour in the hot and cold states. Coloured beads are obtained with compounds of copper, iron, chromium, manganese, nickel, and cobalt; the most characteristic result is for cobalt. Possible results are shown in Table VI.5.

*5. **Test for ammonium ions*** Mix 4–5 mg of the substance with about 0·2 ml (say, 4–5 drops) of sodium hydroxide solution in a semimicro test-tube, introduce a Pyrex filter tube carrying a strip of red litmus paper and place the assembly (Fig. II.29a) in the hot water rack (Fig. II.23). The evolution of ammonia,

Table VI.5 Borax bead reactions

Oxidizing flame	Reducing flame	Metal
1. Green when hot; blue when cold.	Colourless when hot, opaque-red when cold.	Cu.
2. Yellow, hot and cold.	Green, hot and cold.	Fe.
3. Dark yellow when hot, green when cold.	Green, hot and cold.	Cr.
4. Violet (amethyst), hot and cold.	Colourless, hot and cold.	Mn.
5. Blue, hot and cold.	Blue, hot and cold.	Co.
6. Reddish-brown when cold.	Grey when cold.	Ni.

detected by its action upon the reagent paper, indicates the presence of an ammonium salt.

Filter paper moistened with mercury(I) nitrate solution may also be used; this is blackened by ammonia. The mercury(I) nitrate reagent contains an excess of free nitric acid; it is advisable, therefore, to add sodium carbonate solution dropwise and with stirring to about 1 ml of mercury(I) nitrate solution until a slight permanent precipitate is produced. The solution is centrifuged and the centrifugate is employed for the preparation of mercury(I) nitrate paper.

Alternatively, drop-reaction paper treated with 2–3 drops of 5 per cent tannic acid solution and 2–3 drops of 20 per cent silver nitrate solution may be used; it is blackened by ammonia.

Note: Sodium hydroxide is a dangerous substance because of its destructive action upon the eyes. Great care should be taken that the test-tube containing the hot sodium hydroxide solution is directed away from the eyes of the observer and his near neighbours. The solution, when heated directly, has a tendency to 'bump'.

6. Test for nitrate (and/or nitrite) If ammonium is found, transfer the solution from test 5 with the aid of 0·5 ml water to a semimicro boiling tube (or crucible), add 0·5 ml sodium hydroxide solution and evaporate down to a volume of about 0·2 ml – this treatment completely decomposes the ammonium salt. Transfer the residue to a semimicro test-tube, rinse the vessel with 0·5 ml sodium hydroxide solution and add this to the contents of the test-tube. Then add 10 mg of Devarda's alloy (or of aluminium powder or thin foil), introduce a Pyrex filter tube provided with a loose plug of cotton wool at the lower end and containing a strip of red litmus paper or mercury(I) nitrate paper (Fig. II.29*a*), and place the assembly in the hot water rack. If frothing occurs, remove the apparatus from the water bath until the vigorous reaction has subsided.

If ammonium is absent, add about 10 mg of Devarda's alloy (or of aluminium as powder or thin foil) and 0·2–0·3 ml sodium hydroxide solution to the reaction mixture from test 5, introduce the filter tube carrying the reagent paper and proceed as above.

If ammonia is evolved, as detected by its action upon red litmus paper or upon mercury(I) nitrate paper, the presence of a nitrate or nitrite is indicated. Nitrite will also be detected in the reaction with dilute sulphuric acid (see test 7): if nitrite is absent, the presence of nitrate is established.

It must be emphasized that both the mercury(I) nitrate paper test and the tannic acid–silver nitrate paper test are not applicable in the presence of arsenites. Arsenites are reduced under the above conditions to arsine, which also blackens the test papers.

Note: The dangers connected with heating solutions with sodium hydroxide must be re-emphasized here (cf. note to test 5).

*7. **Action of dilute sulphuric acid*** Treat 5–10 mg of the substance in a semimicro test-tube with about 0·3–0·5 ml of M sulphuric acid and note whether any reaction takes place in the cold (indicated on Table VI.6 by *C*). Heat the mixture on a water bath and observe the effect produced. Possible results are shown in Table VI.6.

Table VI.6 Test with dilute sulphuric acid

Observation	Inference
1. Colourless gas is evolved with effervescence; gas is odourless and produces a turbidity when passed into lime water (see Fig. II.30). (*C*)	CO_2 from carbonate or bicarbonate.
2. Nitrous fumes evolved; recognized by reddish-brown colour and odour. (*C*)	NO_2 from nitrite.
3. Yellowish-green gas evolved; suffocating odour, reddens then bleaches litmus paper, also turns starch–KI paper blue; very poisonous. (*C*)	Cl_2 from hypochlorite.
4. Colourless gas evolved with suffocating odour; turns filter paper moistened with $K_2Cr_2O_7$ solution green; decolourizes fuchsin solution.	SO_2 from sulphite.
5. Colourless gas evolved; gives above tests for SO_2; sulphur is deposited in the solution.	SO_2 and S from thiosulphate.
6. Colourless gas evolved; odour of rotten eggs; blackens filter paper moistened with lead acetate solution; turns cadmium acetate paper yellow.	H_2S from sulphide.*
7. Odour of vinegar.	CH_3COOH from acetate.
8. Colourless gas is evolved; rekindles glowing splint.	O_2 from peroxides or peroxo-salts of alkali and alkaline earth metals.
9. Colourless gas evolved; odour of bitter almonds,† highly poisonous.	HCN from cyanide or from soluble hexacyanoferrate(II) or (III).
10. Upon boiling, yellow solution formed and SO_2 (fuchsin solution decolourized, etc.) evolved.	SO_2, from thiocyanate.

* Many sulphides, especially native ones, are not affected by dilute H_2SO_4; some H_2S is evolved by warming with concentrated HCl alone or with a little tin.

† If cyanide is suspected, gently warm (water bath) 5 mg of the mixture with 5 drops of M H_2SO_4 in a semimicro test-tube carrying over the mouth a piece of filter paper moistened at the centre with 1 drop dilute NaOH solution. After 2 minutes, treat the drop on the paper with 1 drop $FeSO_4$ solution, warm and add a drop or two of 2M HCl. A blue colour indicates the presence of a cyanide. For another test, see Section **IV.8**, reaction 1. Mercury(II) cyanide is attacked slowly.

*8. **Action of concentrated sulphuric acid*** Treat 5–10 mg of the substance in a semimicro test-tube with 0·3–0·5 ml concentrated sulphuric acid. If no reaction occurs in the cold, place the tube in the hot water rack. (If chlorate is suspected from the preliminary charcoal reduction test to be present, use not more than 5 mg for this test, as an explosion may result on warming.)

If the substance reacted with dilute sulphuric acid, the addition of the concentrated acid may result in a vigorous reaction and rapid evolution of gas, which may be accompanied by a very fine spray of acid. In such a case it is best to add dilute (M) sulphuric acid from a capillary dropper to another portion of the substance until action ceases, and then to add 1 ml concentrated sulphuric acid.

Results of this test are summarized in Table VI.7.

Table VI.7 Test with concentrated sulphuric acid

Observation	Inference
1. Colourless gas evolved with pungent odour and which fumes in the air; white fumes of NH_4Cl in contact with glass rod moistened with concentrated NH_3 solution; Cl_2 evolved on addition of MnO_2 (reddens then bleaches litmus paper).	HCl from chloride.
2. Gas evolved with pungent odour, reddish colour and fumes in moist air; on addition of MnO_2 increased amount of red fumes with odour of Br_2 (fumes colour filter paper moistened with fluorescein solution red).	HBr and Br_2 from bromide.
3. Violet vapours evolved, accompanied by pungent acid fumes, and often SO_2 and even H_2S.	HI and I_2 from iodide.
4. Yellow gas evolved in the cold with characteristic odour; explosion or crackling noise on warming gently (DANGER!).	ClO_2 from chlorate.
5. 'Oily' appearance of tube in cold; on warming, pungent gas evolved, which corrodes the glass; if moistened glass rod introduced into the vapour, a gelatinous precipitate of hydrated silica is deposited upon it.	HF from fluoride.
6. Pungent acid fumes evolved, often coloured brown by NO_2; colour deepens upon addition of copper turnings (if nitrite absent).	HNO_3 and NO_2 from nitrate.
7. Yellow colouration in cold; upon warming, vigorous reaction, COS (burns with blue flame), SO_2 (decolourizes fuchsin solution) and free S produced.	COS, SO_2 and S from thiocyanate.
8. Colourless gas evolved; burns with a blue flame;* no charring (VERY POISONOUS).	CO from formate.
9. Colourless gas evolved; renders lime water turbid and may also burn with a blue flame; no blackening.	CO and CO_2 from oxalate.

* The burning splint should be introduced into the tube; the application of a flame to the mouth of the tube frequently fails to ignite the gas owing to its dilution with air in the tube.

9. Tests on liquid samples (solutions) If a liquid is supplied for analysis, proceed as follows:

1. Observe the colour, odour, and any special physical properties.
2. Test its reaction to litmus (or equivalent test) paper.
   (*a*) The solution is neutral: free acids, free bases, acid salts, and salts which give an acid or alkaline reaction owing to hydrolysis, are absent.
   (*b*) The solution reacts alkaline: this may be due to the hydroxides of the alkali and alkaline earth metals, to the carbonates, sulphides, hypochlorites, and peroxides of the alkali metals, etc.
   (*c*) The solution reacts acid: this may be due to free acids, acid salts, salts which yield an acid reaction because of hydrolysis, or to a solution of salts in acids.
3. Evaporate a portion of the liquid to dryness on the water bath (use a 5 or 8 ml crucible and stand it over one of the openings in the hot water rack); carefully

smell the vapours evolved from time to time. If a solid residue remains, examine it as detailed above for a solid substance. If a liquid remains, evaporate cautiously on a wire gauze in the fume chamber; a solid residue should then be examined in the usual way. If no residue is obtained, the original liquid consists of some volatile substances which may be water or water containing certain gases or volatile substances, such as CO_2, NH_3, SO_2, H_2S, HCl, HBr, HI, H_2O_2, or $(NH_4)_2CO_3$, all of which can be readily detected by special tests. It is best to neutralize with sodium carbonate and test for anions.

VI.5 TESTING FOR ANIONS IN SOLUTION ON THE SEMIMICRO SCALE The preliminary tests 7 and 8 with dilute sulphuric acid and with concentrated sulphuric acid will have provided useful information as to many anions present. For more detailed information, it is necessary to have a solution containing all (or most) anions free from heavy metal ions. This is best prepared by boiling the substance with concentrated sodium carbonate solution; double decomposition occurs (either partially or completely) with the production of the insoluble carbonates* of the metals (other than alkali metals) and the soluble sodium salts of the anions, which pass into solution. Thus, if the unknown substance is the salt of a bivalent metal M and an acid HA, the following reaction will occur:

$$MA_2 + CO_3^{2-} \rightleftarrows MCO_3\downarrow + 2A^-$$

Sodium ions will accompany the anion A^- in the solution.

Preparation of solution for testing for anions Use analytical grade (sulphate- and chloride-free) sodium carbonate. Boil 200 mg of the finely divided substance or mixture with 2·5 ml of a saturated solution of sodium carbonate (prepared from 0·4 g anhydrous sodium carbonate and 2·5 ml distilled water) for 5–10 minutes in a 10 ml conical flask with a funnel in the mouth to reduce the loss by evaporation. Transfer, with the aid of about 0·5 ml water, to a 4 ml semimicro test-tube and centrifuge. Remove the centrifugate to another test-tube: add 1 ml hot distilled water to the residue, stir, and add the clear washings to the original centrifugate; the total volume should be 3–4 ml. If there is no precipitate formed, the solution is virtually free from heavy metal ions; note this fact for the subsequent testing for cations.

Use the soda extract to carry out the following tests.

1. ***Sulphate test*** To 5 drops of the soda extract in a 4 ml test-tube add dilute hydrochloric acid until acid (test with litmus paper) and then add 3 drops in excess. Place in the hot water rack for 5 minutes to expel carbon dioxide completely, and then add 2–3 drops of barium chloride solution. A white precipitate ($BaSO_4$) shows the presence of sulphate.

2. ***Test for reducing agents*** Acidify 5 drops of the soda extract with dilute sulphuric acid and add 3 drops of dilute sulphuric acid in excess. Add 2–3 drops of dilute potassium permanganate solution (prepared by dilution of 1 drop of

*Certain carbonates, initially formed, are converted into insoluble basic carbonates or into hydroxides.

0·02M $KMnO_4$ with 4 drops of water and mixing well). Bleaching of the permanganate solution indicates the presence of one or more of the following reducing anions: sulphite, thiosulphate, sulphide, nitrite, bromide, iodide, and arsenite. If the permanganate is not decolourized, place the tube in the hot water rack for several minutes and observe the result. If the reagent is bleached only on heating, the presence of oxalate is indicated. A negative test shows the absence of the above anions.

3. Test for oxidizing agents Acidify 5 drops of the soda extract cautiously with concentrated hydrochloric acid and add 2 drops in excess, followed by 3–5 drops of the manganese(II) chloride reagent. Place in the hot water rack for 1 minute. A brown (or black) colouration indicates the presence of nitrate, nitrite, chlorate, or chromate. A negative test indicates the absence of the above oxidizing anions except for small amounts of nitrates or nitrites. If reducing anions have been found, the test is inconclusive.

Note: The reagent consists of a saturated solution of manganese(II) chloride, $MnCl_2.4H_2O$, in concentrated hydrochloric acid. Its action depends upon its conversion by even mild oxidizing agents to a dark-brown coloured $[MnCl_5]^{2-}$ complex ion, in which manganese is tervalent.

4. Tests with silver nitrate solution Sulphide, cyanide, and sulphite interfere in tests with silver nitrate solution, hence if any of these anions was detected in the preliminary test with dilute sulphuric acid, it must be removed first as follows. Acidify 1 ml of the soda extract with dilute acetic acid (use litmus paper) and boil gently in a small conical flask or crucible in the fume cupboard to expel H_2S, HCN or SO_2 (1–2 minutes). It is important that the solution be acid throughout. Centrifuge, if necessary, and allow to cool. If the volume has been reduced appreciably, add water to restore the original volume.

Carry out a separation with this solution, or (if S^{2-}, CN^-, and SO_3^{2-} are absent) with the soda extract, according to the scheme outlined in Table VI.8.

Table VI.8 Separation of certain anions with $AgNO_3$ Acidify 1·0 ml of the solution cautiously with dilute HNO_3 (use litmus paper). Determine the total volume of the acidified solution with the aid of a small measuring cylinder or graduated 2 ml pipette, add one-tenth of the volume of concentrated HNO_3 and stir for 10–15 seconds. Then add a few drops of $AgNO_3$ solution with stirring. If a precipitate forms, place the test-tube or centrifuge tube in the hot water rack and add $AgNO_3$ solution slowly and with stirring until precipitation is complete. Centrifuge and wash with a few drops of M HNO_3.

Residue	Centrifugate
AgCl – white. AgBr – pale yellow. AgI – yellow. [AgSCN – white.] (1)	Add 3–4 drops of $AgNO_3$ solution, then 1–2 drops of 50% KNO_2 solution (prepared from the A.R. solid) and stir. If a white ppt. (AgCl) forms, chlorate is present: continue the addition of $NaNO_2$ solution dropwise until precipitation is complete. [If no ppt. forms, do not add more $NaNO_2$ solution.] Centrifuge, if necessary, and wash with 2 drops of 2M HNO_3 (2).

Residue	Centrifugate
AgCl derived from $AgClO_3$.	Add NaOH solution dropwise and with vigorous stirring until the solution is just neutral to litmus or, better, barely alkaline to nitrazine yellow indicator (3);

Table VI.8 Separation of certain anions with $AgNO_3$ (contd.)

	then add 2–3 drops of dilute acetic acid and 5 drops of $AgNO_3$ solution. Heat to about 80°C in the water bath (4). If a permanent ppt. forms, add more $AgNO_3$ solution until precipitation is complete. Centrifuge, and wash with a few drops of hot water.	
	Residue (5)	**Centrifugate**
	Ag_3PO_4 – yellow. Ag_3AsO_4 – brownish-red. Ag_3AsO_3 – yellow. $Ag(COO)_2$ – white.	Reject.

Notes to Table VI.8: 1. Silver nitrate solution precipitates only AgCl, AgBr, and AgI from a dilute nitric acid solution.
2. Sodium nitrite reduces chlorate to chloride:

$$ClO_3^- + 3NO_2^- \rightarrow Cl^- + 3NO_3^-$$

which is then precipitated with $AgNO_3$ as AgCl.
3. It is essential that the solution be just neutral to litmus or, at most, barely alkaline; the latter will be indicated by a very slight brown opalescence (due to Ag_2O) obtained after shaking. If much brown silver oxide separates, it will redissolve only with difficulty.

The introduction of 1 drop of nitrazine yellow indicator into the solution is to be preferred. (This indicator covers the *p*H range 5·5–7·2 and the colour change is from yellow to blue.) The addition of NaOH solution is continued until the solution just assumes a pale-blue colour.
4. Silver acetate is soluble in hot water and is thus held in solution.
5. In solutions faintly acid with CH_3COOH, PO_4^{3-}, AsO_4^{3-}, AsO_3^{3-} and $(COO)_2^{2-}$ are precipitated by $AgNO_3$.

If CrO_4^{2-} is present (yellow or orange solution) it will be reduced by the $NaNO_2$ treatment and will be precipitated here as green $Cr(OH)_3$. *Chromate* can be readily detected in the soda extract as follows. Acidify 5 drops of the solution with dilute HNO_3 and cool. Add 0·3 ml amyl alcohol and 2 drops of 3 per cent H_2O_2 solution. A blue colouration (chromium pentoxide) in the organic layer confirms chromate.

If a mixture of halides, or of phosphate, arsenate, and arsenite, is suspected, use the methods of separation given in Section **V.18**, 4. The confirmatory tests for the individual anions are collected in Section **VI.6**.

The reactions with silver nitrate solution are intended to act as a guide to the presence of groups of anions, and the table must be interpreted in conjunction with the observations made in the preliminary tests. Arsenite, arsenate, and chromate will be found in the analysis for cations (Section VI.8).

5. ***Tests with calcium chloride solution*** For tests 5 and 6 the soda extract has to be neutralized. To do this, take 0·5 ml of the soda extract in a semimicro boiling tube (or small crucible) and render it faintly acid with dilute nitric acid (use litmus paper). Heat to boiling for about 30 seconds to expel carbon dioxide, etc., allow to cool, then add dilute ammonia solution until just alkaline, and boil for 30 seconds to expel the slight excess of ammonia.

[If a precipitate forms on neutralizing, the presence of sulphides of arsenic, antimony, and tin, as well as salts of amphoteric bases (lead, tin, aluminium, and zinc) is indicated. Note such an observation but discard the precipitate.]

With one half of this solution follow the prescriptions of Table VI.9, keep the other half for test 6.

Table VI.9 Separation of certain anions with $CaCl_2$ Add an equal volume of $CaCl_2$ solution and place in the hot water rack for 5 minutes. Centrifuge and wash with a few drops of hot water.

Residue	Centrifugate
May contain calcium fluoride, oxalate, phosphate, and arsenate. Add 1 ml dilute acetic acid, heat in the hot water rack for 2–3 minutes and centrifuge.	Reject.

Residue	Centrifugate
May be calcium oxalate and calcium fluoride. Extract with 1 ml hot, dilute H_2SO_4 and centrifuge, if necessary. Treat the hot centrifugate or solution with 2 drops 0·004M $KMnO_4$ solution. If the permanganate is decolourized, oxalate is present. If the ppt. is not completely soluble in dilute H_2SO_4, test the original substance for fluoride by the water-film test (Section **VI.6**).	Add NaOH solution dropwise until neutral (use litmus or nitrazine yellow indicator). White ppt. (sometimes separating slowly). Arsenate and/or phosphate present. Treat the suspension or neutral solution with a few drops of $AgNO_3$ solution. A yellow ppt. indicates phosphate: a brownish-red ppt. indicates arsenate or arsenate + phosphate.

6. Test with iron(III) chloride solution A neutralized reagent is needed for this test. As the bench reagent usually contains excess free acid added during its preparation in order to produce a clear solution, the precipitation of the basic acetate may be prevented. Add dilute NH_3 solution dropwise to 1 ml of side-shelf $FeCl_3$ solution until a slight precipitate forms and then centrifuge the mixture. The clear centrifugate is employed in the test.

Treat the other portion of the neutralized soda extract with 5 drops of the reagent.

Reddish-purple colouration indicates thiosulphate.

Reddish-brown colouration, yielding a brown precipitate on dilution and boiling in a semimicro boiling tube, indicates acetate.

Yellowish-white precipitate indicates phosphate.

Blood-red colouration, discharged by $HgCl_2$ solution, indicates thiocyanate.

VI.6 CONFIRMATORY TESTS FOR ANIONS ON THE SEMIMICRO SCALE The tests in the preceding section will indicate the anions present. In general, these should be confirmed by at least one distinctive confirmatory test. The following are recommended. Full experimental details will be found in Chapter IV under the reactions of the anions, the reference to these will be abbreviated as follows: thus (**IV.2**, 7) is to be interpreted as Section **IV.2**, reaction 7. It will of course be realized that the tests in Chapter IV refer to reactions on the macro scale; the student should have no difficulty in reducing these to the semimicro scale once the technique described in Section **II.4** has been acquired. Particular attention is directed to the apparatus shown in Fig. II.30, which is employed in testing for evolved gases with reagent papers and liquid reagents. It is assumed that interfering anions are absent or have been removed as described in Section **VI.7** under *Special tests for mixtures of anions.*

Chloride. Mix 20 mg* of the substance with an equal weight of MnO_2 and 0·5 ml concentrated H_2SO_4, and place in the hot water rack; Cl_2 is evolved (reddens then bleaches litmus paper and also turns potassium iodide–starch paper blue) (**IV.14**, 2).

Bromide. MnO_2 and concentrated H_2SO_4 test (as under Chloride); Br_2 is evolved (**IV.15**, 2) OR dissolve 20 mg of the substance in 0·5 ml water, add 5 drops of dilute HCl, 5 drops of CCl_4, and 2–3 drops of NaOCl solution and shake; reddish-brown colouration of CCl_4 layer (**IV.15**, 5).

Iodide. NaOCl solution, dilute HCl and CCl_4 test (as under Bromide); violet colouration of CCl_4 layer (**IV.16**, 4).

Fluoride. Fit a 4 ml (75 × 10 mm) test-tube with a cork carrying a tube about 8 cm long and of about 3 mm bore: cut a V-shaped groove in the cork. Adjust the tube in the cork so that the lower end is about 2·5 cm from the bottom of the test-tube (cf. Fig. V.1). Place 15–20 mg of the substance and 0·5 ml concentrated H_2SO_4 in the test-tube, dip the glass tube into water so that a film of water almost seals the lower end to a depth of about 5 mm, insert into the test-tube, and place in the hot water rack. The formation of a white film in the water confirms fluoride (**IV:17**, 1).

The zirconium–alizarin-S test (**IV.17**, 6) may be used if oxalate is known to be absent and also in the absence of sulphates, thiosulphates, nitrites, phosphates, and arsenates in quantities greater than that of the fluoride.

Nitrite. Treat 10–20 mg of the substance with 0·5 ml dilute acetic acid, 10 mg thiourea and 3 drops of $FeCl_3$ solution. Red colouration (**IV.7**, 9).

Nitrate. Brown ring test with $FeSO_4$ solution and concentrated H_2SO_4 (**IV.18**, 3) if bromide, iodide, chlorate, and nitrite absent. (i) Dissolve 10–20 mg of the substance in 0·5 ml water. Cautiously add 1 ml concentrated H_2SO_4, mix and cool under running water. Incline the tube and with great care allow about 0·5 ml $FeSO_4$ solution to run slowly down the side of the tube so that it forms a layer above the heavy sulphuric acid (Fig. VI.1). Observe the brown ring at the junction of the two liquids after 1–2 minutes. (ii) Dissolve 10–20 mg of the substance in 0·5 ml water and add about 1 ml $FeSO_4$ solution. Incline the tube and allow 0·5–1 ml concentrated H_2SO_4 to run slowly down the side of the tube to form a layer under the solution (Fig. VI.1), and after 1–2 minutes observe the brown ring at the interface.

Sulphide. Treat 10–20 mg of the substance with 0·5 ml dilute H_2SO_4. Place the tube in the hot water rack and test with lead or cadmium acetate paper (**IV.6**, 1).

Sulphite. Treat 10–20 mg of the substance with 0·5 ml dilute H_2SO_4, place the tube in the hot water rack, and test for SO_2 with $K_2Cr_2O_7$ paper† (Fig. II.29*a*) or with 0·2 ml fuchsin solution (Fig. II.30*b*) (**IV.4**, 9).

Thiosulphate. Dilute H_2SO_4 on solid and liberation of SO_2 ($K_2Cr_2O_7$ test –

---

* The weights and volumes given in the suggested confirmatory tests are very approximate and serve to indicate a reasonable scale for the various operations. They are given solely for the guidance of the student; satisfactory results can, however, be obtained on an appreciably smaller scale.

† For the preparation of $K_2Cr_2O_7$ paper, the dichromate solution should be almost saturated.

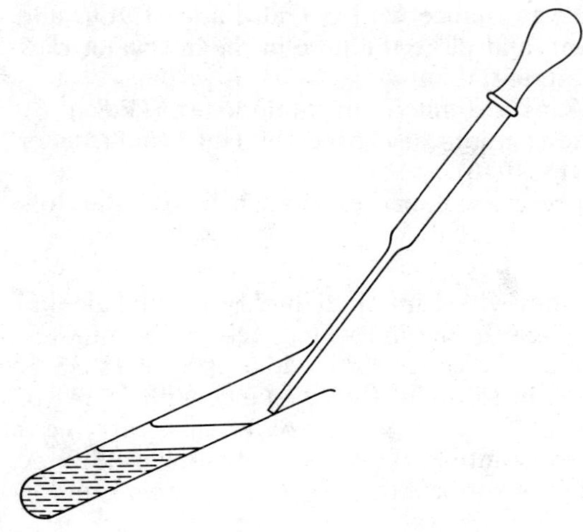

Fig. VI.1

details under Sulphite) and sulphur (**IV.5**, 1).

Sulphate. The $BaCl_2$ solution and dilute HCl test is fairly conclusive.

Further confirmation is obtained as follows. Centrifuge the suspension and remove the supernatant liquid. Add 1 ml water, stir, centrifuge, and discard the centrifugate. Add 3–4 drops of water, stir thoroughly to produce a fairly uniform suspension, transfer the suspension to a small plug of cotton wool in a small ignition tube, introduce 2–3 drops of Na_2CO_3 solution and heat cautiously to redness. Maintain the lower part of the tube at a red heat for 2–3 minutes and allow to cool. Break the tube, transfer the residue as completely as possible to a semimicro test-tube, add 0·5 ml dilute HCl and introduce a Pyrex filter tube carrying a strip of filter paper moistened with lead acetate solution. Place the assembly in the hot water rack. A brown stain on the paper confirms sulphate.

The lead acetate test on the solution and the dissolution of the resulting precipitate of lead sulphate in ammonium acetate solution are also characteristic (**IV.24**, 2).

Carbonate. Treat 10–20 mg of the substance with 0·5 ml dilute H_2SO_4, place the tube in the hot water rack, and test with 0·2–0·5 ml lime water (Fig. II.30*b* or *c*) (**IV.2**, 1).

Hypochlorite. Treat 10–20 mg of the substance with 0·5 ml dilute HCl, place the tube in the hot water rack and test for Cl_2 with potassium iodide–starch paper and with litmus paper (**IV.13**, 2).

Chlorate. The $AgNO_3$–$NaNO_2$ test is conclusive (see Table VI.8, test 5, also **IV.19**).

Chromate. Mix 10–20 mg of the substance with 0·5 ml dilute H_2SO_4, add 0·5 ml amyl alcohol and 0·3 ml of 10-volume H_2O_2; blue colour of alcohol layer (**IV.33**, 4).

Arsenite. Immediate precipitate of As_2S_3 in dilute HCl solution (**III.12**, 1) and absence of precipitate with magnesium nitrate reagent (**III.12**, 3).

Arsenate. Action of H_2S on acid solution (**III.13**, 1), $AgNO_3$ solution test in faintly acetic acid solution (**III.13**, 2), and magnesium nitrate reagent test (**III.13**, 3).

Phosphate. Mix 10–20 mg of the substance with 0·5 ml dilute HNO_3, add 1 ml ammonium molybdate reagent, and place the tube in the hot water rack for a minute or two. Yellow precipitate (**IV.28**, 4).

Cyanide. Prussian blue test (**IV.8**, 4) or ammonium sulphide test (**IV.8**, 1, 6).

Thiocyanate. Iron(III) chloride test; colour discharged by $HgCl_2$ solution or by NaF solution, but not by HCl (**IV.10**, 6).

Borate. Flame test (**IV.23**, 2); or turmeric paper test (**IV.23**, 3); or mannitol–bromothymol blue test (**IV.23**, 8).

Acetate. Mix 20 mg of the substance with 1 ml of ethanol or n-butyl alcohol and 5 drops concentrated H_2SO_4. Heat in the hot water rack for 10 minutes, and pour into 2 ml of Na_2CO_3 solution – characteristic odour of ester (**IV.35**, 3) OR, better, indigo test (**IV.35**, 9). Mix 15 mg of the substance with 15 mg of $CaCO_3$ in a semimicro test-tube, introduce a pressure-filter tube carrying a strip of filter paper moistened with a solution of 5 mg of *o*-nitrobenzaldehyde in 1 ml of NaOH solution. Heat the test-tube strongly. Blue or green stain on paper confirms acetate.

Oxalate. Precipitate with $CaCl_2$ solution in the presence of dilute acetic acid, the precipitate decolourizes a dilute acid solution of $KMnO_4$ (**IV.37**, 3, 4 and **VI.5**, 5); resorcinol test (**IV.37**, 5).

VI.7 SPECIAL TESTS FOR MIXTURES OF ANIONS ON THE SEMIMICRO SCALE The subject is treated fully in Section **IV.45**, but on a macro scale. The student should be able to adapt these to semimicro work. Some typical semimicro separations of mixtures of anions are given below. The quantities are for guidance only and can be reduced, if desired. If the mixture is insoluble in water it is often convenient to employ the soda extract (Section **VI.5**) or the neutralized soda extract (Section **VI.5**, test 5). The special gas testing apparatus of Fig. II.30 will, of necessity, find application here.

*1. **Carbonate in the presence of sulphite*** Treat 20 mg of the mixture with 20 mg of finely powdered $K_2Cr_2O_7$ and 0·5 ml dilute H_2SO_4; place the apparatus (Fig. II.30*b* or *c*) in the hot water rack, and test for CO_2 with lime or baryta water. (The addition of a few milligrams finely divided zinc will generate H_2 and assist in driving the CO_2 out of the tube.)

*2. **Nitrate in the presence of nitrite*** Dissolve 10 mg of the solid in 1 ml of water. Remove 1 drop with a glass stirring rod and 'spot' on KI–starch paper moistened with very dilute H_2SO_4: a blue colour is obtained, due to the iodine liberated by the nitrous acid. Add 20–30 mg sulphamic acid and stir until effervescence ceases (Section **IV.7**, 6). Test for absence of nitrite by 'spotting' on KI–starch paper. Then apply the brown ring test (Section **VI.6**).

*3. **Nitrate in the presence of bromide and iodide*** Treat 10–15 mg of the substance with 1 ml sodium hydroxide solution and 10–15 mg of Devarda's alloy (or of aluminium as powder or as thin foil), place in the hot water rack, and test for ammonia with red litmus paper or mercury(I) nitrate paper (Fig. II.30*a*).

Another procedure is to just acidify 5 drops of the soda extract with dilute acetic acid, and then add the ammoniacal silver sulphate reagent (see Section

IV.45, 3), dropwise until precipitation is complete. Centrifuge. Add excess concentrated H_2SO_4 cautiously to the centrifugate and apply the brown ring test.

*4. **Nitrate in the presence of chlorate*** Test for nitrate as under 3: then acidify with dilute HNO_3 and test for chloride with a few drops of $AgNO_3$ solution.
If chloride is originally present, it may be removed by the addition of saturated silver sulphate solution or of the ammoniacal silver sulphate reagent.

*5. **Chloride in the presence of bromide and iodide*** Dissolve 10–20 mg of the solid in 0·5–1 ml water in a semimicro boiling tube (or use 5–10 drops of the soda extract acidified with dilute HNO_3), add 1–1·5 ml concentrated HNO_3, and boil gently until the bromine and iodine are volatilized. Dilute with 1 ml water, and test for chloride by the addition of a few drops of $AgNO_3$ solution.
Alternatively, repeat the experiment but add 20 mg precipitated PbO instead of the concentrated HNO_3. After the bromine and iodine have been eliminated, add 1 ml water, transfer to a centrifuge tube, centrifuge, and add a few drops of $AgNO_3$ solution to the clear centrifugate: a white precipitate, soluble in dilute NH_3 solution and reprecipitated by dilute HNO_3, indicates chloride.

*6. **Chloride in the presence of iodide (bromide being absent)*** Add excess $AgNO_3$ solution to 0·5 ml of the soda extract acidified with dilute HNO_3; centrifuge and reject the centrifugate. Wash the precipitate with about 0·2 ml dilute NH_3 solution and centrifuge again. Acidify the clear washings with dilute HNO_3; a white precipitate (AgCl) indicates the presence of chloride.

*7. **Chloride in the presence of bromide (iodide being absent)*** Acidify 0·5 ml of the soda extract contained in a semimicro boiling tube with dilute HNO_3, and add an equal volume of concentrated HNO_3. Boil gently until all the bromine is expelled, and then add $AgNO_3$ solution. A white precipitate (AgCl) indicates chloride present.

*8. **Bromide and iodide in the presence of each other*** Dissolve 10–20 mg of the mixture in 0·5–1 ml water, add 5 drops dilute H_2SO_4 and 0·3–0·5 ml CCl_4; or use 0·3 ml of the soda extract acidified with dilute H_2SO_4, add 5 drops dilute H_2SO_4 and 0·5 ml CCl_4. Then introduce dilute NaOCl solution dropwise, shaking after the addition of each drop. A violet colouration of the CCl_4 layer, which appears first, indicates iodide: this subsequently disappears and is replaced by a reddish-brown (or brown) colouration if a bromide is present.

*9. **Thiocyanate, chloride, bromide, and iodide in the presence of each other*** This problem may arise in the tests with $AgNO_3$ solution (see Section **VI.5**, test 4 since AgSCN, AgCl, AgBr, and AgI are precipitated in M HNO_3 solution: interfering anions (e.g. IO_3^- which reacts with SCN^-) are assumed to be absent. The precipitate may be formed, for example, from 0·5 ml of the soda extract and is collected, after centrifuging, in a semimicro centrifuge or test-tube. Add 5–10 drops of water, stir the suspension, and transfer three-quarters of it by means of a capillary pipette to a small crucible; this portion will be used in the tests for Cl^-, Br^-, and I^-, and the remainder for SCN^-.

Thiocyanate test Treat one-quarter of the precipitate with 5–10 drops of 5 per cent solution, heat, and stir in the hot water rack for 3–5 minutes (this dissolves part of the AgSCN). Centrifuge and add to the clear centrifugate one drop dilute HCl and one drop $FeCl_3$ solution. A red colouration indicates thiocyanate present.

If thiocyanate is present, it must be destroyed since it interferes with the tests for the halides. Dry the main precipitate in the crucible by heating gently in an air bath (Fig. II.26), remove the air bath and heat directly to dull redness for one minute or until all the thiocyanate is decomposed, i.e. until blackening of the precipitate and/or burning of sulphur just ceases. Prolonged heating should be avoided. Allow to cool.

Mixtures of halides may now be identified as indicated in 5–8 above; a systematic procedure, which covers the three halides, is given below.

To the residue in the crucible (if thiocyanate is present) or to the remainder of the precipitate transferred to a 5 or 10 ml beaker (if thiocyanate is absent), add 100–150 mg zinc powder (20 mesh) and 10–15 drops dilute H_2SO_4. Allow the reduction to proceed for 5–10 minutes with intermittent stirring: gentle warming for a few seconds may be necessary to start the reaction. Transfer the liquid, with the aid of a few drops dilute H_2SO_4, to a centrifuge tube, centrifuge, and divide the clear centrifugate into three equal parts.

Iodide test Add 4–5 drops CCl_4 and 5 drops of 3 per cent H_2O_2 or 5 drops of 25 per cent $Fe_2(SO_4)_3$ solution. Agitate vigorously and allow to settle. Purple to violet colour of the CCl_4 layer. *Iodide present.*

Bromide test (*a*) If iodide is present, it must be removed by treating one-third of the solution with 5 drops dilute H_2SO_4 and 2 drops of 30 per cent $NaNO_2$ solution (chloride-free). Boil down gently to 3–4 drops and allow to cool. Transfer the solution to a centrifuge tube and test for bromide as under (*b*).

(*b*) If iodide is absent, use one-third of the solution directly. Add an equal volume of concentrated HNO_3, heat in the boiling water bath for 30 seconds, and cool to room temperature with cold water. Add 3–4 drops CCl_4 and stir vigorously with a glass rod. A brown colour in the CCl_4 layer indicates *bromide present.*

Chloride test (*a*) If iodide and/or bromide present, dilute the remaining third of the solution to 0·5 ml, add 1 ml concentrated HNO_3 and boil gently until evolution of bromine ceases. Cool, dilute and test for chloride as in (*b*) below with $AgNO_3$ solution only.

(*b*) If bromide and iodide are absent, add 2 drops each of dilute HNO_3 and $AgNO_3$ solution. A white precipitate (AgCl) indicates *chloride present.*

10. Phosphate, arsenate, and arsenite in the presence of each other This problem may arise in the test with $AgNO_3$ in neutral solution (see Section **VI.5**, test 4). A separation of these ions is necessary, which can be carried out according to the scheme shown in Table VI.10.

Table VI.10 Separation of AsO_3^{3-}, AsO_4^{3-}, and PO_4^{3-} Dilute 5 drops of the soda extract with an equal volume of water, acidify with dilute HCl and render alkaline with dilute NH_3 solution. Centrifuge, if necessary, and discard the ppt. Treat the clear solution with 10–12 drops of $Mg(NO_3)_2$ reagent (or of magnesia mixture) added dropwise and with stirring. Allow to stand for 10 minutes, stirring frequently: centrifuge. Wash the ppt. with 2–3 drops dilute NH_3 solution.

<table>
<tr><th colspan="2">Residue</th><th>Centrifugate</th></tr>
<tr><td colspan="2">May contain $Mg(NH_4)PO_4$, and $Mg(NH_4)AsO_4$ (1).
Dissolve the ppt. in 5–10 drops of dilute HCl, add 1 drop of 10 per cent NH_4I solution, heat to boiling and saturate with H_2S (Fig. II.27). Centrifuge, and saturate again with H_2S to ensure complete precipitation. Wash with a few drops 2M HCl.</td><td rowspan="3">May contain arsenite.
Add 6–8 drops of 3 per cent H_2O_2 solution (to oxidize arsenite to arsenate), heat on water bath for 1 minute, cool, add 5–10 drops of $Mg(NO_3)_2$ reagent (or of magnesia mixture), and allow to stand for 5–10 minutes, stirring frequently.
White ppt. of $Mg(NH_4)AsO_4$.
Arsenite present.
Alternatively, acidify with dilute HCl and pass H_2S. Immediate yellow ppt. of As_2S_3.
Arsenite present.</td></tr>
<tr><th>Residue</th><th>Centrifugate</th></tr>
<tr><td>As_2S_3 – yellow.
Arsenate present.</td><td>Transfer to a small crucible, boil until H_2S is expelled and volume is reduced to 2–3 drops. Transfer, with the aid of a few drops of water, to a centrifuge tube, make alkaline with dilute NH_3 solution and add 5 drops in excess. Add 10 drops of $Mg(NO_3)_2$ reagent (or of magnesia mixture), and allow to stand, with frequent stirring, for 10 minutes.
White ppt. of $Mg(NH_4)PO_4$.
Phosphate present.</td></tr>
</table>

Note to Table VI.10

1. The presence of arsenate is readily detected as follows. Treat a portion of the ppt. with 5 drops $AgNO_3$ solution to which 1 drop of 2M acetic acid has been added. If the ppt. acquires a reddish colour (due to Ag_3AsO_4), arsenate is present. The yellow colour of Ag_3PO_4 is obscured by the red colour of Ag_3AsO_4.

VI.8 PREPARATION OF SOLUTION FOR CATION TESTING ON THE SEMIMICRO SCALE

Preparation of solid samples Since the whole scheme for the analysis of cations depends upon the reactions of ions, it is clear that it is first necessary to get the substance into solution. Water is first tried in the cold and then on warming. If insoluble in water, the following reagents are investigated as solvents in the order indicated: dilute hydrochloric acid, concentrated hydrochloric acid, dilute nitric acid, concentrated nitric acid, and aqua regia (3 volumes of concentrated HCl to 1 volume of concentrated HNO_3). Most substances encountered in an elementary course will dissolve in either water or dilute hydrochloric acid. If concentrated hydrochloric acid has to be used, the solution must be considerably diluted before proceeding with the analysis, otherwise certain cations, such as cadmium and lead, will not be precipitated by hydrogen sulphide. When concentrated nitric acid or aqua regia is employed as the solvent, the solution must be evaporated almost to dryness, a little hydrochloric acid added, the solution evaporated again to small bulk and then diluted with water to dissolve the soluble nitrates (or chlorides). This evaporation is necessary because the nitric acid may react with the hydrogen sulphide subsequently employed in the Group analysis.

To discover the most suitable solvent, treat portions of about 15 mg* of the finely powdered substance successively with 0·3–0·5 ml of water, dilute hydrochloric acid, concentrated hydrochloric acid, dilute nitric acid, concentrated nitric acid, and aqua regia in the order given. Try the solubility first in the cold and then at the temperature of a boiling water bath for water and dilute hydrochloric acid using a semimicro test-tube. For the other solvents, it will be necessary to investigate the solubility in a semimicro boiling tube; if solution does not take place in the cold, warm gently with a semimicro burner. When all the substance has dissolved, transfer to a semimicro test-tube, rinse the boiling tube with a few drops of water, and add the 'rinsings' to the solution. If in doubt as to whether the substance or a portion of the substance has dissolved, evaporate a little of the clear solution on a watch glass.

If the substance dissolves in water, proceed immediately to the test for the metal ions. If the use of dilute hydrochloric acid results in the formation of a precipitate, this may consist of the metals of Group I; the precipitate may either be filtered off and examined for Group I or else the original substance may be dissolved in dilute nitric acid. If concentrated acids are employed for dissolution, the remarks in the first paragraph must be borne in mind.

Oxides, hydroxides, free metals, and simple alloys If a solid substance is found to contain no anions, it may be an oxide, or hydroxide, or a metal or a mixture of metals, or an alloy. Metals and alloys have certain characteristic physical properties; many metals evolve hydrogen on treatment with dilute acids. As a rule, nitric acid must be employed as solvent, and it will then be necessary to remove the excess nitric acid (as already described above) before proceeding to the Group analysis.

When a suitable solvent has been found, prepare the solution for analysis using about 50 mg of the solid: the volume of the final solution should be 1–1·5 ml. Use this solution for the separation of cations into groups according to Section **VI.9**.

VI.9 SEPARATION OF CATIONS INTO GROUPS ON THE SEMIMICRO SCALE Once a solution is produced, the systematic search for cations can be started. As the first step, cations should be separated into groups; later separations within the individual groups must be carried out. When making these separations, the results of preliminary tests must always be kept in mind. If we know for sure that certain cations are present or absent, we can make appropriate simplifications in our separation procedures, which will result in considerable gain of time.

Separation of cations into groups can be carried out according to the instructions of Table VI.11. Note, that this method is suitable only if anions of organic acids, borate, fluoride, silicate, and phosphate are absent. Modifications of the separation scheme in the presence of these anions are described in Section VI.17.

---

* This is most simply estimated by weighing out 90 mg and dividing it into 6 approximately equal parts.

Table VI.11 Separation of cations into Groups on the semimicro scale (anions of organic acids, borate, fluoride, silicate, and phosphate being absent) Add 2 drops (1) of dilute HCl to 1 ml of the clear solution in a 3 ml centrifuge tube (or a 4 ml test-tube). If a ppt. forms, stir and add a further 1–2 drops to ensure complete precipitation. Centrifuge (2): wash the ppt. with a few drops of cold water (3) and add washings to centrifugate.

Residue	**Centrifugate**
The ppt. may contain $PbCl_2$ – white. Hg_2Cl_2 – white. AgCl – white. Group I (silver group) present. Examine by Group Separation Table VI.12 (Section **VI.10**)	This must not give a ppt. with 1 drop of dilute HCl. Add 4 drops of 3% H_2O_2 solution (4), and heat on a water bath for 2–3 minutes. Adjust the HCl concentration to about 0·3M (5). Pass H_2S through the warm solution until precipitation is complete (6). Centrifuge and wash (7).

Residue	**Centrifugate**
The ppt. may contain: HgS – black. PbS – black. Bi_2S_3 – black or dark-brown. CuS – black. CdS – yellow. SnS_2 – yellow. Sb_2S_3 – orange. As_2S_3 – yellow. Groups IIA and IIB (copper and arsenic groups) present. Examine by Group Separation Tables VI.13 and VI.14 (Sections **VI.11** and **VI.12**)	Transfer to a semimicro boiling tube (or crucible or beaker) and boil off H_2S (test with lead acetate paper). Add 3 drops of concentrated HNO_3 and boil to oxidise any iron(II) to iron(III) (8). The volume at this stage should be about 1 ml. Transfer to a semimicro test-tube or centrifuge tube. Add 50–100 mg of solid NH_4Cl (or 0·25–0·50 ml of 20% NH_4Cl solution), heat on a water bath, add concentrated NH_3 solution until alkaline and then 2 drops in excess. Place the tube in the hot water rack for 2–3 minutes, stir and centrifuge. Wash (9).

Residue	**Centrifugate**
The ppt. may contain: $Fe(OH)_3$ – reddish-brown. $Cr(OH)_3$ – green. $Al(OH)_3$ – white. $MnO_2.xH_2O$ – brown. Group IIIA (iron group) present. Examine by Group Separation Table VI.15 Section **VI.13)**	Add 1–2 drops of dilute NH_3 solution, warm and pass H_2S for 1 minute. Centrifuge (10) and wash (11).

Residue	**Centrifugate**
The ppt. may contain: CoS – black. NiS – black. MnS – pink. ZnS – white. Group IIIB (zinc group) present. Examine by Group Separation Table VI.16 (Section **VI.14**)	This must give no further ppt. with H_2S (10). Transfer to a small crucible and acidify with dilute acetic acid (12). Evaporate to a pasty mass [FUME CUPBOARD], allow to cool, add 5–10 drops of concentrated HNO_3 so as to wash down to the centre of the crucible most of the solid adhering to the walls. Heat cautiously until dry and then more strongly until white fumes cease to be evolved (13). Cool. Add 5 drops dilute HCl and 0·5 ml water, warm and stir: transfer the solution with the aid of 0·5 ml water, to a 3 ml centrifuge tube. If the solution is not clear, centrifuge and remove the clear solution to another tube. Add 25 mg solid NH_4Cl (or 0·25 ml 20% NH_4Cl solution), render alkaline with concentrated NH_3 solution, and then add, with stirring, 0·3 ml $(NH_4)_2CO_3$ solution. Keep and stir the mixture in a water bath at 50–60°C (14) for 2–3 minutes. Centrifuge and wash with a few drops of hot water.

Residue	**Centrifugate**
The ppt. may contain:	May contain Mg^{2+}, Na^+, and K^+ (15). Evaporate to a pasty mass in a porcelain

Table VI.11 Separation of cations into Groups on the semimicro scale (anions of organic acids, borate, fluoride, silicate, and phosphate being absent) (contd.)

$BaCO_3$ – white. $SrCO_3$ – white. $CaCO_3$ – white. Group IV (calcium group) present. Examine by Group Separation Table VI.17 or VI.18 (Section **VI.15**).	crucible [FUME CUPBOARD], add 0·5 ml concentrated HNO_3, evaporate cautiously to dryness and then heat until no more white fumes are evolved. White residue. Group V (alkali group) present. Examine by Group Separation Table VI.19 (Section **VI.16**).

Notes to Table VI.11 1. For the sake of uniformity throughout the text, a drop is intended to mean 0·05 ml – the volume of the drop delivered by the common medicine dropper. If the instructions require the addition of 0·5 ml, this quantity can either be measured out in a small measuring cylinder or from a calibrated pipette, or 10 drops can be added directly from a reagent dropper provided, of course, that a drop from the latter does not differ appreciably from 0·05 ml (see, however, Section **II.4**, 3). It is recommended that all droppers be calibrated as follows. The dropper pipette is almost filled with water by alternately compressing and releasing the bulb whilst the capillary end is dipped into a small beaker containing distilled water. The dropper is held vertically over a clean, dry measuring cylinder, the bulb gently pressed and the number of drops counted until the meniscus reaches the 2 ml mark. This process is repeated until two results are obtained which do not differ by more than two drops. A small label, stating the number of drops per ml, should be attached to the upper part of the dropper.

2. It must be remembered that in all operations with the centrifuge, the tube must be counterbalanced with another similar tube containing the same volume of water.

If the substance was completely soluble in dilute HCl, it is evident that no silver or mercury(I) salt is present. When lead is present, the solution may be clear while hot, but $PbCl_2$ is deposited on cooling the solution, due to the slight solubility of this salt in cold water. Lead may be found in Group II, even if it is not precipitated in Group I.

3. It is usually advisable in Group Separations to wash a precipitate with a small volume of a suitable wash solution and to add the washings to the centrifugate. In the present instance cold water or cold, very dilute HCl (say, 0·5M) may be used. The precipitating reagent, diluted 10–100-fold, is generally a suitable wash liquid. Specific directions for washing precipitates will usually be omitted from the present table in order to economize space.

4. The H_2O_2 solution is added to oxidize Sn^{2+} to Sn^{4+}, thus leading ultimately to the precipitation of SnS_2 instead of the somewhat gelatinous SnS. The excess H_2O_2 should preferably be decomposed by boiling before passing H_2S, other-

wise some S may be precipitated; the latter may mislead the unwary student if Group II elements are absent. The subsequent separation of Groups IIA and IIB by means of aqueous KOH is thus rendered more complete since SnS_2 dissolves completely and SnS dissolves only partially in aqueous KOH.

If it is intended to use ammonium polysulphide in the separation of Groups IIA and IIB (by an adaptation of Tables V.14 and V.18 (Sections **V.10** and **V.12**) the addition of H_2O_2 is not essential since $(NH_4)_2S_x$ will oxidize the SnS to SnS_2 and the latter dissolves as the thiostannate, SnS_3^{2-}.

5. It is important that the concentration of HCl be approximately correct, i.e. 0·3M, before passing H_2S: with higher concentrations of acid, lead, cadmium, and tin(II) will be incompletely precipitated; if the acidity is too low, sulphides of Group IIIB (NiS, CoS, and ZnS) may be precipitated. Either of two methods may be employed to adjust the acid concentration.

(*a*) Run in exactly 5·0 ml distilled water from a burette, etc., into a clean, dry conical flask of 10 ml capacity: attach a label to the latter so that the upper edge of the label is in line with the level of the water. Pour out the water.

Transfer the centrifugate from Group I to the calibrated 10 ml conical flask with the aid of a few drops of water. Add concentrated NH_3 solution dropwise (use a capillary dropper) with constant stirring, until the mixture is alkaline. (Ignore any precipitate which may form: this will dissolve when HCl is added or will be converted by the H_2S treatment into the sulphide.) Then add dilute HCl by means of a capillary dropper, with constant stirring, until the mixture is just acid (test with litmus paper by removing a drop with a micro stirring rod). Now add exactly 0·75 ml of 2M HCl (measured from a calibrated dropper or from a 1 ml graduated pipette), and dilute the solution with distilled water to the 5 ml mark.

(*b*) A simple procedure is to use the indicator methyl violet (0·1 per cent aqueous solution or, better, the purchased or prepared indicator paper). The following table gives the colour of the indicator at various concentrations of acid.

Acid concentration	*p*H	Methyl violet indicator
Neutral or alkaline	7+	Violet
0·1M HCl	1·0	Blue
0·25M HCl	0·6	Blue-green
0·33M HCl	0·5	Yellow-green
0·50M HCl	0·3	Yellow

Add 1 micro drop of methyl violet indicator solution and introduce dilute NH_3 solution with constant stirring until the colour of the solution is yellow-green. A blue-green colour is almost but not quite acid enough, yet is acceptable for most analyses: students who find detection of slight colour changes difficult may, indeed, prefer the blue-green colour change. If the indicator paper is available, the thoroughly stirred solution should be spotted with a micro stirring rod on fresh portions of the paper. It is recommended that a comparison solution containing, say, 2 ml of 0·3M HCl and 1 micro drop of indicator be freshly prepared: this will facilitate the correct adjustment of the acidity. A more satisfactory standard is a buffer solution prepared by mixing 1 ml of 2M sodium acetate and 2 ml of 2M HCl: this has a *p*H of 0·5.

6. For the passage of H_2S into the solution, the 'pressure' method detailed in Section **II.3**, 7 (see also Section **II.4**, 12) should be employed. The solution,

contained in a 10 ml conical flask or in a 4 ml test-tube, is heated (the former directly on a wire gauze, the latter in a hot water bath), a capillary delivery tube is inserted and H_2S passed in, whilst slowly shaking the vessel with a swirling motion until precipitation is complete; the latter will be apparent when bubbling either stops altogether or is reduced to a very slow rate of 1–2 bubbles per minute. Saturation is normally complete in 1–2 minutes. The best method of determining whether precipitation is complete is to centrifuge a portion of the solution and to test the centrifugate with H_2S. If only a white precipitate or a suspension of sulphur is obtained, the presence of an oxidizing agent is indicated.

If an oxidizing agent is present (e.g. a permanganate, dichromate or iron(III) ions) as is shown by the gradual separation of a fine white precipitate of sulphur and/or a change in colour of the solution, it is usual to pass SO_2 into the hot solution until reduction is complete, then to boil off the excess SO_2 (conical flask or small beaker or crucible; test with $K_2Cr_2O_7$ paper), and finally to pass H_2S. Arsenates, in particular, are slowly precipitated by H_2S: they are therefore usually reduced by SO_2 to arsenites and then precipitated as As_2S_3 with H_2S, after removal of the excess SO_2 in order to avoid interaction of the latter with H_2S and consequent separation of S. Tin(IV) compounds may be very slightly reduced to the bivalent state by this treatment; the amount of reduction is, however, so small that it may be neglected. The original solution or substance must be tested for the valence state of the arsenic.

The objection to the use of SO_2 is that some sulphuric acid may be formed, especially on boiling, and this may partially precipitate Pb, Sr, and Ba as sulphates. Any precipitate formed in this process should therefore be examined for these cations; $PbSO_4$ is soluble in ammonium acetate solution.

An alternative procedure to be used when arsenate is present, which does not possess the disadvantages associated with SO_2 is to add 0·2 ml concentrated HCl and 2 drops of 10 per cent NH_4I solution; the arsenate is thereby reduced to arsenite and upon saturation of the warm solution with H_2S under 'pressure', the arsenic is completely precipitated as As_2S_3. This reduction may be carried out after the sulphides of the other elements have been precipitated in the presence of 0·3M HCl.

7. The wash liquid is prepared by dissolving 0·1 g of NH_4NO_3 in 2 ml water and treating this solution with H_2S; about 0·2 ml will suffice for the washing. The H_2S must be present in the wash liquid to prevent oxidation of the moist sulphides to sulphates.

8. If iron was originally present in the tervalent state, it will be reduced to the iron(II) ions by H_2S. It must be oxidized to iron(III) with concentrated nitric acid (or with a few drops of saturated bromine water) to ensure complete precipitation with NH_4Cl and dilute NH_3 solution. The original solution must be tested to determine whether the iron is present as Fe^{2+} or as Fe^{3+}.

The nitric acid will simultaneously oxidize any HI, if NH_4I has been used to reduce arsenates, etc.

9. The washing may be made with a little hot water or, better, 2 per cent NH_4NO_3 solution.

10. If the centrifugate is brown or dark-coloured, Ni may be suspected. The dark-coloured solution contains colloidal NiS, which centrifuges with difficulty. It may be acidified with dilute acetic acid and boiled (crucible or semimicro boiling tube) until the NiS has coagulated: this may be added to the Group IIIB precipitate or tested separately for Ni.

11. The wash liquid may consist of 1 per cent NH_4Cl solution to which 1 per cent by volume of ammonium sulphide solution has been added. Oxidation of the moist sulphides to sulphates is thus reduced considerably.
12. The filtrate must be acidified immediately and concentrated to remove H_2S. Ammonium sulphide solution upon exposure to air slowly oxidizes to ammonium sulphate and would then precipitate any Ba or Sr present as $BaSO_4$ or $SrSO_4$. Another reason for acidifying the filtrate from Group IIIB is to prevent absorption of CO_2 from the air with the formation of carbonate ions; the latter would also precipitate the ions of Group IV.
13. The initial centrifugate from Group IIIB will be almost saturated with ammonium salts and this concentration of ammonium ions is higher than is necessary to prevent the precipitation of $Mg(OH)_2$ and it may also lead to incomplete precipitation of the carbonates of Group IV. The latter effect is due to the acidic properties of the ammonium ion:

$$NH_4^+ + CO_3^{2-} \rightleftarrows NH_3 + HCO_3^- ;$$

the concentration of CO_3^{2-} ions upon the addition of $(NH_4)_2CO_3$ would thus be considerably reduced. For these reasons most of the ammonium salts must be eliminated first.

Concentrated NHO_3 decomposes NH_4Cl at a lower temperature than is required for its volatilization:

$$NH_4^+ + HNO_3 \rightarrow N_2O\uparrow + H^+ + 2H_2O$$

Loss by decrepitation and spitting during these operations must be avoided.
14. Ammonium carbonate decomposes appreciably above 60°C

$$2NH_4^+ + CO_3^{2-} \rightarrow 2NH_3\uparrow + CO_2\uparrow + H_2O$$

The digestion also improves the filtering properties of the precipitate.
15. Owing to the slight solubility of $CaCO_3$, $SrCO_3$, and $BaCO_3$ in solutions of ammonium salts, the centrifugate from Group IV, when these metals are present, may contain minute amounts of them. Since the Group IV metals may interfere to a limited extent with the flame tests for Na and K and also the Na_2HPO_4 test for Mg (if employed), it has been recommended that the filtrate from Group IV be heated on a water bath for 2–3 minutes with half a drop each of $(NH_4)_2SO_4$ solution and $(NH_4)_2(COO)_2$ solution; any precipitate which forms is removed by centrifugation and discarded. Owing to the comparatively small concentration of ammonium salts, this is generally unnecessary if the procedure described in Table VI.19 (Section VI.16) is adopted.

VI.10 SEPARATION AND IDENTIFICATION OF GROUP I CATIONS ON THE SEMIMICRO SCALE The separation of Group I cations can be carried out according to the scheme outlined in Table VI.12.

VI.11 SEPARATION OF GROUPS IIA AND IIB AND SEPARATION AND IDENTIFICATION OF GROUP IIA CATIONS When describing these separations in macro scale, two alternative methods have been suggested (see Section **V.10**), one being based on the use of ammonium polysulphide (see Table V.14) and the other on potassium hydroxide (Table V.15). In semimicro scale the potassium hydroxide method is more suitable and this will be presented here. This does not mean however that the ammonium polysulphide

Table VI.12 Separation of Group I cations on the semimicro scale The residue may contain $PbCl_2$, AgCl, and Hg_2Cl_2. Add 1 ml hot water to the ppt., place the tube in a boiling water bath for 1–2 minutes, and stir continuously. Centrifuge rapidly; separate the solution from the residue with a capillary pipette and transfer the clear solution to a centrifuge tube.

Residue	**Centrifugate**
May contain Hg_2Cl_2 and AgCl, and also some undissolved $PbCl_2$. To remove the latter, add 1·0 ml water, place the tube in a boiling water bath for 1 minute; centrifuge and reject the supernatant liquid. Treat the residue with 0·5 ml warm dilute aqueous NH_3, stir and centrifuge.	May contain $PbCl_2$. Add 2 drops ammonium acetate solution and 1 drop of K_2CrO_4 solution. Yellow ppt. of $PbCrO_4$. Lead present.

Residue	**Centrifugate**
Black. $Hg+Hg(NH_2)Cl$. Mercury present.	May contain $[Ag(NH_3)_2]^+$. Add dilute HCl or dilute HNO_3 until acid. White ppt. of AgCl. Silver present.

For explanations consult notes to Table V.13 in Section **V.9**.

method cannot be used in semimicro scale; the student may adapt this method, by studying separation Tables V.14 and V.18 (in Sections **V.10** and **V.12** respectively) and scaling down the separation scheme to semimicro operations.

The scheme recommended for separating Groups IIA and IIB from each other, and for separations within Group IIA is outlined in Table VI.13.

Table VI.13 Separation of Groups IIA and IIB as well as separation of Group IIA cations on the semimicro scale The ppt. obtained with H_2S in the presence of dilute HCl ($\sim$0·3M) may contain the sulphides HgS, PbS, Bi_2S_3, CuS, and CdS, and also As_2S_3, Sb_2S_3, and SnS_2(1). Treat the ppt. with 1·5 ml of 2M KOH solution, and heat in a boiling water bath for 3 minutes with occasional stirring (**CAUTION**: see Note 2). Add 4 drops freshly prepared saturated H_2S water; stir and centrifuge.

Residue	**Centrifugate**
May contain HgS, PbS, Bi_2S_3, CuS, and CdS. Wash the residue once with 0·5 ml water and combine the washings with the first centrifugate. Treat the ppt. with 1–1·5 ml dilute HNO_3; place in a boiling water bath, and heat for 2–3 minutes with stirring. Centrifuge.	May contain metals of Group IIB. (See table VI.14, Section **VI.12**).

Residue	**Centrifugate**
Black: HgS. Wash with 0·5 ml water and discard the washings. Treat the ppt. with 5 drops of M NaOCl solution and 1 drop of dilute HCl. Heat on the	May contain Pb^{2+}, Bi^{3+}, Cu^{2+}, and Cd^{2+}. Add excess concentrated NH_3 solution, and centrifuge.

Residue	**Centrifugate**
May contain $Bi(OH)_3$ and $Pb(OH)_2$. Add 1 ml of NaOH solution, place in the boiling water rack for 2 minutes, and centrifuge.	May contain $[Cu(NH_3)_4]^{2+}$ and $[Cd(NH_3)_4]^{2+}$. If colourless, Cu is absent: test

Table VI.13 Separation of Groups IIA and IIB as well as separation of Group IIA cations on the semimicro scale (contd.)

	Residue	Centrifugate	
water bath for 1 minute. To the clear solution, add 1–2 drops of $SnCl_2$ solution. White ppt., turning grey or black. Hg(II) present.	May be $Bi(OH)_3$. Wash with 0·5 ml water and reject washings. Add 1 ml sodium stannite reagent (3) to the ppt. Immediate blackening of ppt. Bi present. Alternatively, dissolve a little of the ppt. in 1–2 drops dilute HNO_3. Place 1 drop of the solution upon filter paper moistened with cinchonine–potassium iodide reagent. Orange-red spot. Bi present.	May contain $[Pb(OH)_4]^{2-}$. Acidify with dilute acetic acid, add 2 drops of K_2CrO_4 solution. Yellow ppt. of $PbCrO_4$. Pb present.	directly for Cd by passing H_2S for 10 seconds into the ammoniacal solution. Yellow ppt. of CdS. Cd present. If blue, Cu is present. Divide into two unequal parts. Smaller portion (4). Acidify with acetic acid and add 1 drop of $K_4[Fe(CN)_3]$ solution. Reddish-brown ppt. on standing for 2–3 minutes. Cu present. Larger portion. Add KCN solution dropwise, with stirring, until the blue colour is discharged. Pass H_2S for 30–40 seconds. Yellow ppt. of CdS. Cd present.

Notes to Table VI.13: 1. Tin(II) sulphide is not completely soluble in 2M KOH solution. For this reason H_2O_2 is employed in the Group Separation Table VI.11 (Section **VI.9**); any tin(II) is oxidized to tin(IV) leading ultimately to SnS_2, which dissolves readily in warm 2M KOH.
2. Great care should be taken in heating the KOH mixture. The mixture should be stirred constantly with a stirring rod: the face should not be brought directly over the heated tube. Potassium hydroxide solution is a dangerous substance because of its destructive action upon the eyes.
3. The sodium stannite solution is prepared by treating 1–2 drops of $SnCl_2$ solution with NaOH solution dropwise until the initial precipitate of $Sn(OH)_2$ dissolves completely. Cooling is desirable.
4. Alternatively, acidify with dilute H_2SO_4, add a drop or two of $ZnSO_4$ solution, followed by a few drops of ammonium tetrathiocyanatomercurate(II) reagent. A violet precipitate confirms Cu. The precipitate is rendered readily visible by adding a few drops of amyl alcohol and stirring; it collects in and colours the organic layer.

VI.12 SEPARATION AND IDENTIFICATION OF GROUP IIB CATIONS ON THE SEMIMICRO SCALE For this separation the centrifugate from Group IIA is used. The separation scheme outlined in Table VI.14 is therefore linked directly to Table VI in Section **VI.11**. If the ammonium polysulphide method has been adopted for the separation of Groups IIA and IIB, the student should scale down the scheme given in Table V.18 (Section **V.12**) to semimicro scale.

Table VI.14 Separation of Group IIB cations on the semimicro scale The centrifugate from the copper group (Group IIA) may contain AsO_3^{3-}, AsS_3^{3-}, SbO_2^-, SbS_2^-, $[Sn(OH)_6]^{2-}$, SnS_3^{2-} and a little HgS_2^{2-}. Transfer to a small conical flask, add concentrated HCl dropwise and with stirring until the mixture is distinctly acid to litmus paper. Treat with H_2S for 30–60 seconds to ensure complete precipitation of the sulphides. The formation of a precipitate indicates the possible presence of HgS, As_2S_3, Sb_2S_3, and SnS_2. Centrifuge about 2·5 ml of the mixture, remove the supernatant liquid with a dropper pipette and discard it. Transfer the remainder of the mixture in the flask to the centrifuge tube, centrifuge, remove the solution and discard it: wash the residue with a little water, completely remove the washings and reject them (1). Treat the precipitate with 0·5–1 ml concentrated HCl, place in the hot water bath for 2–3 minutes and stir frequently. Centrifuge: remove the centrifugate to a 4 ml test-tube; wash residue with 0·3 ml dilute HCl and add washings to the contents of the test-tube.

<table>
<tr><th colspan="2">Residue</th><th>Centrifugate</th></tr>
<tr><td colspan="2">May contain HgS and As_2S_3. If yellow, only As_2S_3 present. Wash residue with 5 drops of water: discard washings. Treat residue with 0·5 ml dilute NH_3, stir well, and centrifuge.</td><td rowspan="3">May contain Sb^{3+} and Sn^{4+}.
Divide into two parts.
(i) Either – Render just alkaline with concentrated NH_3 solution, add 0·3 g oxalic acid, and pass H_2S for 20–30 seconds.
Orange ppt. of Sb_2S_3.
Sb present.
Or – To 2 drops of the solution on a spot plate, add a minute crystal of $NaNO_2$, stir, and add 2 drops of Rhodamine-B reagent.
Violet colouration.
Sb present.
(ii) Either – Introduce 2 cm clean iron wire or 20 mg iron filings, and heat on a water bath for 3–5 minutes. Centrifuge, and treat the clear centrifugate with 2 drops of $HgCl_2$ solution.
White ppt. of Hg_2Cl_2 or grey ppt. of Hg.
Sn present (2).
Or – Treat 0·2–0·3 ml of the solution with 5–10 mg of Mg powder, add 2 drops of $FeCl_3$ solution, 2–3 drops of 5% tartaric acid solution, 1–2 drops of dimethylglyoxime reagent, and then dil. NH_3 solution until basic.
Red colouration.
Sn present.</td></tr>
<tr><th>Residue</th><th>Centrifugate</th></tr>
<tr><td>If dark-coloured (HgS).
Hg present.
Confirm as in table VI.13, if Hg not found in Group II.A.</td><td>Add dilute HNO_3 until acid.
Yellow ppt. of As_2S_3.
As present.
Confirm thus: centrifuge and reject washings. Dissolve the ppt. in 0·5 ml of warm dilute NH_3, add 0·5 ml of 3% H_2O_2 and heat on a water bath for 3 minutes (to oxidise arsenite to arsenate). Add 4 drops of $Mg(NO_3)_2$ reagent and stir. White, crystalline ppt. of $Mg(NH_4)AsO_4$ Centrifuge and reject centrifugate. Add 2 drops of $AgNO_3$ solution and 1 drop of dilute acetic acid. Briwnish-red residue of Ag_3AsO_4.</td></tr>
</table>

Notes to Table V.14: 1. If Group IIA absent, commence here.
2. If the proportion of Sb is large, a deposit of Sb is found on the iron wire, which tends to slow down the reduction considerably. It is then better to use Mg turnings or powder for the reduction. Alternatively, the solution (containing Sb and Sn) may be added dropwise to the iron wire reacting in dilute HCl.

For further explanations the notes to Table V.19 (Section **V.12**) should be consulted.

VI.13 SEPARATION AND IDENTIFICATION OF GROUP IIIA CATIONS ON THE SEMIMICRO SCALE The precipitation of cations of Group IIIA has been described in the general separation scheme (Table VI.11, Section **VI.9**). The scheme outlined in Table VI.15 commences with the residue obtained after centrifugation.

Table VI.15 Separation of Group IIIA cations on the semimicro scale

Residue	
Wash with a little dilute aq. NH_3 and reject the washings. The ppt. may contain $Fe(OH)_3$, $Al(OH)_3$, $Cr(OH)_3$ and a little $MnO_2.xH_2O$. Transfer the ppt. to a semimicro boiling tube with the aid of 2 ml NaOH solution; add 1 ml 3 per cent H_2O_2 solution or 0·2 g sodium peroxoborate, $NaBO_3.4H_2O$. Boil gently until the evolution of O_2 ceases (about 1 minute). Transfer the mixture with the aid of a little water to a centrifuge tube. Centrifuge.	
Residue	**Centrifugate**
May contain $Fe(OH)_3$ and $MnO_2.xH_2O$. Wash with a few drops of hot water or 2 per cent NH_4NO_3 solution and add washings to *A*. Dissolve the ppt. in 0·5 ml dilute HNO_3 and, if necessary, 2 drops of 3 per cent H_2O_2 solution or 1 drop of saturated H_2SO_3 solution. Warm on water bath to decompose excess H_2O_2. Divide the solution into 2 parts. (i) Add 1 drop of $K_4[Fe(CN)_6]$ solution. Blue ppt. Fe present. The original solution or substance should be tested with $K_4[Fe(CN)_6]$ or KSCN to determine whether Fe^{2+} or Fe^{3+}. (ii) Dilute with 1 ml water, cool, add 10 mg $NaBiO_3$, shake and allow solid to settle. Violet solution of MnO_4^-. Mn present.	May contain $[Al(OH)_4]^-$ and CrO_4^{2-}: the latter is indicated by the yellow colour of the solution. Divide the solution into 2 parts. (i) Either – Acidify with dilute acetic acid and add 1 drop of lead acetate solution. Yellow ppt. of $PbCrO_4$. Cr present. Or – Acidify with dilute HNO_3. Cool: add 0·3–0·5 ml amyl alcohol and 2 drops of 3 per cent H_2O_2 solution. Shake and allow the two layers to separate. Blue colouration of chromium pentoxide in upper layer. Cr present. The blue colour does not last long as the compound is unstable. (ii) Acidify with dilute HCl (litmus paper test) and then render just basic with dilute aq. NH_3, and add 1 drop in excess. Heat on a water bath for 1 minute. White gelatinous ppt. of $Al(OH)_3$. Al present. Confirm Al thus: centrifuge, wash with a few drops of water, dissolve the ppt. in dilute HCl, add 0·3 ml ammonium acetate solution and 1 drop of the 'aluminon' reagent. Mix, allow to stand for 30 seconds, and render alkaline with ammoniacal ammonium carbonate solution. A red ppt. confirms Al.

For explanations to this scheme consult the notes attached to Table V.23 in Section **V.14**.

VI.14 SEPARATION AND IDENTIFICATION OF GROUP IIIB CATIONS ON THE SEMIMICRO SCALE The separation scheme outlined in Table VI.16 commences with the sulphide precipitates obtained according to the prescriptions of the general separation table (Table VI.11 in Section **VI.9**). It is a semimicro adaptation of the hydrochloric acid–hydrogen peroxide method, described in Table V.24 (Section **V.15**).

For explanations to the scheme in Table VI.16 consult the notes to Table V.24 in Section **V.15**.

Table VI.16 Separation of Group IIIB cations on the semimicro scale The ppt. may contain CoS, NiS, MnS, and ZnS. If it is not black, CoS and NiS are absent. Stir the ppt. in the cold with 1 ml of M HCl (1 volume of concentrated acid: 10–12 parts of water) for 1–2 minutes. Centrifuge.

<table>
<tr><th>Residue</th><th colspan="2">Centrifugate</th></tr>
<tr><td rowspan="3">If black, may contain CoS and NiS. Test residue with borax bead.
Blue bead.
Co present.
Add 10–15 drops dilute HCl and 5 drops of M NaOCl solution, stir and place in a hot water bath for 1–2 minutes. Transfer the liquid with the aid of 1 ml water to a semimicro boiling tube and boil gently to expel Cl_2. Divide the solution into 2 parts.
(i) Add 0·5–1 ml amyl alcohol and 50 mg of solid NH_4SCN, and shake.
Blue colouration in the alcohol layer.
Co present.
(ii) Add 1 drop of NH_4Cl solution, render faintly alkaline with NH_3 solution, and add 3–5 drops of dimethylglyoxime reagent.
Red ppt.
Ni present.</td><td colspan="2">May contain Mn^{2+} and Zn^{2+} (plus, possibly, traces of Co^{2+} and Ni^{2+}).
Transfer to a semimicro boiling tube, boil to expel H_2S (test with lead acetate paper), return liquid to semimicro centrifuge tube, cool, add excess NaOH solution (0·5–1 ml) and 4 drops of 3 per cent H_2O_2 solution; heat on a water bath for 3 minutes. Centrifuge.</td></tr>
<tr><th>Residue</th><th>Centrifugate</th></tr>
<tr><td>Dark-coloured. May contain $MnO_2.xH_2O$ (plus traces of $Co(OH)_3$ and $Ni(OH)_2$). Dissolve the ppt. in 0·5 ml dilute HNO_3 and a drop or two of 3 per cent H_2O_2 solution. Warm on a water bath for 2–3 minutes to decompose excess of H_2O_2; cool. Add 50 mg $NaBiO_3$, shake and allow to settle.
Purple solution of MnO_4^-.
Mn present.
Alternatively, dissolve the ppt. in 0·5 ml dilute HNO_3 with the addition of 1–2 drops of 3 per cent H_2O_2 solution. Transfer to a semimicro boiling tube with the aid of 0·5 ml water, and boil to decompose excess H_2O_2. Cool, add 0·5 ml concentrated HNO_3 and 250 mg PbO_2. Boil for 1 minute and allow to stand.
Purple solution of MnO_4^-.
Mn present.</td><td>May contain $[Zn(OH)_4]^{2-}$. Divide into 2 parts.
(i) Pass H_2S.
White ppt. of ZnS.
Zn present.
Use either test (ii) or (iia).
(ii) Just acidify with dilute H_2SO_4, add 5 drops of 0·1 per cent $CuSO_4$ solution, and 5 drops of ammonium tetrathiocyanatomercurate(II) reagent, and stir.
Violet ppt.
Zn present.
(iia) Just acidify with dilute H_2SO_4, add a drop of dilute cobalt acetate or nitrate solution, 0·5 ml ammonium reagent, and stir.
Pale-blue ppt.
Zn present.</td></tr>
</table>

VI.15 SEPARATION AND IDENTIFICATION OF GROUP IV CATIONS ON THE SEMIMICRO SCALE The Group IV cations cannot be precipitated directly from the filtrate of Group IIIB, because of the high concentration of ammonium ions in the solution. The effect of ammonium ions on the subsequent separations of Groups IV and V were described in detail in Section **V.16** and will not be repeated here. The operations described in the general Group Separation Table VI.11 (Section **VI.9**) aim at the removal of the bulk of ammonium ions before precipitating the carbonates of Group IV cations.

As in the macro scale, two methods are recommended for the separation of Group IV cations; the 'sulphate method' is the generally accepted procedure even in semimicro scale, but the 'nitrate method' is equally suitable. The two methods are outlined in Tables VI.17 and VI.18 respectively.

Table VI.17 Separation of Group IV cations with the sulphate method, on the semimicro scale The precipitate may contain $BaCO_3$, $SrCO_3$, and $CaCO_3$. Treat the ppt. with 0·5 ml dilute acetic acid and stir. Place in a hot water bath until the ppt. has dissolved. Dilute with 0·5 ml water. Test 3–4 drops of the hot solution for barium by adding a drop or two of K_2CrO_4 solution. A yellow ppt. ($BaCrO_4$) indicates Ba present.

Ba present. To the remainder of the hot solution add a slight excess of K_2CrO_4 solution (i.e. until the solution acquires an orange tint), and separate the ppt. of $BaCrO_4$ (*C*) by centrifugation. Render the centrifugate alkaline with NH_3 solution, and add excess $(NH_4)_2CO_3$ solution or, better, a little solid Na_2CO_3. Place the tube in the hot water bath. A white ppt. indicates $SrCO_3$ and/or $CaCO_3$. Centrifuge and wash with a little hot water. Dissolve the ppt. in 0·5–1 ml dilute acetic acid, and place the tube in a hot water bath to remove the excess CO_2 (solution *A*).

Ba absent. Discard the portion used in testing for barium, and employ the remainder of the solution (*B*) in testing for strontium and calcium after heating on a water bath for a few minutes to expel CO_2.

<table>
<tr><th>Residue (C)</th><th colspan="2">Solution A or Solution B</th></tr>
<tr><td rowspan="3">Yellow: $BaCrO_4$.
Wash with hot water. Dissolve in a few drops concentrated HCl, evaporate almost to dryness in a small crucible, and apply the flame test.
Green (or yellowish-green flame.
Ba present.
(Use spectroscope, if available.)</td><td colspan="2">Adjust volume to 2 ml (solution D) by evaporation or dilution as necessary.
Either – To 1 ml solution, add 1 ml saturated $(NH_4)_2SO_4$ solution followed by 0·1 g sodium thiosulphate, heat on a water bath for 5 minutes, and allow to stand for a short time. Centrifuge.
Or – To 1 ml solution, add 1 ml triethanolamine and 1 ml saturated $(NH_4)_2SO_4$ solution, heat on a water bath with stirring for for 5 minutes, and allow to stand for 1–2 minutes. Centrifuge.</td></tr>
<tr><th>Residue</th><th>Centrifugate</th></tr>
<tr><td>Largely $SrSO_4$. Wash with a little water. Stir the ppt. with 3–4 drops of water, transfer the suspension by means of a capillary dropper to 1 cm² of quantitative filter paper contained in a 5 ml crucible. Ignite until paper has charred, add 1–2 drops of concentrated HCl, and apply the flame test.
Crimson flame.
Sr present.
(Use spectroscope if available.)</td><td>May contain Ca complex. (If Sr absent, use 1 ml of solution D.) Add a few drops of $(NH_4)_2(COO)_2$ solution (1) and warm on a water bath.
White ppt. of $Ca(COO)_2$.
Ca present.
Confirm by flame test – brick-red flame.
(Use spectroscope, if available.)</td></tr>
</table>

Note to Table VI.17: 1. If triethanolamine has been employed in the separation of Sr and Ca, the addition of a little acetic acid, until faintly acid, may assist the precipitation.

See also notes to Table V.26 in Section **V.16**.

Table VI.18 Separation of Group IV cations with the nitrate method, on the semimicro scale The precipitate may contain $BaCO_3$, $SrCO_3$, and $CaCO_3$. Treat the ppt. with 0·5 ml dilute acetic acid, place in a hot water bath and stir until dissolved. Dilute with 0·5 ml water. Test 3–4 drops of the hot solution for barium by adding a drop or two of K_2CrO_4 solution. A yellow ppt. ($BaCrO_4$) indicates Ba present.

Ba present. To the remainder of the hot solution add a slight excess of K_2CrO_4 solution (i.e. until the solution just assumes an orange tint), and centrifuge the ppt. of $BaCrO_4$ (*C*). Transfer the centrifugate (solution *A*) by means of a capillary dropper to another centrifuge tube; wash the ppt. with 0·5 ml water and combine the washings with solution *A*.

Ba absent. Discard the portion used in testing for barium, and employ the remainder of the solution (*B*) in testing for strontium and calcium after heating on a water bath for a few minutes to expel CO_2.

<table>
<tr><th>Residue (C)</th><th colspan="2">Solution A or Solution B</th></tr>
<tr><td rowspan="3">Yellow: $BaCrO_4$.
Dissolve in a few drops of concentrated HCl, evaporate to dryness in a small crucible and apply the flame test. Green (or yellowish-green) flame.
Ba present.
(Use spectroscope, if available.)</td><td colspan="2">Render alkaline with NH_3 solution, and add excess $(NH_4)_2CO_3$ solution or, better, a little solid Na_2CO_3. Place the tube in a hot water bath. A white ppt. indicates the presence of $SrCO_3$ and/or $CaCO_3$. Centrifuge: discard centrifugate. Wash with 0·5 ml hot water and centrifuge; remove the supernatant liquid as completely as possible with a capillary dropper. Add 4 drops of 83 per cent HNO_3 (1), and cool in a stream of cold water from the tap. Add further 4–5 drop portions of 83 per cent HNO_3, with stirring, from a T.K. dropping bottle until 2 ml (2) are introduced. Stir for 3–4 minutes, and centrifuge.</td></tr>
<tr><td>Residue</td><td>Centrifugate (D)</td></tr>
<tr><td>White: $Sr(NO_3)_2$.
Sr present.
Confirm by flame test. Crimson flame.
(Use spectroscope, if available.)</td><td>May contain Ca^{2+}. Transfer most of the liquid to a semimicro boiling tube or a small crucible, and evaporate almost to dryness (FUME CUPBOARD). Transfer to a centrifuge tube with the aid of 0·5–1 ml water, render alkaline with NH_3 solution, and add excess $(NH_4)_2(COO)_2$ solution (3). Allow to stand in a hot water bath for 2–3 minutes.
White ppt. of $Ca(COO)_2$.
Ca present.
Centrifuge, and confirm Ca in the ppt. by the flame test – brick-red flame.
(Use spectroscope, if available.)
Alternatively, confirm Ca by the $CaSO_4.2H_2O$ (nicroscope) test (4).</td></tr>
</table>

Notes to Table VI.18: 1. The 83 per cent HNO_3 [in which $Sr(NO_3)_2$ is almost insoluble] is prepared by adding 100 g (68·0 ml) concentrated HNO_3 (sp. gr. 1·42: *c.* 70%) to 100 g (66·2 ml) fuming HNO_3 (sp. gr. 1·5: *c.* 95%).
2. The T.K. bottle, charged with 83 per cent HNO_3, should be calibrated. The acid should be added dropwise to a clean 5 ml measuring cylinder until the 2 ml mark is reached and the number of drops counted. It is advisable to place a small label on the T.K. bottle stating the number of drops per millilitre.
3. The addition of dilute acetic acid until faintly acid may assist the precipitation of the $Ca(COO)_2$.
4. Place 1 drop of the centrifugate *D* on a microscope slide and add 1 drop of dilute H_2SO_4. Concentrate by placing the slide on a micro crucible and warming gently until crystallization just commences. Examine the crystals in a microscope (magnification *c.* × 100). Bundles of needles or elongated prisms confirm Ca.

Compare also with notes to Tables V.27 in Section **V.16**.

VI.16 IDENTIFICATION OF GROUP V CATIONS ON THE SEMIMICRO SCALE Apart from a partial separation of magnesium, it is not possible to separate the Group V cations from one another; it is however simple to carry out individual tests for each of them. As we have used ammonia and ammonium salts in the previous separations, we cannot test for ammonium ions in the filtrate of Group IV cations. Test for ammonium ions is to be carried out from the original sample.

To identify Group V cations, follow the instructions of Table VI.19.

Table VI.19 Identification of Group V cations on the semimicro scale Treat the dry residue (contained in a small crucible) with 1 ml water, stir for 1 minute, and transfer with the aid of a further 0·5 ml water to a semimicro centrifuge tube. Centrifuge. (1)

Residue	Centrifugate
Dissolve in a few drops of dilute HCl, and add 1 ml water. Divide into two unequal parts; retain the smaller portion in the centrifuge tube. (i) Larger portion (2). Treat 0·25 ml of 2 per cent oxine solution in 2M acetic acid with 1 ml 2M ammonia solution. Add a little NH_4Cl to the test solution followed by the ammoniacal oxine reagent, and heat in a water bath for 1–2 minutes (the odour of NH_3 should be apparent). Pale-yellow ppt. of Mg 'oxinate'. Mg present. (ii) Smaller portion. To 3–4 drops add 2 drops of 'magneson' reagent, followed by several drops of NaOH solution until alkaline. A blue ppt. confirms Mg.	Divide into two parts (i) and (ii). (i) Add 5–10 drops of uranyl magnesium acetate reagent, shake or stir, and allow to stand for 5 minutes. Yellow crystalline ppt. Na present. Confirm by flame test: persistent yellow flame. (ii) Add 3 drops of sodium hexanitritocobaltate(III) reagent (or 5 mg of the A.R. solid) and 2 drops of dilute acetic acid, warm gently on water bath, and allow to stand for 3 minutes. Yellow ppt. of $K_3[Co(NO_2)_6]$. K present. Confirm by flame test and view through two thicknesses of cobalt glass (3): red (crimson) flame (usually transient).

Test for ammonium by placing 10 mg of the original substance with about 0·5 ml sodium hydroxide solution in a semimicro test-tube (without rim) and attach a 'filter tube' (Fig. II.30). Place a piece of red litmus paper or mercurous nitrate paper in the funnel. Warm on a water bath. Odour of ammonia; red litmus paper turns blue; mercury(I) nitrate paper turns black. Ammonium present.

Notes to Table VI.19: 1. If the residue dissolves completely (or almost completely) in water, dilute the resulting solution (after centrifugation, if necessary) to about 1·5 ml and divide it into three approximately equal parts: (i) Use the major portion to test for Mg with the prepared 'oxine' solution; confirm Mg by applying the 'magneson' test to 3–4 drops of the solution; (ii) and (iii) Test for Na and K respectively, as described in the table.

2. If it is desired to carry out the Na_2HPO_4 test for comparison with the 'oxine' test for Mg, treat the acid solution with a little $NH_4$3l, followed by dilute NH_3 solution until alkaline, add 5–6 drops of Na_2HPO_4 solution. Shake or stir the mixture and allow to stand for 3–5 minutes. A white crystalline precipitate of $Mg(NH_4)PO_4$ indicates Mg. Centrifuge and wash the precipitate with 0·3 ml water: discard the washings. Dissolve the precipitate in 5 drops dilute HCl, warming if necessary. Add 1 drop of the 'magneson' reagent, and then NaOH solution until alkaline. A blue precipitate confirms Mg.

3. See note (4) to Table V.28 in Section **V.17**.

The notes attached to Table V.28 in Section **V.17** may also be helpful.

VI.17 MODIFICATIONS OF SEPARATION PROCEDURES IN THE PRESENCE OF INTERFERING ANIONS The separation of cations, as described in Sections **VI.9** to **VI.16**, is interfered with if anions of certain organic acids, borate, fluoride, silicate, and phosphate are present. During the course of anion testing (Sections **VI.5** and **VI.6**) the presence or absence of these anions has been established. If they are present, the following procedures should be adapted (in the order as they are described):

If **borate** is found to be present, transfer the centrifugate from Group II to a small crucible, heat to expel hydrogen sulphide (do not evaporate to dryness), and allow to cool. Add 2–3 drops concentrated hydrochloric acid and 5–6 drops methanol, and heat on a water bath until the solution is almost evaporated to dryness. Repeat the addition of hydrochloric acid and methanol, and evaporate to dryness on the water bath. If borate is the only interfering ion, dissolve the residue in 2 ml of 2M hydrochloric acid, and continue the analysis for cations. The borate is volatilized as methyl borate (poisonous).

If **oxalate** or **acetate** is found to be present, add to the residue from which borate has been removed (or, if borate is absent, the centrifugate from Group II which has been evaporated almost to dryness) 1 ml concentrated hydrochloric acid and 0·5 ml concentrated nitric acid. Evaporate slowly almost to dryness: use a crucible. Allow to cool, then add 1 ml concentrated hydrochloric acid and 1 ml concentrated nitric acid, and evaporate just to dryness. Dissolve the residue in 2 ml of 2M hydrochloric acid and continue the analysis for cations. The evaporation with $HCl–HNO_3$ mixture will destroy the organic acids and simultaneously remove any **fluoride** which may be present.

If **silicate** is present, evaporate the centrifugate from Group II to dryness, add 1 ml concentrated hydrochloric acid, and evaporate to dryness. Repeat the addition of the acid and evaporation twice. Heat the residue with 2 ml 0·3M hydrochloric acid and centrifugate, and continue the analysis for cations with the solution obtained.

If **phosphate** is present, follow the prescriptions of Table VI.20.

Table VI.20 Removal of Phosphate from Solutions on the semimicro scale Place the centrifugate from Group II in a semimicro boiling tube (or small crucible), boil to expel H_2S, add 1–2 drops concentrated HNO_3 (or 5–10 drops bromine water) and boil gently for 1 minute. Transfer to a 4 ml test-tube with the aid of 0·5 ml water. Add 1 drop of NH_4Cl solution and 2–3 drops of the zirconium nitrate reagent, warm on water bath for 2 minutes and centrifuge. Test for completeness of precipitation by adding a further drop of the zirconium nitrate reagent to the centrifugate; if a ppt. forms, centrifuge again and repeat the process until the addition of 1 drop of the reagent to the clear centrifugate has no visible effect. Heat on a boiling water bath, with stirring, for 1 minute; centrifuge. Wash with a few drops of hot water.

Residue	**Centrifugate**	
Zirconium phosphate. Reject.	Add 50 mg solid NH_4Cl (or 0·25 ml of 20 per cent NH_4Cl solution), heat on a water bath, add concentrated NH_3 solution until alkaline and then 2 drops in excess. Place the tube in the boiling water bath for 3–5 minutes. Centrifuge. Wash with a few drops of hot water or 2 per cent NH_4NO_3 solution.	
	Residue	**Centrifugate**
	Examine for Group IIIA. The excess Zr will be found in the residue after treatment with H_2O_2 and NaOH solution (or with sodium peroxoborate, $NaBO_3.4H_2O$, and boiling), and will accompany any $Fe(OH)_2$, if Fe is present.	Examine for Groups IIIB, IV, and V.

For explanations consult the notes attached to Table V.20 in Section **V.13**.

VI.18 SEPARATIONS BY PAPER AND THIN LAYER CHROMATOGRAPHY. GENERAL INTRODUCTION Separations of certain cations (including some of the 'rarer' elements, discussed in Chapter VII) and anions can be carried out effectively by either paper or thin layer chromatography or both.

In paper chromatography we use filter paper, marketed for this purpose. It comes usually in the form of a 2–5 cm-wide tape, from which a strip of the necessary length can easily be cut. The more modern technique of thin layer chromatography (TLC), makes use of thin sheets of aluminium oxide, silica-gel, cellulose or some other material, supported by a metal sheet or a polymer. Chromatographic thin layers can be prepared in the laboratory from commercially available adsorbents. A thick suspension of these is made with water (usually a 2:1 w/w mixture of water: adsorbent is made up) and this is then spread on a metal plate with a suitable spreader device. Techniques vary from device to device, and the instructions of the manufacturer should be followed whenever thin layer plates are to be prepared. Ready-made thin layer sheets are also available commercially. These contain the active material spread on a plastic support. Thin-layer chromatographic materials, especially ready-made plates, are much more expensive than chromatographic paper, but normally offer faster and sharper separations than the paper. The procedures described in Section **VI.20** can be carried out both on a slow chromatographic paper (e.g. Whatman No. 1) or on a cellulose thin layer (e.g. Whatman cellulose).

In paper or thin layer chromatography a small amount of material (say, an aqueous solution containing a mixture of cations) is placed upon a limited area near the end of a strip of filter paper, or thin layer, and a solvent is allowed to diffuse from the end of the paper or thin layer by capillary action; under suitable conditions and after some time (1–30 hours), the mixture will be found to have migrated from its limited area of application and separated wholly or partially into its components as distinct zones. The zones in the form of spots or bands may be located by the application of appropriate chemical reagents to the paper or by ultraviolet fluorescence. The diffusion of the solvent and the resulting separation into spots or bands is sometimes termed the development of the chromatogram; this term is a little misleading and should not be confused when employed in the above sense with the subsequent identification process by means of which the zones are rendered clearly visible by treatment of the paper or thin layer with various reagents. We will use the phrase 'development of the chromatogram' to signify the treatment of the chromatogram after it has been formed.

To specify the position attained by a substance or ion in a chromatogram, the term R_F was introduced: this is the ratio of the distance travelled by the substance or ion to that of the solvent front, measured from the point of application of the mixture. Figure VI.2 will help to make the definition of R_F clear; in (*a*) the strip of filter paper is shown immersed in the solvent and supported by a glass rod; in (*b*) AB indicates where the spot of the solution was applied at the beginning of the experiment; in (*c*) the position of two bands (rendered visible, for example, by spraying with an appropriate chemical reagent) C and D, and also of the solvent front E, are indicated. The R_F value for the substance or ion at C is x/z and for the substance or ion at D is y/z. It is found that under comparable conditions many ions have characteristic R_F values; separation by paper chromatography is usually possible when the R_F values differ by about 0·1.

The position of AB (usually a pencilled line) is fixed by applying a drop of the solution from a capillary pipette or micro syringe along the line. The distance moved by the substance or ion is generally taken as the distance between the line AB and the 'centre of gravity' of the band; the resulting R_F value does not take into account the width of the band and also some uncertainty may be introduced if the band has diffuse boundaries.

The values of R_F are reasonably constant providing close control of all variables is maintained. It is found, however, that the relative rates of movement are constant with less rigid control, thus enabling one band on a strip to be

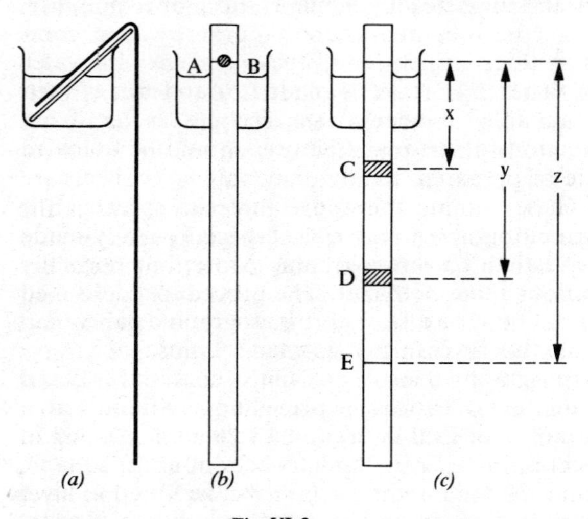

Fig. VI.2

identified by its position relative to known bands. Furthermore, with the large number of 'spot' tests available for the detection of individual inorganic ions, the necessity for an accurate knowledge of R_F values is diminished. If constancy of pure solvents, of temperature and saturation of the atmosphere is maintained, it is found that R_F values are influenced by the following factors among others:

(*a*) The presence of other ions, e.g. the presence of chloride in separations carried out with nitrate solutions.

(*b*) The acidity of the original solution; this may be due to the need for acid in the formation of a complex which is soluble in the organic solvent, to prevent hydrolysis of the salt, etc.

(*c*) The time of running of a strip; sometimes the R_F values increase with the time of running at the outset and this may correspond with a decrease in the rate of movement of the solvent front.

(*d*) The presence and concentration of other cations.

It is not proposed to deal with the theory of paper chromatography here, but it may be pointed out that an important factor is the distribution of the inorganic compound between an organic solvent and water. One way in which the distribution may be varied is by forming complexes of the cations with different organic solvents or mixtures of solvents.

The separation of metals constituting the usual groups in qualitative analysis, and also the isolation of the constituents of mixtures containing some of the less common metals, will be described. A few separations of anions will also be given in outline.

Some advantages of paper and thin layer chromatography may be mentioned:

(i) The procedure is simple and reasonably rapid.

(ii) Costly special apparatus and reagents are not necessary.

(iii) Only small quantities of materials are required.

(iv) The method has proved successful for the separation and detection of groups of metals which are difficult to deal with by routine qualitative analysis, e.g. the separation of the platinum metals from one another, of beryllium from aluminium, of scandium from the rare earths and thorium, and of hafnium from zirconium.

VI.19 APPARATUS AND TECHNIQUE FOR CHROMATOGRAPHIC SEPARATIONS In paper and thin layer chromatography we may use *ascending* or *descending* technique. In the former the solvent moves upwards, while in the latter downwards.

The most convenient method of running thin layer chromatograms is the ascending technique. A large jar or glass tank is used, which is covered with a glass sheet or closed by a cork. The solvent is poured into the bottom of the jar, or, if the vessel is too large, a small beaker or Petri-dish is filled with the solvent and placed on the bottom of the jar. The thin layer plates (2·5 cm wide and 30–45 cm long) are placed in the vessel ensuring that their lower end always reaches into the solvent. Sometimes it is advisable to line the walls of the vessel with strips of filter paper; in other cases the whole tank may be covered with a polythene bag. For the procedures described in Section **VI.20** neither lining of vessels, nor their covering is necessary.

The solvent should be put into the tank about an hour before the thin layer sheet is introduced. The plates are then run either for a fixed time (when the final position of the solvent front must be marked), or up to a fixed distance (12–35 cm), pre-marked with a pencil. The sample (2–5 μl) is taken with a micro syringe (e.g. Hamilton type) or micro pipette (Shandon) and dropped to a point on the lower end of the thin layer. The position of the point must be chosen in such a way that it will not be covered by the liquid solvent when the thin layer plate is placed in it. The location of the starting point should be clearly indicated by a pencil mark. If a wider thin layer plate is used, more chromatograms can be run simultaneously. In this case the individual samples should be placed at least 1·5 cm apart along the starting line.

The chromatograms are then allowed to run at a constant temperature, and then finally taken out of the tank and dried. A hair drier can often be used to speed up the process of drying.

The dried chromatograms can then be examined with an ultraviolet lamp or developed by spraying with suitable reagents or both.

With paper chromatography a descending technique may be more convenient (though an ascending method can of course also be used).

For routine work in qualitative analysis, the apparatus depicted in Fig. VI.3 is inexpensive and efficient. It consists of a glass gas jar or measuring cylinder (say, 50 cm by 7·5 cm), closed with a tight-fitting cork or rubber bung which carries a glass rod to which a glass boat is fused. (The glass boat, with approximate dimensions, is shown in Fig. VI.4.) The boat is suspended about 2·5 cm below the bottom of the cork, and serves as a container for the solvent employed in the separations. Two side arms are fused to the glass rod above the glass boat and these act as supports for a pair of paper strips; only one paper strip is shown in position in the diagram. In order to saturate the atmosphere in the cylinder with respect to the solvent, a layer of the latter is kept at the bottom of the glass cylinder; this will prevent evaporation effects from disturbing the equilibrium of the solvents travelling down the paper. In some separations it is also necessary to control the amount of water vapour in the atmosphere; for this purpose a beaker containing water, saturated salt solution or aqueous acid saturated with the solvent is placed on the bottom of the jar. It is advantageous to maintain fairly constant temperature conditions and for some separations the whole apparatus is kept in a thermostatically controlled ($\pm 1°$) cabinet or bath. The size of the airtight container should be commensurate with the narrow strips of filter paper ordinarily employed, otherwise with volatile solvents it is difficult to maintain the equilibrium of the solvent mixture running in the paper and which is essential for the separation.

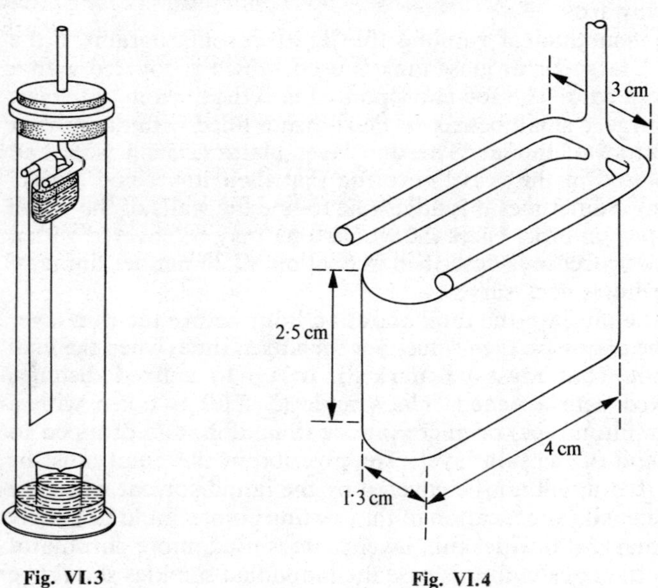

Fig. VI.3 **Fig. VI.4**

The procedure permits the separation of 1–200 μg of ions; under favourable conditions as much as 1 mg (1000 μg) of certain ions may be separated by chromotography. Strips of Whatman No. 1 filter paper, 30–45 cm long and 2·5 cm wide, are cut from the rolls supplied by the manufacturers. A light pencil line is drawn across the width of the paper about 8 cm from the upper edge; this acts as the zero line. A convenient volume (say, 0·05 ml) of the test solution, containing not more than 1 mg of the mixed metals (say, as chlorides) is applied to the centre of the zero line with the aid of a micro pipette; the strip is held horizontally until the liquid is absorbed. The spot usually forms a wet patch about 2·5 cm^2; this is generally allowed to dry completely in the air (*c.* one hour), although in some cases the degree of drying is dependent upon the chemical stability of the salts undergoing separation. The strip is then hung vertically in the cylinder with the upper end immersed in the solvent contained in the glass boat; the uper end of the strip may be held down in the glass boat by means of a glass rod, but with two strips touching in the trough, the capillary and other forces suffice to hold the strips in position without the necessity of weighting. After the solvent has diffused down the paper a sufficient distance to effect a separation (1–24 hours; this may be when the solvent front has travelled within 2·5–5 cm of the lower edge of the paper), the strip is removed from the cylinder. The position of the solvent front should be marked whilst the strip is still wet with solvent since it is sometimes difficult to locate it when dry. For some mixed solvents containing water there may be two solvent fronts, one of the dry solvent (lower area – the position of this is noted) and another wet solvent front (upper area) in which the paper is saturated with water. The solvent is then allowed to evaporate, i.e. the strips are dried by pinning or clipping them over a glass rod clamped or supported at a convenient height. The presence and position of the salts on the strip are detected by spraying with a suitable reagent. The simple type of atomizer shown in Fig. II.48, inserted in a test-tube or small flask, may be used. (If available, it is more satisfactory to employ a commercial form of atomizer operated by a constant pressure device [$\sim 7000 \text{ N m}^{-2}$].) After spraying, the coloured bands may not reach their maximum intensity until several hours have elapsed.

A large number of organic and inorganic reagents have been used. The criteria applied in selecting a reagent for paper chromatography differ from those usually employed in choosing a 'spot' test reagent. It is not necessary for the reagent to be specific for a certain ion, but it is desirable to have one which will give a test for as many ions as possible. The reagents employed include diphenylthiocarbazone (dithizone), rubeanic acid, diphenylcarbazide, alizarin, salicylaldoxime, morin, potassium hexacyanoferrate(II), potassium chromate, ammonium sulphide, and hydrogen sulphide (as the free gas). In many instances mixtures of two or more of these reagents are advantageous.

In order to obtain repeatable results, the compositions of the solvents employed must be constant. In some cases, particularly with water-miscible solvents, small variations in the water content may affect considerably the efficiency of the separation. All the solvents used must be of high purity.

Some of the solvents employed are detailed below:

(i) n-Butyl alcohol saturated with 3M hydrochloric acid. Equal volumes of the alcohol and 3M hydrochloric acid are shaken together; the upper layer is used.

(ii) Acetylacetone, saturated with water; to 7·5 ml of the saturated solvent, 0·05 ml concentrated hydrochloric acid and 2·5 ml dry acetone are added. This solvent is described as acetylacetone saturated with water and containing 0·5 per cent (v/v) of hydrochloric acid (*d* 1·18) and 25 per cent (v/v) of dry acetone.

(iii) Glacial acetic acid containing 25 per cent (v/v) dry methanol.

(iv) Acetone containing 5 per cent (v/v) water and 8 per cent (v/v) hydrochloric acid (*d* 1·18).

(v) Pyridine containing 20 per cent (v/v) water and 1 per cent (w/v) potassium thiocyanate.

(vi) Methanol.

(vii) Methyl ethyl ketone containing 30 per cent (v/v) hydrochloric acid (*d* 1·18).

(viii) Cellosolve (ethylene glycol monoethyl ether) containing 20 per cent (v/v) hydrochloric acid (*d* 1.18).

(ix) Ethyl ether containing 2 per cent (w/v) dry hydrogen chloride and 7·5 per cent (v/v) dry methanol.

(x) Methyl acetate containing 3 per cent (v/v) methanol and 10 per cent (v/v) water.

(xi) Dry n-butyl alcohol containing 40 per cent (v/v) dry methanol.

(xii) 2-Methyl-tetrahydroxyfuran (tetrahydrosylvan) containing 5 per cent (v/v) water and 10 per cent (v/v) nitric acid (*d* 1·42).

(xiii) Pyridine containing 10 per cent (v/v) water.

The following details for the purification of solvents for paper chromatography may be useful:

Acetone. Reflux with solid potassium hydroxide and potassium permanganate for one hour, and fractionate. Use the fraction, b.p. 56° ±0·5°.

Methyl ethyl ketone. As for acetone. Use fraction, b.p. 79·5°–80°.

Methyl n-propyl ketone. As for acetone. Use fraction, b.p. 99°–102°.

Methyl alcohol. Redistil the purest commercial (synthetic) alcohol. B.p. 63°.

n-Butyl alcohol. Dry over anhydrous calcium sulphate, filter, and distil. Use fraction, b.p. 116·5°–117·5°.

Pyridine. The AnalaR brand is satisfactory.

Ethyl acetate. Shake with saturated calcium chloride solution, dry with anhydrous calcium sulphate, filter, and distil. B.p. 76·5–77·5°.

Methyl acetate. As for ethyl acetate. B.p. 56·5–57·5°.

Diethyl ether. Reflux with alkaline potassium permanganate, dry with anhydrous calcium sulphate, filter, and distil. B.p. 35°. The ether is free from peroxides.

2-Methyl-tetrahydroxyfuran (tetrahydrosylvan). As for diethyl ether. B.p. 79°–80°.

VI.20 PROCEDURES FOR SELECTED CHROMATOGRAPHIC SEPARATIONS

A number of procedures for certain cations and anions are given below. First separations within cation groups are described, followed by procedures of special importance. In the latter some of the 'rarer' elements and ions are also included; the reactions of these will be described in Chapter VII. All the procedures are described for descending paper chromatography; they can be readily adapted for ascending thin-layer chromatography on cellulose, with considerably shorter running times. Once these separations have been attempted, the student may study the literature for other, specific procedures.

1. Group I. Ag, Pb, and Hg The metals are present as nitrates in dilute nitric acid. The solution is spotted upon paper and allowed to dry in air for one hour. The solvent consists of n-butyl alcohol, mixed with 5 per cent (v/v) glacial acetic acid, followed by water to turbidity. The separation (elution) is allowed to proceed for 12–16 hours in an atmosphere saturated with the solvent. The strip is removed from the extraction vessel, dried in air, and then sprayed with a 0·05 per cent solution of dithizone in chloroform.

	R_F values	Colour of band
Pb	0·08	Rose pink
Ag	0·16	Orange
Hg(I)	0·85	Pink

The paper chromatogram may also be developed by exposure to hydrogen sulphide gas, but the results are not as satisfactory as with dithizone.

2. ***Group IIA*. *Hg, Pb, Bi, Cu, and Cd*** The metals are present as chlorides in dilute hydrochloric acid. The solvent used is n-butyl alcohol saturated with 3M hydrochloric acid. To obtain a good separation of copper and lead, the strip should be at least 45 cm long. The solvent is allowed to flow for 15–18 hours in an atmosphere saturated with respect to both organic solvent and aqueous phases. After evaporation of the solvent, the strip is sprayed with a solution of dithizone in chloroform (0·05 per cent w/v). Lead gives only a weak colour with dithizone and is best detected by spraying the top portion of the strip with an aqueous solution of rhodizonic acid.

	R_F values	Reagent	Colour of band
Cu	0·20	Dithizone	Purple-brown
Pb	0·27	Rhodizonic acid	Bright blue
Bi	0·59	Dithizone	Purple
Cd	0·77	Dithizone	Purple
Hg(II)	0·81	Dithizone	Pink

A chromatogram (drawn from an actual photograph) is shown in Fig. VI.5*a*.

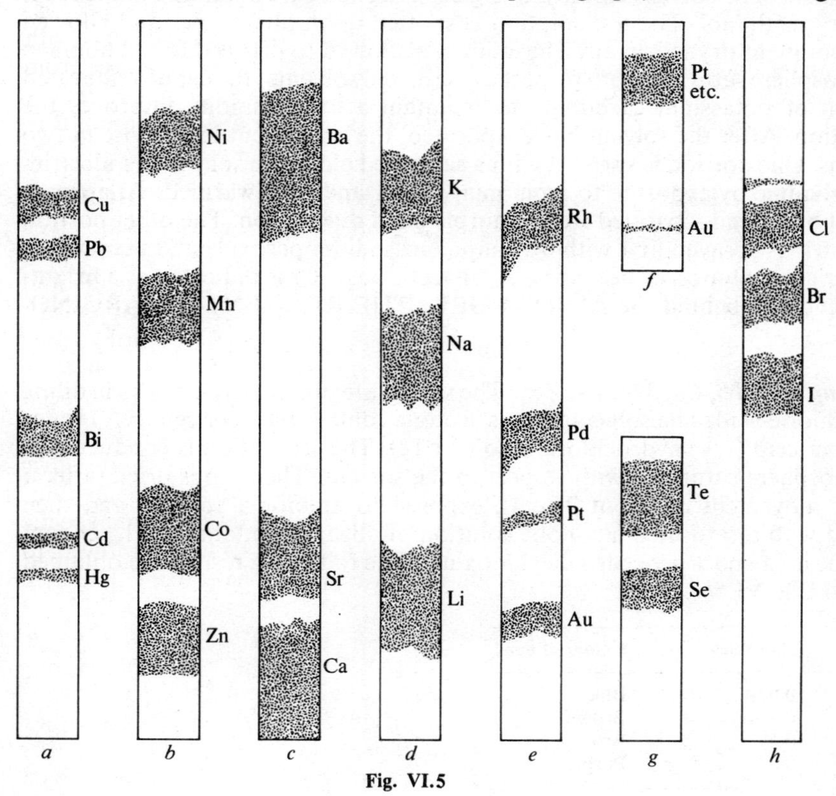

Fig. VI.5

3. Group IIB. As, Sb, and Sn The separation of a mixture of the three elements is a difficult operation. The metals are present as their lower chlorides in dilute (2–4M) hydrochloric acid. The solution is spotted on paper and allowed to dry in air for 15 minutes. The solvent consists of 7·5 ml acetylacetone (b.p. 137°–141°) saturated with water and treated with 0·05 ml concentrated hydrochloric acid and 2·5 ml acetone (sufficient of the last-named to give a clear solution). The separation is allowed to proceed for 1 hour in an atmosphere saturated with respect to a saturated solution of acetylacetone in water; the solvent movement is about 15 cm. The complexes formed are very stable, particularly that with tin (R_F = 1). The strip is removed from the extraction vessel, the solvent is allowed to evaporate for several minutes, and the strip is sprayed (before it is completely dry) with a chloroform solution of dithizone (0·005 per cent w/v), and then allowed to dry thoroughly. The tin is found in the solvent front.

	R_F values	Colour of band
As	0·2	Yellow
Sb	0·5	Red
Sn	1·0	Purple

4. Group IIIA. Fe, Al, and Cr The main difficulty here is the separation of the aluminium and chromium. The metals are present as chlorides in 5M hydrochloric acid. The solvent employed is glacial acetic acid containing 25 per cent (v/v) dry methanol. The test solution is spotted upon filter paper and allowed to evaporate to dryness in air. The elution is allowed to proceed for 12 hours in an atmosphere saturated with respect to the mixed solvents; the use of a saturated solution of potassium carbonate to maintain a low humidity improves the separation. After the solvent has evaporated, the strip is cut lengthwise in two portions. One portion is sprayed with a saturated alcoholic solution of alizarin, made alkaline by exposure to ammonia vapour and then warmed; Al appears as a red band well separated from a purple band due to iron. The other portion of the strip is sprayed first with 0·5M aqueous sodium peroxide and then with a 0·05 per cent solution of benzidine in 2M acetic acid: Cr is indicated as a bright-blue band just behind the Al. (DANGER: THE REAGENT IS CARCINOGENIC.)

5. Group IIIB. Ni, Co, Mn, and Zn The metals are present as chlorides in dilute hydrochloric acid. The solvent used is acetone containing 5 per cent (v/v) water and 8 per cent (v/v) hydrochloric acid (*d* 1.18). The separation is conducted in an atmosphere saturated with respect to the solvent. The strip is dried (after a solvent movement of about 25 cm), exposed to ammonia vapour, and then sprayed with a saturated alcoholic solution of alizarin containing 0·1 per cent rubeanic acid and 1 per cent salicyladoxime. The following results are obtained (cf. also Fig. VI.5*b*).

	R_F values	Colour of band
Ni	0·07	Blue
Mn	0·3	Brown
Co	0·6	Brown
Zn	0·9	Purple

6. Group IV. Ca, Sr, and Ba This is a difficult separation. The metals are present as chlorides. The solvent used is pyridine containing 20 per cent (v/v) water and 1 per cent (w/v) potassium thiocyanate. The elution is allowed to proceed for 5–6 hours in an atmosphere saturated with respect to pyridine and also with a relative humidity of between 65 and 80 per cent; the latter is attained by the use of a saturated aqueous solution of ammonium nitrate or chloride. The solvent front moves about 20 cm. The strip is dried, and then sprayed with a saturated alcoholic solution of alizarin to identify Ca, and with 0·5 per cent (w/v) aqueous sodium rhodizonate (freshly prepared) to reveal Ba and Sr. The R_F values are: Ca, 0·95; Sr, 0·75; Ba, 0·15 (see Fig. VI.5*c*).

7. Group V. Na, K, and Li The metals are present as chlorides in neutral solution. The solvent used is methanol. The position of the alkali metal chloride bands is detected by spraying with 0·1M silver nitrate and saturated fluorescein in 50 per cent alcohol, and then drying the strip. The R_F values are: Li, 0·8; Na, 0·5; K, 0·1 (see Fig. VI.5*d*).

8. Alkaline earths and alkali metals. Be, Ca, Sr, Ba, Mg; Na, K, and Li The experimental details given below permit the separation of Be, Mg, Ca, Sr, and Ba; Mg, Na, and K; Mg, Ca, Na, and K; and of Li, Na, and K. The metals are present as acetates in dilute acetic acid solution. The solvent is a mixture of 80 per cent ethanol and 20 per cent (v/v) 2M acetic acid. The elution is allowed to proceed for about 24 hours. The paper chromatogram, after drying in an air oven at 60°C, is developed by spraying with 0·1 molar aqueous solution of violuric acid (5-isonitroso-barbituric acid) and then dried at 60°C. The R_F values and the colours of the bands are collected below.

Metal	R_F values	Colour of band
Be	0·86	Yellowish-green
Mg	0·76	Yellowish-red
Ca	0·68	Orange
Sr	0·55	Reddish-violet
Ba	0·43	Pale red
Li	0·76	Reddish-violet
Na	0·56	Reddish-violet
K	0·45	Violet

9. Separation of halides. F, Cl, Br, and I The anions are present as their sodium salts. The solvent used is either pyridine containing 10 per cent (v/v) water, or acetone containing 20 per cent (v/v) water. The elution is conducted during 1·5–2 hours in an atmosphere saturated with solvent vapour. After evaporation of the solvent from the strip, the latter is cut longitudinally. One part is sprayed with 0·1M silver nitrate and saturated fluorescein in 50 per cent alcohol, and the other is sprayed with the zirconium–alizarin reagent. Upon drying, the halides are clearly visible as characteristic bands, which when viewed under ultraviolet light appear as dark areas on a fluorescent background. The R_F values are as follows (see also Fig. VI.5*h*):

Ion	Pyridine – 10% H_2O	Acetone – 20% H_2O
Fluoride	0·00	0·25
Chloride	0·24	0·51
Bromide	0·47	0·62
Iodide	0·71	0·77

10. Separations of various anions Using a solvent mixture composed of n-butyl alcohol, pyridine, and 1·5M ammonia solution in the proportions of 2:1:2, the following R_F values are obtained for the sodium or potassium salts: chloride, 0·24; bromide, 0·36, iodide, 0·47; chlorate, 0·42, bromate, 0·25, iodate, 0·09; nitrite, 0·25; nitrate, 0·40; arsenite, 0·19: arsenate, 0·05; phosphate, 0·04; and thiocyanate, 0·56. The positions of the anions may be detected by spraying with ammoniacal silver nitrate; potassium iodide and hydrochloric acid are particularly effective for chlorates, bromates, and iodates. The R_F values provide the basis for the separation of a number of mixtures of anions, e.g. chloride and iodide, bromide or iodide and nitrate.

11. Aluminium and beryllium The metals are present as chlorides in dilute hydrochloric acid. The solvent consists of 80 per cent n-butyl alcohol and 20 per cent hydrochloric acid. After elution overnight, the solvent front moves about 20 cm. The positions of the aluminium and beryllium bands are shown by spraying with an alcoholic solution of 8-hydroxyquinoline. The bands fluoresce under ultraviolet light; the fluorescence intensifies upon drying the paper and exposing to ammonia gas. The R_F values are: Al, 0·03; Be, about 0·3.

This procedure permits the detection of 2 μg of Al in the presence of 300 μg of Be, and vice versa; traces of Be can thus be detected in Al salts and traces of Al in Be salts.

12. Separation of mercury from other metals The procedure enables 1 μg of Hg to be identified in the presence of a large excess of Pb, Cu, Bi, Cd, As, Sb, Fe, Al, Cr, Ni, Co, Mn, and Zn. The metals are present as chlorides. The solvent used is methyl acetate containing 3 per cent (v/v) methanol and 10 per cent (v/v) water.

The test solution should not contain more than 5 per cent (v/v) concentrated hydrochloric acid and must have a $pH < 2$. It is spotted on a paper strip and allowed to evaporate for 10–15 minutes. Diffusion of the solvent takes place in an atmosphere saturated with respect to the vapour of a saturated solution of methyl acetate in water, and the temperature is maintained constant at 22°. The solvent moves sufficiently far in 20–30 minutes to effect a complete separation. After evaporation of the solvent, the strip is made alkaline by exposure to ammonia vapour and then sprayed with a 1 per cent solution of diphenylcarbazide in alcohol. Mercury is indicated by a narrow blue band in the dry solvent front.

13. The platinum metals: Pt, Pd, Rh, Ir, Ru, Os, and Au The metals are present as their chlorides or sodium chloro-salts. The solvent used is methyl ethyl ketone containing 30 per cent (v/v) hydrochloric acid (*d* 1·18); it is freshly prepared. The solution is spotted on the paper strip and allowed to dry thoroughly in the air. The atmosphere in the separation vessel is saturated with respect to the solvent. For the detection of Pt, Pd, Au, and Rh, 0·5M tin(II) chloride in dilute hydrochloric acid is a suitable developing reagent; a mixture of tin(II) chloride and potassium iodide solution is somewhat more sensitive. Iridium is reduced to the colourless iridium trichloride by the tin(II) chloride, but can be reoxidized to the brown tetrachloride by chlorine water. Ruthenium and osmium are detected by spraying with a solution of thiourea in 5M hydrochloric acid; warming is necessary to develop the colours fully.

The R_F values are: Ru, 0·08; Rh+Ir, 0·10; Pd, 0·60; Pt, 0·80; Au, 0·95. The separation of the four elements Rh, Pd, Pt, and Au is comparatively easy (see also Fig. VI.5*e*).

14. Gold from the platinum metals The metals are present as chlorides in hydrochloric acid (concentration $\gg$ 2M). The solvent is diethyl ether containing 2 per cent (w/v) dry hydrogen chloride and 7·5 per cent (v/v) dry methanol. The solution is spotted upon filter paper and the strip dried for one hour only. The separation is conducted in an atmosphere of solvent vapour. The solvent mixture extracts gold in a narrow band in the solvent front, while the platinum metals remain in the original spot; the developing reagent is 0·25M tin(II) chloride in dilute hydrochloric acid. It is easy to detect 1 μg of gold in the presence of more than 100 times the quantity of platinum metals (see also Fig. VI.5*f*).

15. Selenium and tellurium The elements are present as selenite and tellurite in dilute nitric acid solution. The mixture is spotted upon paper and dried thoroughly in the air. The solvent is dry n-butyl alcohol containing 4 per cent (v/v) of dry methanol. The atmosphere in the separation vessel is saturated with respect to the solvent vapour and the relative humidity is also maintained at 50 per cent by means of a saturated solution of calcium nitrate. The solvent is allowed to diffuse 8–10 cm down the strip (*c.* 2 hours). After evaporation of the solvent, the strip is sprayed with 0·5M tin(II) chloride in dilute hydrochloric acid. The tellurium is indicated by a black band (R_F 0·1) and the selenium as an orange band (R_F 0·5). It is possible to detect 1–5 μg of Se in the presence of 1 mg of Te by this method (see also Fig. VI.5*g*).

16. Scandium, thorium and the rare earths The metals are present as nitrates in dilute nitric acid solution. The solvent is 2-methyltetrahydroxyfuran (tetrahydrosylvan) containing 5 per cent (v/v) water and 10 per cent (v/v) nitric acid (*d.* 1·42). The mixture is spotted upon paper and dried thoroughly in the air. The relative humidity inside the extraction vessel is maintained at 80 per cent by means of a saturated solution of ammonium chloride. After the solvent has diffused about 15 cm down the solvent strip, it is allowed to evaporate, and the strip is placed for about 10 minutes in an atmosphere of ammonia vapour. The paper chromatogram is then sprayed with an alcoholic solution of alizarin and finally with 2M acetic acid. The following results are obtained.

	R_F	Colour of band
Th	0·96	Violet-blue
Sc	0·17	Violet
Rare earths	0·00	Violet-blue

Scandium can also be separated from the rare earths by using methyl acetate containing 10 per cent (v/v) water and 5 per cent (v/v) nitric acid (*d* 1·42) as solvent. The strip is subjected to a solvent run of 25 cm. Scandium is then found in a narrow strip (R_F 0·17), but thorium forms a more diffuse band.

17. Uranium in mineral ores An approximately 10 per cent solution of the sample containing 25–50 per cent free nitric acid (by volume) is prepared by a suitable method (e.g. a potassium hydroxide or sodium peroxide fusion, followed by treatment with nitric acid). A portion of the sample (0·05 ml) is spotted upon filter paper in the usual manner and allowed to dry in the air. The solvent consists of 2-methyltetrahydroxyfuran (tetrahydrosylvan) saturated with water and to which sufficient concentrated nitric acid has been added to give a 2·5–10 per cent concentration of nitric acid. The paper strip is removed when the solvent front has moved 5–7 cm beyond the test patch, and dried in a current of warm air. The uranium moves in a narrow band near the solvent front. The paper is sprayed with 1 per cent potassium hexacyanoferrate(II) solution. A brown stain appears in the presence of uranium. The quantity of uranium may be estimated by comparison with standard stains prepared with known amounts (0·1–200 μg) of uranium.

18. Zirconium and hafnium This separation must rank as one of the major achievements of inorganic paper chromatography. The metals are present as zirconyl and hafnyl nitrates; basic nitrates must be absent since these are immobile. The mixture is best prepared by digesting the sample at 80°C with concentrated nitric acid and evaporating the excess acid at the same temperature under reduced pressure; the product gives a clear solution when dissolved in water.

The solvent is prepared by adding slowly 30 ml concentrated nitric acid (*d* 1·42) to 70 ml dichlorotriethylene glycol; the mixture is stable for 2–3 days at room temperature. A total of 150 μg of mixed oxides can be handled with a Whatman No. 1 paper strip, 3 cm wide. The solution (0·02 ml) is applied to the paper and, without drying, is immediately transferred to the extraction vessel. The chromatogram is allowed to run for 18 hours; the solvent moves about 3 cm per hour, and drips from the end of the strip at the end of the period. The wet strip is sprayed with a saturated solution of alizarin in ethanol containing 5 per cent (v/v) 2M hydrochloric acid. The strip is then heated gently, care being taken that it never becomes completely dry; it is resprayed with the reagent, if necessary. The characteristic red lines of the zirconium and hafnium lakes appear slowly against the yellow background of the reagent. Narrow bands are obtained with R_F values of 0·1 for Hf and 0·2 for Zr. It is possible to detect 2 μg of each metal, and also the presence of about 2 per cent of hafnium in commercial zirconium nitrate.

CHAPTER VII REACTIONS OF SOME LESS COMMON IONS

VII.1 INTRODUCTION In the previous chapters the discussions were restricted to those cations and anions which occur most often in ordinary samples. Having studied the reactions, separation, and identification of those ions, the student should now concentrate on the so-called 'rarer' elements. Many of these, like tungsten, molybdenum, titanium, vanadium, and beryllium, have important industrial applications.

The term 'rarer' elements as originally employed in the sense of their comparative rare occurrence and limited availability must now, in a number of cases, be regarded as a misnomer. Large quantities of some of these elements are utilized annually, and the range of their application is slowly but surely widening. A few examples may be mentioned: the use of molybdenum, tungsten, titanium, and beryllium in the steel industry, of tungsten in the manufacture of incandescent lamps, and of titanium and uranium in the paint industry. The interpretation of the term 'rarer' elements, as applied to the elements described in this chapter, is perhaps best accepted in the sense of their comparatively rare occurrence in routine qualitative analysis.

No attempt has been made to give more than a short introduction to the subject; to economize in space, most of the simple equations have been omitted. The elements have been classified, in so far as is possible, in the simple groups with which the student is already familiar, and methods of separation have been briefly indicated. Thus thallium and tungsten are in Group I; molybdenum, gold, platinum, selenium, tellurium, and vanadium in Group II; and beryllium, titanium, uranium, thorium, and cerium in Group III. The presence of vanadium will be revealed by the blue colour and absence of precipitate produced by hydrogen sulphide in acid solution; its actual isolation as sulphide is effected by the addition of acid to the ammonium sulphide solution in Group IIIB. It is hoped that the subject- matter of this chapter will suffice to enable the student to detect the presence of one or two of the less common ions.

VII.2 THALLIUM, Tl (A_r: 204·34) – THALLIUM(I) Thallium is a heavy metal with characteristics reminiscent of lead. It melts at 302·3°C. Thallium metal can be dissolved readily in nitric acid; it is insoluble in hydrochloric acid.

Thallium forms the monovalent thallium(I) and trivalent thallium(III) ions, the former being of greater analytical importance. Thallium(III) ions are less frequently encountered in solutions, as they tend to hydrolyse in aqueous solution, forming thallium(III) hydroxide precipitate. Thallium(I) ions can be oxidized to thallium(III) ions in acid media with permanganate and hexacyanoferrate(III) ions as well as with lead dioxide, chlorine gas, bromine water or aqua regia (but not with concentrated nitric acid). The reduction of thallium(III) ions to thallium(I) is easily effected by tin(II) chloride, sulphurous acid, iron(II) ions, hydroxylamine or ascorbic acid.

Reactions of thallium(I) ions To study these reactions a 0·025M solution of thallium(I) sulphate, Tl_2SO_4, or a 0·05M solution of thallium(I) nitrate, $TlNO_3$, should be used. All these compounds are **HIGHLY POISONOUS**.

1. Dilute hydrochloric acid: white precipitate of thallium(I) chloride, TlCl, sparingly soluble in cold, but more soluble in hot water (compare lead).

2. Potassium iodide solution: yellow precipitate of thallium(I) iodide, TlI, almost insoluble in water; it is also insoluble in cold sodium thiosulphate solution (difference and method of separation from lead).

The spot-test technique is as follows. Place a drop of the faintly acid test solution on a black spot plate or upon a blackened watch glass. Add a drop of 10 per cent potassium iodide solution and, when a precipitate appears, a drop or two of 2 per cent sodium thiosulphate solution. A yellow precipitate is produced.

Sensitivity: 0·6 μg Tl. *Concentration limit:* 1 in 80,000.

The potassium iodide removes mercury as tetraiodomercurate(II) $[HgI_4]^{2-}$, whilst the sodium thiosulphate dissolves lead and silver as complex thiosulphates.

3. Potassium chromate solution: yellow precipitate of thallium(I) chromate, Tl_2CrO_4, insoluble in cold, dilute nitric or sulphuric acid.

4. Hydrogen sulphide: no precipitate in the presence of dilute mineral acid. Incomplete precipitation of black thallium(I) sulphide, Tl_2S, occurs in neutral or acetic acid solution.

5. Ammonium sulphide solution: black precipitate of thallium(I) sulphide, Tl_2S, soluble in mineral acids. The precipitate is oxidized to thallium(I) sulphate, Tl_2SO_4, upon exposure to air.

Owing to the slight solubility of thallium(I) chloride, some of the thallium is also precipitated in Group IIIB (compare lead). It is often, however, precipitated with Group II ions.

6. Sodium hexanitritocobaltate(III) solution: light-red precipitate of thallium(I) hexanitritocobaltate(III), $Tl_3[Co(NO_2)_6]$.

7. Chloroplatinic acid solution: pale-yellow precipitate of thallium(I) hexachloroplatinate(IV) $Tl_2[PtCl_6]$, almost insoluble in water (solubility: 0·06 g ℓ$^{-1}$ at 15°C).

8. Potassium hexacyanoferrate(III) solution: brown precipitate of thallium(III) hydroxide in alkaline solution:

$$Tl^+ + 2[Fe(CN)_6]^{3-} + 3OH^- \rightarrow Tl(OH)_3\downarrow + 2[Fe(CN)_6]^{4-}.$$

A similar result is obtained with sodium hypochlorite, sodium hypobromite or hydrogen peroxide in alkaline solution.

9. Ammonium thiocyanate solution: white precipitate of thallium(I) thiocyanate, TlSCN; the precipitate is dissolved in hot water.

10. Flame test. All thallium salts exhibit a characteristic green colouration when introduced into the colourless Bunsen flame. When examined through the spectroscope only one sharp line can be seen, at 535 nm, in contrast to barium, which exhibits several lines between 510 and 550 nm (cf. Fig. II.5 in Section **II.2**).

VII.3 THALLIUM, Tl (A_r: 204·34) – THALLIUM(III) The general physical and chemical properties of thallium were discussed in Section **VII.2**.

Reactions of thallium(III) ions To study these reactions use a 0·2M solution of thallium(III) chloride, $TlCl_3$.

1. Sodium hydroxide or ammonia solution: brown precipitate of thallium(III) hydroxide, insoluble in excess of the reagent (difference from thallium(I) salts, which give no precipitate), but readily soluble in hydrochloric acid.

2. Hydrochloric acid: no precipitate (difference from thallium(I) salts).

3. Potassium chromate solution: no precipitate (difference from thallium(I) salts).

4. Potassium iodide solution: brownish-black precipitate, probably a mixture of thallium(I) iodide and iodine.

5. Hydrogen sulphide: reduced to thallium(I) with the precipitation of sulphur. If the acid is neutralized, thallium(I) sulphide, Tl_2S, precipitates.

6. Flame test See Section **VII.2**, reaction 10.

Separation. The element may be precipitated as TlCl in Group I, separated from AgCl and Hg_2Cl_2 by solution in boiling water, and from $PbCl_2$ by means of dilute H_2SO_4; the thallium can then be precipitated as TlI by the addition of KI solution. It may also be separated from lead by precipitation as the iodide and treatment with $Na_2S_2O_3$ solution, in which PbI_2 alone is soluble. Useful confirmatory tests are reaction 6 (in Section **VII.2**) and the flame test.

Any thallium not precipitated in Group I as TlCl, will ultimately be found in the precipitate of Group IIIB. It is best to dissolve the Group IIIB precipitate in dilute nitric acid, boil to expel H_2S, add a few drops of sulphurous acid and boil to expel excess SO_2. The resulting solution is then poured into excess sodium carbonate solution when the carbonates of cobalt, nickel, manganese, and zinc are precipitated. The thallium remains in solution as thallium(I) carbonate and can be precipitated as thallium(I) sulphide, Tl_2S, by the addition of ammonium sulphide solution.

VII.4 TUNGSTEN, W (A_r: 183·85) – TUNGSTATE Solid tungsten is a white-coloured metal; the powdered metal is grey. Its melting point is extremely high (3370°C). The metal is insoluble in acids, including aqua regia. To dissolve metallic tungsten, it should be ignited first in a stream of oxygen, and the tungsten trioxide, WO_3, which is formed, can then be fused with solid sodium hydroxide in an iron crucible. The solidified melt will then dissolve in water when tungstate ions, WO_4^{2-}, are formed.

Tungstates form complex acids with phosphoric, boric, and silicic acids; tungstic acid cannot therefore be precipitated from these compounds by hydrochloric acid. The complexes may usually be decomposed by heating with concentrated sulphuric acid, tungstic acid being liberated.

Reactions of tungstate ions To study these reactions use a 0·2M aqueous solution of sodium tungstate $Na_2WO_4.2H_2O$.

1. Dilute hydrochloric acid: white precipitate of hydrated tungstic acid, $H_2WO_4.H_2O$, in the cold; upon boiling the mixture, this is converted into yellow tungstic acid, H_2WO_4, insoluble in dilute acids. Similar results are obtained with dilute nitric and sulphuric acids, but not with phosphoric acid. Tartrates, citrates, and oxalates inhibit the precipitation of tungstic acid. The precipitate is soluble in dilute ammonia solution (distinction from $SiO_2.xH_2O$).

2. Phosphoric acid: white precipitate of phosphotungstic acid, $H_3[PO_4(W_{12}O_{36})].5H_2O$, soluble in excess of the reagent.

3. Hydrogen sulphide: no precipitate in acid solution.

4. Ammonium sulphide solution: no precipitate, but if the solution is afterwards acidified with dilute hydrochloric acid, a brown precipitate of tungsten trisulphide, WS_3, is produced. The precipitate dissolves in ammonium sulphide solution forming a thiotungstate ion, WS_4^{2-}.

5. Zinc and hydrochloric acid If a solution of a tungstate is treated with hydrochloric acid and then a little zinc added, a blue colouration or precipitate is produced. The product is called 'tungsten blue' and has a composition near to the formula W_2O_5.

6. Tin(II) chloride solution: yellow precipitate, which becomes blue upon warming with concentrated hydrochloric acid.

The spot-test technique is as follows. Mix 1–2 drops of the test solution with 3–5 drops of the tin(II) chloride reagent on a spot plate. A blue precipitate or colouration of tungsten blue W_2O_5.

Sensitivity: 5 μg W. *Concentration limit:* 1 in 10,000.

Molybdenum gives a similar reaction. If, however, a thiocyanate is added, the red complex ion $[Mo(SCN)_6]^{3-}$ is formed, and upon the addition of concentrated hydrochloric acid the red colour disappears and the blue colour due to tungsten remains.

The spot test is conducted as follows in the presence of molybdenum.

Place a drop of concentrated hydrochloric acid upon filter or drop-reaction paper and a drop of the test solution in the centre of the spot. A tungstate produces a yellow stain. Add a drop of 10 per cent potassium thiocyanate solution and a drop of saturated tin(II) chloride; a red spot, due to $[Mo(SCN)_6]^{3-}$, is produced, but this disappears when a drop of concentrated hydrochloric acid is added and a blue colour, due to tungsten blue remains.

Sensitivity: 4 μg W. *Concentration limit:* 1 in 12,000.

7. *Iron(II) sulphate solution:* brown precipitate. This turns white upon adding dilute hydrochloric acid, and then yellow upon heating (difference from molybdates).

8. *Silver nitrate solution:* pale yellow precipitate of silver tungstate, soluble in ammonia solution, decomposed by nitric acid with the formation of white hydrated tungstic acid.

9. *$KHSO_4$–H_2SO_4–phenol test (Defacqz reaction)* A little of the solid (or the residue obtained by evaporating a little of the solution to dryness) is heated with 4–5 times its weight of potassium hydrogen sulphate slowly to fusion, and the temperature is maintained until the fluid melt is clear. The cold melt is stirred with concentrated sulphuric acid. Upon adding a few milligrams of phenol to a few drops of the sulphuric acid solution, an intense red colouration is produced (difference from molybdate). A reddish-violet colouration is obtained if hydroquinone replaces the phenol. The test is a highly sensitive one and will detect 2 μg of tungstate.

Heat a few milligrams of the solid unknown with 10–20 mg of potassium hydrogen sulphate and 2 drops of concentrated sulphuric acid in a small porcelain crucible, allow to cool, and add a few milligrams of solid phenol. A red colouration is produced.

10. *Dry test* Microcosmic salt bead: oxidizing flame – colourless or pale-yellow; reducing flame – blue, changing to blood-red upon the addition of a little iron(II) sulphate.

Separation. Tungsten is precipitated in Group I and is associated with Ag in the ammoniacal filtrate of the group separation. The filtrate is almost neutralized with dilute HCl (any precipitate formed being kept in solution by the addition of ammonia solution), and the silver precipitated as AgI by the addition of KI solution. The filtrate is concentrated, acidified with dilute HCl, and reactions 5, 6 or 9 applied.

VII.5 SEPARATION AND IDENTIFICATION OF GROUP I CATIONS IN THE PRESENCE OF THALLIUM AND TUNGSTEN The separation and identification of Hg(I), Ag, W, Pb, and Tl(I) is described in Table VII.1.

VII.6 MOLYBDENUM, Mo (A_r: 95·94) – MOLYBDATE Molybdenum is a silverish-white, hard, heavy metal. In powder form it is grey. It melts at 2622°C. The metal is resistant to alkalis and hydrochloric acid. Dilute nitric acid dissolves it slowly, concentrated nitric acid renders it passive. Molybdenum can readily be dissolved in aqua regia or in a mixture of concentrated nitric acid and hydrogen fluoride.

Molybdenum forms compounds with oxidation numbers +2, +3, +4, +5, and +6. Of these molybdates are the most important (with oxidation number +6). Molybdates are the salts of molybdic acid, H_2MoO_4. This acid tends to polymerize with the splitting off of molecules of water. Thus, the commercial

Table VII.1 Separation and identification of Group I cations in the presence of Tl and W
The precipitate may contain $PbCl_2$, AgCl, Hg_2Cl_2, TlCl, and tungstic acid ($WO_3.xH_2O$). Wash the ppt. on the filter with 2 ml portions of 2M HCl, then 2–3 times with 1 ml portions of cold water, and reject the washings. Transfer the ppt.* to a boiling tube or to a small beaker, and boil with 10–15 ml water. Filter hot.

Residue	Filtrate
May contain Hg_2Cl_2, AgCl, and tungstic acid. Wash the ppt. several times with hot water until the washings give no ppt. with K_2CrO solution; this ensures the complete removal of the Pb and Tl. Pour 5 ml warm, dilute NH_3 solution repeatedly through the filter.	May contain Pb^{2+} and Tl^+; these may crystallize out on cooling. Evaporate to fuming with 2–3 ml concentrated H_2SO_4, cool, dilute to 10–20 ml, cool and filter.

Residue	Filtrate	Residue	Filtrate
If black, consists of $Hg(NH_2)Cl + Hg$. Hg(I) present.	May contain $[Ag(NH_3)_2]^+$ and WO_4^{2-}. Nearly neutralize with dilute HCl, add just enough dilute NH_3 solution to redissolve any ppt. which forms. Add KI solution and filter.	If white – consists of $PbSO_4$. This is soluble in ammonium acetate solution; K_2CrO_4 solution then precipitates yellow $PbCrO_4$, insoluble in 2M acetic acid. Pb present.	May contain Tl^+. Just neutralize with dilute NH_3 solution, and add KI solution. Yellow ppt. of TlI, insoluble in cold $Na_2S_2O_3$ solution. Tl present. Confirm by flame test; intense green flame. (Use spectroscope, if available.)

(From the Filtrate containing $[Ag(NH_3)_2]^+$ and WO_4^{2-}:)

Residue	Filtrate
Pale yellow (AgI). Ag present.	May contain WO_4^{2-}. Evaporate to a small volume, acidify with dilute HCl, add 3 ml $SnCl_2$ solution, boil, add 3 ml concentrated HCl, and heat again to boiling. Blue ppt. or colouration. W present. Confirm by the Defacqz reaction.

* A gelatinous precipitate may also be hydrated silica, which is partially precipitated here from silicates decomposable by acids.

ammonium molybdate is in fact a *heptamolybdate* in which $[Mo_7O_{24}]^{6-}$ ions are present. For the sake of simplicity the formula MoO_4^{2-} will be used in this text whenever molybdates are discussed.

Reactions of molybdates MoO_4^{2-} To study these reactions use a 4·5 per cent solution of ammonium molybdate, $(NH_4)_6Mo_7O_{24}.4H_2O$, which is approximately 0·25M for molybdate MoO_4^{2-}.

1. Dilute hydrochloric acid: white or yellow precipitate of molybdic acid, H_2MoO_4, from concentrated solutions, soluble in excess mineral acid.

2. Hydrogen sulphide With a small quantity of the gas and an acidified molybdate solution, a blue colouration is produced; further passage of hydrogen sulphide yields a brown precipitate of the trisulphide, MoS_3; this is soluble in ammonium sulphide solution to a solution containing a thiomolybdate, $[MoS_4]^{2-}$, from which MoS_3 is reprecipitated by the addition of acids. The precipitation in acid solution is incomplete in the cold; more extensive precipitation is obtained by the prolonged passage of the gas into the boiling solution and under pressure. Precipitation is quantitative with excess hydrogen sulphide at 0° in the presence of formic acid.

3. *Reducing agents,* e.g. zinc, tin(II) chloride solution, colour a molybdate solution acidified with dilute hydrochloric acid blue (probably due to Mo^{3+}), then green and finally brown.

4. *Ammonium thiocyanate solution:* yellow colouration in solution acidified with dilute hydrochloric acid, becoming blood-red upon the addition of zinc or of tin(II) chloride on account of the formation of hexathiocyanatomolybdate(III) $[Mo(SCN)_6]^{3-}$; the latter is soluble in ether. The red colouration is produced in the presence of phosphoric acid (difference from iron).

The spot-test technique is as follows. Place a drop of the test solution and a drop of 10 per cent potassium or ammonium thiocyanate solution upon quantitative filter paper or upon drop-reaction paper. Add a drop of saturated tin(II) chloride solution. A red spot is obtained.

Sensitivity: 0·1 μg Mo. *Concentration limit:* 1 in 500,000.

If iron is present, a red spot will appear initially but this disappears upon the addition of the tin(II) chloride solution (or of sodium thiosulphate solution). Tungstates reduce the sensitivity of the test (cf. Section **VII.4**, reaction 6).

5. *Sodium phosphate solution:* yellow, crystalline precipitate of ammonium phosphomolybdate in the presence of excess nitric acid (cf. phosphates, Section **IV.28**, reaction 4).

6. *Potassium hexacyanoferrate(II) solution:* reddish-brown precipitate of molybdenum hexacyanoferrate(II), insoluble in dilute mineral acids, but readily soluble in solutions of caustic alkalis and ammonia [difference from uranyl and copper(II) hexacyanoferrate(II)s].

7. *α-Benzoin oxime reagent (or 'cupron' reagent)*

$$\{C_6H_5.CHOH.C(=NOH)C_6H_5\}.$$

The molybdate solution is strongly acidified with dilute sulphuric acid and 0·5 ml of the reagent added. A white precipitate is produced.

8. *Potassium xanthate (or potassium ethyl xanthogenate) test* $\{SC(SK)OC_2H_5\}$
When a molybdate solution is treated with a little solid potassium xanthate and then acidified with dilute hydrochloric acid, a red-purple colouration is produced. With large amounts of molybdenum, the compound separates as dark, oily drops which are readily soluble in organic solvents such as benzene, chloroform, and carbon disulphide. The reaction product has been given the formula $MoO_2[SC(SH)(OC_2H_5)]_2$. The test is said to be specific for molybdates, although copper, cobalt, nickel, iron, chromium, and uranium under exceptional conditions interfere. Large quantities of oxalates, tartrates, and citrates decrease the sensitivity of the test.

The spot-test technique, for which the reaction is particularly well adapted, is as follows. Place a drop of the nearly neutral or faintly acid test solution on a spot plate, introduce a minute crystal of potassium xanthate, followed by 2 drops of 2M hydrochloric acid. An intense red-violet colouration is obtained.

Sensitivity: 0·04 μg Mo. *Concentration limit:* 1 in 250,000.

9. Phenylhydrazine reagent ($C_6H_5.NHNH_2$) A red colouration or precipitate is produced when molybdates and an acid solution of phenylhydrazine react. The latter is oxidized by the molybdate to a diazonium salt, which then couples with the excess of base in the presence of the molybdate to yield a coloured compound.

Mix a drop of the test solution and a drop of the reagent on a spot plate. A red colouration appears.

Sensitivity: 0·3 μg Mo. *Concentration limit:* 1 in 150,000.

Alternatively, place a drop of the reagent on drop-reaction paper and immediately add a drop of the test solution. A red ring forms round the spot.

Se, Te, Sb(III), Sn(IV), tungstate, vanadate, and oxalate interfere.

Sensitivity: 0·1 μg Mo. *Concentration limit:* 1 in 300,000.

The reagent consists of a solution of 1 part of phenylhydrazine dissolved in 2 parts of glacial acetic acid.

10. Iron(II) sulphate solution: reddish-brown colour. Upon adding dilute mineral acid, the colour changes to blue; the colour becomes paler and more green upon warming but returns to blue on cooling (difference from tungstate).

11. Dry tests **a.** Microcosmic salt bead: oxidizing flame – yellow to green while hot and colourless when cold; reducing flame – brown when hot, green when cold.

b. Evaporation with concentrated sulphuric acid in a porcelain dish or crucible: a blue mass (containing 'molybdenum blue') is obtained. The blue colour is destroyed by dilution with water.

Separation. Molybdenum appears along with As, Sb, Sn, Au, and Pt in the ordinary process of analysis. Upon acidification of the yellow ammonium sulphide solution extract and boiling with concentrated HCl, MoS_3 appears with the As_2S_3. The mixture of MoS_3 and As_2S_3 is dissolved in HNO_3, the As removed with the magnesium nitrate reagent or with magnesia mixture and the Mo detected in the filtrate by tests 4, 8 or 11**b**.

See also Table VII.4 in Section **VII.14**.

VII.7 GOLD, Au (A_r: 196·97) – GOLD(III) Gold is a heavy metal with its characteristic yellow colour. In powderous form it is reddish-brown. It melts at 1064·8°C.

Gold is resistant against acids, only aqua regia dissolves it when tetrachloroaurate(III), $[AuCl_4]^-$, anions are formed. Gold dissolves slowly in potassium cyanide, when dicyanoaurate(I), $[Au(CN)_2]^-$, anions are formed. From both the monovalent and trivalent form gold can easily be reduced to the metal. Gold(I) compounds are less stable than those of gold(III).

Reactions of gold(III) [tetrachloroaurate(III)] ions To study these reactions use a 0·33M solution of commercial gold(III) chloride, which in fact is hydrogen tetrachloroaurate(III), $H[AuCl_4].3H_2O$.

1. Hydrogen sulphide: black precipitate of gold(I) sulphide, Au_2S (usually mixed with a little free gold), in the cold, insoluble in dilute acids, but largely soluble in yellow ammonium sulphide solution, from which it is reprecipitated by dilute hydrochloric acid. A brown precipitate of metallic gold, together with gold(I) sulphide and sulphur, is obtained upon precipitation of a hot solution; this is also largely dissolved by yellow ammonium sulphide solution.

$$2[AuCl_4]^- + 3H_2S \rightarrow Au_2S\downarrow + 2S\downarrow + 6H^+ + 8Cl^-$$

2. Ammonia solution: yellow precipitate of 'fulminating gold'; this has been formulated as $Au_2O_3.3NH_3 + NH(ClNH_2Au)_2$ but the exact composition is not fully established. The dry substance explodes upon heating or upon percussion.

3. Oxalic acid solution Gold is precipitated as a fine brown powder (or sometimes as a mirror) from cold neutral solutions (difference from platinum and other Group II metals). Under suitable conditions, the gold is obtained in the colloidal state as a red, violet, or blue solution.

$$2[AuCl_4]^- + 3(COO)_2^{2-} \rightarrow 2Au\downarrow + 6CO_2\uparrow + 8Cl^-$$

Similar results are obtained with iron(II) sulphate solution. Reduction also occurs with hydroxylamine and hydrazine salts, and ascorbic acid.

4. Tin(II) chloride solution: purple precipitate, 'purple of Cassius', consisting of an adsorption compound of tin(II) hydroxide, $Sn(OH)_2$, and colloidal gold, in neutral or weakly acid solution. In extremely dilute solutions only a purple colouration is produced. If the solution is strongly acid with hydrochloric acid, a dark-brown precipitate of pure gold is formed.

$$2[AuCl_4]^- + 4Sn^{2+} + 2H_2O \rightarrow 2Au\downarrow + Sn(OH)_2\downarrow + 3Sn^{4+} + 2H^+ + 8Cl^-$$

5. Hydrogen peroxide: the finely divided metal is precipitated in the presence of sodium hydroxide solution (distinction from platinum). The precipitated metal appears brownish-black by reflected light and bluish-green by transmitted light.

$$2[AuCl_4]^- + 3H_2O_2 + 6OH^- \rightarrow 2Au\downarrow + 3O_2\uparrow + 8Cl^- + 6H_2O$$

6. Sodium hydroxide solution: reddish-brown precipitate of gold(III) hydroxide, $Au(OH)_3$, from concentrated solutions. The precipitate has amphoteric properties; it dissolves in excess alkali forming tetrahydroxoaurate(III), $[Au(OH)_4]^-$, ion.

7. p-*Dimethylaminobenzylidene-rhodanine reagent:* (for formula see Silver, Section **III.6**, reaction 11): red-violet precipitate in neutral or faintly acid solution. Silver, mercury, and palladium salts give coloured compounds with the reagent and must therefore be absent.

Moisten a piece of drop-reaction paper with the reagent and dry it. Place a drop of the neutral or weakly acid test solution upon it. A violet spot or ring is obtained.

Sensitivity: 0·1 μg Au. *Concentration limit:* 1 in 500,000.

The reagent consists of a 0·3 per cent solution of *p*-dimethylaminobenzylidene-rhodanine in acetone.

8. Dry test All gold compounds when heated upon charcoal with sodium carbonate yield yellow, malleable, metallic particles, which are insoluble in nitric acid, but soluble in aqua regia. The aqua regia solution should be evaporated to dryness, dissolved in water, and tests 1, 3 or 4 applied.

Separation. Gold is usually detected and determined by dry methods. In the wet way, it is precipitated by H_2S in Group II, dissolved by yellow ammonium sulphide solution, and reprecipitated from the latter by concentrated HCl along with As_2S_3 and MoS_3. Separation from As_2S_3 and MoS_3 is effected by concentrated HNO_3, in which the gold precipitate is insoluble. The gold may then be dissolved in aqua regia and identified by reactions 1, 3 or 4. See also Table VII.2 in Section VII.14.

VII.8 PLATINUM, Pt (*A*$_r$: 195·09) Platinum is a greyish-white, ductile and malleable heavy metal with a density of 21·45 g cm^{-3} and a melting point of 1773°C.

It is a noble metal, not attacked by dilute or concentrated acids, except aqua regia, which dissolves platinum when hexachloroplatinate(IV) ions are formed:

$$3Pt + 4HNO_3 + 18HCl \rightarrow 3[PtCl_6]^{2-} + 4NO\uparrow + 6H^+ + 8H_2O$$

Molten alkalis and alkali peroxides attack platinum, therefore these should never be melted in platinum crucibles. In its compounds, platinum can be mono-, di-, tri-, tetra- and hexavalent, tetravelent platinum being the most important in analytical practice.

Reactions of the hexachloroplatinate(IV) ion [PtCl$_6$]$^{2-}$ To study these reactions use a 0·5M solution of hexachloroplatinic acid [hydrogen hexachloroplatinate(IV)].

1. Hydrogen sulphide: black (or dark brown) precipitate of the disulphide, PtS_2 (possibly containing a little elementary platinum), slowly formed in the cold, but rapidly on warming. The precipitate is insoluble in concentrated acids, but dissolves in aqua regia and also in ammonium polysulphide solution; it is reprecipitated from the last-named solution of thio salt by dilute acids.

$$[PtCl_6]^{2-} + 2H_2S \rightarrow PtS_2\downarrow + 4H^+ + 6Cl^-$$

2. Potassium chloride solution: yellow precipitate of potassium hexachloroplatinate(IV), $K_2[PtCl_6]$, from concentrated solutions (difference from gold). A similar result is obtained with ammonium chloride solution.

3. Oxalic acid solution: no precipitate of platinum (difference from gold). Hydrogen peroxide and sodium hydroxide solution likewise do not precipitate metallic platinum.

4. Sodium formate: black powder of metallic platinum from neutral boiling solutions.

$$[PtCl_6]^{2-} + 2HCOO^- \rightarrow Pt\downarrow + 2CO_2\uparrow + 2H^+ + 6Cl^-$$

5. Zinc, cadmium, magnesium or aluminium: all these metals precipitate finely divided platinum.

$$[PtCl_6]^{2-} + 2Zn \rightarrow Pt\downarrow + 2Zn^{2+} + 6Cl^-$$

6. Hydrazine sulphate: ready reduction in ammoniacal solution to metallic platinum, some of which is deposited as a mirror upon the sides of the tube.

$$[PtCl_6]^{2-} + N_2H_4 + 4NH_3 \rightarrow Pt\downarrow + N_2\uparrow + 4NH_4^+ + 6Cl^-$$

7. Silver nitrate solution: yellow precipitate of silver hexachloroplatinate(IV), $Ag_2[PtCl_6]$, sparingly soluble in ammonia solution but soluble in solutions of alkali cyanides and of alkali thiosulphates.

8. Potassium iodide solution: intense brownish-red or red colouration, due to $[PtI_6]^{2-}$ ions. With excess of the reagent $K_2[PtI_6]$ may be precipitated as an unstable brown solid. On warming, black PtI_4 may be precipitated.

9. Tin(II) chloride solution: red or yellow colouration, due to colloidal platinum, soluble in ethyl acetate or in ether.

To employ this reaction as a spot test in the presence of other noble metals (gold, palladium, etc.), the platinum is fixed as thallium(I) hexachloroplatinate(IV), $Tl_2[PtCl_6]$, which is stable to ammonia solution; upon washing the precipitate with ammonia solution, the thallium complexes with gold, palladium, etc., pass into solution.

Place a drop of saturated thallium(I) nitrate solution upon drop-reaction paper, add a drop of the test solution and then another drop of the thallium(I) nitrate solution. Wash the precipitate with ammonia solution, and add a drop of strongly acid tin(II) chloride solution. A yellow or orange spot remains.

Sensitivity: 0·5 μg Pt. *Concentration limit:* 1 in 80,000.

10. Rubeanic acid (or dithio-oxamide) reagent

$$\left(\begin{array}{c} H_2N{-}\underset{\underset{S}{\|}}{C}{-}\underset{\underset{S}{\|}}{C}{-}NH_2 \end{array}\right):$$

a purplish-red precipitate of the complex

$$\left(Pt \begin{array}{l} \swarrow NH_2{-}C{=}S \\ \quad\quad\quad\quad | \\ \nwarrow S{-}{-}C{=}NH \end{array}\right)_2$$

is formed. Palladium and a large proportion of gold interfere.

Place a drop of the test solution (acid with HCl) upon a spot plate and add a drop of the reagent. A purplish-red precipitate is produced.

Concentration limit: 1 in 10,000.

The reagent consists of a 0·02 per cent solution of rubeanic acid in glacial acetic acid.

11. Dry test All platinum compounds when fused with sodium carbonate upon charcoal are reduced to the grey, spongy metal (distinction from gold).

The residue is insoluble in concentrated mineral acids, but dissolves in aqua regia. The solution is evaporated almost to dryness, dissolved in water, and reactions 1, 2, 5, 9 or 10 applied.

Separation. Platinum is precipitated in Group II as PtS_2. The Group IIB metals are extracted with yellow ammonium sulphide solution and reprecipitated with HCl. The sulphides of As, Sb, Sn, Au, Pt, and Mo are dissolved in aqua regia, the excess of acid evaporated and NH_4Cl solution added. A yellow precipitate of $(NH_4)_2[PtCl_6]$ indicates the presence of Pt. The filtrate is treated with $FeSO_4$ solution; Au is precipitated and is removed. The filtrate is again treated with H_2S to reprecipitate As, Sb, Sn, and Mo as sulphides, which are filtered off. These sulphides are then separated as described under molybdates (Section **VII.6**).

See also Table VII.2 in Section **VII.14**.

VII.9 PALLADIUM, Pd (A_r: 106·4) Palladium is a light-grey metal which melts at 1555°C. Its most interesting physical characteristic is that it is able to dissolve (absorb) hydrogen gas in large quantities.

Unlike platinum, palladium is slowly dissolved by concentrated nitric acid, and by hot concentrated sulphuric acid, forming a brown solution of palladium(II) ions. Palladium can also be dissolved by fusing the metal first with potassium pyrosulphate and then leaching the frozen melt with water. The metal dissolves readily in aqua regia, when both Pd^{2+} and Pd^{4+} (more precisely, $[PdCl_4]^{2-}$ and $[PdCl_6]^{2-}$) ions are formed. If such a solution is evaporated to dryness, the latter loses chlorine so that on treating the residue with water a solution of palladium(II) ions (more precisely: tetrachloropalladium(II) ions $[PdCl_4]^{2-}$) is obtained. Palladium(II) ions are most stable; palladium(III) and (IV) compounds can easily be transformed into palladium(II).

Reactions of palladium(II) ions To study these reactions use a 1 per cent solution of palladium(II) chloride.

1. Hydrogen sulphide: black precipitate of palladium(II) sulphide, PdS, from acid or neutral solutions. The precipitate is insoluble in ammonium sulphide solution.

2. Sodium hydroxide solution: reddish-brown, gelatinous precipitate of the hydrated oxide, $PdO.nH_2O$ (this may be contaminated with a basic salt), soluble in excess of the precipitant.

3. Ammonia solution: red precipitate of $[Pd(NH_3)_4][PdCl_4]$, soluble in excess of the reagent to give a colourless solution of $[Pd(NH_3)_4]^{2+}$ ions. Upon acidifying the latter solution with hydrochloric acid, a yellow crystalline precipitate of $[Pd(NH_3)_2Cl_2]$ is obtained.

4. Potassium iodide solution: black precipitate of palladium(II) iodide, PdI_2, in neutral solution, soluble in excess of the reagent to give a brown solution of $[PdI_4]^{2-}$. In acid solution, black PdI is precipitated.

5. Mercury(II) cyanide solution: white precipitate of palladium(II) cyanide, $Pd(CN)_2$ (difference from platinum), sparingly soluble in dilute hydrochloric acid, readily soluble in potassium cyanide solution and in ammonia solution.

6. α-Nitroso-β-naphthol solution: brown, voluminous precipitate of $Pd(C_{10}H_6O_2N)_2$ (difference from platinum).

The reagent consists of a 1 per cent solution of α-nitroso-β-naphthol in 50 per cent acetic acid.

7. Reducing agents (Cd, Zn or Fe in acid solution, formic acid, sulphurous acid, etc.): black, spongy precipitate of metallic palladium, 'palladium black'. Tin(II) chloride yields a brown suspension containing metallic palladium.

8. Dimethylglyoxime reagent: yellow, crystalline precipitate of palladium dimethylglyoxime, $Pd(C_4H_7O_2N_2)_2$, insoluble in M hydrochloric acid (difference from nickel and from other platinum metals) but soluble in dilute ammonia solution and in potassium cyanide solution (cf. Nickel, Section **III.27**, reaction 8).

Salicylaldoxime reagent also precipitates palladium quantitatively as $Pd(C_7H_6O_2N)_2$ (cf. Copper, Section **III.10**, reaction 10; difference from platinum).

Place a drop of the slightly acid solution on a microscope slide and add a minute crystal of dimethylglyoxime. After some minutes, a yellow precipitate is formed. This is seen to consist of long, very characteristic needles when examined under a microscope (×75).

Platinum does not interfere, but gold and nickel give a similar reaction.

Concentration limit: 1 in 10,000.

A useful spot test utilizes the fact that a suspension of red nickel dimethylglyoxime in water when treated with a neutral or acetic acid solution of a palladium salt yields the yellow palladium dimethylglyoxime, which is sparingly soluble in dilute acids. The test is best performed with dimethylglyoxime paper; the latter is prepared as follows. Immerse drop-reaction paper in a 1 per cent alcoholic solution of dimethylglyoxime, dry, then immerse again in a solution of 0·5M nickel chloride rendered barely ammoniacal. The nickel complex precipitates; wash thoroughly with water, immerse in alcohol and dry.

Place a drop of the neutral or acetic acid test solution upon nickel dimethylglyoxime paper, and almost dry by waving over a flame. Immerse the paper in dilute hydrochloric acid until the surface surrounding the fleck becomes white, and then wash the paper with cold water. A pink to red spot remains, depending upon the quantity of palladium present. The acid-stable palladium dimethylglyoxime at the site of the fleck protects the underlying red nickel dimethylglyoxime from attack by the acid.

Sensitivity: 0·05 μg Pd. *Concentration limit:* 1 in 1,000,000.

9. Mercury(I) chloride (calomel) suspension: reduces palladium(II) ions to the metal

$$Pd^{2+} + 2Cl^- + Hg_2Cl_2\downarrow \rightarrow Pd\downarrow + 2HgCl_2$$

Shake the slightly acid test solution with solid mercury(I) chloride in the cold. The solid acquires a grey colour.

Concentration limit: 1 in 100,000.

Separation. Palladium is precipitated in Group II as PdS; it is insoluble in ammonium polysulphide solution and therefore accompanies the elements of Group IIA. It is ultimately identified in acid solution as the dimethylglyoxime complex.

VII.10 SELENIUM, Se (A_r: 78·96) – SELENITES, SeO_3^{2-} Selenium resembles sulphur in many of its properties. It dissolves readily in concentrated nitric acid or in aqua regia to form selenious acid, H_2SeO_3. In analytical work selenites, SeO_3^{2-}, and selenates, SeO_4^{2-}, are both encountered; selenites being more stable. In this section selenites are dealt with; the reactions of selenates together with dry tests for selenium will be dealt with in Section **VII.10**.

Reactions of selenites To study these reactions use a 0·1M solution of selenious acid, H_2SeO_3 (or SeO_2) or sodium selenite, Na_2SeO_3.

1. Hydrogen sulphide: yellow precipitate, consisting of a mixture of selenium and sulphur, in the cold, becoming red on heating. The precipitate is readily soluble in yellow ammonium sulphide solution.

$$SeO_3^{2-} + 2H_2S + 2H^+ \rightarrow Se\downarrow + 2S\downarrow + 3H_2O$$

2. Reducing agents (sulphur dioxide, solution of tin(II) chloride, iron(II) sulphate, hydroxylamine hydrochloride, hydrazine hydrochloride or hydriodic acid (KI+HCl), zinc or iron): red precipitate of selenium in hydrochloric acid solution. The precipitate frequently turns greyish-black on warming. When solutions in concentrated hydrochloric acid are boiled or evaporated, serious losses of selenium as $SeCl_4$ occur.

3. Copper sulphate solution: bluish-green, crystalline precipitate of copper selenite, $CuSeO_3$, in neutral solution (difference from selenate). The precipitate is soluble in dilute acetic acid.

4. Barium chloride solution: white precipitate of barium selenite, $BaSeO_3$, in neutral solution, soluble in dilute mineral acids.

5. Thiourea test, $CS(NH_2)_2$. Solid or dissolved thiourea precipitates selenium as a red powder from cold dilute solutions of selenites. Tellurium and bismuth give yellow precipitates, whilst large amounts of nitrite and of copper interfere.

Place a little powdered thiourea on quantitative filter paper and moisten it with a drop of the test solution. Orange-red selenium separates out.

Sensitivity: 2 μg Se.

6. Iodides Selenites are reduced by potassium iodide and hydrochloric acid

$$SeO_3^{2-} + 4I^- + 6H^+ \rightarrow Se\downarrow + 2I_2 + 3H_2O$$

The iodine is removed by adding a thiosulphate, and the selenium remains as a reddish-brown powder. Tellurites react under these conditions forming the complex anion, $[TeI_6]^{2-}$, which also has a reddish-brown colour; it is, however, decomposed and decolourized by a thiosulphate, thus permitting the detection of selenium in the presence of not too large an excess of tellurium.

Place a drop of concentrated hydriodic acid (or a drop each of concentrated potassium iodide solution and of concentrated hydrochloric acid) upon drop-reaction paper and introduce a drop of the acid test solution into the middle of the original drop. A brownish-black spot appears. Add a drop of 5 per cent sodium thiosulphate solution to the spot; a reddish-brown stain of elementary selenium remains.

Sensitivity: 1 μg Se (in 0·025 ml). *Concentration limit:* 1 in 25,000.

7. Pyrrole reagent

```
 / CH—CH \
|  ||  ||  |
|  CH  CH  |
|    \ /   |
 \   NH   /
```

Selenious acid oxidizes pyrrole to a blue dyestuff of unknown composition ('pyrrole blue'). Iron salts accelerate the reaction when it is carried out in phosphoric acid solution. Selenic, tellurous and telluric acids do not react under the conditions given below: the test therefore provides a method of distinguishing selenites and selenates.

Place a drop of 0·5M iron(III) chloride solution and 7 drops of syrupy phosphoric acid (sp. gr. 1·75) on a spot plate containing 1 drop of the test solution and stir well. Add a drop of the pyrrole reagent and stir again. A greenish-blue colouration is obtained.

Sensitivity: 0·5 μg Se. *Concentration limit:* 1 in 100,000.

The reagent consists of a 1 per cent solution of pyrrole in aldehyde-free ethanol.

8. Ammonium thiocyanate and hydrochloric acid Selenites are reduced in acid solution to elemental selenium:

$$2SeO_3^{2-} + SCN^- + 4H^+ \rightarrow 2Se\downarrow + CO_2\uparrow + NH_4^+ + SO_4^{2-}$$

Mix 0·5 ml of the test solution with 2 ml of 10 per cent potassium thiocyanate solution and 5 ml of 6M hydrochloric acid, and boil for 30 seconds. A red colouration, due to selenium, is produced.

This sensitive test may be adapted to the semimicro scale: the concentration limit is 1 part in 100,000. The following interfere: arsenic(III), antimony(III), iron(II) and molybdates.

VII.11 SELENIUM, Se (A_r: 78·96) – SELENATES, SeO_4^{2-} The most important properties of selenium were described in Section **VII.10**.

Reactions of selenates To study these reactions use a 0·1M solution of potassium selenate, K_2SeO_4, or sodium selenate, $Na_2SeO_4.10H_2O$.

1. Hydrogen sulphide: no precipitation occurs. If the solution is boiled with concentrated hydrochloric acid, the selenic acid is reduced to selenious acid; hydrogen sulphide then precipitates a mixture of selenium and sulphur.

$$SeO_4^{2-} + 2HCl \rightarrow SeO_3^{3-} + Cl_2\uparrow + H_2O$$

2. Sulphur dioxide: no reducing action.

3. Copper sulphate solution: no precipitate (difference from selenite).

4. Barium chloride solution: white precipitate of barium selenate, $BaSeO_4$, insoluble in dilute mineral acids. The precipitate dissolves when boiled with concentrated hydrochloric acid, and chlorine is evolved (distinction and separation from sulphate).

$$BaSeO_4\downarrow + 2HCl \rightarrow SeO_3^{2-} + Ba^{2+} + Cl_2\uparrow + H_2O$$

5. Dry tests **a.** Selenium compounds mixed with sodium carbonate and heated upon charcoal: odour of rotten horseradish. A foul odour, due to hydrogen selenide, H_2Se, is obtained upon moistening the residue with a few drops of dilute hydrochloric acid. A black stain (due to Ag_2Se) is produced when the moistened residue is placed in contact with a silver coin.

b. Elementary selenium dissolves in concentrated sulphuric acid: green solution, due to the presence of the compound $SSeO_3$. Upon dilution with water, red selenium is precipitated.

Separation. See under Tellurates (Section **VII.13**), and in Section **VII.14**.

VII.12 TELLURIUM, Te (A_r: 127·60) – TELLURITES, TeO_3^{2-} Tellurium is less widely distributed in nature than selenium; both elements belong to Group VI of the periodic system. When fused with potassium cyanide, it is converted into potassium telluride, K_2Te, which dissolves in water to yield a red solution. If air is passed through the solution, the tellurium is precipitated as a black powder (difference and method of separation from selenium). Selenium under similar conditions yields the stable potassium selenocyanide, KSeCN; the selenium may be precipitated by the addition of dilute hydrochloric acid to its aqueous solution.

$$2KCN + Te \rightarrow K_2Te + (CN)_2\uparrow$$

$$2Te^{2-} + 2H_2O + O_2 \rightarrow 2Te\downarrow + 4OH^-$$

$$KCN + Se \rightarrow KSeCN$$

$$SeCN^- + H^+ \rightarrow Se\downarrow + HCN\uparrow$$

Tellurium is converted into the dioxide, TeO_2, by nitric acid. Like sulphur and selenium, it forms two anions, the tellurite, TeO_3^{2-}, and the tellurate, TeO_4^{2-}.

Reactions of tellurites To study these reactions use a 0·1M solution of potassium tellurite, K_2TeO_3, or sodium tellurite, Na_2TeO_3.

1. Hydrogen sulphide: brown precipitate of the disulphide, TeS_2, from acid solutions. The sulphide decomposes easily into tellurium and sulphur, and is readily soluble in ammonium sulphide solution but insoluble in concentrated hydrochloric acid.

2. Sulphur dioxide: complete precipitation of tellurium from dilute (1–5M) hydrochloric acid solutions as a black powder. In the presence of much concentrated hydrochloric acid, no precipitate is formed (difference and method of separation from selenium).

3. Iron(II) sulphate solution: no precipitation of tellurium (difference from selenium). A similar result is obtained with hydriodic acid (KI + HCl).

4. Tin(II) chloride or hydrazine hydrochloride solution or zinc: black tellurium is precipitated.

5. Dilute hydrochloric acid: white precipitate of tellurous acid, H_2TeO_3 (difference from selenite), soluble in excess of the precipitant.

6. Barium chloride solution: white precipitate of barium tellurite, $BaTeO_3$, soluble in dilute hydrochloric acid but insoluble in 30 per cent acetic acid.

7. Potassium iodide solution: black precipitate of TeI_4 in faintly acid solution, dissolving in excess of the reagent to form the red hexaiodotellurate(IV) ion (difference from selenite).

8. Hypophosphorous acid test Both tellurites and tellurates are reduced to tellurium upon evaporation with hypophosphorous acid:

$$TeO_3^{2-} + H_2PO_2^- \rightarrow Te\downarrow + PO_4^{3-} + H_2O$$

$$2TeO_4^{2-} + 3H_2PO_2^- \rightarrow 2Te + 3PO_4^{3-} + 2H^+ + 2H_2O$$

Selenites are likewise reduced to selenium. If, however, the solution of the selenite in concentrated sulphuric acid is treated with sodium sulphite, selenium separates but the tellurite is unaffected; the latter can be detected in the solution after eliminating the sulphur dioxide. Salts of silver, copper, gold, and platinum must be absent for they are reduced to the metal by the reagent.

Mix a drop of the test solution in mineral acid and a drop of 50 per cent hypophosphorous acid in a porcelain micro crucible, and evaporate almost to dryness. Black grains or a grey stain of tellurium are obtained.

Sensitivity: 0·1 μg tellurous acid. *Concentration limit:* 1 in 500,000;
0·5 μg telluric acid; *concentration limit:* 1 in 100,000.

VII.13 TELLURIUM, Te (A_r: 127·60) – TELLURATES, TeO_4^{2-} The most important characteristics of tellurium were described in Section **VII.12**.

Reactions of tellurates To study these reactions use a 0·1M solution of sodium tellurate, Na_2TeO_4.

1. Hydrogen sulphide: no precipitate in hydrochloric acid solution in the cold. In hot acid solution, the tellurate is first reduced to tellurite, and precipitation of the tellurium then occurs (compare Section **VII.12**). Other reducing agents give similar results.

2 Hydrochloric acid: no precipitate in the cold. Upon boiling the solution, chlorine is evolved; tellurous acid, H_2TeO_3, is precipitated upon dilution (distinction from selenium).

$$TeO_4^{2-} + 4HCl \rightarrow H_2TeO_3\downarrow + Cl_2\uparrow + 2Cl^- + H_2O$$

3. Barium chloride solution: white precipitate of barium tellurate, $BaTeO_4$, from neutral solutions; the precipitate is readily soluble in dilute hydrochloric acid and in dilute acetic acid (distinction from selenate).

4. Potassium iodide solution: yellow to red colour, due to hexachlorotellurate(VI) ions, $[TeI_6]^{2-}$, from dilute acid solutions (difference from tellurites).

5. Reducing agents No precipitate is produced in cold solutions with hydrogen sulphide or sulphur dioxide: with hot solutions, or with solutions that have been boiled with hydrochloric acid, precipitates of brown TeS (or Te + S) and of black Te respectively are formed. Tin(II) chloride, hydrazine or zinc in acid solution give black tellurium upon warming.

6. Dry tests **a.** Fusion of any tellurium compound with sodium carbonate upon charcoal: formation of sodium telluride, Na_2Te, which produces a black stain (due to Ag_2Te) when placed in contact with a moist silver coin.

b. Elementary tellurium dissolves in concentrated sulphuric acid: red solution, due to the presence of $TeSO_3$. Upon dilution with water, grey tellurium is precipitated.

Separation. Selenium and tellurium are precipitated in Group II as the yellow Se–S mixture and brown Te–S mixture respectively. Both dissolve in ammonium sulphide solution and are precipitated with As_2S_3 upon the addition of concentrated HCl. They may be identified by the H_2SO_3 tests.

VII.14 SEPARATION AND IDENTIFICATION OF GROUP II CATIONS IN THE PRESENCE OF MOLYBDENUM, GOLD, PLATINUM, PALLADIUM, SELENIUM, AND TELLURIUM The first step in this separation process is to separate cations into Groups IIA and IIB.

Hydrogen sulphide in acid solution precipitates Mo,* Au, Pt, Pd, Se, and Te in addition to the 'common' elements of Group II. Extraction of the group precipitate with ammonium polysulphide solution brings the greater part of the 'rarer' elements (excluding PdS) into Group IIB (arsenic group) but not completely, for appreciable quantities of Mo, Au, and Pt, as well as all the Pd, remain in the Group IIA (copper group) precipitate. The latter three elements and also Pd are therefore also tested for in Group IIA.

For the separation of cations into Groups IIA and IIB the prescriptions of Table VII.2 should be followed.

Table VII.2 Separation of Group II cations into Groups IIA and IIB in the presence of Mo, Au, Pt, Pd, Se, and Te Transfer the Group II precipitate, which has been well washed with M NH_4Cl solution that has been saturated with H_2S, to a porcelain dish, add 5–10 ml ammonium polysulphide solution, heat to 50–60°C and maintain at this temperature for 3–4 minutes with constant stirring. Filter. Wash the precipitate with dilute (1:100) ammonium polysulphide solution.

Residue	Filtrate
May contain HgS, PbS, Bi_2S_3, CuS, CdS, PdS together with Au, Pt, Mo (trace), Sn (trace) as sulphides. Group IIA present.	May contain solutions of the thio-salts of As, Sb, and Sn together with Mo, Au, Pt, Se, and Te. Just acidify by adding concentrated HCl drop by drop (test with litmus paper) and warm gently. A coloured precipitate indicates Group IIB present.

For further explanations consult the notes to Table V.14 in Section **V.10**. For the separation and identification of Group IIA cations, follow Table VII.3.

* For the almost complete precipitation of Mo in Group II, it has been recommended that the solution be first saturated in the cold with hydrogen sulphide, then transferred to a pressure bottle and heated on a water bath.

Table VII.3 Separation and identification of Group IIA cations in the presence of Pt, Au, and Pd The precipitate may contain HgS, PbS, Bi_2S_3, CuS, CdS, and PdS, together with Au and Pt and also Mo sulphide (trace). Transfer to a beaker or porcelain dish, add 5–10 ml dilute HNO_3, boil for 2–5 minutes and filter.

<table>
<tr><th colspan="3">Residue</th><th>Filtrate</th></tr>
<tr><td colspan="3">May contain HgS, Pt, and Au.
Boil with concentrated HCl and a little bromine water, and filter, if necessary, from traces of SnO_2 and $PbSO_4$ which may separate here. Add KCl solution and HCl, and concentrate the solution. Filter.</td><td rowspan="5">May contain Pb, Bi, Cu, Cd, and Pd ions.
Examine for Pb, Bi, Cu, and Cd by Table VII.6 in Section VII.23. After separation of Cu and Cd, acidify the solution with dilute HCl, introduce a few zinc granules and after several minutes filter off any solid and wash with water. Dissolve the ppt. in 2 ml aqua regia, evaporate just to dryness, dissolve the residue in 2M HCl and add dimethylglyoxime reagent.
Yellow ppt.
Pd present.</td></tr>
<tr><th>Residue</th><th colspan="2">Filtrate</th></tr>
<tr><td rowspan="3">Yellow and crystalline $K_2[PtCl_6]$.
Pt present.</td><td colspan="2">May contain $[AuCl_4]$ and $HgCl_2$.
Boil to remove excess acid, render alkaline with NaOH solution, and boil with excess oxalic acid. Filter.</td></tr>
<tr><th>Residue</th><th>Filtrate</th></tr>
<tr><td>Brownish-black or purplish-black.
Au present.</td><td>May contain $HgCl_2$. Add a few drops of $SnCl_2$ solution. White or grey ppt.*
Hg present.</td><td></td></tr>
</table>

* Alternatively, the Di*en*cuprato(II)-sulphate test, Section **III.8**, reaction 8, may be applied to the neutral or slightly ammoniacal solution.

The separation and identification of Group IIB cations can be carried out by following the prescriptions given in Table VII.4.

Table VII.4 Separation and identification of Group IIB cations in the presence of Pt, Au, Se, Te, and Mo Transfer the ppt. to a small conical flask, add 5 ml concentrated HCl, and boil gently for 5 minutes (with funnel in mouth of flask). Dilute with 2–3 ml water, and filter.

<table>
<tr><th colspan="3">Residue</th><th>Filtrate</th></tr>
<tr><td colspan="3">May contain As, Au, Pt, Mo, Se, and Te as sulphides. Dissolve in concentrated HCl + little solid $KClO_3$; concentrate the solution to the crystallization point (use a water bath to reduce loss of Se to a minimum). Filter.</td><td>May contain Sb and Sn as chlorides or complex chloro-acids. Examine by Table V.18 in Section V.12.</td></tr>
<tr><th>Residue</th><th colspan="2">Filtrate</th><td></td></tr>
<tr><td rowspan="3">Yellow $K_2[PtCl_6]$.
Pt present.
Confirm by dissolving in a little hot water and adding KI solution. Red or brownish-</td><td colspan="2">May contain As, Au, Mo, Se, and Te as chlorides or acids. Render alkaline with ammonia solution, add $Mg(NO_3)_2$ reagent or magnesia mixture, allow to stand for 5 minutes with frequent stirring or shaking. Filter.</td><td></td></tr>
<tr><th>Residue</th><th>Filtrate</th><td></td></tr>
<tr><td>White crystalline $Mg(NH_4)AsO_4.6H_2O$.</td><td>May contain Au, Mo, Se, and Te as chlorides or acids. Concentrate to remove ammonia, boil with several millilitres saturated oxalic acid solution, dilute, boil and filter. Extract ppt. with HCl to remove coprecipitated tellurous acid.</td><td></td></tr>
</table>

Table VII.4 Separation and identification of Group IIB cations in the presence of Pt, Au, Se, Te, and Mo (contd.)

red colouration. Alternatively, apply the rubeanic acid test (Section **VII.8**, reaction 10).	As present.	**Residue**	**Filtrate**
		Brownish-black or purplish-black. Au present.	Concentrate with strong HCl on a water bath and, after removing the precipitated KCl, treat with a slight excess of solid Na_2SO_3. Filter.

Residue	**Filtrate**
Red. Se present.	Dilute with an equal volume of water, and add successively a little KI solution and excess of solid Na_2SO_3 whereby the $[TeI_6]^{2-}$ ions are reduced to Te. Filter.

Residue	**Filtrate**
Black. Te present.	Boil with HCl to remove dissolved SO_2 and treat successively with 10 per cent KSCN solution and a little $SnCl_2$ solution. Red colouration, soluble in ether. Mo present. Confirm by potassium xanthate test or by the α-benzoin oxime test. (Section **VII.6**).

If Pt, Au, Se, and Te are known to be absent, the procedure can be simplified. This simplified procedure is given in Table VII.5.

Table VII.5 Separation and identification of Group IIB cations in the presence of Mo Boil the Group ppt. with 5 ml concentrated HCl for 5 minutes, dilute with 2–3 ml water, pass H_2S for 1 minute (to reprecipitate small amounts of As that may have dissolved) and filter.

Residue	**Filtrate**
May contain As_2S_3 (or As_2S_5) and MoS_3. Wash with dilute HCl, followed by water; reject the washings. Warm the ppt. with 3–4 ml 2M NH_3 solution for 3 minutes, and filter.	May contain Sb and Sn as chlorides or complex chloro-acids. Examine by Table V.18 in Section **V.12**.

Residue	**Filtrate**
May contain undissolved MoS_3. Dissolve in concentrated HCl and Br_2 water, and boil to expel Br_2. Test for Mo by the $NH_4SCN–SnCl_2$ reaction.	May contain As and some of the Mo. Identify As as in Table V.18 in Section **V.12** or by the Gutzeit test (Section **III.14**). Identify Mo thus: acidify a portion of the filtrate with dilute HCl and add a little solid potassium xanthate. A red-purple colouration confirms Mo.

If the alternative potassium hydroxide method is employed for the separation of Groups IIA and IIB, the KOH extract may contain As, Sb, Sn, Se, Te, and part of the Mo; the residue may contain, in addition to HgS, PbS, Bi_2S_3, CuS, CdS, and PdS, the gold and platinum partly as sulphides and possibly partly in the form of the free metals. Mo is readily identified by the potassium xanthate or α-benzoin oxime test. The Au and Pt will accompany HgS after extraction with dilute nitric acid: upon dissolution in aqua regia, Pt may be identified as the dimethylglyoxime complex in the presence of M hydrochloric acid after the Pb, Bi, Cu, and Cd have been removed.

VII.15 VANADIUM, V (A_r: 50·94) – VANADATE Vanadium is a hard, grey metal. It melts at 1900°. Vanadium cannot be dissolved in hydrochloric, nitric, or sulphuric acids or in alkalis. It dissolves readily in aqua regia, or in a mixture of concentrated nitric acid and hydrogen fluoride. In its compounds vanadium may have the oxidation numbers +2, +3, +4, +5, and +7, among these +5 is the most common and +4 occurs also frequently. Vanadates contain pentavalent vanadium; these are analogous to phosphates. Vanadic acid, like phosphoric acids, exists in the form of meta-, pyro- and ortho-compounds (HVO_3, $H_4V_2O_7$ and H_3VO_4 respectively). Unlike the salts of phosphoric acid, the metavanadates are the most stable and the orthovanadates the least stable. A solution of an orthovanadate passes on boiling into the metavanadate, the pyro-salt being formed intermediately. In strongly acid solutions dioxovanadium(V) cations VO_2^+ are present. In quadrivalent form vanadium is usually present as the vanadyl ion, VO^{2+}.

Reactions of metavanadates, VO_3^- To study these reactions use a 0·1M solution of ammonium metavanadate, NH_4VO_3, or sodium metavanadate, $NaVO_3$. The addition of some sulphuric acid keeps these solutions stable.

1. Hydrogen sulphide No precipitate is produced in acid solution, but a blue solution (due to the production of quadrivalent vanadium ions) is formed and sulphur separates. Other reducing agents, such as sulphur dioxide, oxalic acid, iron(II) sulphate, hydrazine, formic acid, and ethanol, also yield blue vanadium(IV) (VO^{2+}) ions (cf. Molybdates, Section **VII.6**). The reaction takes place slowly in the cold, but more rapidly on warming.

2. Zinc, cadmium or aluminium in acid solution These carry the reduction still further. The solutions turn at first blue (VO^{2+} ions), then green (V^{3+} ions) and finally violet (V^{2+} ions).

3. Ammonium sulphide solution The solution is coloured claret-red, due to the formation of thiovanadates (probably VS_4^{3-}). Upon acidification of the solution, brown vanadium sulphide, V_2S_5, is incompletely precipitated, and the filtrate usually has a blue colour. The precipitate is soluble in solutions of alkalis, alkali carbonates, and sulphides.

4. Hydrogen peroxide A red colouration is produced when a few drops of hydrogen peroxide solution are added dropwise to an acid (15–20 per cent sulphuric acid) solution of a vanadate; excess hydrogen peroxide should be avoided. The colour is not removed by shaking the solution with ether nor is it affected by phosphates or fluorides (distinction from titanium).

If more hydrogen peroxide is added or the solution is made alkaline or both, the colour changes to yellow.

The red colour is due to the formation of the peroxovanadium(V) cation, VO_2^{3+}:

$$VO_3^- + 4H^+ + H_2O_2 \rightarrow VO_2^{3+} + 3H_2O$$

The yellow colour, on the other hand, originates from the diperoxoortho-vanadate(V) ions, $[VO_2(O_2)_2]^{3-}$, which is formed from the peroxovanadium(V) ions if more hydrogen peroxide is added and the solution is made alkaline:

$$VO_2^{3+} + H_2O_2 + 6OH^- \rightleftarrows [VO_2(O_2)_2]^{3-} + 4H_2O$$

This reaction is reversible; on acidification the solution again turns red.

The spot-test technique is as follows. Mix a drop of 15–20 per cent sulphuric acid and a drop of the test solution either on a spot plate or in a porcelain micro crucible. After a few minutes add 1 drop of 3 per cent hydrogen peroxide solution and then another drop, if necessary. A red to pink colouration appears.

Sensitivity: 2·5 μg V. *Concentration limit*: 1 in 20,000.

Molybdate, chromates, iodides, bromides, cerium(IV) ions and also large amounts of coloured metallic salts reduce the sensitivity of the reaction.

5. Ammonium chloride The addition of solid ammonium chloride to a solution of an alkali vanadate results in the separation of colourless, crystalline ammonium vanadate, NH_4VO_3, sparingly soluble in a concentrated solution of ammonium chloride.

6. Lead acetate solution: yellow precipitate of lead vanadate, turning white or pale yellow on standing; the precipitate is insoluble in dilute acetic acid but soluble in dilute nitric acid.

7. Barium chloride solution: yellow precipitate of barium vanadate (distinction from arsenate and phosphate), soluble in dilute hydrochloric acid.

8. Copper sulphate solution: green precipitate with metavanadates. Pyrovanadates give a yellow precipitate.

9. Mercury(I) nitrate solution: white precipitate of mercury(I) vanadate from neutral solutions.

10. Ammonium molybdate solution No precipitate is produced in the presence of ammonium nitrate and nitric acid, a soluble molybdovanadate being formed (compare phosphate). If the vanadate is mixed with a phosphate, much of the vanadium is coprecipitated with the ammonium phosphomolybdate.

11. Potassium chlorate–p–phenetidine catalytic test Vanadium catalyses the reaction between *p*-phenetidine ($H_2N.C_6H_4.OC_2H_5$) and potassium chlorate; potassium hydrogen tartrate has an activating effect upon the reaction.

Treat 0·5 ml of the test solution in a semimicro test-tube with 0·05 g potassium hydrogen tartrate, 1 ml of the *p*-phenetidine reagent, 1 ml saturated potassium chlorate solution and dilute to 5 ml with distilled water. Immerse the test-tube in a water bath: a violet colour appears within a few minutes.

Lead interferes but is rendered harmless with 100 mg of Na_2SO_4; iron(III) also interferes and is rendered innocuous by adding 50 mg of $NaNO_2$. A similar result is obtained by replacing $KClO_3$ with a saturated solution of $KBrO_3$ but iodide as well as lead and iron(III) interfere.

Sensitivity: 0·001 μg V.

The *p*-phenetidine reagent consists of a 0·1 per cent solution of *p*-phenetidine in 2M hydrochloric acid.

12. Tannin test When a neutral or acetic acid solution of a vanadate is treated with an excess of 5 per cent tannic acid, a deep blue (or blue-black) colouration is obtained. If ammonium acetate is present, a dark-blue (or blue-black) precipitate separates. The precipitate or colouration is destroyed by mineral acids.

13. Iron(III) chloride–dimethylglyoxime test The reaction:

$$VO_3^- + 4H^+ + Fe^{2+} \rightleftarrows VO^{2+} + Fe^{3+} + 2H_2O$$

proceeds from left to right in acid solution and in the reverse direction in alkaline solution. The test for vanadates utilizes the deep-red colouration with dimethylglyoxime given by iron(II) salts (cf. Iron(II), Section **III.21**, reaction 10) and the fact that vanadates are readily reduced to the quadrivalent state by heating with concentrated hydrochloric acid:

$$2VO_3^- + 8HCl \rightarrow 2VO^{2+} + Cl_2\uparrow + 6Cl^- + 4H_2O$$

All oxidizing agents interfere and must be removed.

Evaporate 1 drop of the test solution and 2 drops concentrated hydrochloric acid in a micro crucible almost to dryness. When cold, add a drop of 0·5M iron(III) chloride solution, followed by 3 drops of a 1 per cent alcoholic solution of dimethylglyoxime, and render the mixture alkaline with ammonia solution. Dip a strip of quantitative filter paper or of drop-reaction paper into the solution. The precipitated iron(III) hydroxide remains behind and the red solution of iron(II) dimethylglyoxime diffuses up the capillaries of the paper.

Sensitivity: 1 μg V. *Concentration limit:* 1 in 50,000.

14. Dry test Borax bead: oxidizing flame – colourless (yellow in the presence of much vanadium); reducing flame – green.

Separation. Vanadates are not precipitated by H_2S in acid solution; reduction to the quadrivalent state occurs. With ammonium sulphide solution, the soluble thio-salt is formed from which brown V_2S_5 is precipitated by pouring into 3M H_2SO_4. The precipitate may be dissolved in concentrated HCl, and reactions 4 and 11 applied.

In general, however, the vanadyl ions, VO^{2+}, present in the filtrate from Group II will be largely reoxidized to vanadate by the nitric acid treatment before the precipitation of Group IIIA. If the solution contains the cations of Group IIIA and certain members of later groups, the vanadates of these metals may be precipitated. However, if no other member of Group IIIA is present, vanadium may be incompletely precipitated as ammonium vanadate. It is therefore recommended where vanadium is suspected (pale-blue solution left after the passage of hydrogen sulphide in Group II) that the Group II filtrate be tested for iron with potassium hexacyanoferrate(II) solution. If iron is absent, some iron(III) chloride solution should be added before precipitation of Group IIIA.

See also Table VII.6 in Section **VII.23**.

VII.16 BERYLLIUM, Be (A_r: 9·01) Beryllium is a greyish-white, light but very hard, brittle metal. It dissolves readily in dilute acids. In its compounds beryllium is divalent, otherwise it resembles closely aluminium in chemical properties; it also exhibits resemblances to the alkaline earth metals. The salts react acid in aqueous solution, and possess a sweet taste (hence the name glucinum formerly given to the element). Beryllium compounds are highly poisonous.

Reactions of beryllium ions, Be^{2+} To study these reactions use a 0·1M solution of beryllium sulphate, $BeSO_4.4H_2O$.

1. Ammonia or ammonium sulphide solution: white precipitate of beryllium hydroxide, $Be(OH)_2$, similar in appearance to aluminium hydroxide, insoluble in excess of the reagent, but readily soluble in dilute hydrochloric acid, forming a colourless solution. Precipitation is prevented by tartrates and citrates.

2. Sodium hydroxide solution: white gelatinous precipitate of beryllium hydroxide, readily soluble in excess of the precipitant, forming tetrahydroxoberyllate ion, $[Be(OH)_4]^{2-}$; on boiling the latter solution (best when largely diluted), beryllium hydroxide is reprecipitated (distinction from aluminium). The precipitate is also soluble in 10 per cent sodium hydrogen carbonate solution (distinction from aluminium).

$$Be(OH)_2\downarrow + 2OH^- \rightleftharpoons [Be(OH)_4]^{2-}$$

On the other hand, the precipitate is insoluble in aqueous ethylamine solution whereas aluminium hydroxide dissolves in a moderate excess of the reagent. Precipitation is prevented by tartrates and citrates.

3. Ammonium carbonate solution: white precipitate of basic beryllium carbonate, soluble in excess of the reagent (difference from aluminium). On boiling the solution, the white basic carbonate is reprecipitated.

4. Oxalic acid or ammonium oxalate solution: no precipitate (difference from thorium, zirconium, and cerium).

5. Sodium thiosulphate solution: no precipitate (difference from aluminium).

6. Basic acetate–chloroform test Upon dissolving beryllium hydroxide (reaction 1) in glacial acetic acid and evaporating to dryness with a little water, basic berryllium acetate, $BeO.3Be(CH_3COO)_2$, is produced, which dissolves readily upon extraction with chloroform. This forms the basis of a method for separating beryllium from aluminium, since basic aluminium acetate is insoluble in chloroform. The mixed hydroxides are treated as detailed above.

7. *Quinalizarin reagent:** cornflower-blue colouration with faintly alkaline solutions of beryllium salts. The reagent alone gives a characteristic violet colour with dilute alkali; but this is quite distinct from the blue of the berryllium complex; a blank test will render the difference clearly apparent.

Antimony, zinc, and aluminium salts do not interfere; aluminium should, however, be kept in solution by the addition of sufficient sodium hydroxide; the influence of copper, nickel, and cobalt salts can be eliminated by the addition of potassium cyanide solution; iron salts are 'masked' by the addition of a tartrate but if aluminium salts are also present a red colour is produced. Magnesium salts give a similar blue colour, but beryllium can be detected in the presence of this element by utilizing the fact that in ammoniacal solution the magnesium colour alone is completely destroyed by bromine water.

In adjacent depressions of a spot plate place a drop of the test solution and a drop of distilled water, and add a drop of the freshly prepared quinalizarin reagent to each. A blue colouration or precipitate, quite distinct from the violet colour of the reagent, is obtained.

Sensitivity: 0·15 μg Be. *Concentration limit:* 1 in 350,000.

If magnesium is present, treat a drop of the solution on a spot plate with 2 drops of the reagent and 1 ml saturated bromine water. The original deep-blue colour becomes paler when the bromine is added, but remains more or less permanently blue.

The reagent is prepared by dissolving 0·05 g quinalizarin in 100 ml of 0·1M sodium hydroxide.

8. *para-Nitrobenzene-azo-resorcinol reagent*

CH_3

O_2N—⟨benzene ring⟩—N=N—⟨benzene ring⟩—OH :

OH

orange-red lake with beryllium salts in alkaline solution. Magnesium salts yield a brownish-yellow precipitate; salts and hydroxides of the rare earths, aluminium, and alkaline earths are without influence; the interfering effect of silver, copper, cadmium, nickel, cobalt, and zinc is eliminated by the addition of potassium cyanide solution.

Place a drop of the reagent on drop-reaction paper and into the middle of the resulting yellow area introduce the tip of a capillary containing the test solution so that the latter runs slowly on to the paper. Treat the stain with a further drop of the reagent. The stain is coloured deep orange-red.

Sensitivity: 0·2 μg Be. *Concentration limit:* 1 in 200,000.

The reagent consists of a 0·025 per cent solution of *p*-nitrobenzene-azo-resorcinol in M sodium hydroxide.

9. *Acetylacetone test* Acetylacetone reacts with beryllium salts to yield the

---

* See under Aluminium, Section **III.23**, reaction 10.

complex $Be(C_5H_7O_2)_2$, which possesses a highly characteristic appearance under the microscope.

$$2CH_3.CO.CH_2.CO.CH_3 + Be^{2+} \rightarrow Be(C_5H_7O_2)_2 + 2H^+$$

Place a drop of the test solution on a microscope slide and add a drop of acetylacetone. Crystals separate immediately: these will be found to possess rhombic and hexagonal forms when observed under the microscope (use linear magnification of 75).

Concentration limit: 1 in 10,000.

10. Dry test Upon heating beryllium salts with a few drops of cobalt nitrate solution upon charcoal a grey mass is obtained (difference from aluminium).

Separation. Beryllium is precipitated in Group IIIA. It is ultimately associated with aluminium in solution as tetrahydroxoaluminate and tetrahydroxoberyllate respectively. Upon diluting and boiling, only the $Be(OH)_2$ is precipitated. Alternatively, the quinalizarin test, 7, may be applied to the solution or the basic acetate–chloroform test, 6, to the mixed hydroxides. Beryllium may also be detected in the presence of aluminium by the acetylacetone test, 9; this is specific for Be.

Aluminium and beryllium are separated most satisfactorily by means of 8-hydroxyquinoline (oxine) reagent; aluminium oxinate is precipitated in the presence of an ammonium acetate–acetic acid buffer solution, whereas beryllium oxinate is soluble in acetic acid. For this purpose, the alcoholic solution of the reagent (cf. Magnesium, Section **III.35**, reaction 7) should not be employed as the aluminium complex is slightly soluble in alcohol. The solution of tetrahydroxoaluminate and tetrahydroxoberyllate is just acidified with dilute hydrochloric acid, a slight excess of the oxine solution added, followed by 5 ml saturated ammonium acetate solution. The precipitate of the aluminium com-complex, $Al(C_9H_6ON)_3$, is filtered off. The filtrate is heated nearly to boiling and a slight excess of ammonia solution added. The production of a precipitate (beryllia), usually coloured brown or yellow by adsorbed oxine, indicates the presence of beryllium.

The precipitating reagent consists of a 2 per cent solution of oxine in 2M acetic acid to which ammonia solution is added until a permanent precipitate is produced, and the latter is redissolved by warming.

Another method for separating beryllium and aluminium consists of adding excess of a solution of sodium fluoride to the solution. The complex hexafluoroaluminate, $[AlF_6]^{3-}$, is formed from which the metal is not precipitated as hydroxide by ammonia solution. Beryllium is, however, readily precipitated as the hydroxide under these conditions.

See also note 6 to Table VII.6 in Section **VII.23**.

VII.17 TITANIUM, Ti (A_r: 47·90) – TITANIUM(IV) Titanium is a greyish, hard metal. It can be dissolved in hot, concentrated sulphuric acid and in hydrogen fluoride. Like tin, it is insoluble in concentrated nitric acid, because of the formation of titanic acid on the surface of the metal, which protects the rest of the metal from the acid.

Titanium forms the violet titanium(III), Ti^{3+}, and the colourless titanium(IV), Ti^{4+}, ions. Titanium(III) ions are rather unstable and are readily oxidized to titanium(IV) in aqueous solutions. Titanium(IV) ions exist only in strongly acid solutions; they tend to hydrolyse forming the titanyl ion, TiO^{2+}, first if the acidity of the solution is lowered; later titanium(IV) hydroxide is precipitated. In analytical work titanium(IV) ions are normally encountered.

Reactions of titanium(IV) ions To study these reactions use a 0·2M solution of titanium(IV) sulphate, which is prepared *either* by dissolving the reagent $[Ti(SO_4)_2]$ in 5 per cent sulphuric acid, *or* by fusing titanium dioxide TiO_2 with a 12–15 fold excess of potassium pyrosulphate in a porcelain or platinum crucible.

$$TiO_2 + 2K_2S_2O_7 \rightarrow Ti(SO_4)_2 + 2K_2SO_4$$

The melt, after powdering, is extracted with cold 5 per cent sulphuric acid, and filtered if necessary.

1. Solutions of sodium hydroxide, ammonia or ammonium sulphide All these reagents give a white gelatinous precipitate of orthotitanic acid, H_4TiO_4, or titanium(IV) hydroxide, $Ti(OH)_4$, in the cold; this is almost insoluble in excess reagent, but soluble in mineral acids. If precipitation takes place from hot solution, metatitanic acid, H_2TiO_3, or $TiO.(OH)_2$ is said to be formed, which is sparingly soluble in dilute acids. Tartrates and citrates inhibit precipitation.

2. Water A white precipitate of metatitanic acid is obtained on boiling a solution of a titanic salt with excess water.

3. Sodium phosphate solution: white precipitate of titanium phosphate, $Ti(HPO_4)_2$, in dilute sulphuric acid solution.

4. Zinc, cadmium or tin When any of these metals is added to an acid (preferably hydrochloric acid) solution of a titanium(IV) salt, a violet colouration is produced, due to reduction to titanium(III) ions. No reduction occurs with sulphur dioxide or with hydrogen sulphide.

*5. Cupferron reagent:** flocculent yellow precipitate of the titanium salt, $Ti(C_6H_5O_2N_2)_4$, in acid solution (distinction from aluminium and beryllium). If iron is present, it can be removed by precipitation with ammonia and ammonium sulphide solutions in the presence of a tartrate; the titanium may then be precipitated from the acidified solution by cupferron.

6. Hydrogen peroxide An orange-red colouration is produced in slightly acid solution. The colour is yellow with very dilute solutions. The colouration has been variously attributed to peroxotitanic acid, $HOO\text{-}Ti(OH)_3$ or to peroxodisulphatotitanium(IV) ion $[TiO_2(SO_4)_2]^{2-}$.

Chromates, vanadates, and cerium salts give colour reactions with the reagent and should therefore be absent. Iron salts give a yellow colour with hydrogen peroxide, but this is eliminated by the addition of syrupy phosphoric acid. Fluorides bleach the colour (stable $[TiF_6]^{2-}$ ions are formed), and large amounts of nitrates, chlorides, bromides, and acetates as well as coloured ions

---

* For the preparation of the reagent, see under Iron(III), Section **III.22**, reaction 10.

reduce the sensitivity to the test. A decrease in the intensity of the yellow colouration upon the addition of ammonium fluoride indicates the presence of titanium.

The spot-test technique is as follows. Place a drop of the sulphuric acid test solution on a spot plate and add a drop of 3 per cent hydrogen peroxide. A yellow colouration results.

Sensitivity: 2 μg Ti. *Concentration limit:* 1 in 25,000.

The test is conducted on the semimicro scale as follows. Place 0·5 ml of the test solution in a small test-tube, add 2 drops dilute sulphuric acid and 1 drop of 3 per cent hydrogen peroxide solution. An orange-yellow or orange-red colouration is produced. Introduce a small crystal of ammonium fluoride: the colour disappears.

7. *Chromotropic acid (1,8-dihydroxynaphthalene-3,6-disulphonic acid) reagent*

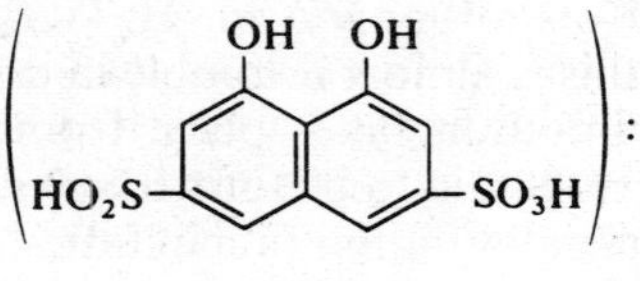

OH OH

HO_2S SO_3H :

reddish-brown colouration with titanium salts in the presence of hydrochloric or sulphuric acid. Appreciable concentrations of nitric acid inhibit the reaction.

Mix a drop of the test solution and a drop of the reagent on drop-reaction paper or upon a spot plate. A reddish-brown (or purplish-pink) spot or colouration results.

Sensitivity: 5 μg TiO_2. *Concentration limit:* 1 in 10,000.

Uranium(VI) (uranyl) and iron(III) salts interfere and yield brown and deep-green colourations respectively; these colours are destroyed by the addition of tin(II) chloride, for uranium(IV) and iron(II) salts do not react with chromotropic acid. Mercury salts give a yellow and silver salts a black stain on drop-reaction paper; the colour due to titanium is, however, still perceptible.

In the presence of uranium(VI) (uranyl) salts and/or iron(III) salts, proceed as follows. Mix a large drop of the test solution on a watch glass with a small quantity of a solution of tin(II) chloride in hydrochloric acid (a large excess is to be avoided) and warm gently (hot plate). Place a drop of the reagent on some drop-reaction paper and then a drop of the clear solution from the watch glass. A reddish-brown (or purplish-pink) spot appears.

The reagent consists of a saturated solution of chromotropic acid. The reagent does not keep well, hence it is preferable to impregnate drop-reaction paper with the reagent solution and allow the paper to dry in the air. The impregnated papers are stable for several months. In use, a drop of the test solution and a drop of M H_2SO_4 are placed upon the impregnated paper; a purplish-pink colour results.

8. *Pyrocatechol reagent*

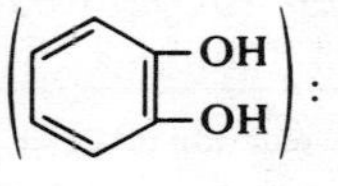

OH

OH :

yellow colouration with neutral or weakly acid (sulphuric acid) solutions of titanium salts. Iron(III), chromium, cobalt, and nickel salts interfere as do also large amounts of free mineral acids; alkali hydroxides and carbonates reduce the sensitivity of the test.

Place a drop of the sulphuric acid test solution on drop-reaction paper impregnated with the reagent. A yellow or yellowish-red spot is obtained.

Sensitivity: 3 μg Ti. *Concentration limit:* 1 in 20,000.

The reagent consists of a freshly prepared 10 per cent aqueous solution of pyrocatechol.

9. Dry test Microcosmic salt bead: oxidizing flame – colourless; reducing flame – yellow whilst hot and violet when cold (this result is obtained more rapidly if a little tin(II) chloride is added). If the bead is heated in the reducing flame with a trace of iron(II) sulphate, it acquires a blood-red colour.

Separation. If sufficient acid is present in the earlier groups to prevent its separation by hydrolysis, titanium is found in Group IIIA. It can be readily detected in the precipitate obtained after treatment with Na_2O_2 by means of the hydrogen peroxide test 6; NaF solution will discharge the colour. It is usually best to fuse the precipitate with 10 times its weight of powdered $K_2S_2O_7$ or $KHSO_4$; the melt, containing the metals as sulphates, is extracted with cold water, and the extract boiled for about 30 minutes. Metatitanic acid separates out. This is filtered off, dissolved in concentrated HCl, and the H_2O_2 test applied.

See also Table VII.6 in Section **VII.23**.

VII.18 ZIRCONIUM, Zr (A_r: 91·22) Zirconium is a grey, very hard metal. It melts at 1860°C. Finely powdered zirconium can be easily ignited in air.

Zirconium metal can be dissolved in aqua regia and in hydrogen fluoride.

Zirconium forms only one important oxide, zirconia ZrO_2, which is amphoteric in character. The normal zirconium salts, like $ZrCl_4$, are readily hydrolysed in solution giving rise chiefly to zirconyl salts, containing the bivalent radical ZrO^{2+}. The zirconates, e.g. Na_2ZrO_3, are best produced from ZrO_2 by fusion methods. Zirconium also readily forms complex ions like hexafluorozirconate(IV) $[ZrF_6]^{2-}$, produced by heating zirconia with potassium hydrogen fluoride.

Ignited zirconium dioxide, or the mineral, is insoluble in all acids except hydrofluoric acid. It is soluble in fused caustic alkalis and in sodium carbonate: the resulting alkali zirconate is practically insoluble in water, being converted into zirconium hydroxide by this solvent. It is therefore best dissolved in hydrochloric acid, and the zirconium hydroxides precipitated by ammonia solution, etc.

Together with zirconium small quantities of hafnium are always present. The two elements cannot be differentiated by classical methods of chemical analysis, but a paper chromatographic separation is possible (see Section **VI.20**).

Reactions of zirconium(IV) [zirconyl ZrO^{2+}] ions To study these reactions use a 0·1M solution of zirconyl nitrate, $ZrO(NO_3)_2.2H_2O$, or zirconyl chloride, $ZrOCl_2.8H_2O$. These solutions must contain some free acid.

1. Sodium hydroxide solution: white, gelatinous precipitate of the hydroxide, $Zr(OH)_4$ (or $ZrO_2.xH_2O$), in the cold, practically insoluble in excess of the reagent (difference from aluminium and beryllium), but soluble in dilute mineral acids (avoid sulphuric acid). With a hot solution of a zirconium salt, a white precipitate of $ZrO(OH)_2$ is obtained; it is sparingly soluble in dilute but soluble in concentrated mineral acids. Tartrates and citrates inhibit the precipitation of the hydroxide.

2. *Ammonia or ammonium sulphide solution:* white, gelatinous precipitate of the hydroxide, $Zr(OH)_4$ (or $ZrO_2.xH_2O$), insoluble in excess of the reagent.

3. *Sodium phosphate solution:* white precipitate of zirconium phosphate, $Zr(HPO_4)_2$ or $ZrO(H_2PO_4)_2$, even in solutions containing 10 per cent sulphuric acid by weight and also tartrates and citrates. No other element forms an insoluble phosphate under these conditions except titanium. The latter element can be kept in solution as peroxotitanic acid by the addition of sufficient hydrogen peroxide solution, preferably of '100-volume' strength, before the sodium phosphate is introduced.

4. *Hydrogen peroxide:* white precipitate of peroxozirconic acid, $HOO\text{-}Zr(OH)_3$, from slightly acid solutions; this liberates chlorine when warmed with concentrated hydrochloric acid. When both hydrogen peroxide and sodium phosphate are added to a solution containing zirconium, the precipitate is zirconium phosphate (see reaction 3).

5. *Ammonium carbonate solution:* white precipitate of basic zirconium carbonate, readily soluble in excess of the reagent, but reprecipitated on boiling.

6. *Oxalic acid solution:* white precipitate of zirconium oxalate, readily soluble in excess of the reagent and also in ammonium oxalate solution (difference from thorium).

7. *Ammonium oxalate solution:* white precipitate of zirconium oxalate, soluble in excess of the reagent (distinction from aluminium and beryllium); the solution gives no precipitate with hydrochloric acid (difference from thorium).

Note: A solution of zirconium sulphate or a zirconium salt solution containing excess sulphate ions does not give a precipitate with either ammonium oxalate or oxalic acid. This is due to the fact that the zirconium is present as the anion $[ZrO(SO_4)_2]^{2-}$, hence sulphuric acid should be avoided in preparing solutions of zirconium salts.

8. *Saturated potassium sulphate solution:* white precipitate of $K_2[ZrO(SO_4)_2]$, insoluble in excess of the reagent. When precipitation takes place in boiling solution, the resulting basic zirconium sulphate is insoluble in dilute hydrochloric acid (difference from thorium and cerium). No precipitate is obtained with sodium sulphate solution.

9. *Phenylarsonic acid reagent* $\{C_6H_5.AsO(OH)_2\}$: white precipitate of zirconium phenylarsonate in the presence of 0·5–1M hydrochloric acid; it is best to boil the solution. Tin and thorium salts must be absent.

The reagent consists of a 10 per cent aqueous solution of phenylarsonic acid.

10. n-Propylarsonic acid reagent

$\{CH_3.CH_2.CH_2.AsO(OH)_2\}$:

white precipitate of zirconium n-propylarsonate in dilute sulphuric acid solution (separation from most other metals including titanium but not tin).

The reagent consists of a 5 per cent aqueous solution of n-propylarsonic acid.

11. Potassium iodate solution In faintly acid solution, a voluminous, white precipitate of basic zirconium iodate is obtained. The precipitate is soluble in warm hydrochloric acid (difference from aluminium).

12. Alizarin-S reagent:* red precipitate in a strongly acid medium. Fluorides discharge the colour because of the formation of the stable hexafluoro-zirconate(IV) ion $[ZrF_6]^{2-}$.

Place a drop of the test solution (which has been acidified with hydrochloric acid) on a spot plate, add a drop of the reagent and a drop of concentrated hydrochloric acid. A red precipitate results.

The reagent consists of a 2 per cent aqueous solution of alizarin-S (sodium alizarin sulphonate).

13. p-*Dimethylaminobenzene-azo-phenylarsonic acid reagent*

$$\left((CH_3)_2N-C_6H_4-N{=}N-C_6H_4-AsO(OH)_2\right):$$

Acid solutions of zirconium salts give a brown precipitate with the reagent. If the test is conducted on filter paper, the brown precipitate remains in the pores of the paper and the excess of the coloured reagent may be washed out with dilute acid.

Impregnate some drop-reaction paper with the reagent and dry the paper. Place a drop of the acid test solution on the paper. Dip the paper into 2M hydrochloric acid at 50–60°C. A brown spot or ring remains.

Sensitivity: 0·1 ug Zr (in M HCl). *Concentration limit:* 1 in 500,000. Free sulphuric acid exceeding M concentration reduces the sensitivity of the test; phosphates, fluorides, and organic acids, which form precipitates or stable complex compounds, either inhibit or retard the reaction; molybdates, tungstates, and salts of titanium or cerium give precipitates, but their interference can be eliminated by mixing the test solution with about an equal volume of concentrated hydrochloric acid, adding some '100-volume' hydrogen peroxide, spotting the mixture on to impregnated drop-reaction paper and finally washing the latter with warm 2M hydrochloric acid. Tin also gives a coloured precipitate, but this may be prevented by the above treatment and omitting the hydrogen peroxide; a brown ring, surrounding a central zone, is obtained.

The reagent is prepared by dissolving 0·1 g of *p*-dimethylaminobenzene-azo-phenylarsonic acid in 5 ml concentrated hydrochloric acid and 100 ml ethanol.

---

* For formula, see under Aluminium, Section **III.23**, reaction 9.

14. Dry tests No characteristic results are obtained with the borax or microcosmic beads nor does zirconium yield a distinguishing flame test.

Separation. Zirconium is precipitated in Group IIIA as $ZrO_2.xH_2O$, if phosphates are absent. It can be readily detected in the residue obtained after treating the Group IIIA precipitate with 20 per cent sodium hydroxide solution and 10 per cent hydrogen peroxide. The residue is dissolved in hydrochloric acid and boiled to expel chlorine. The resulting solution is treated with sodium phosphate solution and hydrogen peroxide, when a white precipitate indicates the presence of zirconium and an orange-yellow colouration the presence of titanium (see reaction 3 above). Alternatively, the alizarin-S test (reaction 12) may be applied to the solution in hydrochloric acid.

See also Table VII.6 in Section **VII.23**.

VII.19 URANIUM, U (A_r: 238·03) Uranium is a grey, heavy metal. It is quite soft, and melts at 1133°C.

Uranium can be dissolved in dilute acids, when uranium(IV) ions, U^{4+}, and hydrogen gas are formed. Uranium(IV) ions can easily be oxidized to the hexavalent state, which is the most stable oxidation state of uranium. At this oxidation state, depending on the *p*H of the solution, two ions can be formed: the uranyl cation, UO_2^{2+}, is stable in acid solutions, while the diuranate anion, $U_2O_7^{2-}$, in alkaline media. The two ions are in equilibrium with each other:

$$2UO_2^{2+} + 3H_2O \rightleftharpoons U_2O_7^{2-} + 6H^+$$

Alkalizing the solution (removing H^+) will shift the equilibrium towards the formation of $U_2O_7^{2-}$, while acidifying it (adding H^+) will shift the equilibrium in the other direction.

In this text only the reactions of uranyl ions will be described.

Reactions of uranyl ions UO_2^{2+} To study these reactions use a 0·1M solution of uranyl nitrate, $UO_2(NO_3)_2.6H_2O$ or of uranyl acetate, $UO_2(CH_3COO)_2.2H_2O$.

1. Ammonia solution: yellow precipitate of ammonium diuranate, insoluble in excess of the reagent, but soluble in ammonium carbonate or sodium carbonate, forming the tricarbonatouranylate(VI) ion:

$$2UO_2^{2+} + 6NH_3 + 3H_2O \rightarrow (NH_4)_2U_2O_7\downarrow + 4NH_4^+$$

$$(NH_4)_2U_2O_7\downarrow + 6CO_3^{2-} + 6H_2O \rightarrow 2[UO_2(CO_3)_3]^{4-} + 2NH_4^+ + 6OH^-$$

No precipitation occurs in the presence of certain organic acids, such as oxalic, tartaric, and citric acids.

2. Sodium hydroxide solution: yellow amorphous precipitate of sodium diuranate, $Na_2U_2O_7$, soluble in ammonium carbonate solution.

3. Ammonium sulphide solution: brown precipitate of uranyl sulphide, UO_2S, soluble in dilute acids and in ammonium carbonate solution.

4. Hydrogen peroxide: pale-yellow precipitate of uranium tetroxide $UO_4.2H_2O$ (sometimes called uranium peroxide), soluble in ammonium carbonate solution with the formation of a deep-yellow solution. Chromium, titanium, and vanadium interfere with this otherwise sensitive test.

5. *Cupferron reagent:* no precipitate (distinction from titanium).

6. *Sodium phosphate solution:* white precipitate of uranyl phosphate UO_2HPO_4, soluble in mineral acids but insoluble in dilute acetic acid. If precipitation is effected in the presence of ammonium sulphate or of ammonium acetate, uranyl ammonium phosphate, $UO_2(NH_4)PO_4$ is precipitated.

7. *Ammonium (or sodium) carbonate solution:* white precipitate of uranyl carbonate, UO_2CO_3, soluble in excess of the reagent forming a clear, yellow solution containing the tricarbonatouranylate(VI) ion (cf. reaction 1).

8. *Potassium hexacyanoferrate(II) solution:* brown precipitate of uranyl hexacyanoferrate(II), $(UO_2)_2[Fe(CN)_6]$, in neutral or acetic acid solutions, soluble in dilute hydrochloric acid (difference from copper). The precipitate becomes yellow upon the addition of sodium hydroxide solution, due to its conversion into sodium diuranate (distinction from copper and from molybdenum).

$$(UO_2)_2[Fe(CN)_6]\downarrow + 2Na^+ + 6OH^- \rightarrow Na_2U_2O_7\downarrow + [Fe(CN)_6]^{4-} + 3H_2O$$

The spot-test technique is as follows. Place a drop of the test solution on drop-reaction paper and add a drop of potassium hexacyanoferrate(II) solution. A brown spot is obtained.

Sensitivity: 0·9 μg U. *Concentration limit:* 1 in 50,000.

Both iron and copper interfere. If, however, potassium iodide solution is added, they are reduced to the non-reactive iron(II) and copper(I) ions; the liberated iodine may be decolourized with sodium thiosulphate solution. Alternatively, the reduction may be carried out with sodium thiosulphate solution alone on a spot plate, the copper acting as a catalyst for the reduction of the iron:

$$2Fe^{3+} + 2S_2O_3^{2-} = 2Fe^{2+} + S_4O_6^{2-};$$
$$2Cu^{2+} + 2S_2O_3^{2-} = 2Cu^+ + S_4O_6^{2-}$$

Place a drop of concentrated potassium iodide solution on a piece of drop-reaction paper and, after the iodide solution has soaked into the paper, add a drop of the acidified test solution. Add a further drop of the potassium iodide solution to complete the reduction, followed by a drop of sodium thiosulphate solution. Then add a drop of potassium hexacyanoferrate(II) solution to the decolourized spot, whereupon a brown ring is obtained.

9. *Fluorescence* Uranium salts, when irradiated by ultraviolet rays (e.g. with a u.v. lamp used for paper chromatography) exhibit a characteristic green fluorescence. The phenomenon is especially surprising if in a dark room a bottle containing solid uranyl nitrate or uranyl acetate is irradiated.

In solution the intensity of fluorescence depends on the *p*H. In acid solutions the fluorescence is strong, but becomes gradually weaker as the *p*H of the solution is raised.

10. *Dry test* Borax or microcosmic salt bead: oxidizing flame – yellow; reducing flame – green.

Separation. Uranium is precipitated in Group IIIA as $(NH_4)_2U_2O_7$. It is most simply separated from $Fe(OH)_3$, $Cr(OH)_3$, and $Al(OH)_3$ by digestion in the cold with a large excess of ammonium carbonate solution. The ammonium diuranate dissolves (see reaction 1 above); upon acidification with HCl and addition of $K_4[Fe(CN)_6]$ solution, a brown precipitate is formed.

See also Table VII.6 in Section **VII.23**.

VII.20 THORIUM, Th (A_r: 232·04) Thorium is a greyish-white, not too hard, heavy metal, which melts at 1750°C. It can be dissolved in concentrated hydrochloric acid or in aqua regia. The quadrivalent Th^{4+} cations are stable in acid solutions.

Reactions of thorium(IV) ions To study these reactions use a 0·1M solution of thorium nitrate $Th(NO_3)_4.4H_2O$ which contains a few per cent free nitric acid also.

1. Ammonia, ammonium sulphide or sodium hydroxide solution: white precipitate of thorium hydroxide, $Th(OH)_4$ or $ThO_2.xH_2O$, insoluble in excess of the reagent, but readily soluble in dilute acids when freshly precipitated. Tartrates and also citrates prevent the precipitation of the hydroxide.

2. Ammonium or sodium carbonate solution: white precipitate of basic carbonate, readily soluble in excess of the concentrated reagent forming the pentacarbonatothorate(IV) anion, $[Th(CO_3)_5]^{6-}$.

3. Oxalic acid solution: white, crystalline precipitate of thorium oxalate, $Th(C_2O_4)_2$ (distinction from aluminium and beryllium), insoluble in excess of the reagent and in 0·5M hydrochloric acid.

4. Ammonium oxalate solution: white precipitate of thorium oxalate, which dissolves on boiling with a large excess of the reagent forming trioxalatothorate(IV) anion, $[Th\{(COO)_2\}_3]^{2-}$, but is reprecipitated upon the addition of hydrochloric acid (difference from zirconium).

5. Saturated potassium sulphate solution: white precipitate of the complex salt potassium tetrasulphatothorate(IV), $K_4[Th(SO_4)_4]$, insoluble in excess of the precipitant, but soluble in dilute hydrochloric acid.

6. Hydrogen peroxide: white precipitate of hydrated thorium heptoxide (thorium peroxide), $Th_2O_7.4H_2O$ in neutral or faintly acid solution.

The compound is not a true peroxide, but an associate compound of thorium dioxide and hydrogen peroxide: $2ThO_2.3H_2O_2.H_2O$.

7. Sodium thiosulphate solution: precipitate of thorium hydroxide and sulphur on boiling (distinction from cerium).

$$Th^{4+} + 2S_2O_3^{2-} + 2H_2O \rightarrow Th(OH)_4\downarrow + 2S\downarrow + 2SO_2\uparrow$$

8. Potassium iodate solution: white, bulky precipitate of thorium iodate, $Th(IO_3)_4$. Precipitation occurs in the presence of 50 per cent by volume of concentrated nitric acid (difference from cerium(III)).

9. Potassium hexacyanoferrate(II) solution: white precipitate of thorium hexacyanoferrate(II), $Th[Fe(CN)_6]$, in neutral or slightly acid solution.

10. Potassium fluoride solution: bulky, white precipitate of thorium fluoride, ThF_4, insoluble in excess of the reagent (distinction and method of separation for aluminium, beryllium, zirconium, and titanium).

11. Saturated sebacic acid solution, $\{COOH.(CH_2)_8COOH\}$*:* white voluminous precipitate of thorium sebacate, $Th(C_{10}H_{16}O_4)_2$ (difference from cerium).

12. m-*Nitrobenzoic acid reagent,* $(NO_2.C_6H_4.COOH)$ Upon addition of excess reagent to a neutral solution of a thorium salt at about 80°, a white precipitate of the salt, $Th(NO_2.C_6H_4.COO)_4$ is obtained (distinction from cerium).

The reagent is prepared by dissolving 1 g of the acid in 250 ml water at 80°, allowing to cool overnight and filtering.

Separation. See under Cerium (Section **VII.22**).

VII.21 CERIUM, Ce (*A*$_r$: 140·12) – CERIUM(III) Cerium is a white-grey, soft metal, which melts at 795°C. In its compounds cerium can be tervalent and quadrivalent, forming cerium(III), Ce^{3+}, and cerium(IV), Ce^{4+}, ions respectively.

Reactions of cerium(III) ions To study these reactions use a 0·1M solution of cerium(III) nitrate $Ce(NO_3)_3.6H_2O$.

1. Ammonia or ammonium sulphide solution: white precipitate of cerium(III) hydroxide, $Ce(OH)_3$ (or $Ce_2O_3.xH_2O$), insoluble in excess of the precipitant, but readily soluble in acids. The precipitate slowly oxidizes in the air, finally becoming converted into yellow cerium(IV) hydroxide, $Ce(OH)_4$ (or $CeO_2.xH_2O$). Sodium hydroxide solution gives a similar result. The precipitation is prevented by tartrates and citrates.

2. Oxalic acid or ammonium oxalate solution: white precipitate of cerium(III) oxalate, insoluble in excess reagent (compare thorium and zirconium), and in very dilute mineral acids.

3. Sodium thiosulphate solution: no precipitate (distinction from thorium and from cerium(IV) ions).

4. Saturated potassium sulphate solution: white, crystalline precipitate, having the composition $Ce_2(SO_4)_3.3K_2SO_4$ in neutral solution and $Ce_2(SO_4)_3.2K_2SO_4.2H_2O$ from slightly acid solution (difference from aluminium and beryllium).

5. Sodium bismuthate This reagent, in the presence of dilute nitric acid, converts cerium(III) ions into cerium(IV) in the cold. A similar result is obtained by heating with ammonium peroxodisulphate or with lead dioxide and dilute nitric acid (1:2). In all cases, the solutions become yellow or orange in colour.

6. Ammonium carbonate solution: white precipitate of cerium(III) carbonate, $Ce_2(CO_3)_3$, nearly insoluble in excess of the precipitant (difference from beryllium, thorium, and zirconium) and insoluble in sodium carbonate solution.

7. Hydrogen peroxide When a cerium(III) salt is treated with ammonia solution and excess hydrogen peroxide is added, a yellowish-brown or reddish-brown precipitate or colouration occurs, due to cerium peroxide gel ($CeO_2.H_2O_2.H_2O$). This is not very stable. Upon boiling the mixture, yellow cerium(IV) hydroxide, $Ce(OH)_4$, is obtained. The test cannot be applied directly in the presence of iron since the colour of iron(III) hydroxide is similar to that of cerium trioxide. The precipitation of iron(III) hydroxide may be prevented by the addition of an alkali tartrate; this, however, reduces the sensitivity of the test for cerium.

The spot-test technique is as follows. Mix a drop of the hot test solution, of '3 per cent' hydrogen peroxide and of dilute ammonia solution in a porcelain micro crucible and warm gently. A yellow or yellowish-brown precipitate or colouration appears.

Sensitivity: 0·35 μg Ce. *Concentration limit:* 1 in 140,000.

8. Ammonium fluoride solution: white, gelatinous precipitate of cerium(III) fluoride, CeF_3, in neutral or slightly acid solution. The precipitate becomes powdery upon standing.

9. Potassium iodate solution: white precipitate of cerium(II) iodate, $Ce(IO_3)_3$, in neutral solution, soluble in nitric acid (difference from cerium(IV) iodate; cf. Thorium, Section **VII.20**, reaction 8).

10. Diammineargentato nitrate (ammoniacal silver nitrate) reagent: This reagent reacts with neutral solutions of cerium(III) salts to form cerium(IV) hydroxide and metallic silver [difference from cerium(IV)]; the former is coloured black by the finely divided silver:

$$Ce(OH)_3\downarrow + [Ag(NH_3)_2]^+ + OH^- = Ce(OH)_4\downarrow + Ag\downarrow + 2NH_3$$

Iron(II), manganese(II), and cobalt(II) ions also give the higher metallic hydroxides and silver, and must therefore be absent.

Mix a drop of the neutral test solution and a drop of the reagent on a watch glass or in a porcelain micro crucible, and warm gently. A black precipitate or brown colouration appears.

Sensitivity: 1 μg Ce. *Concentration limit:* 1 in 50,000.

The reagent is prepared by treating 0·1M silver nitrate with dilute ammonia solution until the precipitate first formed is just redissolved. The reagent should be freshly prepared and discarded after use.

VII.22 CERIUM, Ce (A_r: 140·12) – CERIUM(IV) The most important physical and chemical properties of cerium have been described in Section **IV.22**.

Reactions of cerium(IV) ions To study these reactions use a 0·1M solution of cerium(IV) sulphate, $Ce(SO_4)_2.4H_2O$ or cerium(IV) ammonium sulphate $Ce(SO_4)_2.2(NH_4)_2SO_4.2H_2O$. These solutions should contain a few per cent of free sulphuric acid. Both solutions are orange-yellow. On standing a fine precipitate is sometimes formed; it cannot be filtered easily, the clear solution should rather be decanted before use.

1. Ammonia or sodium hydroxide solution: yellow precipitate of cerium(IV) hydroxide, $Ce(OH)_4$. If the precipitate is warmed with hydrochloric acid, chlorine is evolved and cerium(III) ions are formed.

2. Oxalic acid or ammonium oxalate solution: reduction occurs, more rapidly on warming, to cerium(III) ions, and ultimately white cerium(III) oxalate is precipitated.

3. Saturated potassium sulphate solution: no precipitate (distinction from cerium(III) salts).

4. Reducing agents (e.g. hydrogen sulphide, sulphur dioxide, hydrogen peroxide, and hydriodic acid) These convert cerium(IV) ions into cerium(III).

5. Sodium thiosulphate solution: yellow colour of solution discharged and sulphur precipitated, owing to reduction.

$$2Ce^{4+} + S_2O_3^{2-} + H_2O \rightarrow 2Ce^{3+} + S\downarrow + SO_4^{2-} + 2H^+$$

6. Ammonium fluoride solution: yellow colour of solution is discharged but no precipitate is produced, due to the formation of hexafluorocerate(IV) anions, $[CeF_6]^{2-}$.

7. Potassium iodate solution: white precipitate of cerium(IV) iodate, $Ce(IO_3)_4$, from concentrated nitric acid solution (difference from cerium(III); thorium and zirconium give a similar reaction).

8. Anthranilic acid reagent ($NH_2.C_6H_4.COOH$) Cerium(IV) oxidizes anthranilic acid to a brown compound. Cerium(III) does not react and must be oxidized first with lead dioxide and concentrated nitric acid to cerium(IV); other oxidizing agents cannot be used, since they react with the anthranilic acid. Iron(III) ions inhibit the test and must be masked by the addition of phosphoric acid. The ions of gold and vanadium, as well as chromate ion, react similarly and therefore interfere. Reducing agents must be absent.

Place a drop of the test solution (slightly acid with nitric acid) on a spot plate and add a drop of a 5 per cent solution of anthranilic acid in alcohol. A blackish-blue precipitate appears, which rapidly passes into a soluble product and colours the solution brown.

Concentration limit: 1 in 10,000.

9. Dry test Borax bead: oxidizing flame – dark brown whilst hot and light yellow to colourless when cold: reducing flame – colourless.

Separation. Cerium and thorium salts are precipitated in Group IIIA. They may be separated from the other metals of the group by dissolving the precipitate in dilute HCl and adding oxalic acid solution, when the oxalates of both metals are precipitated. The thorium and cerium may be separated: (*a*) by dissolving the thorium oxalate in a mixture of ammonium acetate solution and acetic acid, cerium oxalate being insoluble under these conditions; (*b*) by treatment with a large excess of hot concentrated ammonium oxalate solution; only the thorium oxalate dissolves (a complex ion being formed), and may be reprecipitated from the resultant solution as oxalate by the addition of hydrochloric acid.

In routine qualitative analysis, it is probably best to boil the mixed oxalates (after washing with 2 per cent oxalic acid solution) with M potassium hydroxide solution, thereby converting them into the hydroxides. The precipitate is separated by filtration, washed with hot water, dissolved in the minimum volume of dilute hydrochloric acid and the solution divided into two parts:

(i) **Thorium.** Neutralize with ammonia solution, and add the *m*-nitrobenzoic acid reagent (Section **VII.20**, reaction 12) or a warm saturated solution of sebacic acid (Section **VII.20**, reaction 11). Alternatively, add hydrochloric acid to the neutral solution until the concentration is about 0·3M, and then sodium pyrophosphate solution. A white precipitate, ThP_2O_7, indicates Th.

(ii) **Cerium.** Identify by the addition of hydrogen peroxide, followed by ammonia solution until the liquid is alkaline. A reddish-brown precipitate indicates Ce (see Section **VII.21**, reaction 7).

VII.23 SEPARATION OF GROUP III CATIONS IN THE PRESENCE OF TITANIUM, ZIRCONIUM, THORIUM, URANIUM, CERIUM, VANADIUM, THALLIUM, AND MOLYBDENUM When in the course of the systematic analysis the separation of Group III cations is to be carried out, the solution should be tested for phosphate. If this test is positive, remove phosphate by the tin(II) chloride method (see Table V.22 in Section **V.13**). Among the less common ions, titanium, zirconium, cerium, thorium, and uranium will be completely precipitated by ammonia solution, and will therefore appear together with the other Group IIIA cations. Vanadium will only partly be precipitated here; some of the vanadium present will be found in the filtrate of the sulphide precipitates of Group IIIB cations. Some of the thallium(I) partly removed with Group I will also appear in this Group; some molybdenum will appear in the filtrate of Group IIIB sulphides.

The precipitation of Group IIIA cations with ammonia solution is carried out according to the procedure given in Table V12. The separation and identification of Group IIIA cations in the presence of some less common ions, as described in Table VII.6 commences with the precipitate obtained with ammonia.

Table VII.6 Separation and identification of Group IIIA cations in the presence of Ti, Zr, Ce, Th, U, and V Dissolve the precipitate in the minimum volume of dilute HCl. Pour the weakly acid solution into an equal volume of a solution which contains 2 ml 30 per cent H_2O_2 and is 2·5M with respect to NaOH. (The latter solution should be freshly prepared.) Boil for 5 minutes, but no longer. Filter and wash the ppt. with hot 2 per cent NH_4NO_2 solution.

Residue	Filtrate
May contain $Fe(OH)_3$, $TiO_2.xH_2O$; $ZrO_2^3.xH_2O$; $ThO_2^2.xH_2O$; $CeO_3^2.xH_2O$ (and some $MnO_2.xH_2O$).	May contain CrO_4^{2-}, $[Al(OH)_4]^-$, VO_3^-, and $U_2O_7^{2-}$. Acidify with dilute HNO_3, add 3–4 ml 0·25M $Pb(NO_3)_2$ solution, followed by 2 g solid ammonium acetate. Stir well, filter and wash with hot water.

Table VII.6 Separation and identification of Group IIIA cations in the presence of Ti, Zr, Ce, Th, U, and V (contd.)

<table>
<tr>
<td rowspan="4">Dissolve in dilute HCl, boil to expel Cl_2 and divide the solution into five parts.
(i) Add KSCN solution (1).
Red colouration.
Fe present.
(ii) If Fe present, add just sufficient H_3PO_4 (2) to mask iron(III) ions, and then H_2O_2. Orange-red colouration, discharged by the addition of solid NH_4F.
Ti present (see Section **VII.17**).
White ppt.
Zr present (3; see also Section **VII.18**).
(iii) Add excess saturated oxalic acid solution (*c.* 10 per cent). White ppt.
Th and/or Ce present.
For separation of Th and Ce, see Section **VII.2**.
(iv) Evaporate to fuming with H_2SO_4 to expel HCl. Cool, dilute, add HNO_3 and a little $NaBiO_3$. Stir and allow to stand.
Purple colouration.
Mn present.</td>
<td>**Residue**</td>
<td colspan="2">**Filtrate**</td>
</tr>
<tr>
<td rowspan="3">May contain $PbCrO_4$ and $Pb(VO_3)_2$ (4).
Dissolve in the minimum volume of hot 2M HNO_3 ($\sim$5–6 ml), thoroughly cool the resulting solution, and transfer to a small separatory funnel. Add an equal volume of amyl alcohol, and a little 6 per cent H_2O_2. Shake well and allow the two layers to separate.
A blue colouration in the upper layer indicates Cr present, and a red to brownish-red colouration (5) in the lower layer indicates V present.
Confirm V by the $KClO_3$–*p*-phenetidine test (Section **VII.15**, reaction 12).</td>
<td colspan="2">May contain Al^{3+}, UO_2^{2+} and excess of Pb^{2+}. Pass H_2S to remove all the Pb as PbS. Filter, wash and boil the filtrate to expel H_2S. Almost neutralise with NH_3 solution, cool and pour into an excess of concentrated $(NH_4)_2CO_3$ solution. Warm for 5 minutes. Allow to stand, filter and wash.</td>
</tr>
<tr>
<td>**Residue**</td>
<td>**Filtrate**</td>
</tr>
<tr>
<td>White: $Al(OH)_3$.
Al present.
Confirm by Thenard's blue test (6).</td>
<td>May contain U, probably as $[UO_2(CO_3)_3]^{4-}$. Evaporate to a small volume (7) acidify with HCl and add $K_4[Fe(CN)_6]$ solution. Brown ppt. of $(UO_2)_2[Fe(CN)_6]$ becoming yellow upon the addition of NaOH solution.
U present.</td>
</tr>
</table>

Notes to Table VII.6: 1. The potassium hexacyanoferrate(III) test for Fe^{3+} is not recommended here when U is present. The NaOH–H_2O_2 separation is not quite quantitative and sufficient U may be present in the precipitate to introduce complications.

2. The addition of phosphoric acid or of sodium phosphate is essential if Zr is to be tested for, even if Fe is absent (cf. Section **VII.18**, reaction 3).

3. If both Ti and Zr are present, the precipitate of zirconium phosphate may be filtered off (best in the presence of a little macerated filter paper, or a Whatman filtration accelerator), and the filtrate treated with Na_2SO_3 or with $Na_2S_2O_3$ solution and warmed. The peroxotitanic acid is reduced and titanium phosphate precipitates. It may be necessary to reduce the acidity of the solution somewhat to precipitate the titanium completely. Zr may also be identified by the alizarin-S reaction.

4. If Group IV metals are present they may be precipitated as vanadates. This is prevented by adding excess $FeCl_3$ solution when iron(III) vanadate is precipitated here.

5. If much Cr and a little V is present, a second extraction with amyl alcohol may be necessary in order to ensure the complete removal of the Cr, and to render the colouration due to V completely visible. The addition of two drops of concentrated nitric acid intensifies the colour.
6. The dry test is preferable to the 'aluminon' reaction, for the latter is not applicable in the presence of Be. If Be is absent, the 'aluminon' test may be applied.
7. A precipitate that separates here may be (basic) beryllium carbonate. It should be filtered off and tested for Be by the basic acetate–chloroform test or by the acetylacetone test (see Section **VII.16**, reactions 6 and 9).

If *thallium* has been found in Group I, some of it may pass into Group IIIB because of the solubility of thallium(I) chloride in water and be precipitated as Tl_2S. It may be readily detected by the green flame colouration, preferably viewed through a hand spectroscope. For the separation of thallium from the rest of Group IIIB cations, commencing with the sulphide precipitate, follow the procedure given in Table VII.7.

Table VII.7 Separation of Tl from the rest of Group IIIB cations Dissolve the precipitate in 2M HNO_3. Expel H_2S by boiling, add H_2SO_3 and expel excess SO_2 again by boiling. Pour the solution into an excess of saturated Na_2CO_3 solution. Filter.

Residue	**Filtrate**
May contain $CoCO_3$, $NiCO_3$, $MnCO_3$ and $ZnCO_3$. Dissolve in 2M HCl, remove CO_2 by boiling, neutralize with NH_3, then add $(NH_4)_2S$. Treat the precipitate according to Tables V.24 or V.25.	May contain Tl^+. Identify by reaction 6 in Section **VII.2** or by flame test (Section **VII.2**, reaction 10). Tl present.

If vanadium and/or molybdenum have been detected in the earlier groups (as indicated by the production of a blue colouration with hydrogen sulphide in acid solution) these elements may also be found in the filtrate from Group IIIB: a violet-red colour points to vanadium and a reddish-brown colouration to molybdenum. Generally, vanadium is removed in Group IIIA as iron vanadate and its presence in the Group IIIB filtrate is therefore unlikely. However, when the filtrate from Group IIIB is acidified with acetic acid and boiled, the production of a brown precipitate will indicate the removal of any residual molybdenum, vanadium or nickel in the form of sulphides. If desired, the precipitate may be dissolved in concentrated nitric acid, evaporated to dryness, and the residue dissolved in hydrochloric acid. Molybdenum is detected by the ammonium thiocyanate or the potassium xanthate test (Section **VII.6**), and vanadium by the $KClO_3$–*p*–phenetidine or hydrogen peroxide reactions (Section **VII.15**).

VII.24 LITHIUM, Li (A_r: 6·94) Lithium is a silver-white metal; it is the lightest metal known (density 0·534 g ml^{-1} at 0°) and floats upon petroleum. It melts at 186°C. It oxidizes on exposure to air, and reacts with water forming lithium hydroxide and liberating hydrogen, but the reaction is not so vigorous as with sodium and potassium. The metal dissolves in acids with the formation of salts. The salts may be regarded as derived from the oxide, Li_2O.

Some of the salts, notably the chloride, LiCl, and the chlorate, $LiClO_3$, are very deliquescent. The solubilities of the hydroxide, LiOH, (113 g ℓ^{-1} at 10°), carbonate, Li_2CO_3, (13·1 g ℓ^{-1} at 13°), the phosphate, Li_3PO_4, (0·30 g ℓ^{-1} at 25°), and the fluoride, LiF, (2·7 g ℓ^{-1} at 18°) are less than the corresponding sodium and potassium salts, and in this respect lithium resembles the alkaline earth metals.

Reactions of lithium ions To study these reactions use a M solution of lithium chloride, LiCl; alternatively dissolve lithium carbonate, Li_2CO_3, in the minimum volume of 2M hydrochloric acid.

1. Sodium phosphate solution: partial precipitation of lithium phosphate, Li_3PO_4, in neutral solutions; the precipitate is more readily obtained from dilute solutions on boiling. Precipitation is almost complete in the presence of sodium hydroxide solution. The precipitate is more soluble in ammonium chloride solution than in water (distinction from magnesium).

Upon boiling the precipitate with barium hydroxide solution, it passes into solution as lithium hydroxide (difference from magnesium).

2. Sodium or ammonium carbonate solution: white precipitate of lithium carbonate, Li_2CO_3, from concentrated solutions and in the presence of ammonia solution. No precipitation occurs in the presence of high concentrations of ammonium chloride since the carbonate-ion concentration is reduced to such an extent that the solubility product of Li_2CO_3 is not exceeded:

$$NH^{+} + CO_3^{2-} \rightleftharpoons NH_3 + HCO_3^{-}$$

3. Ammonium fluoride solution: a white, gelatinous precipitate of lithium fluoride, LiF, is slowly formed in ammoniacal solution (distinction from sodium and potassium).

4. Tartaric acid, sodium hexanitritocobaltate(III) or hexachloroplatinic acid solution: no precipitate (distinction from potassium). A precipitate of lithium hexanitritocobaltate(III) is, however, produced in very concentrated (almost saturated) solutions of lithium salts; interference with the sodium hexanitritocobaltate(III) test for K^{+} is therefore unlikely.

5. Iron(III) periodate test Iron(III) salts react with periodates to yield a precipitate of iron(III) periodate: this precipitate is soluble in excess of the periodate solution and also in excess potassium hydroxide solution. The resulting alkaline solution is a selective reagent for lithium, since it gives a white precipitate, $KLiFe[IO_6]$, even from dilute solutions and in the cold. Sodium and potassium give no precipitate; ammonium salts, all metals of Groups I to IV, and magnesium should be absent.

Place a drop of the neutral or alkaline test solution in a micro test-tube, and add 1 drop of M sodium chloride solution and 2 drops of the iron(III) periodate reagent. Carry out a blank test with a drop of distilled water simultaneously. Immerse both tubes for 15–20 seconds in water at 40–50°C. A white (or yellowish-white) precipitate indicates the presence of lithium; the blank remains clear.

Sensitivity: 0·1 μg Li. *Concentration limit:* 1 in 100,000.

The iron(III) periodate reagent is prepared by dissolving 2 g potassium periodate (KIO_4) in 10 ml freshly prepared 2M potassium hydroxide solution, diluting with water to 50 ml, adding 3 ml of 0·5M iron(III) chloride solution and diluting to 100 ml with 2M potassium hydroxide solution. The reagent is stable.

6. Ammonium carbonate solution (microscope test) Lithium carbonate, when freshly formed, has a characteristic appearance under the microscope.

Place a drop of the concentrated test solution on a microscope slide. Introduce a few minute specks of sodium or ammonium carbonate. Some lithium carbonate crystals are formed immediately. Examine under the microscope (magnification: 200 diameters): the crystals are in the form of either hexagonal stars or plates (compare $CaSO_4.2H_2O$, Section **III.33**, reaction 10).

The cations of the alkaline earth metals and of magnesium must be absent.

Concentration limit: 1 in 10,000.

7. Dry test Flame colouration. Lithium compounds impart a carmine-red colour to the non-luminous Bunsen flame. The colour is masked by the presence of considerable amounts of sodium salts, but becomes visible when observed through two thicknesses of cobalt glass.

The most distinctive test utilizes the spectroscope; the spectrum consists of a beautiful red line at 671 nm.

Separation. In order to separate lithium from the other alkali metals, they are all converted into the chlorides (by evaporation with concentrated hydrochloric acid, if necessary), evaporated to dryness, and the residue extracted with absolute alcohol which dissolves the lithium chloride only. Better solvents are dry dioxan (diethylene dioxide, $C_4H_8O_2$) and dry acetone. Upon evaporation of the extract, the residue of lithium chloride is (*a*) subjected to the flame test, and (*b*) precipitated as the phosphate after dissolution in water and adding sodium hydroxide solution.

VII.25 THE BORAX BEAD TEST IN THE PRESENCE OF LESS COMMON CATIONS The borax bead test has been described in Section **V.2**, 5,

Table VII.8 Borax bead tests in the presence of Mo, Au, W, U, V, Ti, and Ce

Oxidizing flame		Reducing flame		Metal
Hot	Cold	Hot	Cold	
1. Green	Blue	Colourless	Opaque red or brown (1)	Copper
2. Yellowish-brown	Yellow	Green	Green	Iron
3. Yellow	Green	Green	Green	Chromium
4. Violet (amethyst)	Amethyst	Colourless	Colourless	Manganese
5. Blue	Blue	Blue	Blue	Cobalt
6. Violet	Reddish-brown	Grey	Grey	Nickel
7. Yellow	Colourless	Brown	Brown	Molybdenum
8. Rose-violet	Rose-violet	Red	Violet	Gold
9. Yellow	Colourless	Yellow	Yellow-brown	Tungsten (2)
10. Yellow	Pale yellow	Green	Bottle-green	Uranium
11. Yellow	Greenish-yellow	Brownish	Emerald-green	Vanadium
12. Yellow	Colourless	Grey	Pale-violet	Titanium (3)
13. Orange-red	Colourless	Colourless	Colourless	Cerium

where the results obtained with some common metals have been tabulated (cf. Table V.8). Table VII.8 is more comprehensive, and contains results obtainable with molybdenum, gold, tungsten, uranium, vanadium, titanium, and cerium. In some cases the borax bead test is inconclusive, but combined with the microcosmic salt bead test, the appropriate elements can be singled out (see notes).

Notes to Table VII.8: 1. Bright red in the presence of a trace of tin.
2. Microcosmic salt bead test: reducing flame, cold – blue; blood-red when fused with a trace of iron(II) sulphate.
3. Microcosmic salt bead test: reducing flame, hot – yellow, cold – violet; blood-red with a trace of iron(II) sulphate.

CHAPTER VIII AN ABBREVIATED COURSE OF QUALITATIVE INORGANIC ANALYSIS

VIII.1 INTRODUCTION In many universities and colleges there is not enough time allocated in the curriculum to carry out a full study of qualitative inorganic analysis. For such institutions the abbreviated course, described in the present chapter can be recommended. With good preparation and organization such a course can be completed within 24 to 48 hours net laboratory time. It can also be recommended as a course to those whose main interests lie outside chemistry, but who wish to acquire some knowledge of qualitative inorganic analysis.

It is assumed that the student is familiar with the laboratory operations described in Chapter II. First, most important reactions of a limited number of cations and anions should be studied, followed by preliminary tests, testing for anions in mixtures and separation of cations.

VIII.2 REACTIONS OF CATIONS AND ANIONS For the abbreviated study, a selected number of cations and anions only should be studied, and only the most important reactions should be carried out.

The following reactions should be tried in the laboratory:

CATIONS

Pb^{2+}	Section **III.4**, reactions 1, 5, 6, 7, 16a.
Hg_2^{2+}	Section **III.5**, reactions 1, 3, 10.
Ag^+	Section **III.6**, reactions 1, 5, 6, 12.
Hg^{2+}	Section **III.8**, reactions 1, 3, 6, 11.
Bi^{3+}	Section **III.9**, reactions 1, 2, 6, 7, 11.
Cu^{2+}	Section **III.10**, reactions 1, 2, 3, 5, 6, 8, 14.
Cd^{2+}	Section **III.11**, reactions 1, 2, 4, 9.
As^{3+}	Section **III.12**, reactions 1, 2, 3.
As^{5+} (AsO_4^{3-})	Section **III.13**, reactions 1, 2, 3, 4.
Sb^{3+}	Section **III.15**, reactions 1, 2, 8.
Sn^{2+}	Section **III.18**, reactions 1, 2, 3.
Sn^{4+}	Section **III.19**, reactions 1, 3, 4.
Fe^{2+}	Section **III.21**, reactions 1, 2, 3, 4, 6, 7, 8, 9.

Fe^{3+}	Section **III.22**, reactions 1, 2, 3, 4, 6, 7, 11.
Al^{3+}	Section **III.23**, reactions 1, 2, 3, 7, 11.
Cr^{3+}	Section **III.24**, reactions 1, 2, 4, 7a, 9(c).
Co^{2+}	Section **III.26**, reactions 1, 2, 3, 5, 6, 7, 10(b).
Ni^{2+}	Section **III.27**, reactions 1, 2, 3, 6, 7, 8, 11(b).
Mn^{2+}	Section **III.28**, reactions 1, 2, 3, 5, 6, 7, 11.
Zn^{2+}	Section **III.29**, reactions 1, 2, 3, 6, 8, 9, 13.
Ba^{2+}	Section **III.31**, reactions 1, 2, 3, 4, 6, 10.
Sr^{2+}	Section **III.32**, reactions 1, 2, 3, 5, 6, 9.
Ca^{2+}	Section **III.33**, reactions 1, 2, 3, 5, 6, 7, 12.
Mg^{2+}	Section **III.35**, reactions 1, 2, 3, 5, 7, 8, 11.
K^{+}	Section **III.36**, reactions 1, 7.
Na^{+}	Section **III.37**, reactions 1, 4.
NH_4^{+}	Section **III.38**, reactions 1, 2, 7, 10.

ANIONS

CO_3^{2-}	Section **IV.2**, reactions 1, 2.
HCO_3^{-}	Section **IV.3**, reactions 1, 2, 3.
SO_3^{2-}	Section **IV.4**, reactions 1, 2, 4, 5, 8, 9.
$S_2O_3^{2-}$	Section **IV.5**, reactions 1, 2, 3, 6, 8.
S^{2-}	Section **IV.6**, reactions 1, 2, 3, 5, 6.
NO_2^{-}	Section **IV.7**, reactions 1, 2, 7, 10.
SCN^{-}	Section **IV.10**, reactions 1, 2, 3, 6.
Cl^{-}	Section **IV.14**, reactions 1, 2, 3.
Br^{-}	Section **IV.15**, reactions 1, 2, 3, 5, 8.
I^{-}	Section **IV.16**, reactions 1, 2, 4, 7.
F^{-}	Section **IV.17**, reactions 1, 2, 6.
NO_3^{-}	Section **IV.18**, reactions 1, 2, 3, 4.
ClO_3^{-}	Section **IV.19**, reactions 1, 2, 3, 7, 10.
SO_4^{2-}	Section **IV.24**, reactions 1, 2, 5, 6.
PO_4^{3-}	Section **IV.28**, reactions 1, 3, 4, 5, 6.
CrO_4^{2-}	Section **IV.33**, reactions 1, 2, 3, 5, 6.
CH_3COO^{-}	Section **IV.35**, reactions 1, 2, 3, 4, 6.
$(COO)_2^{2-}$	Section **IV.37**, reactions 1, 2, 3, 4.

VIII.3 SYSTEMATIC ANALYSIS. GENERAL CONSIDERATIONS In this abbreviated course of qualitative analysis students should try to analyse dissolved samples provided by the teacher. In the present chapter it is assumed that a solution is to be analysed, which may contain the ions discussed in Section **VIII.2**. If it is felt desirable to carry out analyses of solid samples, the preliminary tests on and dissolution of the sample should be carried out according to Sections **V.I** to **V.3** and **V.5** to **V.7**. In such a case the preliminary tests described under Section **VIII.4** need not be repeated, but the student may continue with Section **VIII.5**, always keeping in mind the results of the preliminary tests.

The systematic analysis of a solution should be started with preliminary tests

(Section **VIII.4**), followed by testing for anions (Sections **VIII.5–VIII.7**), and by separation and identification of the cations present (Section **VIII.8** and **VIII.9**). Some teachers of qualitative inorganic analysis may prefer to start with the separation of cations – in this case a test for phosphate and fluoride has to be carried out before attempting the separation of Group III cations.

Some teachers may prefer not to carry out a complete separation of cations, but to hand out separate unknown mixtures containing cations of one analytical group only. In this case precipitation is made of the particular group reagent (hydrochloric acid, hydrogen sulphide, ammonia, ammonium sulphide or ammonium carbonate) and the precipitate is examined by Group Separation Tables V.12 to V.19 as well as V.23 to V.28 for each group respectively.

VIII.4 PRELIMINARY TESTS ON SOLUTIONS The following preliminary tests need 0·5–1 ml sample. They should be carried out in the order given below.

(i) General observations The following observations and tests should be made on the solution:

1. Observe the colour, odour and any special physical properties.
2. Test its reaction towards litmus paper.
   (*a*) The solution is neutral: free acids, free bases, acid salts, and salts which give an acid or alkaline reaction owing to hydrolysis, are absent.
   (*b*) The solution reacts alkaline: this may be due to the hydroxides of the alkali and alkaline earth metals, to the carbonates, sulphides, hypochlorites, and peroxides of the alkali metals, etc.
   (*c*) The solution reacts acid: this may be due to free acids, acid salts, salts which yield an acid reaction because of hydrolysis, or to a solution of salts in acids.
3. Evaporate the solution to dryness and note the colour and appearance of the residue.

   If no residue remains, then the presence of volatile substances like carbon dioxide, ammonia, sulphur dioxide, hydrogen sulphide, hydrochloric, hydrobromic and hydroiodic acids, hydrogen fluoride, nitric acid, or ammonium salts may only be present.

(ii) Flame colourations To 0·5 ml of the test solution add 0·5 ml concentrated hydrochloric acid. Ignite a *clean* platinum wire in the non-luminous Bunsen flame, and dip the still hot wire into the solution. With the small-size sample adhered to it, place the wire again into the Bunsen flame and note your observation. If in doubt, repeat the procedure several times. Results are shown in Table VIII.1.

Table VIII.1 Flame colourations

Observation	Inference
Persistent golden-yellow flame.	Na.
Violet (lilac) flame (crimson through cobalt-blue glass).	K.
Brick-red (yellowish-red) flame.	Ca.
Crimson flame.	Sr.
Yellowish-green flame.	Ba.
Livid-blue flame (wire slowly corroded).	Pb, As, Sb, Bi, Cu.

The sodium flame masks that of other elements, e.g. that of potassium. Mixtures can be readily detected with a direct-vision spectroscope. A less delicate method is to view the flame through two thicknesses of cobalt-blue glass, whereby the yellow colour due to sodium is masked or absorbed; potassium then appears crimson.

(iii) Test for ammonium ions To 0·5 ml of the test solution add 2–3 ml of 2M sodium hydroxide solution. Boil the solution gently. The evolution of ammonia, detected by its odour, its action upon litmus paper and upon mercury(I) nitrate paper, indicates the presence of an ammonium ion.

In order to avoid the necessity of holding the test-paper (litmus, etc.) in the vapour, the following simple device may be used. A test-tube is fitted with a cork carrying a wide tube (at least half the diameter of the test-tube) about 5 cm long; the bottom of the wide tube should protrude just below the cork (Fig. V.I). A test-paper can then be supported in the wide tube by simply folding it slightly over the upper edge of the glass tube. This device is recommended for all reactions on the macro scale in which evolved gases are identified by means of test-paper.

Note: Sodium hydroxide is a dangerous substance because of its destructive action upon the eyes. Great care should be taken that the test-tube containing the hot sodium hydroxide solution is directed away from the eyes of the observer and of near neighbours.

(iv) Test for nitrate (or nitrite) If ammonium is found, continue boiling until ammonia can no longer be detected by its action upon red litmus paper or upon mercury(I) nitrate paper; it may be necessary to add a further 2–3 ml sodium hydroxide solution. Then add a little zinc dust or aluminium powder or finely powdered Devarda's alloy (Cu, 50 per cent; Al, 45 per cent; Zn, 5 per cent) and warm the mixture gently. Remove the flame as soon as evolution of hydrogen commences and allow the reduction to proceed (the reaction may become vigorous with aluminium powder and cooling with tap water may be necessary to moderate the vigour of the reaction; alternatively, thin aluminium foil may be used). If ammonium is absent, add zinc dust or aluminium powder or Devarda's alloy to the reaction mixture from test 3. If ammonia is evolved, as detected by its action upon red litmus paper or upon filter paper moistened with mercury(I) nitrate solution, then the presence of nitrate or nitrite is indicated.

It must be emphasized that the mercury(I) nitrate paper test for ammonia is not applicable in the presence of arsenites. Arsenites are reduced under the above conditions to arsine, which also blackens mercury(I) nitrate paper. Similar remarks apply to the tannic acid–silver nitrate test (Section **III.38**, reaction 7).

VIII.5 TESTING FOR ANIONS IN SOLUTION Tests for anions should be carried out in the order as given below.

1. Sulphate test To 2 ml of the solution add dilute hydrochloric acid until acid (test with litmus paper) and then add 2 ml in excess. Boil for 1–2 minutes to expel carbon dioxide completely, and then add about 1 ml barium chloride solution. A white precipitate ($BaSO_4$) shows the presence of sulphate. Confirm by test 1 in Section **IV.24** (charcoal reduction of precipitate).

2. Test for reducing agents Acidify 2 ml of the solution with dilute sulphuric acid and add 1 ml dilute sulphuric acid in excess. Add 3–4 drops of 0·004M potassium permanganate solution (prepared by diluting 1 ml of 0·02M $KMnO_4$ to 5 ml) from a dropper. Bleaching of the permanganate indicates the presence of one or more of the following reducing anions: sulphite, thiosulphate, sulphide, nitrite, bromide, iodide, and arsenite. If the permanganate is not decolourized, heat and observe the result. If the reagent is bleached only on heating, the presence of oxalate is indicated. A negative test shows the absence of the above anions.

3. Test for oxidizing agents Treat 2 ml of the solution cautiously with 1 ml concentrated hydrochloric acid, followed by 2 ml of the manganese(II) chloride reagent. A brown (or black) colouration indicates the presence of nitrate, nitrite, chlorate, or chromate. A negative test indicates the absence of the above oxidizing anions except small amounts of nitrates and nitrites. If reducing anions have been found, this test is inconclusive.

Note: The reagent consists of a saturated solution of manganese(II) chloride, $MnCl_2.4H_2O$, in concentrated hydrochloric acid. Its action depends upon its conversion by even mild oxidizing agents to a dark-brown-coloured pentachloromanganate(III) ion $[MnCl_5]^{2-}$.

4. Tests with silver nitrate solution Sulphide and sulphite interfere in the tests with silver nitrate solution, hence must be removed first as follows. Acidify

Table VIII.2 Separation of certain anions with $AgNO_3$ Acidify 10 ml of the solution with dilute acetic acid (use litmus paper). Determine the volume of the acidified solution with the aid of a small measuring cylinder, add one-tenth of the volume of concentrated HNO_3 and stir for 30 seconds. Then add $AgNO_3$ solution with stirring until precipitation is complete. Heat to boiling point, allow the precipitate to settle, cool, and filter. Wash the precipitate with 2–3 ml of M nitric acid. (1)

<table>
<tr><th>Residue</th><th colspan="3">Filtrate</th></tr>
<tr><td rowspan="5">AgCl – white.
AgBr – pale yellow.
AgI – yellow.</td><td colspan="3">Add 1 ml $AgNO_3$ solution, then 5 per cent $NaNO_2$ solution (prepared from the A.R. solid) dropwise and with stirring until precipitation is complete. [If no precipitate forms, do not add more than 0·5 ml $NaNO_2$ solution.] Filter, if necessary, and wash with 2–3 ml of 2M nitric acid (2).</td></tr>
<tr><td>Residue</td><td colspan="2">Filtrate</td></tr>
<tr><td rowspan="3">AgCl derived from $AgClO_3$.</td><td colspan="2">Add NaOH solution dropwise (use a dropper) and with vigorous stirring until neutral to litmus (3), then add 0·5 ml dilute acetic acid, followed by 1 ml $AgNO_3$ solution, and heat to about 80°C (4). If a permanent precipitate forms, add more $AgNO_3$ solution until precipitation is complete. Filter and wash with hot water.</td></tr>
<tr><td>Residue (5)</td><td>Filtrate</td></tr>
<tr><td>Ag_3PO_4 – yellow.
Ag_3AsO_4 – brownish-red.
Ag_3AsO_3 – yellow.
$Ag_2(COO)_2$ – white.</td><td>Discard.</td></tr>
</table>

10 ml of the solution with dilute acetic acid (use litmus paper): boil gently in a conical flask or porcelain dish in the fume cupboard to expel hydrogen sulphide or sulphur dioxide (3–5 minutes). It is important that the solution be acid throughout. Filter, if necessary, and allow to cool. If the volume has been reduced appreciably, add water to restore the original volume. Then carry out the operations described in Table VIII.2, when a number of anions can be separated and/or identified.

Notes to Table VIII.2: 1. $AgNO_3$ solution precipitates only AgCl, AgBr, and AgI from a dilute nitric acid solution, the other silver salts being soluble.
2. $NaNO_2$ solution reduces chlorate to chloride, which is precipitated as AgCl in the presence of $AgNO_3$ solution:

$$ClO_3^- + 3NO_2^- \rightarrow Cl^- + 3NO_3^-$$

In solutions faintly acid with acetic acid, phosphate, arsenate, arsenite, and oxalate are precipitated by $AgNO_3$ solution.
3. It is essential that the solution be just neutral to litmus or, at most, barely alkaline: the latter will be indicated by a very slight brown opalescence (due to Ag_2O) obtained upon stirring or shaking. If much silver oxide separates, it will redissolve only with difficulty.
4. Silver acetate is soluble in hot water and is held in solution.
5. If chromate is present (yellow or orange solution), it will be reduced by the $NaNO_2$ treatment and will precipitate here as green chromium(III) hydroxide. Chromate is readily detected in the solution as follows: acidify (say, 2 ml) with dilute HNO_3, cool, add 1 ml amyl alcohol and 0·5 ml of 3 per cent H_2O_2 solution and stir. A blue colouration (chromium pentoxide) in the organic layer confirms chromate.

If a mixture of the halides, or of phosphate, arsenate or arsenite, is suspected, use Table V.30 or V.31 given in Section **V.18**. The confirmatory tests for individual anions are collected in Section **V.19**. The reactions with silver nitrate solution are intended to act as a guide to the presence of groups of anions, and the table must be interpreted in conjunction with the observations made in the preliminary tests. Arsenite, arsenate, and chromate will also be found in the analysis for cations (Sections **V.12** and **V.14**).

5. ***Test with calcium chloride solution*** For tests 5 and 6 a practically neutral solution is required. This is obtained as follows. Transfer 4–5 ml of the solution to a porcelain dish and render it faintly acid with dilute nitric acid (use litmus paper). Boil for 3–4 minutes, allow to cool, then add dilute ammonia solution until just alkaline;* boil gently for 1 minute to remove any appreciable excess of ammonia. Divide the solution into two equal parts; reserve half for test 6.

Add $CaCl_2$ solution (equal in volume to that of the solution) and a little dilute acetic acid and allow to stand. A white precipitate indicates the presence of oxalate or fluoride or both. Filter off the precipitate and dissolve it by pouring a little hot dilute sulphuric acid into the filter. Treat the hot filtrate with a few drops of potassium permanganate solution. If the permanganate is reduced, oxalate is present.

* If a precipitate forms on neutralizing the solution, the presence of arsenic, antimony, and tin sulphides and possibly salts of amphoteric bases (lead, tin, aluminium, and zinc) is indicated. It shoul be filtered off and rejected.

*6. Test with iron(III) chloride solution** Treat the other portion of the 'neutral' solution from test 5 with aqueous iron(III) chloride solution.

Reddish-purple colouration indicates thiosulphate.

Reddish-brown colouration, yielding a brown precipitate on dilution and boiling, indicates acetate.

Yellowish-white precipitate indicates phosphate.

Blood-red colouration, discharged by mercury(II) chloride solution, indicates thiocyanate.

VIII.6 CONFIRMATORY TESTS FOR ANIONS The tests in the preceding section will indicate the anions present. In general, these should be confirmed by at least one distinctive confirmatory test. The following are recommended. Full experimental details will usually be found in Chapter IV under the reactions of the anions, the reference to these will be abbreviated as follows: thus (**IV.2**, 7) is to be interpreted as Section **IV.2**, reaction 7. It is assumed, of course, that interfering acids are absent or have been removed as described in Section **VIII.7**.

Chloride. Heat solid with concentrated H_2SO_4 and MnO_2; Cl_2 evolved (reddens then bleaches litmus paper and also turns KI–starch paper blue) (**VI.14**, 2).

Bromide. Heat solid with concentrated H_2SO_4 and MnO_2; Br_2 evolved (**IV.15**, 2) or chlorine water (or the equivalent NaOCl solution and dilute HCl) and CCl_4 test; brown colouration of CCl_4 layer (**IV.15**, 5).

Iodide. Chlorine water (or NaOCl solution and dilute HCl) and CCl_4 test; violet colouration of CCl_4 layer (**IV.16**, 4).

Fluoride. Silicon tetrafluoride test (heat with concentrated H_2SO_4 in a test-tube) (**IV.17**, 1); test 8 in **VII.16**; zirconium–alizarin-S test (**IV.17**, 6).

Nitrite. Brown ring test with dilute acetic acid or with dilute H_2SO_4 (**IV.7**, 2) or thiourea test (**IV.7**, 9).

Nitrate. Brown ring test with concentrated H_2SO_4 (**IV.18**, 3), if bromide, iodide, and nitrite absent.

Sulphide. Dilute H_2SO_4 on solid, and action of H_2S on lead or cadmium acetate paper (**IV.6**, 1).

Sulphite. Dilute H_2SO_4 on solid, odour of SO_2 and action of SO_2 on $K_2Cr_2O_7$ paper (**IV.4**, 1), or upon fuchsin solution (**IV.4**, 9); $BaCl_2$–Br_2 water test (**IV.4**, 2).

Thiosulphate. Dilute H_2SO_4 on solid and liberation of SO_2 ($K_2Cr_2O_7$ paper test or fuchsin solution test) and sulphur (**IV.5**, 1).

Sulphate. The $BaCl_2$ solution and dilute HCl test is fairly conclusive. Further confirmation is obtained by reduction of the precipitate ($BaSO_4$) on charcoal to sulphide (test for latter with lead acetate solution) (**IV.24**, 1).

Carbonate. Action of dilute H_2SO_4 on solid, and lime water test (**IV.2**, 1).

Hypochlorite. Action of dilute HCl and test for Cl_2 evolved (**IV.13**, 4).

Chlorate. The $AgNO_3$–$NaNO_2$ test (**IV.19**, 3).

Chromate. Hydrogen peroxide test with amyl alcohol as organic solvent (**IV.33**, 4).

---

*The bench reagent usually contains excess free acid added during its preparation in order to produce a clear solution: this may prevent the precipitation of the basic acetate on boiling. It is therefore recommended that dilute NH_3 solution be added dropwise to the $FeCl_3$ solution until a slight permanent precipitate forms; the precipitate is filtered off. The filtrate may be termed neutral $FeCl_3$ solution.

Arsenite. Immediate precipitate of As_2S_3 in dilute HCl solution (**III.12**, 1) and absence of precipitate with $Mg(NO_3)_2$ reagent (**III.12**, 3).

Arsenate. Action of H_2S on acid solution (**III.13**, 1), $AgNO_3$ solution test in faintly acetic acid solution (**III.13**, 2) and $Mg(NO_3)_2$ reagent test (**III.13**, 3).

Phosphate. Ammonium molybdate test (temperature not above 40°) (**IV.28**, 4).

Cyanide. Prussian blue test (**IV.8**, 4) or ammonium sulphide test (**V.18**, test 9).

Thiocyanate. Iron(III) chloride solution test: colour discharged by $HgCl_2$ solution or by NaF solution, but not by HCl (**IV.10**, 6).

Acetate. Action of ethanol or of n-butyl alcohol and concentrated H_2SO_4 (**IV.35**, 3) or indigo test (**IV.35**, 9) using ordinary test-tube or ignition tube.

Oxalate. The $CaCl_2$ test and decolourization of acidified $KMnO_4$ solution at about 70° is sufficiently conclusive (**IV.37**, 3, 4).

VIII.7 SPECIAL TESTS FOR MIXTURES OF ANIONS The tests described in this section are recommended even for beginners, as the chemistry involved in these is of great educational value. The tests will only briefly be described; full experimental details of the various tests are described in Chapter IV.

The tests should be carried out in the order given below.

1. Carbonate in the presence of sulphite Sulphites, on treatment with dilute sulphuric acid, liberate sulphur dioxide which, like carbon dioxide, produces a turbidity with lime or baryta water. The dichromate test for sulphites is, however, not influenced by the presence of carbonates. To detect carbonates in the presence of sulphites, treat the solid mixture with dilute sulphuric acid and pass the evolved gases through a small wash bottle or boiling tube containing potassium dichromate solution and dilute sulphuric acid. The solution will be turned green and the sulphur dioxide will, at the same time, be completely removed; the residual gas is then tested with lime water in the usual manner.

An alternative procedure is to add a little powdered potassium dichromate to the mixture and then to warm with dilute sulphuric acid; the evolved gas is passed through lime water.

The above method can, of course, be applied in the presence of thiosulphates.

2. Nitrate in the presence of nitrite The nitrite is readily identified in the presence of a nitrate by treatment with dilute mineral acid, potassium iodide and starch paste (or potassium iodide–starch paper), or by means of the thiourea test. The nitrate cannot, however, be detected in the presence of nitrite since the latter gives the brown ring test with iron(II) sulphate solution and dilute sulphuric acid. The nitrite is therefore completely decomposed first by one of the following methods:

(i) boiling with ammonium chloride solution until effervescence ceases;*

(ii) warming with urea and dilute sulphuric acid until evolution of gas ceases;*

(iii) adding a little sulphamic acid to the solution.

The last one is probably the simplest and most efficient method for the removal of nitrite in aqueous solution.

The brown ring test for nitrate can then be applied.

---

* Traces of nitrate are always formed in this reaction.

3. Nitrate in the presence of bromide and iodide The brown ring test for nitrates cannot be applied in the presence of bromides and iodides since the liberation of free halogen with concentrated sulphuric acid will obscure the brown ring due to the nitrate. The solution is therefore boiled with sodium hydroxide solution until ammonium salts, if present, are completely decomposed, and the solution is then cooled under the tap. Powdered Devarda's alloy or aluminium powder or zinc dust is then added and the mixture gently warmed. The evolution of ammonia, detected by its smell, its action upon red litmus paper and upon mercury(I) nitrate paper (see Section **III.38**, reaction 1) indicates the presence of a nitrate.

An alternative method is to remove the halides by precipitation with an almost saturated solution of silver sulphate (nitrate-free) and any excess of the latter with sodium carbonate solution; the nitrate is then tested for in the filtrate in the usual way (see also Section **IV.45**, 3).

4. Nitrate in the presence of chlorate The chlorate interferes with the brown ring test (cf. Section **IV.19**, reaction 1). The nitrate is reduced to ammonia as described under 3; the chlorate is at the same time reduced to chloride which may be tested for with silver nitrate solution and dilute nitric acid.

If chloride is originally present, it may be removed first by the addition of saturated silver sulphate solution.

5. Chloride in the presence of bromide and iodide This procedure involves the removal of the bromide and iodide with potassium or ammonium peroxodisulphate in the presence of dilute sulphuric acid. The free halogens are thus liberated, and may be eliminated either by simple evaporation (addition of water may be necessary to maintain the original volume) or by evaporation at about 80° in a stream of air.

Add solid potassium or ammonium peroxodisulphate to the solution of the mixed halides contained in a conical flask; strongly acidify with dilute sulphuric acid; heat the flask to about 80°, and aspirate a current of air through the solution with the aid of a filter pump (see Fig. IV.2 in Section **IV.45**, 5) until the solution is colourless. Add more solid peroxodisulphate or water as may be found necessary. Test the residual colourless liquid for chloride with silver nitrate solution and dilute nitric acid.

6. Chloride in the presence of iodide (bromide being absent) Acidify the solution with dilute nitric acid, add excess silver nitrate solution, and filter; reject the filtrate. Wash the precipitate with dilute ammonia solution and collect the washings. Add dilute nitric acid to the washings; a white precipitate of silver chloride indicates the presence of chloride.

The separation is based upon the solubility of silver chloride in dilute ammonia solution and the practical insolubility of silver iodide in this reagent.

7. Chloride in the presence of bromide (iodide being absent) Acidify the solution with dilute nitric acid and add an equal volume of concentrated nitric acid. Boil for 5 minutes or until all the bromine is expelled; then add silver nitrate solution. A white precipitate indicates chloride present.

8. Bromide and iodide in the presence of one another The presence of a chloride does not interfere with the reactions described below. Acidify the solution with dilute sulphuric acid and add 1–2 ml carbon tetrachloride; add 1–2 drops dilute sodium hypochlorite solution* (best with a dropper), and shake; a violet colouration in the carbon tetrachloride layer indicates iodide. Continue the addition of the hypochlorite solution drop by drop to oxidize the iodine to iodate (colourless) and shake after each addition. The violet colour will disappear, and a reddish-brown colouration of the carbon tetrachloride layer, due to dissolved bromine (or to bromine chloride BrCl), will be obtained if a bromide is present. If iodide alone is present, the solution will be colourless after the violet colour has disappeared.

*9. **Phosphate in the presence of arsenate*** Both arsenate and phosphate give a yellow precipitate on warming with ammonium molybdate solution and nitric acid, the latter on gentle warming (not above 40°) and the former on boiling. Also both anions give a white precipitate with the magnesium nitrate reagent (or with magnesia mixture). It must also be remembered in connection with the precipitation of Group II that arsenates are only slowly precipitated by hydrogen sulphide in dilute acid solution.

Acidify the solution with dilute hydrochloric acid, pass in sulphur dioxide to reduce the arsenate to arsenite, boil off the excess sulphur dioxide (test with potassium dichromate paper), and pass hydrogen sulphide into the solution to precipitate the arsenic as arsenic (III) sulphide: continue the passage of hydrogen sulphide until no more precipitate forms. Filter, boil off the hydrogen sulphide, and test the filtrate for phosphate by the ammonium molybdate test or with the magnesium nitrate reagent.

An alternative method for the elimination of arsenate is the following. Acidify the solution with dilute hydrochloric acid and then add one-quarter of the volume of concentrated hydrochloric acid (the total volume should be about 10 ml). Add 0·5 ml of 10 per cent ammonium iodide solution, heat to boiling and pass hydrogen sulphide into the boiling solution until precipitation is complete (5–10 minutes). Filter off the arsenic(III) sulphide, and boil off the hydrogen sulphide from the filtrate. Add dilute ammonia solution until alkaline and excess of the magnesium nitrate reagent (or of magnesia mixture). A white precipitate indicates the presence of phosphate.

If the white precipitate of magnesium ammonium phosphate is washed with a little water, and then treated on the filter paper with a little silver nitrate solution containing a few drops of dilute acetic acid, a yellow colouration, due to silver phosphate, is obtained. However, a similar reaction with the white precipitate produced by the magnesium nitrate reagent with a mixture of phosphate and arsenate yields a brownish-red colouration on the white precipitate; this is due to silver arsenate.

It may be pointed out that if arsenite is also present it may be readily detected in the filtrate obtained by treating the original mixture of arsenate, phosphate, and arsenite with the magnesium nitrate reagent; upon acidifying with 2M hydrochloric acid and passing hydrogen sulphide, an immediate yellow precipitate of arsenic(III) sulphide is produced.

---

* The use of dilute sodium hypochlorite solution and dilute acid is far more satisfactory than chlorine water. Alternatively, a 0·05M solution of chloramine-T (14·1 g ℓ^{-1}) may be employed; this is a source of hypochlorous acid (and therefore of chlorine):

$$CH_3.C_6H_4.SO_2.N.ClNa + H_2O \rightarrow CH_3.C_6H_4SO_2.NH_2 + Na^+ + OCl^-$$

VIII.8 SEPARATION AND IDENTIFICATION OF CATIONS IN SOLUTION The preliminary tests and tests for anions may have indicated the presence of certain elements (e.g. As, Cr, Mn) which are normally identified as cations. This information should always be kept in mind when the separation and identification of cations is to be attempted.

Before describing the general scheme for the separation of the cations into groups, the student should take note of the following facts:

1. The analysis should not be conducted with large quantities of the substance because much time will be spent in filtering the precipitates and difficulty may be experienced in washing and dissolving them. It is therefore recommended that about 5–20 ml be employed for the analysis. After a little experience the student will be able to judge from the relative sizes of the precipitates the relative quantities of the various components present in the mixture.

2. The tests must, in the first place, be carried out in the order given. A group reagent will separate its particular group from those which follow it and not from those which precede it. Thus hydrogen sulphide in the presence of 0·3M hydrochloric acid will separate Group II from Groups IIIA, IIIB, IV, and V, but does not separate Group II from Group I. It is most important, therefore, that one group should be completely precipitated before precipitation of the next group is attempted, otherwise the group precipitate will be contaminated by metals from the preceding groups and misleading results will be obtained.

The separation of cations into groups should be carried out according to the procedure given in Table VIII.3. Separations within individual groups should be carried out according to the instructions given in the Group Separation Tables in Chapter V, to which references are given in Table VIII.3. The notes, following the table contain much useful information and should be read *before* the separation into groups is attempted.

Table VIII.3 Separation of cations into groups (anions of organic acids, fluoride, and phosphate being absent) Add a few drops of dilute HCl to the cold solution. If a ppt. forms, continue adding dilute HCl until no further precipitation takes place. Filter (1) and wash the ppt. with a little water: add washings to filtrate (2).

<table>
<tr><th>Residue</th><th colspan="3">Filtrate</th></tr>
<tr><td rowspan="6">The ppt. may contain:
$PbCl_2$ – white.
Hg_2Cl_2 – white.
AgCl – white.
Group I present.
Examine by Group Separation Table V.13 (Section V.9).</td><td colspan="3">This must give no further precipitate with a few drops of dilute HCl. Add 1 ml of 3 per cent H_2O_2 solution (3). Adjust the HCl concentration to 0·3M (4). Heat to boiling and pass H_2S through the solution until precipitation is complete (5). Filter and wash (6).</td></tr>
<tr><th>Residue</th><th colspan="2">Filtrate</th></tr>
<tr><td rowspan="4">The ppt. may contain:
HgS – black.
PbS – black.
Bi_2S_3 – black or dark brown.
CdS –</td><td colspan="2">Test a small portion with H_2S to be certain that precipitation of Group II is complete. Boil down to about 10 ml in a porcelain dish and thus ensure that all H_2S has been removed (test with lead acetate paper). Add 1–2 ml of concentrated HNO_3 and boil to oxidise any iron(II) to iron(III) (7). Add 1–2 g of solid NH_4Cl, heat to boiling, add dilute NH_3 solution until mixture is alkaline and then 1 ml in excess, boil for 1 minute and filter immediately. Wash (8).</td></tr>
<tr><th>Residue</th><th>Filtrate</th></tr>
<tr><td>The ppt. may</td><td>Add 2–3 ml dilute NH_3 solution, heat, pass H_2S (under 'pressure') for 1 minute. Filter (9) and wash (10).</td></tr>
</table>

Table VIII.3 Separation of cations into groups (anions of organic acids, fluoride, and phosphate being absent) (contd.)

		Residue	**Filtrate**
yellow. CuS – black. SnS_3 – yellow. Sb_2S_3 – orange. As_2S_3 – yellow. Groups IIA and IIB present. Examine by Group Separation Tables V.14 to V.19 (Sections **V.10** to **V.12**).	contain: $Fe(OH)_3$ – reddish-brown. $Cr(OH)_3$ – green. $Al(OH)_3$ – white. $MnO_2 \cdot xH_2O$ – brown. Group IIIA present. Examine by Group Separation Table V.23 (Section **V.14**).	The ppt. may contain: CoS – black. NiS – black. MnS – pink. ZnS – white. Group IIIB present. Examine by Group Separation Tables V.24 or V.25 (Section **V.15**).	This must give no further ppt. with H_2S (9). Transfer to a porcelain dish and acidify with dilute acetic acid (11). Evaporate to a pasty mass [FUME CUPBOARD], allow to cool, add 2–3 ml concentrated HNO_3 so as to wash the solid around the walls to the centre of the dish and heat cautiously until the mixture is dry. Then heat more strongly until no more white fumes are evolved (12). Cool. Add 3 ml dilute HCl and 10 ml water: warm and stir to dissolve the salts. Filter, if necessary. Add 0·25 g solid NH_4Cl (or 2·5 ml 10 per cent NH_4Cl solution), render alkaline with concentrated NH_3 solution and then add, with stirring, $(NH_4)_2CO_3$ solution in slight excess. Keep, and stir the mixture in a water bath at 50–60°C for 3–5 minutes (13). Filter and wash with a little hot water.

Residue	**Filtrate**
The ppt. may contain: $BaCO_3$ – white. $SrCO_3$ – white. $CaCO_3$ – white. Group IV present. Examine by Group Separation Tables V.26 or V.27 (Section **V.16**).	May contain Mg^{2+}, Na^+ and K^+ (14). Evaporate to a pasty mass in a porcelain dish [FUME CUPBOARD], add 2 ml concentrated HNO_3, evaporate cautiously to dryness and then heat until white fumes cease to be evolved. White residue. Group V present. Examine by Group Separation Table V.28 (Section **V.17**).

Notes to Table VIII.3: This table and these notes which follow are intended for beginners and in consequence the various operations are given in somewhat greater detail than in Table V.12 (Section **V.8**).

1. If the solution contains chloride ions, it is evident that no silver or mercury(I) salt is present. When lead is present, the solution may be clear while hot, but $PbCl_2$ is deposited upon cooling the solution, due to the slight solubility of the salt in cold water. Lead may be found in Group II, even if it is not precipitated in Group I.
2. It is usually advisable in group separations to wash a precipitate with a small

volume of a suitable wash solution and to add the washings to the filtrate. In the present instance cold water or cold, very dilute HCl (say, 0·5M) may be used. The precipitating reagent, diluted 10 to 100-fold, is generally a suitable wash liquid. Specific directions for washing precipitates will usually be omitted from the present table in order to economize space.

3. The H_2O_2 solution is added to oxidize Sn^{2+} to Sn^{4+}, thus leading ultimately to the precipitation of SnS_2 instead of the somewhat gelatinous SnS. The excess H_2O_2 should be decomposed by boiling before passing H_2S, otherwise some S may be precipitated. The subsequent separation of Groups IIA and IIB by means of aqueous KOH is thus rendered more complete, since SnS_2 dissolves entirely and SnS dissolves only partially in aqueous KOH.

If it is intended to use ammonium polysulphide in the separation of Groups IIA and IIB, the addition of H_2O_2 is not essential since the $(NH_4)_2S_x$ will oxidize the SnS to SnS_2 and the latter dissolves as the thiostannate, $(NH_4)_2SnS_3$.

4. It is important that the concentration of the HCl be approximately correct, i.e. 0·3M, before passing H_2S: with higher concentrations of acid, lead, cadmium, and tin(II) will be incompletely precipitated. If the acidity is too low, sulphides of Group IIIB (NiS, CoS, and ZnS) may be precipitated.

Either of two methods may be employed to adjust the acid concentration.

(*a*) Concentrate the solution (if necessary) to a volume of 10–15 ml, cool. Add concentrated NH_3 solution dropwise from a dropper (the commercial 'medicine dropper' is satisfactory), with constant stirring, until the mixture is just alkaline. Introduce dilute HCl dropwise (use a dropper) until the mixture is just acid (use litmus paper). Then add 2·0 ml of 3M HCl (measured from a graduated pipette or calibrated dropper) and dilute the solution to a volume of 20 ml with distilled water.

(*b*) An alternative procedure is to make use of the indicator methyl violet (0·1 per cent aqueous solution or, better, the purchased or prepared indicator paper). The following table gives the colour of the indicator at various concentrations of acid:

Acid concentration	*p*H	Methyl violet indicator
Neutral or alkaline	7+	Violet
0·1M HCl	1·0	Blue
0·25M HCl	0·6	Blue-green
0·33M HCl	0·5	Yellow-green
0·5M HCl	0·3	Yellow

Add 1 drop of methyl violet indicator solution and introduce dilute HCl or dilute NH_3 solution (as necessary) dropwise and with constant stirring until the colour of the solution is yellow-green: a blue-green colour is almost but not quite acid enough, yet it is acceptable for most analyses. (If the indicator paper is available, the thoroughly stirred solution should be spotted on fresh portions of the paper.) It is recommended that a comparison solution containing, say, 10 ml of 0·3M HCl and 1 drop of indicator be freshly prepared; this will facilitate the correct adjustment of the acidity. A more satisfactory standard is a buffer solution prepared by mixing 5 ml of M sodium acetate, 5 ml of 2M HCl, and 5 ml of water; this has a *p*H of 0·5.

5. For the passage of H_2S into the solution, the latter is placed in a small

conical flask (one of 50 ml capacity is suitable) or in a boiling tube and the 'pressure' method used as detailed in Section **II.3**, 7. Heat the solution almost to boiling and pass in H_2S, whilst slowly shaking the flask with a swirling motion, until precipitation is complete: the latter will be apparent when bubbling either stops altogether or is reduced to a very slow rate of 1–2 bubbles per minute. Saturation is normally reached in 2–5 minutes. The best method of determining whether precipitation is complete is to filter off a portion of the solution and test the filtrate with H_2S. If only a white precipitate or suspension of sulphur is obtained, the presence of an oxidizing agent is indicated.

If an oxidizing agent is present (e.g. permanganate, dichromate or iron(III) ions) as is shown by the gradual separation of a fine white precipitate of sulphur and/or a change in colour of the solution, it is usual to pass SO_2 into the hot solution until reduction is complete, then to boil off the excess SO_2 (test with $K_2Cr_2O_7$ paper), and finally to pass H_2S. Arsenates, in particular, are slowly precipitated by H_2S; they are therefore usually reduced by SO_2 to arsenites and then precipitated as As_2S_3 with H_2S, after prior removal of the excess SO_2 in order to avoid interaction of the latter with H_2S and the consequent separation of S. Tin(IV) ions may be very slightly reduced to tin(II) by this treatment; the amount of reduction is, however, so small that it may be neglected. The original solution or substance must be tested for the valence state of the arsenic.

The objection to the use of SO_2 is that some sulphuric acid may be formed, especially upon boiling, and this may partially precipitate Pb, Sr, and Ba as sulphates. Any precipitate formed in this process should accordingly be examined for these cations: $PbSO_4$ is soluble in ammonium acetate solution.

An alternative procedure to be borne in mind when arsenate, etc., is present, which does not possess the disadvantages associated with SO_2 and is perhaps more expeditious, is to add 2–3 ml concentrated HCl and 0·5 ml of 10 per cent NH_4I solution. The arsenate is thereby reduced to arsenite and upon saturation of the hot solution with H_2S under 'pressure', the arsenic is completely precipitated as As_2S_3. This reduction can be carried out after the sulphides of the other elements have been precipitated in the presence of 0·3M HCl.

6. The wash liquid is prepared by dissolving 0·5 g of NH_4NO_3 in 10 ml water and treating this solution with H_2S; about 5 ml will suffice for the washing. The H_2S must be present in the wash liquid to prevent oxidation of some of the moist sulphides to sulphates.

7. If the iron was originally present as iron(III), it will be reduced to iron(II) by H_2S. It must be oxidized to the tervalent state (1–2 ml saturated Br_2 water may also be used) in order to ensure complete precipitation with NH_4Cl and dilute NH_3 solution. The original solution must be tested to determine whether the iron is present as Fe^{2+} or as Fe^{3+}.

8. The washing may be made with a little hot water or, better, with 2 per cent NH_4NO_3 solution.

9. If the filtrate is brown or dark-coloured, Ni may be suspected. The dark-coloured solution contains colloidal NiS, which runs through the filter paper. It may be acidified with acetic acid and boiled until the NiS is coagulated: this may either be added to the Group IIIB precipitate or tested separately for Ni. As a general rule the addition of macerated filter paper (e.g. in the form of a portion of a Whatman filtration accelerator or ashless tablet) to the suspension before filtration will lead to a clear or colourless filtrate.

10. The wash liquid may consist of 1 per cent NH_4Cl solution containing 1 per

cent by volume of ammonium sulphide solution. Oxidation of the moist sulphides to soluble sulphates is thus considerably lessened.

11. The filtrate must be immediately acidified and concentrated to remove H_2S. Ammonium sulphide solution on exposure to air slowly oxidizes to ammonium sulphate and would then precipitate any barium or strontium present as $BaSO_4$ and $SrSO_4$. Another reason for acidifying the filtrate from Group IIIB is to prevent the absorption of CO_2 from the air with the formation of carbonate ions; the latter would also precipitate the ions of Group IV.

12. The initial filtrate from Group IIIB will be almost saturated with ammonium salts. This concentration of ammonium ions is higher than is necessary to prevent the precipitation of $Mg(OH)_2$ and it may also lead to incomplete precipitation of the carbonates of Group IV metals. The latter effect is due to the acidic properties of ammonium ion:

$$NH_4^+ + CO_3^{2-} \rightleftharpoons NH_3 + HCO_3^-;$$

the concentration of CO_3^{2-} ions upon the addition of $(NH_4)_2CO_3$ would thus be considerably decreased, and incomplete precipitation of Group IV may occur. For these reasons most of the ammonium salts must be eliminated first.

Concentrated HNO_3 decomposes NH_4^+ ions at a lower temperature than is required for its volatilization:

$$NH_4^+ + HNO_3 \rightarrow N_2O\uparrow + H^+ + 2H_2O.$$

Loss by decrepitation and spurting during this operation must be avoided.

13. Ammonium carbonate decomposes appreciably above 60°C:

$$2NH_4^+ + CO_3^{2-} \rightarrow 2NH_3\uparrow + CO_2\uparrow + H_2O.$$

The digestion also improves the filtering properties of the precipitate.

14. Owing to the slight solubility of $CaCO_3$, $SrCO_3$, and $BaCO_3$ in solutions of ammonium salts, the filtrate from Group IV will, when these metals are present, contain minute amounts of the ions of the alkaline earth metals. Since the Group IV metals may interfere to a limited extent with the flame tests for Na and K and also the Na_2HPO_4 test for Mg (if employed), it has been recommended that the filtrate from Group IV be heated with a little (say, 1 ml) of $(NH_4)_2SO_4$ solution and $(NH_4)_2(COO)_2$ solution and filtered from any precipitate which forms. Owing to the comparatively small concentration of ammonium salts, this is generally unnecessary if the procedure described in Table V.28 (Section V.17) is adopted.

VIII.9 MODIFICATIONS IN THE PRESENCE OF ANIONS OF ORGANIC ACIDS, FLUORIDE, AND PHOSPHATE Anions of organic acids, fluoride, and phosphate interfere with the separation of cations after Group II has been precipitated. All these ions will cause the metals of Groups IIIB, IV, and also magnesium to precipitate when the solution is made ammoniacal to precipitate Group IIIA. It is essential that, if not done before, the solution should be tested for acetate, oxalate, fluoride, and phosphate at this stage. If any of these ions are found to be present, proceed with their removal in the order given below.

(i) If *acetate and/or oxalate* ions are present, evaporate the filtrate from Group II to dryness, when some carbon may be liberated and the organic matter decomposed. By repeated evaporation with concentrated nitric acid, oxidize the black residue completely. The residue must not be heated too strongly, as iron(III) oxide, chromium(III) oxide, and aluminium oxide may be rendered sparingly soluble in hydrochloric acid. Treat the residue as described under (ii).

(ii) If *fluoride* is present, the residue of (i), or the filtrate of Group II, must be evaporated almost to dryness with concentrated hydrochloric acid, and the procedure repeated once or twice. Hydrogen fluoride volatilizes and is thus removed.

(iii) If *phosphate* is present, follow the procedure given in Table VIII.4. It must be emphasized however that even if phosphate has been shown to be present, its separation need not be carried out if no precipitate is obtained in Group IIIA (with ammonia and ammonium chloride) and also if it is known that Group IV cations and magnesium are absent.

Table VIII.4 Removal of phosphate ions before precipitating the Group III cations Boil the filtrate from Group II until free from H_2S, add a few drops concentrated HNO_3 (or 1–2 ml bromine water) and boil gently for 1 minute. Test a small portion for phosphate with ammonium molybdate and nitric acid, and a further portion for the presence of Groups IIIA, IIIB, IV, or Mg by the addition of NH_4Cl and NH_3 solution. If both tests are positive, proceed as follows.

Adjust the volume of the solution to 10 ml (1). Add 0·5–1 g solid NH_4Cl, stir until dissolved, then add the zirconium nitrate reagent slowly and with stirring until precipitation is complete (2): a large excess of the reagent must be avoided. Heat the contents of the test-tube or small conical flask to boiling and stir with a glass rod to prevent bumping. Filter through a Whatman No. 32 filter paper (3): wash the ppt. with a little hot water and combine the washings with the filtrate.

Residue	Filtrate	
Zirconium phosphate. Reject.	Test if all the phosphate has been precipitated by the addition of a drop of the zirconium nitrate reagent: if no ppt. forms, all the phosphate has been removed. Add about 0·5 g solid NH_4Cl, heat to boiling, add a slight excess of dilute NH_3 solution (i.e. until the odour of ammonia is permanent in the boiling solution), boil for 2–3 minutes and filter.	
	Residue	**Filtrate**
	Examine for Group IIIA. The excess Zr will be found in the residue after treatment with H_2O_2 and NaOH solution (or with sodium perborate, $NaBO_3.4H_2O$, and boiling), and will accompany any Fe, if present.	Examine for Groups IIIB, IV, and V.

Notes to Table VIII.4: 1. It is essential that the acidity with respect to hydrochloric acid should not exceed M, otherwise a turbid supernatant liquid is obtained and the removal of phosphate is not quite complete. Test with methyl violet or any other suitable indicator (brilliant cresyl blue, cresol red, etc.).

2. It is important that the excess of the zirconium nitrate reagent should not exceed 25 per cent, otherwise a turbid supernatant liquid will be obtained; this turbidity cannot be removed by filtration or centrifugation. It is best, therefore, to add the zirconium nitrate solution slowly and with stirring until precipitation appears complete, heat just to boiling, filter, and test the filtrate with the reagent, etc.

3. The addition of half a Whatman filtration accelerator (or a little filter paper pulp) assists filtration; the precipitate must be washed thoroughly with hot water.

IX APPENDIX

IX.1 RELATIVE ATOMIC MASSES OF THE ELEMENTS

Element	Symbol	Atomic number	Relative atomic mass (1971)	Log of relative atomic mass
Actinium	Ac	89	227	2·3560
Aluminium	Al	13	26·9815	1·4311
Americium	Am	95	243	2·3856
Antimony	Sb	51	121·75	2·0855
Argon	Ar	18	39·948	1·6015
Arsenic	As	33	74·9216	1·8746
Astatine	At	85	210	2·3222
Barium	Ba	56	137·34	2·1378
Berkelium	Bk	97	247	2·3927
Beryllium	Be	4	9·01218	0·9548
Bismuth	Bi	83	208·9806	2·3201
Boron	B	5	10·81	1·0338
Bromine	Br	35	79·904	1·9026
Cadmium	Cd	48	112·90	2·0527
Calcium	Ca	20	40·08	1·6029
Californium	Cf	98	251	2·3997
Carbon	C	6	12·011	1·0796
Cerium	Ce	58	140·12	2·1465
Cesium	Cs	55	132·9055	2·1235
Chlorine	Cl	17	35·453	1·5497
Chromium	Cr	24	51·996	1·7160
Cobalt	Co	27	58·9332	1·7704
Copper	Cu	29	63·546	1·8031
Curium	Cm	96	247	2·3927
Dysprosium	Dy	66	162·50	2·2109
Einsteinium	Es	99	254	2·4048
Erbium	Er	68	167·26	2·2234
Europium	Eu	63	151·96	2·1817
Fermium	Fm	100	257	2·4099
Fluorine	F	9	18·9984	1·2787
Francium	Fr	87	223	2·3483
Gadolinium	Gd	64	157·25	2·1966
Gallium	Ga	31	69·72	1·8434
Germanium	Ge	32	72·59	1·8609
Gold	Au	79	196·9655	2·2944
Hafnium	Hf	72	178·49	2·2516
Helium	He	2	4·00260	0·6023
Holmium	Ho	67	164·9303	2·2173

Element	Symbol	Atomic number	Relative atomic mass (1971)	Log of relative atomic mass
Hydrogen	H	1	1·0080	0·0035
Indium	In	49	114·82	2·0600
Iodine	I	53	126·9045	2·1035
Iridium	Ir	77	192·92	2·2854
Iron	Fe	26	55·847	1·7470
Krypton	Kr	36	83·80	1·9232
Lanthanum	La	57	138·9055	2·1427
Lawrencium	Lr	103	257	2·4099
Lead	Pb	82	207·2	2·3164
Lithium	Li	3	6·941	0·8414
Lutetium	Lu	71	174·97	2·2427
Magnesium	Mg	12	24·305	1·3857
Manganese	Mn	25	54·9380	1·7399
Mendelevium	Md	101	256	2·4082
Mercury	Hg	80	200·59	2·3023
Molybdenum	Mo	42	95·94	1·9820
Neodymium	Nd	60	144·24	2·1591
Neon	Ne	10	20·179	1·3049
Neptunium	Np	93	237·0482	2·3748
Nickel	Ni	28	58·71	1·7687
Niobium	Nb	41	92·9064	1·9680
Nitrogen	N	7	14·0067	1·1463
Nobelium	No	102	254	2·4048
Osmium	Os	76	190·2	2·2792
Oxygen	O	8	15·9994	1·2041
Palladium	Pd	46	106·4	2·0269
Phosphorus	P	15	30·9738	1·4910
Platinum	Pt	78	195·09	2·2902
Plutonium	Pu	94	244	2·3874
Polonium	Po	84	210	2·3222
Potassium	K	19	39·102	1·5922
Praesodymium	Pr	59	140·9077	2·1489
Prometheum	Pm	61	145	2·1614
Protactinium	Pa	91	231·0359	2·3637
Radium	Ra	88	226·0254	2·3542
Radon	Rn	86	222	2·3464
Rhenium	Re	75	186·2	2·2700
Rhodium	Rh	45	102·9055	2·0124
Rubidium	Rb	37	85·4678	1·9318
Ruthenium	Ru	44	101·07	2·0046
Samarium	Sm	62	150·4	2·1772
Scandium	Sc	21	44·9559	1·6528
Selenium	Se	34	78·96	1·8974
Silicon	Si	14	28·086	1·4485
Silver	Ag	47	107·868	2·0329
Sodium	Na	11	22·9898	1·3615
Strontium	Sr	38	87·62	1·9426
Sulphur	S	16	32·06	1·5060
Tantalum	Ta	73	180·9479	2·2576
Technetium	Tc	43	98·9062	1·9952
Tellurium	Te	52	127·60	2·1059
Terbium	Tb	65	158·9254	2·2012
Thallium	Tl	81	204·37	2·3104
Thorium	Th	90	232·0381	2·3656
Thulium	Tm	69	168·9342	2·2277
Tin	Sn	50	118·69	2·0744

Element	Symbol	Atomic number	Relative atomic mass (1971)	Log of relative atomic mass
Titanium	Ti	22	47·90	1·6803
Tungsten	W	74	183·85	2·2645
Uranium	U	92	238·029	2·3766
Vanadium	V	23	50·9414	1·7071
Xenon	Xe	54	131·30	2·1183
Ytterbium	Yb	70	173·04	2·2381
Yttrium	Y	39	88·9059	1·9489
Zinc	Zn	30	65·38	1·8154
Zirconium	Zr	40	91·22	1·9601

IX.2 REAGENT SOLUTIONS AND GASES The reagents are listed alphabetically. One asterisk * indicates that a reagent has a limited stability and should not be kept for longer than 1 month. Two asterisks** indicate that the reagent should be prepared freshly and discarded after use. Reagents with no asterisk can be kept for at least one year after preparation.

Albumin solution.* Dissolve 0·1 g albumin in 20 ml water to obtain a colloid solution.

Acetic acid (concentrated, 'glacial'). The commercial concentrated or glacial acetic acid is a water-like solution with a characteristic smell, having a density of 1·06 g cm^{-3}. It contains 99·5% (w/w) of CH_3COOH (1·06 g CH_3COOH per ml) and is approximately 17·6 molar. When cooling to 0°C, the reagent freezes forming ice-like crystals, hence the name 'glacial'. It has to be handled with care.

Acetic acid (1:1 or 9M). To 50 ml water add 50 ml glacial acetic acid and mix.

Acetic acid (30% v/v). Dilute 30 ml glacial acetic acid, CH_3COOH, with water to 100 ml.

Acetic acid (2M). Dilute 114 ml glacial acetic acid with water to 1 litre.

Acetone. The pure solvent is a clear colourless liquid with a characteristic odour. Its specific gravity is 0·79 g cm^{-3} and it boils at 56·2°C.

Acetylacetone. $CH_3.CO.CH_2.CO.CH_3$. The commercial reagent is a colourless, sometimes yellowish liquid with a density of 0·97 g cm^{-3}. It boils at 137°C.

Alizarin (saturated solution in ethanol). To 2 g alizarin, $C_{14}H_8O_4$, add 10 ml ethanol and shake. Use the clear solution for the tests.

Alizarin–hydrochloric acid solution. Mix 19 ml saturated alcoholic solution of alizarin and 1 ml of 2M hydrochloric acid.

Alizarin–rubeanic acid–salicylaldoxime reagent (for paper chromatography).** Dissolve 20 g alizarin, $C_{14}H_8O_4$, 0·1 g rubeanic acid, $NH_2.CS.CS.NH_2$, and 1 g salicylaldoxime, $C_6H_4CH(NOH)OH$, in 100 ml of 96% ethanol.

Alizarin-S (2%). Dissolve 2 g alizarin-S (sodium alizarinsulphonate), $C_{14}H_7O_4.SO_3Na.H_2O$, in 100 ml water.

Alizarin-S (0·1%). Dissolve 0·1 g alizarin-S (sodium alizarinsulphonate), $C_{14}H_7O_4.SO_3Na.H_2O$, in 100 ml water.

Aluminium chloride (0·33M). Dissolve 80·5 g aluminium chloride hexahydrate, $AlCl_3.6H_2O$, in water and dilute to 1 litre.

Aluminium sulphate (0·17M). Dissolve 107·2 g aluminium sulphate hexadecahydrate, $Al_2(SO_4)_3.16H_2O$, in water and dilute to 1 litre. Alternatively, dissolve 158·1 g aluminium potassium sulphate icosytetrahydrate (potash alum), $K_2SO_4.Al_2(SO_4)_3.24H_2O$, in water and dilute to 1 litre.

Aluminon (0·1 %).** Dissolve 0·1 g aluminon (tri-ammonium-aurine-tri-carboxylate), $C_{22}H_{14}O_9$, in 100 ml water.

p-**Amino-dimethylaniline hydrochloride** (1 %).* Dissolve 1 g *p*-amino-NN-dimethylaniline dihydrochloride, $H_2N.C_6H_4.N(CH_3)_2.2HCl$, in 100 ml water.

Ammonia solution (concentrated). The commercial concentrated ammonia solution is a water-like liquid with a characteristic smell, owing to the evaporation of ammonia gas. It has a density of 0·90 g cm^{-3}, contains 58·6% (w/w) NH_3 (or 0·53 g NH_3 per ml), and is approximately 15·1 molar. It should be handled with care, wearing eye protection. Direct smelling of the solution should be avoided. The solution should be kept far apart from concentrated hydrochloric acid to avoid the formation of ammonium chloride fumes.

Ammonia solution (1:1, approx. 7·5M). To 500 ml water add 500 ml concentrated ammonia solution and mix.

Ammonia solution (2M). Dilute 134 ml concentrated ammonia solution with water to 1 litre.

Ammonia solution (1·5M). Dilute 100 ml concentrated ammonia solution, to 1 litre.

Ammonia solution (2·5%). Dilute 5 ml concentrated ammonia with water to 100 ml.

Ammonia solution (0·1M). Dilute 5 ml 2M ammonia with water to 100 ml.

Ammonia solution (0·02M). Dilute 1 ml 2M ammonia with water to 100 ml.

Ammonium acetate (10M). Dissolve 77 g ammonium acetate, CH_3COONH_4, in water and dilute the solution to 100 ml.

Ammonium carbonate (M).* Dissolve 96·1 g ammonium carbonate, $(NH_4)_2CO_3$, in water and dilute to 1 litre. Aged reagents must be boiled up and cooled before use to remove ammonium carbaminate from the solution.

Ammonium carbonate (concentrated).* Shake 20 g ammonium carbonate, $(NH_4)_2CO_3$, with 80 ml water. Allow to stand overnight and filter. Prepare the reagent freshly.

Ammonium chloride (saturated). To 4 g ammonium chloride add 10 ml water. Warm the mixture on a water bath, and use the clear, supernatant liquid of the cooled mixture as a reagent.

Ammonium chloride (20%). Dissolve 20 g ammonium chloride, NH_4Cl, in water and dilute to 100 ml.

Ammonium chloride (M). Dissolve 53·5 g ammonium chloride in water and dilute the solution to 1 litre.

Ammonium chloride (1 %). Dissolve 1 g ammonium chloride, NH_4Cl, in water and dilute to 100 ml.

Ammonium chloride–ammonium sulphide wash solution. Dissolve 1 g ammonium chloride in 80 ml water, add 1 ml of M ammonium sulphide and dilute the solution to 100 ml.

Ammonium fluoride (0·1M). Dissolve 3·7 g ammonium fluoride, NH_4F, in water and dilute to 1 litre.

Ammonium iodide (10%). Dissolve 10 g ammonium iodide, NH_4I, in water and dilute to 100 ml.

Ammonium metavanadate (0·1M). Dissolve 11·7 g ammonium metavanadate, NH_4VO_3, in 100 ml of M sulphuric acid and dilute with water to 1 litre.

Ammonium molybdate (0·25M for molybdenum).* Dissolve 44·2 g ammonium molybdate, $(NH_4)_6Mo_7O_{24}.4H_2O$, in a mixture of 60 ml concentrated ammonia and 40 ml water. Add 120 g ammonium nitrate and after complete dissolution dilute the solution to 1 litre. Before use, the solution, to which the reagent is added, must be made acid by adding nitric acid.

Ammonium molybdate (0·025M).* Dilute 1 ml of 0·25M ammonium molybdate with water to 10 ml.

Ammonium molybdate–quinine sulphate reagent.* Dissolve 4 g finely powdered ammonium molybdate, $(NH_4)_6Mo_7O_{24}.4H_2O$, in 20 ml water. Add to this solution, with stirring, a solution of 0·1 g quinine sulphate, $C_{20}H_{24}N_2O_2.H_2SO_4.2H_2O$, in 80 ml concentrated nitric acid.

Ammonium nitrate (M). Dissolve 80 g ammonium nitrate, NH_4NO_3, in water and dilute to 1 litre.

Ammonium nitrate (2%). Dissolve 20 g ammonium nitrate, NH_4NO_3, in water and dilute the solution to 1 litre.

Ammonium nitrate wash solution.** Dissolve 0·5 g ammonium nitrate, NH_4NO_3, in 10 ml water and saturate the solution with hydrogen sulphide gas.

Ammonium oxalate (0·17M or 2·5%). Dissolve 25 g ammonium oxalate monohydrate, $(COONH_4)_2.H_2O$, in 900 ml water and after complete dissolution dilute to 1 litre.

Ammonium oxalate (0·1M). Dissolve 14·2 g ammonium oxalate monohydrate, $(COONH_4)_2.H_2O$, in water and dilute to 1 litre.

Ammonium peroxodisulphate (M).* Dissolve 22·8 g ammonium peroxodisulphate, $(NH_4)_2S_2O_8$, in water and dilute to 100 ml.

Ammonium peroxodisulphate (0·1M).* Dissolve 22·8 g ammonium peroxodisulphate, $(NH_4)_2S_2O_8$, in water and dilute to 1 litre.

Ammonium polysulphide (M). To 1 litre ammonium sulphide solution (M) add 32 g sulphur, and heat gently until the latter dissolves completely and a yellow solution is formed. The formula of the reagent, $(NH_4)_2S_x$, where x is approx. 2.

Ammonium sulphate (saturated). Mix 45 g ammonium sulphate, $(NH_4)_2SO_4$, and 50 ml water. Heat the mixture on a water bath for 3 hours and allow to cool. Use the clear supernatant liquid for the tests.

Ammonium sulphate (M). Dissolve 132 g ammonium sulphate, $(NH_4)_2SO_4$, in water and dilute to 1 litre.

Ammonium sulphide (M).* Saturate 500 ml ammonia solution (2M) with hydrogen sulphide, until a small sample (1 ml) of the solution does not cause any precipitation in magnesium sulphate (M) solution. Then add 500 ml ammonia solution (2M). Store the solution in a well-stoppered bottle. The solution must be colourless. Yellow or orange colour indicates that considerable amounts of polysulphide are present in the solution.

Ammonium tartrate (6M).* Dissolve 50 g tartaric acid, $C_4H_6O_6$, in 50 ml water and add 50 ml concentrated ammonia solution. The solution contains an excess of ammonia.

Ammonium tetrathiocyanatomercurate(II) (0·3M).** Dissolve 9 g ammonium thiocyanate, NH_4SCN, and 8 g mercury(II) chloride, $HgCl_2$, in water and dilute to 100 ml.

Ammonium thiocyanate (0·1M). Dissolve 7·61 g ammonium thiocyanate, NH_4SCN, in water and dilute to 1 litre.

Ammonium thiocyanate (saturated solution in acetone).* Shake 1 g ammonium thiocyanate, NH_4SCN, with 5 ml acetone, and use the clear supernatant liquid for the tests.

Amyl acetate, $CH_3COOC_5H_{11}$. The commercial reagent is a clear colourless liquid with a pleasant odour. It has a density of 0·87 g ml^{-1} and boils at 138°C.

(iso)-Amyl alcohol (3-methylbutan-1-ol). The pure commercial solvent is a colourless liquid with a characteristic odour. It has a density of 0·81 g ml^{-1}. It should be kept in a metal box or cupboard.

Aniline. The commercial product is an almost colourless oily liquid, which darkens on longer standing. Its density is 1·02 g ml^{-1}.

Aniline sulphate (1%).* Dissolve 1 g aniline sulphate $(C_6H_5NH_2)_2.H_2SO_4$ in 100 ml of water.

Anthranilic acid (5%).* Dissolve 0·5 g anthranilic acid $NH_2.C_6H_4.COOH$ in 10 ml of ethyl alcohol.

Antimony(III) chloride (0·2M).* Dissolve 45·6 g antimony(III) chloride $SbCl_3$ in 200 ml of 1+1 hydrochloric acid and dilute the solution to 1 litre. Alternatively, take 29·2 g antimony(III) oxide Sb_2O_3, dissolve it in 250 ml of 1:1 hydrochloric acid by heating the mixture on a water bath. After cooling dilute the solution to 1 litre.

Aqua regia.** To 3 volumes of concentrated hydrochloric acid add 1 volume of concentrated nitric acid. Mix and use immediately, the solution does not keep at all.

Arsenic trioxide (arsenious acid) (0·1M for arsenic; **POISON**). Boil 9·89 g arsenic trioxide with 500 ml water until complete dissolution. Cool the solution and dilute with water to 1 litre.

Ascorbic acid (0·05M.** Dissolve 0·9 g ascorbic acid in 100 ml water.

Barium chloride (0·25M). Dissolve 61·1 g barium chloride dihydrate, $BaCl_2.2H_2O$, in water and dilute to 1 litre.

Barium nitrate (0·25M). Dissolve 65·3 g barium nitrate, $Ba(NO_3)_2$, in water and dilute to 1 litre.

Baryta water (saturated). Shake 5 g barium hydroxide octahydrate, $Ba(OH)_2.8H_2O$, with 100 ml water. Allow to stand for 24 hours. Use the clear supernatant liquid for the tests.

Benzene. The commercial solvent is a clear, colourless liquid with a characteristic odour. It has a specific weight of 0·878 g cm^{-3} and boils at 80°C. It is highly inflammable.

Benzidine acetate (0·05%).* Dissolve 0·05 g benzidine (4-4′-diaminodiphenyl), $H_2N.C_6H_4.C_6H_4.NH_2$, in 100 ml of 2M acetic acid. (DANGER: THE REAGENT IS CARCINOGENIC.)

Benzidine hydrochloride (0·5%).* Dissolve 0·5 g benzidine (4-4′-diaminodiphenyl), $H_2N.C_6H_4.C_6H_4.NH_2$, in 2 ml of 1:1 hydrochloric acid and dilute the solution with water to 100 ml. (DANGER: THE REAGENT IS CARCINOGENIC.)

α-Benzoin oxime (5% in alcohol.** Dissolve 5 g benzoin oxime, $C_6H_5.CH(OH).C(NOH).C_6H_5$, in 95% ethanol and dilute with the solvent to 100 ml.

Beryllium sulphate (0·1M). Dissolve 17·7 g beryllium sulphate tetrahydrate, $BeSO_4.4H_2O$ in water and dilute to 1 litre.

Bismuth nitrate (0·2M). To 500 ml water add cautiously 50 ml concentrated nitric acid, HNO_3. Dissolve 97·0 g bismuth nitrate pentahydrate, $Bi(NO_3)_3.5H_2O$, in this mixture, and dilute with water to 1 litre.

Bromine water (saturated).** Shake 4 g (or 1 ml) liquid bromine with 100 ml water. Ensure that a slight excess of undissolved bromine is left at the bottom of the mixture. The solution keeps for 1 week. When handling bromine, exercise utmost care; use rubber gloves and eye protection always.

Bromothymol blue (0·04%). Dissolve 40 mg bromothymol blue in 100 ml ethanol.

n-Butyl alcohol (butan-1-ol). The commercial solvent is a clear, colourless liquid with a characteristic odour. It has a density of 0·81 g cm^{-3} and boils at 116°C.

Cacothelyne (0·25%).* Dissolve 0·25 g cacothelyne (nitrobruciquinone hydrate, $C_{21}H_{21}O_7N_3$) in 100 ml water.

Cadmium acetate (0·5M). Dissolve 13·7 g cadmium acetate dihydrate, $Cd(CH_3COO)_2.2H_2O$, in water and dilute to 100 ml.

Cadmium carbonate suspension (freshly precipitated).** To 20 ml of 0·5M cadmium acetate add 20 ml of 0·5M sodium carbonate solution. Allow the precipitate to settle and wash 4–5 times with water by decantation.

Cadmium sulphate (0·25M). Dissolve 64·13 g tricadmium sulphate octahydrate $3CdSO_4.8H_2O$ in water and dilute to 1 litre.

Calcium chloride (0·5M). Dissolve 109·5 g calcium chloride hexahydrate, $CaCl_2.6H_2O$, in water and dilute to 1 litre. If the solid reagent is not pure enough, weigh 50 g calcium carbonate into a large porcelain dish, add slowly 100 ml of 1:1 hydrochloric acid, waiting with the addition of a new portion until effervescence ceases, then evaporate the mixture to dryness and dissolve the residue in water. Dilute finally to 1 litre.

Calcium nitrate (saturated). Mix 80 g calcium nitrate tetrahydrate, $Ca(NO_3)_2.4H_2O$, and 20 ml water. Heat the mixture on a water bath until complete dissolution. Allow to cool.

Calcium sulphate (saturated). Shake 0·35 g calcium sulphate monohydrate, $CaSO_4.H_2O$, with 100 ml water. Allow the mixture to stand for 24 hours and use the clear supernatant liquid for the tests.

Carbon dioxide gas. This gas can be obtained from a Kipp apparatus using calcium carbonate pieces (broken marble) and 1:1 hydrochloric acid. The gas should be washed with concentrated sulphuric acid.

Carbon tetrachloride. The commercial reagent is a colourless liquid with a characteristic smell. It has a high density (1·59 g cm^{-3}) and a high refractive index (1·46). Although it is not inflammable, it should be kept together with other organic solvents in a metal box or cupboard.

Cellosolve $C_2H_5O.CH_2.CH_2OH$ (2-ethoxyethanol, ethylene glycol monoethyl ether). The commercial solvent is a clear liquid with a characteristic odour. It has a density of 0·93 g cm^{-3} and boils at 134°C.

Cerium(III) nitrate (0·1M). Dissolve 43·5 g cerium(III) nitrate hexahydrate, $Ce(NO_3)_3.6H_2O$, in 100 ml 2M nitric acid and dilute with water to 1 litre.

Cerium(IV) sulphate (0·1M). Dissolve 40·4 g of cerium(IV) sulphate tetrahydrate, $Ce(SO_4)_2.4H_2O$, or 63·6 g cerium(IV) diammonium sulphate dihydrate, $Ce(SO_4)_2.2(NH_4)_2SO_4.2H_2O$, in a cold mixture of 500 ml water and 50 ml concentrated sulphuric acid, and dilute the solution with water to 1 litre.

Chloramine-T (0·05M).** Dissolve 14·1 g chloramine-T, $CH_3.C_6H_4.SO_2.N.NaCl.3H_2O$, in water and dilute to 1 litre.

Chlorine gas. This gas is available commercially in steel cylinders and should be taken from these. Alternatively, it can be produced in a Kipp apparatus (with glass joints only) from calcium hypochlorite (bleaching powder) and 1:1 hydrochloric acid. It is advisable to mix the bleaching powder with some gypsum powder, making the mixture wet, to produce lumps of the reagent. After drying these lumps can be placed into the Kipp apparatus.

Chlorine water (about 0·1M).** Saturate 200 ml of water with chlorine gas.

Chloroform. The commercial reagent is a colourless liquid with a characteristic smell. Chloroform vapour should not be inhaled; it causes drowsiness. It has a high density (147 g cm^{-3}) and a high refractive index (1·44). Although it is not inflammable, it should be kept together with other organic solvents in a metal box or cupboard.

Chromic acid (25%). Pour 25 ml chromosulphuric acid cautiously, under constant stirring, into 70 ml water. After cooling dilute the solution with water to 100 ml.

Chromium(III) chloride (0·33M). Dissolve 88·8 g chromium(III) chloride hexahydrate, $CrCl_3.6H_2O$, in water and dilute to 1 litre.

Chromium(III) sulphate (0·167M). Dissolve 110·4 g chromium(III) sulphate pentadecahydrate, $Cr_2(SO_4)_3.15H_2O$, in water and dilute to 1 litre.

Chromosulphuric acid (concentrated). To 100 g potassium or sodium dichromate ($K_2Cr_2O_7$ or $Na_2Cr_2O_7$) add 1 litre concentrated sulphuric acid. Stir the mixture occasionally and keep it in a stoppered vessel. Because of its strong oxidizing and dehydrating properties handle the solution with the greatest care (eye glasses, rubber gloves). Its cleansing action is slow but effective; best results are obtained if the mixture is left overnight in the vessel to be cleaned. Used portions should be poured back into the stock of the mixture. As the brownish colour changes slowly to green, discard the solution.

Chromotropic acid (saturated).** To 0·1 g chromotropic acid (1,8-dihydroxy-naphthalene-3,6-disulphonic acid; sodium salt) $C_{10}H_6O_8S_2Na_2$ add 5 ml water, mix thoroughly, and use the pure supernatant liquid for the tests.

Cinchonine–potassium iodide reagent (1%).* Dissolve 1 g cinchonine in a mixture of 99 ml water and 1 ml dilute nitric acid (2M) by boiling, let the mixture cool and dissolve 1 g potassium iodide, KI, in the solution. The reagent is stable for 2 weeks.

Cobalt(II) acetate (0·1M). Dissolve 2·5 g cobalt(II) acetate tetrahydrate, $(CH_3COO)_2Co.4H_2O$, in 100 ml water.

Cobalt(II) chloride (0·5M). Dissolve 119 g cobalt(II) chloride hexahydrate, $CoCl_2.6H_2O$, in water and dilute to 1 litre.

Cobalt(II) nitrate (0·5M). Dissolve 146 g cobalt(II) nitrate hexahydrate, $Co(NO_3)_2.6H_2O$, in water and dilute to 1 litre.

Cobalt(II) thiocyanate (10%).** Dissolve 1 g cobalt acetate tetrahydrate $CO(CH_3COO)_2.4H_2O$ in 5 ml water. In a separate vessel dissolve 1 g ammonium thiocyanate, NH_4SCN, in 5 ml water, and mix the two solutions.

Copper(II) acetate–benzidine solution.** *Solution I:* Dissolve 3 g copper(II) acetate monohydrate in $(CH_3COO)_2Cu.H_2O$ in 100 ml water. *Solution II:* Dissolve 1 g benzidine, $H_2N.C_6H_4.C_6H_4.NH_2$, in 100 ml 2M acetic acid. The two solutions should be kept separately; equal volumes of each should be mixed immediately before use. (DANGER: THE REAGENT IS CARCINOGENIC.)

Copper(I) chloride reagent (M).* Dissolve 9·9 g copper(I) chloride, CuCl, in a mixture of 60 ml water and 40 ml concentrated hydrochloric acid. Heating accelerates dissolution. Add to the solution strips of bright copper, which should be left there.

Copper(II) sulphate (0·25M). Dissolve 62·42 g copper sulphate pentahydrate, $CuSO_4.5H_2O$, in water and dilute to 1 litre.

Copper(II) sulphate (0·1%). Dissolve 0·1 g copper sulphate pentahydrate, $CuSO_4.5H_2O$, in 100 ml water.

Copper sulphide suspension.** Dissolve 0·12 g copper sulphate pentahydrate $CuSO_4.5H_2O$ in 100 ml water, add 5 drops of 2M ammonia solution and introduce hydrogen sulphide gas until the solution becomes cloudy. The suspension must be freshly prepared.

Cupferron (2%).** Dissolve 2 g cupferron, $C_6H_5.N(NO)ONH_4$, in 100 ml water. The solution does not keep well. Addition of 1 g ammonium carbonate, $(NH_4)_2CO_3$, enhances the stability.

Diammineargentatonitrate (0·1M).** To 10 ml 0·1M silver nitrate add 2M ammonia solution, until the precipitate first formed is just redissolved. Discard the unused reagent, because on standing it forms silver azide which may cause serious explosions.

Diammonium hydrogen phosphate (0·5M).* Dissolve 66 g diammonium hydrogen orthophosphate, $(NH_4)_2HPO_4$, in water and dilute to 1 litre.

Diazine green (0·01%). Dissolve 0·01 g diazine green (Janus green, Colour Index 11050) in 100 ml water.

Di-*en*-cuprato(II) sulphate (0·125M).* To 10 ml copper sulphate (0·25M) add 5 ml ethylenediamine and dilute with water to 20 ml. The solution has a dark-blue colour.

Diethyl ether. The commercial product is a clear, colourless, mobile liquid, which is highly inflammable. It has a density of 0·71 g cm^{-3}. It should be kept in a metal box or cupboard.

p-**Dimethylaminobenzylidene–rhodanine** (0·3% in acetone).** Dissolve 30 mg commercial *p*-dimethylaminobenzylidene–rhodanine reagent in 10 ml acetone.

p-**Dimethylaminobenzene-azo-phenyl-arsonic acid** (0·1%).** Dissolve 0·1 g 4′-dimethylaminoazobenzene-4-arsonic acid, $(CH_3)_2N.C_6H_4.N:N.C_6H_4AsO(OH)_2$, in 5 ml concentrated hydrochloric acid, and dilute with ethanol to 100 ml.

Dimethyl glyoxime (1%). Dissolve 1 g dimethyl glyoxime, $CH_3.C(NOH).C(NOH).CH_3$, in 100 ml ethanol.

β,β′-Dinaphthol (0·05%).* Dissolve 0·05 g di-β-naphthol, $(C_{10}H_6OH)_2$, in 100 ml concentrated sulphuric acid.

Dinitro-*p*-diphenylcarbazide (0·1%).** Dissolve 0·1 g *pp*′-dinitro-*sym*.-diphenyl carbazide, $CO(NH.NH.C_6H_4.NO_2)_2$, in 100 ml ethanol.

Dioxan (1,4-dioxan, diethylene dioxide), $C_4H_8O_2$. The commercial solvent is a clear, colourless liquid, with a characteristic odour. It has a density of 1·42 g cm^{-3}. The liquid freezes at 11·5°C and boils at 101°C.

Diphenylamine (0·5%).* Dissolve 0·5 g diphenylamine, $(C_6H_5)_2NH$, in 85 ml concentrated sulphuric acid, and dilute the solution with the greatest care with water to 100 ml.

Diphenylcarbazide (1% alcoholic solution).** Dissolve 0·10 g *sym*.diphenylcarbazide, $(C_6H_5.NH.NH)_2CO$, in 10 ml of 96% ethanol. The solution decomposes rapidly if exposed to air.

Diphenylcarbazide (0·2% solution with acetic acid).** Dissolve 0·2 g *sym*. diphenyl carbazide, $(C_6H_5.NH.NH)_2CO$, in 10 ml glacial acetic acid and dilute the mixture to 100 ml with ethanol.

Dipicrylamine (1%).** Boil a mixture of 0·2 g dipicrylamine (hexanitrodiphenylamine), $C_6H_2(NO_2)_3NH$, 4 ml of 0·5M sodium carbonate and 16 ml water. Allow to cool and filter. Use the filtrate for the tests.

Diphenylthiocarbazone (dithizone) (0·005%).* Dissolve 5 mg diphenylthiocarbazone (dithizone), $C_6H_5.N:N.CS.NH.NH.C_6H_5$, in 100 ml carbon tetrachloride or chloroform.

α,α′-Dipyridil reagent.* Dissolve 0·01 g of the solid reagent in 0·5 ml 0·1M hydrochloric acid. Alternatively, dissolve 0·01 g of the reagent in 0·5 ml ethanol. The reagent is very expensive, and it is worthwhile to use the minimum amount required.

Disodium hydrogen arsenate (0·25M). Dissolve 76 g disodium hydrogen arsenate, $Na_2HAsO_4.7H_2O$, in water and dilute the solution to 1 litre.

Disodium hydrogen phosphate (0·033M). Dissolve 12 g disodium hydrogen orthophosphate dodecahydrate, $Na_2HPO_4.12H_2O$, *or* 6 g disodium hydrogen orthophosphate dihydrate, $Na_2HPO_4.2H_2O$, *or* 4·2 g anhydrous disodium hydrogen orthophosphate, Na_2HPO_4, in water and dilute to 1 litre.

Dithizone. See diphenylthiocarbazone.

Ethanol (ethyl alcohol) (96%). The commercial product contains about 95–96% C_2H_5OH and has a density of 0·81 g cm^{-3}. It is inflammable and should be kept in a metal box or cupboard.

Ethanol (80%). To 80 ml 96% ethanol add 16 ml water.

Ethylamine (M). Dissolve 4·5 g ethylamine, $C_2H_5NH_2$, in water and dilute to 100 ml. Alternatively, dilute 7 ml of 70% ethylamine solution with water to 100 ml.

Ethylenediamine (diamino-ethane). The commercial product is a liquid with a density of 0·90 g cm^{-3}.

Ferron (0·2%).* Dissolve 0·2 g ferron (8-hydroxy-7-iodoquinoline-5-sulphonic acid), $C_9H_6O_4NSI$, in 100 ml water.

Fluorescein reagent. To 1 g fluorescein, $C_{20}H_{12}O_5$ (Colour Index 45350), add a mixture of 50 ml ethanol and 50 ml water. Shake, allow to stand for 24 hours, and filter the solution.

Formaldehyde solution (40%). Commercial formaldehyde solution contains about 40 g HCHO per 100 ml. It normally contains some methanol as stabilizer. It has a characteristic, pungent odour. If kept in cold for longer times, solid paraldehyde separates from the solution.

Formaldehyde (10%).* Dilute 10 ml of 40% formaldehyde solution (formalin) with 30 ml water.

Formaldehyde (4%)–**sodium carbonate** (0·05M) **reagent.**** To 50 ml water add 10 ml 0·5M sodium carbonate and 10 ml 40% formaldehyde solution, and dilute the mixture to 100 ml. The reagent decomposes on standing.

Formaldoxime (2·5%).* Dissolve 2·5 g formaldoxime hydrochloride, $(CH_2{:}NOH)_3.HCl$, in 100 ml water.

Fuchsin (0·1%). Dissolve 0·1 g fuchsin (magenta, Colour Index 42500) in 100 ml water.

Fuchsin (0·015%). Dissolve 0·015 g fuchsin (magenta, Colour Index 42500) in 100 ml water.

α-Furil dioxime (10%).** Dissolve 1 g α-furil dioxime, $C_{10}H_8O_4N_2$, in 10 ml alcohol.

Gallocyanine (1%).** Dissolve 1 g gallocyanine, $C_{15}H_{12}O_5N_2$, in 100 ml water.

Gold(III) chloride (0·33M for gold). Dissolve 13·3 g sodium tetrachloroaurate(III) dihydrate, $NaAuCl_4.2H_2O$, in water and dilute to 100 ml. Alternatively, dissolve 6·57 g gold metal in 50 ml aqua regia, evaporate the solution to dryness, and dissolve the residue in 100 ml water.

Hexachloroplatinic(IV) acid (26% or 0·5M). Dissolve 2·6 g hexachloroplatinic(IV) acid, $H_2[PtCl_6].6H_2O$, in 10 ml water.

Hydrazine sulphate (saturated).* To 2 g hydrazine sulphate, $N_2H_4H_2SO_4$, add 5 ml water and saturate the solution by vigorous shaking.

Hydrochloric acid (concentrated, 'fuming'). The commercial concentrated hydrochloric acid is a water-like solution with a characteristic smell, and is 'fuming' owing to the evaporation of hydrogen chloride gas. It has a density of 1·19 g cm^{-3}, contains 36·0% (w/w) HCl (or 0·426 g HCl per ml), and is approximately 11·7 molar. The reagent should be stored far away from concentrated ammonia to prevent the formation of ammonium chloride fumes. It should be handled with care, using eye protection.

Hydrochloric acid (6M, 1:1). To 500 ml water add 500 ml concentrated hydrochloric acid, and let the solution cool to room temperature.

Hydrochloric acid (3M). To 500 ml water add 265 ml concentrated hydrochloric acid, and dilute with water to 1 litre.

Hydrochloric acid (2M). Pour 170 ml concentrated hydrochloric acid into 800 ml water under constant stirring, and dilute with water to 1 litre.

Hydrochloric acid (0·5 M). Dilute 4·5 ml concentrated hydrochloric acid with water to 100 ml.

Hydrochloric acid (0·1M). Dilute 20 ml 0·5M hydrochloric acid with water to 100 ml.

Hydrogen peroxide (concentrated).* Commercial concentrated hydrogen peroxide (so called '100 volume') solution contains 30% (w/w) of H_2O_2. Sometimes it contains small amounts of sulphuric acid or organic material as stabilizer. It should be handled with care.

Hydrogen peroxide (10%).* To 30 ml concentrated (30%) hydrogen peroxide add 60 ml water and mix.

Hydrogen peroxide (3%, approx. M).* Dilute 100 ml concentrated ('100 volume') hydrogen peroxide with water to 1 litre.

Hydrogen sulphide gas. This gas can be obtained from a Kipp apparatus, using solid iron(II) sulphide, FeS, and 1:1 (6M) solution of hydrochloric acid. The gas can be washed by bubbling it through water.

Hydrogen sulphide (saturated solution, about 0·1M).** Saturate 250 ml water with hydrogen sulphide gas, obtained from a Kipp apparatus. The solution contains approximately 4 g H_2S per litre. The solution will keep for about 1 week.

Hydroxylamine hydrochloride (10%).* Dissolve 10 g hydroxylammonium chloride, $NH_2OH.HCl$, in water and dilute to 100 ml.

8-Hydroxyquinoline (5%).* Dissolve 5 g 8-hydroxyquinoline, C_9H_7ON, in a mixture of 90 ml water and 10 ml of M sulphuric acid. The reagent is stable for several months.

8-Hydroxyquinoline (2%, in acetic acid).* Dissolve 2 g 8-hydroxyquinoline, C_9H_7ON, in 100 ml 2M acetic acid.

8-Hydroxyquinoline (1% in alcohol).* Dissolve 1 g 8-hydroxyquinoline, C_9H_7ON, in 100 ml of 96% ethanol.

Hypophosphorous acid (50%). The commercial reagent contains 49–53% pure HPO_2 and has a density of 1·2 g cm^{-3}. Dilute the solution with 4 volumes of water for the tests.

Indigo solution (1%). Dissolve 0·1 g indigo, $C_{16}H_{10}O_2N_2$ (Colour Index 73000), in 10 ml concentrated sulphuric acid.

Indole (0·015%). Dissolve 15 mg indole, $C_6H_4.NH.CH.CH$, in 100 ml ethanol.

Iodine (potassium tri-iodide) (0·05M). Dissolve 12·7 g iodine, I_2, and 25 g potassium iodide, KI, in water, and dilute to 1 litre. Do not handle iodine with a metal or plastic spatula: use a spoon made of porcelain or glass.

Iodine (potassium tri-iodide) (0·005M). Dilute 10 ml 0·05M iodine solution with water to 100 ml.

Iron(III) chloride (0·5M). Dissolve 135·2 g iron(III) chloride hexahydrate in water, add a few millilitres concentrated hydrochloric acid if necessary, and dilute with water to 1 litre. If the solution turns dark, add more hydrochloric acid.

Iron(III) periodate reagent.* Dissolve 2 g potassium periodate, KIO_4, in 10 ml freshly prepared 2M potassium hydroxide solution. Dilute with water to 50 ml, add 2 ml of 0·5M iron(III) chloride solution and dilute with 2M potassium hydroxide to 100 ml.

Iron(II) sulphate (saturated).** To 1 g iron(II) sulphate, $FeSO_4.7H_2O$, add 5 ml water. Shake well in the cold. Decant the supernatant liquid, which should be used for the tests. The reagent should be prepared freshly.

Iron(II) sulphate (0·5M).* Dissolve 139 g iron(II) sulphate heptahydrate, $FeSO_4.7H_2O$, or 196 g iron(II) ammonium sulphate hexahydrate (Mohr's salt, $FeSO_4.(NH_4)_2SO_4.6H_2O$) in a cold mixture of 500 ml water and 50 ml M sulphuric acid, and dilute the solution with water to 1 litre.

Iron(III) sulphate (25%). Heat 5 g iron(III) sulphate, $Fe_2(SO_4)_3$, with 15 ml water. After complete dissolution and cooling dilute with water to 20 ml.

Iron(III) thiocyanate (0·05M). Dissolve 1·35 g iron(III) chloride hexahydrate, $FeCl_3.6H_2O$, and 2 g potassium thiocyanate, KSCN, in water and dilute the solution to 100 ml.

Lanthanum nitrate (0·1M). Dissolve 4·33 g lanthanum nitrate hexahydrate, $La(NO_3)_3.6H_2O$, in 100 ml water.

Lead acetate (0·25M). Dissolve 95 g lead acetate trihydrate, $Pb(CH_3COO)_2.3H_2O$, in a mixture of 500 ml water and 10 ml glacial acetic acid, CH_3COOH, and dilute the solution with water to 1 litre.

Lead acetate (0·0025M). Dilute 1 ml of 0·25M lead acetate solution with water to 100 ml.

Lead nitrate (0·25M) Dissolve 82·8 g lead nitrate, $Pb(NO_3)_2$, in water and dilute to 1 litre.

Lime water (saturated).* Shake 5 g calcium hydroxide, $Ca(OH)_2$, with 100 ml water. Allow to stand for 24 hours. Use the clear supernatant liquid for the tests.

Lithium chloride (M). Dissolve 42·4 g anhydrous lithium chloride, LiCl, or 60·4 g lithium chloride monohydrate $LiCl.H_2O$ in water and dilute to 1 litre.

Magnesia mixture (0·5M for magnesium). Dissolve 102 g magnesium chloride hexahydrate, $MgCl_2.6H_2O$, and 107 g ammonium chloride, NH_4Cl, in water, dilute to 500 ml, add 50 ml concentrated ammonia and dilute the solution to 1 litre with water.

Magnesium chloride (0·5M). Dissolve 101·7 g magnesium chloride hexahydrate, $MgCl_2.6H_2O$, in water and dilute the solution to 1 litre.

Magnesium nitrate reagent (ammoniacal) (0·5M for magnesium). Dissolve 128 g magnesium nitrate hexahydrate, $Mg(NO_3)_2.6H_2O$, and 160 g ammonium nitrate, NH_4NO_3, in water, add 50 ml concentrated ammonia and dilute with water to 1 litre.

Magnesium sulphate (M). Dissolve 246·5 g magnesium sulphate heptahydrate, $MgSO_4.7H_2O$, in water and dilute the solution to 1 litre.

Magnesium sulphate (0·5M). Dissolve 123·2 g magnesium sulphate heptahydrate, $MgSO_4.7H_2O$, in water and dilute to 1 litre.

Manganese(II) chloride (saturated). Heat 7 g manganese(II) chloride tetrahydrate, $MnCl_2.4H_2O$, with 3 ml water on a water bath until complete dissolution. Allow to cool and use the clear supernatant liquid for the tests.

Manganese(II) chloride (saturated in HCl). To 7 g manganese(II) chloride tetrahydrate, $MnCl_2.4H_2O$, add 5 ml concentrated hydrochloric acid. Shake, allow to stand for 24 hours and use the clear, supernatant liquid for the tests.

Manganese(II) chloride (0·25M). Dissolve 49·5 g manganese(II) chloride tetrahydrate, $MnCl_2.4H_2O$, in water and dilute to 1 litre.

Manganese(II) nitrate–silver nitrate reagent.* Dissolve 2·87 g manganese(II) nitrate hexahydrate, $Mn(NO_3)_2.6H_2O$, in 40 ml water. Add a solution of 3·55 g silver nitrate, $AgNO_3$, in 40 ml water and dilute the mixture to 100 ml. Neutralize the solution with 2M sodium hydroxide until a black precipitate is starting to form. Filter and keep the solution in a dark bottle.

Manganese(II) nitrate–silver nitrate–potassium fluoride reagent.* To 100 ml manganese(II) nitrate–silver nitrate reagent add a solution of 3·5 g potassium fluoride in 50 ml water. Boil, filter off the dark precipitate in cold, and use the clear solution for the tests.

Manganese(II) sulphate (saturated). Shake 4 g manganese(II) sulphate tetrahydrate, $MnSO_4.4H_2O$, with 6 ml water. Allow to stand for 24 hours. Use the clear supernatant liquid for the tests.

Manganese(II) sulphate (0·25M). Dissolve 55·8 g manganese(II) sulphate tetrahydrate, $MnSO_4.4H_2O$, in water and dilute to 1 litre.

Mannitol (10%).* Dissolve 10 g mannitol, $C_6H_{14}O_6$, in water, and neutralize the solution with 0·01M sodium hydroxide against bromothymol blue indicator, until the colour of the solution just turns to green.

Mercury(I) chloride (calomel) suspension.* To 5 ml of 0·05M mercury(I) nitrate add 1 ml of 2M hydrochloric acid. Wash the precipitate 5 times with 10 ml water by decantation. Suspend the precipitate finally with 5 ml water.

Mercury(II) chloride (5% in alcohol). Dissolve 1 g mercury(II) chloride, $HgCl_2$, in 20 ml ethanol.

Mercury(II) chloride (saturated). Shake 7 g mercury(II) chloride, $HgCl_2$, with 100 ml water. Allow to stand for 24 hours and use the clear supernatant liquid for the tests.

Mercury(II) chloride (0·05M). Dissolve 13·9 g mercury(II) chloride, $HgCl_2$, in water and dilute to 1 litre.

Mercury(I) nitrate (0·05M).* Dissolve 28·1 g mercury(I) nitrate dihydrate, $Hg_2(NO_3)_2.2H_2O$, in a cold mixture of 500 ml water and 10 ml concentrated nitric acid. Dilute with water to 1 litre. Add 1 drop of pure mercury metal to prevent oxidation.

Mercury(II) nitrate (0·05M). Dissolve 17.1 g mercury(II) nitrate monohydrate, $Hg(NO_3)_2.H_2O$, in water and dilute the solution to 1 litre.

Mercury(II) sulphate reagent (0·2M). To 40 ml water add cautiously 10 ml concentrated sulphuric acid under constant stirring. Dissolve in this mixture 2·2 g mercury(II) oxide.

Methanol (methyl alcohol). The commercial product is a clear, colourless liquid with a characteristic odour. It boils at 64·7°C and has a density of 0·79 g cm^{-3}. It is **POISONOUS** (causing blindness and ultimately death) and should therefore never be tasted.

Methyl acetate, $CH_3COO.CH_3$. The commercial solvent is a liquid with a pleasant odour. It has a density of 0·93 g cm^{-3}.

2-Methyltetrahydroxyfuran (tetrahydrosylvan), $C_5H_{10}O$. The commercial solvent is a clear liquid with a density of 0·85 g cm^{-3} and a boiling range of 79–80°C.

4-Methyl-1:2-dimercapto benzene (dithiol reagent).** Dissolve 0·2 g 4-methyl-1:2-dimercapto benzene in 100 ml of 1 per cent hydroxide. If tin(IV) ions are to be tested, add 0·5 g thioglycollic acid to the solution.

Methylene blue (0·1 %). Dissolve 0·1 g methylene blue (Colour Index 52015) in 100 ml water.

Methyl-ethyl ketone CH_3-CO-C_2H_5 (butan-2-one). The commercial product is a clear liquid with a characteristic odour. It has a density of 0·803 g cm^{-3} and a boiling point of 80°C.

Nessler's reagent.* Dissolve 10 g potassium iodide in 10 ml water (solution *a*). Dissolve 6 g mercury(II) chloride in 100 ml water (solution *b*). Dissolve 45 g potassium hydroxide in water and dilute to 80 ml (solution *c*). Add solution *b* to solution *a* dropwise until a slight permanent precipitate is formed, then add solution *c*, mix, and dilute with water to 200 ml. Allow to stand overnight and decant the clear solution, which should be used for the test.

Nickel chloride (0·5M). Dissolve 119 g nickel chloride hexahydrate, $NiCl_2.6H_2O$, in water and dilute to 1 litre.

Nickel sulphate (0·5M). Dissolve 140 g nickel sulphate reagent (approx. composition $NiSO_4.7H_2O$) in water and dilute to 1 litre.

Nitrazine yellow (0·1 %). Dissolve 0·1 g of nitrazine yellow (2,4-dinitro-benzeneazo)-1-naphthol-4,8-disulphonic acid, disodium salt, (Colour Index 14890) in 100 ml water.

Nitric acid (concentrated). The commercial concentrated nitric acid is a water-like solution with a density of 1·42 g cm^{-3}. It contains 69·5 % (w/w) HNO_3 or 0·99 g HNO_3 per millilitre. It is approx. 15·6 molar. Nitrous fumes make the somewhat decomposed reagent reddish-brown. It should be handled with utmost care, wearing gloves and eye protection.

Nitric acid (83 %). Mix 68·0 ml concentrated (70 %) nitric acid with 66·2 ml fuming (95 %) nitric acid.

Nitric acid (1:1 or 8M). To 500 ml water add cautiously 500 ml concentrated nitric acid.

Nitric acid (2M). Pour 128 ml concentrated nitric acid into 500 ml water, and dilute to 1 litre.

Nitric acid (0·4M). Dilute 20 ml of 2M nitric acid with water to 100 ml.

***o*-Nitrobenzaldehyde** (0·5 %).* Dissolve 5 mg of 2-nitrobenzaldehyde, $NO_2.C_6H_4.CHO$, in 1 ml of 2M sodium hydroxide.

***p*-Nitrobenzene-azo-α-naphthol** (0·5 %).* Dissolve 0·5 g 4-(4-nitrophenylazo)-1-naphthol (magneson II), $NO_2.C_6H_4.N{:}N.C_{10}H_6.OH$, in a mixture of 10 ml 2M sodium hydroxide and 10 ml water. Dilute the solution to 100 ml.

***p*-Nitrobenzene-azo-chromotropic acid** (0·005 %).* Dissolve 5 mg 4-(4-nitro-phenylazo)-chromotropic acid, sodium salt (Chromotrope 2B, Colour Index 16575) in 100 ml of concentrated sulphuric acid.

***p*-Nitrobenzene-azo-resorcinol** (0·5 %).* Dissolve 0·5 g 4-nitrobenzene-azo-resorcinol [4-(4-nitrophenylazo)-resorcinol or magneson], $NO_2.C_6H_4.N{:}N.C_6H_2(OH)_2CH_3$, in a mixture of 10 ml 2M sodium hydroxide and 10 ml water. Dilute the solution with water to 100 ml.

p-**Nitrobenzene-diazonium chloride reagent.**** Dissolve 1 g 4-nitroaniline, $NO_2.C_6H_4.NH_2$, in 25 ml 2M hydrochloric acid and dilute with water to 160 ml. Cool, add 20 ml of 5 per cent sodium nitrite solution and shake until the precipitate dissolves. The reagent becomes turbid on keeping but can be employed again after filtering.

p-**Nitrobenzene-azo-orcinol** (0·025%).* Dissolve 25 mg 4-(4-nitrophenylazo)-orcinol, $NO_2.C_6H_4.N{:}N.C_6H_2(OH)_2.CH_3$, in 50 ml of 2M sodium hydroxide, and dilute the solution with water to 100 ml.

m-**Nitrobenzoic acid** (saturated).* Dissolve 1 g 3-nitrobenzoic acid, $NO_2.C_6H_4.COOH$, in 250 ml water by heating the mixture on a water bath. Allow to cool and filter.

Nitron reagent (5%).* Dissolve 5 g nitron ($C_{20}H_{16}N_4$) in 100 ml 2M acetic acid.

4-Nitrophthalene-diazoamino-azobenzene (0·02%).** Dissolve 0·02 g Cadion 2B, $NO_2.C_{10}H_6.N{:}N.NH.C_6H_4.N{:}N.C_6H_5$, in 100 ml ethanol to which 1 ml 2M potassium hydroxide is added.

α-Nitroso-β-naphtol (1%). Dissolve 1 g 1-nitroso-2-naphtol, $C_{10}H_6(OH)NO$, in 100 ml of 1:1 acetic acid. Instead of acetic acid 100 ml ethanol or acetone may be used.

Osmium tetroxide (1%). Dissolve 1 g osmium tetroxide (osmic acid), OsO_4, in 10 ml M H_2SO_4 and dilute with water to 100 ml. The solution is available commercially.

Oxalic acid (saturated). Shake 1 g oxalic acid dihydrate, $(COOH)_2.2H_2O$, with 10 ml water, and use the clear supernatant liquid for the tests.

Oxalic acid (0·5M). Dissolve 63 g oxalic acid dihydrate, $(COOH)_2.2H_2O$, in water and dilute to 1 litre.

Oxalic acid (2%). Dissolve 2 g oxalic acid dihydrate, $(COOH)_2.2H_2O$, in 100 ml water.

Oxine. See 8-hydroxyquinoline.

Palladium chloride (1%). Dissolve 1 g palladium chloride, $PdCl_2$, in water and dilute to 100 ml.

Perchloric acid (concentrated). The concentrated perchloric acid solution is a clear, colourless liquid with a density of 1·54 g cm^{-3} and contains 60% (w/w) $HClO_4$ (or 0·92 g $HClO_4$ per ml). It is approximately 9·2 molar.

Perchloric acid (2M). Dilute 216 ml concentrated perchloric acid with water to 1 litre.

o-**Phenanthroline** (0·1%). Dissolve 0·1 g 1,10-phenanthroline hydrate, $C_{12}H_8N_2.H_2O$, in 100 ml water.

p-**Phenetidine hydrochloride** (2%).* Dissolve 2 g *p*-phenetidine, $C_2H_5.O.C_6H_4.NH_2$, in 5 ml concentrated hydrochloric acid and dilute with water to 1 litre (complete dissolution may take place only after dilution).

p-**Phenetidine** (0·1%).* Dissolve 0·1 g *p*-phenetidine, $C_2H_5OC_2H_4NH_2$, in 100 ml 2M hydrochloric acid.

Phenolphthalein (0·5%). Dissolve 0·5 g phenolphthalein, $C_6H_4(COOH).CH(C_6H_4OH)_2$, in 100 ml ethanol.

Phenylarsonic acid (10%).** Shake 1 g phenylarsonic acid, $C_6H_5AsO(OH)_2$, with 10 ml water.

Phenylhydrazine reagent.* Dissolve 1 g phenylhydrazine, $C_6H_5.NH.NH_2$, in 2 ml glacial acetic acid.

Phosphomolybdic acid (5%).* Dissolve 0·5 g dodecamolybdophosphoric acid icosytetrahydrate, $H_3PO_4.12MoO_3.24H_2O$, in 10 ml water. The solution does not keep well.

Phosphoric acid (concentrated). Concentrated phosphoric acid is a viscous, clear, colourless liquid with a high density (1·75 g cm^{-3}). It contains 88 per cent (w/w) H_3PO_4 (or 1·54 g H_3PO_4 per ml) and is approximately 16 molar.

Phosphoric acid (M). Dilute 63·7 ml concentrated phosphoric acid with water to 1 litre.

Picrolonic acid (saturated).* Shake 1 g picrolonic acid, $C_{10}H_7O_5N_4$, with 5 ml water, and use the clear supernatant liquid for the tests.

Potassium acetate (M). Dissolve 98·1 g potassium acetate, CH_3COOK, in water and dilute to 1 litre.

Potassium antimonate (0·2M). Dissolve 10·5 g potassium antimonate, $K[Sb(OH)_6]$, in water and dilute the solution to 200 ml. Alternatively, dissolve 6·47 g antimony(V) oxide, Sb_2O_5, in 100 ml 1:1 hydrochloric acid and dilute the solution to 200 ml. Neither solution keeps well.

Potassium benzoate (0·5M). Dissolve 80·1 g potassium benzoate, C_6H_5COOK, in water and dilute to 1 litre.

Potassium bromate (saturated). To 7 g potassium bromate, $KBrO_3$, add 100 ml water and shake. Allow to stand for 24 hours and use the clear, supernatant liquid for the tests.

Potassium bromate (0·1M). Dissolve 16·7 g potassium bromate, $KBrO_3$, in water and dilute to 1 litre.

Potassium bromide (0·1M). Dissolve 11·9 g potassium bromide, KBr, in water and dilute to 1 litre.

Potassium chlorate (saturated). To 7 g potassium chlorate, $KClO_3$, add 100 ml water and shake. Allow to stand for 24 hours and use the clear supernatant liquid for the tests.

Potassium chlorate (0·1M). Dissolve 12·6 g potassium chlorate, $KClO_3$, in water and dilute the solution to 1 litre.

Potassium chloride (saturated). To 30 g potassium chloride add 70 ml water and heat on a water bath until all the solid dissolves. Pour the hot solution into a 100 ml reagent bottle and allow to cool. Some solid KCl will crystallize on cooling. Use the clear liquid for the reactions.

Potassium chloride (M). Dissolve 74·6 g potassium chloride, KCl, in water and dilute to 1 litre.

Potassium chromate (saturated). To 60 g potassium chromate add 100 ml water and heat on a water bath. After complete dissolution allow to cool. Use the clear supernatant liquid for the tests.

Potassium chromate (0·1M). Dissolve 19·4 g potassium chromate, K_2CrO_4, and dilute the solution to 1 litre.

Potassium cyanate (0·2M). Dissolve 16·2 g potassium cyanate, KOCN, in water and dilute the solution to 1 litre.

Potassium cyanide (10%).** **(POISON)** Dissolve 1 g potassium cyanide, KCN, in 10 ml water. Discard the solution immediately after use.

Potassium cyanide (M).* **(POISON)** Dissolve 0·61 g potassium cyanide, KCN, in 10 ml water. Discard the reagent immediately after use.

Potassium cyanide (0·1M).* **(POISON)** Dissolve 0·61 g potassium cyanide in water and dilute to 100 ml. The solution should not be stored for long because it takes up carbon dioxide from the air releasing hydrogen cyanide gas

$$2CN^- + CO_2 + H_2O \rightarrow 2HCN\uparrow + CO_3^{2-}$$

so the solution smells like hydrogen cyanide.

Potassium dichromate (0·1M). Dissolve 29·4 g potassium dichromate, $K_2Cr_2O_7$, in water and dissolve to 1 litre.

Potassium fluoride (0·1M). Dissolve 5·81 g anhydrous potassium fluoride, KF, in water and dilute to 1 litre.

Potassium hexacyanocobaltate(III) reagent (4%). Dissolve 4 g potassium hexacyanocobaltate(III) (potassium cobalticyanide) $K_3[Co(CN)_6]$ and 1 g potassium chlorate, $KClO_3$, in water and dilute to 100 ml.

Potassium hexacyanoferrate(II) (potassium ferrocyanide) (0·025M). Dissolve 10·5 g potassium hexacyanoferrate(II) trihydrate, $K_4[Fe(CN)_6].3H_2O$, in water and dilute to 1 litre.

Potassium hexacyanoferrate(III) (potassium ferricyanide) (0·033M).* Dissolve 10·98 g potassium hexacyanoferrate(III), $K_3[Fe(CN)_6]$, in water and dilute to 1 litre.

Potassium hydroxide (2M). To 112 g potassium hydroxide, KOH, add 50 ml water. Stir the mixture several times, until all the solid dissolves. The mixture warms up considerably. Allow to cool and dilute with water to 1 litre.

Potassium iodate (0·1M). Dissolve 21·4 g potassium iodate, KIO_3, in 500 ml hot water. Allow to cool and dilute to 1 litre.

Potassium iodide (6M).** Dissolve 5 g potassium iodide, KI, in 5 ml water.

Potassium iodide (10%). Dissolve 1 g potassium iodide, KI, in water and dilute to 10 ml.

Potassium iodide (0·1M). Dissolve 16·6 g potassium iodide, KI, in water and dilute to 1 litre.

Potassium nitrate (0·1M). Dissolve 10·1 g potassium nitrate, KNO_3, in water and dilute the solution to 1 litre.

Potassium nitrite (saturated).** To 2 g potassium nitrite, KNO_2, add 1 ml water and shake in the cold. Use the clear supernatant liquid for the test.

Potassium nitrite (50%).** Dissolve 5 g potassium nitrite, KNO_2, in 5 ml water. The solution decomposes within a few hours.

Potassium nitrite (0·1M).** Dissolve 0·9 g potassium nitrite, KNO_2, in water and dilute to 100 ml.

Potassium periodate (saturated).* To 0·1 g potassium periodate add 20 ml water and heat on a water bath until dissolution. Allow to cool and use the clear supernatant liquid for the tests.

Potassium permanganate (0·02M). Dissolve 3·16 g potassium permanganate, $KMnO_4$, in water and dilute to 1 litre.

Potassium permanganate (0·004M). Dilute 4 ml 0·02M potassium permanganate solution with water to 20 ml.

Potassium peroxodisulphate (0·1M).* Dissolve 27·0 g potassium peroxodisulphate $K_2S_2O_8$, in water and dilute to 1 litre.

Potassium selenate (0·1M). Dissolve 22·1 g potassium selenate, K_2SeO_4, in water and dilute to 1 litre.

Potassium sulphate (saturated). Shake 10 g potassium sulphate, K_2SO_4, with 90 ml water. Allow to stand for 24 hours and use the clear, supernatant liquid for the tests.

Potassium tellurite (0·1M). Dissolve 25·4 g potassium tellurite, K_2TeO_3, in water and dilute to 1 litre.

Potassium thiocyanate (10%). Dissolve 1 g potassium thiocyanate in 10 ml water.

Potassium thiocyanate (0·1M). Dissolve 9·72 g potassium thiocyanate,

KSCN, in water and dilute to 1 litre.

Propylarsonic acid (5%).** Dissolve 0·5 g propylarsonic acid, $CH_3CH_2CH_2AsO(OH)_2$, in 10 ml water.

Pyridine. The pure solvent is a clear, colourless liquid with a characteristic odour. It has a specific gravity of 0·98 g cm^{-3} and boils at 113°C.

Pyrocatechol (10%).* Dissolve 10 g catechol (*o*-dihydroxybenzene), $C_6H_4(OH)_2$, in water and dilute to 100 ml.

Pyrogallol (10%).* Dissolve 0·5 g pyrogallol, $C_6H_3(OH)_3$, in 5 ml water. The reagent decomposes slowly.

Pyrrole (1%).* Dissolve 1 g pyrrole, C_4H_5N, in 100 ml of (aldehyde-free) ethanol.

Quinaldic acid (1%).* Dissolve 1 g quinaldic acid, $C_{10}H_7O_2N$, in 5 ml 2M sodium hydroxide and dilute with water to 100 ml.

Quinalizarin (0·05%, in NaOH).* Dissolve 0·05 g quinalizarin in a mixture of 50 ml water and 5 ml 2M sodium hydroxide. Dilute the solution with water to 100 ml.

Quinalizarin (0·05%) in pyridine.* Dissolve 0·01 g quinalizarin (1,2,5,8-tetrahydroxyanthraquinone) in 2 ml pyridine, and dilute the solution with acetone to 20 ml.

Quinalizarin (0·02%).* Dissolve 0·02 g quinalizarin (1,2,5,8-tetrahydroxyanthraquinone) in 100 ml ethanol.

Rhodamine-B (0·01%). Dissolve 0·01 g rhodamine-B, $C_{28}H_{31}N_2O_3CL$ (Colour Index 45170), in 100 ml water. For the preparation of a more concentrated solution see Section **III.15**, reaction 7.

Rochelle salt. See sodium potassium tartrate.

Rubeanic acid (0·5%).** Dissolve 0·5 g rubeanic acid [or dithio-oxamide] $NH_2.CS.CS.NH_2$ in 95 per cent ethanol and dilute the solvent to 100 ml. The solution decomposes rapidly and should be prepared freshly each time.

Rubeanic acid (0·02%).** Dissolve 20 mg rubeanic acid (dithio-oxamide), $NH_2.CS.CS.NH_2$, in 100 ml glacial acetic acid. The reagent should be prepared freshly.

Salicylaldoxime (1%).* Dissolve 1 g salicylaldoxime, $C_6H_4.CH(NOH).OH$, in 5 ml cold ethanol (96%) and pour the solution dropwise into 95 ml water at a temperature not exceeding 80°C. Shake the mixture until clear, and filter if necessary.

Sebacic acid (saturated). Mix 5 g sebacic acid $(C_4H_8COOH)_2$ with 95 ml water and shake. Allow to stand for 24 hours and use the clear supernatant liquid for the tests.

Selenious acid (0·1M). Dissolve 11·1 g selenium dioxide, SeO_2, in water and dilute to 1 litre.

Silver nitrate (20%). Dissolve 2 g silver nitrate, $AgNO_3$, in water and dilute to 10 ml.

Silver nitrate (0·1M). Dissolve 16·99 g silver nitrate, $AgNO_3$, in water, and dilute to 1 litre. Keep the solution in a dark bottle.

Silver sulphate (saturated, about 0·02M). To 1 g silver sulphate, Ag_2SO_4, add 100 ml water and shake. Allow to stand for 24 hours, and use the clear supernatant liquid for the tests. Alternatively, heat the mixture on a water bath until complete dissolution, and allow to cool.

Silver sulphate (ammoniacal) (0·25M).** Dissolve 7·8 g silver sulphate, Ag_2SO_4, in 50 ml 2M ammonia solution, and dilute with water to 100 ml. The reagent should be prepared freshly and discarded after use. Aged reagents might cause serious explosions.

Sodium acetate (saturated). Dissolve 53 g sodium acetate trihydrate, $CH_3COONa.3H_2O$, in 70 ml hot water. Allow to cool and dilute to 100 ml.

Sodium acetate (2M). Dissolve 272 g sodium acetate trihydrate, $CH_3COONa.3H_2O$, in water and dilute to 1 litre.

Sodium arsenate (0·1M). Dissolve 31·2 g disodium hydrogen arsenate heptahydrate, $Na_2HAsO_4.7H_2O$, in water to which 10 ml concentrated hydrochloric acid was added. Dilute the solution to 1 litre.

Sodium arsenite (0·1M). Dissolve 13·0 g sodium meta-arsenite, $NaAsO_2$, in water and dilute to 1 litre.

Sodium azide–iodine reagent.* Dissolve 3 g sodium azide, NaN_3, in 100 ml 0·05M iodine (sodium tri-iodide) solution.

Sodium carbonate (saturated, about 1·5M). Dissolve 4 g anhydrous sodium carbonate, Na_2CO_3, in 25 ml water.

Sodium carbonate (0·5M). Dissolve 53 g anhydrous sodium carbonate, Na_2CO_3, in water and dilute to 1 litre.

Sodium carbonate (0·05M). Dilute 10 ml 0·5M sodium carbonate with water to 100 ml.

Sodium chloride (M). Dissolve 58·4 g sodium chloride, NaCl, in water and dilute to 1 litre.

Sodium chloride (0·1M). Dissolve 5·84 g sodium chloride, NaCl, in water and dilute to 1 litre.

Sodium citrate (0·5M).* Dissolve 147 g sodium citrate dihydrate, $Na_3C_6H_5O_7.2H_2O$ in water and dilute to 1 litre.

Sodium disulphide (2M).* First prepare 100 ml 2M sodium sulphide. Add to this liquid 8 g finely powdered sulphur, and heat a water bath for 2 hours. Stir the mixture from time to time. Let the yellow solution cool and filter if necessary. The composition of the product is uncertain; the formula Na_2S_x is often used and the substance called sodium polysulphide. With the amounts of reagents mentioned x is approximately equal to 2.

Sodium dithionite (0·5M).** Dissolve 10·5 g sodium dithionite dihydrate, $Na_2S_2O_4.2H_2O$, in water and dilute to 100 ml. The solution should be freshly prepared.

Sodium ethylenediamine tetra-acetate (5%). Dissolve 5 g sodium ethylenediamine tetra-acetate dihydrate, $[CH_2N(CH_2COOH).CH_2.COONa]_2.2H_2O$ (Na_2EDTA), in water and dilute to 100 ml.

Sodium fluoride (0·1M). Dissolve 4·2 g sodium fluoride, NaF, in water and dilute to 1 litre.

Sodium formate (M).* Dissolve 68 g sodium formate, H.COONa, in water and dilute to 1 litre.

Sodium hexafluorosilicate (0·1M).* Dissolve 18·8 g sodium hexafluorosilicate, $Na_2[SiF_6]$, in water and dilute to 1 litre.

Sodium hexanitritocobaltate(III) (0·167M).* Dissolve 6·73 g sodium hexanitritocobaltate(III) (sodium cabaltinitrite) in 100 ml water.

Sodium hydrogen carbonate (10%). Dissolve 10 g sodium hydrogen carbonate in 80 ml cold water, by shaking. Dilute the solution to 100 ml.

Sodium hydrogen carbonate (0·5M). Dissolve 42·0 g sodium hydrogen carbonate, $NaHCO_3$, in water and dilute to 1 litre.

Sodium hydrogen tartrate (saturated).* To 10 g sodium hydrogen tartrate monohydrate, $C_6H_5O_6Na.H_2O$, add 100 ml water and shake. Allow to stand for 24 hours and use the clear supernatant liquid for the tests.

Sodium hydroxide (concentrated). To 5 g solid sodium hydroxide, NaOH, add 5 ml water and mix. Dissolution is slow first but becomes rapid as the mixture gets hotter. Use the cold liquid for the tests.

Sodium hydroxide (20 %, 5M). Dissolve 20 g sodium hydroxide in 30 ml water, after dissolution dilute the solution to 100 ml. Keep the solution in a plastic bottle.

Sodium hydroxide (2M). To 80 g solid sodium hydroxide, NaOH, add 80 ml water. Cover the mixture in the beaker with a watch-glass and mix its content from time to time. The heat liberated during this process ensures a quick dissolution. Allow the solution to cool and dilute with water to 1 litre.

Sodium hydroxide (0·01M). Dilute 0·5 ml 2M sodium hydroxide with water to 100 ml. The solution should be prepared freshly.

Sodium hypobromite (0·25M).** To 50 ml freshly prepared bromine water add 2M sodium hydroxide dropwise, until the solution becomes colourless (about 50 ml is needed).

Sodium hypochlorite (M).** This reagent is available commercially as a solution, or can be prepared in the laboratory by saturating 2M sodium hydroxide solution with chlorine gas. This operation must be carried out in a fume cupboard.

Sodium hypophosphite (0·1M).** Dissolve 0·88 g sodium hypophosphite, NaH_2PO_2, in water and dilute to 100 ml.

Sodium metaphosphate.** Heat 1 g ammonium sodium hydrogen phosphate (microcosmic salt), $NH_4NaHPO_4 . 4H_2O$, in a porcelain crucible until no more gases are liberated. Dissolve the cold residue in 50 ml water. The reagent must be prepared freshly.

Sodium metavanadate (0·1M). Dissolve 12·2 g sodium metavanadate, $NaVO_3$, in 100 ml M sulphuric acid and dilute with water to 1 litre.

Sodium nitrite (5 %).** Dissolve 5 g sodium nitrite, $NaNO_2$, in water and dilute to 100 ml.

Sodium nitroprusside reagent. Rub 0·5 g sodium nitroprusside dihydrate, $Na_2[Fe(CN)_5NO].2H_2O$, in 5 ml water. Use the freshly prepared solution.

Sodium 1-nitroso-2-hydroxy-naphthalene-3,6-disulphonate (1 %).* Dissolve 1 g sodium 1-nitroso-2-hydroxynaphthalene-3,6-disulphonate [Nitroso R-salt, $C_{10}H_4(OH)(SO_3Na)_2NO$] in 100 ml of water.

Sodium oxalate (0·1M). Dissolve 13·4 g sodium oxalate in water and dilute to 1 litre.

Sodium peroxide (0·5M).** Dissolve 3·9 g sodium peroxide, Na_2O_2, in water and dilute to 100 ml. The solution must be prepared freshly.

Sodium phosphite (0·1M).** Dissolve 2·16 g sodium phosphite pentahydrate, $Na_2HPO_3.5H_2O$, in water and dilute the solution to 100 ml. The reagent should be prepared freshly.

Sodium potassium tartrate (10 %).* Dissolve 10 g sodium potassium tartrate tetrahydrate, $KNaC_4H_4O_6.4H_2O$ (Rochelle salt or Seignette salt), in water and dilute to 100 ml.

Sodium potassium tartrate (0·1M).* Dissolve 28·2 g sodium potassium tartrate tetrahydrate, $KNaC_4H_4O_6.4H_2O$ (Rochelle salt or Seignette salt), in water and dilute to 1 litre.

Sodium pyrophosphate.** Heat 1 g disodium hydrogen phosphate,

$Na_2HPO_4.2H_2O$, in a porcelain crucible until no more water is liberated. Dissolve the cold residue in 50 ml water. The reagent must be prepared freshly.

Sodium rhodizonate (0·5%).* Dissolve 0·5 g sodium rhodizonate (rhodizonic acid, sodium salt $C_6O_6Na_2$) in 100 ml water. The solution decomposes rapidly.

Sodium salicylate (0·5M). Dissolve 80 g sodium salicylate, $C_6H_4(OH)COONa$, in water and dilute to 1 litre.

Sodium selenate (0·1M). Dissolve 36·9 g sodium selenate decahydrate, $Na_2SeO_4.10H_2O$, in water and dilute to 1 litre.

Sodium selenite (0·1M). Dissolve 17·3 g sodium selenite, Na_2SeO_3, in water and dilute to 1 litre.

Sodium silicate (M).* Dilute 200 ml commercial (30%) water glass solution with water to 1 litre.

Sodium succinate (0·5M). Dissolve 135 g sodium succinate hexahydrate, $(CH_2COONa)_2.6H_2O$, in water and dilute to 1 litre.

Sodium sulphate (0·1M). Dissolve 32·2 g sodium sulphate decahydrate, $Na_2SO_4.10H_2O$, in water and dilute to 1 litre.

Sodium sulphide (2M). To 16 g solid sodium hydroxide, NaOH, add 20 ml water. Cover the beaker with a watch glass and shake the mixture gently, until complete dissolution. Dilute the mixture with water to 100 ml. Take 50 ml of this solution and saturate with hydrogen sulphide gas. Then add the rest of sodium hydroxide solution to the latter.

Sodium sulphite (0·5M).* Dissolve 6·3 g anhydrous sodium sulphite, Na_2SO_3, or 12·6 g sodium sulphite heptahydrate, $Na_2SO_3.7H_2O$, in water and dilute to 100 ml.

Sodium tellurate (0·1M). Dissolve 27·4 g sodium tellurate dihydrate, $Na_2TeO_4.2H_2O$, in water and dilute to 1 litre.

Sodium tellurite (0·1M). Dissolve 22·2 g sodium tellurite, Na_2TeO_3, in water and dilute to 1 litre.

Sodium tetraborate (0·1M). Dissolve 38·1 g sodium tetraborate decahydrate (borax), $Na_2B_4O_7.10H_2O$, in water and dilute to 1 litre.

Sodium tetrahydroxo stannate(II) reagent (0·125M).** To 2 ml 0·25M tin(II) chloride add 2M sodium hydroxide solution under vigorous shaking, until the precipitate just dissolves (about 2 ml of the latter is needed). The reagent decomposes rapidly, it has to be prepared freshly each time.

Sodium tetraphenyl boron (0·1M).** Dissolve 3·42 g sodium tetraphenyl boron, $Na[B(C_6H_5)_4]$, in 100 ml water. The solution does not keep well.

Sodium thiosulphate (0·5M). Dissolve 124·1 g sodium thiosulphate pentahydrate, $Na_2S_2O_3.5H_2O$, in water and dilute to 1 litre.

Sodium thiosulphate (0·1M). To 2 ml 0·5M sodium thiosulphate add 8 ml water and mix.

Sodium tungstate (0·2M). Dissolve 65·97 g sodium tungstate dihydrate, $Na_2WO_4.2H_2O$, in water and dilute to 1 litre.

Starch solution.* Suspend 0·5 g soluble starch in 5 ml water, and pour this into 20 ml water, which has just ceased to boil. Mix. Allow to cool, when the solution becomes clear.

Strontium chloride (0·25M). Dissolve 66·7 g strontium chloride hexahydrate, $SrCl_2.6H_2O$, in water and dilute to 1 litre.

Strontium nitrate (0·25M). Dissolve 52·9 g strontium nitrate, $Sr(NO_3)_2$, in water and dilute to 1 litre.

Strontium sulphate (saturated). Shake 0·1 g strontium sulphate, $SrSO_4$, with

100 ml water and allow to stand for at least 24 hours. Use the clear supernatant liquid for the tests.

Sulphur dioxide gas. This gas is available commercially in a liquefied state in aluminium canisters, from which it can be taken. Alternatively it can be produced from sodium sulphite and 1:1 sulphuric acid. The solid reagent should be placed into a round-bottomed flask, which can be heated. The acid is kept in a funnel with a stopcock, inserted into one opening of the flask. By adding some sulphuric acid to the solid and by gentle heating, sulphur dioxide gas is coming through the second opening of the flask, which can be washed in concentrated sulphuric acid.

Sulphur dioxide (sulphurous acid), saturated aqueous solution.** Bubble sulphur dioxide gas through a thin glass tube into 200 ml water, until the solution becomes saturated. The gas can be taken from a cylinder or can be generated from solid sodium hydrogen sulphite, $NaHSO_3$, and dilute sulphuric acid (M). The solution should be well stoppered.

Sulphuric acid (concentrated). The commercial reagent is a colourless, oil-like liquid of a high density ($1{\cdot}84\ g\ cm^{-3}$). It contains 98% (w/w) H_2SO_4 (or 1·76 g H_2SO_4 per ml) and is approximately 18 molar. The reagent must be handled with utmost care, wearing rubber gloves and eye protection. When diluting, concentrated sulphuric acid must be poured slowly into water (*and never vice versa*), under stirring and, if necessary, cooling. Traces of the reagent have to be removed immediately from the skin or from clothing by washing with large amounts of water.

Sulphuric acid (3:2, approx. 12M). To 40 ml water add cautiously 60 ml concentrated sulphuric acid, under constant stirring. If the mixture becomes hot, set aside for a while to cool, and then continue mixing.

Sulphuric acid (8M). To 500 ml water add slowly, under constant stirring, 445 ml concentrated sulphuric acid. (If the mixture becomes too hot, wait for 5 minutes before resuming the dilution.) Finally, when the mixture is cool again, dilute with water to 1 litre.

Sulphuric acid (3M). To 50 ml water add, under constant stirring, 16·6 ml concentrated sulphuric acid, with greatest caution. When cool, dilute the solution to 100 ml.

Sulphuric acid, (M). Pour 55·4 ml concentrated sulphuric acid slowly into 800 ml cold water, under vigorous stirring. Wear protective glasses and rubber gloves during this operation. Finally, dilute the cold solution with water to 1 litre.

Tannic acid (5%).* Dissolve 5 g tannic acid, $C_{76}H_{52}O_{46}$, in 100 ml water and filter the solution.

Tartaric acid (M).* Dissolve 15 g tartaric acid, $C_4H_6O_6$, in water and dilute to 100 ml. The solution does not keep for very long (max. 1–2 months).

Tartaric acid (0·1M).* Dissolve 15 g tartaric acid, $C_4H_6O_6$, in water and dilute to 1 litre.

Tartaric acid–ammonium molybdate reagent.** Dissolve 15 g tartaric acid, $C_4H_6O_6$, in 100 ml 0·25M ammonium molybdate reagent.

Tetrabase (1%).* Dissolve 0·1 g tetrabase (4,4′-tetramethyldiamino-diphenyl-methane) $[(CH_3)_2NC_6H_4]_2$ in 10 ml chloroform.

Tetrahydroxysylvan. See 2-Methyltetrahydroxyfuran.

Tetramethyldiamino-diphenylmethane (0·5%).** Dissolve 0·5 g tetrabase (4,4′-tetramethyldiaminodiphenyl methane), $[(CH_3)_2N.C_6H_4]_2CH_2$, in a mixture of 20 ml glacial acetic acid and 80 ml 96% ethanol.

Thallium(III) chloride (0·2M). Dissolve 45·7 g thallium(III) oxide, Tl_2O_3, in 50 ml of 1:1 hydrochloric acid. Evaporate the solution to dryness and dissolve the residue in water. Dilute the solution to 1 litre.

Thallium(I) nitrate (0·05M). Dissolve 13·32 g thallium(I) nitrate, $TlNO_3$, in water and dilute to 1 litre.

Thallium(I) sulphate (0·025M). Dissolve 12·62 g thallium(I) sulphate, Tl_2SO_4, in water and dilute to 1 litre.

Thiourea (10%).* Dissolve 1 g thiourea, $CS(NH_2)_2$, in 10 ml water. The reagent decomposes slowly and hydrogen sulphide is formed; an aged reagent therefore should not be used for the tests.

Thiourea (5% in 5M hydrochloric acid).* To 50 ml water add 43 ml concentrated hydrochloric acid. Allow to cool. Dissolve in this mixture 5 g thiourea, $CS(NH_2)_2$, and dilute the solution with water to 100 ml.

Thorium nitrate (0·1M). Dissolve 58·8 g thorium nitrate hexahydrate, $Th(NO_3)_4.6H_2O$, in a mixture of 500 ml water and 25 ml concentrated nitric acid, and dilute the solution to 1 litre.

Tin(II) chloride (saturated).* Shake 2·5 g tin(II) chloride dihydrate, $SnCl_2.2H_2O$, in 5 ml concentrated hydrochloric acid. Allow the solid to settle and use the clear solution for the tests.

Tin(II) chloride (0·25M).* Dissolve 56·5 g tin(II) chloride dihydrate, $SnCl_2.2H_2O$, in a cold mixture of 100 ml concentrated hydrochloric acid and 80 ml water, then dilute with water to 1 litre. To prevent oxidation keep a small piece of granulated tin metal at the bottom of the solution.

Tin(IV) chloride reagent** (for removal of phosphate). Dissolve 5 g tin(IV) chloride, $SnCl_4$, in 5 ml water. Prepare the reagent freshly.

Titanium(IV) chloride (10%). To 90 ml of 1:1 hydrochloric acid add 10 ml liquid titanium tetrachloride, $TiCl_4$, and mix.

Titanium(III) sulphate (15%).** The commercial titanium(III) sulphate solution contains about 15 per cent $Ti_2(SO_4)_3$ and also free sulphuric acid. It has a violet colour. On standing it is slowly oxidized to colourless titanium(IV) sulphate.

Titanium(IV) sulphate (0·2M). Dilute 33 ml 15% (w/v) commercial titanium(IV) sulphate solution with M sulphuric acid to 100 ml.

Titan yellow (0·1%). Dissolve 0·1 g Titan yellow (Clayton yellow, Colour Index 19540) in 100 ml water.

Triethanolamine $N(CH_2.CH_2.OH)$. The commercial liquid is a viscous, dense liquid with a density of 1·12 g cm^{-3}.

Uranyl acetate (0·1M). Dissolve 42·4 g uranyl acetate dihydrate, $UO_2(CH_3COO)_2.2H_2O$, in a mixture of 200 ml water and 30 ml concentrated acetic acid. After dissolution dilute the solution with water to 1 litre.

Uranyl magnesium acetate reagent.* Dissolve 10 g uranyl acetate dihydrate, $UO_2(CH_3COO)_2.2H_2O$, in a mixture of 6 ml glacial acetic acid and 100 ml water (solution *a*). Dissolve 33 g magnesium acetate tetrahydrate, $Mg(CH_3COO)_2.4H_2O$, in a mixture of 100 ml glacial acetic acid and 100 ml water (solution *b*). Mix the two solutions, allow to stand for 24 hours, and filter.

Uranyl nitrate (0·1M). Dissolve 50·2 g uranyl nitrate hexahydrate, $UO_2(NO_3)_2.6H_2O$, in water and dilute to 1 litre.

Uranyl zinc acetate reagent.* Dissolve 10 g uranyl acetate dihydrate, $UO_2(CH_3COO)_2.2H_2O$, in a mixture of 5 ml glacial acetic acid and 20 ml water, and dilute the solution to 50 ml (solution *a*). Dissolve 30 g zinc acetate dihydrate, $Zn(CH_3COO)_2.2H_2O$, in a mixture of 5 ml glacial acetic acid and 20 ml water, and dilute with water to 50 ml (solution *b*). Mix the solutions, add 0·5 g sodium chloride, NaCl, allow to stand for 24 hours and filter.

Violuric acid (0·1M).* Dissolve 1·6 g violuric acid (5-iso-nitroso-barbituric acid), $C_4H_3O_4N_3$, in 100 ml of water.

Zinc acetate (1 %). Dissolve 1 g zinc acetate dihydrate, $Zn(CH_3COO)_2.2H_2O$, in water and dilute to 100 ml.

Zinc hexammine hydroxide. To 25 ml 0·5M zinc nitrate add 12·5 ml of 2M potassium hydroxide. Filter, wash the precipitate with water, and dissolve the precipitate off the filter with 15 ml 1:1 ammonia. Pour the liquid on the filter several times, until complete dissolution.

Zinc nitrate (0·5M). Dissolve 149 g zinc nitrate hexahydrate, $Zn(NO_3)_3.6H_2O$, in water and dilute to 1 litre.

Zinc nitrate (0·1M). Dilute 1 ml 0·5M zinc nitrate with water to 5 ml.

Zinc sulphate (0·25M). Dissolve 72 g zinc sulphate heptahydrate, $ZnSO_4.7H_2O$, in water and dilute to 1 litre.

Zirconium nitrate reagent (for phosphate separations). Heat 10 g commercial zirconium nitrate, $ZrO.(NO_3)_2.2H_2O$, and 100 ml 2M nitric acid to boiling under constant stirring. Allow to cool and set aside for 24 hours. Decant the clear liquid and use this for the tests.

Zirconyl chloride (0·1M). Dissolve 32·2 g zirconyl chloride octahydrate, $ZrOCl_2.8H_2O$, in 100 ml 2M hydrochloric acid and dilute with water to 1 litre.

Zirconyl chloride (0·1 %). Dissolve 0·1 g zirconyl chloride octahydrate, $ZrOCl_2.8H_2O$, in 20 ml concentrated hydrochloric acid and dilute with water to 100 ml.

Zirconyl nitrate (0·11M). Dissolve 26·2 g zirconyl nitrate dihydrate, $ZrO(NO_3)_2.2H_2O$, in 100 ml 2M nitric acid and dilute with water to 1 litre.

IX.3 SOLID REAGENTS

Aluminium powder, Al
Aluminium sheet, Al
p-Amino-N′N-dimethyl aniline, $H_2N.C_6H_4.N(CH_3)_2$
Ammonium acetate, CH_3COONH_4
Ammonium carbonate, $(NH_4)_2CO_3$
Ammonium chloride, NH_4Cl
Ammonium fluoride, NH_4F
Ammonium nitrate, NH_4NO_3
Ammonium peroxodisulphate, $(NH_4)_2S_2O_8$
Ammonium sodium hydrogen phosphate (microcosmic salt), $NH_4NaHPO_4.4H_2O$
Ammonium thiocyanate, NH_4SCN
Arsenic(III) oxide, As_2O_3 (**POISON**)
Benzoic acid, C_6H_5COOH
Cadmium metal, Cd (granulated)
Calcium carbonate, $CaCO_3$
Calcium chloride anhydrous, $CaCl_2$
Calcium fluoride, CaF_2
Calcium hydroxide, $Ca(OH)_2$

Chromotropic acid (sodium salt), $C_{10}H_6O_8S_2Na_2$
Cobalt(II) acetate, $Co(CH_3COO)_2.4H_2O$
Copper foil, Cu
Copper sheet or coin, Cu
Copper turnings, Cu
Devarda's alloy (50% Cu, 45% Al, 5% Zn)
Dimethyl glyoxime, $CH_3C(NOH)C(NOH).CH_3$
Diphenylamine, $(C_6H_5)_2NH$
Disodium hydrogen phosphate, $Na_2HPO_4.2H_2O$
Hydrazine sulphate, $(N_2H_4).H_2SO_4$
Hydroquinone (quinol), $C_6H_4(OH)_2$
Hydroxylamine hydrochloride (hydroxylammonium chloride), $HO.NH_3.Cl$
Iron (filings and nails), Fe
Iron wire, Fe
Iron(II) ammonium sulphate, $FeSO_4.(NH_4)_2SO_4.6H_2O$
Iron(II) sulphate, $FeSO_4.7H_2O$
Lead acetate, $(CH_3COO)_2Pb.3H_2O$
Lead dioxide, PbO_2
Litmus (test) paper
Magnesium metal (powder), Mg
Magnesium metal (turnings), Mg
Manganese dioxide, MnO_2
Mannitol, $C_6H_{14}O_6$
Mercury metal, Hg
Mercury(II) oxide, HgO (red)
Microcosmic salt (see ammonium sodium hydrogen phosphate)
Oxalic acid, $(COOH)_2.2H_2O$
Paraffin wax
Phenol, C_6H_5OH
Platinum sheet or foil, Pt
Potassium benzoate, C_6H_5COOK
Potassium bromate, $KBrO_3$
Potassium bromide, KBr
Potassium chlorate, $KClO_3$
Potassium chromate, K_2CrO_4
Potassium cyanide, KCN (**POISON**)
Potassium dichromate, $K_2Cr_2O_7$
Potassium fluoride, KF
Potassium hydrogen sulphate, $KHSO_4$
Potassium hydrogen tartrate, $KHC_4H_4O_6$
Potassium iodide, KI
Potassium nitrate, KNO_3
Potassium nitrite, KNO_2
Potassium periodate, KIO_4
Potassium permanganate, $KMnO_4$
Potassium peroxodisulphate, $K_2S_2O_8$
Potassium xanthate (ethyl potassium xanthate, $C_2H_5O(CS)SK$)
Resorcinol, m-$C_6H_4(OH)_2$
Salicylic acid, $C_6H_4(OH)COOH$
Silica (precipitated), SiO_2

Silver sheet or coin, Ag
Silver sulphate, Ag_2SO_4
Soda lime
Sodium acetate, $CH_3COONa.3H_2O$
Sodium bismuthate, $NaBiO_3$
Sodium carbonate (anhydrous), Na_2CO_3
Sodium chloride, NaCl
Sodium citrate, $Na_3C_6H_5O_7.2H_2O$
Sodium dihydroxytartrate osazone $(NaO.CO.C{:}N.NH.C_6H_5)_2$
Sodium dithionite, $Na_2S_2O_4.2H_2O$
Sodium fluoride, NaF
Sodium formate, HCOONa
Sodium hexafluorosilicate, $Na_2[SiF_6]$
Sodium hexanitritocobaltate(III), $Na_3[Co(NO_2)_6]$
Sodium hydrogen carbonate, $NaHCO_3$
Sodium hypophosphite, NaH_2PO_2
Sodium nitrite, $NaNO_2$
Sodium nitroprusside, $Na_2[Fe(CN)_5NO].2H_2O$
Sodium oxalate (Sörensen's salt), $(COONa)_2$
Sodium perborate, $NaBO_3.4H_2O$
Sodium perchlorate, $NaClO_4+H_2O$, (the commercial solid product contains about 75 per cent of $NaClO_4$)
Sodium peroxide, Na_2O_2
Sodium phosphite, $Na_2HPO_3.5H_2O$
Sodium potassium tartrate, COOK.CHOH.CHOH.COONa, (Rochelle salt or Seignette salt).
Sodium salicylate, $C_6H_4(OH)COONa$
Sodium succinate, $(CH_2COONa)_2.6H_2O$
Sodium sulphate (anhydrous), Na_2SO_4
Sodium sulphite (anhydrous), Na_2SO_3
Sodium tetraborate (Borax), $Na_2B_4O_7.10H_2O$. The anhydrous (fused) sodium tetraborate, $Na_2B_4O_7$, is even more suitable for the borax bead tests
Sodium thiosulphate, $Na_2S_2O_3.5H_2O$
Starch (potato-starch or soluble starch)
Sulphamic acid, $H_2N.SO_3H$
Sulphur, S, sublimed ('flowers of sulphur')
Tartaric acid, $C_4H_6O_6$[NOOC.CH(OH).CH(OH).COOH]
Thiourea, $(NH_2)_2CS$
Tin, granulated, Sn
Turmeric (test) paper
Urea, $H_2N.CO.NH_2$
Uranyl acetate, $UO_2.(CH_3COO)_2.2H_2O$
Uranyl nitrate, $UO_2(NO_3)_2.6H_2O$
Zinc powder, Zn
Zinc, granulated, Zn

IX.4 SOLUBILITIES OF SALTS AND BASES IN WATER AT 18°C*

	K	Na	Li	Mg	Ba	Sr	Ca	Zn	Pb	Ag
Cl	32·95	35·86	77·79	55·81	37·24	51·09	73·19	203·9	1·49	$1{\cdot}4\times10^{-4}$
Br	65·86	88·76	168·7	103·1	103·6	96·52	143·3	478·2	0·97	1×10^{-3}
I	137·5	177·9	161·5	148·2	201·4	169·2	200·0	419·0	0·08	$3{\cdot}0\times10^{-7}$
F	92·56	4·44	0·27	0·0076	0·16	0·012	0·0016	0·005	0·06	195·4
NO_3	30·34	83·97	71·43	74·31	8·74	66·27	121·8	117·8	51·66	213·4
ClO_3	6·6	97·16	313·4	126·4	35·42	174·9	179·3	183·9	150·6	12·25
BrO_3	6·38	36·67	152·5	42·86	0·8	30·0	85·17	58·43	1·3	0·59
IO_3	7·62	8·33	80·43	6·87	0·05	0·25	0·25	0·83	0·002	0·004
SO_4	11·11	16·83	35·64	35·43	$2{\cdot}3\times10^{-4}$	0·011	0·20	53·12	0·0041	0·55
CrO_4	63·1	61·21	111·6	73·0	$3{\cdot}8\times10^{-4}$	0·12	0·4	—	2×10^{-5}	0·0025
$(COO)_2$	30·27	3·34	7·22	0·03	0·0086	0·0046	$5{\cdot}6\times10^{-4}$	$7{\cdot}9\times10^{-4}$	0·0001	0·0035
CO_3	108·0	19·39	1·3	0·1	0·0023	0·0011	0·0013	—	0·0001	0·003
OH	142·9	116·4	12·04	0·001	3·7	0·77	0·17	0·0005	0·01	0·002 (?)

* The solubilities are expressed in grams of anhydrous salt dissolved by 100 ml of water.

IX.5 LOGARITHMS

	0	1	2	3	4	5	6	7	8	9	1 2 3 4	5	6 7 8 9
10	0000	0043	0086	0128	0170	0212	0253	0294	0334	0374	4 9 13 17 4 8 12 16	21 20	26 30 34 38 24 28 32 37
11	0414	0453	0492	0531	0569	0607	0645	0682	0719	0755	4 8 12 15 4 7 11 15	19 19	23 27 31 38 22 26 30 33
12	0792	0828	0864	0899	0934	0969	1004	1038	1072	1106	3 7 11 14 3 7 10 14	18 17	21 25 28 32 20 24 27 31
13	1139	1173	1206	1239	1271	1303	1335	1367	1399	1430	3 7 10 13 3 7 10 12	16 16	20 23 26 30 19 22 25 29
14	1461	1492	1523	1553	1534	1614	1644	1673	1703	1732	3 6 9 12 3 6 9 12	15 15	18 21 24 28 17 20 23 26
15	1761	1790	1818	1847	1875	1903	1931	1959	1987	2014	3 6 9 11 3 5 8 11	14 14	17 20 23 26 16 19 22 25
16	2041	2068	2095	2122	2148	2175	2201	2227	2253	2279	3 5 8 11 3 5 8 10	14 13	16 19 22 24 15 18 21 23
17	2304	2330	2355	2380	2405	2430	2455	2480	2504	2529	3 5 8 10 2 5 7 10	13 12	15 18 20 23 15 17 19 22
18	2553	2577	2601	2625	2648	2672	2695	2718	2742	2765	2 5 7 9 2 5 7 9	12 11	14 16 19 21 14 16 18 21
19	2788	2810	2833	2856	2878	2900	2923	2945	2967	2989	2 4 7 9 2 4 6 8	11 11	13 16 18 20 13 15 17 19
20	3010	3032	3054	3075	3096	3118	3139	3160	3181	3201	2 4 6 8	11	13 15 17 19
21	3222	3243	3263	3284	3304	3324	3345	3365	3385	3404	2 4 6 8	10	12 14 16 18
22	3424	3444	2364	3483	3502	3522	3541	3560	3579	3598	2 4 6 8	10	12 14 15 17
23	3617	3636	3655	3674	3692	3711	3729	3747	3766	3784	2 4 6 7	9	11 13 15 17
24	3802	3820	3838	3856	3874	3892	3909	3927	3945	3962	2 4 5 7	9	11 12 14 16
25	3979	3997	4014	4031	4048	4065	4082	4099	4116	4153	2 3 5 7	9	10 12 14 15
26	4150	4166	4183	4200	4216	4232	4249	4265	4281	4298	2 3 5 7	8	10 11 13 15
27	4314	4330	4346	4362	4378	4393	4409	4425	4440	4456	2 3 5 6	8	9 11 13 14
28	4472	4487	4502	4518	4533	4548	4564	4579	4594	4609	2 3 5 6	8	9 11 12 14
29	4624	4639	4664	4669	4683	4698	4713	4728	4742	4757	1 3 4 6	7	9 10 12 13
30	4771	4786	4800	4814	4829	4843	4857	4871	4886	4900	1 3 4 6	7	9 10 11 12
31	4914	4928	4942	4955	4969	4983	4997	5011	5024	5038	1 3 4 6	7	8 10 11 12
32	5051	5065	5079	5002	5105	5119	5132	5145	5159	5172	1 3 4 5	7	8 9 11 12
33	5185	5198	5211	5224	5237	5250	5263	5276	5289	5302	1 3 4 5	6	8 9 10 12
34	5315	5328	5340	5353	5366	5378	5391	5403	5416	5428	1 3 4 5	6	8 9 10 11
35	5441	5453	5465	5478	5490	5502	5514	5527	5539	6551	1 2 4 5	6	7 9 10 11
36	5563	5575	5587	5599	5611	5623	5635	5647	5658	5670	1 2 4 5	6	7 8 10 11
37	5682	5694	5705	5717	5729	5740	5752	5763	5775	5786	1 2 3 5	6	7 8 9 10
38	5798	5809	5821	5832	5843	5855	5866	5877	5888	5899	1 2 3 5	6	7 8 9 10
39	5911	5922	5933	5944	5955	5966	5977	5988	5999	6010	1 2 3 4	5	7 8 9 10
40	6021	6031	6042	6053	6064	6075	6085	6096	6107	6117	1 2 3 4	5	6 8 9 10
41	6128	6138	6149	6160	6170	6180	6191	6201	6212	6222	1 2 3 4	5	6 7 8 9
42	6232	6243	6253	6263	6274	6284	6294	6304	6314	6325	1 2 3 4	5	6 7 8 9
43	6335	6345	6355	6365	6375	6385	6395	6405	6415	6425	1 2 3 4	5	6 7 8 9
44	6435	6444	6454	6464	6474	6484	6493	6503	6513	6522	1 2 3 4	5	6 7 8 9
45	6532	6542	6551	6561	6571	6580	6590	6599	6609	6618	1 2 3 4	5	6 7 8 9
46	6628	6637	6646	6656	6665	6675	6684	6693	6702	6712	1 2 3 4	5	6 7 7 8
47	6721	6730	6739	6749	6758	6767	6776	6785	6794	6803	1 2 3 4	5	5 6 7 8
48	6812	6821	6830	6839	6848	6857	6866	6875	6884	6893	1 2 3 4	4	5 6 7 8
49	6902	6911	6920	6928	6937	6946	6955	6964	6972	6981	1 2 3 4	4	5 6 7 8
50	6990	6998	7007	7016	7024	7033	7042	7050	7059	7067	1 2 3 3	4	5 6 7 8

LOGARITHMS

	0	1	2	3	4	5	6	7	8	9	1	2	3	4	5	6	7	8	9
51	7076	7084	7093	7101	7110	7118	7126	7135	7143	7152	1	2	3	3	4	5	6	7	8
52	7160	7168	7177	7185	7193	7202	7210	7218	7226	7235	1	2	2	3	4	5	6	7	7
53	7243	7251	7259	7267	7275	7284	7292	7300	7308	7316	1	2	2	3	4	5	6	6	7
54	7324	7332	7340	7348	7356	7364	7372	7380	7388	7396	1	2	2	3	4	5	6	6	7
55	7404	7412	7419	7427	7435	7443	7451	7459	7466	7474	1	2	2	3	4	5	5	6	7
56	7482	7490	7497	7505	7513	7520	7528	7536	7543	7551	1	2	2	3	4	5	5	6	7
57	7559	7566	7574	7582	7589	7597	7604	7612	7619	7627	1	2	2	3	4	5	5	6	7
58	7634	7642	7649	7657	7664	7672	7679	7686	7694	7701	1	1	2	3	4	4	5	6	7
59	7709	7716	7723	7731	7738	7745	7752	7760	7767	7774	1	1	2	3	4	4	5	6	7
60	7782	7789	7796	7803	7810	7818	7825	7832	7839	7846	1	1	2	3	4	4	5	6	6
61	7853	7860	7868	7875	7882	7889	7896	7903	7910	7917	1	1	2	3	4	4	5	6	6
62	7924	7931	7938	7945	7952	7959	7966	7973	7980	7987	1	1	2	3	3	4	5	6	6
63	7993	8000	8007	8014	8021	8028	8035	8041	8048	8055	1	1	2	3	3	4	5	5	6
64	8062	8069	8075	8082	8089	8096	8102	8109	8116	8122	1	1	2	3	3	4	5	5	6
65	8129	8136	8142	8149	8156	8162	8169	8176	8182	8189	1	1	2	3	3	4	5	5	6
66	8195	8202	8209	8215	8222	8228	8235	8241	8248	8254	1	1	2	3	3	4	5	5	6
67	8261	8267	8274	8280	8287	8293	8299	8306	8312	8319	1	1	2	3	3	4	5	5	6
68	8325	8331	8338	8344	8351	8357	8363	8370	8376	8382	1	1	2	3	3	4	4	5	6
69	8388	8395	8401	8407	8114	8420	8426	8432	8439	8445	1	1	2	2	3	4	4	5	6
70	8451	8457	8463	8470	8476	8482	8488	8494	8500	8506	1	1	2	2	3	4	4	5	6
71	8513	8519	8525	8531	8537	8543	8549	8555	8561	8567	1	1	2	2	3	4	4	5	5
72	8573	8579	8585	8591	8597	8603	8609	8615	8621	8627	1	1	2	2	3	4	4	5	5
73	8633	8639	8645	8651	8657	8663	8669	8675	8681	8686	1	1	2	2	3	4	4	5	5
74	8692	8698	8704	8710	8716	8722	8727	8733	8730	8745	1	1	2	2	3	4	4	5	5
75	8751	8756	8762	8768	8774	8779	8785	8791	8797	8802	1	1	2	2	3	3	4	5	5
76	8808	8814	8820	8825	8831	8837	8842	8848	8854	8859	1	1	2	2	3	3	4	5	5
77	8865	8871	8876	8882	8887	8893	8899	8904	8910	8915	1	1	2	2	3	3	4	4	5
78	8921	8927	8932	8938	8943	8949	8954	8960	8965	8971	1	1	2	2	3	3	4	4	5
79	8976	8982	8987	8993	8998	9004	9009	9015	9020	9025	1	1	2	2	3	3	4	4	5
80	9031	9036	9042	9047	9053	9058	9063	9069	9074	9079	1	1	2	2	3	3	4	4	5
81	9085	9090	9096	9101	9106	9112	9117	9122	9128	9133	1	1	2	2	3	3	4	4	5
82	9138	9143	9149	9154	9159	9165	9170	9175	9180	9186	1	1	2	2	3	3	4	4	5
83	9191	9196	9201	9206	9212	9217	9222	9227	9232	9238	1	1	2	2	3	3	4	4	5
84	9243	9248	9253	9258	9263	9269	9274	9279	9284	9289	1	1	2	2	3	3	4	4	5
85	9234	9299	9304	9309	9315	9320	9325	9330	9335	9340	1	1	2	2	3	3	4	4	5
86	9245	9350	9355	9360	9365	9370	9375	9380	9385	9390	1	1	2	2	3	3	4	4	5
87	9395	9400	9405	9410	9415	9420	9425	9430	9435	9440	0	1	1	2	2	3	3	4	4
88	9445	9450	9455	9460	9465	9469	9474	9479	9484	9489	0	1	1	2	2	3	3	4	4
89	9494	9499	9504	9509	9513	9518	9523	9528	9533	9538	0	1	1	2	2	3	3	4	4
90	9542	9547	9552	9557	9562	9566	9571	9576	9581	9586	0	1	1	2	2	3	3	4	4
91	9590	9595	9600	9605	9609	9614	9619	9624	9628	9633	0	1	1	2	2	3	3	4	4
92	9638	9643	9647	9652	9657	9661	9666	9671	9675	9680	0	1	1	2	2	3	3	4	4
93	9685	9689	9694	9699	9703	9708	9713	9717	9722	9727	0	1	1	2	2	3	3	4	4
94	9731	9736	9741	9745	9750	9754	9759	9763	9768	9773	0	1	1	2	2	3	3	4	4
95	9777	9782	9786	9791	9795	9800	9805	9809	9814	9818	0	1	1	2	2	3	3	4	4
96	9823	9827	9832	9836	9841	9845	9850	9854	9859	9863	0	1	1	2	2	3	3	4	4
97	9868	9872	9877	9881	9886	9890	9894	9899	9903	9908	0	1	1	2	2	3	3	4	4
98	9912	9917	9921	9926	9930	9934	9939	9943	9948	9952	0	1	1	2	2	3	3	4	4
99	9956	9961	9965	9969	9974	9978	9983	9987	9991	9996	0	1	1	2	2	3	3	3	4

IX.6 ANTILOGARITHMS

	0	1	2	3	4	5	6	7	8	9	1	2	3	4	5	6	7	8	9
·00	1000	1002	1005	1007	1009	1012	1014	1016	1019	1021	0	0	1	1	1	1	2	2	2
·01	1023	1026	1023	1030	1033	1035	1038	1040	1042	1045	0	0	1	1	1	1	2	2	2
·02	1047	1050	1052	1054	1057	1059	1062	1064	1067	1069	0	0	1	1	1	1	2	2	2
·03	1072	1074	1076	1079	1081	1084	1086	1089	1091	1094	0	0	1	1	1	1	2	2	2
·04	1096	1099	1102	1104	1107	1109	1112	1114	1117	1119	0	1	1	1	1	2	2	2	2
·05	1122	1125	1127	1130	1132	1135	1138	1140	1143	1146	0	1	1	1	1	2	2	2	2
·06	1148	1151	1153	1156	1159	1161	1164	1167	1169	1172	0	1	1	1	1	2	2	2	3
·07	1175	1178	1180	1183	1186	1189	1191	1194	1197	1199	0	1	1	1	1	2	2	2	3
·08	1202	1205	1208	1211	1213	1216	1219	1222	1225	1227	0	1	1	1	1	2	2	2	3
·09	1230	1233	1236	1239	1242	1245	1247	1250	1253	1256	0	1	1	1	1	2	2	2	3
·10	1259	1262	1265	1268	1271	1274	1276	1279	1282	1285	0	1	1	1	1	2	2	2	3
·11	1288	1291	1294	1297	1300	1303	1306	1309	1312	1315	0	1	1	1	2	2	2	2	3
·12	1318	1321	1324	1327	1330	1334	1337	1340	1343	1346	0	1	1	1	2	2	2	2	3
·13	1349	1352	1355	1358	1361	1365	1368	1371	1374	1377	0	1	1	1	2	2	2	3	3
·14	1380	1384	1387	1390	1393	1396	1400	1403	1406	1409	0	1	1	1	2	2	2	3	3
·15	1413	1416	1419	1422	1426	1429	1432	1435	1439	1442	0	1	1	1	2	2	2	3	3
·16	1445	1449	1452	1455	1459	1462	1466	1469	1472	1476	0	1	1	1	2	2	2	3	3
·17	1479	1483	1486	1489	1493	1496	1500	1503	1507	1510	0	1	1	1	2	2	2	3	3
·18	1514	1517	1521	1524	1528	1531	1535	1538	1542	1545	0	1	1	1	2	2	2	3	3
·19	1549	1552	1556	1560	1563	1567	1570	1574	1578	1581	0	1	1	1	2	2	3	3	3
·20	1585	1589	1592	1596	1600	1603	1607	1611	1614	1618	0	1	1	1	2	2	3	3	3
·21	1622	1626	1629	1633	1637	1641	1644	1648	1652	1656	0	1	1	2	2	2	3	3	3
·22	1660	1663	1667	1671	1675	1679	1683	1687	1690	1694	0	1	1	2	2	2	3	3	3
·23	1698	1702	1706	1710	1714	1718	1722	1726	1730	1734	0	1	1	2	2	2	3	3	4
·24	1738	1742	1746	1750	1754	1758	1762	1766	1770	1774	0	1	1	2	2	2	3	3	4
·25	1778	1782	1786	1791	1795	1799	1803	1807	1811	1816	0	1	1	2	2	2	3	3	4
·26	1820	1824	1828	1832	1837	1841	1845	1849	1854	1858	0	1	1	2	2	3	3	3	4
·27	1862	1866	1871	1875	1879	1884	1888	1892	1897	1901	0	1	1	2	2	3	3	3	4
·28	1905	1910	1914	1919	1923	1928	1932	1936	1941	1945	0	1	1	2	2	3	3	4	4
·29	1950	1954	1959	1963	1968	1972	1977	1982	1986	1991	0	1	1	2	2	3	3	4	4
·30	1995	2000	2004	2009	2014	2018	2023	2028	2032	2037	0	1	1	2	2	3	3	4	4
·31	2042	2046	2051	2056	2061	2065	2070	2075	2080	2084	0	1	1	2	2	3	3	4	4
·32	2089	2094	2099	2104	2109	2113	2118	2123	2128	2133	0	1	1	2	2	3	3	4	4
·33	2138	2143	2148	2153	2158	2163	2168	2173	2178	2183	0	1	1	2	2	3	3	4	4
·34	2188	2193	2198	2203	2208	2213	2218	2223	2228	2234	1	1	2	2	3	3	4	4	4
·35	2239	2244	2249	2254	2259	2265	2270	2275	2280	2286	1	1	2	2	3	3	4	4	4
·36	2291	2296	2301	2307	2312	2317	2323	2328	2333	2339	1	1	2	2	3	3	4	4	5
·37	2344	2350	2355	2360	2366	2371	2377	2382	2388	2393	1	1	2	2	3	3	4	4	5
·38	2399	2402	2410	2415	2421	2427	2433	2438	2443	2449	1	1	2	2	3	3	4	4	5
·39	2455	2460	2466	2472	2477	2483	2489	2495	2500	2506	1	1	2	2	3	3	4	5	5
·40	2512	2518	2523	2529	2535	2541	2547	2553	2559	2564	1	1	2	2	3	4	4	5	5
·41	2570	2576	2582	2588	2594	2600	2606	2612	2618	2624	1	1	2	2	3	4	4	5	5
·42	2630	2636	2642	2649	2655	2661	2667	2673	2679	2685	1	1	2	2	3	4	4	5	6
·43	2692	2698	2704	2710	2716	2723	2729	2735	2742	2748	1	1	2	3	3	4	4	5	6
·44	2754	2761	2767	2773	2780	2786	2793	2799	2805	2812	1	1	2	3	3	4	4	5	6
·45	2818	2825	2831	2838	2844	2851	2858	2864	2871	2877	1	1	2	3	3	4	5	5	6
·46	2884	2891	2897	2904	2911	2917	2924	2931	2936	2944	1	1	2	3	3	4	5	5	6
·47	2951	2958	2965	2972	2979	2985	2992	2999	3006	3013	1	1	2	3	3	4	5	5	6
·48	3020	3027	3034	3041	3048	3055	3062	3069	3076	3083	1	1	2	3	4	4	5	6	6
·49	3090	3097	3105	3112	3119	3126	3133	3141	3148	3155	1	1	2	3	4	4	5	6	6

ANTILOGARITHMS

	0	1	2	3	4	5	6	7	8	9	1	2	3	4	5	6	7	8	9
·50	3162	3170	3177	3184	3192	3199	3206	3214	3221	3228	1	1	2	3	4	4	5	6	7
·51	3236	3243	3251	3253	3258	3273	3281	3289	3296	3304	1	2	2	3	4	5	5	6	7
·52	3311	3319	3327	3334	3342	3350	3357	3365	3373	3381	1	2	2	3	4	5	5	6	7
·53	3388	3396	3404	3412	3420	3428	3436	3443	3451	3459	1	2	2	3	4	5	6	6	7
·54	3467	3475	3483	3491	3499	3508	3516	3524	3532	3540	1	2	2	3	4	5	6	6	7
·55	3548	3556	3565	3573	3581	3589	3597	3606	3614	3622	1	2	2	3	4	5	6	7	7
·56	3631	3639	3648	3656	3664	3673	3681	3690	3698	3707	1	2	3	3	4	5	6	7	8
·57	3715	3724	3733	3741	3750	3758	3767	3776	3784	3793	1	2	3	3	4	5	6	7	8
·58	3802	3811	3819	3828	3837	3846	3855	3864	3873	3882	1	2	3	4	4	5	6	7	8
·59	3890	3899	3908	3917	3926	3936	3945	3954	3963	3972	1	2	3	4	5	5	6	7	8
·60	3981	3990	3999	4009	4018	4027	4036	4046	4055	4064	1	2	3	4	5	6	6	7	9
·61	4074	4083	4093	4102	4111	4121	4130	4140	4150	4159	1	2	3	4	5	6	7	8	9
.62	4169	4178	4188	4198	4207	4217	4227	4236	4246	4256	1	2	3	4	5	6	7	8	9
·63	4266	4276	4285	4295	4305	4315	4325	4335	4345	4355	1	2	3	4	5	6	7	8	9
·64	4365	4375	4385	4395	4406	4416	4426	4436	4446	4457	1	2	3	4	5	6	7	8	9
·65	4467	4477	4487	4498	4508	4519	4529	4539	4550	4560	1	2	3	4	5	6	7	8	9
·66	4571	4581	4592	4603	4613	4624	4634	4645	4656	4667	1	2	3	4	5	6	7	9	10
·67	4677	4688	4699	4710	4721	4732	4742	4753	4764	4776	1	2	3	4	5	7	8	9	10
·68	4786	4797	4808	4819	4831	4842	4853	4864	4875	4887	1	2	3	4	6	7	8	9	10
·69	4898	4909	4920	4932	4943	4955	4966	4977	4989	5000	1	2	3	5	6	7	8	9	10
·70	5012	5025	5035	5047	5058	5070	5082	5093	5105	5117	1	2	4	5	6	7	8	9	11
·71	5129	5140	5152	5164	5176	5188	5200	5212	5224	5236	1	2	4	5	6	7	8	10	11
·72	5248	5260	5272	5284	5297	5309	5321	5333	5346	5358	1	2	4	5	6	7	9	10	11
·73	5370	5383	5395	5408	5420	5433	5445	5458	5470	5493	1	3	4	5	6	8	9	10	11
.74	5495	5508	5521	5534	5546	5559	5572	5585	5598	5610	1	3	4	5	6	8	9	10	12
·75	5623	5636	5649	5662	5675	5689	5702	5715	5728	5741	1	3	4	5	7	8	9	10	12
·76	5754	5768	5781	5794	5808	5821	5834	5848	5861	5875	1	3	4	5	7	8	9	11	12
·77	5888	5902	5916	5929	5943	5957	5970	5984	5998	6012	1	3	4	5	7	8	10	11	12
·78	6026	6039	6053	6067	6081	6095	6109	6124	6138	6152	1	3	4	6	7	8	10	11	13
·79	6166	6180	6194	6209	6223	6237	6252	6266	6281	6295	1	3	4	6	7	9	10	11	13
·80	6310	6324	6339	6353	6368	6383	6397	6412	6427	6442	1	3	4	6	7	9	10	12	13
·81	6457	6471	6486	6501	6516	6531	6546	6561	6577	6592	2	3	5	6	8	9	11	12	14
·82	6607	6622	6637	6653	6668	6683	6699	6714	6730	6745	2	3	5	6	8	9	11	12	14
·83	6761	6776	6792	6808	6823	6839	6855	6871	6887	6902	2	3	5	6	8	9	11	13	14
·84	6918	6934	6950	6966	6982	6998	7015	7031	7047	7063	2	3	5	6	8	10	11	13	15
·85	7079	7096	7112	7129	7145	7161	7178	7194	7211	7228	2	3	5	7	8	10	12	13	15
·86	7244	7261	7278	7295	7311	7328	7345	7362	7379	7396	2	3	5	7	8	10	12	13	15
·87	7413	7430	7447	7464	7482	7499	7516	7534	7551	7568	2	3	5	7	9	10	12	14	16
·88	7586	7603	7621	7638	7656	7674	7691	7709	7727	7745	2	4	5	7	9	11	12	14	16
·89	7762	7780	7798	7816	7834	7852	7870	7889	7907	7925	2	4	5	7	9	11	13	14	16
·90	7943	7962	7980	7998	8017	8035	8054	8072	8091	8110	2	4	6	7	9	11	13	15	17
·91	8128	8147	8166	8185	8204	8222	8241	8260	8279	8299	2	4	6	8	9	11	13	15	17
·92	8318	8337	8356	8375	8395	8414	8433	8453	8472	8492	2	4	6	8	10	12	14	15	17
·93	8511	8531	8551	8570	8590	8610	8630	8650	8670	8690	2	4	6	8	10	12	14	16	18
·94	8710	8730	8750	8770	8790	8810	8831	8851	8872	8892	2	4	6	8	10	12	14	16	18
·95	8913	8933	8954	8974	8995	9016	9036	9057	9078	9099	2	4	6	8	10	12	15	17	19
·96	9120	9141	9162	9183	9204	9226	9247	9268	9290	9311	2	4	6	8	11	13	15	17	19
·97	9333	9354	9376	9397	9419	9441	9462	9484	9506	9528	2	4	7	9	11	13	15	17	20
·98	9550	9572	9594	9616	9638	9661	9683	9705	9727	9750	2	4	7	9	11	13	16	18	20
·99	9772	9795	9817	9840	9863	9886	9908	9931	9954	9977	2	5	7	9	11	14	16	18	20

IX.7 CONCENTRATED ACIDS AND BASES

	Symbol	Unit	Reagent						
			HCl	HNO_3	H_2SO_4	CH_3COOH	$HClO_4$	H_3PO_4	$NH_4OH(NH_3)$
Relative molecular mass of active ingredient	M	g mol^{-1}	36·46	63·02	98·08	60·03	100·47	98·00	35·04
Percentage concentration (w/w)	P		36·0	69·5	96·0	99·5	60·0	88·0	58·6
Grams active ingredient per cm^3		g cm^{-3}	0·426	0·985	1·76	1·055	0·924	1·54	0·527
Molarity	M	mol ℓ$^{-1}$	11·7	15·6	18·0	17·6	1·21	15·7	15·1
Density	d	g cm^{-3}	1·19	1·42	1·84	1·06	1·54	1·75	0·90
Specific volume		cm^3 g^{-1}	0·84	0·70	0·54	0·94	0·65	0·57	1·11
cm^3 concentrated reagent per litre molar solution			85·1	64·0	55·6	56·9	108·6	63·6	66·5
Factor $F = \frac{M}{10pd}$	F		0·0851	0·0639	0·0556	0·0568	0·1086	0·0636	0·0665

Rule of dilution: If v cm^3 M molar solution has to be prepared, the V/cm^3 volume of concentrated reagent needed is

$$V = \frac{v.M.\text{M}.}{10p.d} = v.\text{M}.F$$

IX.8 PERIODIC TABLE OF THE ELEMENTS The relative atomic masses have been rounded off to a maximum of 4 significant figures. An alphabetical list of the elements with precise relative atomis masses and their logarithms is to be found in Section **IX.1**.

1 H 1·008																	2 He 4·003
3 Li 6	4 Be 9·012											5 B 10·81	6 C 12·01	7 N 14·01	8 O 16·00	9 F 19·00	10 Ne 20·18
11 Na 22·99	12 Mg 24·31											13 Al 26·98	14 Si 28·09	15 P 30·97	16 S 32·06	17 Cl 35·45	18 Ar 39·94
19 K 39·10	20 Ca 40·08	21 Sc 44·96	22 Ti 47·90	23 V 50·94	24 Cr 52·00	25 Mn 54·94	26 Fe 55·85	27 Co 58 93	28 Ni 58·71	29 Cu 63·55	30 Zn 65·38	31 Ga 69·72	32 Ge 72·59	33 As 74·92	34 Se 78·96	35 Br 79·90	36 Kr 83·80
37 Rb 85·47	38 Sr 87·62	39 Y 88·91	40 Zr 91·22	41 Nb 92·91	42 Mo 95·94	43 Tc (97)	44 Ru 101·1	45 Rh 102·9	46 Pd 106·4	47 Ag 107·9	48 Cd 112·41	49 In 114·8	50 Sn 118·7	51 Sb 121·8	52 Te 127·6	53 I 126·9	54 Xe 131·3
55 Cs 132·9	56 Ba 137·3	57 La 138·9	72 Hf 178·5	73 Ta 180·5	74 W 183·9	75 Re 186·2	76 Os 190·2	77 Ir 192·22	78 Pt 195·1	79 Au 197·0	80 Hg 200·6	81 Tl 204·4	82 Pb 207·2	83 Bi 209·0	84 Po (209)	85 At (210)	86 Rn (222)
87 Fr (223)	88 Ra 226·0	89 Ac 227·03															
				58 Ce 140·1	59 Pr 140·9	60 Nd 144·2	61 Pm 145	62 Sm 150·4	63 Eu 152·0	64 Gd 157·3	65 Tb 158·9	66 Dy 162·5	67 Ho 164·9	68 Er 167·2	69 Tm 168·9	70 Yb 173·0	71 Lu 175·0
				90 Th 232·0	91 Pa 231·0	92 U 238·0	93 Np 237·0	94 Pu (244)	95 Am (243)	96 Cm (247)	97 Bk (247)	98 Cf (251)	99 Es (254)	100 Fm (257)	101 Md (258)	102 No (259	103 Lr (260)

Table for Appendix IX.8

INDEX

Abbreviations:
AC: in abbreviated course
CHR: chromatographic, *or* by chromatography
CT: confirmatory test for
D: detection of
I: idendification of
ip: in the presence of
MS: on micro scale
PT: preliminary test for
R: reactions of
S: separation of
SM: on semimicro scale
T: tabulation, *or* table of